Nutritional Quality Management of Forages in the Himalayan Region

Rajan Katoch

Nutritional Quality Management of Forages in the Himalayan Region

Springer

Rajan Katoch
Biochemistry Laboratory,
Department of Genetics & Plant Breeding
CSKHPKV
Palampur, Himachal Pradesh, India

ISBN 978-981-16-5436-7 ISBN 978-981-16-5437-4 (eBook)
https://doi.org/10.1007/978-981-16-5437-4

This Springer imprint is published by the registered company Springer Nature Singapore Pte Ltd.
The registered company address is: 152 Beach Road, #21-01/04 Gateway East, Singapore 189721, Singapore

Foreword

Livestock rearing is one of the most important pursuits of Himalayan people where natural pastures and grasslands are the mainstay for livestock in the region. Agriculture and livestock farming in Himalayan communities are interwoven with the intricate fabric of cultural and religious beliefs along with economical perspectives. Though rearing of livestock is a major occupation of the rural households of the Himalayan region, the productivity of livestock is still low due to the unavailability of quality forages. With increasing grazing pressure and expanding human population, the natural forage resources are continuously degrading at a faster pace resulting in the shortage of fodder. Limited availability of quality feeds is still a major constraint behind low livestock productivity among smallholder farmers in the Himalayan region. It is a concerning fact that the area under natural feed resources is continuously decreasing worldwide. The volatile prices of concentrated feeds and overdependency on low-quality feeds are also responsible for poor livestock productivity.

I would appreciate and congratulate Dr. Rajan Katoch for bringing out this book on *Nutritional Quality Management of Forages in the Himalayan Region* housing information on different forage resources, nutritional quality of forages, niche-based nutritive forage species, varietal improvement of different species for nutritionally rich forages, nonconventional forages, and modem biotechnological intervention for quality improvement of forages. The book presents a valuable resource of information on forages for researchers and scientists in public and private institutions, dairy industry, policy planners, animal nutritionists, and students. I hope that the information in the book would be helpful to farmers in one way or the other for improving the livestock productivity with the provision of quality forages culminating in the improvement of their socioeconomic status.

Harinder Kumar Chaudhary

Vice-Chancellor,
CSK Himachal Pradesh
Agriculture University, Palampur
Palampur, Himachal Pradesh, India

Foreword

The Himalayan regions have rich diversity in climate, landforms, species, ethnicity, natural resource availability, and agricultural practices. This region supports the significant number of livestock population. However, in recent times unsystematic exploitation of natural feeding resources including permanent pastures and rangelands has resulted in their degradation throughout the region. Although livestock rearing is a major occupation of rural households of the Himalayan region, the productivity of livestock is still low due to limited availability of quality forages. The continuous hike in the prices of commercial feeds, lack of in-depth knowledge of management practices of existing fodder resource, limited information on fodder crop varieties with high productivity and nutritional quality, and overreliance on low-quality feeds are still major constrains in achieving sustainable livestock productivity. In the past, the main priority was to produce high yielding forage crops, but now with the understanding of the importance of forage quality in livestock production, researchers have focused their attention on the development of nutritionally superior forages and biofortified varieties.

I would like to place on record my appreciations to Dr. Rajan Katoch for the comprehensive work on the improvement of nutritional quality of forage crops for the benefit of the farming community. The book *Nutritional Quality Management of Forages in the Himalayan Region* covers different aspects of forages which are to be considered for ensuring quality feed for sustainable livestock production. The book contains important information on different forage genetic resources, nutritional quality of forages, niche-based nutritive forage species, varietal improvement of different forage species, and modem biotechnological intervention for quality improvement of the forages. The work will have global attention on forage resources and their quality for livestock feeding and will suggest models for quality improvement of forages worldwide in similar topographies.

T. R. Sharma

Deputy Director General,
Indian Council of Agricultural Research, Ministry of Agriculture
and Farmers Welfare, Govt. of India
New Delhi, India

Preface

In the Himalayan region natural pastures, grasslands and forests are the main resources for the livestock feeding. Livestock rearing is an integral part of farming system in a region, playing an important role in the economy of the rural households. Almost every rural family maintains a number of livestock of one kind and other for sustenance. In the present scenario, the livestock productivity is under tremendous pressure of acute shortage of quality fodder. Due to the shortage of cultivated fodder, the majority of livestock meet their feed requirements from natural grasslands and pastures, resulting in their continuous deterioration. Inadequate fodder supply, mounting prices of feed concentrates, improper management practices, over-dependency on low quality feeds, shifting of cropping patterns have now emerged as major constraints for improving livestock productivity. With the development of superior quality forages, livestock productivity could be enhanced many folds; however, the farmers need to be acquainted with improved fodder varieties and their nutritional attributes. Based on availability, farmers could then prioritise the feed or fodder of adequate nutritive quality with desirable effects on animal health and productivity.

The present book *Nutritional Quality Management of Forages in the Himalayan Region* presents comprehensive information on the nutritional quality management of forages in the biodiversity-rich Himalayan region. It includes information about the nutritional quality of a range of forage grasses, legumes, tree fodder and non-conventional forage resources. The findings and data are based on more than two decades of research on nutritional quality of different temperate grasses, fodder trees, legumes and non-conventional forage resources. The book is especially relevant to the readers interested in understanding forage quality and suggesting models for quality improvement of forages in similar topographies of the world.

The aim of bringing out this manuscript is to disseminate scientific knowledge on different forage resources, constraints affecting forage availability, factors affecting forage quality, nutritional and anti-nutritional components in forages, nutritional quality estimation of forages, nutritional quality of forage different resources, post-harvest processing and conservation, management practices for availing quality fodder, methods for improving the nutritive value of low quality feeds, crop residues in animal feeding, non-conventional fodder resources and biotechnological approaches for forage quality improvement among other. The in-depth knowledge on forage quality would be useful for improving the livestock production and socio-economic status of rural households in the Himalayan region. The book presents a valuable resource of information on nutritional quality of forages for researchers and scientists, animal nutritionists, dairy industry, policy planners and students.

Palampur, Himachal Pradesh, India Rajan Katoch, Ph.D

Acknowledgements

The author thankfully acknowledges the encouragement, support and facilities provided by Prof. H.K. Chaudhary, Vice-Chancellor, CSKHPKV, Palampur HP, India for bringing out this publication.

The author is highly thankful to Dr. T. R. Sharma (Deputy Director General, Crop Sciences) ICAR, GOI, New Delhi, India for his guidance, encouragement and support for publishing this book.

The author is thankful to Dr. A.K. Roy (Project coordinator, AICRP, FC&U) and Dr. R. K. Aggarwal (Principal Scientist), IGFRI, Jhansi for their orientation and technical support for bringing out this dedicated book on nutritional quality of forages.

The author thankfully acknowledges the support by AICRP (FC&U), ICAR, New Delhi for conducting experimental trials and studies on the nutritional quality of different forage crops of the Himalayan region. The forage breeders, scientists and researchers engaged in forage development programme are also acknowledged for providing germplasm and varieties for conducting different studies.

Lastly, I express my deep gratitude towards my parents for their incessant motivation, encouragement and blessings for bringing out this monumental book on forage quality.

Contents

1 **Overview** . . . 1
1.1 Understanding Forage Quality . . . 5
1.2 Influence of Forage Quality on Livestock Productivity . . . 8
1.3 Diversity in Forage Resources of the Himalayan Region . . . 10
1.4 Nutritional Diversity in Forage Resources . . . 11
1.5 Regional Imbalances and Demand and Supply of Forages . . . 13
1.6 Insight of Livestock Production and Productivity . . . 15
1.6.1 Indian Perspective . . . 15
1.6.2 Himalayan Perspective . . . 17
References . . . 19

2 **Impact of Forage Availability on Livestock Rearing and Socioeconomic Aspects** . . . 21
2.1 Livestock Rearing: An Indispensable Component of Hill Agriculture . . . 21
2.2 Importance of Livestock Production in Rural Economy of Himalayan Region . . . 23
2.2.1 Contribution of Livestock to National and Regional Economy . . . 24
2.2.2 Socio-economic Status of Farmers Depending on Livestock-Based Agriculture . . . 27
2.3 Constraints in Achieving Optimal Livestock Productivity . . . 28
2.3.1 Feeds and Fodder Shortage . . . 28
2.3.2 Poor Livestock Extension Services . . . 28
2.3.3 Poor Credit and Marketing Facilities . . . 28
2.3.4 Poor Veterinary Services . . . 29
2.3.5 High Prices of Commercial Feeds . . . 29
2.3.6 Poor Animal Breeding Facilities . . . 29
2.3.7 Gender Monoply in Livestock Activities . . . 29
2.3.8 Poor Animal Management System . . . 31
2.3.9 Lack of Knowledge of Anti-Nutrients in Different Forage Species . . . 31
2.4 Conventional Practices of Livestock Feeding . . . 32

2.4.1 Grazing in Natural Pastures and Grasslands 32
2.4.2 Dry Fodder Feeding 33
2.4.3 Dry Fodder Mixture with Green Fodder 33
2.4.4 Concentrate Feeds 36
2.5 Nutritionally Compromised Fodder Resources for Livestock Feeding 39
2.6 Scope of Non-conventional Forages in Livestock Feeding 42
References 45

3 Forage Resources in Himalayan Region 49
3.1 Grasslands, Pastures and Rangelands of Himalayan Region 50
3.1.1 Grasslands and Pastures of North-West Himalayan Region 50
3.1.2 Grasslands and Pastures of Eastern Himalayas 52
3.1.3 Rangelands of the Indian Himalayan Region 53
3.1.4 Grassland Distribution Pattern 53
3.2 Grasses Predominant in Himalayan Region 58
3.2.1 Tall Fescue (*Festuca arundinacea* Schreb.) 58
3.2.2 Perennial Rye Grass (*Lolium perenne* L.) 59
3.2.3 Setaria (*Setaria anceps* Schumach.) 59
3.2.4 Guinea Grass (*Panicum maximum* Jacq.) 59
3.2.5 Orchard Grass (*Dactylis glomerata* L.) 60
3.2.6 Napier Grass (*Pennisetum purpureum* Schumach.) . . . 60
3.2.7 Napier-Bajra Hybrid (NB Hybrid) 60
3.2.8 Kentucky Bluegrass (*Poa pratensis* L.) 61
3.2.9 Bahia Grass (*Paspalum notatum* Flügge) 61
3.2.10 *Tripsacum laxum* Nash. 61
3.2.11 Mission Grass [*Pennisetum polystachion* (L.) Schultes] 61
3.2.12 Big-Leaf Bristle Grass or Buffalo Grass [*Setaria palmifolia* (J. Koenig) Stapf.] 62
3.2.13 *Arundinella nepalensis* Trinius 62
3.2.14 *Festuca rubra* L. sensu lato 62
3.2.15 Spear Grass (*Heteropogon contortus* (L.)) 62
3.2.16 Blue Panic Grass (*Panicum antidotale* Retz.) 64
3.3 Common Forage Legumes of Himalayan Region 64
3.3.1 Berseem (*Trifolium alexandrium* L.) 64
3.3.2 White Clover (*Trifolium repens* L.) 64
3.3.3 Alfalfa (*Medicago sativa* L.) 65
3.3.4 Cowpea (*Vigna unguiculata* L.) 65
3.3.5 Rice Bean [*Vigna umbellata* (Thunb.) Ohwi & H. Ohashi] 65
3.3.6 Birdsfoot Trefoil (*Lotus corniculatus* L.) 66
3.3.7 Red Clover (*Trifolium pratense* L.) 66
3.3.8 Grass Pea (*Lathyrus sativus* L.) 66

3.3.9 Siratro [*Macroptilium atropurpureum* (DC.) Urb.] . . . 66
3.3.10 Perennial Soybean [*Neonotonia wightii* (Wight & Arn.) J.A. Lackey] 67
3.3.11 Centro (*C. pubescens* Benth.) 67
3.3.12 Sweet Clover [*Melitotus officinalis* (L.) Lam] 67
3.3.13 Shaftal or Persian Clover (*Trifolium resupinatum* L.) 67
3.3.14 Moth Bean [*Vigna aconitifolia* (Jacq.) Marechal] 68
3.3.15 Perennial Horse Gram (*Macrotyloma axillare* var. axillare) 68
3.3.16 Faba Bean (*Vicia faba* L.) 68
3.3.17 Guar Bean (*Cymopsis tetragonaloba* [L.] Taub.) 68
3.3.18 *Macroptilium lathyroides* (L.) Urban. 69
3.3.19 *Macrotyloma uniflorum* Lam. 69
3.3.20 Himalayan Gram (*Cicer microphyllum* Benth.) 69
3.3.21 *Caragana vesicolor* Benth. 69
3.3.22 *Astragulus alpinus* L. 70
3.3.23 *Desmodium heterocarpon* (L.) DC 70
3.4 Cereal Fodder 70
3.4.1 Fodder Maize (*Zea mays* L.) 70
3.4.2 Oat (*A. sativa* L.) 70
3.4.3 Sorghum (*Sorghum bicolor* L.) 72
3.4.4 Barley (*Hordeum vulgare* L.) 72
3.4.5 Pearl Millet (*Pennicetum glaucum* L.) 72
3.4.6 Indian Barnyard Millet [*Echinochloa frumentacea* (Link) Ridl.] 73
3.4.7 Finger Millet [*Eleusine coracana* (L.) Gaertn] 73
3.4.8 Foxtail Millet [*Setaria italica* (L.) Beauv.] 73
3.5 Range Forages 75
3.5.1 Bermuda Grass (*C. dactylon* Pers.) 75
3.5.2 Nut Grass (*Cyperus rotundus* L.) 75
3.5.3 Para Grass (*Brachiaria mutica* (Forssk.) Stapf) 75
3.5.4 Dallis Grass (*Paspalum dilatatum* Poir.) 76
3.5.5 Cogon Grass (*Imperata cylindrica* (L) P. Beauv) 76
3.5.6 Other Important Range Forages 76
3.6 Fodder Trees and Shrubs of the Himalayan Region 76
3.7 Crop Residues 78
3.8 Other Feed Resources in Himalayan Region 80
3.9 Forage Productivity in Himalayan Region 81
3.9.1 Forage Resources and Their Productivity in North-West Himalayas 81
3.9.2 Forage Resources and Productivity in Eastern Himalayan Region 83

3.10 Grazing Patterns and Maintenance of Pastures and Grasslands in Himalayan Region . . . 84
3.11 Multiple Cropping Systems Influencing Forage Productivity and Nutritional Value . . . 86
References . . . 88

4 Constraints Affecting Forage Availability in Himalayan Region . . . 93
4.1 Land Scarcity . . . 93
4.2 Undulating Topography . . . 94
4.3 Erratic Environmental Conditions . . . 95
4.4 Climate Change . . . 95
4.5 Increasing Population of Unproductive Animals . . . 96
4.6 Uncontrolled Pasture Grazing . . . 96
4.7 Invasion of Alien Species in Grasslands and Pastures . . . 96
4.8 Collapse of Traditional Agro-Forestry Practices . . . 96
4.9 Lack of Management and Conservation Practices . . . 97
4.10 Lack of Improved Varieties . . . 97
4.11 Lack of Knowledge of Non-conventional Fodder . . . 97
4.12 Poor Credit and Marketing Facilities . . . 98
References . . . 98

5 Factors Influencing Forage Nutritional Quality . . . 99
5.1 Plant Factors Affecting Forage Quality . . . 99
5.1.1 Stage of Maturity . . . 99
5.1.2 Anti-Quality Factors . . . 104
5.1.3 Species Difference . . . 104
5.1.4 Plant Anatomy . . . 105
5.1.5 Variety (Cultivar) . . . 105
5.1.6 Cellular Composition and Metabolism . . . 106
5.2 Animal Perspective of Forage Quality . . . 106
5.2.1 Palatability . . . 106
5.2.2 Intake . . . 107
5.2.3 Digestibility . . . 107
5.3 Environmental Factors Affecting the Forage Quality . . . 107
5.3.1 Temperature . . . 107
5.3.2 Water Stress . . . 108
5.3.3 Photoperiod . . . 109
5.3.4 Solar Radiation . . . 109
5.3.5 Shading . . . 109
5.3.6 Soil Properties . . . 110
5.3.7 Soil Fertility/Fertilizer Application . . . 110
5.3.8 Altitude . . . 112
5.4 Seasonal Variations and Quality of Forages . . . 115
5.4.1 Seasonal Dynamics in Nutritional Quality of Major Forage Grasses . . . 116

5.4.2 Seasonal Dynamics in Nutritional Constituents of Major Fodder Trees 123
5.4.3 Seasonal Alterations in Anti-Nutritional Constituents of Fodder Trees of Mid-Himalayas 129
5.5 Effect of Foliar Diseases and Insect Pests on Quality of Forages 138
5.5.1 Effect of Foliar Disease and Insect on Quality of Forage Cowpea 139
5.5.2 Effect of Powdery Mildew Disease on Forage Quality of Oats 141
References 144

6 Forage Genetic Resources (FGR) 151
6.1 Forage Genetic Resource Conservation 152
6.2 Conservation and Maintenance of Forage Genetic Resources . . . 153
6.3 Forage Genetic Resources and Varietal Improvement 159
References 165

7 Approaches for Nutritional Quality Improvement in Forages 167
7.1 Forage Research: National and Regional Scenario 168
7.2 Need for Nutritional Quality Improvement of Forages 169
7.2.1 Targets for Nutritional Enhancement of Forages 170
7.3 Breeding for Improvement in Forage Quality 173
7.3.1 Introduction and Selection 177
7.3.2 Hybridization 178
7.3.3 Interspecific and Polyploidy Breeding 179
7.3.4 Mutation Breeding 181
7.3.5 Apomixis 181
7.3.6 Wide Hybridization 182
7.4 Techniques for the Development of Improved Forage Crop Varieties 186
7.4.1 Embryo Rescue 187
7.4.2 Callus Culture 187
7.4.3 Meristem Culture 188
References 189

8 Nutritional and Anti-Nutritional Constituents in Forages 193
8.1 Nutritional and Anti-Nutritional Constituents of Forages and Their Importance in Animal Nutrition 193
8.1.1 Nutritional Components 194
8.1.2 Anti-Nutritional Constituents 202
References 221

9 Nutritional Quality Estimation of Forages 225
9.1 Scheme for Forage Quality Analysis 225
9.2 Destructive Methods for Forage Quality Estimation 226
9.2.1 The Weende Scheme 226
9.2.2 The Van Soest System 226
9.3 Nondestructive Methods for Forage Quality Estimation 228
9.3.1 Near-Infrared Reflectance Spectroscopy (NIRS) Technique 228
9.3.2 Fourier-Transform Infrared Spectroscopy (FTIR) 230
9.4 Forage Quality Indices 232
9.5 Estimation of Forage Quality Components 236
9.5.1 Moisture 236
9.5.2 Estimation of Dry Matter (DM) 237
9.5.3 Estimation of Crude Protein 240
9.5.4 Estimation of Nonprotein Nitrogen in Feed Samples 241
9.5.5 Estimation of Urea Nitrogen in Feed Samples 242
9.5.6 Determination of Fiber Content in Forages 243
9.5.7 Estimation of Cell Wall Constituents 246
9.5.8 Estimation of Carbohydrate Content in Forages 251
9.5.9 Determination of Digestibility of Forages 254
9.5.10 Mineral Analysis of Forages 258
9.5.11 Estimation of Flavanols 259
9.5.12 Estimation of Chlorophyll Content in Forages 260
9.5.13 Estimation of Antioxidant Activity 261
9.6 Estimation of Anti-Nutritional Factors in Forages 261
9.6.1 Estimation of Phenols 262
9.6.2 Estimation of Tannin Content Using *FOLIN-DENIS* Reagent 263
9.6.3 Determination of Condensed Tannins (Proanthocyanidins) 264
9.6.4 Estimation of Oxalate Content 264
9.6.5 Estimation of Saponins Content 265
9.6.6 Estimation of Cyanogenic Glycosides 265
9.6.7 Estimation of Nitrates and Nitrites 267
9.6.8 Estimation of Nitrate-N 268
9.6.9 Estimation of Trypsin Inhibitor 268
9.6.10 Estimation of Mimosine Content in Forages 269
9.6.11 Estimation of Alkaloids 269
9.6.12 Determination of Phytic Acid Content 270
9.7 Estimation of Quality of Grasslands and Pastures 271
9.7.1 Determination of Quality of Grasslands and Pastures Using Remote Sensing Technology 273
References 276

10 Nutritional Quality of Major Forage Grasses of Himalayan Region . . . 279
10.1 Range Forages . . . 279
10.1.1 Guinea Grass (*Panicum maximum* Jacq.) . . . 279
10.1.2 Bermuda Grass (*Cynadon dactylon* Pers.) . . . 280
10.1.3 Cogon Grass [*Imperata cylindrica* (L.) Raeuschel] . . . 280
10.1.4 Dallis Grass (*Paspalum dilatatum* Poiret) . . . 281
10.1.5 Para Grass [*Brachiaria mutica* (Forssk.) Stapf] . . . 281
10.1.6 Guria Grass (*Chrysopogon fulvus* (Spreng.) Chiov.) . . . 281
10.1.7 Marvel Grass [*Dichanthium annalatum* (Forssk.) Stapf] . . . 282
10.1.8 Creeping Bluegrass (*Bothriochloa insculpta* (Hochst. ex A. Rich.) A. Camus) . . . 282
10.1.9 Dinanath Grass (*Pennesitum pedicellatum* Trin.) . . . 282
10.1.10 Congo Grass (*Brachiaria ruziziensis* Germ. & Evrard) . . . 283
10.1.11 Alpine Timothy (*Phleum alpinum* L.) . . . 283
10.1.12 Kentucky Bluegrass (*Poa pratensis* L.) . . . 283
10.1.13 Bahia Grass [*Paspalum notatum* Flueggé] . . . 283
10.1.14 Mission Grass [*Pennisetum polystachion* (L.) Schultes] . . . 284
10.1.15 Big-Leaf Bristle Grass or Buffalo Grass [*Setaria palmifolia* (J.Koenig) Stapf] . . . 284
10.1.16 Spear Grass [*Heteropogon contortus* (L.)] . . . 284
10.1.17 Bristle Grass, Rough Bristle Grass and Bristly Foxtail [*Setaria verticillata* (L.) P. Beauv] . . . 285
10.1.18 Blue Panic (*Panicum antidotale* Retz) . . . 285
10.1.19 Rhode Grass (*Chloris gayana* Kunth.) . . . 285
10.1.20 Buffel Grass or Anjan Grass (*Cenchrus ciliaris* L.) . . . 286
10.2 Cultivated Forage Grasses of the Himalayan Region . . . 286
10.2.1 Tall Fescue (*Festuca arundinacea* Schreb) . . . 286
10.2.2 Perennial Rye Grass (*Lolium perenne* L.) . . . 288
10.2.3 Golden Timothy (*Setaria* spp.) . . . 288
10.2.4 Orchard Grass/Cock's Foot (*Dactylis glomerata* L.) . . . 289
10.2.5 Elephant Grass (*Pennisetum purpureum* Schumach.) . . . 289
10.2.6 Bajra × Napier Hybrids (BN Hybrids) . . . 289
10.3 Cereal Fodders . . . 290
10.3.1 Fodder Maize (*Zea mays* L.) . . . 290
10.3.2 Oats (*Avena sativa* L.) . . . 290
10.3.3 Sorghum [*Sorghum Bicolour* (L.) Moench] . . . 291
10.3.4 Barley (*Hordeum vulgare* L.) . . . 291
10.3.5 Pearl Millet (*Pennisetum glaucum* (L.) . . . 292
10.3.6 Indian Barnyard Millet [*Echinochloa frumentacea* (Link) Ridl.] . . . 292

10.3.7 Finger Millet [*Eleusine coracana* (L.) Gaertn] 292
10.3.8 Foxtail Millet [*Setaria italica* (L.) Beauv.] 293
10.4 Studies on Forage Grasses of the Himalayan Region 293
10.4.1 Comparative Biochemical Profile of Tall Fescue (*Festuca arundinacea* Schreb.) Varieties Grown in the Himalayan Region 293
10.4.2 Productivity and Biochemical Attributes of Different Maize Varieties 297
10.4.3 Studies on Quality Parameters and Production Economics of Maize Under Different Use Patterns . . . 300
References 306

11 Forage Legumes in Himalayan Region 309
11.1 Cultivated Forage Legumes of the Himalayan Region 310
11.1.1 Alfalfa (*Medicago sativa* L.) 310
11.1.2 Berseem (*Trifolium alexandrium* L.) 313
11.1.3 White Clover (*Trifolium repens* L.) 314
11.1.4 Red Clover (*Trifolium pratense* L.) 314
11.1.5 Birdsfoot Trefoil (*Lotus corniculatus* L.) 315
11.1.6 Caribbean Stylo [*Stylosanthes hamata* (L.) Taub.] . . . 315
11.1.7 Sainfoin (*Onobrychis vicifolia* L.) 316
11.1.8 Sweet Clover (*Melitotus officinalis* L.) 316
11.1.9 Shaftal or Persian Clover (*Trifolium resupinatum* L.) . . 316
11.1.10 Moth Bean (*Vigna aconitifolia* L.) 317
11.1.11 Butterfly Pea (*Clitoria ternatea* L.) 317
11.1.12 Phasey Bean [*Macroptilium lathyroides* (L.) Urban.] . . 318
11.2 Dual-Purpose Legumes 318
11.2.1 Cowpea (*Vigna unguiculata* L.) 318
11.2.2 Garden Pea (*Pisum sativum* L.) 319
11.2.3 Grass Pea (*Lathyrus sativus* L.) 319
11.2.4 Common Bean (*Phaseolus vulgaris* L.) 320
11.2.5 Rice Bean (*Vigna umbellata* (Thunb.) Ohwi and Ohashi) 320
11.3 Studies on Forage Legumes in the Himalayan Region 320
11.3.1 Rice Bean: A Potential Forage Legume 320
11.3.2 Effect of Altitude on Nutritional Profile of White Clover 343
11.3.3 Effects of Foliar Diseases and Insect Pests on Quality Parameters of Legumes 345
References 349

12 Fodder Trees in Mitigating Forage Demand 355
12.1 Fodder Trees: An Important Forage Resource in the Himalayan Region 355
12.1.1 Fodder Trees as Nutritional Feeding Resource 356
12.1.2 Fodder Trees as Dry Season Supplement 357

12.1.3 Multiple Usages of Fodder Trees 357
12.2 Tree Fodder: Alternative Source of Quality Fodder 357
12.3 Nutritional Composition of Fodder Trees of Himalayan Region 359
12.4 Anti-nutritional Components of Fodder Trees 368
12.5 Forages Availability During Lean Period 378
References 381

13 Prospects of Nonconventional Feed Resources of Himalayan Region 383
13.1 Nonconventional Feed Resources in Livestock Feeding 383
13.2 Sources of NCFR 384
13.2.1 Fruit and Vegetable Waste 384
13.2.2 Agro-Industrial by-Products (AIBP) 385
13.2.3 Wood-Derived Products 385
13.2.4 Weeds 385
13.3 Nutritive Value of NCFR 386
13.3.1 Fruits and Vegetable Wastes 386
13.3.2 Crop by-Products as NCFR 387
13.3.3 Microalgae 388
13.3.4 Weeds 389
13.4 Constraints in Using Nonconventional Feeds 394
13.5 Improving Quality and Digestibility of NCFR 395
13.5.1 Supplementation 395
13.5.2 Chemical Treatment 395
13.5.3 Ensiling 395
13.5.4 Feed Block Technology 395
13.5.5 Removal of Anti-nutritional Factors 396
13.6 Using NCFR in Livestock Feeding 396
References 397

14 Quality and Digestibility of Crop Residues 399
14.1 Crop Residues in Livestock Feeding 399
14.2 Nutritive Value of Common Crop Residues 400
14.2.1 Rice Straw 400
14.2.2 Wheat Straw 401
14.2.3 Barley and Oat Straw 401
14.2.4 Maize Stover 402
14.2.5 Sorghum Stover 402
14.2.6 Legume Straw 402
14.3 Constraints in Utilization of Crop Residues as Sole Livestock Feed 403
14.3.1 Lignin Content 403
14.3.2 Silica Content 404
14.3.3 Fiber Content 404

14.3.4 Protein Content and Digestibility 404
14.4 Improving Quality and Digestibility of Crop Residues 404
14.4.1 Physical Treatment 406
14.4.2 Chemical Treatment 408
14.4.3 Biological Treatment 437
14.4.4 Exogenous Enzymatic Treatment 456
14.4.5 Supplementation 457
References 458

15 Biotechnological Techniques for Nutritional Quality Improvement in Forages 465
15.1 Molecular Approaches for Forage Nutritional Quality Improvement 466
15.1.1 Genetic Engineering 466
15.1.2 Cisgenesis/Intragenesis 467
15.1.3 RNA Interference (RNAi) and Antisense RNA Technologies 467
15.1.4 Genome Editing Technologies 468
15.1.5 QTLs and Identification of Candidate Genes 468
15.2 Nutritional Quality Enhancement of Forage Crops with New Biotechnological Techniques 470
15.2.1 Transfer of Nutritionally Potential Genes 470
15.2.2 Engineering Metabolic Pathways 474
15.3 Genomic *In Situ* Hybridization (GISH) and DNA-Based Genetic Marker Technology 478
15.3.1 Genomic *In Situ* Hybridization (GISH) 478
15.3.2 DNA-Based Marker Technology/Linkage Mapping 480
15.4 Prospects of Gene Editing Techniques in Nutritional Quality Improvement of Forage Crops 481
References 486

16 Lignin: Possible Manipulations in Forages 493
16.1 Lignin and Plant Growth Development 493
16.1.1 Lodging Resistance to Plants 494
16.1.2 Lignin and Plant Stress Adaptation 494
16.1.3 Plant Drought and Salt Stress Tolerance 495
16.1.4 Plant Temperature Stress Adaptability 495
16.2 Factors Affecting Lignification in Plants 496
16.2.1 Species Difference 496
16.2.2 Temperature 496
16.2.3 Moisture Stress 497
16.2.4 Light Intensity and Quality 497
16.2.5 Plant Anatomy and Morphology 498
16.3 Lignin Content and Forage Digestibility 498

16.4 Lignin Biosynthesis 499
16.5 Lignin Biosynthetic Pathway 501
16.5.1 Shikimic Acid Pathway 501
16.5.2 Phenylpropanoid Pathway 502
16.5.3 Monolignol Biosynthesis Pathway 503
16.5.4 Polymerization of Monolignols to Form H, G, and S Lignin 507
16.6 Intricacies of Lignin Biosynthesis in Plant Cell Wall 507
16.7 Manipulation of Lignin Biosynthesis in Forages 509
16.7.1 RNAi-Mediated Manipulation of Lignin Biosynthesis 511
16.7.2 Antisense RNA-Mediated Manipulation of Lignin Biosynthesis 513
16.7.3 Lignin Biosynthetic Pathway Editing Techniques 515
References 520

17 Postharvest Processing and Preservation of Forages 529
17.1 Postharvest Preservation of Forages 529
17.1.1 Haymaking 530
17.1.2 Silage Making 533
17.1.3 Fodder Chaffing 537
17.1.4 Bale Silage 539
17.1.5 Baleage or Haylage 539
17.2 Packaging and Storage of Preserved Forages 539
17.2.1 Baling of Hay/Straw 539
17.2.2 Pelleting 541
17.2.3 Densified Complete Feed Block (DCFB) 542
17.3 Transporting and Storage Practices 545
References 546

18 Effect of Harvest and Storage on Forage Quality 547
18.1 Nutrient Losses During Harvest 548
18.1.1 Carbohydrate Losses 548
18.1.2 Protein Losses 548
18.1.3 Vitamin Losses 550
18.1.4 Other Nutrient Losses 550
18.2 Nutrient Losses Due to Weathering 551
18.3 Losses During Mowing and Conditioning 551
18.4 Nutrient Losses During Baling and Chopping 552
18.5 Changes in Forage Quality During Storage 552
18.5.1 Quality Loss During Inside Storage 552
18.5.2 Quality Loss During Outside Storage 554
References 554

19 Challenges and Opportunities in Forage Production and Livestock Management in the Himalayan Region 557
19.1 Challenges for Forage Production and Livestock Management 558
19.1.1 Land Constraints 558
19.1.2 Uncontrolled Grazing of Natural Pasture Resources 558
19.1.3 Changing Climatic Conditions 559
19.1.4 Lack of Planting Material 559
19.1.5 Inefficient Forage Resource Management 559
19.1.6 Inefficient Livestock Management Practices 560
19.1.7 Lack of Strong Policy Support 560
19.2 Opportunities for Improving Forage and Livestock Production 560
References 567

Perspective 569

Conclusion 577

Glossary 585

About the Author

Rajan Katoch is currently working in the Department of Genetics and Plant Breeding at CSK Himachal Pradesh Krishi Vishvavidyalaya, Palampur. He has more than two decades of research and teaching experience in the field of Biochemistry and Molecular Biology. He is actively involved in research on nutritional evaluation and quality improvement of forages and headed more than ten research projects as Principal Investigator funded by prestigious government organizations. He has received several prestigious awards and honors including Young Scientist Award, Award from CSIR for excellence in biochemistry, Group Study Exchange (GSE) fellowship from South America (Brazil), ICAR fellowship, and DBT Crest Award from Government of India. He has also received "appreciation awards" from the United States Department of Agriculture (USDA) for his exemplary work on rice bean protease inhibitor as a transgene for resistance. He is a fellow and member of various professional societies and secretary of the Indian Society of Agricultural Biochemists (PC). He has published over 150 papers in high-impact peer-reviewed international and national journals and has authored more than ten successful books with prestigious publishers.

Abbreviations

°C	Degree Celsius
μg	Microgram
μL	Microliter
ACU	Adult Cattle Unit
ADF	Acid detergent fiber
ADIN	Acid detergent insoluble N
ADL	Acid detergent lignin
AICRP (FCU)	All India Coordinated Research Project on (Fodder Crops and Utilization)
amsl	Above mean sea level
ADF-N	ADF-nitrogen
BN hybrid	Bajra-Napier hybrid
CC	Cell content
CF	Crude fiber
Chl	Chlorophyll
CL	Cellulose
cm	Centimeter
CP	Crude protein
CT	Condensed tannins
CWD	Cell wall digestibility
DDM	Digestible dry matter
DFY	Dry fodder yield
DM	Dry matter
DMD	Dry matter digestibility
DMI	Dry matter intake
DMY	Dry matter yield
DOMD	Digestible organic matter in dry matter
EE	Ether extract
FAO	Food and Agriculture Organization
FGR	Forage genetic resource
FLD	Front Line Demonstration
g	Gram
GDP	Gross domestic product

GFY	Green fodder yield
GVA	Gross value added
HC	Hemicellulose
HT	Hydrolysable tannins
ICAR	Indian Council of Agricultural Research
IGFRI	Indian Grassland and Fodder Research Institute
IHR	Indian Himalayan Region
IVDMD	*In vitro* dry matter digestibility
kg	Kilogram
km	Kilometer
KVK	Krishi Vigyan Kendra
L	Liter
LAB	Lactic acid bacteria
m	Meter
M	Molarity
ME	Metabolizable energy
mg	Milligram
mL	Milliliter
N	Normality
NBPGR	National Bureau of Plant Genetic Resources
NCFR	Nonconventional feed resources
NDF	Neutral detergent fiber
NDFCP	Neutral detergent fiber crude protein
NFC	Non-fibrous carbohydrate
NFE	Nitrogen-free extract
NGB	National Gene Bank
NIRS	Near infrared spectroscopy
nm	Nanometer
NPN	Non-protein nitrogen
NSC	Nonstructural carbohydrates
NT	Net tannins
NVI	Nutritive value index
OM	Organic matter
OMD	Organic matter digestibility
PAU	Punjab Agricultural University
PGR	Plant genetic resource
ppm	Parts per millions
q/ha	Quintal per hectare
QI	Quality index
QTL	quantitative trait locus
RFQ	Relative feed quality
RFV	Relative feed value
RS	Reducing sugar
SAU	State Agriculture Universities
SCFA	Short chain fatty acids

SP	Simple phenols
SSR	Simple sequence repeats
t/ha	Tons per hectare
TC	Total carbohydrates
TDN	Total digestible nitrogen
TF	Total flavonoids
TNAU	Tamil Nadu Agricultural University
TNC	Total nonstructural carbohydrates
TP	Total phenols
TT	Total tannins
WSC	Water soluble carbohydrates

Overview

1

The Hindukush–Karakoram–Himalaya (HKKH) region extends up to 3500 km from north to south with Afghanistan in the west and Myanmar in the east. It includes Nepal and Bhutan as a whole along with the mountainous regions from Afghanistan, Bangladesh, China, Myanmar, Pakistan, and India; thus traversing across eight countries.

The Asia-Pacific region is known for having one of the largest, highest, and densely populated mountain systems in the world. The Himalayas, Karakoram, and Hindu-Kush range of mountains extend from Namcha Barwa in the east to the borders of Afghanistan in the west. This mountain system has a series of parallel and converging ranges forming one of the highest mountain regions in the world. The Hindu-Kush Himalayan region extends up to 3500 km, covering an area of about 4.30 million square km from north to south with Afghanistan in the west and Myanmar in the east. It includes eight countries including Nepal and Bhutan as a whole along with the mountainous ranges of Afghanistan, Bangladesh, China, Myanmar, Pakistan, and India. The region is further geographically divided into two different subregions, in which one region has mountainous terrain with extreme changes in altitude and the other region has less mountainous ranges and a more plateau-type landscape (Sharma et al. 2019). Due to unique geographical landscape, the Hindu-Kush Himalayan region is one of the most dynamic, complex, and fragile mountain ecosystems of the world. The region exhibits great diversity in climate, landforms, species, ethnicity, bio-resource availability, and agricultural practices. The Himalayan region also serves as an enormous base for different bio-resources. The region is also known for great diversity of forage resources on which large and diverse livestock population in the region thrive comfortably. The forage resources of Himalayan region include range forages, grasslands and pastures, cultivable forages, trees, and shrubs. Figure 1.1 depicts the extent of the Hindu-Kush Himalayan region.

The Indian Himalayan region (IHR) covering an area of about 5 lakh square km forms the northern boundary of the country. It stretches over 2500 km from Jammu

R. Katoch, *Nutritional Quality Management of Forages in the Himalayan Region*,
https://doi.org/10.1007/978-981-16-5437-4_1

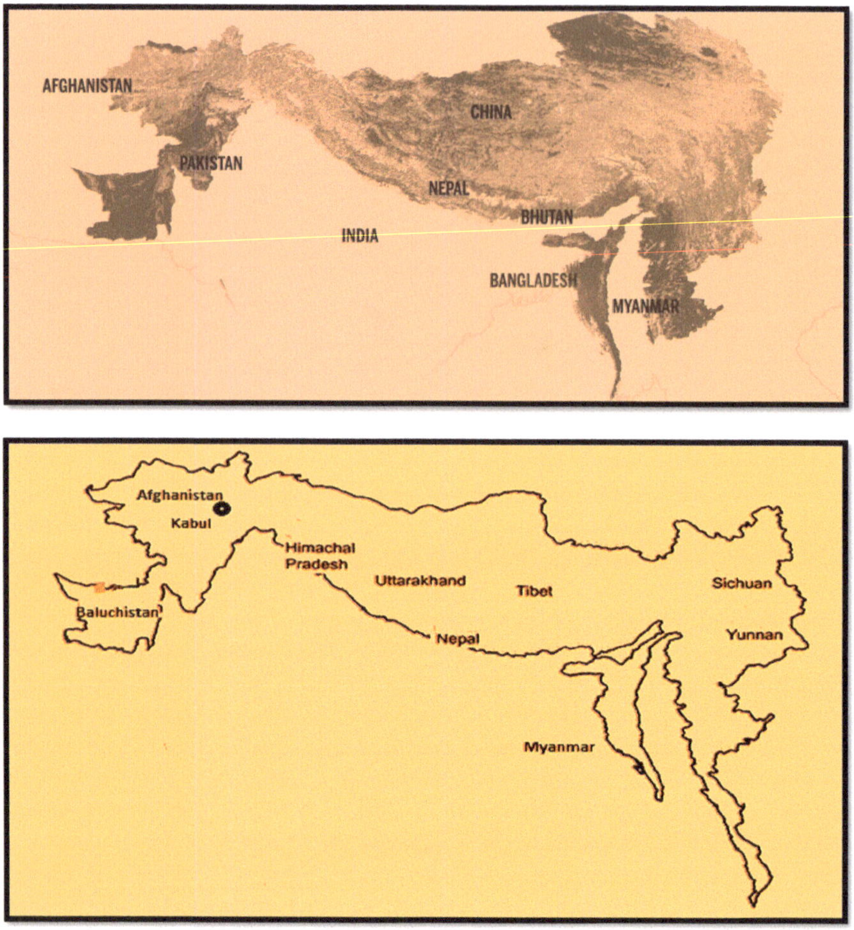

Fig. 1.1 The Hindu-Kush Himalayan region

and Kashmir in the west to Arunachal Pradesh in the east. Longitudinally, Himalayas are also classified as Shivaliks flat summits (600 to 1200 m msl altitude), Middle Himalayas (average altitude 3000 m amsl), Greater Himalayas (average altitude 5200 m amsl,) and Trans-Himalayas (average altitude 4500 m amsl). The Himalayas separate Indian subcontinent from Tibetan Plateau which covers the mountainous ranges of the countries like Pakistan, India, Bhutan, China, and Nepal.

The Indian Himalayan region is spread across 12 Indian states (Jammu and Kashmir, Uttarakhand, Himachal Pradesh, Arunachal Pradesh, Manipur, Meghalaya, Mizoram, Nagaland, Sikkim, Tripura, two districts of Assam (Dima Hasao and Karbi Anglong), and two districts of West Bengal (Darjeeling and Kalimpong). Approximately 50 million people reside in the Indian Himalayan region alone. It starts from the Shivalik foothills and the range extends up to Tibetan plateau in the north. The region occupies strategic position of entire northern boundary (northwest to northeast) of the country and touches almost all international borders (seven countries) with India. The region covers about 16% of India's total geographical area and most of the area in the region is covered by snow, glaciers in higher Himalayas, and forests cover mid-Himalayan region (Fig. 1.2).

Production of food crops, horticultural and cash crops, and livestock are three integral components of mountain farming. The farming in Himalayan region has more of traditional practices than of modern interventions but is incomplete without livestock farming. Therefore, livestock farming is the most important pursuit of the Himalayan people (Figs. 1.2–1.5). Almost every rural family maintains livestock of one kind and other in the region. Livestock along with dairy animals play an important role in hill agriculture and improve economy mainly through milk, meat, and wool production. Farmers involved in rearing dairy animals also keep bullocks to plough fields and sell them to supplement incomes. The whole system is referred as dairy-manure-draught cattle production system.

There are three systems of grazing in the Himalayan region, i.e., sedentary, semi-migratory, and migratory. The sedentary system is practiced by the farmers living in or around subalpine pastures. Under semi-migratory system, the livestock reared below the subalpine areas are taken to subalpine and alpine areas for grazing during summer and are brought back during winter. Under migratory systems of livestock rearing, the nomadic tribes practice a continuous transhumance activity along with their livestock. As soon as the snow melts in these pastures, large herds of sheep, goat, and cattle are taken there for grazing; when it starts getting colder, their movement starts from high hills to low hills.

In the mountains, natural pastures and grasslands are the main feeding resource to livestock in the Himalayan region. Fodder trees are also one of the important feed resources as they substitute stall-feeding to livestock by providing green fodder during lean periods. Stovers and straws are also important feeding resources in the Himalayan region.

There are increasing global concerns regarding the attainment of food and nutritional security not only for humans but also for livestock (Wheeler and Von Braun 2013). The current trends in global livestock production indicate rapid increase in livestock population and consumption of livestock products. This has resulted in a huge gap in demand and supply of feed to livestock and livestock products to human population. The gap is expected to increase further and the agriculture-based countries will be under tremendous pressure for meeting feeding requirements of human as well as livestock population. Since the economic and sociocultural values of livestock in any agrarian country cannot be overlooked, the comprehensive efforts are required (1) to increase livestock productivity, (2) to satisfy teeming demands for livestock products, and (3) to improve the socioeconomic status of the farmers.

Similar to the value of food crops in human nutrition, forage resources are vital for livestock for meeting their feed as well as nutritional requirements. The major forage resources include permanent pastures and grasslands, forage grasses and legumes, tree fodder and commercial feed concentrates. It is a concerning fact that

Fig. 1.2 (**a**, **b**) Cut and carry system of forages for livestock feeding; (**c**, **d**, and **e**) Grazing in Himalayan region

the area under natural feed resources is continuously decreasing worldwide. Moreover, the volatile prices of concentrated feeds and overdependency on low-quality feeds are responsible for their poor performance due to which global livestock productivity is declining. Therefore, this is the high time for the concerted efforts for the attainment of food as well as nutritional security to livestock.In Himalayan region, livestock farming in combination with crop farming has significant importance in making life of rural households more sustainable. On the whole, despite making significant contributions to rural livelihoods, the productivity of livestock in the Himalayan region is low. The region supports enormous resource base, however, in recent times the unsystematic exploitation of available feed resources and uncontrolled grazing have resulted in environmental imbalance and degradation of natural resources throughout the region including permanent pastures and rangelands which further leads to the acute shortage of fodder in the Himalayan region. The concern for forage production is aggravated in the region due to land scarcity where farmer prefers cultivation of food crops instead of forage crops. Moreover, lack of in-depth knowledge of management practices and about existing fodder resources, lack of information on fodder varieties having higher production and excellent nutritional quality, and the overreliance on low-quality feeds are other major constraints in achieving the sustainable livestock production in the region.

Apart from forage availability, forage quality is one another major factor affecting livestock production and productivity in the Himalayan region. In the past, the main priority was to produce high-yielding forage crops, but now with the understanding of importance of forage quality in livestock production, researchers have focused their attention on the development of nutritionally superior forage crop varieties. With the development of superior-quality forages, livestock productivity could be enhanced many folds. Farmers need to be acquainted with the knowledge of nutritional composition and presence of anti-nutritional factors in livestock feed. Based on availability, farmers could prioritize the feed or fodder with highest nutritive quality and less or no ill effects on animal health and their productivity.

The development of improved forage varieties suitable for individual priorities for different regions in the Himalayan region needs to be accelerated. Forage improvement either through conventional approaches or through modern plant breeding and biotechnological approaches is the possible avenue for the breeding of improved forage cultivars. The plant genetic resource (PGR) management, including germplasm conservation, exploration, documentation, and utilization of elite cultivars, has great impact on the development of improved forage cultivars (Fig. 1.3).

1.1 Understanding Forage Quality

Forage quality is based on the ability of livestock to digest and utilize nutritional constituents provided by the forages. Alternatively, forage quality is also defined as a function of total nutrient content in herbage, intake potential, and partitioning of nutrients in livestock body (Adesogan et al. 2009). Forage nutritive value and forage quality are two different terms which are used interchangeably to represent potential of forages for meeting nutritional requirements of livestock. Forage nutritive value is usually calculated by measuring the amount of total digestible nutrients and crude

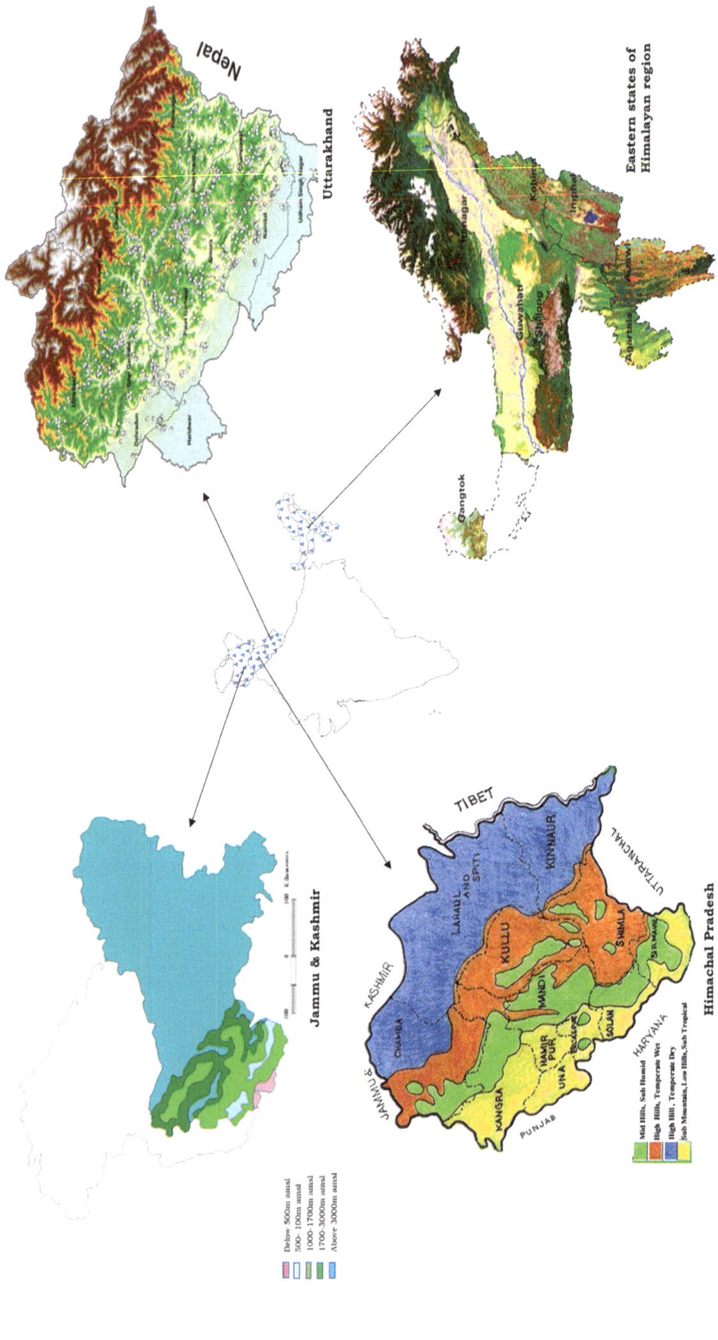

Fig. 1.3 The Indian Himalayan region

protein content in forages while forage quality is a broader term that includes nutritive value of forages as well as their intake. Forage nutritional value depends on the amount of proteins and digestible carbohydrates and additionally ash, lignin, cellulose, crude fiber, and some other plant chemical compounds are also considered as indicators of forage quality (France et al. 2000). Dry matter digestibility and forage intake are the main indexes for determining the quality of forages (El-Waziry 2007; Rhodes and Sharrow 1990). Acid detergent fiber (ADF) provides a good measure of forage digestibility because ADF contains cellulose and lignin. The dry matter digestibility decreases with the increase in lignin content (Van Soest 1994). Nitrogen content and ADF are two important factors in determining metabolizable energy requirements of livestock (Belyea et al. 1993). In practice, animal performance also reflects the quality of forages. Therefore, forage quality can be determined by measuring the production and quality of livestock products.

The quality of forages has many determinants *viz.*, palatability, forage intake, digestibility, and the presence of anti-nutritional factors. Palatability is the key factor influencing the quality of forages. Forage palatability is influenced by texture, leafiness, fertilization, dung or urine patches, moisture content, pest infestation, or compounds. The dietary intake by livestock depends upon the quantity of palatable forage consumed. Intake is higher when quality and palatability are higher. Intake of available energy primarily depends on the level of plant cell wall constituents. In general, energy intake and protein determine the animal performance and productivity. Available energy and protein content in forages are often low relative to animal requirements. Forage digestibility is defined as the extent to which nutrients from forages are absorbed and assimilated as they pass through the digestive tract. Immature leafy tissues are 60 to 70% digestible while mature, stemmy tissues are less than 50% digestible.

Livestock performance is the ultimate test of quality of forages, especially when forages are fed alone. It can be influenced by several factors associated with either the plants or the animals. Failure to give proper consideration to any of these factors results in low animal performance, which in turn decreases the livestock production and productivity. Different anti-nutritional compounds in forages can lower animal performance, by causing sickness or even death. These compounds include tannins, nitrates, alkaloids, cyanoglycosides, estrogens, and mycotoxins. The presence and/or severity of these elements in forages depend on environmental conditions, animal sensitivity as well as the presence of the poisonous plant species (including weeds) in growing area. High-quality forages should not contain high levels of anti-quality components.

In any pasture land, each forage species does not have the same nutritive value as different plant characteristics directly or indirectly influence the quality of forage. Forage maturity is the most important factor for determining the amount of available energy and protein content in forages which is available to livestock. In addition to this, environmental factors such as temperature and photoperiod also have significant impact on quality of forages.

The quality of forages has historically been determined by evaluating different plant characteristics such as color, leafiness, and texture. These factors are still important in assessing the forage quality but if used solely, they provide inaccurate measure of forage quality as they are more subjective and difficult to standardize. In order to feed the livestock having different feed requirements or to formulate supplement diets, nutritional analysis of forages is mandatory for having a better estimate of different nutrients and digestibility. Nutritional analysis combined with the physical evaluation is useful in providing first assessment of relative potential of a forage crop in having significant impact on animal performance. In addition to this, the knowledge of factors affecting the quality of forages also help in making appropriate selection of forages and supplements to fulfill livestock requirements.

1.2 Influence of Forage Quality on Livestock Productivity

Forages are essential for the successful operation of animal production system. This is more relevant to livestock which are heavily dependent upon forages as feeding resources for their health and productivity. Forages are an economical source of nutrients for animal production and also have significant role in maintaining sustainability of environment. Although the role of forages in animal production could vary depending upon regional preferences for livestock and forages, climate and resources, their importance in the success of livestock production is evident. With increasing human population and urbanization, the area under forage crops cultivation is declining, therefore, forage availability to livestock has become a major concern in making livestock production more sustainable especially in the developing countries.

India has the largest and diverse livestock population in the world, yet livestock productivity is comparatively lower than countries holding smaller number of livestock. The unavailability of feed providing essential nutrients for the growth and development of the livestock has put livestock production at the stake of available forage resources. The provision of nutritional feed has further aggravated due to continuously shrinking cultivated area under forage crops, increasing prices of feed concentrates, and mounting demands of human population. Presently, the country is facing acute shortages of green fodder, dry fodder, and commercial feed concentrates. It is speculated that the demand will reach 1170 million tonnes for green forage and 650 million tonnes for dry forage with 152 million tonnes for concentrate feeds in the next 20 to 25 years.

In Indian Himalayan region, the availability of green fodder in Western Himalayan region, Upper Gangetic Plains, and Eastern Plateau is more than 60% of actual requirement. In the hilly state like Himachal Pradesh, the total annual requirement of green and dry fodder is about 62.06 million tonnes and 198.38 million tonnes, respectively, whereas the total availability of green and dry fodder in the state is 31.45 million tonnes and 52.98 million tonnes, respectively. The above data suggest that availability of feed and fodder is essential for improving livestock productivity. This is critical particularly during the lean periods when availability of green fodder is very low.

Appropriate nutrition to livestock is also essential to sustain higher livestock production as well as productivity. A balanced nutrition has the potential of improving end-product quality that ultimately increases economic viability of livestock operation (Makkar 2016). Usually, the quality of forages is determined by measuring their intake and digestibility. Intake of available energy and digestibility are primarily the functions of amount of fiber content and cell-wall constituents in forages. US Dairy Forage Research Center (USDFRC) has estimated that a 10% increase in fiber digestibility would result in an annual $350 million increase in milk and beef production in USA (Reddy et al. 2005). Forages with high ash and ether extract are favorable to provide extra nutrients needed for livestock. Else than nutritional parameters, anti-nutrients are not preferred which otherwise inhibit various vital functions of livestock affecting their health and productivity.

Anti-quality compounds in forages reduce the economic viability of livestock production system (Allen and Segarra 2001). The presence of anti-quality compounds in forages reduces dry matter intake, low dry matter digestibility, and nutritional imbalances in animals. Depending on the body size, lactating cows require 11% CP and 60% TDN. Sometimes the forage species like tall fescue (*Festuca arundinacea*) could be associated with endophyte fungus *Neotyphodium coenophialum* and could produce toxic alkaloids (Stuedemann and Hoveland 1988). The losses in beef cattle industry in USA due tall fescue toxicosis have been estimated to be over $600 million annually (Fribourg et al. 1991).

Profitable livestock production always requires provision of feed of excellent nutritional quality and digestibility. The nutrient composition (net energy, protein, vitamins, minerals, etc.), chemical–physical characteristics of its fiber (fragility of the cell walls), acceptability (palatability) and their associative interactions with other dietary ingredients influence health and productivity of livestock. Nutritive value, especially crude protein and digestibility, is associated with the passage rate of forage through gastrointestinal tract. Forages of low digestibility and protein have slower passage rates, physically limiting their intake. Forage intake is also affected by animal body size, performance level, health, genotype, and social hierarchy. The rumen has an upper limit on its physical capacity. As the rate of fiber digestion decreases, the amount of slowly digestible organic matter (OM) in the rumen increases.

The protein content in forages varies with forage species, nitrogen fertility and maturity. Legumes forage species have high protein content than grasses and cool-season grasses have high CP content than warm-season grasses. Protein supplementation is only needed when the forage cannot provide required levels to a cow for maintaining a desired level of production. When nutritious forages are scarce during lean period, one cannot expect productivity of livestock to be high. During lean period, dry fodder and nonconventional feeds could provide energy and nutrition, but not to the extent comparable to green forages.

1.3 Diversity in Forage Resources of the Himalayan Region

The Himalayan region supports diverse habitats, species, communities, and ecosystems. Livestock in the Himalayan hills largely depend upon natural pastures and grasslands. In the mid-hills of Himalayan region, about half of the total feeds are mainly grass and tree fodder. Around two-thirds to three-fourths of the fodder requirement are met from the forest in mid-hills and 26 to 43% in lower hills (Singh 1999). Natural grasslands and pastures are prime assets of the Himalayan region, which contribute large percentage of forage biodiversity in the region. One of the surveys indicated that a sizeable number of grass species in grasslands of the Himalayan region belong to the Poaceae family (Singh et al. 2008). The grasslands cover large number of herbaceous species, where majority are considered for forage purposes. Total grassland area in the Himalayan region covers warm temperate grasslands, cool temperate grassy slopes, subalpine meadows and thaches, and alpine meadows and steppes.

Rangelands in Himalayan region are the one prime source on which pastoralists are dependent. Pastoralists travel from one place to another with their herd of sheep, goats, horses, and cattle. *Trifolium* species (red and white clover), *Festuca arundinacea* (tall fescue), *Setaria sphacelata* (golden timothy), *Cynadon dactylon* (Bermuda grass), *Dactylis glomerata* (orchard grass), *Paspalum dilatatum* (dallisgrass) and *Panicum maximum* (Guinea grass) are some of the dominating range forages in grasslands and pastures of the Himalayan region. Cultivated forage crops are least-prioritized forages in the region. The region is scarce for cultivable land where farmers prefer to cultivate food crops for human sustainability. However, forage crops are cultivated in some plain areas, wastelands, and bunds in and around the fields. Majority of the range forages are now cultivated in these regions. The major cultivated forages include alfalfa (*Medicago sativa*), berseem (*Trifolium alexandrinum*), stylosanthes (*Stylosanthes hamata*), sainfoin (*Onobrychis viciifolia*), fodder maize (*Zea mays*), fodder sorghum (*Sorghum bicolor*), pearl millet (*Pennisetum glaucum*), barley (*Hordeum vulgare*), teosinte (*Zea mexicana*), and oats (*Avena sativa*).

The Himalayan region is enriched with diversity of multipurpose trees and shrubs. The foliage from these resources provides an important feed resource to livestock in the region. Tree foliage is also considered as feed insurance to farmers during dry period. Around 84 major fodder trees and 40 shrubs in mid-hills of North-West Himalayan region are enlisted as important feed resources for constituting major proportion of livestock feed (Radotra et al. 2015). The major fodder tree species include *Robinia pseudoacacia*, *Quercus leucotrichophora*, *Albizia lebbeck*, *Albizia chinensis* and *Bambusa* spp. *Grewia optiva*, *Morus alba*, *Salix* species, *Bauhinia variegata* and *Leucaena leucocephala* among others.

The residues of cereal and leguminous crops are crucial feed resources during the period of green fodder unavailability in the Himalayan region. Straws and stovers of maize, sorghum, rice, wheat, pearl millet, barley, and oats, and hay of alfalfa, stylo, setaria, guinea grass, rhode grass, etc. are important feed resources during lean period of fodder availability. Feeds prepared after processing *viz.*, silages and haylages are used to some extent in the region. Apart from the conventional feeds, fruits and vegetable wastes, kitchen wastes, industrial wastes (pomaces, brewer's spent grains, molasses, and bagasse) are other supplementary feed resources of the region (Table 1.1).

1.4 Nutritional Diversity in Forage Resources

The Himalayan region has wide range of fodder resources. Natural niches of forages *viz.*, grasslands, pastures and rangelands fully flourish in the mountains during rainy and spring seasons. These contain good amount of dry matter and crude protein. The crude protein in alpine pastures of the state of Sikkim varies from 7.00 to 11.69%, while crude fiber content varies from 11.71 to 28.45%. Nutritional diversity in pasture herbages in lower, mid and upper hills of Himachal Pradesh has also been investigated. The mean crude protein content in herbage was estimated as 4.36% of DM. The NDF and ADF content ranged between 69.25 to 75.62% and 43.12 to 48.12% of DM in lower hills. Similarly, in mid-hills of Himachal Pradesh, the crude protein in pasture herbage was 8.27 to 10.13%, with 72.45% and 38.97% mean NDF and ADF content, respectively. In the upper hills of Himachal Pradesh, the mean crude protein content in forages was 10% with 62.65% NDF and 37.31% ADF content (Radotra et al. 2015).

In the parts of Himalayan region, the livestock are usually left out for grazing for 2–3 months of the year in natural grasslands and pastures. The grasslands and pastures are dominated by grasses, which tend to have very low nutritional quality in comparison to legumes and commercial feed concentrates. Generally, forage grasses contain less than 10% of crude protein content and more than 60% of NDF content. Forage grasses having NDF content less than 50% are considered of good nutritional quality and more than 60% are considered as low in nutritional value. The presence of lignin is another important factor that influences the quality and digestibility of forage grasses. Forage grasses tend to have more lignin and silica content than legumes which inhibit the use of other nutritional constituents in the animal rumen. Biochemically, lignin in grasses is more esterified to hemicellulose and is more recalcitrant due to higher proportion of syringyl subunits that indicate suppressed degradability of grasses than forage legumes (Singh et al. 2018).

Forage legumes provide better nutrition than cereal forages/grasses and lead to higher forage intake and improved animal performance. The symbiotic association between legumes and *rhizobia* provides the plant with an ample supply of N and it is one of the reasons why crude protein (CP) content of legumes is higher than cereals/grasses. Forage legumes contain around 15 to 20% CP content with less than 50% NDF and less than 40% ADF. Legumes also contain high digestible energy than grass/cereal forages. The cell wall of forage legumes contains low hemicellulose content and more pectin content in comparison to grasses, thus have high digestibility. Most of the digestible energy in legumes comes from cell solubles. The proportion of digestible energy obtained from NDF varies from 20 to 40% for legumes

Table 1.1 Common forage species in Himalayan region

Forage grasses	Family	Forage legumes	Family	Forage trees	Family
Andropogon pulmis	Poaceae	*Crotalaria medicaginea*	Fabaceae	*Albizia chinensis*	*Mimosaceae*
Arthraxon lanceolatus	Poaceae	*Desmodium laxiflorum*	Fabaceae	*Albizia lebeck*	*Mimosaceae*
Arthraxon prionodes	Poaceae	*Desmodium microphyllum*	Fabaceae	*Bambusa* spp.	*Poaceae*
Arundinella bengalensis	Poaceae	*Medicago sativa*	Fabaceae	*Bauhinia variegata*	*Fabaceae*
Capillipedium parviflorum	Poaceae	*Trifolium* spp.	Fabaceae	*Berberis ulcina*	*Berberidaceae*
Chrysopogon gryllus	Poaceae	**Cereal forages**	**Family**	*Capparis spinosa*	*Capparidaceae*
Chrysopogon montanus	Poaceae	Fodder maize (*Zea mays*)	Poaceae	*Galium aparine*	Rubiaceae
Cyperus monocephalus	Poaceae	Oat (*Avena sativa*)	Poaceae	*Grewia optiva*	*Tiliaceae*
Cyperus nutans	Poaceae	Sorghum (*Sorghum bicolor*)	Poaceae	*Leucaena leucocephala*	*Fabaceae*
Cyperus rotundus	Poaceae	Pearl millet (*Pennicetum glaucum*)	Poaceae	*Morus alba*	*Moraceae*
Cymbopogon martini	Poaceae	Barley (*Hordeum vulgare*)	Poaceae	*Morus serrata*	*Moraceae*
Cynodon dactylon	Poaceae	Teosinte (*Zea Mexicana*)	Poaceae	*Populus balsamifera*	*Salicaceae*
Dactylis glomerata	Poaceae	**Weeds**	**Family**	*Robinia pseudoacacia*	*Fabaceae*
Dichanthium aristatum	Poaceae	*Pouzolzia zeylanica*	Urticaceae	*Salix tetrasperma*	*Salicaceae*
Festuca arundinacea	Poaceae	*Rubia manjith*	Rubiaceae	*Salix alba*	*Salicaceae*
Heteropogon contortus	Poaceae	*Rubus ellipticus*	Rosaceae	*Toona sinensis*	*Meliaceae*
Lolium multiflorum	Poaceae	*Rumex dentatus*	Polygonaceae	*Robinia pseudoacacia*	*Fabaceae*
Paspalum notatum	Poaceae	*Rumex hastatus*	Polygonaceae	**Other forage species**	**Family**
Pennisetum orientale	Poaceae	*Rumex nepalensis*	Polygonaceae	*Artemisia nilagirica*	Asteraceae
Pennisetum clandestinum	Poaceae	*Rumex obtusilolius*	Polygonaceae	*Bidens biternata*	Asteraceae
Saccharum spontaneum	Poaceae	*Spermadictyon sauveolens*	Rubiaceae	*Erigeron bonariensis*	Asteraceae
Sorghum halepense	Poaceae	*Thalictrum foliolosum*	Ranunculaceae	*Barleria cristata*	Acanthaceae
Saccharum bengalensis	Poaceae	*Salvia hians*	Lamiaceae	*Bupleurum hamiltonii*	Apiaceae
Panicum antidotale	Poaceae	*Urtica dioica*	Urticaceae	*Boehmeria platyphylla*	Urticaceae
Themeda anathera	Poaceae	*Bidens pilosa*	Asteraceae	*Panicum miliaceum*	Poaceae

(60 to 80% from cell solubles) and from 50 to 80% for grasses (20 to 50% from cell solubles) (Buxton and Redfearn 1997).

Fodder trees provide forage and nutrition to livestock during lean period of the year. Tree foliage contains significant amount of essential nutrients, which are required to maintain the growth of intestinal microflora. The tree foliage has diversified nutritional composition and provides significant amount of crude protein (15 to 29%). Makkar and Becker (1997) recorded 8.00 to 25.90% crude protein content in fodder trees in the foothills of Himalayas. Crude fiber content in tree foliage ranges from 11.00 to 42.00% which forms great nutritional diversity, whereas NDF and ADF vary from 22.00 to 70.00% and 12.00 to 60.00%, respectively. Foliage of trees and shrubs is usually fed as a supplement to crop by-products or grass and it is regarded as high milk-producing feed resource with high palatability.

1.5 Regional Imbalances and Demand and Supply of Forages

In Indian Himalayan region, livestock rearing is an integral part of mixed farming. Sedentary, semi-migratory, and migratory systems of livestock rearing are prevalent in Indian Himalayan region. Although, livestock rearing has significant role in economy of the Himalayan region, forage cultivation has remained almost neglected as less than 1% of arable land is under fodder cultivation. This is because of higher number of farmers having marginal and small landholdings in the region. Besides grazing and tree foliage, the major local forage resource is crop residues, which again are too inadequate to sustain livestock production in the region. Due to less cultivated area under forage crops and overdependence on crop residues, the productivity of livestock in the Himalayan region is low.

India primarily is an agrarian country where livestock production is integrated with agricultural farming practices. Over the last couple of decades, the demand for major livestock commodities has increased rapidly, leading to unprecedented growth in the livestock sector. Today the livestock sector constitutes one of the major engines of agricultural growth in the country. Livestock sector has approximately 26% share in total agricultural GDP of the country which equals to about 4.11% of National GDP (Anonymous 2012). Livestock constitutes a significant source of income for nearly 70% of rural households, who derive between 14 and 40% of their income.

On the whole, despite making significant contribution in sustaining the rural livelihoods and national economy, the productivity of livestock is on lower side as compared to the other countries. The wide gap in the demand and supply of feeds in terms of either quality or quantity is one of the critical reasons for low livestock productivity in the country. Presently, there is a net deficit of 35.60% green fodder, 10.95% dry crop residues and 44% concentrate feeds in the country (IGFRI Vision 2050 2015). The percentage proportion of fodder scarcity will increase further continuously and will reach 1012 million tonnes of green fodder and 631 million tonnes of dry forage by the year 2050 (Table 1.2).

Table 1.2 Demand and supply deficit of fodder in India (million tonnes)

Year	Demand		Supply		Net deficit		Deficit (%)	
	Dry fodder	Green fodder	Dry fodder	Green fodder	Dry fodder	Green fodder	Dry fodder	Green fodder
2020	530.50	851.30	467.60	590.40	62.85	260.90	11.85	30.65
2030	568.10	911.60	500.00	687.40	68.07	224.20	11.98	24.59
2040	594.90	954.80	524.40	761.70	70.57	193.00	11.86	20.22
2050	631.00	1012.70	547.70	826.00	83.27	186.60	13.20	18.43

Source: Ghosh et al. (2016)

The pattern of forage deficit varies in different parts of the country. For instance, the green fodder availability in Western Himalayas, Upper Gangetic Plains and Eastern Plateau and Hilly Zones is more than 60% of the actual requirement. In Trans-Gangetic Plains, feed availability is between 40 and 60% of the requirement and in the remaining zones the figure is below 40%. In case of dry fodder, the availability is over 60% in the Eastern Himalayas, Middle Gangetic Plains, Upper Gangetic Plains, East Coast Plains, and Hilly Zones. In Trans-Gangetic Plains, Eastern Plateau and Hills, and Central Plateau and Hills, the availability is in the range of 40–60%, while in the remaining zones of the country the availability is around 40% (Bhagmal et al. 2009). The only way to meet the fodder needs is to look for increased productivity per unit land area and also through integration of fodder crops in the cropping system (Fig. 1.4).

In the states of Himachal Pradesh, Jammu and Kashmir, and Uttarakhand, there is an overall deficit of 24.90% in green fodder availability. The green fodder deficit in Jammu and Kashmir and Uttarakhand is 53% and 55.50%, respectively. Small ruminants form a major part of livestock in Himachal Pradesh and they are mostly dependent on grazing lands. Nomadism and semi-nomadism are also in practice in major part. The demand is also fulfilled by top feeds and migration to nearby states (Roy et al. 2019a, b).

In North East Zone, major source of green fodder availability is cultivated land (17,803.90 thousand tonnes) followed by forests (3510.00 thousand tonnes) and pasture land (1035.00 thousand tonnes) (Roy et al. 2019a, b). There exists an overall deficit of 23.10% of green fodder in North-East Himalayan region. However, some of the states such as Mizoram and Arunachal Pradesh are having surplus green fodder. States which are most deficit in green fodder availability are Tripura (58%) followed by Sikkim (48.50%) and Meghalaya (44.60%). The situation in North-East Himalayan states is quite different than the rest of the country as the major livestock commodity in demand is meat with milk as secondary product.

There is vast availability of wasteland and forestland which are not utilized for grazing due to difficult terrain (Roy et al. 2019a, b). In North East Zone, forests are the major sources of dry fodder followed by cereal crop residues and kitchen/horticultural/top feed/farm waste. There exists an overall surplus of 14.00% of dry fodder in NEH region and except Assam all other states in the region are having surplus dry fodder. In Himachal Pradesh, Jammu and Kashmir, and Uttarakhand

(a)

(b)

Fig. 1.4 (**a**, **b**): Wheat straw supply from Punjab to Himachal Pradesh, India

states, 55.90% of dry fodder is available in surplus amount (Roy et al. 2019a, b). Though the Indian Himalayan states have surplus dry fodder, it is not fully utilized due to hilly and difficult terrains, thereby creating a deficit in the availability of dry fodder in these states. Cosiderable quantity of straw is imported from the neighboring states every year (Roy et al. 2019a, b).

1.6 Insight of Livestock Production and Productivity

1.6.1 Indian Perspective

Livestock is a source of subsidiary income for many families in Himalayan region especially the resource poor who maintain few heads of animals. The livestock provide food items such as milk, meat and eggs for human consumption. In Indian context, according to 20th livestock census, total livestock population in the country is 536.76 million. The livestock population in rural and urban areas are 514.11 million (95.78%) and 22.65 million (4.22%), respectively (Anonymous 2019a).

Compared with previous census, the total livestock population in rural and urban area have increased by 4.56% and 11.19%, respectively. The bovine population in India is 303.76 million of which 193.46 million are cattle. The number of total sheep and goats are 74.26 million and 148.88 million, respectively. India has population of 9.06 million pigs, 851.81 million ducks and chicken, 0.25 million camels, 85 thousand yaks, 0.39 million mithun, and 0.34 million horses and ponies. In comparison to 2012 livestock census, there is 4.80% increase in total livestock population in the country.

Taking individual account of percent increase in the population of livestock, there is 1.30% increase in total bovine population, 1.10% increase in total buffalo population, 14.10% increase in total sheep population, 10.10% increase in goat population and 12.03% decrease in total pig population, and 16.80% increase in poultry population in the country (Fig. 1.5 and Table 1.3).

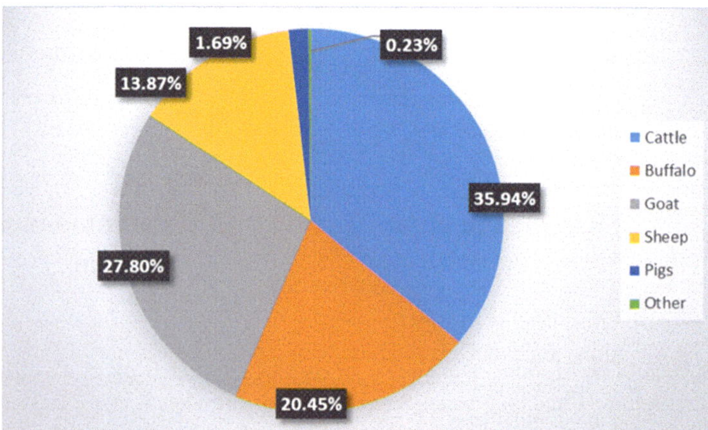

Fig. 1.5 Percentage share of livestock population in India. (Source: Anonymous 2019a)

Table 1.3 Status of livestock population in India

Livestock species	Population (millions)
Cattle	193.46
Buffaloes	109.85
Sheep	74.26
Goats	148.88
Pigs	9.06
Mithun	0.39
Yaks	0.06
Horses and Ponies	0.34
Mules	0.08
Donkeys	0.12
Camels	0.25

Source: 20th livestock census (Anonymous 2019a)

Table 1.4 India's ranking in livestock sector (population and production)

Rank	Sector
1st	Total livestock population, total milk production, total buffalo population, total carabeef production, total goat milk production, total bovine population
2nd	Cattle population, goat population, bristle production (a pig industry by-product), aquaculture, goat meat production
3rd	Egg production, sheep population, fisheries production
5th	Poultry production, meat production
6th	Poultry meat production
8th	Duck production
9th	Camel population, wool production

According to the Livestock Production Statistics of India-2019, the total milk production is around 187.75 million tonnes per year. The milk production has shown 6.50% increase during 2019. Uttar Pradesh (16.30%), Rajasthan (12.60%), Madhya Pradesh (8.30%), Andhra Pradesh (8.0%), and Gujarat (7.70%) are the major milk-producing states in the country (Anonymous 2019b). Together they contribute nearly 60% of the total livestock production in the country. The total egg production 103.32 billion numbers during 2019, which has shown 8.50% increase in comparison to the previous year. Andhra Pradesh (19.10%), Tamil Nadu (18.20%), Telangana (13.20%), West Bengal (8.30%), and Haryana (5.90%) are the leading states in egg production in the country. Together they contribute nearly 65% of the total egg production in the country. The total meat production in India is 8.11 million tonnes, of which the poultry has about 50% share (4.06 million tonnes). Uttar Pradesh (15.10%), Maharashtra (12.60%), West Bengal (10.20%), Andhra Pradesh (9.6%), and Telangana (9.30%) are the major meat-producing states in the country. The total wool production in the country is 40.42 million kgs. In comparison to past years, the total wool production has been declined by 2.50%. Rajasthan (35.90%), Jammu and Kashmir (18.90%), Telangana (10.20%), Karnataka (7.60%), and Gujarat (5.60%) are the leading states in total wool production in the country. Table 1.4 presents the status of Indian livestock sector in the world. On an average, the exotic cows, crossbred cows, indigenous cows, nondescript cows, indigenous buffaloes, and nondescript buffaloes produce 11.67, 7.85, 3.85, 2.50, 6.34, and 4.35 kg milk/day, respectively.

1.6.2 Himalayan Perspective

Livestock production is considered the engine of mountain economy as a source of food security in the form of milk and meat, providing supplementary income for rural households and reducing the vulnerability of livelihoods. The major sources of herbage for livestock mainly in hilly regions are pastures and grasslands. The pastoralists of the Himalayan region move from one place to another and livestock are fed with herbages of pastures and grasslands and sometimes tree foliage. The livestock sector contributes about 25% of agricultural gross domestic product and provides nutritionally rich food to many people in both rural and urban areas. The subsequent paragraphs present the status of livestock population in Indian Himalayan region:

- Cattle are most common followed by goats, buffaloes, and sheep in mixed crop farming systems in the Central and Eastern Himalayas. The state of Uttarakhand has 47.94 millions of livestock whereas the total poultry population is 46.41 million. Total livestock population includes 4.97 million exotic/crossbred cattle, 20.06 million indigenous cows, 9.87 million buffaloes, 1.27 million exotic/crossbred sheep, 13.67 million goats, and 27.10 thousand pigs. Milk production has major share in livestock production of Uttarakhand. In the year 2015, Uttarakhand state produced 331 hundred tonnes of milk from indigenous cows and 826 hundred tonnes from buffaloes.
- Total livestock and poultry population in the state of Himachal Pradesh was 4.41 million and 13.41 million, respectively. In the state, the cattle population was 1.826 million, buffaloes 0.646 million, goats 1.108 million, and sheep 0.791 million (Anonymous 2019b). Further 2.48 million pigs, 8.85 million horses/ponies, 0.020 million mules, 0.0480 million donkeys and 1.94 thousand yaks and camels have been recorded. Total milk produced during 2019–2020 was 1.460 million tonnes of which maximum production has been recorded from crossbred cows (0.877 million tonnes) and least from goats (0.05191 million tonnes). Wool and meat production were 1503.15 thousand kgs and 4.60 tonnes, respectively (Department of Economics and Statistics, HP).
- Total livestock and poultry population in the state of Jammu and Kashmir was 83.25 and 73.66 million, respectively. The bovine population was 35.91 million and yak population 5.47 million which was highest in the country. The state has second place with respect to population of horses/ponies and mules in the country. The percent share of total of horses and ponies in Jammu and Kashmir was 23.13 and 18.59 against the total population of 6.25 and 1.96 lakh, in the country, respectively (Anonymous 2019a).
- Northeast states, Assam, Manipur, Meghalaya, Mizoram, Nagaland, Tripura, and Sikkim have 18.92, 0.55, 2.03, 0.35, 0.55, 0.27 and 1.31 million of total livestock population and 4.67, 5.89, 5.37, 2.04, 2.83, 0.58 and 4.16 millions of poultry birds, respectively. The status of livestock productivity in the Indian Himalayan states has been presented in Table 1.5.

In the Himalayan region, the majority of farmers operate mixed farming system in which rearing of livestock is integrated with food crop production. Different crops provide feed and fodder, livestock provide meat, milk and milk products for subsistence and as a source of cash income. Natural pastures and grasslands are the major feed resources for the Himalayan livestock. Fodder trees, cultivated forages and crop residues are other important feed resources. However, natural pastures and grasslands are overexploited and overgrazed in the Himalayan region, and therefore the total area under these feed resources is decreasing at a faster pace. Furthermore,

Table 1.5 Livestock productivity in the Indian Himalayan states during 2018–2019

Himalayan state	Milk production (Million tonnes)	Egg production (Lakh number)	Wool production (Thousand kg)	Meat production (Thousand tonnes)
Jammu and Kashmir	2.54	2348.84	7629.28	91.61
Himachal Pradesh	1.46	1007.00	1475.00	4.60
Uttarakhand	1.79	4531.83	564.07	29.18
Assam	0.88	5014.61	–	50.40
Manipur	0.085	1053.24	–	28.05
Meghalaya	0.086	1090.36	–	45.25
Mizoram	0.025	415.19	–	16.11
Nagaland	0.072	374.74	–	32.28
Tripura	0.18	2759.56	–	47.82
Sikkim	0.060	54.56	–	3.72

Source: Anonymous (2019b)

the landholdings of Himalayan people are small and fragmented, consisting mostly of marginal uplands which leave little space for cultivation of forage crops. These circumstances have resulted in acute shortage of fodder to livestock in Himalayan region. The inadequacy of fodder, rising prices of feed concentrates, improper management practices, overutilization of low-quality feeds in livestock feeding are other major constraints in achieving sustainable livestock production in the Himalayan region. In Himalayan region, multifaceted efforts are required for improving the availability of quality fodder for improving production and productivity of livestock, ultimately affecting the socio-economic status of the farmers of the region.

References

Adesogan AT, Newman Y, Vendramini JM, Sollenberger L (2009) Defining forage quality. US Department of Agriculture, UF/IFAS Extension Service, University of Florida, Gainesville, FL, pp 1–5

Allen IG, Segarra E (2001) Anti-quality components in forage: overview, significance, and economic impact. J Range Manag 54:409–412

Anonymous (2012) Department of Animal Husbandry and Dairying. 19th Livestock Census

Anonymous (2019a) Department of Animal Husbandry and Dairying. 20th Livestock Census

Anonymous (2019b) Basic livestock production statistics. Department of Animal Husbandry and Dairying. Govt. of India, New Delhi. www.dahd.nic.in/

Belyea RL, Steevens B, Garner G, Whittier JC, Sewell H (1993) Using NDF and ADF to balance diets. Agricultural Publication, G., p 3161

Bhagmal, Singh KA, Roy A, Ahmed S, Malaviya D (2009) Forage crops and grasses. In: Handbook of agriculture. ICAR, New Delhi, pp 1353–1417

Buxton DR, Redfearn DD (1997) Plant limitations to fiber digestion and utilization. J Nutr 127: 814–818

El-Waziry M (2007) Nutritive value assessment of ensiling or mixing Acacia and Atriplex using *in vitro* gas production technique. Res J Agric Biol Sci 3:605–614

France J, Theodorou M, Lowman R, Beever D (2000) Feed evaluation for animal production. Feeding systems and feed evaluation models. CABI, Wallingford, pp 12–20

Fribourg HA, Gwinn KD, Chestnut AB, Oliver JW, Waller JC, Reynolds JH, Conger BV (1991) Tall fescue and the fungal endophyte. Tennessee Farm Home Sci 160:10–29

Ghosh PK, Palsaniya DR, Srinivasan R (2016) Forage research in India: issues and strategies. Agric Res J 53:1–12

IGFRI (2015) IGFRI Vision 2050. Indian Grassland and Fodder Research Institute, Jhansi

Makkar HPS (2016) Animal nutrition in a 360-degree view and a framework for future R&D work: towards sustainable livestock production. Anim Prod Sci 56(10):1561–1568

Makkar HPS, Becker K (1997) Potential of *J. curcas* seed meal as a protein supplement to livestock feed; constraints to its utilization and possible strategies to overcome constraints. In: Giibitz GM, Mittelbach M, Trabi M (eds) Biofuels and industrial from Jatropha curcas. DBV Graz, pp 190–205

Radotra S, Dev I, Ahmad S, Kannan A (2015) Pasture and forages in North Western Himalayan region: current status and future strategies. In: Conference: status and future prospects of animal production system in Himalayan Region, pp 1–11

Reddy MSS, Chen F, Shadle G, Jackson L, Aljoe H, Dixon RA (2005) Targeted down-regulation of cytochrome P_{450} enzymes for forage quality improvement in alfalfa (*Medicago sativa* L.). Proc Natl Acad Sci U S A 102:16573–16578

Rhodes BD, Sharrow SH (1990) Effect of grazing by sheep on the quantity and quality of forage available to big game in Oregon's Coast Range. J Range Manag 43(3):235–237

Roy AK, Agarwal R, Bhardwaj N, Mishra A, Mahanta SK (2019a) Revisiting national forage demand and availability scenario. In: Indian fodder scenario: redefining state wise status. AICRP on Forage Crops and Utilization, ICAR-IGFRI, Jhansi, pp 1–21

Roy AK, Malaivya DR, Kaushal P (2019b) Breeding strategies to improve fodder legumes with special emphasis on clover and medics. In: Roy MM, Malaviya DR, Yadav VK, Singh T, Sah RP, Vijay D, Radhakrishna A (eds) Proceedings: The 23rd International Grassland Congress (Sustainable use of grassland resources for forage production, biodiversity and environmental protection) New Delhi, India November 20–24, 2015. Range Management Society of India, Jhansi, pp 327–332

Sharma ED, Molden A, Yuba R, Khatiwada R, Zhang L, Singh SP, Yao Y, Wester P (2019) Introduction to the Hindu Kush Himalaya Assessment. In: Wester P, Mishra A, Mukherji A, Shrestha A (eds) The Hindu Kush Himalaya Assessment. Springer, Cham, pp 1–16. https://doi.org/10.1007/978-3-319-92288-1_1

Singh R (1999) Smallholder dairy farming initiatives: Success and failure of milk cooperatives in the HKH. In: Paper presented at the International Symposium on Livestock in Mountain/Highland Production Systems: Research and Development Challenges into the Next Millennium, 7–10 December 1999, Pokhara, Nepal

Singh V, Gaur RD, Bohra BA (2008) Survey of fodder plants in mid-altitude Himalayan rangelands of Uttarakhand, India. J Mt Sci 5:265–278

Singh T, Ramakrishnan S, Mahanta SK, Tyagi VC, Roy AK (2018) In: Edvan RL, Santos EM (eds) Tropical forage legumes in India: status and scope for sustaining livestock production. IntechOpen, London. https://doi.org/10.5772/intechopen.81186

Stuedemann JA, Hoveland CS (1988) Fescue endophyte: history and impact on animal agriculture. J Prod Agric 1:39–44

Van Soest PJ (1994) Nutritional ecology of the ruminant. Cornell University, Ithaca, p 476

Wheeler T, Von Braun J (2013) Climate change impacts on global food security. Science 341:508–513

Impact of Forage Availability on Livestock Rearing and Socioeconomic Aspects

2

Livestock farming makes multifaceted contribution to socio-economic development of rural households. As an important source of income and employment, livestock farming helps in improving socioeconomic status of rural households (Birthal et al. 2002). Livestock farming sector also has scope for income-generating opportunities, especially for marginal and small farmers and landless laborers having around 70% of total livestock of the country. Livestock is also important as savings and investments for poor households and also ensures economic security through multiple ways. In mixed farming system, its importance goes beyond direct food production. Livestock production directly depends on availability of fodder both in terms of quality and quantity, thus feed availability is considered as one of the important determinants of livestock productivity.

2.1 Livestock Rearing: An Indispensable Component of Hill Agriculture

The socio-economic status of farmers is predicted from different agricultural practices including livestock rearing. The rearing of livestock is considered as the prime source of income for farmers in the Himalayan region (Fig. 2.1). The Himalayan region is home to number of breeds of cattle, small ruminants, fowl, pigs, equine species, and others. India, being the second most populated country in the world, has upper hand in livestock production too. The major contribution to total livestock population is by bovine animals (cows and buffaloes) which are mainly used for milching, while bullock used for ploughing, horse and mule for transportation in hilly region, poultry for egg production, sheep and goat for wool in hilly region and meat production, are other major contributors in total livestock production of the country.

In India, according to Livestock census 2019 (Anonymous 2019), total number of milch animals was around 125.34 million having an increase of 6% over the

R. Katoch, *Nutritional Quality Management of Forages in the Himalayan Region*, https://doi.org/10.1007/978-981-16-5437-4_2

Fig. 2.1 (a–e) Livestock rearing practices in Himalayan region

previous census. The country has 109.85 million buffaloes compared to 108.70 million in the 2012 census. The goat population was 148.88 million, which showed an increase of 10.10% over the previous census. Total number of sheep in the country were 74.26 million, showing an increase of 14.10% over the previous enumeration.

In the Himalayan region, livestock farming is an indispensable component of the life of rural households. It is the source of income for people of the Himalayan region which includes dairy products from milk and dung as manure to crops. Livestock ensures socio-economic security to people of the Himalayan region during emergency situations. In central and eastern Himalayan region, cattle are prevalent, followed by goats, buffaloes, and sheep (Rao and Saxena 1994). Under mixed farming system, landholdings are small and livestock production is a critical supplement to total farm income. Livestock population in Assam (10.77%), Sikkim (7.96%) and Meghalaya (7.41%) has increased gradually over the years (Anonymous 2012). Livestock population in the state of Himachal Pradesh increased from 4.98 million in the year 1982 to 5.22 million till year 2007. According to 20th livestock census, total livestock population in Himachal Pradesh is decreased by 8.95% which equals to 0.441 million. The buffalo and cattle population was 0.64 and 1.82 million, respectively. Population of sheep and goat were 0.79 and 1.18 million, respectively. Number of yaks were 1.94 thousand in the state while horses/ponies and mules were 0.08 lakh and 0.20 lakh, respectively (Table 2.1). Uttarakhand, another state in Himalayan region, had 2 million cattle including 0.497 million total exotic/cross-bred cows, 0.151 million indigenous cows and 0.987 million buffaloes whereas sheep, goats and pigs contributed 0.57%, 1.01%, and 0.19%, respectively, to total livestock population in the state (Anonymous 2012). The state of Jammu and Kashmir had 4.30 million cattle, 0.738 million

Table 2.1 Livestock population in the state of Himachal Pradesh

Livestock	Population
Cattle	18.26 lakh
Buffaloes	6.46 lakh
Sheep	7.91 lakh
Goats	11.08 lakh
Pigs	0.02 lakh
Horses/ponies	0.08 lakh
Mules	0.20 lakh
Donkeys	0.48 lakh
Camels	30
Yaks	1.94 thousand
Mithun	0
Total livestock (2019)	**44.11 lakh**
Total livestock (2012)	**48.44 lakh**
% decrease	**8.95%**

Source: Anonymous (2019)

buffaloes, 0.328 million sheep, 0.198 million goats, 54.49 thousand yaks, 2431 pigs, and 0.140 million horses and ponies. The state of Jammu and Kashmir also has the highest number of yaks in India.

In the Himalayan region, every family rears either a cow or a buffalo, irrespective of their economic viability. Although, the number of livestock varies from household to household, still they are the main sources of milk and other by-products.

2.2 Importance of Livestock Production in Rural Economy of Himalayan Region

In Himalayas, a large proportion of livestock is raised under mixed farming system. This farming system has key role in maintaining socio economic status of the farmers. The livestock are used not only for milching purpose but also used as a security asset during scarcity of life resources. Cattle have their own contribution in socio-cultural status in hill farming system and in context of religious beliefs. With large number of people moving away from agriculture to other sectors of livelihood, majority of Himalayan farmers opt livestock rearing with crop farming as an option for their livelihood. Livestock rearing in Himalayan region contribute to

(a) **Income:** Livestock is a source of subsidiary income for rural households through direct and indirect ways. They provide income through the sale of animals and/or through selling milk, meat, eggs, and other animal products. They are also considered as a form of savings and insurance during emergency situations and also provide manure, draft power, and transport services. Being a source of wealth, livestock not only contribute to social status but may also facilitate access to financial services, both in formal and informal markets.

Livestock are generally kept close to home and thereby require few labor inputs. A small unit of livestock could be reared by women while managing other household activities and sale of their products provide them financial assistance and plays a role in women empowerment.

(b) **Employment**: A major proportion of total population in Himalayan region depends on agriculture for their livelihoods. But agriculture being seasonal in nature could provide employment for some period in a year. The landless and less land people depend upon livestock for utilizing their labor during lean agricultural season.

(c) **Food**: The livestock products such as milk, meat, and eggs are an important source of animal food to the members of the livestock owners. Milk further extends its value as dozens of milk products, namely curd, clarified butter (ghee), butter, cheese, buttermilk, lassi, etc. Other than milk products, meat from sheep and goats is the prime food source in most of the Himalayan region.

(d) **Social and financial security**: Livestock rearing offers security to their owners in terms of food and economic source. The families with meager agricultural land but with sufficient number of livestock have better life-thriving resources. Gifting animals during marriages and other ceremonies is still a common practice in the Himalayan region. Animals are also used in different socioreligious functions.

(e) **Draft**: The bullocks are essential for different agricultural practices in the Himalayan region. Farmers, especially small and marginal of the region, depend on bullocks for ploughing, carting, and transporting both inputs and outputs. While some use horses and mules for transporting materials in upper tracts of hills.

2.2.1 Contribution of Livestock to National and Regional Economy

Livestock has significant contribution in main food supply, nutritional security to humans generating income and employment, providing insurance to farmers during emergency situation, soil productivity, agricultural diversification, and sustainable agricultural production (Saxena et al. 2020). Livestock farming has been recognized as one of the key sectors in achieving the ambitious goal of doubling the farmer's income by the year 2022 (Das and Kumar 2017). The demand for livestock products will be high in near future due to human population growth, increasing urbanization, and rising income. Therefore, this sector has enormous opportunities to increase farmer income and accelerate the pace of poverty reduction (Saxena et al. 2020). Economists have emphasized on growing importance of livestock in the gross domestic product (GDP) (Birthal and Taneja 2006, 2012). The gross value added (GVA) from livestock sector contributed 4.10% to the total GVA and 27.50% to the agricultural GVA in 2017–2018 (Saxena et al. 2020). The vast and diverse population of livestock (11.54% of world population) and poultry animal (3.10% of total world population) resources play crucial role in improving the socioeconomic status of rural households in India. The vast animal population with adaptability to wide range of agro-climatic conditions in the country is a vital asset and provides scope for diversified agriculture.

Dairy farming is the mainstream of livestock rearing in India. Milk production in India remained stagnant during the period between 1950 and 1970; since then milk production is increasing continuously (Birthal and Taneja 2006). India is the largest producer of milk in the world with 187.75 million tonnes and per capita availability of 394 g/day (Singh 2020). Indigenous cows provide 3.73 milk kg/day whereas indigenous buffaloes provide 6.19 kg milk/day. Buffaloes and cows are important milch species contributing 49% and 47%, respectively, in total milk production in the country. Goat milk contributes only 3% to the total milk production in the country (Singh 2020). Milk and milk products contribute only 1.86% in total export of livestock products from India.

Next to milk production, meat production has major share in value terms. Bovine meat has been a major export commodity with around 70% share in the total livestock exports from India (Saxena et al. 2020). The total meat production in India is 8.11 million tonnes. The meat from poultry (~50%), buffaloes (~19%), and goats (~13.35%) have maximum share in total meat production. Livestock products account for about one-fifth of the global trade of agricultural products. Despite holding large number of livestock in the world, India does not have a significant contribution in global livestock trade. During the year 2017–2018, India's total import of livestock and livestock products was worth 101.24 billion while the worth of India's total export of livestock and livestock products was 457.76 billion (Singh 2020). During the year 2018–2019, India exported approximately 1.99 million tonnes of livestock products worth Rs. 30,632 crores (Singh 2020). The export of meat had large share in total export of livestock products from India. Buffalo meat contributed around 19.30% to the total export of animal products from India. Around 1.23 million tonnes of buffalo meat, worth Rs. 25,168 crores, exported from India in the year 2018–2019. The goat sector contributes Rs. 14,453 crores to the agricultural economy of the country through meat (Rs. 6851 crores), milk (Rs. 4588 crores), skin (Rs. 648 crores), etc., which accounts for around 8% of GDP from livestock this sector (Singh 2020). Sheep and goats are major wool-producing livestock. India ranks sixth among clean wool producer countries and ninth among greasy wool producers. The wool and woollen textiles industry in the country is small in size and widely scattered. India has third largest sheep population in the world producing around 45 million kg of raw wool, and accounts for 3.10% of total world wool production. The wool production is not sufficient to meet the total requirement of raw wool for textiles industry in the country. The bulk of Indian wool is of coarse quality and is used mostly in handmade carpet industry. About 80% of good-quality indigenous wool is used for manufacturing carpets, which alone is a major contributor to exports earning.

The Himalayan region has the major contribution to wool industry in the country, while milk production also contributes exceptionally in the total productivity. Milk production in Himachal Pradesh was 877.97, 131.04, 399.41, and 51.91 thousand tonnes from cross-bred cow, indigenous cow, buffalo, and goat, respectively. During

Table 2.2 Livestock productivity in Himachal Pradesh (2019–2020)

Livestock products	Quantity production
Crossbred cow milk production	877.97 (thousand tonnes)
Indigenous cow milk production	131.04 (thousand tonnes)
Buffalo milk production	399.41 (thousand tonnes)
Goat milk production	51.91 (thousand tonnes)
Total milk production	**1460.15 (thousand tonnes)**
Wool production	1503.15 (thousand kg)
Meat production	4.60 tonnes

Fig. 2.2 (a–e): Pastoralism in Himalayan region

the year 2019–2020, meat production was 4.60 tonnes whereas wool production was 1503.15 thousand kgs (Table 2.2).

Trends in average daily milk yield have been examined species-wise in Himachal Pradesh. The average cow milk yield in the state increased from 1.38 kg to 2.54 kg during the past two decades at an annual growth rate of 2.72%. For buffaloes, the growth rate has been observed negative.

Dairy farming is the main activity and source of livelihood of livestock-dependent farming communities in the mountains. It involves with natural resources base—forest/rangelands, croplands, livestock breeds, feeding, health management, marketing and consumption of the product. Milk, curd, ghee, cheese, butter, and others are the mainstream products from dairy industries.

Pastoralists completely depend on livestock for their family income in the Himalayan region (Fig. 2.2). Bakarwals are transhumance pastoralists mainly depending on goats, sheep, and horses. Gujjars are semi-sedentary and sedentary population depending mainly on cattle, buffalo, sheep, and horses for their livelihood. Chopans and Gaddies mainly depend on sheep for their economy. Changpas are nomadic and rear pashmina goats, sheep, and yaks. Wool production in the country mainly relies on Himalayan region for quality fiber. Pashmina and Angora wool in Jammu and Kashmir is a valuable source of fiber to the textile industry of Himalayan region. These goats and sheep are generally used for wool than meat. The share of poultry in meat production is the highest.

Livestock asset is more equitably distributed than land. The regression analysis of relationship between poverty and livestock income has shown that growth in livestock sector contributes more toward poverty reduction. These results indicate that livestock could be used as an effective tool for reducing rural poverty. The declining trends in livestock holdings among landless households which constitute 32% of rural households, is an emerging issue that needs to be tackled urgently by providing adequate common property resources to ensure them sufficient livelihood opportunity.

2.2.2 Socio-economic Status of Farmers Depending on Livestock-Based Agriculture

Livestock rearing is the part of agriculture sector where farmers contribute to the economic growth of the country. The ownership of the livestock is more evenly distributed with landless laborers and marginal farmers owning the bulk of livestock. In the hill state of Himachal Pradesh, average family size of farmers having marginal farms is large. Literacy rate is higher in males for marginal and small farms in Himachal Pradesh. Females in marginal farms are slightly higher in literacy than small farm holding. The gender-wise comparison across the farm categories indicates low educational status of females as compared to their male counterparts (Pathania and Sharma 2016). The household income from different sources is approximately 0.20 million in the overall farm situation, which is found to increase with the size of holding. The total value of livestock inventory is approximately Rs. 0.36 lakh, Rs. 0.41 lakh, and Rs. 0.37 lakh on marginal, small, and overall farms, respectively (Pathania and Sharma 2016).

Education is a basic key factor for the development of an individual, society, and the nation as a whole. In a survey conducted in Champawat district of Uttarakhand, the educational status of nonmembers and members of dairy cooperatives in a survey shows that the level of education of nonmembers (84.66%) was higher than the education level of members (79.13%). The percentage of family members having education as graduation and above was 7.14% for members and 7.80% for nonmembers. The percentage of farmers involved in dairy as main occupation was 13.30% and 16.70% for members and nonmembers, respectively, and in case of dairy as subsidiary occupation the respective percentage for members and nonmembers were 86.70% and 83.30 (Dubey et al. 2014).

2.3 Constraints in Achieving Optimal Livestock Productivity

Animal husbandry has been a critically important subsector of agriculture. Livestock farming is not only an indispensable component of agriculture but is also the most suitable production system that has enormous potential to improve the socioeconomic status of the large percentage of the rural population. It is the only source of cash income to a vast majority of landless people who maintain small stocks of cattle on the fodder resources available on common forest lands. Apart from constraints associated with fodder production, livestock production and productivity also face other challenges which hinder optimum production in hilly areas. Some of the constraints in livestock productivity are:

2.3.1 Feeds and Fodder Shortage

The unavailability or irregularity in fodder supply is one of the most critical components for low livestock productivity and production in the region. There are many factors responsible for low feed availability to livestock in the region. The diversion of grazing lands for non-pasture uses is one of the crucial factors for low availability of feeds to livestock in the Himalayan region. Natural pastures are main feed resources to Himalayan livestock. Most of the natural pastures are neither defined nor marked on ground. In the absence of such designation and accompanying land records and maps, the grazing lands are gradually being put to other land-use system, causing reduction in the extent of already fragmented grazing lands. Moreover, topography and land availability, less preference for fodder crops under cultivation, harsh environmental condition, lack of knowledge on modern technologies, etc. have hindered fodder production in the Himalayan region. The underfeeding of livestock has resulted in late maturity, high calf and adult mortality, poor lifetime performance, and infertility.

2.3.2 Poor Livestock Extension Services

Poor livestock extension services are also one of the important factors for low livestock productivity in the region. Farmers are not aware of recent development in the areas of animal nutrition, particularly improved utilization of existing feed resources, strategically supplementation of roughage-based diet, use of additives for improving the animal health, production, reproduction, and feeding of colostrum to newly born calves.

2.3.3 Poor Credit and Marketing Facilities

Farmers depend on credit facilities of agriculture department and subsidies provided by the government. Poor economic status of farmers, credit of private level at high interest rates, and unavailability of credit are other major constraints to livestock sector development for the smallholders in the region. Farmers also face problems at the market level. Lack of granted marketing prospect for livestock products and unfair prices further add to vagaries faced by the farmers. Lack of efficient marketing systems and policies to supply necessary inputs for basic agricultural activities are the other hurdles.

2.3.4 Poor Veterinary Services

Difficult terrain and remote areas of the Himalayan region unfortunately have negative impact on services provided by veterinary hospitals. These terrains make the distance between healthcare centers and delivery systems too far to provide emergency and necessary services. The veterinary workers generally provide different services like first aid, deworming, vaccination, mineral mixture, and information on cattle feed to the farmers which are limited in the Himalayan region.

2.3.5 High Prices of Commercial Feeds

Pastures and grasslands in the hilly region have meager maintenance costs. Majority of farmers in hilly areas are self-sustainable for fodder production. But only fodder availability cannot meet the nutritional requirements of livestock. The farmers need to feed the commercial feeds like rice and wheat brans, feed pellets, oil cakes, etc. to provide protein and other nutrition to livestock. Involvement of high cost to supplement the livestock feeding with concentrate feed is difficult for the farmers of the region.

2.3.6 Poor Animal Breeding Facilities

The cattle/buffalo development programs suffer in the Himalayan region due to genetic potential of bulls. Bull populations with potential ability to develop superior calves in hilly regions are very low. Farmers often resort to natural breeding and move from one place to another. The remote areas in Himalayan region are deprived of breeding procedure as veterinary services for improved breeding process fail to reach farmers in the remote places.

2.3.7 Gender Monoply in Livestock Activities

Women play crucial role in maintaining the livestock in the Himalayan region (Fig. 2.3). Most of the livestock activities are incomplete without the assistance of women. Their activities vary widely ranging from care of animals, grazing and fodder collection, cleaning of animal sheds to processing milk and livestock

Fig. 2.3 (**a**–**d**): Gender monoply in livestock rearing in Himalayan region

products. In livestock management, indoor jobs like milching, feeding, cleaning, etc. are done by women in most of the families. In Nepal, women contribute nearly 70% of the work in raising livestock. Women account for most of the total employment in dairy production. Depending upon the economic status, women perform the tasks of collecting fodder and collecting and processing dung (Tulachan 1994).

Women also prepare cooking fuel by mixing dung with twigs and crop residues. Local women generally spent most of their time (8 to 10 h/day) in drudgery by walking long distances to collect fuelwood and fodder from trees, shrubs and grasses available in forests. Moreover, women also play significant role in maintaining health and hygiene conditions of livestock.

Mountain women farmers are not familiar about appropriate techniques as well as merits and demerits of livestock rearing practices. Generally, women are excluded from extension, marketing, credit, and other activities which are critical to increasing livestock productivity and income. In Uttarakhand, women devote nearly 170 h/month on average (nearly 5 h/day) to different dairy activities ranging from feed items to value addition of milk products and men spend only 31 h/month (less than 1 h/day). As much as 85% of women's farm work time is devoted to dairy production (Tulachan et al. 2002).

2.3.8 Poor Animal Management System

In the traditional Himalayan animal management system, different livestock species are housed in one large room on the ground floor of a two- or three-storied building, generally without ventilation or appropriate floor space. Due to mixed housing of different livestock species, outbreaks of contagious diseases cause heavy mortality. Pneumonia, enteritis, tuberculosis, enterotoxemia, and gastrointestinal parasitism are important diseases in goats and sheep, resulting in mortality and high morbidity. In cattle, the predominant diseases are the foot-and-mouth disease, hemorrhagic septicemia, hemoprotozoan diseases, ectoparasites, mastitis, gastrointestinal parasitism, warts, and laminitis. The surplus grain and concentrated feeds, as well as kitchen rejects, are generally supplied to the animals, but the total supplied feedstuff fall far short of the animals' actual requirements. Grazing on depleted natural ranges in private or community holdings is the conventional way of feeding the animals. The total grazing hours are frequently restricted by poor weather. Supplementary feeding is primarily by means of dry grasses. The whole situation leads to malnutrition, poor growth, and low production, and makes the animals susceptible to various diseases.

2.3.9 Lack of Knowledge of Anti-Nutrients in Different Forage Species

The presence of anti-nutrients in forages reduces their nutritive value and can affect the economic viability of livestock production system. Anti-quality in forages may have evolved as structural components of the plant or as secondary metabolites for plant defense mechanisms. Anti-quality components could result in mineral deficiencies, toxicities, or nutritional imbalances. The origin of anti-quality attributes can be phytochemicals in plant tissues or structural inhibitors in leaf and stem arrangement. Specific chemical inhibitors of quality can result from plant metabolism or microbes living in plants. Anti-quality factors in forages may reduce dry matter intake, limit dry matter digestibility, or cause nutritional imbalances. Anti-quality factors also act as toxins which shut down vital systems in animals, result in abnormal reproduction, disturb endocrine or neurological function, cause genetic aberrations, or suppress immune function, and in severe cases lead to diseases and death. The economic impact of anti-quality factors on individual flocks and herds can be devastating when the result is a large loss in production, reproduction, morbidity, or mortality. Economic consequences can also be severe if the loss is but a single animal with high economic value. For example, the presence of the endophyte fungus *Neotyphodium coenophialum* confers stress tolerance to the plant but alkaloids produced by fungus cause several disorders in animals. Therefore, farmers need to be educated through camps about anti-nutrients in common forage species of area and their effects on animal health and production.

2.4 Conventional Practices of Livestock Feeding

Livestock production is the highly practiced occupation in Indian Himalayan region. Marginal farmers constitute the core of the livestock production in India. Approximately, 80% of large ruminants and over 85% of other livestock are owned by the marginal farmers in the country. Livestock production in India is predominantly based on grazing on natural grasslands and pastures and crop residues. Concentrate feeds are also used to supplement the dry fodder diets but not on large scale. Supplementation of dry fodder with green fodder is useful to enhance the efficiency of livestock production system including reducing the emission of greenhouse gases and reducing the dependence on expensive concentrate feeds and lowering production costs. Majority of the farmers in the Himalayan region are smallholders and they depend on diversification of agriculture to reduce risks and derive economic benefits from the limited resources they possess.

2.4.1 Grazing in Natural Pastures and Grasslands

Pastoralism is one of the important livestock feeding practices in the Himalayan region. It is a subsistence pattern in which people make their living by domesticating large herds of animals. It is an effective means of exploiting marginal environments (Bhasin 1988). Pastoralists are the members of caste or ethnic groups with a strong traditional association with livestock-keeping, where a substantial proportion of group derives over 50% of household consumption from livestock products or their sale and where over 90% of animal consumption is from natural pasture or browse and where households are responsible for the full cycle of livestock breeding. Pastoralism in the Himalayan region is based on transhumant practices and involves movements of pastoralists from elevation to high elevation to take advantage of seasonally available pastures (Bhasin 1988). During the summer season, when the snow melts in the higher alpine regions, Himalayan pastoralists move up to these areas to graze their animals. Migratory pastoralism is common throughout the Himalayas and, from west to east, some of the herding communities in the region include the goat- and sheepherding Bakkarwals of Jammu and Kashmir; the buffalo herding Gujjars in Kashmir, parts of Himachal Pradesh, and Uttarakhand; goat- and sheepherding Gaddis, Kanets, Kaulis, and Kinnauras in the state of Himachal Pradesh; the sheepherding Bhotias of Uttarakhand; and yak herding Sherpas of Khumbu, Nepal, and less well-known communities in the mountainous ranges of Bhutan, Sikkim, and Arunachal Pradesh. All of these herders continue a long-standing tradition of migrating up to the alpine pastures of the high Himalayas for the summer and descending to the low-lying Himalayan foothills in the winter. Pastoralists like Gaddis, Bakkarwals, Gujjars, and Chopans majorly have livestock as their source of income. Some pastoralists are nomadic, while some are sedentary and semi-sedentary. They generally solely depend on grasslands and pastures to feed the livestock. In the state of Himachal Pradesh, feeding livestock with traditional practices is affected by seasons. Livestock are grazed in local pasturelands for 7 months of the year, i.e., May to November. With the advancement of winter season, fodder scarcity occurs in November. From November to April stall-feeding at household level and local-grazing in fields are practiced. During summer (April to September) sheep and goats are grazed at local high-altitude pasturelands by transhumant pastoralists. The arrival of winter is marked by seasonal migration of sheep and goat owners to low-altitude areas. Small livestock are reared in three different systems like extensive, semi-intensive, and intensive (Thakur et al. 2012). Major pastoralist groups in Indian Himalayan region and their seasonal activities have been enlisted in Table 2.3.

Extensive system of feeding is the one in which lambs are allowed to graze in rangelands and stubble grazes on croplands or forestlands during the monsoons. This system is prevalent in transhumant migratory pastoralists where they go long way to graze their livestock. In semi-intensive system, animals are allowed to graze on the common property resources or cultivable/fallow land for 8 to 12 h per day and later supplemented with concentrates. Trees and shrubs are lopped or pruned and often contribute substantial amount of leaf fodder during lean period. The intensive system includes complete stall-feeding on cultivated fodders, crop residues, concentrates, compounded feeds and grazing on developed pastures combined with stall-feeding (Fig. 2.4).

2.4.2 Dry Fodder Feeding

Once the season of grazing is over, farmers feed the livestock in their sheds (Stall-feeding). Stall-feeding is practiced with different feeding habits. One of the common practices of Himalayan region is feeding of livestock with crop residues. Crop residues include cereal straw, maize and sorghum stover, sugarcane etc. Crop residues are available in large quantities after harvest of grains and are important feed resource during the unavailability of green fodder. Apart from crop residues, hay forms the next set of feed to livestock. Crop residues or hay are fed to livestock in the region at least once a day.

2.4.3 Dry Fodder Mixture with Green Fodder

Stall-feeding of animals is done according to the availability of fodder crops and resources in particular cropping season (Pathania and Sharma 2016) (Table 2.4). Green fodder is the prime source of nutrition to livestock. Farmers in hilly regions are totally dependent on the locally available feed resources like oak tree leaves and

Table 2.3 Major pastoralist groups in Indian Himalayan region and their seasonal activities

Pastoralist group	Location		Activities
Bakarwals	Jammu and Kashmir		Move to Jammu and Punjab plains in winter times and to Kishtwar and other higher alpine valleys of Kashmir Himalayas in summer months
Gujjars	Jammu, Himachal Pradesh & Uttaranchal		Winter times are spent in the regions of Jammu, Punjab, lower districts of HP & UP, Saharanpur regions, and to the areas adjoining Rajaji National Park. They migrate to higher (non-alpine) regions of Himachal Pradesh and Uttaranchal in summer times
Changpas	Southeast Ladakh		Their migration cycle is around the various high-altitude pastures of Rupshu plains in Changthang region of Ladakh
Gaddis	Kangra and Dharamsala regions of Himachal Pradesh, parts of UP and Punjab		Occupy lower districts of HP during winter months and occupy Lahaul and Dhauladhar pastures in summer months

(continued)

Table 2.3 (continued)

Pastoralist group	Location		Activities
Bhotias	Upper regions of Garhwal & Kumaon of Uttaranchal		Occupy lower districts of Uttaranchal like Dehradun, Bhabar valley in winter months and move to higher pastures of Garhwal and Kumaon Himalayas toward Nanda Devi, Gwaldam, Mana pastures, and adjoining regions
Bhuttias	North district of Sikkim		Alpine regions of Lachung and Lachen valleys of the north district of Sikkim and move to lower forest below Mangan in the summer times
Monpas	Tawang and West Kameng district of Arunachal Pradesh		Higher reaches of East Kameng and Tawang district in summer season of Arunachal Pradesh and migrate to low lands around Tawang in the winter months
Kinnauras	Kinnaur district of Himachal Pradesh		In summer, sheep and goat flocks are driven to higher parts inside HP and in winter the flocks are driven to the foothills of Uttaranchal and HP

unclassified grasses grown in the forest area for feeding their animals round the year (Meena et al. 2008).

In the state of Uttarakhand, 95.30% farmers feed their livestock with oak tree leaves and grasses without chaffing, 92% offer fodder tree, grasses cut from forest,

Fig. 2.4 Grazing as a conventional practice in the Himalayan region

grazing in nearby forest and hay around the year and 86.70% offer only oak tree leaves throughout the year, while 85.70% farmers feed vegetable residues to milch animals and calf. Dry fodder solely fed to livestock can lead to health problems. To avoid these issues, dry fodder is mixed with green fodder to improve palatability and digestibility. Green fodder includes forage grasses and forage legumes. Apart from pastures and grassland forages, cultivated green forages are fed mixing with dry fodder while tree foliage contains good amount of nutrients but also contains anti-nutrients like tannins and phenols. In majority of the rural areas in India, chopped green fodder like maize, oats, sorghum, berseem, alfalfa is mixed with chopped straw/stover (wheat, paddy, maize, sorghum) in different proportions on dry matter basis (Figs. 2.5 and 2.6).

Table 2.4 Fodder availabilty for stall-feeding in Himachal Pradesh (% of total cropped area)

Season/crop	Marginal farmers	Small farmers
Kharif		
Maize	21.75%	22.75%
Vegetables	0.60%	0.25%
Fodder crops	25.38%	23.88%
Subtotal	47.73%	46.88%
Rabi		
Wheat	48.34%	48.38%
Vegetables	0.30%	0.24%
Fodder crops	3.63%	4.50%
Subtotal	52.27%	53.12%
Total cropped area (thousand ha)	0.331	0.800
Cropping intensity	191.11	187.97

Source: Pathania and Sharma (2016)

Fig. 2.5 Harvesting green fodder for stall-feeding

2.4.4 Concentrate Feeds

Owing to continuously building population pressure, the available land for forage production is declining day by day. Insufficient feed availability has compromised the genetic potential of livestock. In such circumstances, utilization of crop residues in livestock feeding is indispensable to meet the needs of livestock; however, the major constraint in the utilization of these crop residues is high ligno-cellulosic contents and poor nutritive value which again cannot support the maintenance of nutrient requirement of the animals. The utilization of low-quality feedstuffs more efficiently could contribute in enhancing livestock productivity in hilly regions (Beigh et al. 2017). To supplement low-quality feedstuffs and make them more

Fig. 2.6 Stall-feeding practices in Himalayan region. (**a**) Green fodder; (**b**) Dry fodder; (**c**) Dry fodder and green fodder mixture; (**d**) and (**e**) Cattle feeding

palatable and nutritionally superior, concentrate feeds are used by the farmers. Numerous reports in the literature indicate substantial increases in live weight gain by supplementing cattle consuming low digestibility forages with concentrate feeds (Lee et al. 1987; Hennessy et al. 1995).

The forage is generally enriched with addition of concentrate feeds such as oilseed cakes (Khal), feed pellets (Bovino feed), feed prepared from cotton seeds (Binola), rice or wheat bran, maize or wheat flour mixed with essential nutrients and minerals (Taramira and Choker), and broken seeds of food legumes. Oilseed cakes are the residues remaining after the removal of the greater part of the oil from oilseeds. The residues are rich in protein and most are valuable feeds for livestock. Based on method of processing, the cakes are classified into (a) *ghani*-pressed, (b) expeller-pressed, and (c) solvent-extracted. Of these, *ghani*-pressed cakes contain the maximum amount of ether extract, while solvent-extracted contain traces of oil. Conversely, the protein content is higher in solvent-extracted cakes and lowest in the *ghani* cakes. Oilseed cakes are in general very good sources of protein and about 95% of the nitrogen is present as true protein. Wheat bran consists of the outermost layer of the seed along with some flour. It is flaky, reddish brown material. Wheat bran is quite palatable and is well known for its ability to prevent constipation because of its swelling and water-holding capacities. Wheat bran has a high capacity to absorb water and swell because of its fiber and non-starch carbohydrates and so it has a bulk effect in the colon, giving it laxative properties. Rice bran consists of the fibrous outer layer of the grain, some hull, chipped grain, and calcium carbonate which is added during milling process. Raw rice bran contains 13 to 19% oil which is removed by solvent-extraction process leading to production of deoiled rice bran. It has high crude protein content ranging from 13 to 16% and TDN from 55 to 65%. It also contains better proportions of amino acids, particularly lysine and methionine, compared to other cereal grains, including maize and wheat. The phosphorus content is high (1.30%) and calcium content is low.

Though nutritive value of concentrate feeds is higher, but if the proportion of concentrate in the diet is too high, it increases the cost of livestock feeding. In North-West Himalayan region, farmers mix different grains, chopped straws, and brans and store them for long time to feed livestock once in a day to provide adequate nutrition. Broken wheat grains, broken oats, broken maize grains, paddy husk, and wheat and rice bran are mixed to prepare a healthy and nutritious feed (Fig. 2.7). Homemade feed is used for improving the nutritional quality of diet in which crop residues are also one of the feed components.

2.5 Nutritionally Compromised Fodder Resources for Livestock Feeding

Livestock productivity depends on many factors, among which forage quality is of prime importance. Forage with high nutritive value is considered one with high crude protein, high palatability, high ash, moderate ether extract content, low crude fiber content, low lignin, other cell wall constituents, and lesser anti-nutrients affecting the health and livestock productivity. Green fodder or forage is highly palatable for animal consumption, but unavailability of green forage at all times and circumstances is the limiting factor. Fresh green forages are obtained majorly from cultivated forage crops, pastures and grasslands, tree fodder, shrubs, and weeds. These forages are considered high in nutritive value with appropriate crude protein and crude fiber content. During lean season months when there is scarcity of green forage, farmers use hay, crop residue and silage for feeding the livestock.

Hay is prepared from grass, legumes, or other herbaceous plants which have been cut, dried, and stored for feeding. It could be used for feeding when or where there is not enough pasture or rangeland on which to graze an animalor when grazing is not feasible due to weather especially during winter or when pasture lush by itself would be too rich for the health of the animal.

In severe winters at high altitudes of Himalayan region, livestock are fed indoors for long periods. Traditionally, large areas of grass are cut for hay in the humid temperate zone, despite only moderate climatic conditions for drying. In lowland arable areas, silage from sown pasture or cereals has replaced hay feeding slowly.

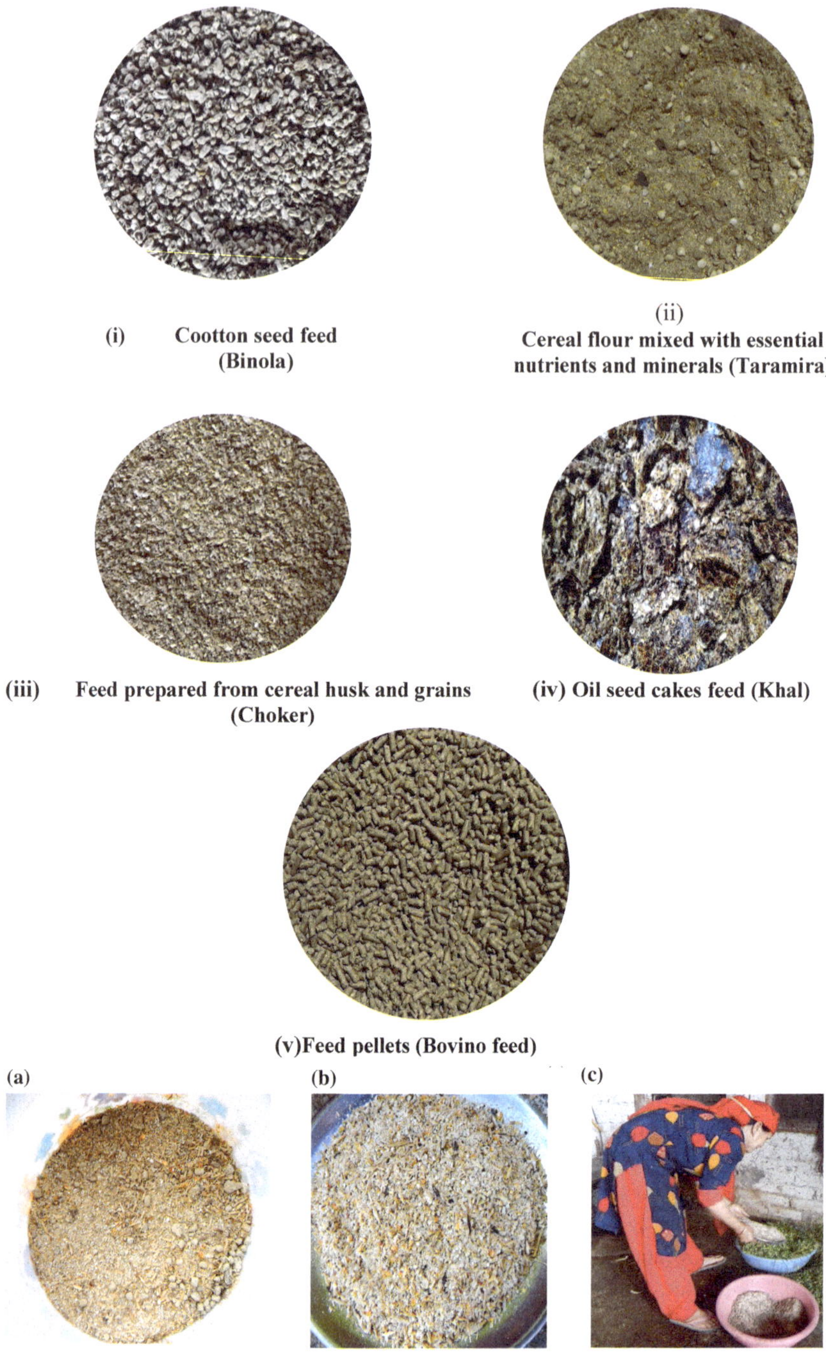

Fig. 2.7 (**a**) (i–v): Commercial concentrate feeds; (**b**) Homemade feed concentrates; and (**c**) Mixing of concentrate feed with fodder for livestock feeding

Commonly used plants for hay include mixtures of grasses such as rye grass (*Lolium* species), timothy, brome, fescue, Bermuda grass, orchard grass, and other species, depending on the region. Hay may also include legumes, such as alfalfa (lucerne) and clovers (red, white, and subterranean). Legumes in hay are ideally cut pre-bloom. Other pasture forbs are also sometimes a part of the mix, though these plants are not necessarily desired as certain forbs are toxic to some animals. High-quality legume hays like clovers, alfalfa and berseem made in dry warm climates are usually of high feeding value, but grass hays in more humid climates and from tropical herbage are of lower quality. In stall-fed production systems, as in dairying, the role of hay and straw is relatively simple. They provide bulk dry matter in the ration and provide some energy and protein content.

Silage plays important role in providing nutrition especially when green fodder is unavailable. It is the end product of fermenting a high moisture crop (40 to 80% moisture). The purpose of making silage is to preserve forage nutrients for feeding. This is accomplished by the conversion (by fermentation) of plant sugars to organic acids. Wastage of the plant is less in preparing silage as the whole plant is being consumed for ensiling, which is an important consideration with coarse, stemmy forages. Silage can be stored in less space than dry fodder of same quantity. It furnishes high-quality forage in any season of the year at relatively low expenses. During acute shortage of green fodder in summers, silage could complement the demand. Grass silage preserves 85% or more of feed value of the crop, whereas hay-making preserves significantly less percentage of nutrients. It is the most economical form in which the whole stalk of grasses can be processed and stored. On the other hand, a considerable part of this crop is wasted during the course of feeding in dry condition even if it is of good quality. During the monsoon months, it becomes difficult with dry grasses for making hay, therefore preserving the fodder as silage provides better option.

The silage prepared from fodder maize is an important feed source for livestock during the scarcity of green fodder. Maize silage is also a good way to secure the crops as it is possible to turn a maize crop damaged by frost, rain, or drought into maize silage. The popularity of maize silage is due to several factors. It is a consistent source of palatable and high-energy forage for all classes of livestock, including dairy cattle, beef cattle, sheep, and goats. It is one of the most high-yielding forage crops, requires less labor (since it is harvested in a single operation) and is generally less costly to produce than other forage crops. Maize silage contains about 25 to 35% dry matter and is rich in fiber (15 to 27%), with a highly variable starch content (18 to 37%). It is low in lignin and lipids (about 2 to 3%). The organic matter digestibility of maize silage ranges from 62 to 76%, depending on the growing conditions and on the stage of growth at harvest (Barrière et al. 2003). In India, baby corn husk silage offered as a total mixed ration with wheat straw and concentrate mixture, to male buffalo calves, resulted in higher nutrient digestibility, voluntary intake and nitrogen retention than for fresh baby corn husks (Bakshi and Wadhwa 2012).

Crop residues are the materials left in an agricultural field after the crop is harvested. Crop residues, comprising mainly of straw (wheat, oat, rice, etc.) and stover (maize, sorghum, millets, etc.) obtained after harvesting the crops, form an abundantly available feedstock. Cereal crop residues are of low feeding value and have low voluntary intakes. Major constraint to utilize the straws by livestock is the cross-linking of cell wall polysaccharides with the resistant biopolymer lignin [up to 15%], which prevents their microbial biodegradation in the rumen (Mahesh and Mohini 2013) and thereby decreasing the overall nutritive worth for livestock. Despite this, they constitute and continue to be an important feed resource for sustainable livestock production. The crop residues have due importance in livestock feeding during the unavailability of green fodder. Straws of legume crops have generally better nutritive value, forage quality, and thus are nutritionally superior to cereal straws. Stovers have better nutritional quality than straws with respect to intake and organic matter (Walli 2004). Straws and stovers are generally used to feed low-producing animals or can be used as a source of bulk in the high producers' ration to fulfill their appetite, can help correct physically effective fiber shortage for milk fat synthesis in high concentrate feeding systems, and may beneficially provide additional heat increment during cold stress conditions. Nutritive quality of crop residues could be improved with various treatments like physical, chemical and biological treatment. Among physical treatment, high temperature/steam pressure treatment has been found to improve digestibility of crop residues. Alkali and ammoniation of crop residues can improve protein and fiber content which enhances the digestibility. Lignocellulolytic microorganisms like white rot fungi can effectively reduce lignin content and improve fiber quality of crop residues.

2.6 Scope of Non-conventional Forages in Livestock Feeding

Different forages are used for livestock feeding in Himalayan region, still there are other resources that are not common in feeding practices, however, they have comparatively better nutritional quality as compared with common forages of the region. Some species like *Rumex*, *Moringa oliefera*, chicory, and *Azolla* are good in nutritionally desirable traits albeit also with anti-nutritional constituents. These plants have better options for inclusion in feeding resources in Himalayan region. A brief description of these plants is given below.

***Rumex* spp.** *Rumex* spp. belongs to Polygonaceae family and grows commonly at higher altitudes (900 to 4000 m amsl) on moist as well as dry slopes, under shades, and even in plains. It is a weed in pastures. *R. nepalensis* plant contains moderate levels of crude protein (Humaira et al. 2014), but crude protein of leaf of young plants is over much higher than 30% (Abbasi et al. 2015). Although, the tannin and mimosine content are slightly higher, it possesses higher levels of Ca and P than white clover and willow leaves (Wangchuk 2015). The *Rumex* spp. contains low level of tannins (5 to 10%) which protects protein and allows a high efficiency of feed utilization by the animal (Seresinhe and Pathirana 2003). Mimosine content is generally below 0.38%, which is roughly 3.80 g kg^{-1} DM (Wangchuk 2015). There exists scope of inclusion of these species in feeding practices after region-specific trials to study their effects on animal health and production (Fig. 2.8).

Fig. 2.8 (**a**) *Rumex nepalensis* (**b**) *Rumex obtusifolius* (**c**) *Rumex hastatus*

***Moringa oliefera* (Drumstick tree)** *Moringa oliefera* is a tropical plant originated from the southern hills of the Himalayas and was introduced in many tropical and subtropical areas, largely by migrant Asian populations (Bosch 2004; Orwa et al. 2009). *Moringa* leaves are a good source of digestible protein, digestible OM, and energy for livestock and therefore a valuable protein supplement. In addition, they are a valuable mineral supplementation when minerals are limited in feeds. The palatability of moringa forage has been reported to be only average. *Moringa* leaves could successfully supplement low-quality forage diets and improve animal performance. *Moringa* leaves are typically fed fresh to livestock however they have been ensiled alone, or in mixtures with Napier grass or sugarcane, to increase the nutritive value of the silage (Mendieta-Araica et al. 2011). Its leaves have been evaluated as a protein supplement for medium- or low-producing dairy cows fed with low-quality forages. *Moringa* oilseed cake, the by-product of oil extraction is rich in a wide range of vitamins (β-carotene, ascorbic acid, vitamin B_1, B_6, and niacin) (Reyes Sanchez 2004) as well as flavonoids (quercetin and kaempferol), which are known to be more potent antioxidants (Yang et al. 2006). *Moringa* leaves may thus be used as an antioxidant feed (Makkar et al. 2007). They have a relatively high concentration of lipids (5 to 6%, up to 10% DM) with an important proportion (33 to 45%) of α-linolenic acid (C18:3n-3) (Moyo et al. 2011; Olaofe et al. 2013). Its seed cake defatted with a solvent is a high-protein ingredient containing about 60% DM of protein, with limited amounts of fiber (NDF 9% DM) and fat (less than 1%). Its amino acid profile is rich in sulfur-containing amino acids but very poor in lysine. Feeding *Moringa* leaves had variable effects on DM intake and milk yield but did not change milk composition. The leaves offered as the sole forage (fresh or ensiled) and molasses gave the same results as low-quality forage supplemented with concentrate. Use of moringa leaves for feeding poultry, pigs, and fish is feasible but only in limited amounts due to the presence of fiber and anti-nutritional factors (Fig. 2.9).

Chicory (*Cichorium intybus*) Chicory is a perennial herb (not a legume) with a deep root system, making it drought-persistent. It could be a good complement in a forage seed mix and gives stable and persistent hay. It has a relatively high content of

Fig. 2.9 (**a**) Moringa foliage and (**b**) Tree

Fig. 2.10 Chicory (*Cichorium intybus)*

lysine in the leaves, but the crude protein content is lower compared to the legumes. Its forage is highly palatable to livestock. Chicory inclusion to growing pigs has shown to favor lactobacillus in pigs' intestinal microflora, which can give a prebiotic effect. This is because of the high content of dietary fiber (40%). Massoud et al. (2009) reported that chicory contains 14.70 and 4.65% of CP in leaves and roots, respectively. Leaves contain 70.71% of total carbohydrates and 7.80% of total soluble sugars. Chicory is also known for its anti-parasitic properties owing to the presence of bioactive compounds, such as tannins or sesquiterpene lactones. Studies with forage chicory offered at levels more than 80% of the diet [based on dry matter (DM) intake] to parasitized lambs have reported anthelminthic effects in animals. Chicory by itself is not suitable for hay-making, however it has been successfully used for silage production for many years, and is suitable for pit silage and wrapped haylage. Seeds, leaves, and roots of chicory are successfully used as supplements in small ruminant diets without compromising animal performance (Fig. 2.10).

Fig. 2.11 Azolla (*A. pinnata*)

Table 2.5 Nutritional composition of *Rumex* spp. *Moringa*, Chicory and *Azolla*

Nutrients	*Rumex* spp.			Moringa	Chicory	Azolla
	R. nepalensis	*R. obtusilolius*	*R. hastatus*			
	Composition (% DM)					
Crude protein	31.20	30.15	22.60	24.30	14.70	22.48
Ether extract	–	–	–	5.40	3.68	4.50
Crude fiber	12.20	15.20	13.23	13.60	16.78	14.70
Total ash	6.60	8.50	18.74	10.30	10.91	17.34
NDF	41.00	42.50	43.90	28.30	40.21	54.85
ADF	26.30	28.41	33.10	19.30	31.20	36.47

Azolla (A. pinnata) *Azolla* is an aquatic, free-floating fern belonging to the family Salviniaceae. *Azolla* has been well documented as feed to livestock which shows that it is a good source of protein with almost all essential amino acids required for animal nutrition. Cultivation of *Azolla* is easy with high productivity and good nutritive value and can be used as a beneficial fodder supplement by various researchers (Prabina and Kumar 2010). *Azolla* is rich in crude protein or ionic matter like iron, calcium, magnesium, phosphorus, copper, manganese, etc. apart from appreciable amount of vitamins A and B_{12}. Livestock can easily digest *Azolla* due to its high protein content and low lignin content (Anitha et al. 2016) (Fig. 2.11 and Table 2.5).

The socio-economic status of rural households in Himalayan region is dependent on livestock and crop farming. An understanding of the socio-economic status of the hill farmers and its determinants could help in accelerating effective transfer of technology. The utilization of forage resources based on season or period of the year and appropriate feeding practices could enhance livestock production and the status of livelihood of the farmers dependent on livestock.

References

Abbasi AM, Shah MH, Khan MA (2015) Wild edible vegetables of Lesser Himalayas : ethnobotanical and nutraceutical aspects, vol 1, pp 1–360

Anitha KC, Rajeshwari YB, Prasanna SB, Shilpashree J (2016) Nutritive evaluation of azolla as livestock feed. J Exp Biol Agric Sci 4:670–674

Anonymous (2012) Department of Animal Husbandry and Dairying. 19th Livestock Census

Anonymous (2019) Department of Animal Husbandry and Dairying. 20th Livestock Census

Bakshi MPS, Wadhwa M (2012) Nutritional evaluation of baby corn husk—a new feed resource for livestock. Indian J Anim Sci 82(12):1548–1550

Barrière Y, Guillet C, Goffner D, Pichon M (2003) Genetic variation and breeding strategies for improved cell wall digestibility in annual forage crops. A review. Anim Res 52(3):193–228

Beigh YA, Ganai AM, Ahmad HA (2017) Prospects of complete feed system in ruminant feeding: a review. Vet World 10:424–437

Bhasin V (1988) Himalayan ecology. In: Transhumance and Social Organisation: Gaddis of Himachal Pradesh. Kamla-Raj Enterprises, Delhi

Birthal PS, Taneja VK (2006) Livestock sector in India: opportunities and challenges. In: Presented at the ICAR-ILRI workshop on 'Smallholder livestock production in India' held during January 24–25, 2006 at NCAP, New Delhi

Birthal PS, Taneja VK (2012) Operationalizing the pro-poor potential of livestock: issues and strategies. Indian J Anim Sci 82:441–447

Birthal PS, Joshi PK, Kumar A (2002) Assessment of livestock priorities for livestock sector in India. In: Policy paper (15) NCAP. ICAR, New Delhi

Bosch CH (2004) *Moringa oleifera* Lam. In: Grubben GJH, Denton OA (eds) PROTA (Plant Resources of Tropical Africa/Resources végétales de l'Afrique tropicale), Wageningen, Netherlands

Das VK, Kumar AG (2017) Drivers of farmers' income: the role of farm size and diversification. In: Working paper 013, Indira Gandhi Institute of Development Research, Mumbai

Dubey LR, Pali HR, Datta KK (2014) Socio-economic profile of members and non-members of dairy co-operatives in Champawat district of Uttarakhand. Asian J Dairy Food Res 33(4): 263–266

Hennessy DW, Kohun PJ, Williamson PJ, Brown DA, Nolan JV (1995) The effect of nitrogen and protein supplementation on feed intake, growth and digestive function of steers with different *Bos indicus, Bos taurus* genotypes when fed a low quality grass hay. Aust J Agr Res 46:1121–1136

Humaira S, Rahmatullah Q, Shahid I, Mirza FQ (2014) Seasonal availability and palatability of native flora of Santh Saroola Kotli Sattian, Rawalpindi, Pakistan. Afr J Plant Sci 8:92–102

Lee GJ, Hennessy DW, Nolan JV, Leng RA (1987) Responses to nitrogen and maize supplements by young cattle offered low-quality pasture hay. Aust J Agr Res 38:195–207

Mahesh MS, Mohini M (2013) Biological treatment of crop residues for ruminant feeding: a review. Afr J Biotechnol 12:4221–4231

Makkar HPS, Francis G, Becker K (2007) Bioactivity of phytochemicals in some lesser-known plants and their effects and potential applications in livestock and aquaculture production systems. Animal 1(9):1371–1391

Massoud MI, Amin WA, Elgindy AA (2009) Chemical and technological studies on chicory (*Cichorium intybus* L) and its applications in some functional food. J Adv Agric Res 14(3): 735–756

Meena H, Ram H, Singh SK, Sahoo A, Mahapatra RK, Rasool TJ (2008) Livestock husbandry scenario at high altitude Kumaon Himalaya. Livestock Res Rural Dev 19(11):7

Mendieta-Araica B, Spörndly R, Reyes-Sánchez N, Spörndly E (2011) Moringa (*Moringa oleifera*) leaf meal as a source of protein in locally produced concentrates for dairy cows fed low protein diets in tropical areas. Livest Sci 137(1/3):10–17

Moyo B, Masika PJ, Hugo A, Muchenje V (2011) Nutritional characterization of Moringa (*Moringa oleifera* lam.) leaves. Afr J Biotechnol 10(60):12925–12933

Olaofe O, Adeyeye EI, Ojugba S (2013) Comparative study of proximate, amino acids and fatty acids of *Moringa oleifera* tree. Elixir Appl Chem 54:12543–12554

Orwa C, Mutua A, Kindt R, Jamnadass R, Anthony S (2009) Agroforestree database: a tree reference and selection guide version 4.0. World Agroforestry Centre, Kenya

Pathania MS, Sharma A (2016) Economic analysis of milch animals in Jaisinghpur tehsil of district Kangra. Himachal J Agric Res 42(1):15–23

Prabina BJ, Kumar K (2010) Dried Azolla as a nutritionally rich cost effective and immuno-modulatory feed supplement for broilers. Asian Australas J Anim Sci 5:20–22

Rao KS, Saxena KG (1994) In: Singh B, Singh MP (eds) Sustainable development and rehabilitation of degraded village lands in Himalaya, Dehradun

Reyes Sanchez N (2004) Marango: cultivation and utilization in animal feeding. Guia tecnica N°5, Universidad Nacional Agraria, Por un desarollo agrario integral y sostenible, p. 24

Saxena R, Khan M, Choudhary B, Kanwal V (2020) The trajectory of livestock performance in India: a review. J Dairy Sci 72. https://doi.org/10.33785/IJDS.2019.v72i06.001

Seresinhe T, Pathirana KK (2003) Forage tannins in ruminant nutrition. Trop Agric Res Ext 6

Singh A (2020) Livestock production statistics of India—2019. https://doi.org/10.13140/RG.2.2.32034.86721

Thakur D, Sharma A, Katoch S, Chander M, Mane B, Sharma P (2012) High altitude livestock farming: a participatory appraisal in Himachal Pradesh, India. Indian J Anim Sci 82:644–648

Tulachan PM (1994) Gender differences in livestock production management in the Chitwan District of Nepal. J Farm Syst Res Ext 4(3):121–135

Tulachan PM, Jabbar MA, Saleem M (2002) Smallholder dairy in mixed farming systems of the Hindu Kush-Himalayas. Report number: system study report. ICIMOD, ILRI, Kathmandu, Addis Ababa

Walli TK (2004) Straws as important feed resource under sustainable crop-dairy production system. Indian Dairyman 56:35–43

Wangchuk K (2015) Himalayan dock (*Rumex nepalensis*): the flip side of obnoxious weed. J Anim Sci Technol 57:34

Yang RY, Tsou SCS, Lee TC, Chang L-C, Kuo G, Lai PY (2006) Moringa, a novel plant rich in antioxidants, bioavailable iron, and nutrients. Am Chem Soc Symp Ser 925(17):224–239

3 Forage Resources in Himalayan Region

The fundamental characteristic of any mountain agroecosystem in the world is extreme internal variability and complexity with localized micro-ecosystems and habitats for diverse flora and fauna. The Himalayan region is an important hot spot of biological diversity among other mountain systems of the world. Agriculture along with animal production system is a principal occupation and source of livelihood for more than 60% of population lives in the Himalayan region (Sharma et al. 1999). Natural grasslands and pastures are important ecological resources on which a huge population of livestock is directly dependent upon their feeding requirements. These natural resources also support a great diversity of flora such as trees and shrubs, range grasses, and legumes, which have great importance from the prospect of livestock feeding. Fodder trees and shrubs are another important forage resource for livestock in the Himalayan region. The importance of the fodder trees and shrubs could be understood from the fact that they provide green fodder to livestock when other resources of green fodder are not available.

The cultivated forage crops also meet some percentage of total feeding requirements of livestock in the region, however due to land constraints forage crops are less prioritized by farmers in comparison to the food crops. Inspite of occupying place for cultivation in main farming lands, farmers grow these crops on bunds to support livestock farming but to a lesser extent. The residues from main food crops are another feed resource to livestock in the Himalayan region. Crop residues including straw, stovers, haulms, wines and tops serve as lean season feed resources. These crop residues are normally nutritionally compromised resources that bulk the available feed during the period of fodder scarcity. Major forage resources of the region are:

R. Katoch, *Nutritional Quality Management of Forages in the Himalayan Region*,
https://doi.org/10.1007/978-981-16-5437-4_3

3.1 Grasslands, Pastures and Rangelands of Himalayan Region

Grasslands have been defined as land covered with graminoid vegetation having less than 10% tree and shrub cover (House and Hall 2000). Grasslands represent largest single component of Earth covering approximately 117 million square km of vegetated lands (Pathak and Dagar 2016). The plants belong to the families Poaceae, Cyperaceae, Fabaceae, and Asteraceae and are present in higher proportion in grasslands. Grasslands are considered as a backbone of livelihoods for all the pastoral communities (White et al. 2000). In addition, they are niche for different wild herbivores and faunal groups for their breeding, migration, and wintering (Rahmani 2006). Due to continuous increase in human and livestock population in the last few decades, abrupt changes in land-use pattern, and transformation of traditional pastoral practices, most of the grasslands are degrading and resulting in the loss of biodiversity including decline in forage species, loss of pastoral livelihoods, and desertification (Faber-Langendoen and Jose 2010).

3.1.1 Grasslands and Pastures of North-West Himalayan Region

The North-West Himalayan region includes the Indian states of Jammu and Kashmir, Himachal Pradesh, and Uttarakhand. About 15.18 million ha of land in the state of Himachal Pradesh is under grasslands and pastures which is 35.44% of the total geographical area. Besides barren and uncultivable land (3.30% of the area), cultivable wastelands (2.58% of the area) are also utilized for livestock grazing. The field bunds and terrace risers comprising 1.64% of total area are also a source of forage for livestock. The productive area under grasslands and pastures in the states of Jammu, Kashmir, and Ladakh is 3.50, 13.20 and 5.08%, respectively (Radotra et al. 2015). In North-Western Himalayan region, the state of Himachal Pradesh consists of four zones ranging from low hills with subtropical areas to high hill dry temperate zones. These zones are dominated by different forage species.

- Low hill subtropical zone includes the areas of Kangra, Hamirpur, Una Bilaspur, Mandi and Sirmaur districts of the state. The area under grasslands in this zone is 15.38% of total area in Himachal Pradesh (Radotra et al. 2015). The dominant species of grasses in this zone include *Saccharum spontaneum* (Kai), *Imperata cylindrica* (Chhiz), *Chrysopogon gryllus* (Chhota Dhaulu), *Sorghum halepense* (Baroo), *Dichanthium annulatum* and *Eragrostis* sp.
- Mid-hill zone of Himachal Pradesh covers the areas of Chamba, Kangra, Hamirpur, Mandi, Bilaspur, Shimla, Kullu, Kinnaur and Sirmour districts. The area under grassland in this zone is 12.83% of total grasslands of Himachal Pradesh and grasses such as *Chrysopogon montanus* (Bara Dhaulu), *Cymbopogon martini* (Makora), *Cynodon dactylon* (Doob), *Paspalum notatum* (Kodri), *Pennisetum clandestinum* (Kikyue) and *Pennisetum orientale* (Jhaan), *Lotus corniculatus* and *Trifolium* sp. are dominant forage legumes in mid-Hill zone of Himachal Pradesh.

Table 3.1 Agro-ecological zones of Indian North-West Himalayas and the predominant forage grasses and legumes

Zones	Grasses	Legumes
Subtropical zone	*Chloris gayana, Cenchrus ciliaris, Chrysopogon fulvus, Dicanthium annulatum, Heteropogon contortus, C. setigerus, Paspalum notatum, Avena sativa, Paspàlum notatum, Lolium perenne, Festuca arundinacea, Setaria anceps, Cenchrus ciliaris, Dactylis glomerata, Chloris gayana, Apluda mutica, Pleum pretense, Panicum maximum, Heteropogon controtus, Bracharia brizantha*, Bajra-napier hybrids	*Trifolium alexandrinum, Stylosanthus hamata, Stylosanthus scabra, Macrotyloma axillare, Nenoatonia wightii*
Intermediate zone	*Chrysopogon fulvus, Chloris gayana, Dactylis glomerata, Dicanthium annulatum, Festuca arundinacea, Lolium perenne, Heteropogon contortus, Setaria* spp., *Avena sativa*	*Trifolium alexandrinum, Stylosanthus hamata, Macroptelium atropupreum*
Temperate zone	*Avena sativa, Dactylis glomerata, Festuca arundinacea, Lolium perenne, Phleum pratense, Bromus unioloides, Phalaris* spp., *Poa pratensis, Lolium multiflorum, Agrostis* spp.	*Trifolium pratense, Trifolium repens, Onobrychis viciifolia, Medicago sativa, Trifolium alexandrinum*
Cold arid zone	*Festuca arundinacea, Avena sativa, Phalaris* spp., *Dactylis glomerata*	*Medicago sativa, Medicago falcata, Lotus corniculatus, Astragalus* spp., *Caragana* spp., *Melilotus officinalis, Cicer microphyllum*

- Third agroecological zone includes high hill areas of Chamba, Kangra, Kullu, Mandi, Shimla, Solan, Sirmou, and Kinnaur districts. This zone is characterized by high hills and temperate wet climate. This zone has about 18% of total grassland area and the dominant grasses are *Dichanthium aristatum* (Angleton grass), *Pennisetum orientale* (Bari Jhaan grass), *Andropogon pulmis* (Creeping bent grass), *Heteropogon contortus* (Spear grass), *Festuca* sp. (Fescue grass) and *Dichanthium annulatum* (Marvel grass). *Lotus corniculatus, Trifolium* sp., and *Medicago denticulata* are some of the important forage legumes of this region.
- Cold arid zone or temperate dry zone covers north and northeastern part of Chamba and Kullu districts, high hills and valley of Lahaul-Spiti, Kinnaur, and Shimla districts. These temperate grasslands are infested with weeds and unpalatable plant species. The important forage grasses of this zone include *Festuca arundinacea, Avena sativa, Phalaris* spp., and *Dactylis glomerata*. The important forage species adaptable to the grasslands of different agroclimatic zones of North-Western Himalayan region are presented in Table 3.1.

3.1.2 Grasslands and Pastures of Eastern Himalayas

The Eastern Himalayan region is surrounded by China in the north, Bhutan in the west, Bangladesh and Myanmar in the south. This Himalayan region has great variation with respect to temperature, humidity, and topography. Important forages adaptable to the grasslands of different agroclimatic zones of Eastern Himalayan region have been presented in Table 3.2.

Table 3.2 Predominant forage grasses and legumes in agro-ecological zones of eastern Indian Himalayas

Zone	Distribution	Major fodder species
The Himalayan Hills	Northern tip of the state of West Bengal-Darjeeling and the further northern territory of Sikkim	*Allium prattii, Carex nivalis, Carex nigra, Elymus nutans, Festuca valesiaca, Heracleum* sp. *Juncus* sp. *Kobresia duthiei, Kobresia nepalensis, Kobresia pygmaea, Kobresia capillifolia, Phleum alpinum, Selinum tenuifolium, Rheum acuminatum*
North East Hills	Karbi Anglong and Cachar Hills in Assam and the entire territories of Arunachal Pradesh, Nagaland, and Meghalaya	*Bambusa longispiculata, Brachiaria mutica, Cynodon dactylon, Imperata cylindrica, Saccharum spontaneum Saccharum officinarum, Seteria palmifolia, Thysanchaena maxima, Themada caudata*
Southern Hills	Manipur, Tripura, and Mizoram	*Brachiaria mutica, Calopogonium mucunoides, Cenchrus ciliaris, Cynodon dactylon, Dichanthium annulatum, Medicago sativa, Melilotus alba, Pennisetum purpureum, Phalaris minor, Sorghum halepense, Vicia sativa*
Lower Brahmaputra	Assam (Barpeta, Dhubri, Nagaon, Darrang, Kamrup, Goalpara, Kokrajhar, and Sonitpur)	*Avena sativa, Brachiaria mutica, Chrysopogon zizanioides, Coix lacryma, Eleusine indica, Miscanthus fuscus, Panicum antidotale, Panicum maximum, Pennisetum pedicellatum, Pennisetum purpureum, Saccharum ravennae, Setaria palmifolia, Setaria sphacelata, Sorghum halepense, Themeda villosa, Tripsacum andersonii, Vigna umbellate, Vigna unguiculata, Zea mays var.mexicana*
Upper Brahmaputra	Districts of Lakhimpur, Sibsagar, Cachar, Dibrughar, Jorhat, and Karimganj (all in Assam) and Jalpaiguri and Koch Bihar in West Bengal	Common fodder species of Assam

3.1.3 Rangelands of the Indian Himalayan Region

Rangelands are the extensive area of land that is occupied by native herbaceous or shrubby vegetation and grazed by domestic or wild herbivores. Globally, over 100 million pastoralists depend on more than five billion hectares of rangelands for meeting their needs for sustaining livelihoods. Due to geographic extent and resources of rangelands, the proper management of grasslands is essential. The traditional rangeland management practices are sustainable, but increasing pressure on land and inappropriate management and development policies are now resulting in rapid degradation of rangelands.

High-altitude mountain regions of India are dominated by rangelands, with Ladakh area at the northern tip of the Indian subcontinent in Jammu and Kashmir as a typical example. The changing altitude and variation in climatic conditions make agriculture highly vulnerable in the Himalayan region. The lifestyle of the Himalayan people is somewhat traditional and depends on livestock rearing that relies on rangeland foraging. These rangelands are important source of forages and meet up to 50 to 60% of the total forage requirement.

Rangelands of Uttarakhand Himalaya have predominance of *Themeda anathera*, *Cassia mimisoides*, *Digitaria stricta*, *Pennisitum flaccidum* and *Setaria homonyma* grasses. The productivity of these rangelands depends on annual seasons. During rainy season, productivity is comparatively high than summer season. In the rangelands of the state of Jammu and Kashmir, *Lolium arundinaceum*, *Dactylis glomerata*, *Poa alpina*, and *Medicago* spp. are dominant forage species. The important range forage species in the rangelands of Eastern Himalayan region are *Apluda mutica, Arun dodonax, Brachiaria mutica, Cyrtococcum longipes, Cliotoria ternatea, Cyrtococcum accresecns, Dichanthium annulatum Panicum maximum, Setaria italica, Panicum auritum, Paspalidium punctatum Pennisetum orientale, Saccharum longisetosum, Stylosanthes seabrana and Stylosanthes hamata.* The important forages in rangelands of Himachal Pradesh and Uttarakhand include *Apluda mutica, Brachiaria mutica, Cenchrus ciliaris, Clitoria ternatea, Desmanthus virgatu, Hetropogan contortus, Panicum maximum, Poa alpine, Stylosanthes hamata, Medicago sativa,* and *Stylosanthes seabrana*.

3.1.4 Grassland Distribution Pattern

Grasslands occupy approximately 24% of geographical area in India which are spread across the different biogeographic regions and exhibit diverse ecological functions. Major types of grasslands in the country are (1) Alpine moist meadows of the Greater Himalayas; (2) Alpine arid pastures or trans-Himalayan grasslands; (3) Hill-side grasslands in mid-elevation ranges of Himalayas; (4) "***Chaurs***" of Himalayan foothills; (5) Wet alluvial ("***Terai***") grasslands of Gangetic and Brahmaputra flood plains; (6) "***Phumdi***" grasslands of Manipur; (7) "***Banni***" and "***Vidis***" of Gujarat; (8) Savannas of western and peninsular India; (9) Plateau and

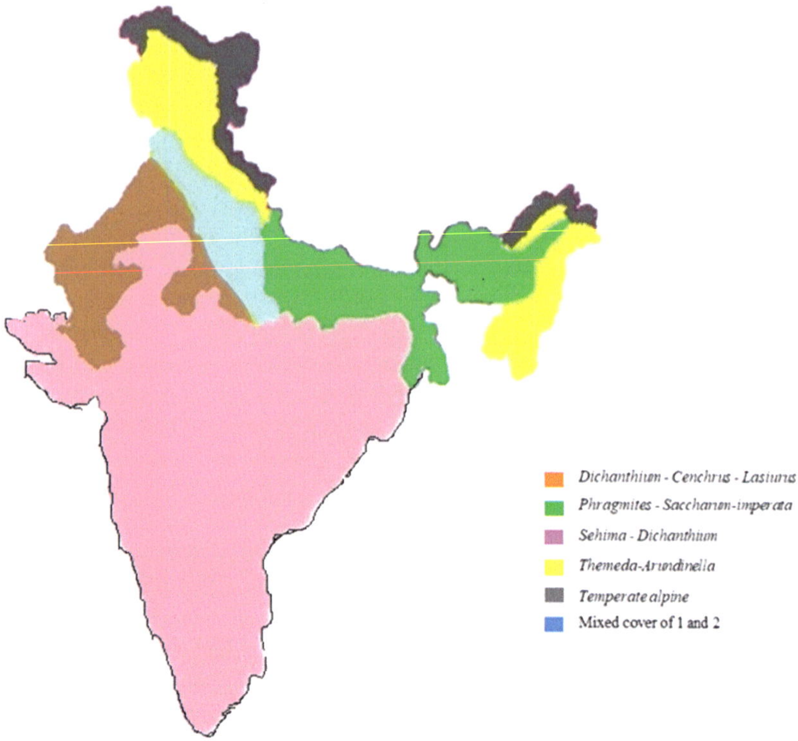

Fig. 3.1 Indian Grasslands

valley grasslands in Satpuras and Maikal hills; (10) Dry grasslands of Andhra Pradesh and Tamil Nadu; and (11) "***Shola***" grasslands of Western Ghats.

Besides alpine grasslands of the Himalayan region and grasslands in the high hills of Western Ghats, the majority of grasslands have originated as a result of anthropogenic activities (Singh et al. 1983). Dabadghao and Shankarnarayan (1973) identified five broad types of grass cover in India. The nature of these grass covers is primarily governed by climatic factors and especially by latitudes resulting in the occurrence of *Sehima - Dichanthium* cover in tropical region, *Dichanthium - Cenchrus-Lasiurus, Phragmites - Saccharum - Imperata,* and *Themeda-Arundinella* grass covers in subtropical region, and temperate alpine grass cover in temperate region.

Altitude separates *Themeda-Arundinella* cover (restricted to the northern hills) from *Phragmites - Saccharum-imperata* (in moist to wet habitats) and *Dichanthium-Cenchrus - Lasiurus* (in relatively dry habitats) types which occur in plains. Each of these grasslands has its own characteristics which differentiate them from one another (Fig. 3.1).

3.1.4.1 *Sehima: Dichanthium* type

The *Sehima-Dichanthium* grass cover is lying approx. between 8° and 28°N and 68° and 87°E (Malaviya et al. 2015). This type of grassland covers whole of peninsular India (dry subhumid zone except Nilgiris). The thorny bushes in this grassland are *Acacia catechu*, *Mimosa hamata*, *Zizyphus nummularia* and sometimes fleshy *Euphorbia*, along with *Anogeissus latifolia*, *Acasia pendula*, *Soymida febrifuga* and other species. *Sehima* grows easily on gravel and the cover may be 87% of the area on ground and *Dichanthium* flourishes on level soils and may cover up to 80% of the ground. The grasslands of arid regions are mainly located in Rajasthan, part of Kutch and Saurashtra in India.

3.1.4.2 *Dichanthium - Cenchrus - Lasiurus* Type (Semi-Arid Zone)

This type of grass cover lies between 23° and 32°N and 68° and 80°E. This type of grass cover extends to the northern portion of Gujarat, Rajasthan (excluding Aravallis), Western Uttar Pradesh, Delhi State, and Punjab with 0.43 million sq. km. Major forage resources are *Acacia senegal, Calotropis gigantea, Calotropis procera, Coreopsis. auriculata, Cenchrus ciliaris, Cenchrus setigerus, Cymbopogon jawarancusa, Cynodon dactylon, Dactyloctenium sindicum, Desmostachya bipinnata, Dichanthium annulatum, Eleusine compressa, Laesiurus sindicus, Sporobolus marginatus, Prosopis cineraria, Salvadora oleoides and Zizyphus nummularia* (Malaviya et al. 2015).

3.1.4.3 *Phragmitis - Saccharum - Imperata* Type (Moist Subhumid Zone)

This grass cover lies between 26° and 32°N and 74° and 96°E. This grassland covers the Ganga alluvial plain in Northern India. Predominant forage grass species in this grassland include *B. pertusa, Cynodon dactylon, Imperata cylindrica, Saccharum arundinaceum, Saccharum spontaneum, Phragmites karka, Desmostachya bipinnata, Barbarea intermedia, Vitevaria zizanioides, Imperata cylindrica, Chrysopogoaciculatus, Panicum notatum, and Dichanthium annulatum* and they are frequently observed in transition zones. The common trees and shrubs in the region are *Acacia nilotica, Acacia latifolia, Butea monosperma, Phoenix sylvestris,* and *Zizyphus nummularia* (Malaviya et al. 2015).

3.1.4.4 *Themeda - Arundinella* type

Themeda - Arundinella grass cover lies between 29° and 37°N, and between 73° and 81°E. This grass cover extends to the humid mountain regions and moist subhumid areas of Assam, Manipur, West Bengal, Uttarakhand, Himachal Pradesh, and Jammu and Kashmir. The principal forage species of this grass cover are *Arundinella benghalensis, Alnus nepaolensis, Bothriochloa intermedia, Chrysopogon fulvus, Cymbopogon jwarancusa, Cynodon dactylon, Heteropogon contortus, Themeda anathera, Euloliopsis binata,* and *Ischaemum barbatum.* Associated perennial species are *Achyranthes mutica, Arundinella khaseana, Pennisetum flaccidum, Chloris,* and *Desmostachya* (Malaviya et al. 2015).

3.1.4.5 Temperate Alpine Type

Temperate alpine grass cover lies between 29° and 37°N and between 73° and 81°E. The grassland spreads across the Himalayan states and temperate areas of Nagaland, Manipur, and Western Ghats having an altitude of above 2000 m above mean sea level. The principal perennial species are *Agropyron conaliculatum, Bromus ramosus, Chrysopogon gryllus, Dactylis glomerata, Danthonia cachemyriana, Eragrostis nigra, Festuca lucida, Phleum alpinum, Carex nubigena, Poa pratensis* and *Stipa concinna*. Other important grass species are *Poa alpina*, among others (Malaviya et al. 2015).

Grasslands and other grazing resources occupy a very important position in mountain agriculture. The mountainous regions of the country have different types of grassland communities based on their altitude, slope, aspect, and rock strata. They occur as multispecies communities over large tracts of land as well as intermixed with shrubs or as undergrowth in open to moderately dense forest. The major grasslands in the Himalayan region include (Fig. 3.2):

(a) **Warm temperate grasslands**: These are the ranges from 1500 to 2500 m amsl, present in North-Western, Western, and Central Himalayan regions, especially on south and south-eastern slopes. These grasslands are characterized by extensive grassy slopes with scattered trees and shrubs. The grass cover in these areas falls under *Themeda-Arurtdinella* type. Such grasslands are locally known as "ghasnis" in the state of Himachal Pradesh and hills of Uttarakhand (Dabadghao and Shankarnarayan 1973). The slopes are dominated by *Imperata cylindrica* and *Cymbopogon distans*.

(b) **Cool temperate grassy slopes**: Slopes having more than 45° angle with thin soil in the cool temperate and subalpine zone do not favor the tree growth and generally support herbaceous or grassland vegetation. These grasslands are dominated by *Chrysopogon gryllus, Andropogon tristis, Themeda anathera, Themeda tremula, Erianthus rufipilus, Miscanthus nepalensis, Brachypodium sylvaticum, Bromus unioloides* and several other grasses and sedges, and are also rich in terrestrial orchid species like *Satyrium nepalense, Herminium lancium*, and *Habenaria intermedia*.

(c) **Subalpine meadows and Thaches:** Thaches are blanks within the cool temperate and subalpine forests due to the activity of migratory grazers. These grasslands are covered with herbaceous plants such as *Origanum vulgare, Taraxacum officinale, Ranunculus hirtellus, Rumex nepalensis, Anemone rivalaris, Senecio chrysantbemoides*, and *Anapbalis cunelfolia*. Majority of the species are unpalatable and weedy in nature.

(d) **Alpine meadows:** Alpine meadows are located above the tree line on the southern face of Greater Himalayan region having altitude more than 3000 m amsl and up to 5200 m amsl. Major grasses in these areas are *Poa, Festuca, Bromus, Briza, Calamagrostis, Agrostis*, etc.

(e) **Steppe formations of trans-Himalayas**: Indian Trans-Himalaya (ITH), located beyond Greater Himalaya, is one of the important feed resources to local and migratory pastoral communities for livestock grazing. These grasslands are

Fig. 3.2 Grasslands and pastures of the Himalayan region

located in Ladakh in Jammu and Kashmir, Lahul, Spiti and Kinnaur districts of Himachal Pradesh, Nilang, Niti, Malla Johar, Malla Darma, and Lapthal areas of Uttarakhand, and Tso Lhamu plateau of Sikkim. This region is characterized by sparse or treeless vegetation, often dominated by scrub, desert steppe, or mixed herbaceous vegetation (Mediterranean type of vegetation). These grasslands also have the dominance of short grasses which can be classified as steppe according to the global classification of grasslands. Major forage grass species include *Elymus nutans*, Stipa species, *Agropyron*, *Poa*, *Calamagrostis*, and *Festuca* along with thorny bushes such as *Caragana*, *Astragalus*, *Lonicera*, and also *Juniperus*, *Ephedra* and several herbs of Asteraceae, Saxifragaceae, Caryophyllaceae and Scrophulariaceae families.

3.2 Grasses Predominant in Himalayan Region

In the Himalayan region, majority of forage grass resources are present in grasslands, pastures, and rangelands. Other than naturally available forage grasses, the Himalayan region also has rich diversity in cultivated forage grasses. Rye grass, tall fescue, *Setaria*, Napier grass and orchard grass are important cultivated forages. *Arundinella hookeri, Andropogon tristis, Poa* spp., *Chrysopogon gryllus, Stipa concinna, Cymbopogon* spp., *Bothriochloa* spp., *Desmodium* spp., and *Agrostos micrantha* are among dominant forage grasses of grasslands and pastures of the Himalayan region. A brief description of important grasses of the Himalayan region has been given below:

3.2.1 Tall Fescue (*Festuca arundinacea* Schreb.)

Tall fescue is a cool-season hardy perennial forage grass. It is a bunchgrass grown for pasture, hay and silage. Having Europe as center of origin, this grass is also adapted to the lower areas of Himalayas where the season is too hot for other grasses and in the area that is too cold in the winter for the warm-season grasses. It is one of the best temperate perennial grasses for supplying nutritious forage in late spring and early summer. Tall fescue could produce 30 to 40 t/ha and 5.0 t/ha of green and dry fodder, respectively. Tall fescue has comparatively better nutritive quality and desirable agronomic traits. Nutritive value of tall fescue is similar to that of other perennial, cool-season grasses. It has 9.33 to 12.54% CP content, 1.36 to 3.19% soluble carbohydrates, 62.40 to 71.40% NDF, 26.90 to 41.70% ADF, 2.20 to 10.0% ADL content and 47.00 to 56.00% digestibility. The high lignin content in tall fescue is associated with low digestibility of the grass. Sometimes, the grazing on this grass has the associated problem due to the presence of endophytic fungus (*Neotyphodium uncinatum*). Endophyte toxicity has been reported to cause problems in reproduction, low weaning weights, and poor gains of grazing stocker animals.

3.2.2 Perennial Rye Grass (*Lolium perenne* L.)

Perennial rye grass is a cool-season perennial bunchgrass native to Europe, temperate Asia, and North Africa. It is widely distributed throughout the world. The growing nature of the perennial rye grass can change depending on environmental conditions. It resembles annual rye grass (*Lolium multiflorum* Lam.), although perennial rye grass has more leaves in lower parts of the plant canopy. It provides highly digestible pasture in comparison to other perennial grass species. The grass has good nutritional quality and digestibility with appreciable level of other essential nutrients. Because of its better nutritional quality, perennial forage-type rye grass is fed to lactating dairy cows on pasture and also to other classes of livestock. It has 9.43 to 12.09% crude protein content, 37.24 to 40.36% ADF, 55.31 to 58.52% NDF, and 49.24 to 53.27% total digestible nutrient (TDN) (Surmen et al. 2013). Perennial rye grass is a valuable grass in temperate regions worldwide including the Himalayan region. The grass also has high total usable carbohydrates, however nutritional composition depends largely on maturity and soil fertility. It produces good-quality forage with fairly a good amount of digestible dry matter with 12 to 15% crude protein and can be used for pasture, hay, silage, turf, and conservation purposes. Perennial rye grass often is mixed with orchard grass. Mixing rye grass with orchard grass improves swathing and drying, thus reducing mechanical harvesting problems often encountered with pure stands of both forage grasses.

3.2.3 Setaria (*Setaria anceps* Schumach.)

The grass thrives well in areas with an annual rainfall ranging from 1000 to 1500 mm and can survive long, hot, and dry seasons. The grass grows well at 20 to 25°C. It is more cold-tolerant than most of tropical and subtropical grasses. The grass can be used as green cut fodder, silage, and hay with 50 to 70 t/ha green fodder and 7 to 8 t/ha dry fodder productivity. The green foliage has 11.50% CP, 75.50% NDF, and 40.80% ADF, whereas hay has 4.00% CP, 78.0% NDF, and 56.00% ADF. The grass is considered as low-quality roughage usually too coarse and can be supplemented with urea-molasses feeding. *Setaria* contains oxalate, which is an anti-nutritional component. Level of oxalic acid varies with the cultivar and stage of growth. It is very palatable at young but becomes less palatable with the advancement in maturity. The setaria grass has two to three times higher productivity, extended green period and better nutritional quality as compared with other grasses.

3.2.4 Guinea Grass (*Panicum maximum* Jacq.)

Guinea grass is a perennial bunchgrass that is grown for pasture, cut-and-carry, silage, and hay. Guinea grass as understorey grass in agroforestry system has high production potential. It is also suitable for rangelands receiving 900 to 1500 mm rainfall, although it can also survive in areas receiving less than 400 mm rainfall. The crop has been adopted well by the farmers in India because of its multicut nature and high yield of green fodder. It is a fast-growing and leafy grass palatable to livestock with a good nutritional value. The grass could produce up to 40 to 60 t/ha of dry matter. The grass has CP content ranging from 5.50 to 13.50% (Aganga and Tshwenyane 2004) and *in vitro* DM digestibility of around 66% (Santos et al. 2008).

3.2.5 Orchard Grass (*Dactylis glomerata* L.)

Orchard grass is grown for pasture and hay in all temperate countries and is also cultivated in temperate regions of tropical and subtropical areas. The grass is also grown for ground cover and lawns. Orchard grass with its deep root system is also useful for checking soil erosion. This grass has crude protein content between 10 to 15%. The protein content in grass declines as the maturity advances. The high protein content in grass is counterbalanced by 25 to 30% fiber content that is often higher than that of other grasses at the same stage of maturity. The grass is widely used for hay-making and pastures because of its high yield and sugar content, which makes it sweeter than most other temperate grasses. The orchard grass stays green for longer period during summer in comparison to other grasses.

3.2.6 Napier Grass (*Pennisetum purpureum* Schumach.)

Napier grass is also known as elephant grass. It grows well at sea level and up to 2000 m and grows best in high-rainfall areas. The deep root system allows the grass to survive during drought period. Elephant grass has low protein content (about 10%). The grass is rich in fiber and depending on the stage of maturity, NDF content varies from 55 to 75% DM. It provides good hay if cut at early stage. Elephant grass can cause nitrate poisoning in cattle when used as the sole component of the diet (Cook et al. 2005). The grass is extremely palatable at early growth stages. However, it becomes coarse and unpalatable as the maturity progresses. The grass is rather low in energy and protein content, due to high cell wall constituents. It is usually made into good-quality silage without additives.

3.2.7 Napier-Bajra Hybrid (NB Hybrid)

NB hybrid, an interspecific hybrid between bajra and Napier grasses, is developed to combine the high-quality and faster growth traits of bajra and deep root system habit of Napier grass. NB hybrid has large and green leaves, sheaths are softer, margins are less serrated, and the herbage is highly digestible and palatable. It contains 8 to 11% CP content, 27 to 30% CF, 60 to 65% NDF, 39 to 41% ADF, 9 to 13% lignin content, and 8.20 to 10.70% ash content. It is juicier and succulent at all the stages of growth. It is less fibrous and more acceptable. The oxalate content of some of the varieties may be high. It can withstand drought for a short period and regenerates with rains but is susceptible to frost.

3.2.8 Kentucky Bluegrass (*Poa pratensis* L.)

Commonly known as Kentucky bluegrass, smooth meadow grass or common meadow grass, is a perennial grass species native to Europe, North Asia, and the mountains of Algeria and Morocco. The grass is best suited for early season grazing. Grazing on this grass is best suited to maintain a stand height of 2 to 6 in. It contains 12.88% crude protein content, 54.67% NDF, 30.27% ADF, 2.98% Ca, 2.80% K, 0.29% P, and 0.26% Mg (Tenikecier and Ates 2018). The grass has good drought tolerance and avoids drought by going dormant during dry period. The grass is suited to well-drained, fertile soils in moist or humid region.

3.2.9 Bahia Grass (*Paspalum notatum* Flügge)

Bahia grass is a perennial rhizomatous grass, notable for its prominent, dual V-shaped inflorescence (6 cm long). The grass grows from sea level to an altitude of 2000 m and in regions with annual rainfall from 750 mm to 2000 mm. The grass is very tolerant to drought and to a lesser extent to salinity when well established. It is known to survive 20 to 36 days of flooding. The yield is 3 to 20 t/ha but its nutritive value decreases quickly after the seedling stage and it is recommended to keep a 6-week cutting interval to maintain forage quality. It can withstand heavy grazing pressure.

3.2.10 *Tripsacum laxum* Nash.

It is a robust, strongly rhizomatous, tufted, and leafy perennial grass that can form large bunches. The grass is cultivated primarily for fodder in cut-and-carry systems. The grass provides several environmental benefits, notably against soil erosion and development of pests and diseases (Cook et al. 2005). It is a warm-season grass that grows from sea level up to an altitude of 1800 m, at temperatures ranging from 18 to 30°C. It has 14.20% crude protein, 66.90% NDF, 37.00% ADF, 2.50% EE, and 10.70% ash content during early stage. Palatability varies with age, genotype, and soil fertility. It does better under good soil moisture but can withstand short droughts. It is intolerant to water-logging, flooding, heavy grazing, or frequent cutting.

3.2.11 Mission Grass [*Pennisetum polystachion* (L.) Schultes]

The grass is native to India and tropical Africa and distributed in different parts of the world. The grass thrives in difficult conditions and is adaptable. It prefers high rainfall but tolerates short drought periods. In India, total annual yield is around 3.40 t green fodder/ha/year. It provides valuable fodder, can be directly grazed, or cut to be used as hay. Its nutritive value and palatability are high before seedling stage and reduce at subsequent stages.

3.2.12 Big-Leaf Bristle Grass or Buffalo Grass [*Setaria palmifolia* (J. Koenig) Stapf.]

The grass is a robust perennial tufted grass growing up to 2.50 to 3 m in height and often forming large clumps. Big-leaf bristle grass is used as forage and fodder grass as it is appreciated by all livestock. It is valuable for ground cover and soil rehabilitation in stream bank stabilization. Big-leaf bristle grass is a very hardy species that is found in tropical and subtropical areas of Africa, America, and India, in areas where rain falls during summer. The grass has a relatively poor nutritive value with protein content lower than 10% (Fig. 3.3).

3.2.13 *Arundinella nepalensis* Trinius

The grass has a very wide range of distribution. It is an evergreen, clump-forming grass with short, scaly rhizomes; it produces slender to stout, erect culms usually 90 to 200 cm tall, occasionally to 300 cm.

3.2.14 *Festuca rubra* L. sensu lato

Festuca rubra is also known as red fescue or creeping red fescue. It is widespread across much of the Northern Hemisphere and in the Himalayan region. It can tolerate variations in the temperature and climate. It has 10 to 13.50% crude protein content, 50 to 55% NDF, 30 to 35% ADF, 2 to 2.90% Ca, 2.50 to 2.90% K, 0.20 to 0.30% P, and 0.20 to 0.30% Mg (Tenikecier and Ates 2018).

3.2.15 Spear Grass (*Heteropogon contortus* (L.))

The grass is mainly used as fodder when it is young. It can be grazed or cut for hay or silage. Once flowering occurs, its overall nutritive value declines. It should then not be grazed or cut for fodder after flowering (Soromessa 2011). The grass is seldom cultivated and usually found wherever grasslands are periodically burned. The yield is very variable (0.50 to 8.70 t/ha) and depends on soil fertility and rainfall. The grass does not withstand heavy grazing and may become dominated by other grasses (Cook et al. 2005). Crude protein levels, even in the very young grass, rarely rise

Fig. 3.3 The portfolio of some important forage grasses in the Himalayan region. In reading order (**a**) Tall fescue; (**b**) Rye grass; (**c**) Setaria grass; (**d**) Guinea grass; (**e**) Orchard grass; (**f**) Napier grass; (**g**) Napier-Bajra hybrid grass; (**h**) Kentucky bluegrass; (**i**) Bahia grass; (**j**) *Tripsacum laxum;* (**k**) Mission grass; (**l**) Big-leaf bristle grass or buffalo grass; (**m**) *Arundinella nepalensis;* (**n**) Spear grass; (**o**) *Festuca rubra;* (**p**) Blue panic grass

above 10%, ranges between 4 and 6% for much of growing season and between 2 and 3% during dry season. Protein digestibility is also low. Intake and digestibility decrease rapidly with age of material and stage of growth. The grass is highly palatable at early stages but becomes less nutritive as it matures.

3.2.16 Blue Panic Grass (*Panicum antidotale* Retz.)

The grass is vigorous, and tufted perennial in nature (Surhone et al. 2010). The grass can be sown alone or with other grasses such as buffel grass (*C. ciliaris*) in pastures. At later stages of the growth, it is recommended to graze the grass rotationally and heavily in order to maintain its nutritive value. The grass can yield 10 to 50 t of fresh matter/ha. Hay yields vary from 2.50 to 6 t DM/ha under rainfed conditions to about 5 t/ha under irrigated conditions (FAO 2011). It has a large range of CP 8 to 10% crude protein content, 40 to 45% ADF, 8 to 10% ash, and 0.8 to 1.60% EE content. The NDF content varies between 55 and 85% depending on the stage of maturity.

3.3 Common Forage Legumes of Himalayan Region

3.3.1 Berseem (*Trifolium alexandrium* L.)

Egyptian clover or berseem is one of the important members of genus *Trifolium*. Berseem is one of the oldest cultivated clovers, domesticated in Egypt and later introduced into many other parts of the world. It is now widespread in the different regions of South and West Asia. This potential fodder crop is cultivated over two million hectares of area in India with 65 to 100 t/ha green fodder production. Berseem has been recognized as a potential fodder crop due to its multicut nature providing fodder for a long duration (November to May), high quality of green fodder with high digestibility (up to 65%) and palatability. Berseem is also established as one of the best *Rabi* (winter season) fodder crops in North West Zone, Hill Zone, and part of Central and Eastern Zone of India. It supports milk production in lactating buffaloes, cows, and crossbred cattle. The additional advantage of berseem cultivation is the ability of substantial improvement in soil fertility. It is also commonly grown in mixtures with brassica and sometimes with oat to have high quality in early cuts. Since the crop does not tolerate close and frequent grazing, the cut-and-carry system is adopted to provide green feed to stall-fed cattle and buffaloes.

3.3.2 White Clover (*Trifolium repens* L.)

White clover is a common component in cool-season perennial legume. In warmer areas, it behaves as a summer-growing annual. White clover is recognized as one of the nutritious forage legumes due to its excellent feeding value, but has underlying problem of its bloat-causing characteristics. Fodder from white clover is more digestible than other forage legumes owing to low structure fiber and high voluntary intake at equivalent digestibility. Apart from its usage as fodder crop, white clover forage crop is also valuable in rejuvenating permanent pasture lacking forage legumes in temperate regions and cold subtropical areas. The crop could be used for haylage, hay, soil improvement, and land reclamation. However, the green forage yield is less as compared to red clover and lucerne.

3.3.3 Alfalfa (*Medicago sativa* L.)

Alfalfa (*Medicago sativa* L.) also known as Lucerne, Purple Medic, Trefoil, and "Queen of Forages" is a perennial flowering plant mainly cultivated for fodder. It is widely cultivated as the most important forage legume in temperate areas of the world. Lucerne is native to South West Asia and is considered as one of the excellent animal feeds due to high protein content, high intake potential and digestibility. Being perennial in nature, it provides nutritious fodder throughout the year with 60 to130 t/ha green fodder production. The fodder obtained from alfalfa is of high nutritive value, which supports lactation. The rapid ruminal degradability of alfalfa increases the voluntary feed intake.

3.3.4 Cowpea (*Vigna unguiculata* L.)

Cowpea, a native to West Africa is cultivated in tropics as well as subtropics across the world. It is a quick-growing legume and highly valued for human food and livestock feed (dual-purpose legume). Its dual-purpose character makes it a very attractive crop for increasing productivity and profitability of mixed farming system. The vines, leaves, and sometimes pods from cowpea are often used as a potential feed source. It also has great potential to be grown as mixed crop with maize, sorghum, and millets to produce an ideal cereal–legume fodder mixture. The fodder from cowpea is utilized for green-feeding, hay-making, and ensiling with other crops.

3.3.5 Rice Bean [*Vigna umbellata* (Thunb.) Ohwi & H. Ohashi]

Rice bean belongs to subgenus Ceratotropis genus *Vigna*. Rice bean is a tropical-to-temperate grain legume grown for food and fodder. The vegetative parts of the crop are fed fresh or made into hay or as crop residue and the seeds are used as concentrates. Despite having high yield potential, i.e., 15 to 30 t/ha green fodder, the crop has not been given due attention in the past (Katoch et al. 2007a, b). Generally, late maturing and photosensitive cultivars are used for fodder production with average green fodder yield ranging from 25 to 35 t/ ha. Rice bean has the ability to withstand biotic and abiotic stresses, possesses drought tolerance, resistant to many pests and diseases, and can be grown in marginal soils to provide fodder during scarcity periods. It is often grown as intercrop or mixed crop with maize. Rice bean hay increases DM intake, digestible protein, and N retention along with positive effect on rumen microflora resulting in increased volatile fatty acid production (Chanthakhoun and Wanapat 2010; Chanthakhoun et al. 2011). Crude protein content in rice bean seeds and leaves varies from 14 to 27%. Rice bean is rich in limiting amino acids such as methionine and tryptophan and also valine, tyrosine, and lysine (Mohan and Janardhan 1994). Rice bean seeds which are occasionally used as livestock feed also have excellent nutritional quality with appreciable level of minerals and vitamins (Katoch 2013).

3.3.6 Birdsfoot Trefoil (*Lotus corniculatus* L.)

It is a perennial self-generating legume grown in wild and naturalized populations of temperate regions where it is used as fodder crop. In some parts of the world, birdsfoot trefoil was introduced to improve pasture for increasing animal production due to non-bloating and highly palatable forage. This crop is well adapted to stress conditions as compared to alfalfa.

3.3.7 Red Clover (*Trifolium pratense* L.)

Red clover is a short-lived perennial best-suited for hay and silage. It is also suited best for short crop rotations and pastures that can be reseeded regularly. Red clover is widely cultivated in temperate regions of the world and provides 40 to 60 t/ha green fodder. It has high importance for its forage quality. It is widely grown as a fodder crop and valued for its nitrogen fixation, which increases soil fertility (Katoch et al. 2007c).

3.3.8 Grass Pea (*Lathyrus sativus* L.)

Grass pea is a dual-purpose legume. It is one of the preferred legume seeds in low fertility soils and arid areas because of its outstanding tolerance to dry or flooding condition. However, it contains a toxic component, which may cause paralysis if consumed in excessive amounts. The legume can be grown as a sole crop, in intercropping systems or in mixtures. Grass pea is grown in rotation after barley or sometimes after a legume crop, such as pea or chickpea, which has been sown in April and harvested in July. It possesses high levels of proteins, carbohydrates, and minerals but is deficient in some essential sulfur-containing amino acids.

3.3.9 Siratro [*Macroptilium atropurpureum* (DC.) Urb.]

Siratro is a productive and palatable forage legume. It is mainly sown in mixed legume-grass pastures. It is used for cut-and-carry systems and hay, though its twining habit makes harvesting difficult. Siratro is also used as a fallow crop (after lowland rice). Siratro is mainly sown in mixed stands with erect or tussock-forming grasses such as *Chloris gayana*, *Cenchrus ciliaris*, *Panicum maximum*, *Setaria sphacelata*, or with legumes that require similar growing conditions such as *Campanula rotundifolia*, *Desmodium intortum*, *Macrotyloma axillare*, Neonotonia *wightii* and *Stylosanthes* species.

3.3.10 Perennial Soybean [*Neonotonia wightii* (Wight & Arn.) J.A. Lackey]

Perennial soybean is mainly used as pasture or hay and is valued for its drought resistance and high yields. Perennial soybean is grazed rotationally during the warm wet months and continuously during winter period. It is cut at the flowering stage, dried down to 12% moisture, and then stored as hay. In mixed stands, perennial soybean and grasses such as *Digitaria eriantha*, *Pennisetum purpureum*, and *Megathyrsus maximus* can be ensiled together for making good-quality silage.

3.3.11 Centro (*C. pubescens* Benth.)

It is a vigorous twining, trailing, and climbing perennial forage legume. After 4 to 8 months of growth, centro forms a dense cover of 40 to 45 cm high and is able to fix N ranging from 120 to 270 kg/ha. It is then commonly cultivated with grasses mainly *Panicum maximum*, *Chloris gayana*, and *Pennisetum purpureum*), which benefit from N enrichment of soil and thus gives high yield. The annual green matter yield is around 5 to 14 t/ha but yield to the extent of 40 t/ha has also been recorded. The grass has average dry matter yields of about 3 t/ha in mixed pastures (Teitzel and Chen 1992).

3.3.12 Sweet Clover [*Melitotus officinalis* (L.) Lam]

Melitotus officinalis is native to Europe and Asia. Sweet clover can be used as pasture or livestock feed when properly cured. Sweet clover can be used for hay, silage, or pasture. It has less palatability because of bitter taste due to coumarin in the plant tissues. It has good palatability for cattle, sheep, and horses.

3.3.13 Shaftal or Persian Clover (*Trifolium resupinatum* L.)

Persian clover is an annual, prostrate, or semi-erect branched legume, 20 to 60 cm high and similar to berseem. The high moisture content makes the legume unsuitable for ensiling. In temperate regions, the legume is sown during autumn for livestock feeding during spring season. Persian clover is a valuable addition to a barley–rye grass mixture as it can reduce the need for mineral fertilizers, and improve mid-season yield and *in vitro* digestibility (Thompson and Stout 1997).

3.3.14 Moth Bean [*Vigna aconitifolia* (Jacq.) Marechal]

Moth bean is an annual herbaceous trailing legume that has outstanding tolerance to dry conditions. It is an underutilized multipurpose legume that can provide both palatable and drought-resistant, hot-season pasture, and hay for livestock and seeds for human consumption. Though annual of only 5 months duration, it could serve as a pasture legume as it reseeds itself if allowed to become well established before grazing. The green matter yield ranges from 37 to 50 t/ha. The yield of hay has been reported to be 7.50 to 10 t/ha (Brink and Jansen 2006).

3.3.15 Perennial Horse Gram (*Macrotyloma axillare* var. axillare)

Perennial horse gram is a trailing or climbing vine with stems usually growing up to 3.50 m in height. It is also an underutilized multipurpose legume. The crop has significant value with the prospects of livestock and human consumption. The plant has a bitter taste which may cause the animal to reject intake until they get used to its taste.

3.3.16 Faba Bean (*Vicia faba* L.)

Faba bean is a multipurpose legume used for both food and fodder (hay, silage, and straw). *Vicia faba* is an upright annual forage legume that can grow to a height of 1.50 to 2 m. Two subspecies are commonly grown (i) *Vicia faba* var. major for human consumption, though beans can be fed to livestock and (ii) *Vicia faba* var. minor which produces smaller seeds and is used mainly for livestock feeding. The crop is suitable for crop rotation with winter or summer cereals. The fresh faba bean forage has good nutritional quality with protein content ranging from 14 to 20%.

3.3.17 Guar Bean (*Cymopsis tetragonaloba* [L.] Taub.)

It is also known as cluster bean, and siam bean. In addition to the human nutrition, this crop has significant value in livestock feeding. It is an erect, bushy annual herbaceous legume that reaches up to a height of 3 m. Guar bean is not very suitable for grazing due to its hairy leaves and unpalatability. Guar is sometimes grazed to reduce bloat in livestock. Palatability improves after cutting and wilting. The best time for cutting guar for fodder is during flowering and early pod formation. Guar meal is an important nutritious feed resource to livestock.

3.3.18 *Macroptilium lathyroides* (L.) Urban.

This forage legume is also known as phasey bean, phasemy bean, wild bean, wild bush bean and wild pea bean. It is an erect branching legume that reaches up to a height of 0.6 to 1 m. It is usually annual but it is occasionally biennial or a short-lived perennial. It is mainly used as pioneer forage that rapidly provides feed in pastures. As an N-fixing legume, phasey bean is used for green manure and is a valuable cover crop in rotations. It can be cut or grazed, but does not tolerate sustained heavy cutting or grazing which can greatly reduce or even eliminate seeding. DM yield to the extent of 13 t/ha after 90 days of sowing could be obtained from the crop.

3.3.19 *Macrotyloma uniflorum* Lam.

It is also known as horse gram, horse grain, kulthi bean and poor man's pulse. The twining habit enables climbing. Under normal growing conditions, it can produce more than 6 t/ha DM. This legume is grown as a pulse, for deferred grazing or as a fodder crop for dry season feed for livestock. Like its perennial relative, *Macrotyloma axillare*, it is often not relished by cattle on first introduction, but then well eaten following a period of adjustment. It has value as a pioneer legume or sometimes a regenerating annual in permanent pasture.

3.3.20 Himalayan Gram *(Cicer microphyllum* Benth.)

It is locally known as Kukunnory or wild gram, which is closely related to cultivated chickpea (*Cicer arietinum*). It is the only wild species of genus *Cicer* that occurs in the cold arid deserts of the western Himalayas from Afghanistan to Tibet and western Nepal at an altitude of about 3300 to 4800 m amsl (Kour et al. 2011). Whole plant is used as fodder for increasing milk production and has medicinal properties.

3.3.21 *Caragana vesicolor* Benth.

This legume is the dominant shrub species of Tibetan plateau and in Trans-Himalayan grasslands (Kumar et al. 2016). This thorny shrub species is confined to West and North-West Himalayan regions of India, China, Nepal, Afghanistan, and Pakistan. The dense bushy nature, thick rootstock, and stunted growth with cushion-like appearance are its adaptations to cold arid conditions. It has many local names such as Trama (Western Ladakh) and Dam or Dama (Uttarakhand), this legume is among the most important and highly preferred fodder and fuelwood above 4000 m in the Trans-Himalayan region.

3.3.22 *Astragulus alpinus* L.

The forage legume is also known as alpine milkvetch. It is a perennial legume with a matted growth habit. This plant grows in subalpine and alpine climates, often in moist areas, such as woodlands and meadows around streams and lakes. It is an important feed resource to wild animals inhabiting the Himalayan region.

3.3.23 *Desmodium heterocarpon* (L.) DC

It is an aggressively creeping, stoloniferous perennial legume, producing dense stands. The stems are much-branched, especially at their base and usually become quite woody. The legume has comparatively low nutritional quality than other forage legumes (Fig. 3.4).

3.4 Cereal Fodder

3.4.1 Fodder Maize (*Zea mays* L.)

Maize produces good-quality herbaceous fodder with high palatability. Maize must be harvested at early and full bloom stage to get maximum benefit of the forage. Maize crop yields around 0.16 to 0.20 tonnes of green fodder per acre during rainy season and 0.1 to 0.12 tonnes of fodder per acre during summer season. Maize fodder could also be preserved as silage. Silage is the product from a series of processes by which cut forage of high moisture content is fermented to produce a stable feed that resists further breakdown in anaerobic storage. It should have 30% dry matter to be ensiled successfully. Maize grain is an important feed component of dairy cattle and poultry feed.

3.4.2 Oat (*A. sativa* L.)

Oat is grown in winter season in India and is a multipurpose crop for grain, pasture, forage, or as a rotation crop. It is a palatable, succulent, and nutritious crop. Oat requires a long and cool season for its growth; therefore, it is successfully grown in different regions of the world. In temperate climates, oats are spring, winter, and/or autumn sown depending on regional climatic conditions, crop rotation requirements, end use, and other farming practices. In slightly warmer regions, spring-type oats are sown in autumn to avoid summer heat and drought.

Fig. 3.4 The portfolio of some important forage legume of Himalayan region. In reading order (**a**) Berseem; (**b**) White clover; (**c**) Alfalfa; (**d**) Cowpea; (**e**) Rice bean; (**f**) Birdsfoot trefoil; (**g**) Red clover; (**h**) *Lathyrus;* (**i**) Siratro; (**j**) Perennial soybean; (**k**) Centro; (**l**) Sweet clover; (**m**) Persian clover; (**n**) Moth bean; (**o**) Butterfly pea; (**p**) Perennial horse gram; (**q**) Faba bean; (**r**) Guar bean; (**s**) *Macroptilium lathyroides*; (**t**) *Cicer microphyllum;* (**u**) *Macrotyloma uniflorum*; (**v**) *Caragana vesicolor*; (**w**) *Astragulus alpines*; (**x**) *Desmodium heterocarpon*

3.4.3 Sorghum (*Sorghum bicolor* L.)

Sorghum is a fast-growing, warm-weather annual that can provide fodder during lean periods. Sorghum is best suited to slightly warm, fertile soils whereas cool, wet soils generally limit its growth. The crop tolerates drought relatively well, though adequate fertility and soil moisture maximize sorghum yields. Sorghum is used both for grain and fodder. While some varieties are grown solely for grain, others have been developed for forage production, and some varieties are dual-purpose (Harada et al. 2000). The young plants contain an alkaloid, which releases hydrocyanic, or prussic acid, when hydrolyzed. When the crop is cut and field-cured or ensiled, the hydrocyanic acid degrades (2 to 3 weeks after ensiling) which greatly reduces toxicity.

3.4.4 Barley (*Hordeum vulgare* L.)

Barley is a major cereal crop primarily grown for its grain. Also yields valuable forage that can be grazed, cut for hay or silage while plant is still green, or cut after grain harvest as straw (Göhl 1982). Barley has fast establishment and winter production. It is best suited to late planting situations where its quick early growth under cold conditions is an advantage over other forage cereals. Barley provides valuable forage during drought periods or when the barley crop has suffered frost damage that has hindered grain production. Until the late 1990s, forage barley varieties had rarely been selected for improved forage quality and variety selection was based solely upon yield and other agronomic characteristics (Surber et al. 2011). When barley forage is directly grazed or cut at early stages, it could be harvested later for grain without significant decrease in grain yield.

3.4.5 Pearl Millet (*Pennicetum glaucum* L.)

Pearl millet is an erect annual grass with a profuse root system. It is mainly grown to produce hay, silage, green-chop, pasture and standover feed grazed directly other than for grains. Pearl millet stover is the part of the plant that remains after grain harvest and is a fibrous by-product with comparative low nutritive value. Pasture can be grazed 40 to 50 days after seeding, when it is about 40 to 50 cm high. Regular rotational grazing or mowing is advisable, because it increases the productive season by preventing heading and maintains high nutritive value. Pearl millet intended for hay should be cut at the boot stage. Two to three could be taken after cutting intervals of 3 to 4 weeks.

3.4.6 Indian Barnyard Millet [*Echinochloa frumentacea* (Link) Ridl.]

Barnyard millet is also an important source of fodder in the Himalayan region. It has probable Indian origin. It is a staple cereal in areas where climatic and edaphic conditions are unsuitable for rice cultivation. In India, barnyard millet is grown in the Himalayan region from the north to the Deccan plateau in the south. It is generally cultivated in hill slopes and undulating fields of hilly, tribal, or marginal areas. Barnyard millet has a wide adaptation capacity and can grow up to an altitude of 2000 m amsl during summer season (Gupta et al. 2009). It is a very fast-growing grass. Seed germinates quickly, axillary shoots develop already in the second week after emergence, and flowering starts 2 to 3 weeks later. Barnyard millet leaves are broad and the plant picks up good growth in short time thus producing voluminous fodder. Its fodder has high palatability and can be used for making hay or silage. It has a higher straw yield and fodder value even at multiple cuttings (Bandyopadhyay 2009). The fodder yield is about 6.30 tons/ha (Vanniarajan et al. 2018).

3.4.7 Finger Millet [*Eleusine coracana* (L.) Gaertn]

Finger millet is a variety of millet grown in the arid parts of Africa and Asia. Finger millet straw is used as fodder for cattle, sheep, and goats. It is a highly adaptable crop and has grown up to 2400 m above mean sea level in higher elevations in the Himalayas. Finger millet is a fast-growing cereal crop that reaches maturity within 3 to 6 months (Dida and Devos 2006). Under rainfed conditions, finger millet straw yield ranges between 2 and 3 t/ha and can reach 10 t/ha while under irrigated conditions (Sampath 1986). Finger millet should be cut at vegetative stage (around 80 days after planting) for making good-quality hay or silage (Madibela and Modiakgotla 2004). When cut two or three times during the growing season, it can produce green matter to the extent of 33 t/ha (Sampath 1986).

3.4.8 Foxtail Millet [*Setaria italica* (L.) Beauv.]

Foxtail millet is an annual, erect, fast-growing, leafy, and tufted grass that can reach up to 220 cm in height. Foxtail millet grows quickly and maturity occurs within less than 90 days. It does not regrow well after cutting or grazing. Foxtail millet may yield 15 to 20 t green matter/ha and 3.5 t hay/ha. The OMD and ME values for the hay and stover are much lower. The low digestibility of mature foxtail millet and other related stovers and straws is mainly due to the high proportion of coarse cell wall material that induces an incomplete and delayed access of the rumen microorganisms (Wilman et al. 1999) (Fig. 3.5).

Fig. 3.5 The portfolio of some important cereal forages of the Himalayan region. In reading order (**a**) Fodder maize; (**b**) Oat; (**c**) Sorghum; (**d**) Barley; (**e**) Pearl millet; (**f**) Indian barnyard millet; (**g**) Finger millet; (**h**) Foxtail millet

3.5 Range Forages

Rangelands are the major herbaceous sources for livestock feeding in the Himalayan region. The region has number of ranges dominated by range forages that cover large areas. Tall fescue, *Setaria*, orchard grass, and guinea grass are some range grasses that are now being domesticated for forage cultivation.

3.5.1 Bermuda Grass (*C. dactylon* Pers.)

Bermuda grass produces fodder of moderate nutritional quality. It is highly tolerant to drought and heavy grazing and therefore extremely present in rangelands of Himalayan region. It is also used for cut-and-carry, hay, and deferred feed. It has profound spreading ability. It is used to form dense swards, lawns, and turf with DM yield of about 5 to 15 t/ha. Bermuda grass is suitable for all ruminant species as pasture, hay, and haylage. OM digestibility for the fresh forage is comprised between 45 and 65% and slightly lower for the hay. HCN poisoning is the problem that has been reported when grazed Bermuda grass stands too soon after N fertilizer application. Bermuda grass hay could be fed to early lactating ewes as the sole forage, provided adequate energy and protein are supplied by a concentrate.

3.5.2 Nut Grass (*Cyperus rotundus* L.)

The names nut grass and nut sedge shared with the related species *Cyperus esculentus* are derived from its tubers having strong resemblance with nuts. *Cyperus rotundus* is a perennial grass that reaches a height of up to 140 cm. Young plants initially form white, fleshy rhizomes, up to 25 mm in dimension, in chains. Some rhizomes grow upward in the soil then form a bulb-like structure from which new shoots and roots grow. Rhizomes grow horizontally or downward and form dark reddish-brown tubers or chains of tubers.

3.5.3 Para Grass (*Brachiaria mutica* (Forssk.) Stapf)

Para grass is a semiaquatic, palatable and good-quality forage grass. It is particularly suited to poorly drained, swampy, and flooded areas. Para grass can be grazed, used in cut-and-carry systems or can be used for making hay or silage. It can be used for erosion control on river banks and steep slopes. The grass produces 5 to 12 t DM/ha/year. A comparative study based on *in sacco* measurements on 25 feed resources inferred that *Brachiaria mutica* has a good nutritive value and could be included in livestock diet to increase their productivity (Sharma and Sharma 2006).

3.5.4 Dallis Grass (*Paspalum dilatatum* Poir.)

The grass is a strongly tufted, leafy, sod-forming perennial grass. It is very palatable up to the pre-flowering stage. The palatability of this grass declines with maturity. It has more palatability than tall fescue (*Festuca arundinacea*) and Bahia grass (*Paspalum notatum*) (Nada and Sawamura 1985). The grass can be used for making good-quality silage provided that it is cut before flowering and remains free from paspalum ergot. The grass does not have high nutritional value; supplementation with maize silage, cereal grains, and protein or nitrogen sources is a preferred practice.

3.5.5 Cogon Grass (*Imperata cylindrica* (L) P. Beauv)

This grass is a component of native pastures and grasslands. The grass could be used at young stage, at 15 to 25 cm height. The grass has relatively poor nutritional quality with protein content less than 7% of DM. It is palatable only if cut frequently. Mature leaves are sharp and irritating which are not preferred by livestock.

3.5.6 Other Important Range Forages

Themeda anethera, Cassia mimisoides, Digitaria stricta, Pennisitum flaccidum, Setaria homonyma, Panicum maximum, Setaria italica, Arun dodonax, Achyranthes mutica, Cyrtococcum accresecns, Cleomella longipes, Dichanthium annulatum, Panicum auritum, Paspalidium punctatum Pennisetum orientale, Saccharum longisetosum, and *Brachiaria mutica* are other grasses of Himalayan region. The other range legumes naturally dominating in the Himalayan region include *Stylosanthes seabrana, Stylosantheshamata, Cliotoria ternatea,* and *Medicago* spp. (Fig. 3.6).

3.6 Fodder Trees and Shrubs of the Himalayan Region

Fodder trees and shrubs are important feed resources that could supply energy, protein, and other nutrients to livestock (Papachristou and Papanastasis 1994; Kemp et al. 2001). They are alternative sources of fodder during lean periods in the Himalayan region. Fodder trees remain lush green or less affected by seasonal dry conditions due to extensive root systems enabling them to extract water and nutrients from deep soil profile. These trees have good nutrient profile with high feeding value, longevity, and capacity for alleviating acute feed shortages when all other fodder sources have been exhausted (Katoch 2009; Katoch et al. 2017). The

Fig. 3.6 The portfolio of some important range forages in the Himalayan grasslands and rangelands. In reading order (**a**) Bermuda grass; (**b**) Nut sedge; (**c**) Cogon grass; (**d**) *Themeda* spp.; (**e**) Dallis grass; (**f**) Kentucky bluegrass; (**g**) Bahia grass; (**h**) Guatemala grass; (**i**) Mission grass; (**j**) Big-leaf bristle grass; (**k**) River grass; (**l**) Red fescue; (**m**) Spear grass; (**n**) Blue panic grass; (**o**) *Stylosanthes hamata*; (**p**) *Clitoria ternatea* (**q**) *Medicago tranculata*

presence of anti-nutritional components like tannins and phenols inhibit the sole use of tree fodder as livestock feed. Balaraman (1981) reported 44 fodder trees from Sikkim Himalaya for livestock feeding and many of these are common in other parts of Himalayan region.

In Himalayan region, farmers maintain naturally regenerating tree species, particularly on edges of terraced agriculture fields. This reduces the pressure from forest resources in terms of fodder and fuelwood along with site improvement. Around 84 major fodder trees species and 40 shrubs are used for livestock feeding in middle hills of Himachal Pradesh (Radotra et al. 2015).

In Uttarakhand hills, farmers maintain naturally regenerating tree species, particularly on the edges of terraced agriculture fields without any significant input of manpower. This reduces the pressure from forest resources in terms of fodder and fuelwood along with site improvement. Important forage trees of Himalayan region have been enlisted in Table 3.3.

The temperate species *Salix* and *Populus* spp. have been used as direct browsing feed for livestock during summer and autumn droughts (Moore et al. 2003; McWilliam et al. 2005; Pitta et al. 2005). Hedgerows and windbreaks (i.e., shelterbelts) are primarily used to enclose the fields and meadows (Baudry et al. 2000) in addition to controlling soil erosion (Nerlich et al. 2013), while foliage is browsed for feeding livestock. Leucaena (*Leucaena leucocephala*) is another nutritive tree foliage used for livestock feeding in Himalayan region.

In Garhwal Himalaya, *Grewia oppositifolia, Morus serrata, Bauhinia variegata, Quercus leucotrichophora, Quercus floribunda, Ilex dipyrena, Oogeinia oojeinensis, Boehmeria rugulosa, Morus alba* and *Celtis australis* are common species that are used as fodder. C*eltis australis, Morus serrata, Oogeinia oojeinensis, Acer* spp., *Ficus* spp., *Machilus duthei* are the main fodder trees that are lopped during summer.

3.7 Crop Residues

Crop residues are mainly fibrous materials that are by-products of cultivated crops. Due to the intensive crop production, these residues form a high percentage of the total mass produced annually. Rice, wheat, oat, and barley straw are the main crop residues of the region.

Rice straw is one of the major crop residues worldwide. Rice is harvested manually and threshed by hand and occasionally by bullock treading. Hand-beaten straw is of better quality and easier to store. Storage is usually in circular or rectangular stacks in the open field. Wheat is the second most important food crop. After harvesting wheat, straw is heaped either under cover or in mud-plastered waterproof heaps. Though oat and barley straw are also important feeding resources, they are less preferred as compared to rice and wheat straw. Maize produces 18 to 20 t/ha of stover. Maize cobs are hand-harvested and the stover thereafter is dried in the field. The residue from vegetable crops such as brassicas, legumes (haulms) like soybean, rice bean, and common beans are also used in the livestock feeding.

Table 3.3 Major forage resource of the Himalayan region

Family name	Plant name	Plant part used	Family name	Plant name	Plant part used
Forage grasses		**Forage shrubs and trees**			
Poaceae	*Avena sativa*	Whole plant	*Salicaceae*	*Salix alba*	Leaves and twigs
	Hordeum vulgare	Straw and seed		*Salix tetrasperma*	-do-
	Triticum aestivum	Straw		*Salix elegans*	-do-
	Panicum millaceum	Straw and seed		*Salix sclerophylla*	-do-
	Bromus spp.	Whole plant		*Salix daphnoides*	do
	Elymus spp.	-do-		*Populus balsamifera*	Leaves
	Digitaria sanguinalis (L.) Scop.	-do-		*Populus euphratica*	Leaves
	Festuca rubra	-do-	*Elaeagnaceae*	*Hippophaea rhamnoides*	Fruits and leaves
	Poa annua	-do-		*Elaeagnus angustifolius*	Fruits and leaves
	Poa pratensis	-do-	*Saxifragaceae*	*Ribes alpestre*	Berries and leaves
	Trisetum spp.	-do-		*R. orientale*	-do-
	Setaria spp	-do-	*Moraceae*	*Morus alba*	Leaves
	Phragmites australis	-do-		*Morus serrata*	Leaves
	Pennisetum spp.	-do-	*Polygonaceae*	*Fagopyrum tataricum*	Seeds and leaves
Forage legumes				*Fagopyrum esculentum*	-do-
Leguminosae	*Medicago sativa*	Whole plant		*Poligonum aviculare*	Whole plant
	Medicago falcata	-do-	*Chenopodiaceae*	*Chenopodium album*	Leaves
	Medicago media	-do-		*Chenopodium botrys*	-do-
	Lotus corniculatus	-do-		*Kochia stellaris*	-do-

(continued)

Table 3.3 (continued)

Family name	Plant name	Plant part used	Family name	Plant name	Plant part used
	Astragalus spp.	Leaves	*Brassicaceae*	*Capsella bursa-pastoris*	-do-
	Cicer microphyllum	Pods and leaves		*Lepidium* spp.	Leaves
	Lathyrus sativus	-do-	*Asteraceae*	*Artemisia siversiana*	Whole plant
	Lathyrus humilis	-do-		*Cichorium intybus*	-do-
	Melilotus officinalis	Whole plant		*Lactuca* spp.	-do-
Other forages				*Taraxacum* spp.	-do-
Cyperaceae	*Carex* spp.	-do-	*Convolvulaceae*	*Convolvulus arvensis*	-do-
	Cyperus spp.	-do-	*Leguminosae*	*Robinia pseudoacacia*	Pods and leaves
	Kobresia spp.	-do-	*Rosaceae*	*Rosa webbiana*	Fruits
Amaranthaceae	*Amaranthus caudatus*	Grains		*Potentilla fruticosa*	Whole plant
Iridaceae	*Iris lactea*	Whole plant	*Berberidaceae*	*Berberis ulcina*	Fruits and leaves
			Capparidaceae	*Capparis spinosa*	Fruits and leaves
			Apiaceae	*Prangos pabularia*	-do-

3.8 Other Feed Resources in Himalayan Region

Besides forage grasses and legumes, root crops, *Brassica* spp., and vetches are also used as supplementary feed sources to livestock. Due to early bulking capacity and short duration, these are often grown as catch crops. Aquatic weeds are also collected from lakes and other water bodies and fed to the animals (Radotra et al. 2015). Majority of aquatic plants have little, if any, value as human food and a practical utilization of this resource is as fodder for livestock. Fruit and vegetable wastes contain more than 7.0% crude protein and are suitable for maintenance requirement of ruminant animals. The vegetable wastes contain more protein as compared with fruit wastes. Vegetable wastes like cabbage, cauliflower, baby corn husk, and radish leaves are used as feed in upper Himalayan regions. Potato haulms and pea vines are also fed to livestock where crop flourishes well and occupies larger areas in upper regions of Himachal Pradesh.

3.9 Forage Productivity in Himalayan Region

3.9.1 Forage Resources and Their Productivity in North-West Himalayas

Grasslands, rangelands, and forests are major sources of forage for livestock in the Himalayan region. Traditionally, the pastures in subalpine and alpine Himalayan regions had been the important source of fodder availability, however, due to extremes in climate, poor management, and constant grazing, the grasslands have become less productive. Nomadic graziers (Khadwals, Gaddis, and Gujjars) use subalpine and alpine pastures for rearing their livestock. Due to ever-increasing demand for animal products, the livestock population has increased manifold, resulting in grazing pressure on pastures and loss of edible species. Forage resources and their productivity in North-West Himalayan region have been discussed below.

3.9.1.1 Major Forage Resources and Productivity in Himachal Pradesh

Livestock production is an integral part of the mixed farming system in the hilly areas. It is crucial for the well-being of farmers in hills. Himachal Pradesh is an agrarian state where crop farming, animal husbandry, and pastoralism are intimately integrated and play significant role in sustaining the rural livelihoods. Owing to marginal and submarginal nature of land holdings of the farmers, the forage resource base in the state is multidimensional. The contribution of livestock to income and employment generation is quite significant and second after the crop production. The contribution of livestock to the GDP of the state is about 40% emanating from milk and other livestock products.

Although, livestock farming plays a significant role in the economy of rural households, yet area under forage cultivation remained almost negligible. The area under permanent pastures and grasslands is 36% of the total geographical area. In mid- and high hills, the land constraints do not allow the farmers to go for cultivation of the forage crops resulting in gap between demand and supply of feed resources to livestock. This striking imbalance in requirement and availability of fodder has declined the livestock productivity in the state. The area under forages and forage productivity in Himachal Pradesh has been summarized in Table 3.4.

3.9.1.2 Major Forage Resources and Productivity in Uttarakhand

Uttarakhand, an important state in Indian Himalayan region, is characterized by mixed crop–livestock systems in smallholder farming. Diversity in livestock composition is the characteristic feature of mainland of Uttarakhand. Climatic conditions are suitable, ranging from subtropical to temperate and alpine, and provide suitability for livestock farming. A large proportion of livestock is raised under mixed

Table 3.4 Area under production and productivity of forages in Himachal Pradesh

Source	Area (Million ha)	Production (Million tonnes)	Productivity (t/ha)
Cultivated fodder crops	0.28	0.76	9.00
Permanent pastures and grasslands	15.03	36.07	8.00
Cultivable wastelands	1.18	1.42	4.00
Barren and uncultivable lands	6.54	5.88	3.00
Misc. tree crops and groves	0.67	0.41	2.00
Area under bunds/terraces	0.57	0.85	5.00
Total	**24.27**	**45.39**	

(Source: http://himachal.nic.in/economics/reports/StateAbstract/2013_A1b.pdf)

Table 3.5 Fodder availability components in Uttarakhand

Sources	Production (Million tonnes)	Percentage
A. Grasses (fresh and dry)	3.89	25.82
B. Fodder leaves	3.96	27.67
C. Agriculture sector	6.65	46.51
Total green fodder	**14.30**	
D. Sugarcane tops	1.71	25.79
E. Seasonal fodder crops and potato leaves	1.62	24.25
F. Straws, stalks, and stovers	3.32	49.96
Total dry fodder production	**6.65**	

Source: Uttaranchal Livestock Development Board (2015)

cropping system. The region possesses a huge natural resource base for livestock farming that includes common property resources and grasslands. Temperate grasslands occupy the mainland of Uttarakhand. During summer, farmers migrate to higher hills along with their cattle around for 4 months and move downhills during winters. In low hills, the fodder is available from seasonal fodder crops, grasslands and pastures, fodder trees, and crop residues.

In Uttarakhand hills, fodder trees, shrubs, and grazing in the forests are the main sources for the livestock feed including use of agricultural residue (Singh and Sundriyal 2009). In mid-hills of Uttarakhand, about 30 to 50% of total animal feed, mainly grass and tree fodder, is produced from forests and grasslands (Singh and Naik 1987). In parts of Garhwal and Kumaon Himalaya (Uttarakhand), the standing biomass of grasses increases with increasing altitude up to 3750 m (Dabadghao and Shankarnarayan 1973).

While the growing season in the temperate region generally begins in April, the subalpine and alpine grasslands start sprouting during June to July. Thus, biomass production in these grasslands is lower than in tropical grasslands (Ram et al. 1989) due to the shorter growing season. Different sources contributing to fodder availability have been presented in Table 3.5.

3.9.1.3 Major Forage Resources and Productivity in Jammu and Kashmir

Livelihood of population of Jammu and Kashmir state revolves around agriculture, horticulture, and allied sectors. These sectors constitute the mainstay of the state's economy. Pastures and grasslands are the major feed resource of Jammu and Kashmir but during winter season, the entire landscape gets covered with snow and the animals cannot graze freely in grasslands. This period lasts from November to March. From April to October besides grazing following forage resources are used. Fodder trees and shrubs contribute 8 to 60% of total fodder supply in the state of Jammu and Kashmir. Over 200 species of fodder trees and shrubs have been identified in different agroecological zones. Farmers use fodder trees as a feed source during the unavailability of green fodder. *Celtis australis*, *Salix alba*, and *Robinia pseudoacaia* are the major fodder trees. Foliage of *Robinia* and *Salix* is not much relished by the livestock because of their bitter taste. Rice and maize are the major crops of the area and paddy straw and maize stover are extensively used to feed the animals.

More than 340 thousand hectares of area is under orchards and there is immense scope for introduction of fodder crops as intercrops. Green herbage productivity in alpine pastures of Jammu and Kashmir varies in different regions. The productivity in Daomarg and Duksum areas of the Jammu and Kashmir state is 4.70 t/ha and 29.10 t/ha, respectively (Radotra et al. 2015). The pastoralists mainly depend on grasslands and rangelands for grazing. Bakarwal, Gujjars, Chopans, Gaddis, and Changpas are the major pastoralists of the state which depend on livestock for their livelihood.

3.9.2 Forage Resources and Productivity in Eastern Himalayan Region

The Eastern Himalayas extend along the north and south faces of the Himalayan range from the Kali Gandaki Gorge in central Nepal eastward through Tibet and India's Sikkim state, Bhutan, India's Arunachal Pradesh, and northernmost Myanmar. In Eastern Himalayan region, among annual forage crops, oats are cultivated as fodder crops. Forage crops like maize, teosinte, jobstear, dinanath grass, and soybean are successfully grown during pre-*kharif* and *kharif* season and while oat, pea, forage mustard, Japanese rape, Chinese cabbage, and turnip could be grown during *rabi*. They could produce an average of 25 to 50 t/ha green forage. *Festuca arundinacea* and *Lolium multiflorum* provide 41.50 and 35.60 tonnes of green fodder /ha/year. Paddy straw and maize stover are the major crop residues of this region (IGFRI, Jhansi vision 2050 2015).

Tree leaves constitute 25 to 60% of total green fodder fed to livestock in the hills of Sikkim and Nepal. *Brassaiopsis hainia* and *Litsea polyantha* are adapted between 600 and 1700 m and same adaptability is found in *Garuga pinnata*. Important tree species used for fodder purpose are *Artocarpus lakoocha, Camptomeris leucaene, Bombus Mori, Brassaiopsis hainia, Litsea polyantha, Salix* spp. *Helianthera ligustrina, Ficus auriculata, Ficus nerifolia, Bauhania varigeta, Grewia alastica,*

Table 3.6 Productivity of important forage grasses in Eastern Himalayas

Tropical grasses	Productivity (t/ha/year)	Temperate grasses	Productivity (t/ha/year)
Bajra × Napier hybrid	222.80	*Dactylis glomerata*	26.70
Bracharia ruziensis	140.50	*Festuca arundinacea*	41.50
Erianthus longisetosum	42.10	*Lolium multiflorum*	35.60
Erianthus sikkimensis	47.00	*Lolium perenne*	21.20
Napier × Bajra hybrid	158.80		
Paspalum dilatatum	64.20		
Pennisetum polystachyon	61.10		
Setaria palmifolia	45.40		
Setaria sphacelata cv. *Nandi*	105.10		
Thysanlaena agrostis	73.60		
Tripsacum laxum	94.00		

Source: (IGFRI, Jhansi, vision 2050)

and *Bahunia purpurea.* The productivity of fodder grasses grown in Eastern Himalayas has been given in Table 3.6.

3.10 Grazing Patterns and Maintenance of Pastures and Grasslands in Himalayan Region

In the Himalayan region, livestock plays a crucial role both at the household level as well as the state level for overall socioeconomic development. The main sources of fodder availability, in the Himalayan region, for the livestock are the abundantly available grazing lands. Pastoralists view livestock as symbols of wealth as they provide milk, small stock meat, and wool and also provide security against unpredictable climatic conditions. They also rear indigenous animal breeds, retaining and maintaining rich genetic diversity. The grazing lands are utilized under following traditional livestock rearing/grazing systems:

(a) **Transhumance system:** In this system, grazing is cyclical beginning at the end of winter with flocks and herds leaving lowland grazing areas where permanent villages are located and moving to mountain pastures for grazing during summer. Latitudinal transhumance occurs in tropical climates where the cyclic movement is under the influence of alternating wet and dry seasons.

(b) **Nomadic system:** This pastoralist system is based on extensive movement of herds and flocks in search of forage, led by human family units with no permanent home base.

(c) **Semi-sedentary system:** This pastoralist system is based on a village permanently occupied by women and children where livestock herds and flocks are tended by men. They are absent from the village for extended periods of time in search of forage.

(d) **Sedentary systems:** This system is managed at a particular location by resident management. This system may be managed by either or both extensive and intensive grazing managements and can include rangeland, pastureland, cropland, and forestland within grazing system (Allen et al. 2011).

Transhumance is the movement of herds to exploit seasonal availability of pastures. Transhumant pastoralists migrate continuously from one place to another. The migratory routes are well defined. In this system, grazing areas are demarcated for different groups. They migrate from summer pastures to winter pastures with their flocks having some sort of living arrangement at both the places.

In the hills of Himalayan region, the transhumant pastoralists follow a cyclical migratory pattern from cool highland valleys in summer to warmer lowland valleys in winter. In terms of ecological adaptations, the two most significant factors for transhumance are seasonal severity of winters, associated with the presence of territorial use of highland and lowland pastures.

Nomadism is viable in the extreme hot and cold. Changpas are nomads of Changthang in Ladakh region who raise herds of yaks and flocks of sheep and pashmina goats. However, their animals are not raised on any cultivated fodder crops but survive exclusively on natural pastures.

The nomadic Dokpas (Tibetan pastoralists) of Trans-Himalayan ecosystems are a unique example of survival in drought and cold-stricken landscapes through mobile livestock production systems. They rear yaks, dzos (cow yak hybrid), sheep, and goats (pashmina type) in the plateaus and meadows. With adaptation to the harsh climatic conditions, Dzumsa (traditional institution in North Sikkim) and the Dokpas have indigenously regulated and managed the grazing regimes of the Greater Himalayas, transition zones, and the Trans-Himalayan meadows. The Dzumsa specifies dates of movement of herds from one location to another as per the lunar calendar which is applicable to all herders round the year. Dzumsa strongly controls the rotational grazing for rangeland resource management and regeneration.

The sedentary system is practiced by the farmers living between an altitude of 1000 and 2500 m amsl. In this zone, all cultivable lands are utilized for cultivation of cereals, while animals roam free in forests and subalpine pastures for grazing. The grazing continues for about 10 months in a year. During the months of December and January, crop residues and conserved tree leaf fodder are fed to the livestock.

Semi-sedentary system of livestock rearing is practiced by the farmers living below 1000 m altitude. The farmers of this zone generally depend on professional graziers who take livestock to pastures and forests in subalpine and alpine zones during summer. The livestock is brought back during autumn and is stall-fed on crop residues and tree fodder during winter season.

Since the lower altitude grazing lands are limited in area and the livestock population in these areas far exceeds the carrying capacity, the practice of summer migration to the higher altitude alpine meadows has become necessary to sustain the number of livestock. Agropastoralists in the western and central Himalayan regions keep more cattle than they really need because of easy access to free grazing areas and their inability to dispose or cull the population due to religious sentiments. Uncontrolled grazing on the steeper slopes reduces water-holding capacity and compaction reduces the permeability of the soil. Continuous grazing also creates channels or paths on hill slopes which remove huge quantities of soil during rains.

Overgrazing in mid- and high hills has resulted in decrease in forage grass cover and increase in the unpalatable alien species such as *Rhododendron anthopogon*, *Berberis* spp., *Euphorbia wallichii*, *Euphorbia longifolia*, *Iris kumaonensis*, *Lantana*, *Eupatorium*, *Parthenium*, and others.

Despite the fact that domestic animals are an integral part of agropastoral ecosystems and that grazing-based animal husbandry is the mainstay of the economy in many parts of the Himalayas, no studies and policy guidelines are available for optimal use of grazing resources. Plantation of trees under agroforestry system and round the year production of fodder could be the best option for the agropastoralists.

3.11 Multiple Cropping Systems Influencing Forage Productivity and Nutritional Value

Multiple cropping is a form of ecological intensification, which is highly sustainable when two or more crops are grown at the same time or in a sequence. Intercropping forage legumes with cereals or grasses is one of the viable options offering potential for increasing forage productivity and nutritional quality. The inclusion of forage legumes in low-input grassland mixtures is vital to improve biomass production, forage quality, and ultimately soil fertility. The improvement of forage quantity and quality is crucial for improved animal performance that is a goal of all livestock farmers. Mixing legumes and grasses serves as the forage supplementary alternative since pure grasses or cereals provide comparatively poor-quality fodder due to their inherent lower crude protein content.

Legume species like *Trifolium* species whose CP levels remain higher even during the driest seasons, while simultaneously partitioning the fixed nitrogen to the companion grasses, are highly recommended for grass–legume mixtures. Mixing legumes with grasses increases the crude protein content of the herbage mixture relative to that of grass monocultures. Grasses have lower crude protein content usually lower during the winter season (dry) therefore, forage legumes could be incorporated in sole grass stands to increase forage quantity and quality during dry seasons.

Inclusion of forage legumes is a better alternative to supplement the nutritive value of natural grasses. The inclusion of forage legumes in cropping sequence improves the quality in terms of proteins and minerals and helps to economize the use of nitrogenous fertilizers. Legume species like *Lespedeza cuneata* could be utilized for production and baling of hay during summer and autumn for utilization during winter or early spring as its fiber content increases with advancing maturity.

Table 3.7 Multiple cropping systems

Cropping sequence for year-round fodder production	
Napier × Bajra hybrid + cowpea – berseem	
Maize + cowpea – MP Chari + cowpea – berseem + Japanese rape	
MP Chari + cowpea – berseem + Japanese rape	
Cowpea – MP Chari + cowpea – berseem + Japanese rape	
Napier × Bajra hybrid + cowpea – berseem – cowpea	
Intensive forage sequences recommended for different regions	
(i) North zone	
(a)	Maize + cowpea – sorghum + cowpea (2 cuts) – berseem + mustard
(b)	Sudan grass + cowpea – maize + cowpea – turnip – oats (2 cuts)
(c)	Hybrid Napier or Setaria interplanted with cowpea in summer and berseem in winter
(d)	Teosinte + cowpea (2 cuts) – carrot – oats + mustard (2 cuts)
(ii) Western and Central zone	
(a)	Bajra + guar (Clusterbean) (2 cuts) – Annual Lucerne (6 cuts)
(b)	MP Chari + cowpea (2 cuts) – maize + cowpea - teosinte + cowpea (2 cuts)
(c)	Hybrid Napier or Guinea or Setaria grass interplanted with cowpea in summer + berseem in winter (8–9 cuts/year)
(d)	Hybrid Napier or Guinea or Setaria grass interplanted with Lucerne (8–9 cuts/year)
(iii) Eastern zone	
(a)	Maize + cowpea – teosinte + rice bean (2 cuts) – berseem + mustard (3 cuts)
(b)	M.P. Chari + cowpea – Dinanath grass (2 cuts) – berseem + mustard (3 cuts)
(c)	Para grass + *C. pubescens* (8–9 cuts/year)
(d)	Hybrid Napier or Setaria grass interplanted with Subabul or common Sesban (*Sesbania sesban*) (9–10 cuts/year)
(iv) Southern zone	
(a)	Sorghum + cowpea (3 cuts) – maize + cowpea – maize + cowpea
(b)	Hybrid Napier or Guinea or Setaria grass interplanted with Lucerne (8–9 cuts) or hybrid Napier + Subabul/Sesbania (9–11 cuts/year)
(c)	Sudan grass + cowpea (3 cuts) – M.P. Chari + cowpea (3 cuts)
(d)	Para grass + Centro (*C. pubescens*) (9–11 cuts/year)

Source: Bhagmal et al. (2009)

Trifolium repens is a promising forage, which partition fixed nitrogen to the companion nonlegume plant and remain palatable throughout the growing seasons. Various studies have been conducted to investigate the effect of intercropping of legumes with grasses. Intercropping of alfalfa and corn-rye revealed higher forage production performance with net increase of 9.52% and 34.81% in DM yield, 42.13% and 16.74% in CP yield, 25.94% and 69.99% in degradable dry matter yield, and 16.96% and 5.50% in degradable CP yield than rotation and alfalfa sole cropping systems, respectively (Zhang et al. 2015). Multiple cropping system involving berseem with hybrid Napier in spring and intercropping the inter-row spaces of the grass with cowpea during summer after the final harvest of berseem are the promising cropping sequences for fodder availability for longer period of year (Table 3.7). The perennial grasses like *Setaria* and guinea grass could be intercropped with the inclusion of berseem with lucerne for improving productivity and nutritional value.

Mixed farming is indispensable component of Himalayan region, where livestock farming and crop farming are interlinked with socioeconomic values and cultural beliefs. Natural grasslands and pastures are the important feed resources to livestock. The other important feed resources of the region are cultivated fodder, crop residues and concentrates. Tree forages and crop residues are specially used as fodder during forage scarcity condition in Himalayan region. Due to increasing livestock population, the attainment of feed as well as nutritional security to livestock is one of the major concerns for achieving sustainable livestock productivity in Himalayan region. The adoption of modern practices and effective management like multiple cropping, timely harvest of fodder crops for enhanced nutrition, crop residues management and other improved agronomic practices could have visible impact on improving fodder availability to the livestock.

References

Aganga AA, Tshwenyane S (2004) Potentials of guinea grass (*Panicum maximum*) as a forage crop in livestock production. Pak J Nutr 3:1–4

Allen VG, Batello C, Berretta EJ, Hodgson J, Kothmann M, Li X, McIvor J, Milne J, Morris C, Peeters A, Sanderson M (2011) An international terminology for grazing lands and grazing animals. Grass Forage Sci 66:2–28

Balaraman N (1981) Feeds and fodder resources for livestock in Sikkim. Research bill no. 13 ICAR, research complex for NEH region, Sikkim Centre, Gangtok

Bandyopadhyay BB (2009) Evaluation of barnyard millet cultivars for fodder yield under single and double cut treatments at higher elevation of hills. Agric Sci Dig 29:66–68

Baudry J, Bunce RGH, Burel F (2000) Hedgerows: an international perspective on their origin, function and management. J Environ Manage 60:7–22

Bhagmal, Singh KA, Roy A, Ahmed S, Malaviya D (2009) Forage crops and grasses. In: Handbook of agriculture. ICAR, New Delhi, pp 1353–1417

Brink M, Jansen PCM (2006) *Vigna aconitifolia* (Jacq.) Maréchal. In: Brink M, Belay G (eds) PROTA (plant resources of tropical Africa/Ressources végétales de l'Afrique tropicale), Wageningen

Chanthakhoun V, Wanapat M (2010) Effect of legume (*Phaseolus calcaratus*) hay supplementation on rumen cellulolytic bacterial populations in swamp buffaloes investigated by the real-time PCR technique. J Anim Vet Adv 9:1654–1659

Chanthakhoun V, Wanapat M, Wachirapakorn C, Wanapat S (2011) Effect of legume (*Phaseolus calcaratus*) hay supplementation on rumen micro-organisms, fermentation and nutrient digestibility in swamp buffalo. Livest Sci 140(1/3):17–23

Cook BG, Pengelly BC, Brown SD, Donnelly JL, Eagles DA, Franco MA, Hanson J, Mullen BF, Partridge IJ, Peters M, Schultze-Kraft R (2005) Trop forages. CSIRO, DPI&F (Qld), CIAT and ILRI, Brisbane

Dabadghao PM, Shankarnarayan KA (1973) The grass cover of India. Indian Council of Agricultural Research, New Delhi, p 714

Dida MM, Devos KM (2006) Finger millet. In: Chittarajan K (ed) Genome mapping and molecular breeding in plants, vol 1. Springer-Verlag, Berlin, pp 333–344
Faber-Langendoen D, Jose C (2010) World grasslands and biodiversity patterns. NatureServe, Arlington, VA, p 24
FAO (2011) Grassland index: a searchable catalogue of grass and forage legumes. FAO, Rome
Göhl (1982) Division de Production et Santé Animale. FAO, Rome
Gupta A, Mahajan V, Kumar M, Gupta HS (2009) Biodiversity in the barnyard millet (*Echinochloa frumentacea* link, Poaceae) germplasmin India. Genet Resour Crop Evol 56:883–889
Harada H, Yoshimura Y, Sunaga Y, Hatanaka T (2000) Variations in nitrogen uptake and nitrate-nitrogen concentration among sorghum groups. Soil Sci Plant Nutr 46(1):97–104
House JI, Hall DO (2000) Tropical savannas and grasslands. In: Mooney H, Roy J, Saugier B (eds) Terrestrial global productivity: past, present, and future. Academic, San Diego, CA, pp 363–400
IGFRI (2015) IGFRI Vision 2050. Indian Grassland and Fodder Research Institute, Jhansi
Katoch R (2009) Tree fodder: an alternate source of quality fodder in Himachal Pradesh. Range Manag Agrofor 30:16–24
Katoch R (2013) Nutritional potential of ricebean (*Vigna umbellata*): an underutilized legume. J Food Sci 78:8–16
Katoch R, David B, Priya KN, Bandhari JC (2007a) Fodder yield and nutritional attributes of ricebean (*Vigna umbellata*). Range Manag Agrofor 28:260–262
Katoch R, Sood S, Kumar N, Bandhari JC (2007b) Fodder production and nodulation efficiency of different rice bean genotypes. Forage Res 33:73–77
Katoch R, Chand U, Bandana (2007c) Red clover: present status and future needs. Forage Res 33: 1–8
Katoch R, Singh SK, Tripathi A, Kumar N (2017) Effect of seasonal variation in biochemical composition of leaves of fodder trees prevalent in the mid-hill region of Himachal Pradesh. Range Manag Agrofor 38:234–240
Kemp PD, Mackay AD, Matheson LA, Timmins ME (2001) The forage value of poplars and willows. Proc N Z Grassl Assoc 63:115–120
Kour K, Sangwan PL, Khan I, Koul S, Sharma SN, Kitchlu S, Bani S (2011) Alcoholic extract of *Cicer microphyllum* augments Th1 immune response in normal and chronically stressed Swiss albino mice. J Pharm Pharmacol 63(2):267–277
Kumar A, Adhikari B, Rawat G (2016) *Caragana versicolor* Benth. (Fabaceae), a keystone species of high conservation concern in the Hindu Kush Himalayan region. Curr Sci 111:985–987
Madibela OR, Modiakgotla E (2004) Chemical composition and *in vitro* dry matter digestibility of indigenous finger millet (*Eleusine coracana*) in Botswana. Livest Res Rural Dev 16(4):26
Malaviya DR, Roy AK, Kaushal P (2015) Conservation of grassland plant genetic resources through people participation. In: Roy MM, Malaviya DR, Yadav VK, Singh T, Sah RP, Vijay D, Radhakrishna A (eds) The 23rd International Grassland Congress (Sustainable use of grassland resources for forage production, biodiversity and environmental protection) New Delhi, India 20–24 November 24, 2015. Range Management Society of India, Jhansi, pp 318–326
McWilliam EL, Barry TN, López-Villalobos N (2005) Organic matter digestibility of poplar (*Populus*) and willow (*Salix*) forage trees and it's *in vitro* prediction. J Sci Food Agric 85: 1098–1104
Mohan VR, Janardhan K (1994) Chemical composition and nutritional evaluation of raw seeds of six rice bean varieties. J Indian Bot Soc 73:259–263
Moore KM, Barry TN, Cameron PN, Lopez-Villalobos N, Cameron DJ (2003) Willow (*Salix* sp.) as a supplement for grazing cattle under drought conditions. Anim Feed Sci Technol 104:1–11
Nada Y, Sawamura H (1985) Seasonal productivity of tall fescue, Dallis grass and Bahia grass grown in pure pastures, and rotational grazing combining these pastures in Kyushu district. J Japan Soc Grassl Sci 30:428–433

Nerlich K, Graeff-Hönninger S, Claupein W (2013) Agroforestry in Europe: a review of the disappearance of traditional systems and development of modern agroforestry practices, with emphasis on experiences in Germany. Agr Syst 87:475–492

Papachristou TG, Papanastasis VP (1994) Forage value of Mediterranean deciduous woody fodder species and its implication to management of silvo-pastoral systems for goats. Agr Syst 27:269–282

Pathak PS, Dagar J (2016) Indian grasslands and their management. In: Grassland: a global resource perspective. IGFRI, Jhansi, pp 3–36

Pitta DW, Barry TN, Lopez-Villalobos N, Kemp PD (2005) Effects on ewe reproduction of grazing willow fodder blocks during drought. Anim Feed Sci Technol 120:217–234

Radotra S, Dev I, Ahmad A, Kannan A (2015) Pasture and forages in North Western Himalayan region: current status and future strategies. In: Proceedings of the status and future prospects of animal production system in Himalayan Region

Rahmani A (2006) A note for the task force on deserts and grasslands of India. Submitted to Planning Commission, Government of India

Ram J, Singh JS, Singh SP (1989) Plant biomass, species diversity and net primary production in central Himalayan high altitude grassland. J Ecol 77:456–468

Sampath SR (1986) Scope for using small millets as forage in India. In: Seetharam A, Riley KW, Harinarayana G (eds) Small millets in global agriculture. Proceedings of the 1st International small millets workshop Bangalore, India, October 29-November 2, 1986

Santos EM, Zanine AM, Ferreira DJ, Oliveira JS, Penteado DSC, Pereira OG (2008) Activated inoculant improve Tanzania-grass (*Panicum maximum*) silage. Arch Zootec 57:35–42. (Eng. Abstr)

Sharma K, Sharma SD (2006) Chemical composition and *in sacco* degradibility of some fodder tree leaves, shrub leaves and herbaceous plants. Indian J Anim Sci 76:538–541

Sharma CM, Baduni NP, Nautiyal DP (1999) Socio-economic strategies for environmental conservation in rural Garhwal Himalaya. Ecol Environ Conserv 5:227–231

Singh V, Naik DG (1987) Western Himalayas: environment, problems and development. In: Pangtey YPS, Joshi SC (eds) Animal draught power in the mid-altitude Himalayan villages. Gyanodaya Prakashan, Nainital

Singh N, Sundriyal RC (2009) Fuelwood, fodder consumption and deficit pattern in central Himalayan village. Nat Sci 7:85–88

Singh JS, Lauenroth WK, Milchunas DG (1983) Geography of grassland ecosystems. Prog Phys Geo 7:46–80

Soromessa T (2011) *Heteropogon contortus* (L.) P. Beauv. Ex Roem. & Schult. Record from Protabase. In: Brink M, Achigan-Dako EG (eds) PROTA (plant resources of tropical Africa/ Ressources végétales de l'Afrique tropicale), Wageningen, Netherlands

Surber L, Abdel-Haleem H, Martin J, Hensleigh P, Cash D, Bowman J, Blake T (2011) Mapping quantitative trait loci controlling variation in forage quality traits in barley. Mol Breed 28(2): 189–200

Surhone LM, Tennoe MT, Henssonow SF (2010) Panicum antidotale. VDM Verlag Dr. Mueller AG & Co. Kg

Surmen M, Yavuz T, Albayrak S, Cankaya N (2013) Forage yield and quality of perennial ryegrass (*Lolium perenne* L.) lines in the black sea coastal area of Turkey. Turk J Field Crops 18:40–45

Teitzel JK, Chen CP (1992) In: Mannetje LT, Jones RM (eds) *Centrosema pubescens* Benth. Record from Proseabase. PROSEA (Plant Resources of South-East Asia) Foundation, Bogor

Tenikecier H, Ates E (2018) Chemical composition of six grass species (Poaceae sp.) from protected forest range in northern Bulgaria. Asian J Appl Sci 11:56. https://doi.org/10.3923/ajaps.2018.71.75

Thompson DJ, Stout DG (1997) Mixtures of Persian clover with Italian ryegrass or barley-Italian ryegrass for annual forage. Can J Plant Sci 77:579–585

Vanniarajan C, Anand G, Kanchana S, Arun Giridhari V, Renganathan VG (2018) A short duration high yielding culture—barnyard millet ACM 10145. Agric Sci Dig A Res J 8:123–126

White R, Murray S, Rohweder M (2000) Grassland ecosystems. World Resources Institute, Washington, DC

Wilman D, Lian W, Qin HW, Ji YL (1999) Intake and digestibility of diets derived from stovers and straws compared with lucerne hay and sweet potato haulm. J Agric Sci 132(2):207–213

Zhang B, Yin Y, Xie J, Li Z, Yang G (2015) Legume-cereal intercropping improves forage yield, quality and degradability. PLoS One 10(12):e0144813. https://doi.org/10.1371/journal.pone.0144813

4 Constraints Affecting Forage Availability in Himalayan Region

Livestock contributes enormously to food and nutritional security apart from livelihood security to the rural households all over the world. In the Himalayan region, the rearing of livestock is one of the important pursuits of rural households. The agriculture and livestock farming are interlinked *via* cultural, religious beliefs and economical perspectives. Though rearing of livestock is a major occupation of the rural households of the Himalayan region, the productivity of livestock is low.

While discussing about livestock farming in the Himalayan region, there are many constraints which affect the sustainable improvement of livestock productivity. Livestock in the mountains depend predominantly on open grazing or stall-feeding. In later, usually the by-products of agricultural produce like wheat straw, paddy straw, hay, tree foliage are mixed with green fodder or sometimes with feed concentrates. Degradation of forest in the high-pressure mid-hills has reduced natural fodder availability to livestock. Moreover, continuous increase of feed concentrate prices, lack of management practices, and information on fodder varieties excelling in productivity and quality and excessive utilization of low-quality feeds have emerged as major hurdles in sustainable livestock productivity in the Himalayan region. The main constraints which influence fodder availability and affect livestock productivity in Himalayan region (Fig. 4.1) have been discussed.

4.1 Land Scarcity

Majority of the population in the Himalayan region revolve around agriculture and allied activities for their livelihood. The wealth and properties of small and marginal farmers are associated with size of land holdings. The majority of the farmers in Himalayan region are small and marginal, holding small pieces of land. Scarcity of land is one of the major constraints for low agriculture and forage production in the region. The per capita available cultivable land in areas across the Indian Himalayas

R. Katoch, *Nutritional Quality Management of Forages in the Himalayan Region*,
https://doi.org/10.1007/978-981-16-5437-4_4

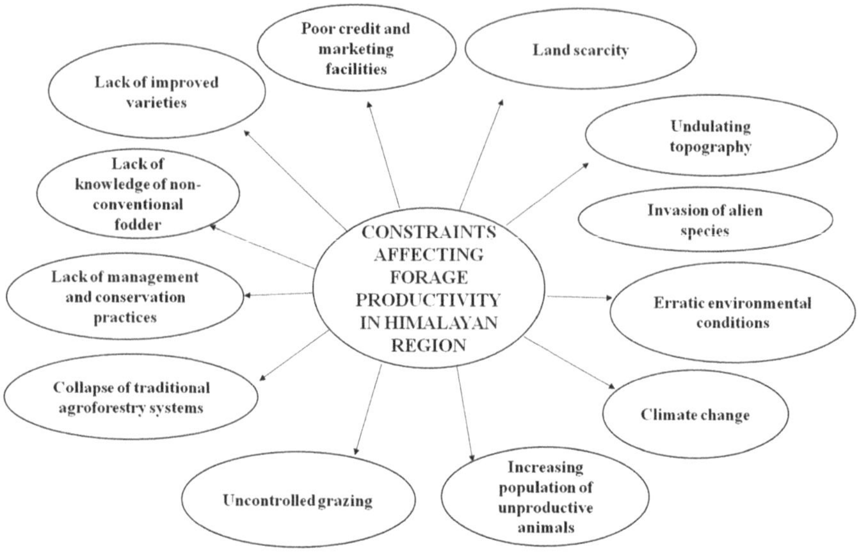

Fig. 4.1 Constraints affecting forage productivity in Himalayan region

is mostly insufficient to sustain livelihoods. Owing to land constraints, small and marginal farmers prioritize food crops over forage crops.

In the last few decades, the area under the natural pastures and grasslands in the Himalayan region has decreased to a considerable extent. For example, the available grazing area in subalpine and alpine pastures of Jammu and Kashmir has decreased from 0.15 ha/animal in 1977 to 0.10 ha/animal in 1982 (Misri 2003) and is continuously decreasing further (Qureshi et al. 2007). Similarly in Himachal Pradesh, the average grazing pressure and grazing intensity in the state were 1.26 ACU/ha (Adult cattle unit with average body weight of 350 kg) and 0.79 ACU/ ha, respectively (Dev et al. 2006). Overexploitation, overstocking, and poor management are the main reasons for depleting natural pastures and grasslands in the Himalayan region.

The increased pressure of human population is more focused on growing food crops rather than forage crops, therefore, the area under the forage crops is continuously shrinking. Moreover, the division in the families has further divided large land holdings into smaller land holdings, resulting in increase in the number of marginal and small farmers in the Himalayan region ultimately affecting the fodder availability to the livestock.

4.2 Undulating Topography

The Himalayan region has a mountainous profile that splits the environment into different altitudinal and agro-ecological zones. Lower plains of the Himalayan region have less sloppy areas while the topographic hindrance increases with increasing altitude. In Indian Himalayas, more than 21% of land has more than 30% slope, whereas 30.70% of land has slope between 8 and 30% (Partap 1999). Moreover, heavy snowfall and rainfall in higher altitudes make the area prone to landslides. The combination of a steep topography and high rainfall events pattern results in the absence of adapted measures and leads to intensive erosion to loss of soil nutrients leading to lower productivity.

4.3 Erratic Environmental Conditions

The Himalayan region has huge variation in the micro-climates. The high mountains in the Himalayan region are covered by snow for over half a year. Scanty rainfall during the peak growth period in the region hampers the seed germination and growth phase of the crop and sometimes, heavy rainfall during flowering and multiplication stage reduces the seed formation efficiency. The region also experiences water scarcity not only due to rainfall but also due to lack of water storage facility. The harsh environmental conditions of the region have direct influence on forage availability.

4.4 Climate Change

Climate change is not a sole problem of Himalayan region but a worldwide phenomenon. Climate change has direct influence on quality and availability of fodder. Livestock production systems rely upon a wide variety of feed sources which include grazing of native forage and browse species, planted pasture, and feed products from arable land. Due to climate change, the length of the growing season and climate suitability has been affected in all farming systems due to the limitation of moisture availability or extreme heat. For crops suitable for snow-bound areas, plants require low temperature to flourish but prolonged higher temperature than optimum temperature reduce or alter the forage availability and quality (Hopkins and Del Prado 2007).

The climate change has a direct influence on forage availability due to

- changes in herbage growth brought about by changes in atmospheric carbon dioxide concentrations and temperature,
- changes in composition of pastures, such as changes in the ratio of grasses to legumes,
- changes in herbage quality with changing concentrations of water-soluble carbohydrates and N at given dry matter yields,
- greater incidences of biotic and abiotic stresses and changes in rainfall patterns.

4.5 Increasing Population of Unproductive Animals

Livestock population is increasing due to large-scale rearing of animals for meeting different human requirements. Due to religious beliefs and other regional considerations, the population of unproductive cattle is continuously increasing. These cattle are abandoned due to loss of their productivity with ageing. Maintenance of excessive number of poorly fed and low-yielding animals has contributed to the deprivation of forage resources which has further widened the gap between demand and supply of feed to the livestock.

4.6 Uncontrolled Pasture Grazing

Grasslands and pastures are major forage resources of Himalayan region and heavy grazing on these resources has resulted in decline in total biomass availability. The grazing pattern has further raised many problems in these pastures. The grasslands and natural pastures in outer hills are used for the grazing of local livestock. Around 30 pastoral communities persist in hilly or arid/semi-arid regions in northern and western parts of India whereas about 20 communities in temperate/hilly regions depend on grazing-based livestock production (Roy and Singh 2013).

Transhumance pastoralism and nomadic pastoralism in Himalayan region largely depend on grazing lands and rarely practice stall-feeding. The mid-hill grasslands are protected only for about 2 to 3 months during monsoon and are grazed during the rest of the year. Sub-alpine and alpine pastures of higher Himalayas are grazed from the month of April to October, which is the active period for the growth of vegetation. Due to overgrazing, the perennial and palatable grass species become less dominant in natural grasslands and pastures and give place to obnoxious weeds.

4.7 Invasion of Alien Species in Grasslands and Pastures

Several studies have suggested that alien species cause a decline in species richness of native plant communities in natural pastures. This phenomenon is linked with the ability of alien species to become dominant in a plant community outside its native range and to replace native species. Alien plant species (exotic weeds) *viz., Lantana, Parthenium*, and *Eupatorium* have invaded most of the grazing lands in the Himalayan region severely affecting their productivity.

4.8 Collapse of Traditional Agro-Forestry Practices

Traditional agro-practices involved intimate integration of tree component with the crops. The tree component is used to fulfill the small wood and fodder requirements, especially of the small and marginal farmers. Due to intensification of agriculture, cultivation of high-yielding crops and implementation of policies having less incentive for forage tree farming, have limited, tree component from the agri-practices, resulting in pressure on the already degraded grazing lands. It therefore necessitates the revival of age-old practice of integrating multiuse tree component in the agri-practices through supportive policies and reforms.

4.9 Lack of Management and Conservation Practices

Crop management practices have considerable importance in determining the productivity and quality of grasslands. The presence of inferior and unproductive forage species, lack of fertilizer application, absence of forage legumes, inappropriate cutting intervals and indiscriminate grazing are some of the contributing factors for poor productivity and quality of grasslands. There exists a wealth of indigenous knowledge pertaining to proper utilization and management of natural forage resource base. The farmers in mountainous regions generally follow conventional and traditional approaches of fodder production. Farmers use traditional varieties of forage crops which are low-yielding and susceptible to pests and diseases. Lack of knowledge on mineral and nutritional constituents of forages leads the farmers to cultivate nutritionally compromised forage crops.

Varieties released by different institutes are often specified for particular locations and there are meager efforts for popularization of variety for different regions. Forage crops in Himalayan region are harvested manually without use of any machinery. The topographic constraints restrict the farmers in using large machineries for cultivation and harvesting of forages. Processing and conservation techniques like preparation of silages and haylage are rarely practiced for livestock rearing in Himalayan region. Poor infrastructure facilities for fodder storage are one of the major constraints for fodder conservation for lean period. The lack of knowledge on fodder-based technologies and conservation approaches largely impacts fodder production in the Himalayan region.

4.10 Lack of Improved Varieties

Many aspects related to forage breeding, plant genetic resources, plant-protection measures, forage quality, palatability and seed production have not been adequately addressed in the Himalayan region.

4.11 Lack of Knowledge of Non-conventional Fodder

The Himalayan region is the house of diverse plant species supporting the existence of large number of animal species. Besides providing a diverse resource for a large number of grazing animals, a large food base in terms of natural grasslands and natural pastures, wide food resource base in the form of non-conventional feed resources is still unrecognized. Weeds, kitchen wastes, vegetable wastes, and fruit peels are some of the non-conventional but nutritious feeds. The use of non-conventional feed resources in livestock feeding could vitalize the livestock productivity in the region. These fodder resources are better alternative to conventional feeds.

4.12 Poor Credit and Marketing Facilities

There are many schemes which facilitate farmers for agricultural activity, but their high interest rates sometimes dissuade farmer from availing their benefits. Lack of efficient marketing systems for basic agriculture activities has also aggravated the problems in fodder production.

Livestock-based farming in the Himalayan region has become a challenging aspect. For improving grasslands and pastures, holistic management of these resources is required by promoting sustainable interaction between grasslands, livestock, and grazing communities. The adoption of new management practices by farming communities both for fodder and livestock production would be supportive for the efficient livestock management in the Himalayan region.

References

Dev I, Misri B, Pathania MS (2006) Forage demand and supply in western Himalaya: a balance sheet for Himachal Pradesh. Indian J Anim Sci 76:720–726

Hopkins A, Del Prado A (2007) Implications of climate change for grassland in Europe: impacts, adaptations and mitigation options: a review. Grass Forage Sci 62:118–126

Misri B (2003) Improvement of sub-alpine and alpine Himalayan pastures. Research Centre, Indian Grassland and Fodder Research Institute, Regional Station, Palampur HP, Palampur

Partap T (1999) Sustainable land management in marginal mountain areas of the Himalayan region. Mountain Res Dev 19(3):251–260

Qureshi RA, Ghufran MA, Gilani SA, Sultana K, Ashraf M (2007) Ethnobotanical studies of selected medicinal plants of Sudhan Gali and Ganga Chotti hills, district Bagh, Azad Kashmir. Pak J Bot 39:2275–2283

Roy AK, Singh J (2013) Grasslands in India: problems and perspectives for sustaining livestock and rural livelihoods. Trop Grassl Forrajes Trop 1:240–243

Factors Influencing Forage Nutritional Quality 5

Forage quality generally represents the total amount of nutrients, which could be derived by livestock from forages. Forage quality is affected by a number of factors, which may be categorized into three groups *viz.,* plant factors, animal factors, and environmental factors. Understanding of interrelationship among these factors with forage quality would help in making appropriate selection of forages and supplements that would match animal daily requirements and result in economically favored livestock performances. Different factors affecting forage quality have been discussed in the chapter.

5.1 Plant Factors Affecting Forage Quality

5.1.1 Stage of Maturity

Maturity stage is one of the crucial factors that affect quality of forages. Generally, all forages are highly succulent at early growth stages; however, with the advancement in maturity, succulence of forages decreases considerably due to changes in composition of cell wall, accumulation of lignin and other polyphenolic compounds. In forages, the digestibility and CP content decrease, whereas fiber and lignin content increase with advancing maturity (Fig. 5.1) (Kilcher and Troelsen 1973; Jung and Allen 1995; Khorasani et al. 1997).

Lignin in forages is indigestible to ruminal microflora and impedes the digestibility of other nutritional constituents in forages (Jung and Allen 1995). The negative effects of association of starch and neutral detergent fiber (NDF) could also result in reduced digestibility of forages. Advancing maturity of cereal forages results in high starch content that may decrease ruminal fluid pH of livestock and results in low microbial population in rumen and subsequent decrease in NDF digestibility (Calsamiglia et al. 2002). The protein content in forages is usually high at early growth stages, maximum at the end of vegetative stage and decline at later stages

R. Katoch, *Nutritional Quality Management of Forages in the Himalayan Region*, https://doi.org/10.1007/978-981-16-5437-4_5

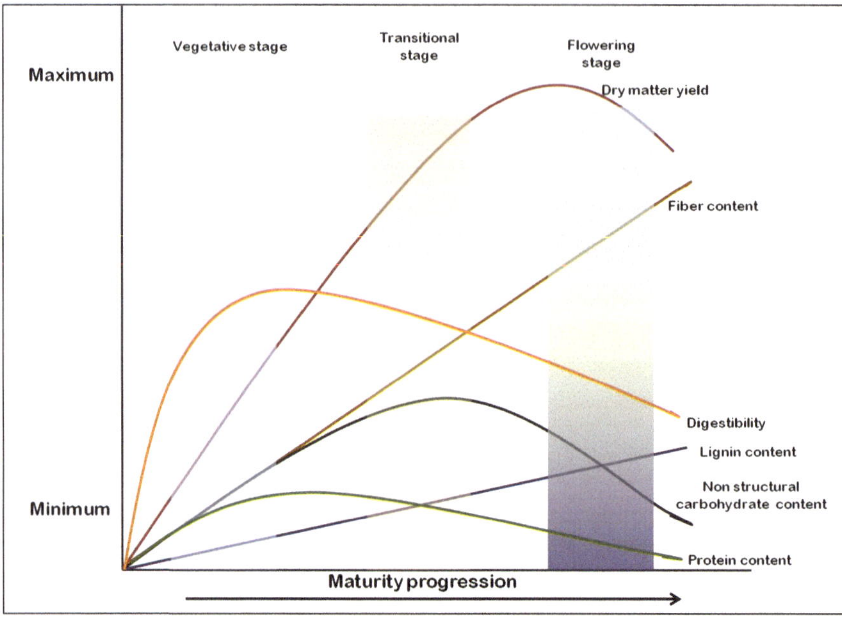

Fig. 5.1 Effect of maturity on quality of forage grasses and legumes

(Mero and Uden 1998). Decrease in CP content with maturity is related to dilution of nitrogen in leaves with increasing dry matter content (Verma and Mishra 1999). The immature growth stage is accompanied by high mitotic activity and needs more nitrogen for growth, thus resulting in high CP content at early growth stages (Ryan and Bormann 1982). The relocation of nitrogen from mature leaves to perennial tissues before abscission also contributes to low protein content in mature leaves (Menino et al. 2007). The decrease in CP content with advancing maturity for different forages has also been reported (Subba et al. (1994), Singh et al. (2010), Baloyi et al. (2013), and Ncube et al. (2015)).

Fatty acids composition in forages is strongly influenced by maturity of forages (Clapham et al. 2006; Khan et al. 2011a). In photosynthetic tissues, fatty acids are present in membrane lipids, predominantly in thylakoid membranes of chloroplasts, which on average contain 20% lipids. In a number of forage species, a strong positive correlation has been observed between chlorophyll content and total fatty acid content (Dierking et al. 2010). In addition to stage of maturity, (Khan et al. 2011b), low leaf/stem ratio as well as senescence (Mishra and Sangwan 2008; Yang and Ohlrogge 2009) is attributed to low fatty acid content in forages.

Various studies have confirmed maturity as important factor influencing the nutritional quality of forages. Tall fescue grass has high 11.81% CP content and 55.20% IVDMD with low NDF (63.65%), ADF (48.22%) and hemicellulose (15.46%) contents during early growth stages (Katoch et al. 2012). With advancing maturity, the CP content and digestibility of tall fescue grass reduce to the extent of

9.47 and 47.95%, respectively, whereas NDF, ADF and hemicellulose increase to the extent of 70.31%, 52.11%, and 18.20%, respectively. CP content of alfalfa decreases from 20.20 to 17.60%, whereas in timothy grass CP content decreases from 11.10 to 7.90% with advancing maturity. The fiber content i.e., NDF and ADF in timothy grass increases from 68.70 to 72.80% and 38.00 to 40.70%, respectively with advancement of maturity. Alfalfa and timothy grass have high digestibility at late bud stage and pre-bloom stage, however, when they reach early bloom and full head stages, digestibility of both forages decreases (Yu et al. 2003).

Maturity also has significant influence on anti-nutrients content in forages. During early facet of growth, there is a rapid rise in oxalate content in forages followed by a decline with advancing maturity (Davis 1981). Oxalate content in grasses could be managed by adjusting harvesting period (Rahman et al. 2009; Katoch et al. 2012).

Maturity also influences the mineral content and composition as rapid uptake of minerals by plants usually occurs during early stages of growth. At later stages of maturity, DM content of forages generally increases more rapidly than mineral uptake resulting in decrease in total mineral content (Fleming 1973). Phosphorus and potassium decrease markedly with increasing maturity while calcium is less affected by the stage of maturity (Ammerman et al. 1982). The level of other micronutrients like cobalt, copper, iron, molybdenum, and zinc is also affected by advancing maturity. Effect of maturity on different quality components of alfalfa has been presented in Table 5.1.

The study conducted by our group also revealed that stage of maturity affects the nutritional composition of *Leucana* foliage (Table 5.2) (Unpublished data). In the *Leucana* foliage at three different maturity stages, crude protein content in tender, premature and mature *Leucaena* leaves varied from 25.70 to 34.85%, 25.33 to 32.64% and 21.84 to 28.65%, respectively (Table 5.2). Total soluble carbohydrates content in tender, premature and mature leaves ranged from 45.24 to 48.51%, 39.83 to 41.72%, and 35.88 to 38.59%, respectively.

In tender *Leucaena* foliage, the NDF content varied from 28.89 to 33.89%, whereas ADF content from 19.90 to 23.73%. In premature *Leucaena* leaves, NDF and ADF content increased as compared to tender leaves. The NDF and ADF content varied from 32.47 to 37.36% and 23.59 to 26.11%, respectively, in premature leaves. Mature *Leucaena* leaves had high NDF and ADF content ranging from 36.91 to 41.51% and 28.46 to 34.67%, respectively. Among all stages of maturity, tender foliage had lowest NDF and ADF content. In tender and premature leaves, the ADL content varied from 3.58 to 6.75% and 4.21 to 8.23%, respectively.

In mature leaves, the ADL content varied from 8.30 to 11.20% (Table 5.2). There was direct relationship between lignin content and maturity of *Leucaena* foliage. Similar observation was also recorded for cellulose content, where tender leaves has low cellulose content in comparison to mature leaves. Hemicellulose content decreased with maturity. Ash content varied from 6.45 to 10.94% at different stages of *Leucaena* foliage. Highest ash content was observed in mature leaves (8.77 to 10.94%) followed by premature foliage (8.09 to 9.14%) and tender leaves (6.45 to 7.94%) (Table 5.2).

Table 5.1 Effect of maturity on nutritional quality of alfalfa

Stage of maturity		Digestible dry matter	CP	ADF	NDF	Relative feed value index
		(% dry weight)				
Vegetative		>69	>22	<25	<34	>188
Bud stage		67	20–22	25–31	34–41	166
Early bloom		62	18–19	32–36	42–46	131
Late bloom		60	16–17	37–40	47–50	115
Seed stage		<58	<16	>41	>50	<108

Total phenolic content in tender leaves varied from 12.15 to 13.60%, whereas simple phenols varied from 4.45 to 6.10%. In premature *Leucaena* leaves, total phenols are in lower proportions as compared to tender leaves. The total phenolic and simple phenolic content varied from 10.35 to 12.69% and 4.24 to 6.11%, respectively, in premature leaves.

In mature *Leucaena* leaves, total phenols and simple phenols varied from 10.19 to 12.45% and 4.24 to 5.95%, respectively. Total tannins content in tender, premature and mature *Leucaena* leaves ranged from 7.33 to 8.41%, 4.85 to 7.37%, 5.23 to 7.21%, respectively. The mimosine content in tender *Leucaena* leaves was from 1.87 to 3.61%, whereas in premature and mature leaves the mimosine content was in the range of 1.53–3.48% and 1.47–2.84%, respectively. The study revealed that crude protein and total carbohydrates content decrease with maturity, while fiber content and cell wall components increase with maturity. Ash content also increase with maturity whereas anti-nutritional factors like total phenols, simple phenols, total tannins and mimosine content reduce with advancement in maturity of *Leucaena* foliage (Table 5.2).

Table 5.2 Effect of maturity stages on nutritional quality of *Leucaena* foliage

Leaf maturity stage		CP%	TC%	NDF%	ADF%	HC%	ADL%	CL%	Ash%	TP%	SP%	TT%	Mimosine%
Tender		25.70–34.85	45.24–48.51	28.89–33.89	19.90–23.73	8.18–10.21	3.58–6.75	15.71–17.38	6.45–7.94	12.15–13.60	4.45–6.10	7.33–8.41	1.87–3.61
Premature		25.33–32.64	39.83–41.72	32.47–37.36	23.59–26.11	8.74–11.97	4.21–8.23	16.04–17.88	8.09–9.14	10.35–12.69	4.24–6.11	4.85–7.37	1.53–3.48
Mature		21.84–28.65	35.88–38.59	36.91–41.51	28.46–34.67	6.38–8.45%	8.30–11.20	18.81–23.51	8.77–10.94	10.19–12.45	4.24–5.95	5.23–7.21	1.47–2.84

Source: Katoch et al. (Unpublished data)

5.1.2 Anti-Quality Factors

Anti-nutritional factors reduce the nutritional quality of forages. These factors are synthesized due to normal metabolism and express their adverse effect on health and productivity of livestock. These factors in forages are generally disliked by animals because of their bitter taste, poor palatability and digestibility. If the forages having high anti-nutritional content are consumed in large amount, these factors not only affect animal productivity but also cause toxicity, which can lead to abnormal reproduction, disturbed neuro-endocrine functions and even cause death of livestock in severe cases (Allen and Segarra 2001). Livestock fed with forages with high nitrate cannot complete the conversion of nitrate into ammonia, as a result nitrates accumulate in body and convert hemoglobin into methemoglobin, which does not have capacity to bind and carry oxygen to different vital organs and may cause death due to asphyxiation.

The presence of saponins in forages such as alfalfa are characterized by their bitter taste and can affect livestock health and production through erythrolysis, inhibition of smooth muscle activity, low microbial fermentation and nutrient absorption (Sen et al. 1998). Alkaloids also reduce the quality of forages and affect livestock performance. For example, tryptamine and tyramine alkaloids produced by *Phalaris* are responsible for phalaris staggers and sudden death syndrome in livestock (Edgar 1994). Cyanogenic glycosides liberate hydrocyanic acid on hydrolysis with β-glucosidase enzyme in gastrointestinal tract of animals. Excessive consumption of cyanide ions inhibits the action of cytochrome oxidase enzyme in respiratory chain and leads to death due to unavailability of oxygen and energy (Montgomery 1980; Lehmann et al. 1991; Crush and Caradus 1995). The presence of estrogenic compounds (Phytoestrogens) such as formononetin reduces animal fertility (Woclawek-Potocka et al. 2013). Anti-nutrients in forages can have serious economic impact on health and productivity of livestock (Allen and Segarra 2001).

5.1.3 Species Difference

Forages include diverse species belonging to different plant families. Poaecae family includes forage grasses, while leguminosae family includes forage legumes. The protein content of legumes is higher than grasses. Due to their interaction with nitrogen-fixing bacteria, legumes have high protein content. Legumes have low NDF and ADF content than grasses and digest faster than grasses. Differences in digestibility between grasses and legumes have been well documented. Grasses have high cell wall constituent content and less digestibility than legumes (Waldo and Jorgensen 1981; Buxton and Russel 1988; Buxton and Brasche 1991). Legumes generally have high calcium, magnesium, potassium, copper, zinc and cobalt content than grasses, while grasses are rich in manganese and silica when grown under similar conditions (Fleming 1973; Minson 1990).

Differences also exist between forage grasses and legumes in mineral uptake when grown under same climatic and soil conditions. Tall fescue has high calcium and magnesium and low potassium content than orchard grass (*Dactylis glomerata*) and canary grass (*Phalaris arundinacea*) (Odom et al. 1980). *Setaria*, *Cenchrus* and *Pennisetum* grasses show intermediate oxalate accumulation while *Rumex* and *Oxalis* have high oxalate accumulation. Majority of commonly consumed forages accumulate oxalate content, typically around 1.0 to 2.0% (Rahman and Kawamura 2011).

5.1.4 Plant Anatomy

Structure and composition of cells constituting tissues of forage plants differ markedly. Plant cells are differentiated and aggregated into multiple tissues that perform different and multiple functions within the plant. All plant cells have cell walls but in mesophyll cells, they are thin and easily digestible. Other cells such as vascular xylem and phloem cells undergo secondary thickening. The epidermal cells form waxy layers on leaves that restrict rapid digestion in the rumen. Lignin deposition occurs primarily in tissues that have function in water conduction, mechanical support, or protection. Sclerenchyma and xylem cells are generally considered indigestible, whereas mesophyll cells are almost completely digestible. Epidermal cells and parenchyma bundle sheath cells in C_4 grasses differ greatly in digestibility. Though proportions of xylem and sclerenchyma cells are similar in warm- and cool-season grasses, parenchyma bundle sheath and the number of epidermal cells are high in warm-season grasses (Twidwell et al. 1990). The anatomy of stem is similar for warm- and cool-season grasses and both contain high proportions of sclerenchyma and xylem cells which are highly lignified. Digestibility of stem parenchyma varies among forages and decreases with maturity of tissue (Akin 1989; Grabber et al. 1992). Temperate and tropical forage legumes show similar anatomies with large areas of highly digestible mesophyll tissue. Unlike forage grasses, the legume stem parenchyma does not become lignified with maturity and is almost completely degradable and accounts for high voluntary intake of forage legumes (Demarquilly and Jarrige 1974).

5.1.5 Variety (Cultivar)

Development of superior varieties is mainstream discipline of forage breeding program. After decades of breeding forages for yield, attention has been on development of varieties. Variety or cultivar also affects forage quality. For example, among two cultivars (Nandi and Kazungula) of *Setaria*, Nandi contains low oxalate, while Kazungula has high oxalate content. Cultivars with high number of leaves may also have high oxalate content than varieties with low proportion of leaves, since leaves normally contain high oxalate content than stems. It shows that forage quality differs among cultivars and attention should be given to cultivars having excellent nutritional quality. Therefore, breeding commercially viable high-quality cultivars is the one of the major objectives of forage breeding program.

5.1.6 Cellular Composition and Metabolism

The essential biochemical processes are carried out within specific cells and tissues which have specific functions. For example, photosynthesis primarily occurs within specialized parenchyma cells. These mesophyll cells contain chloroplasts that capture energy from sunlight and enzymes necessary for conversion of carbon dioxide into carbohydrates. Photosynthetic enzymes account for about 50% of total proteins in plants. In addition to proteins and amino acids, the cytoplasm contains sugars, storage polysaccharides, minerals and other essential nutrients. The biochemical composition of cytoplasm varies with cell type and function and nearly all cytoplasmic constituents are completely digestible (Van Soest 1994).

Cell wall is made up of cellulose fibrils embedded in a matrix of hemicellulose, pectic substances and lignin. Cell wall composition varies with plant species and maturity. The highly digestible primary cell wall is formed next to the middle lamella during cell elongation and is mainly composed of polysaccharides including cellulose, hemicellulose, pectic compounds and glycoproteins. The strands of cellulose, consisting of polymers of β-D-glucose molecules are laid down in mesh-like layers. These microfibers are then cross-linked with lignin resulting in the formation of a rigid wall that is very resistant to rumen microbes. The digestibility of plant cell walls varies considerably among different cell types and tissues (Moore and Hatfield 1994).

From a nutritional perspective, plant cell walls comprise most of the fiber content of forages, which is relatively resistant to digestion due to interlinking with lignin. The indigestible fiber limits the energy value of most of the forages (Buxton and Redfearn 1997; Moore et al. 2017). Reduction in the total fiber content or improving its degradability is the most effective way of increasing digestible energy of forages.

5.2 Animal Perspective of Forage Quality

Forage quality can also be analyzed by assessing different parameters through *in vitro* and *in vivo* techniques. These parameters give a picture of forage nutritional value from animal perspective which includes forage palatability, intake capacity and digestibility.

5.2.1 Palatability

Animals prefer one forage crop over another based on the smell, feel and taste. Palatability is affected by texture, leafiness, fertilizer application, moisture content, pest infestation, or the level of different nutrients. Forage palatability is based on a combination of many aspects such as plant species, fertilization, maturity and the level of nutritional as well as anti-nutritional constituents. High-quality forages generally have high palatability.

5.2.2 Intake

Animals must consume adequate quantity of forages in order for better performance. Typically higher the palatability and quality of forages, higher the voluntary intake of forages by livestock. Forages having good texture and taste without any foul smell are generally accepted by animals and therefore have high voluntary intake.

5.2.3 Digestibility

In animals, there is variation in absorption and assimilation of nutrients from the ingested forages as they pass through gastrointestinal tract. Immature, leafy tissues may be digested to the extent of 80 to 90%, whereas stemmy material has digestibility less than 50%.

5.3 Environmental Factors Affecting the Forage Quality

Environmental factors are considered as one of the major factors influencing growth and development as well as nutritional quality of forages. Effects of different environmental factors on nutritional quality of different forages have been discussed under subsequent subheads.

5.3.1 Temperature

Among various environmental factors, temperature has major influence on quality of forages. The ambient temperature is required for optimum growth and development of plants. For cool-season forages, the optimal growth temperature is about 20 °C whereas 30 to 35 °C is for warm-season forages. The marked effect of temperature on forage growth, development and chemical composition was first demonstrated in studies conducted on perennial rye grass (*Lolium perenne* L.). High ambient temperature during growth affects dry matter digestibility, morphological development, physiological age and composition of cell wall. At high temperature, dry matter digestibility of forages decreases. The difference in digestibility of forages due to high-temperature stress is mainly due to decrease in digestibility of stem (Deinum and Dirven 1975). Temperate forage grasses show decline in the digestibility for each 1 °C increase in growth temperature (Wilson and Minson 1980). A 1 °C increase in temperature decreases the digestibility of temperate forages by 0.30 to 0.70% with minor effect on CP content.

Another temperature effect on cell wall digestibility is independent of lignin content in plant cell wall. This effect is probably connected with organization characteristics of cell wall. It has been reported that high temperature causes a slight increase in cell wall content, but a strong decline in cell wall digestibility of leaf blades, leaf sheaths, and stems. Because reactions determining synthesis of cell wall constituents are enzyme-mediated, it is likely that temperature influences cell wall composition and linkages between them. Cool-season and warm-season grasses also respond differently to temperature with respect to individual cell wall constituents (Ford et al. 1979). For each percentage increase in fiber and lignin content there is approximately 4% decrease in digestibility.

Temperature also influences fatty acid composition in forages. A decrease in temperature during growth induces enzymatic desaturation of fatty acids to regulate membrane fluidity and to provide an optimal environment for photosynthesis and maintain transmembrane transport (Allakhverdiev 2009). On the other hand, high growth temperature decreases the number of C18:3n-3 fatty acids and increases the number of C16:0 and C18:2n-6 fatty acids (Clapham et al. 2006).

High temperature during growth also changes the anatomical features of plants. In tall fescue, grown under a temperature regime of 20/18 °C, epidermis, mesophyll, parenchyma bundle sheath and phloem are completely removed during digestion, leaving a residue of vascular tissue, sclerenchyma and cuticle. However, at 30/27 °C, only slight digestion of mesophyll occurs, whereas remainder of leaf remains relatively intact (Akin et al. 1987). Histological reactions for lignin showed formation of large amounts of lignified vascular tissue at high temperature. The analysis for phenolic acids reveals significant quadratic relationships between temperature and content of *p*-coumaric and ferulic acids. Effects of temperature on nutritional quality attributes of forages have also been investigated in *Lespedeza cuneata* and *Lotus pedunculatus*. The condensed tannins content in foliar tissue of both forage legumes increases with the increase in temperature during growth (Fales 1986; Lees et al. 1995).

5.3.2 Water Stress

Moisture stress alters the physiology of forages in ways that have profound influence on their quality. Typically, drought results in an improvement in forage quality as new growth is inhibited, nutrients are concentrated and growth and development of plant decreases (Grant et al. 2014). Neutral detergent fiber (NDF) content is consistently affected by water stress. In rye grass and orchard grass cultivars, it has been observed that water stress increases the nutritional value by increasing CP content and digestibility of NDF content (Jensen et al. 2003). Photosynthetic rates are usually less affected by water stress in comparison to respiration rates, resulting in increase in digestible soluble sugars content. Nitrate toxicity is of particular concern in drought-stressed sorghum. The reduction of nitrate to ammonia is inhibited causing accumulation of nitrate within the plant tissues, particularly in stem tissues (Mayland et al. 2007; Gleadow et al. 2016).

5.3.3 Photoperiod

Photoperiod has significant effect on growth and nutritional quality of forages. Long photoperiods improve the quality of forages because of higher photosynthetic activity, which in turn increases the total soluble sugar content. Each 1 h increase in day length increases the digestibility to about 0.20% along with increase in yield (Deinum et al. 1981). In long-day photoperiod, shoot-to-root ratios generally increase whereas leaf/stem ratios decrease. Juan et al. (1993) reported that alfalfa grown under 13 h photoperiod has higher leaf/stem ratio than 16 h photoperiod. Despite a low leaf/stem ratio, forages adapted to higher latitudes generally have excellent quality than forages from lower latitudes. The enhanced growth during long days often reduces the CP content in forage legumes. Perennial forages adapted to high latitudes have high growth rates with a short life cycle to ensure that plants reach reproductive development before the end of growing season.

5.3.4 Solar Radiation

The amount of solar radiation received at the earth's surface is strongly influenced by growing season and latitude. Forage legumes at higher latitudes have longer day lengths but the intensity of sunshine is less than lower latitudes. Conversely, forage legumes from lower latitude experience shorter day lengths, but greater sunshine intensity. High light intensity increases the digestibility of forage legumes with a reduction in protein content. Plants grown under low irradiance levels may have higher nitrate content and consequently low CP content. Radiation quality can also affect digestibility of forage plants.

5.3.5 Shading

Shading has both direct and indirect effects on quality of forages in that it alters the chemical composition as well as morphological development and yield. Crude protein content is usually higher in leaves and stems from the top of plant canopies than from bottom. Shading slows the growth rate by which it has greater effect on forage yield and quality. Crude protein content is more responsive to shading than other quality characteristics. Kephart et al. (1992) and Kephart and Buxton (1993) reported that after imposing 63% shade on five perennial forage grasses reduces yield by 43% and specific leaf weight by 24%, but reduce NDF content by 3%, cell wall lignin by 4%, and increase digestibility of the herbage by 5%. Kephart and Buxton (1993) reported that 63% shade increases N content by 26% and the responses were generally greater for leaf blades than for stems. Increased content of nitrogenous compounds from shading is usually at the expense of soluble carbohydrates (Wilson and Wong 1982). The changes in CP content in legumes due to shading are generally lesser in grasses. The soluble nonprotein N of alfalfa is increase, whereas soluble protein N decrease with shading (Walgenbach and Marten 1981). Shading may increase silica content of herbage, which reduce herbage digestibility. Studies have indicated that shading also decrease cell wall content of forages. Hemicellulose content is less affected by shading than cellulose and lignin content. Digestible dry matter content of forages grown under shade is higher than full sunlight (Samarakoon et al. 1990). Forage quality undergoes diurnal variation with highest quality in mid-afternoon. Water-soluble carbohydrate content in forages is lowest before sunrise and highest in the afternoon. The digestibility of alfalfa is 1.60% units higher in the late afternoon than early morning (Lechtenberg et al. 1971). Additionally, protein tends to be degraded at night through proteolysis followed by remobilization of nitrogen and protein synthesis during daylight. This results in fluctuation in protein content in forages (5 to 15%).

5.3.6 Soil Properties

Soil properties are important in determining forage quality through its characteristics such as water-holding capacity, soil aeration and nutrient availability. In general, unfavorable soil conditions affect the pasture production by creating a poor environment for root growth and limiting water and nutrient uptake. The unfavorable soil conditions affect forage quality by altering growth patterns. Alfalfa has higher forage quality, lower lignin content and higher crude protein content on heavy clay loams in comparison to sandy soils.

The mineral content of forages is greatly influenced by quantity and the availability of minerals in the soil as in case of many minerals, only a small portion of total mineral content is available for uptake by plants. The mineral composition of soil is affected by the origin of the parent rock, losses that occur through leaching and surface erosion and applications of fertilizers and waste materials. Availability of minerals in the soil and uptake by plants are dependent on factors such as soil pH, moisture and drainage conditions.

Soil pH also has direct impact on availability of certain trace elements in forages. A change in mineral content of herbage due to altering soil pH is therefore dependent on soil type. Molybdenum and selenium content in forages increase as soil pH increases from acid to neutral or alkaline conditions. Decrease in manganese and cobalt content in forages has been observed due to relative increase in soil pH. Macro-minerals are less affected by soil pH in comparison to micro-minerals.

5.3.7 Soil Fertility/Fertilizer Application

The availability of soil nutrients affects the productivity and quality of forages. Of all the nutrients, N has the most dramatic impact on plant growth whether it is supplied through mineralization of soil organic matter, symbiotic bacteria, or through applications of N fertilizer. In intensively managed situations involving monoculture grasses, application of N can significantly enhance yields. Much of the N absorbed by plants is incorporated into amino acids and proteins. Nitrogen is absorbed and transported primarily as nitrate (NO_3^-) in grasses, before being reduced to ammonium (NH_4^+) in leaves and incorporation into amino acids and protein.

When growth of grasses is restricted by drought, cold stress, limited radiation, or mineral deficiency, NO_3^- can accumulate to toxic levels for livestock. Application of N fertilizer to grasses reduces soluble carbohydrate content and increases water content. At very high rates, it can cause an imbalance of energy and protein that results in poor animal performance, reproductive problems and other metabolic disorders (Messman et al. 1991).

Applications of N fertilizer generally have little effect on forage digestibility; though there are some evidences that indicate that the rate of fiber digestion could be increased by N fertilization, especially in situations where N content in forage tissue is low prior to fertilizer application (Messman et al. 1991). Increases in herbage digestibility also have been reported because of N fertilization of warm-season grasses, which normally have low N content (George and Hall 1983). This response occurs because forage N is limiting for ruminal microbial growth under low-N situations. N fertilization may modify herbage digestibility by altering leaf/stem ratio. Since nitrogen application increases the vegetative growth, there are chances of low mineral content in forages due to dilution effect.

Most of phosphorus in animals is present in bones, where it functions as a storage pool to maintain balance of phosphorus in blood (Minson 1990). Requirements for P are similar for forages and animals that consume them. The P content in forages is generally adequate for animal performance. The exception is for soils that are very low in available phosphorus, where deficiencies can cause variety of symptoms, including low intake, poor performance, breeding disorders, bone-chewing, and rickets (Minson 1990).

Grasses are generally lower in phosphorus content than legumes. The cool-season grasses tend to have higher phosphorus content than warm-season grasses. There is no clear evidence that soil phosphorus content has any effect on intake or digestibility.

Potassium is required in higher amounts by plants than animals. High soil potassium is related to excess potassium uptake in cool-season grasses, wherein potassium interferes with uptake of Ca, Mg, and Na, resulting in an imbalance and grass tetany in animals.

The effect of nitrogen (N) fertilization on forage mineral content depends on the soil mineral supply. When the soil mineral supply is adequate, uptake of certain minerals may increase in response to N fertilization due to increase in the leaf-to-stem ratio. Phosphorus fertilization generally increases herbage phosphorus when applied to soils deficient in phosphorus. Potassium fertilization can decrease forage magnesium content, especially when applied at high rates (Fleming 1973). The

Table 5.3 Effect of fertilizer application on nutritional quality of forages

Crop	Fertilizer	Parameters	Reference
Stylosanthes scabra	25 kg N/ha + 20 kg P/ha	Increase in leaf crude protein content (21.00% CP)	Sodeinde et al. (2009)
Rhodes grass	Urea 100 kg N/ha, + FYM 5 t/ha + chicken manure 3ton/ha	Increase in crude protein (20.67%) and ash content (4.00%)	Yossif and Ibrahim (2013)
Rhodes grass	FYM 5 t/ha + chicken manure 3 t/ha	Decrease in crude fiber content (29.05%)	Yossif and Ibrahim (2013)
Fodder maize	Ammonium sulphate nitrate (26%N + 14%S)- 46 kg N/ha	Increase in crude protein content (39.57% leaf CP); Decrease in crude fiber content (14.65%)	Amin (2011)
Fodder maize	80 kg N ha^{-1}	Threefold increase in CP content	Eltelib et al. (2006)
Rhodes grass	80 kg N ha^{-1}	Increase in crude protein content (15.00%)	Keftasa (1990)
Bermuda grass	224 kg N ha^{-1}	Increase in crude protein content (78%) CP	Kering et al. (2011)
Bermuda grass	448 kg ha^{-1}	Decrease in NDF content (63%); decrease in ADF content (12.50%)	Kering et al. (2011)
Cowpea forage	30 kg N ha^{-1}	Increase in crude protein content (7.66%)	Hasan et al. (2010)

effect of fertilizer application on quality of important forage species has been presented in Table 5.3.

5.3.8 Altitude

The effects of altitude on growth, development and biochemical composition of plants could be seen from sea level to high land areas. With an increase in altitude, precipitation, light and radiation intensity increases while temperature and gravity decrease. At low altitude areas, the rate of synthesis of different cell wall constituents is high which ultimately affects forage digestibility (Sarwar et al. 1991; Buxton and Fales 1994). At higher altitudes, cell content (highly digestible carbohydrates) produced from photosynthesis accumulates due to low temperature sensitivity of photosynthesis process (Nelson and Moser 1994). Thus, forages at high elevations have excellent nutritional quality than those from low elevations (Deinum et al. 1981). NDF, ADF and ADL content in Desho grass (*Pennisetum pedicellatum*) decreases with the increase in altitude (Asmare et al. 2017). In *Brachiaria* grass, CP and ash content increase with the increase in altitude (Wassie et al. 2018). Chlorophyll *a* and b contents in plants decrease with the increase in altitude. This reflects an adaptive response of plants to avoid oxidative damage by reducing light

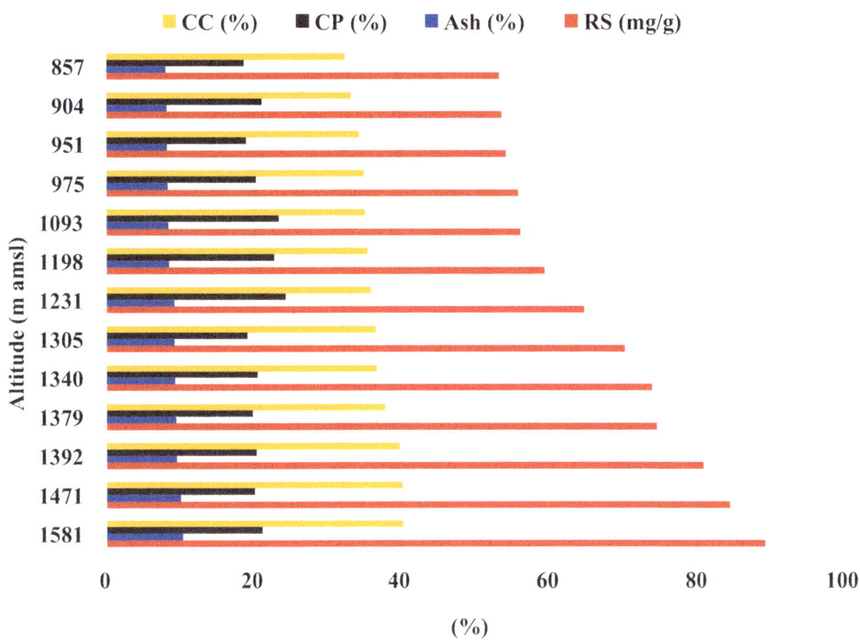

Fig. 5.2 Variation in crude protein (CP), reducing sugars (RS), ash and cell content in white clover with altitude

absorption thereby preventing the production of reactive oxygen species. NDF, ADF, potassium, phosphorus, calcium and magnesium contents in orchard grass decrease with the increase in altitude (Tenikecier and Ates 2020). Nutritional composition of clovers and sweet pea varies with the elevation of their growing area (Ates and Tekeli 2011). Katoch and Tripathi (2020) reported variation in nutritional composition of white clover with varying altitudes (857 and 1581 m amsl) in mid-Himalayan region. CP content revealed variation from 18.66 to 24.83% in white clover at different elevations. White clover plants from lower elevations had low protein content. Variations in ADF and NDF content have also been recorded with altitude. White clover plants from lower altitude had higher ADF, NDF content (49.73% and 67.53%, respectively) as compared to higher altitude (46.60% and 59.06%, respectively). White clover from higher altitudes revealed higher levels of cell content (40.37%) as compared to those from lower altitudes (32.47%) (Fig. 5.2). The cell content was positively correlated with altitude in white clover. The reducing sugar content in white clover plants varied from 53.32 to 89.08 mg/g of dry matter at different altitudes. White clover plants from higher altitude had higher reducing sugar content indicating importance of carbohydrate reserve for winter survival at higher altitude (Boller and Nosberger 1983). Low temperature also initiates transformation of starch into sugars for different metabolic processes in plants.

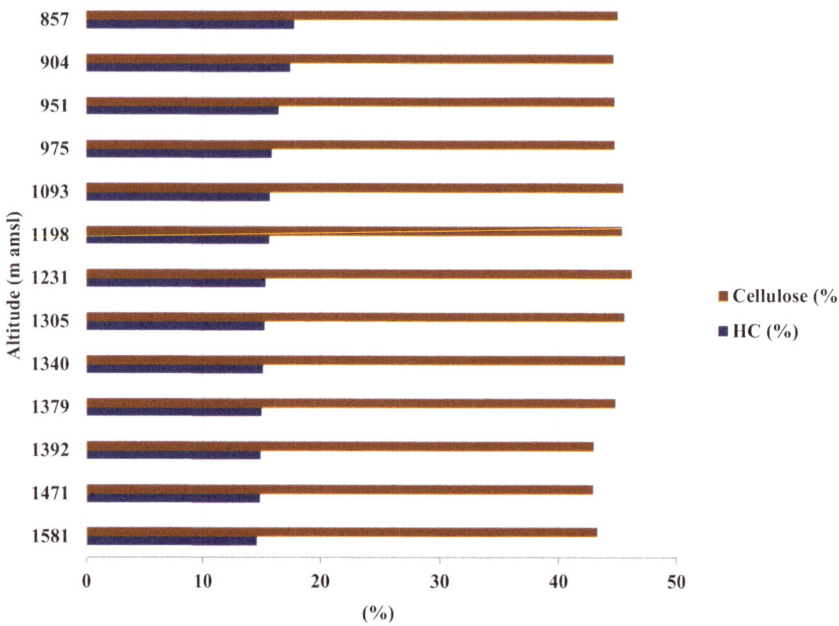

Fig. 5.3 Variation in cellulose and hemicellulose contents in white clover with altitude

At higher altitude, white clover plants have high mineral content (10.46%) which gradually decreases with decrease in elevation. This may be attributed to high soil moisture content as well as to increased activity of phosphate solubilizing bacteria in plant rhizosphere (Selvakumar et al. 2009). The level of cell wall constituents in white clover plants also varies with altitude (Fig. 5.3). ADL content in white clover plants increased from 1.30 to 4.73%, while hemicellulose content increased from 14.57 to 17.80% at lower altitudes. This may be associated with increased partitioning of cellulose and hemicelluloses content for lignification (Fig. 5.4).

Altitude also has significant effect on the level of polyphenolic compounds in forages. With the increase in elevation of the growing area, level of total phenols, simple phenols and tannin contents in white clover plants reduced from 1.84 to 0.82%, 0.81 to 0.31%, and 1.03 to 0.51%, respectively. Reverse trend was observed for hydrocyanic acid (HCN) content. HCN content increases in white clover plants proportionally with the increase in altitude from 0.360 to 1.044 mg/g. The higher levels of polyphenolic compounds in white clover at lower altitudes may be due to increased activities of enzymes involved in the formation of shikimic acid pathway intermediates for alleviation of climatic stresses. The elevated level of HCN in white clover at higher altitude is the probable way of desisting insect pest attack (Stochmal and Oleszek 1997). At higher altitudes, UV-B radiation also increases the rate of hydrocyanic acid accumulation in plants (Lindroth et al. 2000). Although, white clover growing at higher altitudes have better nutritional proportions in comparison to lower altitudes, but HCN content may exceed the

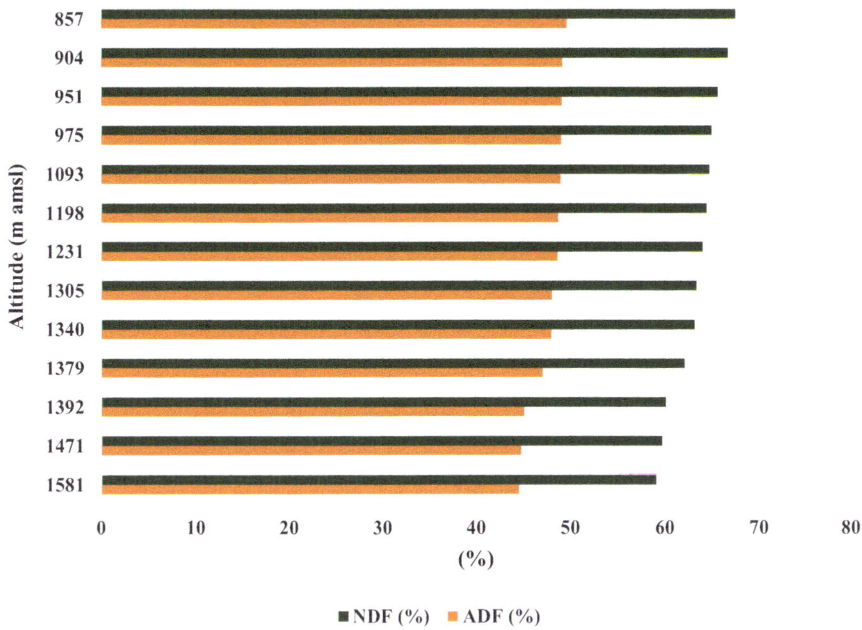

Fig. 5.4 Variation in fiber composition in white clover with altitude

normal level (0.37 mg HCN/g of dry matter; Lehmann et al. 1991). Therefore, excessive feeding of white clover from higher altitudes need regular monitoring to avoid the adverse effect of cyanide-poisoning in livestock (Fig. 5.5).

5.4 Seasonal Variations and Quality of Forages

Studies have revealed that nutritional composition of forages is not uniform throughout the year and changes with seasonal variations affecting performance of livestock (Evans 1989; Ramirez et al. 2001; Aganga and Tshwenyane 2003; Singh and Todaria 2012; Al-Masri 2013). The knowledge of alterations in the level of different nutrients in forages at different time intervals is essential for determining appropriate lopping period for forages in a defined agroclimatic region (Verma and Mishra 1999).

Grasses have high crude protein content during early growing season. The fiber content of pastures changes in the growing season due to change in growth stage and temperature (Sharpe and Rayburn 2019). During early growing season, grasses on an average has 8 to 10% CP content with approximately 60 to 65% digestibility. However, the CP content decrease during dry season. Singh (2002) reported that the level of oxalate content in Napier-Bajra hybrid is not uniform during the growth season. Tropical forages generally have high oxalate content than forage species grown in temperate regions, which implies a relationship between temperature and

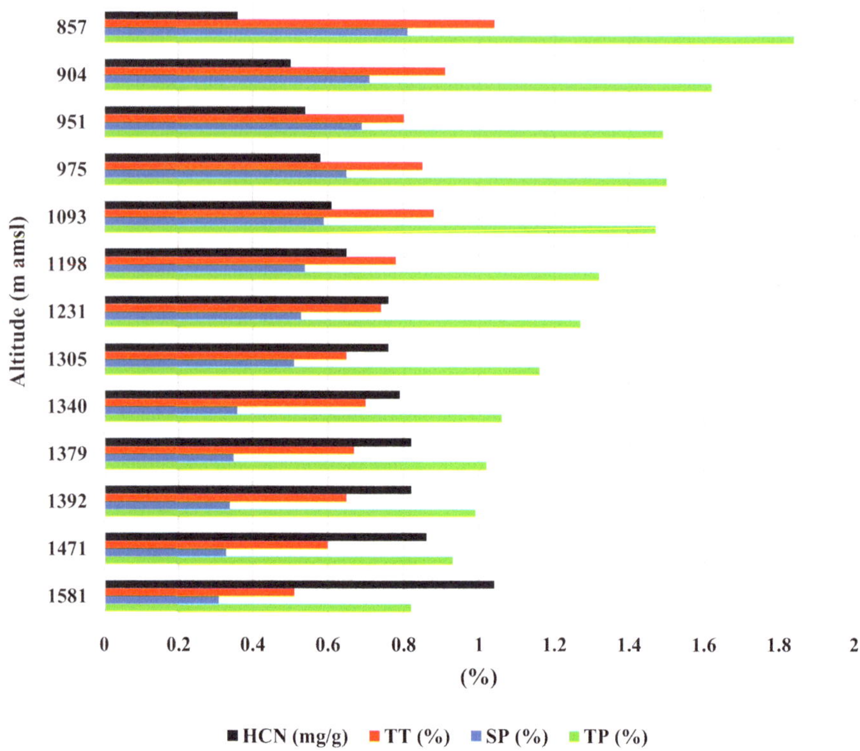

Fig. 5.5 Variation in total phenols (TP), simple phenols (SP), total tannins (TT), and hydrocyanic acid of white clover with varying altitude

oxalate accumulation in plants. Many environmental effects associated with seasonal variation also affect the nutritional composition of forages.

5.4.1 Seasonal Dynamics in Nutritional Quality of Major Forage Grasses

Seasonal variation affects the nutritional quality of forage grasses. Katoch et al. (2012) investigated the effect of seasonal variation on nutritional composition of annual grasses *viz.*, Tall fescue and *Setaria* and the results of their study have been discussed under subsequent subheads.

5.4.1.1 Seasonal Variation in Biochemical Composition of Tall Fescue

- Biochemical constituents of four tall fescue varieties *viz.*, Hima-1, Hima-4, Hima-5 and EC-178182 revealed variation throughout the year (Katoch et al. 2012) (Fig. 5.6a, b).
- *Green fodder yield and dry matter content*: The green fodder yield was higher in all four tall fescue varieties in the month of July as compared to the month of March. The green fodder yield in Hima-5 increased from 46.80 q/ha in the month of March to 75.60 q/ha in the month of July (Fig. 5.6a). DM content in tall fescue was highest during the month of July (26.89%) in comparison to the month of March (22.77%). Variety, Hima-1 had highest dry matter content during March (23.50%) and April (24.47%). In variety Hima-4, the dry matter content was 26.27% during June and increased to 27.10% in the month of July.
- *Crude protein and fiber content*: CP content was highest during April (11.95%) and minimum during the month of July (9.47%). Variety, EC-178182 exhibited 12.54% (March), 12.25% (April) and 11.08% (May) crude protein content. Varieties, Hima-1, Hima-4, and Hima-5 have higher crude protein content during the months of June and July. Tall fescue had highest NDF content in the month of July (70.31%). Minimum level of NDF content was observed during the month of March (63.65%). The variety Hima-5 exhibited low NDF content during March, April, May, and June. In variety Hima-1, the NDF content was lowest during April (66.26%) that increased to 69.40% in the month of July. Tall fescue has high ADF content in the month of July (52.11%) and minimum in March (48.16%). The variety, Hima-1 was observed with minimum ADF content during all cutting intervals. Variety Hima-4 hads minimum ADF content during the month of March (46.60%) that increased to 51.93% during the month of July. Variety, Hima-5 exhibited minimum ADF content during March (50.93%), April (50.93%), May (50.60%), and June (50.40%) and increased to 53.06% during the month of July. During the months of June and July, variety EC-178182 revealed highest ADF content (50.66% and 53.06%, respectively).
- *Hemicellulose content and digestibility*: Tall fescue has highest hemicellulose content (18.20%) in the month of July and lowest during March (15.45%). Variety, Hima-1 exhibited high level of hemicellulose content during April (19.20%), May (19.73%) June (19.46%), and July (19.06%). Among these four varieties, Hima-5 had least hemicellulose content. The dry matter digestibility of tall fescue was high during the month of March (55.20%) and declined during subsequent months. Among these varieties, EC-178182 exhibited highest dry matter digestibility (Fig. 5.6b).

Tall fescue grass has high protein content as well as protein digestibility in the month of April and minimum fiber content in March. The decrease in dry matter digestibility with advancing maturity of tall fescue may be attributed to lignification of cell wall. The variety Hima-4, revealed nutritional superiority over other varieties for dry matter content, Hima-1 for crude protein and ADF, Hima-5 for NDF and hemicellulose and EC-178182 for *in vitro* dry matter digestibility. The results of the study

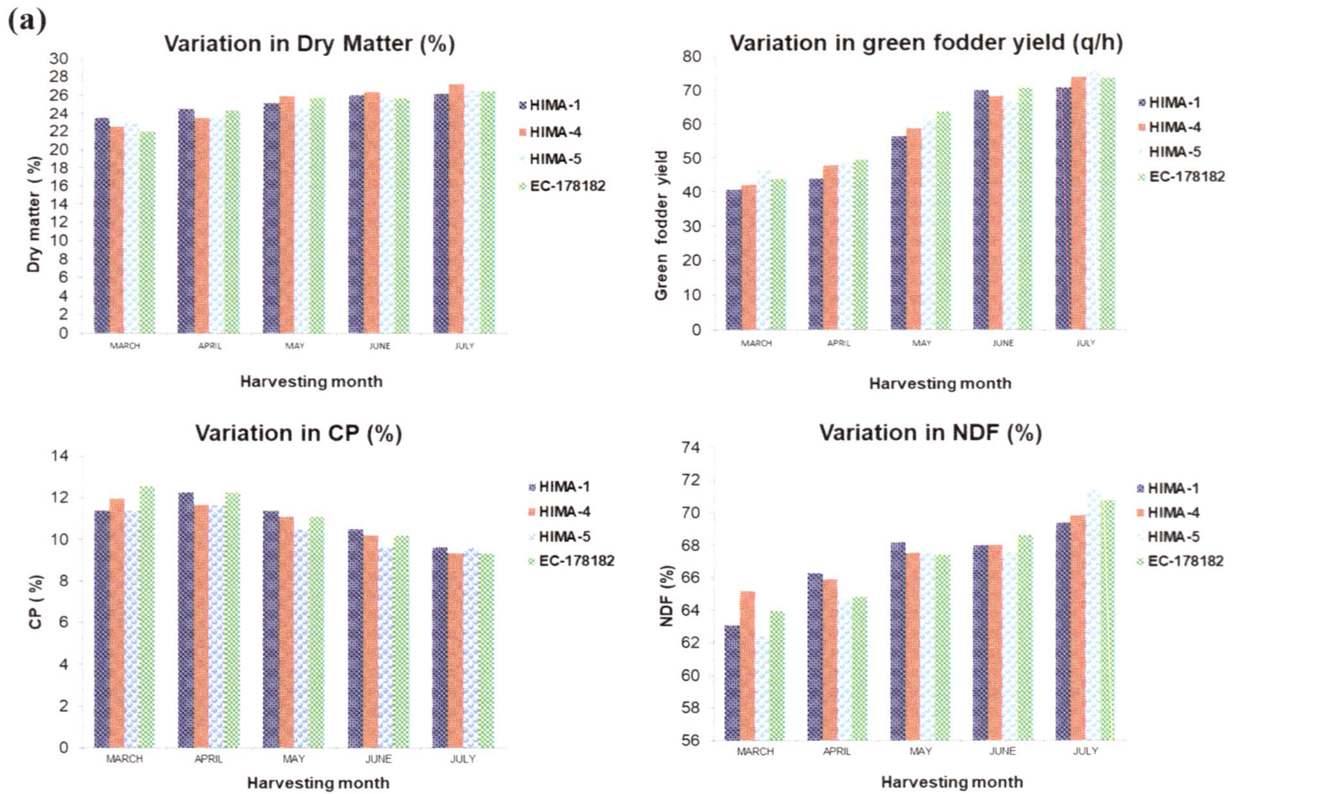

Fig. 5.6 (**a** and **b**): Effect of seasonal variation on forage quality in tall fescue

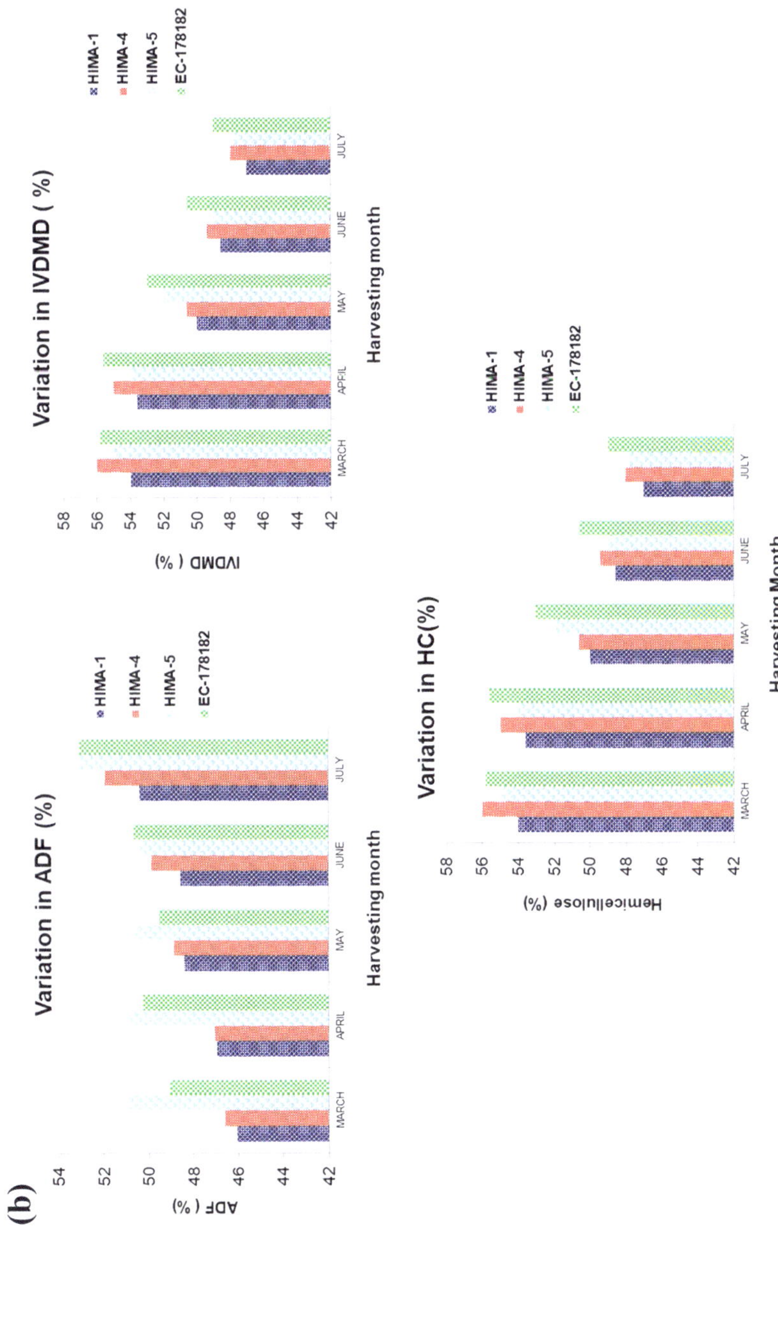

Fig. 5.6 (continued)

revealed that the quality of tall fescue is highly influenced by seasonal variations and has high nutritive value during March and April in sub-temperate Himalayan region.

5.4.1.2 Seasonal Variation in Biochemical Composition of *Setaria* Grass

Setaria grass also exhibits seasonal variations in biochemical composition. A study was conducted with three *setaria* cultivars *viz.*, PSS-1, S-20, and S-92 to understand variations in biochemical parameters with seasonal variations throughout the year (Katoch et al. 2012). The observations of the study have been discussed below and presented in Fig. 5.7a, b.

- *Green fodder yield and dry matter content*: The green fodder yield from *setaria* grass varieties was highest during the month of August (100.51 q/ha) and minimum during October 70.28 q/ha. The variety, S-92 had the highest green fodder production (107.80 q/ha) in the month of August. The green fodder yield in all three cultivars was highest in the month of August and declined in the months of September and October (Fig. 5.7a). *Setaria* grass had highest dry matter content during the month of October (25.81%) while lowest in the month of May (23.86%). In variety S-92, dry matter content was highest in the months of May (24.80%), July (23.83%), and October (26.06%). Varieties S-92 and S-20 has maximum dry matter content during the month of October while PSS-1 during the months of June and August.
- *Crude protein and fiber content*: Crude protein in *Setaria* grass was highest during May (10.88%) and minimum during October (5.05%). Varieties PSS-1 and S-20 has high CP content during the months of May, June, July, August and September while the variety S-92 in the month of October (5.25%) (Fig. 5.7a). NDF (Neutral detergent fiber) content in *Setaria* grass was maximum during October (71.71%) and minimum during May (60.13%). S-92 has highest ADF content during months of June (47.86%), August (48.46%) and October (51.26%) clippings (Fig. 5.7a). Variety S-20 has high ADF content during May (45.60%) and July (47.93%) (Fig. 5.7b). During September, the variety PSS-1 has high ADF content. Among these *Setaria* cultivars, PSS-1 exhibited minimum ADF in the months of May, June, July, and October. During the month of August, variety S-20 has minimum ADF content.
- *Hemicellulose and oxalate content*: Hemicellulose content in *Setaria* grass varied from 14.87 to 16.20% and 20.47 to 22.13% during May and October, respectively. Variety PSS-1 revealed high hemicellulose content during clipping of grass in June (17.13%), July (18.66%), September (20.66%) and October (22.13%). The hemicellulose content is high during October (21.11%) and minimum in the month of May (15.10%). Oxalate content in *Setaria* grass during the month of May varied from 4.72 to 4.97% and during October from 2.61 to 2.71%. Variety S-92 exhibited lowest oxalate content during May. The variety PSS-1 had minimum oxalate content in the month of June and variety S-20 during August (Fig. 5.7b).
- *In vitro dry matter digestibility*: IVDMD of *Setaria* grass is also affected by seasonal variations. The digestibility of *Setaria* grass was high during month of

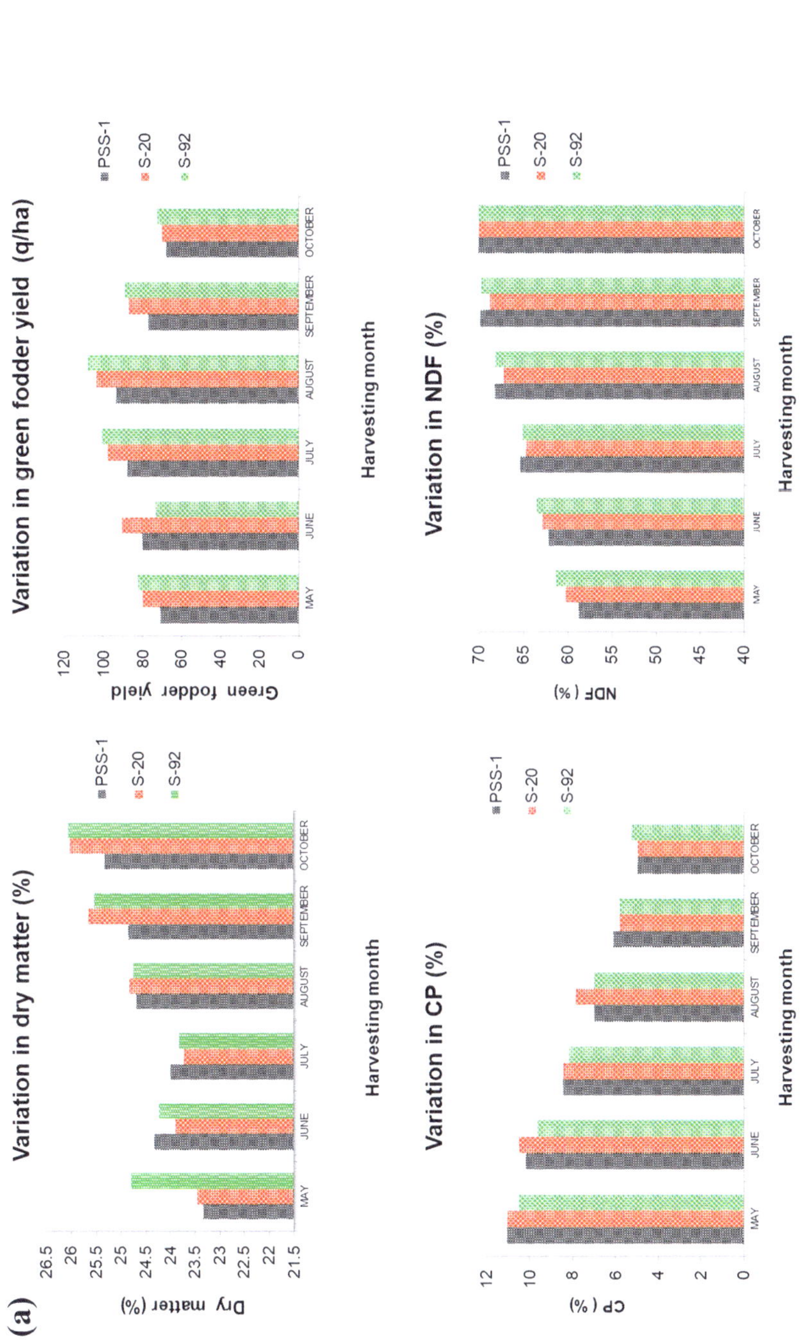

Fig. 5.7 (**a** and **b**): Seasonal variation in biochemical composition of *Setaria* grass

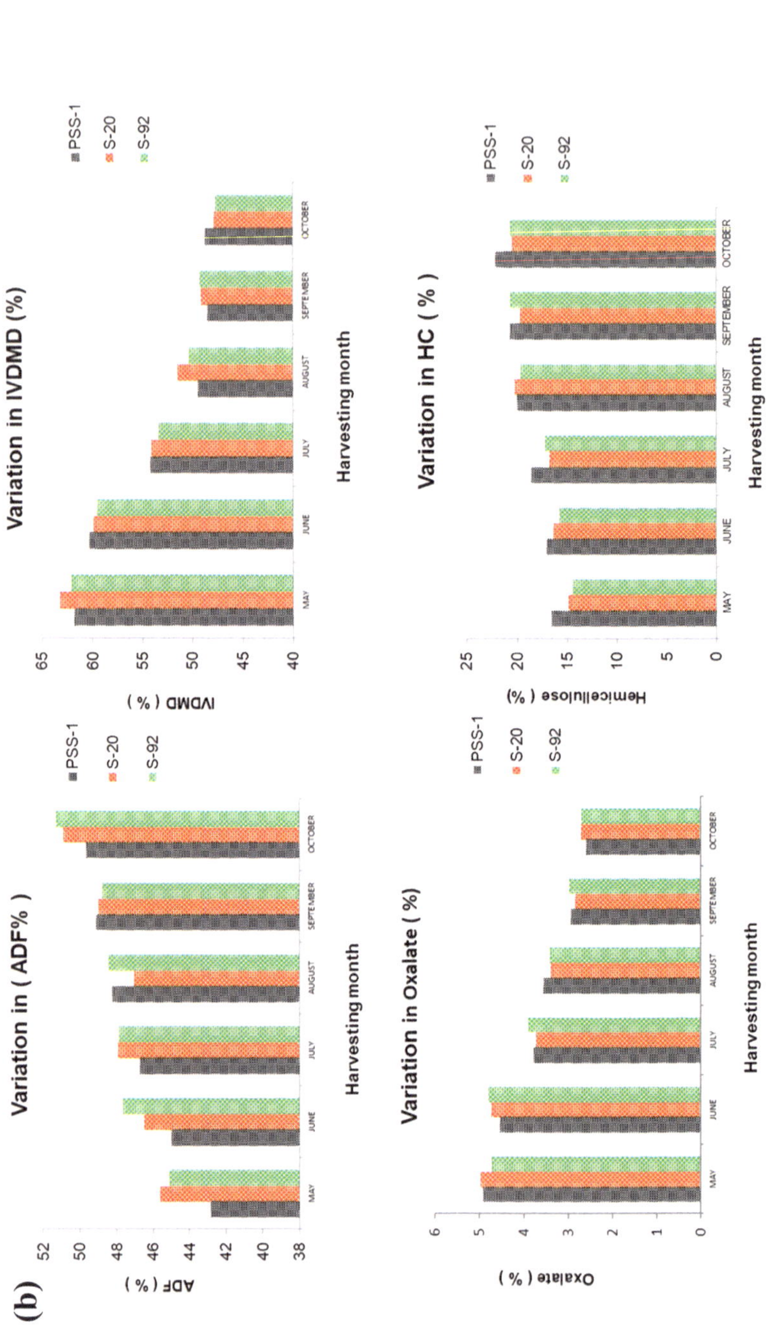

Fig. 5.7 (continued)

May (62.34%) and minimum during October (48.04%). Variety S-20 had maximum digestibility during the months of May (63.13%) and August (51.40%), while PSS-1 during June (60.26%), July (54.20%) and October (48.70%). Variety S-92 had maximum dry matter digestibility in the month of September (49.20%) (Fig. 5.7b).

In *Setaria* grass growing in sub-temperate conditions, nutrients are in balanced proportion during the months of May and July. However, the oxalate content is also simultaneously high, which is rather nutritionally undesirable. Harvesting of *Setaria* grass in the month of August could maintain a better balance in nutrients and anti-nutrients components. Further, harvesting grass after the month of September, when the oxalate content is low, is not appropriate period as grass has high fiber and low protein content.

5.4.2 Seasonal Dynamics in Nutritional Constituents of Major Fodder Trees

Trees are important feed components of livestock diet and play an important role in the meeting the feed as well as nutritional requirements of livestock in areas where few or no alternatives for the availability of forages are feasible. Fodder trees are now recognized as important components of animal feeding, as suppliers of protein and other essential nutrients. Fodder trees and shrubs have always played a role in feeding livestock during adverse climatic conditions especially during winter season/lean period, where available grazing material is not sufficient to meet the maintenance requirements of livestock. Tree fodders contain high levels of crude protein and minerals and many species have high levels of digestibility, however, anti-nutrients could be a problem in some tree species. Seasonal variations during active growth period are important determinants of quality of forages including fodder trees.

Seasonal changes in the nutritional composition of major fodder trees of the Himalayan region have been studied by various researchers. The study conducted by Nabi et al. (2017) reported seasonal variation in nutritional profile of *Populus deltoides*, *Populus nigra*, *Robinia pseudoacacia*, *Salix alba* and *Salix fragilis* foliage. The mean CP content in tree foliage in spring season was 13.33%. Highest CP content was observed in *Robinia pseudoacacia* foliage (18.73%) and lowest in *Populus deltoides* (11.52%). During summer mean CP content in tree foliage was 10.87%. *Robinia pseudoacacia* had highest CP content (18.58%) whereas *Populus deltoides* had lowest CP content (8.80%). During autumn season, tree foliage had 10.05% mean crude protein content. Highest CP content was observed in *Robinia pseudoacacia* foliage (17.01%) and minimum (7.73%) in *Populus deltoides* foliage. The mean crude fiber content in tree foliage was minimum during spring (15.00%) with highest value recorded for *Populus nigra* 15.92% and lowest for *Robinia*

Table 5.4 Effect of seasonal variations on crude protein and crude fiber in tree foliage

	Crude protein (%)			Crude fiber (%)		
Tree foliage	Spring	Summer	Autumn	Spring	Summer	Autumn
Populus deltoides	11.52	8.80	7.73	15.89	19.63	19.92
Populus nigra	11.73	8.93	7.92	15.92	19.65	19.93
Robinia pseudoacacia	18.73	18.58	17.01	12.91	13.16	14.11
Salix alba	12.42	9.07	8.87	15.14	16.53	17.26
Salix fragilis	12.27	8.97	8.71	14.80	16.15	17.24
Mean	13.33	10.87	10.05	15.00	17.02	17.77

Table 5.5 Effect of seasonal variations on ether extract and total ash contents in tree foliage

	Ether extract (%)			Total ash (%)		
Tree foliage	Spring	Summer	Autumn	Spring	Summer	Autumn
Populus deltoides	2.28	3.03	3.88	13.22	13.36	13.80
Populus nigra	2.30	3.09	3.90	13.23	13.40	13.82
Robinia pseudoacacia	4.86	5.10	5.12	8.91	9.51	9.98
Salix alba	3.64	4.14	4.34	12.63	12.94	13.39
Salix fragilis	3.62	4.12	4.31	12.61	12.90	13.37
Mean	3.34	3.96	4.31	12.12	12.42	12.87

pseudoacacia (12.91%). Highest nutritive profile of tree foliage in terms of crude fiber content was observed during autumn season (17.77%). *Populus nigra* (19.93%) and *Robinia pseudoacacia* (14.11%) had maximum and minimum crude fiber content, respectively, during autumn season (Table 5.4).

Ether extract was lowest during spring (3.34%) and increased in advancing seasons. The foliage from *Robinia pseudoacacia* (4.86%) and *Populus deltoids* (2.28%) has highest and lowest ether extract in spring season. In tree foliage, the ether extract was highest during autumn season (4.31%). *Robinia pseudoacaia* (5.12%) and *Populus deltoids* foliage (3.38%) has highest and lowest ether extract content, respectively, during autumn season (Table 5.5). Total ash content in tree foliage also increased in successive growth seasons. The total ash content was minimum during spring season (12.12%). Highest ash content was observed in *Populus nigra* foliage (13.23%) and lowest in *Robinia pseudoacacia* foliage (8.91%) during spring season. In summer, *Populus nigra* foliage had highest ash content (13.36%).

Seasonal variations in biochemical composition of common fodder trees of mid-hills of the Himachal Pradesh state were also investigated by Katoch et al. (2017). The study revealed remarkable variation in nutritional constituents of tree foliage throughout the year. The major outcomes of the study have been presented below:

- Crude protein content in *Bauhinia variegata* varied from 10.20 to 18.08% throughout the year, where highest CP content was observed during the month of March (Fig. 5.8a). In *Bauhinia* foliage, ether extract varied from 2.62 to 4.50% (Fig. 5.8b). The highest ether extract content was observed during the month of May. *Bauhinia* foliage had highest ash content (6.50%) and crude fiber content (32.38%) during September and January, respectively (Fig. 5.8c, d). NDF and ADF in foliage varied from 58.00 to 67.00% and 47.20 to 56.00% from March to January, respectively (Fig. 5.8e, f).

- *Arratocarpus lakoocha* leaves had highest CP content (22.88%) during the month of March (Fig. 5.8a). The ether extract content in *Arratocarpus lakoocha* varied from 1.00 to 1.60%, where highest content was observed during the month of May (Fig. 5.8b). *Arratocarpus lakoocha* leaves revealed highest ash content (13.00%) and crude fiber content (29.83%) during November and January, respectively (Fig. 5.8c, d). The NDF (63.80%) and ADF content (49.20%) in *Arratocarpus lakoocha* leaves were highest during the month of January (Fig. 5.8e, f).
- The foliage from *Albizzia chinensis* revealed highest crude protein content (13.41%) during month of March (Fig. 5.8a). The variation in ether extract content in *Albizzia chinensis* foliage ranged from 1.20 (November) to 3.50% (May) (Fig. 5.8b). Ash content was highest (7.90%) in *Albizzia chinensis* foliage during September (Fig. 5.8c). Highest crude fiber content in *Albizzia chinensis* foliage (34.73%) was observed during January (Fig. 5.8d). The fiber content (NDF and ADF) in *Albizzia chinensis* foliage varied from 60.20 to 69.60% and 37.80 to 51.00% from March to January, respectively (Fig. 5.8e, f).
- During the month of March, crude protein content was highest in *Albizzia lebbek* foliage (19.13%) (Fig. 5.8a). Ether extract was highest (4.90%) during May and July (Fig. 5.8b). The ash content was highest (8.40%) in the month of July (Fig. 5.8c). Highest crude fiber content in *Albizzia lebbek* foliage (32.50%) was observed during January (Fig. 5.8d). The NDF (65.00%) and ADF contents (52.20%) were high during the month of January (Fig. 5.8e, f).
- *Morus alba* foliage revealed highest CP content (22.77%) (Fig. 5.8a) in the month of May, highest ether extract content was observed in *Morus alba* foliage (4.30%) during the month of March (Fig. 5.8b). Ash content were higher (16.90%) in the month of September (Fig. 5.8c). The variation in crude fiber content in *Morus alba* foliage ranged from 13.30 (March) to 20.03% (January) (Fig. 5.8d). *Morus alba* foliage had highest NDF content (59.80%) and ADF content (46.20%) during January (Fig. 5.8e, f).
- Variation in crude protein content of *Grewia oppositifolia* ranged from 9.50 (January) to 18.92% (May) (Fig. 5.8a). In *Grewia* foliage, ether extract varied from 2.50 to 4.00%. The highest ether extract was observed during the month of September (Fig. 5.8b). *Grewia* foliage had highest ash content (9.60%) and crude fiber content (27.00%) during July and January, respectively (Fig. 5.8c, d). The fiber content (NDF and ADF) varied from 55.00 to 62.60% and 38.00 to 51.20% from March to January, respectively (Fig. 5.8e, f).

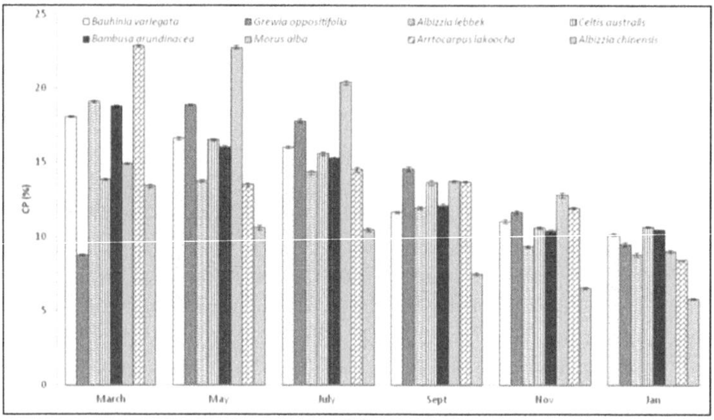

(a)

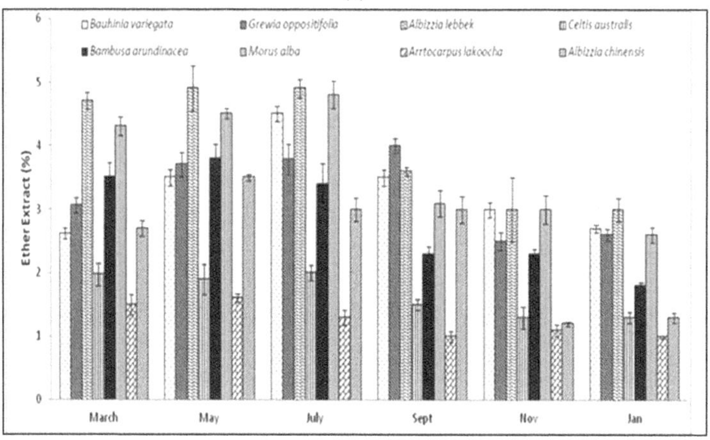

(b)

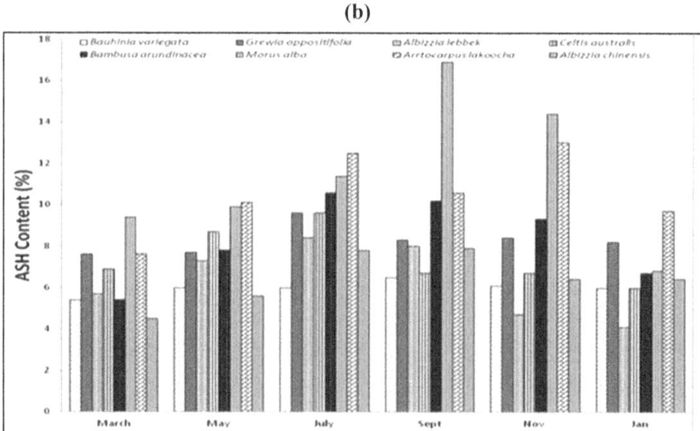

(c)

Fig. 5.8 (**a** to **g**): Effect of seasonal variation on biochemical constituents of fodder trees

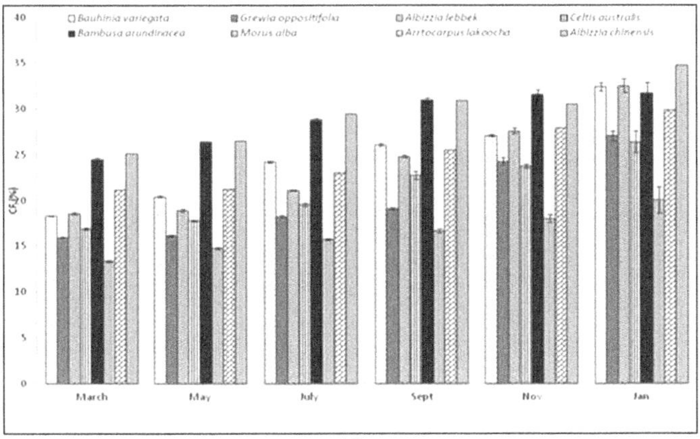

(d)

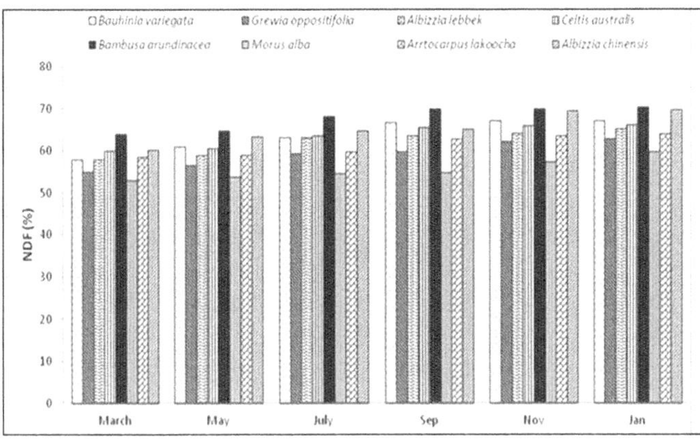

(e)

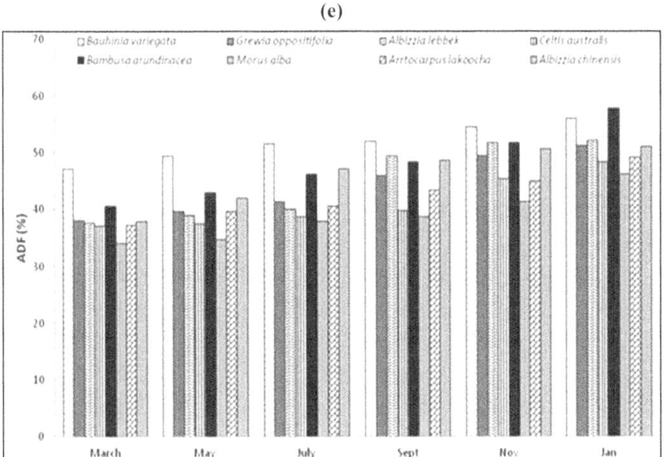

(f)

Fig. 5.8 (continued)

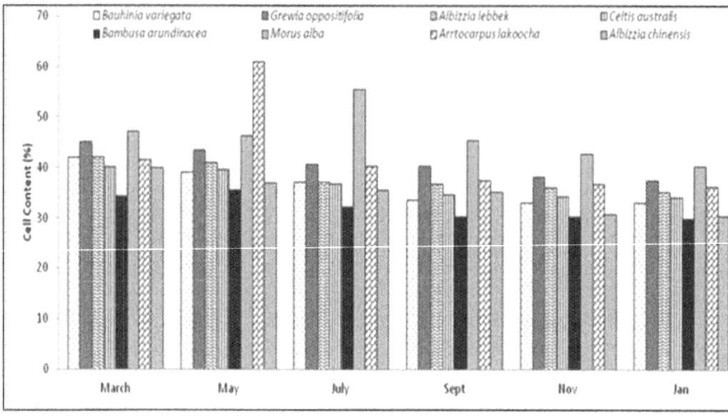

(g)

Fig. 5.8 (continued)

- *Bambusa arundinacea* foliage revealed variation in crude protein content from 10.40 (November) to 18.81% (March) (Fig. 5.8a). Highest ether extract content was observed in *Bambusa arundinacea* foliage (3.80%) during May (Fig. 5.8b). *Bambusa* leaves had high ash content (10.60%) in the month of July (Fig. 5.8c). The variation in crude fiber content ranged from 24.47 (March) to 31.73% (January) (Fig. 5.8d). The variation in NDF and ADF content ranged from 63.80 to 70.20% and 40.60 to 57.80% from March to January, respectively (Fig. 5.8e, f).
- *Celtis australis* foliage had highest crude protein content (16.53%) during the month of May (Fig. 5.8a). The variation in ether extract of foliage ranged from 1.30% (January) to 2.00% (July) (Fig. 5.8b). Highest ash (9.60%) and crude fiber content (26.36%) in *Celtis australis* foliage were observed during July and January, respectively (Fig. 5.8c, d). The NDF and ADF contents were highest during the month of January (Fig. 5.8e, f).
- In the month of March, higher content was observed in *Morus alba* (47.00%) followed by *Grewia oppsitifolia* (45.00%), *Bauhinia variegata* and *Albizzia lebbek* (42.00%) whereas minimum value was observed in *Bambusa arundinacea* foliage (34.20%). *Arratocarpus lakoocha* revealed significant increase in cell content up to 61.00% during the month of May. The cell content was highest in *Morus alba* foliage (55.40%) and lowest in *Bambusa arundinacea* foliage (32.00%) during the month of July. The highest cell content during the month of November was recorded in *Morus alba* foliage (42.60%) whereas *Bambusa arundinacea* had the least value (30.20%). During January, *Morus alba* (40.20%) foliage had the highest cell content. The cell content was highest in *Arratocarpus lakoocha* (61.00%) during the month of May, whereas lower values were observed during September and November in *Bambusa* foliage (30.20%) (Fig. 5.8g).

Nutritional quality of *Leuceana leucocephala* is also affected by variations in seasonal conditions. Young foliage of *Leucaena* has highest CP content (32.10%) during January, that decreases to 20.70% during July (Akingbade et al. 2001). *Leucaena* foliage, has lowest ash content in the month of February (7.30%) and highest is in July (10.40%). The higher mineral components of the forage during the rainy season compared with dry season may be attributed to the availability of the essential nutrient elements in the soil, ability of the forages to utilize the favorable climatic and edaphic growth conditions necessary for high-quality forage production (Akingbade et al. 2001).

One of the studies conducted by our group (Unpublished data) revealed seasonal variations affecting the nutritional composition of *Leucana* foliage. Tender, premature and mature *Leucaena* foliage revealed variation from 25.70 to 34.85%, 25.33 to 32.64%, and 21.84 to 28.65%, respectively, throughout the year. Highest CP content was observed during the month of April and lowest during January. Crude protein content was high in young leaves and decreased with the advancement in maturity. High carbohydrate content in forages is a nutritionally desirable characteristic as they provide instant source of energy to livestock. Level of total soluble carbohydrates in *Leucaena* foliage was highest during winter season. This may ascribe the importance of carbohydrate reserve for survival in harsh winter conditions. During spring season, when leaves are able to get more sunshine hours, they start accumulating carbohydrates for their survival during dry period which is characterized by low temperature and low water availability in the Himalayan region. Similar to CP content, the level of soluble carbohydrates decreased with advancing maturity of *Leucaena* foliage. This could be attributed to formation of complex wall polysaccharides and their probable utilization in lignification of cell wall that is an important process associated with plant maturity. Seasonal variation also affects the ash content in tender, premature, and mature *Leucaena* leaves. Tender leaves had significantly ($p < 0.05$) highest ash content from the month of November till February whereas in premature and mature leaves, the ash content was significantly highest during the month of January. During winter season, high ash content in leaf tissue reflects the reduced utilization of minerals in metabolic reactions (Tables 5.6, 5.7 and 5.8).

5.4.3 Seasonal Alterations in Anti-Nutritional Constituents of Fodder Trees of Mid-Himalayas

Seasonal variations also affect the level of anti-nutritional constituents in forages (Onyeonagu and Asiegbu 2013). It has been reported that alkaloid, phytate, saponins, tannins, HCN and other anti-nutrients are in higher amounts during rainy season in comparison to dry season in tree foliage. Akingbade et al. (2001) reported changes in NDF and ADF content of *L. leucocephala* tree foliage from 39.90% (February) to 24.60% (July) and 26.70% (February) to 19.30% (July). The mimosine content of young leaves of *L. leucocephala* was lowest in the month of March (0.80%) that increased to (2.90%) in the month of December. A seasonal

Table 5.6 Seasonal variation in the nutritive quality of tender leaves of *Leucaena leucocephala*

Month	CP	TC	NDF	ADF	HC	ADL	CL	Ash	TP	SP	TT	Mimosine
	(%) Content											
Nov	27.43^{i}	48.51^{a}	29.23^{i}	20.93^{h}	8.30^{gh}	4.48^{h}	16.45^{ef}	7.83^{a}	13.60^{a}	6.10^{a}	7.50^{fg}	3.61^{a}
Dec	26.25^{j}	45.24^{j}	28.89^{j}	20.36^{i}	8.53^{f}	3.91^{i}	16.45^{ef}	7.86^{a}	13.26^{d}	5.93^{b}	7.33^{h}	2.55^{e}
Jan	25.70^{k}	45.71^{h}	28.75^{j}	19.90^{j}	8.85^{e}	3.58^{j}	16.32^{f}	7.94^{a}	12.96^{e}	4.73^{e}	8.22^{b}	2.50^{f}
Feb	29.62^{f}	45.46^{i}	31.16^{g}	21.62^{f}	9.54^{c}	5.90^{e}	15.71^{g}	7.86^{a}	12.67^{g}	4.51^{fg}	8.16^{bc}	3.51^{b}
Mar	31.38^{c}	46.67^{g}	32.44^{d}	22.60^{e}	9.84^{b}	6.09^{d}	16.51^{e}	7.07^{c}	12.77^{f}	5.11^{d}	7.66^{e}	2.98^{c}
Apr	34.85^{a}	47.68^{d}	32.70^{c}	22.93^{d}	9.77^{b}	6.21^{cd}	16.72^{d}	6.97^{cd}	13.22^{d}	5.42^{c}	7.80^{d}	2.86^{d}
May	32.52^{b}	47.77^{c}	33.48^{b}	23.27^{c}	10.21^{a}	6.51^{b}	16.76^{d}	6.87^{cd}	13.48^{b}	5.38^{c}	8.10^{c}	2.42^{g}
June	30.29^{d}	47.84^{c}	33.89^{a}	23.73^{a}	10.16^{a}	6.75^{a}	16.97^{c}	6.58^{ef}	13.46^{b}	5.05^{d}	8.41^{a}	1.87^{k}
Jul	29.94^{e}	46.99^{f}	32.70^{c}	23.54^{b}	9.17^{d}	6.33^{c}	17.20^{b}	6.75^{de}	12.37^{h}	4.53^{f}	7.84^{d}	1.97^{j}
Aug	29.61^{f}	46.94^{f}	32.27^{e}	22.94^{d}	9.33^{d}	5.89^{e}	17.05^{bc}	6.45^{f}	12.15^{j}	4.45^{g}	7.70^{e}	2.08^{i}
Sept	28.52^{g}	47.23^{e}	31.36^{f}	22.91^{d}	8.45^{fg}	5.53^{f}	17.38^{a}	6.97^{cd}	12.27^{i}	4.75^{e}	7.51^{f}	2.18^{h}
Oct	28.18^{h}	47.92^{b}	29.61^{h}	21.42^{g}	8.18^{h}	4.87^{g}	16.55^{e}	7.38^{b}	13.37^{c}	5.95^{b}	7.41^{gh}	2.95^{c}

Table 5.7 Seasonal variation in the nutritive quality of premature leaves of *Leucaena leucocephala*

	CP	TC	NDF	ADF	HC	ADL	CL	Ash	TP	SP	TT	Mimosine
Month	(%) Content											
Nov	26.62[g]	41.72[a]	33.75[h]	24.10[g]	9.65[i]	5.94[g]	18.16[c]	8.59[c]	12.69[a]	6.11[a]	6.58[d]	3.48[a]
Dec	26.22[h]	39.90[h]	33.60[i]	23.96[h]	9.64[i]	4.86[h]	19.11[b]	8.91[b]	12.43[b]	5.49[de]	6.94[c]	2.46[d]
Jan	25.33[i]	39.83[h]	32.47[j]	23.73[i]	8.74[j]	4.21[i]	19.52[a]	9.14[a]	11.96[d]	4.59[h]	7.37[a]	2.06[e]
Feb	26.26[h]	40.15[g]	35.40[f]	23.59[j]	11.88[b]	6.48[f]	17.11[h]	8.97[b]	11.57[e]	4.24[i]	7.33[a]	3.28[b]
Mar	27.64[e]	40.48[f]	35.94[e]	23.97[h]	11.97[a]	7.93[c]	16.04[j]	8.64[c]	11.55[ef]	5.45[e]	6.10[f]	2.77[c]
Apr	32.64[a]	40.94[d]	36.27[d]	24.95[e]	11.32[c]	8.09[b]	16.86[i]	8.22[d]	11.50[f]	5.97[b]	5.53[g]	2.43[d]
May	30.25[b]	40.89[d]	36.40[c]	25.60[c]	10.80[f]	8.22[a]	17.38[g]	7.97[f]	12.25[c]	5.22[g]	7.03[b]	2.07[e]
June	29.98[b]	40.80[e]	37.36[a]	26.11[a]	11.25[d]	8.23[a]	17.88[d]	7.72[g]	12.39[b]	5.77[c]	6.62[d]	1.53[i]
Jul	28.85[c]	39.90[h]	36.97[b]	25.87[b]	11.10[e]	8.12[b]	17.74[e]	7.39[i]	10.42[g]	5.23[g]	5.19[h]	1.78[h]
Aug	28.31[d]	39.87[h]	36.45[c]	25.30[d]	11.15[e]	7.87[c]	17.43[g]	7.63[h]	10.35[h]	5.36[f]	4.99[i]	1.98[f]
Sept	27.86[e]	41.10[c]	35.42[f]	24.94[e]	10.48[g]	7.16[d]	17.78[e]	8.09[e]	10.36[h]	5.51[d]	4.85[j]	1.93[g]
Oct	27.20[f]	41.26[b]	34.84[g]	24.57[f]	10.26[h]	6.96[e]	17.62[f]	8.25[d]	12.42[b]	5.97[b]	6.45[e]	2.42[d]

Table 5.8 Seasonal variation in the nutritive quality of mature leaves of *Leucaena leucocephala*

	CP	TC	NDF	ADF	HC	ADL	CL	Ash	TP	SP	TT	Mimosine
Month	(%) Content											
Nov	22.42^{j}	37.94^{b}	38.65^{g}	32.22^{f}	6.43^{f}	8.94^{i}	23.28^{c}	9.51^{e}	12.40^{b}	5.95^{a}	6.45^{d}	2.84^{a}
Dec	21.97^{k}	38.59^{a}	37.86^{h}	30.97^{g}	6.89^{d}	8.30^{j}	22.67^{d}	9.90^{d}	11.92^{f}	5.40^{c}	6.52^{d}	2.12^{d}
Jan	21.84^{l}	38.50^{a}	37.23^{j}	30.61^{h}	6.62^{e}	8.34^{j}	22.26^{e}	10.94^{a}	11.32^{j}	4.91^{f}	6.42^{d}	2.21^{c}
Feb	27.85^{b}	36.96^{f}	36.91^{k}	28.46^{k}	8.45^{a}	9.66^{g}	18.81^{h}	10.29^{b}	11.39^{i}	4.56^{h}	6.83^{c}	1.55^{g}
Mar	27.25^{c}	37.49^{d}	37.23^{j}	29.55^{j}	7.68^{bc}	10.22^{e}	19.33^{g}	10.10^{c}	11.44^{h}	4.24^{i}	7.21^{ab}	2.41^{b}
Apr	28.65^{a}	37.78^{c}	37.65^{i}	29.85^{i}	7.80^{b}	10.45^{d}	19.40^{g}	9.94^{cd}	11.85^{g}	5.13^{e}	6.72^{c}	1.72^{f}
May	26.81^{d}	37.92^{b}	39.82^{d}	32.20^{f}	7.62^{c}	10.54^{c}	21.66^{f}	9.41^{e}	12.36^{c}	5.24^{d}	7.12^{ab}	2.09^{d}
June	24.25^{e}	37.22^{e}	41.51^{a}	34.67^{a}	6.84^{d}	11.20^{a}	23.47^{a}	9.17^{f}	12.45^{a}	5.34^{c}	7.11^{b}	1.47^{h}
Jul	23.86^{f}	36.83^{g}	40.81^{b}	34.24^{b}	6.57^{e}	10.97^{b}	23.27^{c}	8.96^{g}	10.81^{k}	5.08^{e}	5.73^{e}	1.54^{g}
Aug	23.53^{g}	35.88^{j}	40.36^{c}	33.78^{c}	6.58^{e}	10.27^{e}	23.51^{a}	8.77^{h}	10.19^{l}	4.96^{f}	5.23^{f}	1.82^{e}
Sept	23.09^{h}	36.16^{i}	39.63^{e}	33.25^{d}	6.38^{f}	9.92^{f}	23.33^{bc}	8.78^{h}	12.07^{d}	4.82^{g}	7.24^{a}	1.86^{e}
Oct	22.91^{i}	36.34^{h}	39.27^{f}	32.84^{e}	6.43^{f}	9.41^{h}	23.43^{ab}	9.38^{e}	11.94^{e}	5.51^{b}	6.43^{d}	2.14^{d}

*In Tables 5.6, 5.7, and 5.8, each value represents mean of variation in nutritional constituent during 2 experimental years. In the same column, significant differences ($p \leq 0.05$) of nutritional constituents according to CRD (Completely Randomized Design) are indicated by different superscripts. Different superscripts represent the significant values whereas same superscripts represent that their values are statistically at par ($p \leq 0.05$)

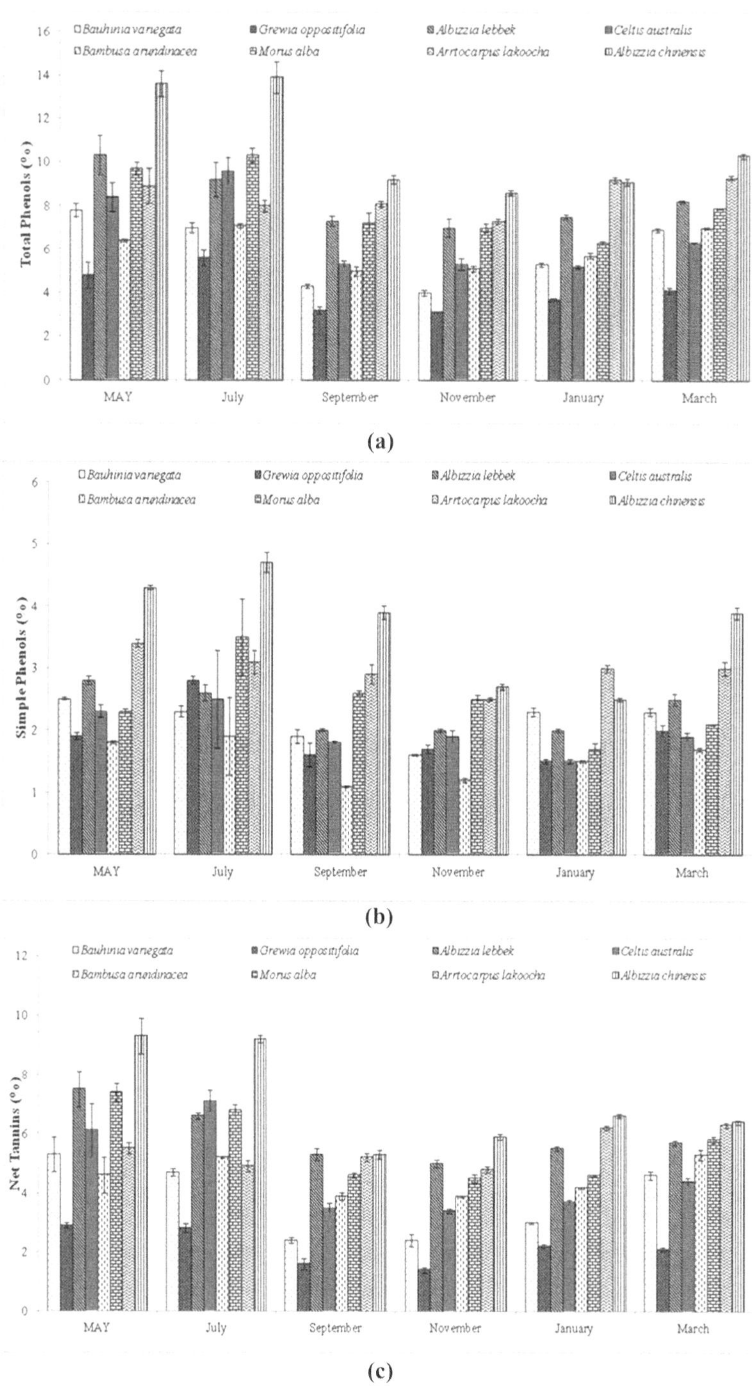

Fig. 5.9 (**a** to **d**): Effect of seasonal variation in anti-nutritional constituents of fodder trees in mid-Himalayas

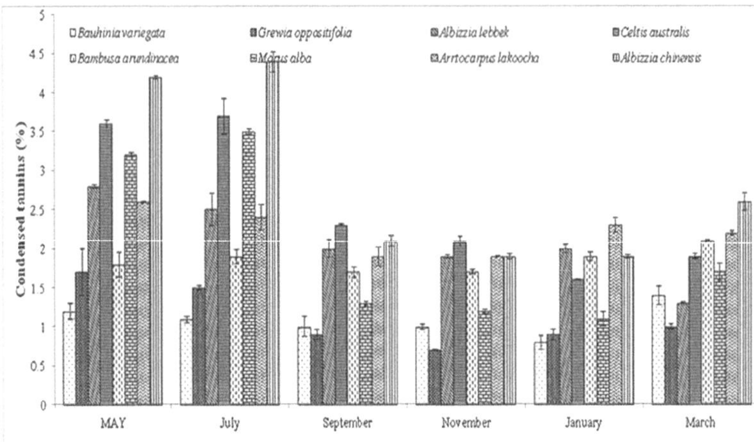

(d)

Fig. 5.9 (continued)

pattern of increase in total phenolic content during summer season and decrease in winter season in tree foliage has been reported by Rana et al. (2006) and Amaral et al. (2010). Condensed tannins start to accumulate during winter season and reach highest level during summer season (Rana et al. 2006).

Seasonal variations in anti-nutritional constituents of fodder trees of mid-Himalayas have been investigated by Katoch (2019). The important results of the study have been discussed Fig. 5.9a–d:

- During the month of May, foliage of *Bauhinia variegata* has significantly been higher level of total phenols (7.80%), thereafter from July to November there was a significant decline in the total phenolic content from 7.00 to 4.00%. The level of simple phenols in *Bauhinia variegata* foliage was highest during the month of May (2.50%) and lowest during the month of November (1.60%). The foliage had significantly high net tannins content during the months of May to July. *Bauhinia variegata* foliage lopped during month of March, has high condensed tannins (1.40%) which reduced during subsequent months till the month of January (0.80%). During the months of May and July, leaves of *Bauhinia variegata* had high hydrolysable tannins content (4.10 and 3.60%, respectively).
- *Arratocarpus lakoocha* foliage has low level of total phenols (7.30%) during November which subsequently increased from the month of January to March (9.20 to 9.30%). The simple phenolic content in *Arrtocarpus lakoocha* foliage was high during the month of May (3.40%) and low during the month of November (2.50%). *Arrtocarpus lakoocha* foliage has low net tannin content during months of July and November. In the month of May, the foliage has higher levels of condensed tannins (2.60%) thereafter, there was reduction in condensed tannins content (1.90%) during the months of September and November. The

foliage from *Arrtocarpus lakoocha* revealed high levels of hydrolysable tannins (3.90 and 4.10%) during the months of January and March.

- During May–July, *Albizia chinensis* foliage has high level of total phenols. The decrease in total phenolic content was observed in the month of November (8.60%). In the month of July, the foliage exhibited high level of simple phenols (4.70%), thereafter, there was a significant reduction in content from November to January (2.70 to 2.50%). High level of net tannins was observed during the months of May and July (9.30 and 9.20%, respectively) thereafter, net tannin content decreased during the month of September (5.30%). *A. chinensis* had higher level of condensed tannins during the months of May and July (4.20 and 4.40%, respectively), which decreased during the months of November and January. Higher level of hydrolysable tannins were observed in the month of May (5.10%) whereas, lowest level (2.20%) was recorded during month of September.
- In the period between the months of May and July, *Albizia lebbek,* foliage high total phenolic content was observed whereas lowest total phenolic content was observed during November (8.60%). *Albizia lebbek* leaves had high simple phenolic content during the months of May and July. Reduction in simple phenolic content was observed from September to January. *Albizia lebbek* foliage had high levels of net tannins during the month of May (7.50%), thereafter reduction in the net tannin content was observed during the month of November (5.00%). Higher condensed tannins (2.80 and 2.50%) were observed during months of May and July, respectively, whereas low levels of condensed tannins was observed during the month of March (1.30%). *Albizia lebbek* foliage harvested during the months of May and March revealed high levels of hydrolysable tannins (4.70 and 4.40%, respectively), whereas lower level was observed during the months of September and November.
- *Morus alba* foliage has high total phenolic content from the month of May to July. The lowest total phenolic content was observed during January (6.30%). The simple phenols content (3.50%) were higher during July. Reduction in simple phenolic content was observed during January (1.70%). The foliage had high net tannins content during the months of May to July. However, consistent reduction in tannins content was observed up to the month of January. *Morus alba* had high condensed tannins content during May and July (3.20 and 3.50%, respectively), which decreased during the months of November and January. High hydrolysable tannins content was observed during May and March (4.20 and 4.10%, respectively), while low level was observed between July and November.
- *Grewia oppositifolia* foliage has higher total phenolic content during the months of May (4.80%) and July (5.60%). However, reduction in total phenols was observed from September to November. The foliage had higher level of simple phenols (2.80%) during July while lowest simple phenolic content was observed during January (1.50%). *Grewia oppositifolia* foliage had high net tannins content during May to July, which reduced during the months from September to November. The foliage harvested during the months of May and July had high condensed tannins content (1.70 and 1.50%, respectively). Thereafter, there was a

reduction in subsequent months where lowest level was observed during November (0.70%). *Grewia* foliage revealed high hydrolysable tannins content (1.30%) during July and January, whereas reduction (up to 0.70%) was observed during September and November.

- *Celtis australis* foliage has high total phenolic content during May–July. The lowest total phenolic content was observed in the month of January (5.20%). The foliage also had high simple phenolic content during the months of May and July. However, low simple phenolic content was observed during the month of January (1.50%). *Celtis australis* had high net tannins content during the months of May to July. Higher levels of condensed tannins was observed during the months of May and July (3.60 and3.70%, respectively). After the month of July, leaves of *Celtis australis* had low level of condensed tannins. During July, leaves of *Celtis australis* exhibited high amount of hydrolysable tannins (3.40%) however, reduction in hydrolysable tannins content were recorded in the months of September and November (1.20 and 1.40%, respectively).
- During the months of March and July, *Bambusa arundinacea* has high total phenolic content (7.00 and 7.10%, respectively), however there was reduction in content during September and November (5.00 and 5.10%, respectively). The leaves have high simple phenolic content during the months of May and July. Thereafter, reduction in simple phenolic content was observed from September to November. In the month of July, foliage had high net tannins content (5.20%) which reduced in subsequent months to 3.90%. The foliage showed high condensed tannins content during the month of March (2.10%) however, during September and November, the total phenolic content was low. The foliage revealed high hydrolysable tannins content in the months of July and March (3.30 and 3.20%, respectively), whereas low level of hydrolysable tannins content (2.20%) was observed in the months of September and November.
- Foliage from *Grewia oppsitifolia* and *Bauhinia variegata* had comparatively lower levels of anti-nutrients. Lopping of these trees from the month of September to December provides better quality foliage for livestock feeding. *Grewia oppsitifolia* may be lopped preferably in the month of November, while the month of October is favourable for lopping *Bauhinia variegata*. During the months of September and November, *Morus alba* foliage had low levels of anti-nutrients, therefore, *Morus alba* could be lopped between September and November for feeding livestock.
- The anti-nutrients in *Leucaena* foliage also revealed variation with seasonal changes. In tender and premature *Leucaena* foliage, the higher values of NDF and ADF were recorded during the month of June whereas the lower values were recorded during the month of January. In mature *Leucaena* leaves, the NDF and ADF content were comparatively higher during the month of June. High fiber content in June reflects the increased synthesis of fiber components in supportive tissues. In tender and premature leaves, highest ADL content was observed

during January and lowest in the month of June. In mature leaves, the ADL content in foliage varied from 8.30 to 11.20%. The higher value of ADL was observed during the month of December and lowest during June. Hemicellulose and cellulose content in leucaena foliage also varied significantly with seasonal variation. In tender and premature *Leucaena* leaves, total and simple phenolic contents were recorded highest during the month of November while in mature leaves the contents were high during June and November, respectively. The lowest tannins content in tender leaves was observed during the month of December whereas in premature and mature *Leucaena* leaves during August and September, respectively. High polyphenolic content in tender Leucaena leaves could be explained by the reason that any damage at this stage is more destructive than the later stages of growth, therefore the high amount of polyphenolic compounds in tender leaves seems to be an adaptive process to avoid any damage by harsh winter season. Mimosine content varies in the different plant parts of *Leucaena* tree where it may reach up to 12% in the growing tips (Kumar 2003), whereas in young pods, seeds, and green stems, it can reach up to 5% (Ghosh and Bandhopadhyay 2007). Mimosine content in *Leucaena* foliage is therefore dependent upon seasonal variations and plant maturity. In *Leaucaena* foliage, mimosine content is high during winter season (November to February) and low during rainy season (June to September). Tender *Leaucaena* leaves had highest mimosine content followed by premature and mature leaves.

- The above study revealed seasonal dynamics in nutritional profile and mimosine content of *Leucaena* foliage. A number scale system (1 to 12) was used to grade variation in the level of a particular nutrient during different months (Table 5.9). The numbers were assigned in increasing order (1 to 12) representing the variation in the content of nutrient from highest to lowest in the particular month whereas in case of anti-nutrients, numbers were assigned in decreasing order (12 to 1) representing variation in anti-nutrients ranged from highest to lowest during a particular month. The grading revealed that the tender *Leucaena* leaves had maximum balance between the nutritional and anti-nutritional constituents from the month of September to February. In premature leaves, the variability in the balance of nutrients and anti-nutrients was observed during different months where highest nutritive value was observed during January–April. Considering variations in nutritional constituents of tender, premature and mature *Leucaena* leaves, the cumulative grading revealed good nutritional quality of *Leucaena* foliage between the months from January and April. During the period between the months of August and December, the *Leucaena* foliage has moderate balance between nutritional and anti-nutritional constituents.

Table 5.9 Cumulative grading showing seasonal variations in nutritional constituents in *Leucaena* foliage

Month	Tender leaves	Premature leaves	Mature leaves	Cumulative grading	Ranking
Jan	60	67	53	56	1
Feb	72	73	48	69	2
Mar	80	76	62	73	3
Apr	84	74	62	73	4
May	96	84	87	86	9
June	99	98	104	103	10
Jul	82	81	79	84	8
Aug	74	78	87	77	6
Sept	67	63	91	79	7
Oct	64	74	89	79	7
Nov	67	82	83	76	5
Dec	64	79	71	77	6

5.5 Effect of Foliar Diseases and Insect Pests on Quality of Forages

Depending on edaphic and environmental conditions, forage quality is affected by different disease causing organisms and insects, which either directly or indirectly alter nutritional value of forage. Forage quality is therefore highly dynamic, changing over the course of the growing season as plants pass through stages of morphologic development, and are impacted by diseases and insects.

Insects infestation could significantly alter both the yield and quality of forages (Buxton and Fales 1994). Insect feeding generally lowers the quantity of forage but, has variable impacts on forage quality. Insects which preferentially feed on leaves such as the alfalfa weevil reduce leaf biomass and consequently, the leaf:stem ratio. The overall plant development is stunted, including stem growth, however, nutritive value does not change appreciably unless plants are severely defoliated. Insects that feed on assimilates, such as the potato leafhopper cause a reduction in stem growth thereby increasing the leaf:stem ratio and nutritive value. Insects usually feed on leaves and young stems of legumes that have direct effects on growth and loss of high quality leaf tissues. Alfalfa weevil removes the highly digestible mesophyll cells prior to the first harvest which decreases forage quality and yield due to lower leaf digestibility and reduced photosynthesis.

Foliar diseases induce synthesis of undesirable constituents and cause defoliation in plants reducing their quality. Diseased plants typically have lower digestibility and non-structural carbohydrate content than in healthy plants, with variable differences in crude protein content. Stem rust, caused by *Puccinia graminis* reduces digestibility of orchard grass, presumably because of the increased resistance of leaf mesophyll tissue to degradation by ruminal microorganisms (Edwards et al. 1981).

Digestibility of smooth brome grass decreases up to 1.20 g kg^{-1} for each 1% increase in diseased area of leaves. Diseases of red clover including clover rot and powdery mildew and in berseem, root rot and leaf blight hamper yield and nutritional quality of fodder. Anthracnose and leaf spot in cowpea, powdery mildew in oats are the major foliar diseases influencing forage quality. Insect pests like stem borer in maize, leaf minor in cowpea and rice bean, cyst nematode in clover, white grubs and termite in other crops damage foliage and other plant parts. The chlorosis and later dropping of foliage after insect attack reduces photosynthetic area and damage cells that ultimately affect forage quality.

5.5.1 Effect of Foliar Disease and Insect on Quality of Forage Cowpea

Cowpea (*Vigna unguiculata* [L.] Walp.) is an important component in mixed farming systems and is valued for its potential to produce high levels of fodder for livestock in addition to seed grain for human consumption. It is drought-tolerant crop, could be grown on relatively poor soils, and fixes nitrogen, thereby improves soil fertility. This forage legume is also prone to a number of diseases and insect pests causing severe losses in yield and quality of green fodder. High incidence of diseases profoundly reduces both yield and quality, whereas insects typically reduce yield more prominently than nutritional quality.

Cowpea is either grown as sole crop or intercropped with maize mainly as *kharif* crop in low-lying regions and mid-hill areas. Wet and humid environmental conditions during the cropping season predispose the crop to the attack of various fungal pathogens and insect pests which cause substantial yield losses. In one of our study the cowpea seeds were treated with imidacloprid 70 Ws @ 5g/kg seed + carbendazim @ 2 g/kg seed followed by foliar sprays of imidacloprid 17.80 SL @ 0.30ml/L at 15 days interval for the management of insects and alternate foliar spray of mancozeb and metalaxyl + mancozeb @ 2.50g/L at 10 and 15 days interval for the management of incidence of diseases (protected condition). The quality attributes in cowpea foliage were compared with untreated control (unpublished data).

The effect of foliar diseases and insect pests on quality and yield attributes of fodder cowpea plants both under protected and unprotected conditions has been presented in Table 5.10. Unprotected cowpea plants showed high wilt and root rot incidence (14.50 to 37.20%), leaf blight severity (15.80 to 44.70%), and YMV incidence (1.90 to 10.30%). Under protected conditions, the incidence of these diseases varied from 8.30 to 21.80%, 6.00 to 19.90%, and 0.60 to 3.00%, respectively. Number of sucking pests were higher (0.60 to 5.20 insects/leaf) from unprotected cowpea plants as compared to protected ones (0.20 to 2.90 insects/leaf).

Table 5.10 Effect of foliar diseases and insect pests on forage quality in cowpea

Treatment	Quality characters														
	DM	CP %	NDF %	ADF %	Hemicellulose %	Chl a (mg/g)	Chl b (mg/g)	Total Chl (mg/g)	Ash %	K (ppm)	Ca (ppm)	IVDMD %	ADF-N (CP %)	CHO (g/100 g)	Phenolics (g/100 g)
Unprotected crop															
After 30 days of sowing	31.10	19.50	60.50	51.60	8.92	0.40	0.35	0.73	4.47	158.50	318.30	64.23	5.23	4.13	0.47
After 60 days of sowing	37.80	15.30	63.20	55.30	7.90	0.43	0.28	0.71	3.97	116.70	252.80	60.98	5.96	3.37	0.59
Protected crop															
After 30 days of sowing	31.70	20.40	58.70	50.40	8.42	0.49	0.36	0.86	4.61	162.40	290.90	66.58	4.34	4.40	0.40
After 60 days of sowing	38.50	18.60	61.700	53.50	8.20	0.44	0.35	0.79	4.02	132.10	269.40	62.36	5.67	3.97	0.50

Green fodder yield was 16.50% higher in protected (60.20 q/ha) as compared to unprotected crop (51.70 q/ha).

In 60-day-old cowpea plants, DM, NDF, ADF and phenolics contents were higher as compared to 30-day-old cowpea plants both under protected and unprotected conditions. After 30 days of sowing, the DM content varied from 31.10 in unprotected cowpea plants to 31.70% in protected cowpea plants. The crude protein content was 19.50% in unprotected cowpea plants and 20.40% in protected cowpea plants after 30 days of sowing. NDF content was 58.70% in protected cowpea plants in comparison to 60.50% in unprotected one. ADF content also revealed variations from 50.40 in protected cowpea plants to 51.60% in unprotected plants after 30 days of sowing. The chlorophyll content (a, b, and total) was also on higher end in protected cowpea plants (0.49, 0.36, and 0.86mg/g) as compared to unprotected cowpea plants (0.40, 0.35, and 0.73 mg/g), after 30 days of sowing. The ash content (4.61%), K (162.40 ppm), IVDMD (66.58%) and CHO (4.40 g/100g) were higher in protected cowpea plants in comparison to 4.47 %, 158.50 ppm, 258.8 ppm, 64.23%, and 4.13 g/100g, in unprotected cowpea plants.

In the 60-day-old unprotected cowpea plants, DM and CP content revealed difference from 37.80 to 38.50% and 15.30 to 15.66% in comparison to protected conditions. The NDF and ADF decreased from 63.21 to 61.78 and 55.32 to 53.53% under protected conditions as compared to unprotected plants. Chlorophyll contents (a, b, and total) was also high in protected (0.44, 0.35, and 0.79 mg/g) as compared to unprotected cowpea plants (0.43, 0.28, and 0.71 mg/g). It was also observed that the ash (4.02%), K (132.1 ppm), Ca (269.20 ppm), IVDMD (62.32 %) and CHO (3.97 g/100g) contents were higher in protected cowpea plants in comparison to 3.97 %, 116.7 ppm, 258.8 ppm, 60.98%, and 3.37 g/100g, in unprotected conditions. The results of the study clearly indicated that incidence of diseases and insect pests has direct influence on the quality of cowpea forage.

5.5.2 Effect of Powdery Mildew Disease on Forage Quality of Oats

Oat (*Avena sativa* L.) locally known as Jai is one of the major fodder crops mainly grown as *rabi* season crop in North, Central and West Zone of the country and as *kharif* season crop in dry temperate zone of North-Western Himalayas. Oat has sixth position in world cereal production after wheat, rice, maize, sorghum and barley. It is the most preferred feed by animals and its straw is soft and superior to wheat and barley. The incidence of foliar diseases has significant impact on yield and quality attributes of oat fodder.

The relationship between powdery mildew disease incidence and quality attributes of oat fodder was studied in five oat varieties (Unpublished data). One set of five varieties was protected with fungicides *i.e.*, alternate foliar sprays of hexaconazole (0.1%) and propiconazole (0.01%) against powdery mildew, whereas, another set served as control (Table 5.11).

Table 5.11 Effect of powdery mildew disease on forage quality of oat

Treatment	Powdery mildew severity (%)		CP %		Quality characters: Chlorophyll a (mg/g)		Chlorophyll b (mg/g)		Total chlorophyll (mg/g)		NDF %		ADF%		Dry matter yield		CP yield
	Year		Year		Year		Year		Year		Year		Year		Year		
	I	II	I	II	I	II	I	II	I	II	I	II	I	II	I	II	
T_1 Kent (sprayed)	(16.80)	(14.10)	9.30	10.20	0.95	1.32	0.31	0.48	1.28	1.80	64.00	60.80	49.40	51.40	87.10	88.30	8.10
T_2 KENT (UNSPRAYED)	(39.20)	(55.70)	9.10	7.29	0.80	0.75	0.31	0.29	1.11	1.04	64.60	63.00	52.00	53.20	68.10	81.00	6.20
T_3 PLP-1 (SPRAYED)	(15.20)	(9.90)	10.50	9.33	0.73	1.37	0.50	0.40	1.77	1.77	63.40	59.30	50.20	50.60	61.30	73.80	6.40
T_4 PLP-1 (UNSPRAYED)	(36.30)	(31.50)	8.20	7.58	0.80	0.87	0.28	0.29	1.06	1.16	63.80	62.20	53.40	55.80	50.80	79.50	4.20
T_5 OL-9 (SPRAYED)	(16.40)	(13.30)	10.20	9.33	1.20	1.29	0.42	0.37	1.65	1.66	62.00	62.40	48.20	54.50	85.40	88.80	8.70
T_6 OL-9 (UNSPRAYED)	(47.70)	(47.90)	7.30	6.70	0.77	0.92	0.26	0.30	1.06	1.22	67.40	61.00	51.80	55.40	64.70	76.90	4.70
T_7 OL-125 (SPRAYED)	(13.30)	(12.50)	10.80	10.22	1.43	1.40	0.47	0.39	1.91	1.80	62.60	60.00	48.60	50.40	87.70	91.40	9.70
T_8 OL-125 (UNSPRAYED)	(46.30)	(44.90)	7.60	8.16	0.82	0.79	0.31	0.33	1.11	1.12	64.00	61.00	50.00	56.40	66.50	76.60	5.20
T_9 OS-6 (SPRAYED)	(11.90)	(11.50)	9.90	8.75	1.35	1.30	0.41	0.43	1.77	1.73	64.80	58.20	48.80	50.60	82.90	85.90	8.20
T_{10} OS-6 (unsprayed)	(45.40)	(51.70)	7.30	7.29	0.76	0.78	0.27	0.34	1.03	1.12	64.60	62.20	50.60	53.50	65.60	75.60	4.80
CD (5%)	3.60	3.80	0.80	0.75	0.20	–	0.10	–	0.20	–	–	–	–	–	13.60		1.50
CV %	7.30	–	5.00		12.50		11.90		7.90		–		–		11.00		13.10

- In protected oat varieties, the minimum disease incidence was recorded in variety OS-6 (11.90%) and maximum in variety Kent (16.80%), whereas under unprotected condition the variety, PLP-1 showed high degree of resistance against the disease. The variety Kent exhibited low degree of resistance against powdery mildew during the first year of experimentation. During second year of experimentation, the disease incidence was again minimum in PLP-1 (9.90%) and variety Kent (14.10%) was highly susceptible among others under protected conditions. In unprotected conditions the incidence of disease was minimum in variety PLP-1(31.50%).
- The results revealed that the incidence of powdery mildew disease reduced the dry matter yield and CP yield in oat varieties. Significant reduction in the level of different quality attributes in oat fodder was observed under unprotected conditions. During the first year, the variation in different quality attributes *viz.*, crude protein (9.30 to 10.80%), chl a (0.73 to 1.43 mg/g), chl b (0.31 to 0.47 mg/g), total chlorophyll (1.28 to 1.91 mg/g), NDF (62.00 to 64.80%), and ADF (48.20 to 50.20%) was observed in chemically protected oat varieties.
- In unprotected oat varieties the crude protein, chl a, chl b, total chlorophyll, NDF, and ADF contents varied from 7.30 to 9.10%, 0.76 to 0.82 mg/g, 0.26 to 0.31 mg/g, 1.03 to 1.11 mg/g, 63.80 to 64.60%, and 50.00 to 53.40%, respectively. The comparison in the results of first year of experimentation revealed reduction in the level of different quality attributes in oat varieties after the incidence of powdery mildew disease especially under unprotected conditions.
- Crude protein, chl a, chl b, total chlorophyll, NDF, and ADF content in protected oat varieties varied from 8.75 to 10.33%, 1.29 to 1.40 mg/g, 0.30 to 0.48 mg/g, 1.73 to 1.80%, 58.20 to 62.40%, and 50.40 to 55.40, respectively, during the second year of experimentation. Variation in crude protein, chl a, chl b, total chlorophyll, NDF, and ADF content in unprotected oat varieties were in range of 6.70 to 8.16%, 0.74 to 0.92 mg/g, 0.29 to 0.34 mg/g, 1.03 to 1.22 mg/g, 60.00 to 63.00%, and 53.20 to 55.40% during the second year of experimentation.
- The reduction in crude protein content was 29.62% (OL-9) and 28.52% (Kent) after disease incedence during first and second years of experimentation, respectively. Chlorophyll is the most important class of pigments that involved in photosynthesis, the process by which light energy is converted to chemical energy in the form of carbohydrates. With the incidence of powdery mildew disease, the chlorophyll a, chlorophyll b, and total chlorophyll reduced to 42.65%, 44%, and 41.88%, respectively, during first year and 43.57%, 39.58%, and 42.22%, respectively, during the second year of experimentation. NDF and ADF content are negatively associated with the digestibility of forages. The incidence of disease increased NDF and ADF content of oat varieties [except OS-6 (First year) and OL-9 (Second year). The NDF and ADF content in diseased plants increased to 8.01% and 6.94% during first year and 4.66% and 9.31% during second year, respectively. Due to the incidence of powdery mildew disease, the dry matter yield from fodder oat varieties reduced to 24.17% and 16.19%, during first and second years of experimentation.

- The study indicated that incidence of powdery mildew disease reduced productivity as well as its nutritional quality in different oat varieties. Level of crude protein, crude fiber, chlorophyll content, neutral detergent fiber and acid detergent fiber contents was also affected with the incidence of powdery mildew disease.

Forages are the most important and economic source of nutrients to livestock. Studies have revealed that the quality of forages is affected by many factors, which could be categorized into plant-based, animal-based, and environmental factors. Stage of maturity is the one of the important factors, which directly or indirectly affects the quality of forages. Generally, maturity has negative relation with the quality of forages. Livestock productivity is the ultimate test of forage quality. High-quality forages have high digestibility and palatability. Different environmental factors have also been considered as major determinants of forage quality. Growth stages and seasonal variations have impact on the nutritional profiles of different forages species in Himalayan region.

References

Aganga AA, Tshwenyane SO (2003) Feeding values and anti-nutritive factors of forage tree legumes. Pak J Nutr 2:170–177

Akin DE (1989) Histological and physical factors affecting digestiferulic and sinapic acids. Agron J 74:424

Akin DE, Fales SL, Rigsby LL, Snook ME (1987) Temperature effects on leaf anatomy, phenolic acids, and tissue digestibility in tall fescue. Agron J 79:271–275

Akingbade AA, Nsahlai IV, Morris CD (2001) Seasonal variation in forage quality and mimosine contents of two varieties of *Leucaena leucocephala*. Afr J Range Forage Sci 18:131–135

Allakhverdiev SI (2009) Regulatory roles inphotosynthesis of unsaturated fatty acids in membrane lipids. In: Wada H, Murata N (eds) Lipids in photosynthesis: essential and regulatory functions. Springer, Dordrecht, pp 265–282

Allen IG, Segarra E (2001) Anti-quality components in forage: overview, significance, and economic impact. J Range Manage 54:409–412

Al-Masri MR (2013) An in vitro nutritive evaluation of *Medicago arborea* as affected by growth stage and cutting regimen. Lives Res Rural Dev 25:77

Amaral JV, Valentao P, Andrade P, Martins R (2010) Phenolic composition of hazelnut leaves: influence of cultivar, geographical origin and ripening stage. Sci Hortic 126:306–313

Amin MEMH (2011) Effect of different nitrogen sources on growth, yield and quality of fodder maize (*Zea mays* L.). J Saudi Soc Agric Sci 10:17–23

Ammerman CB, Moore JE, Henry PR, Miller SM, Martin FG (1982) Effect of age and sample preparation on mineral content of bermudagrass hay. J Dairy Sci 65:1329–1333

Asmare B, Demeke S, Ejeta T, Firew T, Haile A, Wamatu J (2017) Effects of altitude and harvesting dates on morphological characteristics, yield and nutritive value of desho grass (*Pennisetum pedicellatum* Trin.) in Ethiopia. Agric Nat Resour 51:148–153. https://doi.org/10.1016/j.anres.2016.11.001

Ates E, Tekeli AS (2011) Change of some morphological and forage quality properties depends on different pasture aspects in sweet pea (*Lathyrus odoratus* L.). Roman J Grassl Forage Crops 3: 31–38

Baloyi BM, Ayodele VI, Addo-Bediako A (2013) Effects of leaf harvest on crude protein and mineral contents of selected early maturing lines of lablab (*Lablab purpureus*). Afr J Agric Res 8:449–453

Boller BC, Nosberger J (1983) Effects of temperature and photoperiod on stolon characteristics, dry matter partitioning and nonstructural carbohydrate concentration of two white clover ecotypes. Crop Sci 23:1057–1062

Buxton DR, Brasche MR (1991) Digestibility of structural carbohydrates in cool season grass and legume forages. Crop Sci 31:1348–1351

Buxton DR, Fales SL (1994) Plant environment and quality. In: Fahey GC Jr, Collins M, Mertens DR, Moser LE (eds) Forage quality, evaluation and utilization. Am Soc Agron Monogr Series, Madison, WI, pp 155–199

Buxton DR, Redfearn DD (1997) Plant limitations to fiber digestion and utilization. J Nutr 127: 814S–818S

Buxton DR, Russel JR (1988) Lignin constituents and cell-wall digestibility of grass and legume stems. Crop Sci 28:553–558

Calsamiglia S, Ferret A, Devant M (2002) Effects of pH and pH fluctuations on microbial fermentation and nutrient flow from a dual-flow continuous culture system. J Dairy Sci 85: 574–579

Clapham W, Foster J, Neel J, Fedders J (2006) Fatty acid composition of traditional and novel forages. J Agric Food Chem 53:10068–10073. https://doi.org/10.1021/jf0517039

Crush JR, Caradus JR (1995) Cyanogenesis potential and iodine content in white clover (*Trifolium repens* L.) cultivars. N Z J Agric Res 38:309–316

Davis AM (1981) The oxalate, tannin, crude fiber, and crude protein composition of young plants of some *Atriplex* species. J Range Manage 34:329–331

Deinum B, Dirven JGP (1975) Climate, nitrogen and grass. 6. Comparison of yield and chemical composition of some temperate and tropical grass species grown at different temperatures. Neth J Agric Sci 23:69–82

Deinum B, De Beyer J, Nordfeldet PH, Kornher A, Ostgard O, Bogaert GV (1981) Quality of herbage at different latitudes. Neth J Agric Sci 29:141–150

Demarquilly C, Jarrige R (1974) The comparative nutritive value of grasses and legumes. Page 33 in Vaxtodling 28. In: Proc. 5th gen. Mtg Eur Grassl fed, Uppsala

Dierking RM, Kallenbach RL, Roberts CA (2010) Fatty acid profiles of orchard grass, tall fescue, perennial ryegrass, and alfalfa. Crop Sci 50:391–402

Edgar JA (1994) Toxins in temperate grasses-implications and solutions. N Z J Agric Res 37:341–347

Edwards MT, Sleper DA, Loegering WQ (1981) Histology of healthy and digested orchardgrass leaves subjected to digestion in rumen fluid. Crop Sci 21:341–343

Eltelib AH, Hamad AM, Ali EE (2006) Effect of nitrogen and phosphorus fertilization on growth, yield and quality of fodder maize. J Agron 5:515–518

Evans JR (1989) Photosynthesis and nitrogen relationships in leaves of C-3 plants. Oecologia 78:9–19

Fales SL (1986) Effects of temperature on fiber content, composition, and in vitro digestion kinetics of tall fescue. Agron J 78:963–966

Fleming GA (1973) Mineral composition of herbage. In: Butler GW, Bailey RW (eds) Chemistry and biochemistry of herbage. Academic Press, New York, pp 529–566

Ford CW, Morrison IM, Wilson JR (1979) Temperature effects on lignin, hemicellulose and cellulose in tropical and temperate grasses. Aust J Agr Res 30:621–634

George JR, Hall KE (1983) Herbage quality of three warm-season grasses with nitrogen fertilization. Iowa State J Res 58:247–259

Ghosh MK, Bandhopadhyay S (2007) Mimosine toxicity—a problem of *Leucaena* feeding in ruminants. Asian J Anim Vet Adv 2:63–73

Gleadow M, Ottman MJ, Kimball BA, Wall GW, Pinter PJ Jr, Morte RLL, Leavitt WS (2016) Drought-induced changes in nitrogen partitioning between cyanide and nitrate in leaves and stems of sorghum grown at elevated CO_2 are age dependent. Field Crop Res 185:97–102

Grabber JH, Ralph J, Pell A, Hatfield RD (1992) Synthesis of guaiacyl and syringyl lignins and their effects on fiber digestion. Agron Abstracts. ASA-CSSA-SSSA, Madison, WI, p 181

Grant K, Kreyling J, Dienstbach LFH, Beierkuhnlein C, Anke J (2014) Water stress due to increased intra-annual precipitation variability reduced forage yield but raised forage quality of a temperate grassland. Agric Ecosyst Environ 186:11–22

Hasan M, Akbar M, Khandaker Z, Mostafizar R (2010) Effect of nitrogen fertilizer on yield contributing character, biomass yield and nutritive value of cowpea forage. Bangladesh J Anim Sci 39:83–88

Jensen KB, Waldron BL, Asay KH, Johnson DA, Monaco TA (2003) Forage nutritional characteristics of orchard grass and perennial ryegrass at five irrigation levels. Agron J 95: 668–675

Juan NA, Sheaffer CC, Barnes DK (1993) Temperature and photoperiod effects on multifoliolate expression and morphology of alfalfa. Crop Sci 33:573–578

Jung HG, Allen MS (1995) Characteristics of plant cell walls affecting intake and digestibility of forages by ruminants. J Anim Sci 73:2774–2790

Katoch R (2019) Seasonal alterations in anti-nutritional factors of common fodder trees of mid Himalayas. Range Manag Agrofor 40(1):143–149

Katoch R, Tripathi A (2020) Effect of altitude on nutritional profile of white clover (*Trifolium repens* L.). Range Manag Agrofor 41(1):173–177

Katoch R, Thakur M, Kumar N (2012) Effect of morphological stage and clipping intervals on quality and digestibility of tall fescue (*Festuca arundinacea* Schreb.) and Setaria (*Setaria anceps* Stapf.). Range Manag Agrofor 33(2):185–192

Katoch R, Singh SK, Tripathi A, Kumar N (2017) Effect of seasonal variation in biochemical composition of leaves of fodder trees prevalent in the mid-hill region of Himachal Pradesh. Range Manag Agrofor 38(2):234–240

Keftasa D (1990) Effect of management practices on Rhodes grass and Lucerne pasture with especial references to development stage at cutting and associated change in nutritional quality. Institute of Agriculture Research, Kulumsa Research Center, Ethiopia

Kephart KD, Buxton DR (1993) Forage quality responses of C3 and C4 perennial grasses to shade. Crop Sci 33:831–837

Kephart KD, Buxton DR, Taylor SE (1992) Growth of C3 and C4 perennial grasses in reduced irradiance. Crop Sci 32:1033–1038

Kering M, Guretzky J, Funderburg E, Mosali J (2011) Effect of nitrogen fertilizer rate and harvest season on forage yield, quality, and macronutrient content in Midland Bermuda grass. Comm Soil Sci Plant Analys 42:1958–1971

Khan NA, Cone JW, Fievez V, Hendriks WH (2011a) Stability of fatty acids during wilting of perennial ryegrass (*Lolium perenne* L.): effect of bruising and environmental conditions. J Sci Food Agric 91:1659–1665

Khan NA, Cone JW, Pellikaan WF, Khan MA, Struik PC, Hendriks WH (2011b) Changes infatty acid content and composition in silage maize during grain filling. J Sci Food Agric 91:1041–1049

Khorasani GR, Jedel PE, Helm JH, Kennelly JJ (1997) Influence of stage of maturity on yield components and chemical composition of cereal grain silages. Can J Anim Sci 77:259–267

Kilcher MR, Troelsen JE (1973) Contribution and nutritive value of the major plant components of oats through progressive stages of development. Can J Plant Sci 53:251–256

Kumar R (2003) Anti-nutritive factors, the potential risks of toxicity and methods to alleviate them. http://www.fao.org/docrep/003/t0632e/T0632E10.html

Lechtenberg VL, Holt DA, Youngberg HW (1971) Diurnal variation in non structural carbohydrates, in vitro digestibility, and leaf to stem ratio of alfalfa. Agron J 63:719–724

Lees GL, Gruber MY, Suttill NH (1995) Condensed tannins in sainfoin. II.Occurrence and changes during leaf development. Can J Bot 73:1540–1547

Lehmann J, Meister E, Gutzwiller A, Jans F, Charles JP, Blum J (1991) Should one use white clover (*Trifolium repens* L.) varieties rich in hydrogen cyanide? Revue Suisse d' Agric 23:107–112

Lindroth LR, Hofman RW, Campbell BD, McNabb CW, Hunt YD (2000) Population differences in *Trifolium* repens L. response to ultraviolet-B radiation: foliar chemistry and consequences for two lepidopteran herbivores. Oecologia 122:20–28

Mayland HF, Cheeke PR, Majak W, Goff JP (2007) Forage-induced animal disorders. In: Barnes RF, Nelson CJ, Moore KJ, Collins M (eds) Forages: the science of grassland agriculture. Blackwell Publishing Professional, Ames, IA, pp 687–707

Menino MR, Carranca C, De Varennes A (2007) Distribution and remobilization of nitrogen in young non-bearing orange trees grown under Mediterranean conditions. J Plant Nutr 30:1083–1096

Mero RN, Uden P (1998) Promising tropical grasses and legumes as feed resources in Central Tanzania. VI Nitrogen balance in growing bulls consuming tropical herbaceous forage legumes. Anim Feed Sci Technol 72(3–4):387–396

Messman MA, Weiss WP, Erickson DO (1991) Effects of nitrogen fertilization and maturity of bromegrass on in situ ruminal digestion kinetics of fiber. J Anim Sci 69:1151–1161

Minson DJ (1990) Forage in ruminant nutrition. Academic Press, New York

Mishra S, Sangwan RS (2008) Changes in fattyacid composition of polar lipids associated withgrowth and senescence in leaves of *Catharanthus roseus*. Afr J Plant Sci 2:34–37

Montgomery RD (1980) Cyanogens. In: Liener IE (ed) Toxic constituents of plant foodstuffs. Academic Press, New York, pp 143–160

Moore KJ, Hatfield RD (1994) Carbohydrates and forage quality. In: Fahey GC Jr (ed) Forage quality, evaluation, and utilization. Am Soc Agron, Madison, WI, pp 229–280

Moore KJ, Archontoulis SV, Lenssen AW (2017) Simple models for describing ruminant herbivory. In: Shields VDC (ed) Herbivore. In Tech, Rijeka. ISBN 978-953-51-4901-9

Nabi S, Qaisar KN, Khan PA, Rather SA, Nabi B (2017) Seasonal nutrient profile of some preferred fodder tree species of Kashmir Valley. Int J Chem Stud 6:1311–1314

Ncube S, Saidi PT, Halimani TE (2015) Effect of stage of growth and processing methods on the nutritional content of *Acacia angustissima* leaf meal harvested for broiler feeding. Lives Res Rural Dev 27:89

Nelson DJ, Moser LE (1994) Plant factors affecting forage quality. In: Fahey GC Jr, Collins MC, Mertens DR, Moser LE (eds) Forage quality, evaluation and utilization. Am Soc Agron, Madison, WI, pp 115–154

Odom JW, Haaland RL, Hoveland CS, Anthony WB (1980) Forage quality response of tall fescue, orchard grass and *Phalaris* to soil fertility level. Agron J 72:401–402

Onyeonagu CC, Asiegbu JE (2013) Harvest frequency effect on plant height, grass tiller production, plant cover and percentage dry matter production of some forage grasses and legumes in the derived savannah, Nigeria. Afr J Agric Res 8:608–617

Rahman M, Kawamura O (2011) Oxalate accumulation in forage plants: some agronomic, climatic and genetic aspects. Asian Aust J Anim Sci 24:439–448

Rahman MM, Ishii Y, Niimi M, Kawamura O (2009) Change of oxalate form in pot-grown napier grass (*Pennisetum purpureum* Schumach) by application of calcium hydroxide. Grassl Sci 55: 18–22

Ramirez RG, Haenlein GFW, Nunez-Gonalez MA (2001) Seasonal variation of macro and trace mineral contents in 14 browse species that grow in northeastern mexico. Small Rum Res 39: 153–159

Rana KK, Wadhwa M, Bakshi MPS (2006) Seasonal variations in tannin profile of tree leaves. Asian Aust J Anim Sci 19:1134–1138

Ryan DF, Bormann FH (1982) Nutrient resorption in northern hardwood forests. BioScience 32: 29–32

Samarakoon SP, Shelton HM, Wilson JR (1990) Voluntary feed intake by sheep and digestibility of shaded *Stenotaphrum secundatum* and *Pennisetum clandestinum* herbage. J Agric Sci Camb 114:143–150

Sarwar M, Firkins JL, Eastridge ML (1991) Effects of replacing NDF of forage with soyhulls and corn gluten feed for dairy heifers. J Dairy Sci 74:1006–1015

Selvakumar G, Joshi P, Nazim S, Mishra PK, Bisht JK, Gupta HS (2009) Phosphate solublization and growth promotion by *Pseudomonas fragi* CS11RH1 (MTCC 8984) a psychrotolerant bacterium isolated from a high altitude Himalayan rhizosphere. Biologia 64:239–245

Sen S, Makkar HPS, Becker K (1998) Alfalfa saponins and their implication in animal nutrition. J Agric Food Chem 46:131–140

Sharpe P, Rayburn E (2019) Forage yield and its determination. In: Horse pasture management, pp 105–119

Singh A (2002) A note on seasonal variations in oxalate content of Napier bajra hybrid. Indian J Anim Nutr 19:282–284

Singh B, Todaria NP (2012) Nutrients composition changes in leaves of *Quercus semecarpifolia* at different seasons and altitudes. Ann Forest Res 55:189–196

Singh B, Bhatt BP, Prasad P (2010) Altitudinal variation in nutritive value of adult-juvenile foliage of *Celtis australis* L.: a promising fodder tree species of central Himalaya, India. J Am Sci 6: 108–112

Sodeinde FG, Aderinola OA, Amao SR, Ayinde TL, Akinola JO (2009) Fertilizer effect on yield and composition of shrubby Stylo in the Nigerian derived savanna zone. Bull Anim Health Prod Afr 58:273–277

Stochmal A, Oleszek W (1997) Changes of cyanogenic glucosides in white clover (*Trifolium repens L.) during growing season*. J Agric Food Chem 45:4333–4336

Subba DB, Tamang PM, Tamang BB (1994) Seasonal variation in the proximate principles of some common tree fodder in the eastern hills of Nepal. Vet Rev 10(1):23–26

Tenikecier H, Ates E (2020) Effect of altitude on morphological and nutritive characteristics of orchard grass (*Dactylis glomerata* L.) collected from natural flora of Ganos Mountain in Thrace region, Turkey. Range Manag Agrofor 40:286–292

Twidwell EK, Johnson KD, Patterson JA, Cherney JH, Bracker CE (1990) Degradation of switchgrass anatomical tissue by rumen microorganisms. Crop Sci 30:1321–1328

Van Soest PJ (1994) Nutritional ecology of the ruminant. Cornell University Press, Ithaca, NY

Verma KS, Mishra VK (1999) Foliage nutrient content and lopping time of agroforestry tree species in the Western Himalayan region of India. Agr Syst 42:283–290

Waldo DR, Jorgensen NA (1981) Forages for high animal production: nutritional factors and effects of conservation. J Dairy Sci 64:1207–1229

Walgenbach RP, Marten GC (1981) Release of soluble protein and nitrogen in alfalfa. II. Influence of shading. Crop Sci 21:859–862

Wassie WA, Tsegay BA, Wolde AT, Asmare B (2018) Evaluation of morphological characteristics, yield and nutritive value of *Brachiaria* grass ecotypes in northwestern Ethiopia. Agric Food Secur 7:89. https://doi.org/10.1186/s40066-018-0239-4

Wilson JR, Minson DJ (1980) Prospects for improving the digestibility and intake of tropical grasses. Trop Grassl 14:253–257

Wilson JR, Wong CC (1982) Effects of shade on some factors influencing nutritive quality of green panic and siratro pastures. Aust J Agr Res 33:937–950

Woclawek-Potocka I, Mannelli C, Boruszewska D, Kowalczyk-Zieba I, Wasniewski T, Skarzynski DJ (2013) Diverse effects of phytoestrogens on the reproductive performance: cow as a model. Int J Endocrinol 2013:650984. https://doi.org/10.1155/2013/650984

Yang Z, Ohlrogge JB (2009) Turnover of fattyacids during natural senescence of *Arabidopsis*, *Brachypodium* and switchgrass and in *Arabidopsis* oxidation mutants. Plant Physiol 150:1981–1989

Yossif AM, Ibrahim YM (2013) Effect of fertilizers (Urea, Farmyard and Chicken Manure) on growth and yield of rhodes grass (*Chloris Gayana* L. Knuth.). Univ J Plant Sci 1:85–90

Yu P, Christensen D, McKinnon JJ, Markert JD (2003) Effect of variety and maturity stage on chemical composition, carbohydrate and protein subfractions, *in vitro* rumen degradability and energy values of timothy and alfalfa. Can J Anim Sci 83:279–290

Forage Genetic Resources (FGR) 6

Forages are the important feed resource to livestock. A desirable forage crop variety has excellent yield, quality characteristics and superiority in other traits. For the varietal improvement and building up the germplasm resource of different forage species, the maintenance and conservation of forage genetic resources is of prime importance. Different international and national agencies have now realized the importance of forage genetic resources in breeding forage cultivars with superior traits. Major forage grass collections in CGIAR (The Consultative Group on International Agricultural Research) are maintained by the Centro Internacional de Agricultura Tropical (CIAT) and the International Livestock Research Institute (ILRI). CIAT focuses on tropical grasses and legumes with emphasis on species adapted to acid infertile soils. ILRI has a broad collection covering both tropical lowland and highland grasses. The forage grasses have also been conserved at EMBRAPA (Brazilian Agricultural Research Corporation, Brazil), CSIRO (Commonwealth Scientific and Industrial Research Organisation, Australia), and USDA (The United States Department of Agriculture, USA). In India, NBPGR (National Bureau of Plant Genetic Resources), New Delhi and IGFRI (Indian Grassland and Fodder Research Institute), Jhansi are leading institutes for forage conservation, research and development. These institutes have collections of forage germplasm from diverse locations in the country as well as introduced germplasm from other countries. In the last two decades, attention has been given to indigenous forage germplasm as they are invaluable resources for present and future forage breeding programs. The collection of diverse and superior germplasm and their conservation is an important aspect of forage improvement program (Singh et al. 2008; Tyagi et al. 2008).

R. Katoch, *Nutritional Quality Management of Forages in the Himalayan Region*,
https://doi.org/10.1007/978-981-16-5437-4_6

6.1 Forage Genetic Resource Conservation

Forage genetic resources include grasses, legumes, and other herbaceous species. Grasslands are among the largest biomes in the world and are a resource for sustaining productivity of pastures for intensive livestock production in all environments. They are a major reserve of biodiversity, a cheap and an important form of *in situ* conservation of plant genetic resources including forage species.

The continuous genetic erosion of forage diversity is due to changes in land-use pattern from natural pastures and rangelands to crop production, overgrazing and degradation of invasive weed species in natural pastures and grasslands. In the last two decades, invasion of *Lantana camara*, *Ageratum conyzoides*, *Eupatorium adenophorum*, *Eupatorium odoratum*, *Bidense pilosa*, *Partheneum hysterophorus*, *Polygonum molli* and *Cystissus scipasriushas* has resulted in enormous loss to plant genetic biodiversity including forages (Rana et al. 2005; Kohli et al. 2006).

Agricultural biodiversity is an essential component for making agro-ecosystem more sustainable. The Indian agrobiodiversity has been divided into five agroclimatic zones. Each agroclimatic zone has distinct agroecosystem with unique gene pools and consists of landraces, primitive forms and wild relatives of different crops including forage species.

Forage genetic resources (FGR) are the basic requirement in the development of new forage varieties. Therefore, maintenance of genetic diversity is essential for broadening the genetic base for forage breeding programs. The extensive collection, conservation, cataloguing of forage germplasm and its characterization are prerequisites for efficient forage breeding programs. India has large reservoir of biodiversity in forage resources, due to its geographical position and the diversity of its agroclimatic conditions.

The farmers in the Himalayan region maintain high levels of crops diversity by rotation of crops in time and space together with adoption of both mono- and mixed cropping practices. The crop diversity in Himalayan region is mainly due to ecological differentiation and intensive natural and human selection exercised by various ethnic communities over centuries with time-to-time introduction of exotics.

Among forages, the Indian Himalayan region has rich genetic diversity of forage grasses and legumes. The major temperate grasses and legumes which offer diversity include *Agrostis*, *Agropyron*, *Dactylis*, *Elymus*, *Festuca*, *Lolium*, *Phalaris* and *Phleum* among grasses while legumes include *Astragalus*, Lespodeza, *Lotus*, *Medicago*, *Melilotus*, and *Trifolium* (Roy et al. 2009). In case of forage legumes, the diversity is extensive. The extent of diversity present in natural populations of *Medicago sativa* and *Medicago falcata* in Ladakh is a classical example of natural variations (Rana et al. 2012).

India's cultural diversity also plays a significant role in enriching its diversity by introducing new crops (Sahay et al. 2016). This results in identification of several promising types that could adapt to harsh environmental and degraded soil conditions and thereby, ensure feed and nutritional quality as well as provide additional incomes to resource-poor farmers of remote, tribal, hilly and other difficult areas of the country (Singh et al. 2009). The wild relatives of crops are considered as large reservoirs of genes for stress tolerance and quality characteristics have been utilized for genetic enhancement of forages. Table 6.1 presents wealth of India in forage genetic resources distributed in 15 agroclimatic zones of the country (Singh et al. 2009).

6.2 Conservation and Maintenance of Forage Genetic Resources

Different international and national agencies are involved in the maintenance and conservation of forage genetic resources. The international organizations such as Livestock Research Institute (ILRI, www.ilri.org), CIAT (Centro Internacional de Agricultura Tropical, www.ciat.cgiar.org), ICARDA (International Center for Agriculture Research in the Dry Areas), CSIRO (Commonwealth Scientific and Industrial Research Organisation), IGER (Institute of Grassland and Environmental Research) and USDA (United States Department of Agriculture) are major forage germplasm repositories. Forage germplasm diversity in these organizations is part of a Consultative Group of International Agricultural Research (CGIAR) coordinated activity in plant genetic resources. The ILRI Gene bank conserves more than 18 thousand accessions of forages from over 1000 species (www.ilri.org). CIAT gene bank holds 35,898 accessions of beans from 44 species of the genus *Phaseolus* from 109 countries and 23,139 forage accessions belonging to 668 different species of grasses and legumes from 72 countries that have been introduced over the span of 30 years (www.ciat.cgiar.org).

The PGR activities in India were initially executed by the Division of Plant Introduction at Indian Agricultural Research Institute (IARI), New Delhi, under the aegis of the Indian Council of Agricultural Research (ICAR). This was later upgraded as the National Bureau of Plant Genetic Resources (NBPGR, www.nbpgr.ernet) in 1976. Since its inception in the year 1976, ICAR-NBPGR has been nodal organization in India for survey, exploration, collection, characterization, documentation and conservation of plant genetic resources (PGR) including forages.

IGFRI (Indian Grassland and Fodder Research Institute) has mandate of collection, evaluation, characterization, documentation and conservation of forage genetic resources. A national-level program, AICRP (FCU) (All India Coordinated Research Projects on Forage Crops and Utilization), was started in the year 1970. Under this program research activities on forage crops were carried out through 22 coordinated centers located all over the country. The mandates of AICRP on Forage Crops and Utilization are (1) Coordinate multilocation testing program at the national level with a view to identify appropriate varieties and production technologies for different agroecological conditions; (2) Coordinate and monitor research related to problems of national and regional importance; (3) Conduct strategic and applied research for boosting production and productivity of forages (Arable, rainfed and irrigated forage

Table 6.1 Major forage genetic resources distributed in different agro-climatic zones in India

Sr. no.	Agroclimatic zone/regions	Subzones/subregions	Prominent forage genetic resources
1.	Western Himalayan Region	Jammu and Kashmir, Himachal Pradesh, Uttarakhand Hills	*Medicago* spp., *Arundinella nepalensis*, Chrysopogon, *Dactylis glomerata*, *Eleusine*, *Echinochloa*, *Festuca*, *Zea mays*, Kikui *grass*
2.	Eastern Himalayan Region	Sikkim, Arunachal Pradesh, Meghalaya, Nagaland, Manipur, Tripura, Mizoram, Assam, Jalpaiguri, and Cooch Bihar district of West Bengal	Rice bean, maize, range grasses, *Brachiaria*, broom grass, and lablab bean
3.	Lower Gangetic Plains	Basin plains, central alluvial plains, alluvial coastal plains, and *Rarh* plains	Rice bean, guinea grass, coix, and range grasses
4.	Middle Gangetic Plains	12 districts of Eastern Uttar Pradesh and 27 districts of Bihar plains	Maize, cowpea, rice bean, *Pennisetum pedicellatum*, and coix.
5.	Upper Gangetic Plains	Central, South-Western, and North-Western Uttar Pradesh	Maize, sorghum, cowpea, Senji, *Dichanthium*, sehima, and *Heteropogon*
6.	Trans-Gangetic Plains	Punjab, Haryana, Delhi, Chandigarh, and Sri Ganganagar district of Rajasthan	Guar, maize, bajra, berseem, lucerne, guinea grass, sorghum, and cowpea
7.	Eastern Plateau and Hills	(i) Subregion of Wainganga, Madhya Pradesh, eastern hills, and Orissa inland; (ii) Orissa northern, Madhya Pradesh, eastern hills and plateau; (iii) North and eastern Chota Nagpur hills and plateau; (iv) Chota Nagpur south, West Bengal hills and plateau; and (v) Chhattisgarh and south-western Orissa hills	Cowpea, rice bean, *Pennisetum pedicellatum*, guinea grass, *Dichanthium* spp., and *Atylosia*
8.	Central Plateau and Hills	46 districts of Uttar Pradesh, Madhya Pradesh, and Rajasthan	Maize, cowpea, rice bean, *P. pedicellatum*, *Coix*, *Atylosia*, sorghum, bajra, guar, *Cenchrus*, range grasses, and legumes
9.	Western Plateau and Hills	Maharashtra, parts of Madhya Pradesh, and one district of Rajasthan	Maize, sorghum, *Dichanthium* spp., pearl millet, *Dichanthium carzacosum*, *Vicia*, cowpea, rice bean, Cenchrus, range grasses, and legumes
10.	Southern Plateau and Hills	35 districts of Andhra Pradesh, Karnataka and Tamil Nadu	Small millet, *Heteropogon*, *Dichanthium*, *Sehima*, and *Stylosanthes* sp.
11.	East Coast Plains and Hills	(i) Coastal Orissa; (ii) North-Coastal Gujarat; (iii) South-Coastal Andhra Pradesh; (iv) North-Coastal Tamil Nadu; (v) Thanjavur; and (vi) South-Coastal Tamil Nadu	Cowpea, rice bean, guinea grass, coix, small millet, sorghum, *Heteropogon*, *Dichanthium*, and *Stylosanthes* sp.

(continued)

Table 6.1 (continued)

Sr. no.	Agroclimatic zone/regions	Subzones/subregions	Prominent forage genetic resources
12.	West Coast Plains and Hills	Western coast of Tamil Nadu, Kerala, Karnataka, Maharashtra and Goa	Congo, signal grass, *Paspalum*, *panicum*, *Digitaria*, *Brachiaria*, *Iseilemalaxum*, *Isilemia*, and *Vicia*
13.	Gujarat Plains and Hills	19 districts of Gujarat	Lucerne, sorghum, small millet, pearl millet, chioori, range grasses, and legumes
14.	Western Dry Region	Nine districts of Rajasthan	Guar, moth, cowpea, sorghum, pearl millet, and *Cenchrus* spp.
15.	Island Region	Territories of the Andaman and Nicobar Islands and Lakshadweep	

Source: Singh et al. (2009)

crops, range grasses and legumes); and (4) Function as a major service center for exchange of scientific information and research materials related to forages.

At national level, a network approach is being followed for conservation of plant genetic resources. The network consists of National Gene Bank situated at NBPGR headquarters, New Delhi and is primarily responsible for conservation of collection on long-time basis to support regeneration or restoration activities of active collections. The Gene bank functions through a comprehensive management system which is complemented by the 10 regional stations of NBPGR and the 59 National Active Germplasm Sites (NAGS). The ICAR-National Bureau of Plant Genetic Resources (NBPGR), New Delhi is the nodal organization in India for exchange, quarantine, collection, conservation as National Gene Bank (NGB), evaluation and the systematic documentation of plant genetic resources. The Indian NGB has 12 modules with a capacity to hold around one million accessions. The present base collection holdings in NGB are 3,96,189 (www.nbpgr.ernet.in). NBPGR has introduced a total of 13,181 forage germplasm during the last 33 years. A total of 10,049 samples in sorghum and 20,085 samples in pearl millet have been introduced for dual purpose. In oats, genetic resources of grain and fodder type comprising of 1827 accessions have also been procured (Joshi et al. 2009).

Forage genetic resources provide basic material for the improvement of forage crops through breeding to ensure food and nutritional security to livestock vis-à-vis to rapidly increasing human population. Conservation and utilization of these genetic resources are important components of establishment of germplasm collections in gene bank. The germplasm introduced/imported into India is being multiplied, characterized, evaluated and conserved at different NBPGR regional stations located in different agroclimatic zones and also at IGFRI (Indian Grassland and Fodder Research Institute, www.igfri.res.in), Jhansi, Uttar Pradesh. Different released varieties are also conserved in long-term storage at National Gene Bank, NBPGR, New Delhi (Table 6.2).

The Indian Grassland and Fodder Research Institute (IGFRI), Jhansi is maintaining a diverse forage germplasm collection in its mid-term storage module

Table 6.2 Forage germplasm holding at NBPGR, New Delhi

	Crop group	Source	Status	
Germplasm holding	All crops	–	Species	Accessions
			1584	3,96,189
	Millets and forages	–		
Germplasm exchange division introductions	Forages	From Australia, Italy, USA, UK, Brazil, Germany, Egypt, Ethiopia, Bulgaria, Philippines, Singapore, Costa Rica, Zimbabwe, Japan, Russia, and New Zealand		13,181
Long-term Storage	Forage crops	–	206	5594

Source: PGR Portal NBPGR, New Delhi (www.nbpgr.ernet)

(MTS). As a National Active Germplasm Site (NAGS) for forage crops, IGFRI holds approximately 10,116 germplasm accessions of more than 70 genera of cereal forages, range legumes, and range grasses (Sahay et al. 2018) (Table 6.3).

There are two approaches that are mainly used for the conservation of the plant germplasm including forage germplasm, namely *in situ* conservation and *ex situ* conservation.

In situ conservation involves maintenance of genetic variation at the location where it is encountered, either in the wild or in traditional farming. The three main *in situ* techniques on forages may be broadly defined as genetic reserve, grasslands, and rangelands (natural) and on farm.

- Genetic reserve conservation approach involves management and monitoring of genetic diversity in a particular, natural location. This technique is the most appropriate for the management of wild relatives. The genetic reserves are appropriate for forages producing orthodox seeds (i.e., species with seeds that can be maintained in long-term storage by a combination of reduction in moisture content and storage at low temperature) and recalcitrant seeds (i.e., species with seeds that cannot be stored long term). Genetic reserves also permit conservation of multiple wild relatives in a single reserve and allow continued evolution of the species. However, the disadvantages are that the conserved material is not immediately available for human exploitation. Moreover, if the management regime is minimal, little characterization or evaluation data may be available for further exploration in breeding program.
- Pastures, grasslands, and rangelands are the natural resources where *in situ* conservation is majorly relying upon. The grasslands are conserved under the supervision of institutes or organizations to control erosion of genetic diversity of plant species.

Table 6.3 Forage crops germplasm conservation at IGFRI, Jhansi

Crop (Scientific name)	No. of accessions
Cereal fodders	3437
Barley (*Hordeum vulgare*)	
Bajra (*Pennisetum glaucum*)	
Finger millet (*Eleucine indica*)	
Jowar (*Sorghum bicolor*)	
Maize (*Zea mays*)	
Oat (*Avena sativa*)	
Cultivated legumes	2682
Berseem (*Trifolium alexandrinum*)	
Common bean (*Phaseolus vulgaris*)	
Cowpea (*Vigna unguiculata*)	
Guar (*Cyamopsis tetragonoloba*)	
Lablab (*Lablab purpureus*)	
Lucerne (*Medicago* spp.)	
Range legumes	746
Centrosema (*Centrosema* spp.)	
Clitoria (*Clitoria ternatea*)	
Desmanthus (*Desmanthus virgatus*)	
Desmodium (*Desmodium* spp.)	
Gambhari (*Calotropis ensiformis*)	
Horsegram (*Macrotyloma uniflorum*)	
Jackbean (*Canavalia virosa*)	
Jointvetch (*Aeschynomene* spp.)	
Lathyrus (*Lathyrus sativa*)	
Neotonia (*Neotonia wightii*)	
Red clover (*Trifolium pratens*)	
Rice bean (*Vigna umbellata*)	
Rhyncosia (*Rhynchosia* spp.)	
Sainfoin (*Onobrychis vicifolia*)	
Shaftal (*Trifolium resupinatum*)	
Siratro (*Macroptilium atropurpureum*	
Stylosanthes (*Stylosanthes hamata*, spp.)	
Subabul (*Leucaena leucocephala*)	
Swordbean (*Canavalia gladiata*)	
White clover (*Trifolium repens*)	
Wild groundnut (*Arachis pintoi*, spp.)	
Wild pea (*Pisum* spp)	
Zornia (*Zornia falcata*)	
Range grasses	3175
Bothriocloa (*Bothriochloa pertusa*)	
Buffel grass (*Cenchrus* spp.)	
Dhawalu (*Chrysopogon fulvus*)	
Digiteria (*Digiteria bentzii*)	

(continued)

Table 6.3 (continued)

Crop (Scientific name)	No. of accessions
Goat grass (*Aegilops tauschii*)	
Guinea grass (*Panicum maximum*)	
Iselema (*Iselema laxum*)	
Job's tears (*Coix lacryma-jobi*),	
Jhonson grass (*Sorghum halepense*)	
Marvel grass (*Dichanthium annulatum*)	
Napier grass (*Pennisetum purpureum*)	
Orchard grass (*Dactylis glomerata*)	
Panicum (*Panicum* spp.)	
Para grass (*Bracharia* spp.)	
Paspalum (*Paspalum scrobiculatum*)	
Pennisetum (*Pennisetum* spp.)	
Prairie grass (*Bromus unioloides*)	
Rhodes grass (*Chloris gayana*)	
Rye grass (*Lolium perenne*)	
Sen grass (*Sehima nervosum*)	
Spear grass (*Heteropogon contortus*)	
Setaria (*Setaria spacelata*)	
Sewan grass (*Lasiurus indicus*)	
Tall fescue (*Festuca arundinacea*)	
Teosinte (*Zea mays americana*)	
Themeda (*Themeda arundinacea*)	
Timothy (*Phleum pratense*)	
Vetiveria (*Vetiveria zizanoides*)	
Unidentified	76
Total	10,116

Source: Sahay et al. (2019)

Modern-day breeding methods have largely taken the place of traditional farming and selection methods, which has led to the loss of many old landraces. There is a need to conserve the diversity found both in the wild relatives and in the old varieties of domesticated species. On-farm conservation involves the maintenance of crop by farmers within traditional agricultural systems. The advantage of on-farm conservation is that it ensures the maintenance of ancient landraces and those wild species dependent on traditional agriculture. However, the landraces may yield less than their modern counterparts and may be less popular with farmers.

Ex situ conservation involves the maintenance of genetic variation away from its original location. Samples of populations of species, subspecies, cultivars, breeds, or varieties are collected and conserved either as living collections of plants in field, gene banks, botanic gardens and arboreta or as samples of seed, tubers, tissue explants, pollen, or DNA maintained under special artificial conditions. *Ex situ* collection and storage of seeds or embryos is the most convenient and widely used method of genetic conservation. This technique involves samples being collected from individuals or populations and then transferred to a gene bank for storage under controlled conditions, usually at sub-zero temperatures.

Conservation of forage genetic resources is mainly carried out through two types of collections. One collection of seed samples for long-term conservation, which is known as base collection. Base collections are maintained at −18 to −20 °C, to ensure seed viability for maximum possible time period. The moisture content of seed to be stored as base collections is between 3% and 7% depending on the species. Second, collection of seed samples for immediate use, termed as active collection. Active collection is maintained in conditions that ensure at least 65% viability for 10 to 20 years. The moisture content of seeds to be stored as active collections is between 3% and 8% for seeds having poor storability and between 7% and 11% for seeds having good storability, depending on the temperature used for storage. These collections are conserved in different types of storage facilities. Depending on the duration of storage, three basic types of storage are recognized.

- *Short-term storage (STS)*: The period for short-term storage of seeds is from 1 year up to 18 months. It requires a cool and dry atmosphere (20–22 °C and 45–50% RH) where the seeds could be conveniently stored for 1 to 2 years without much loss in their viability.
- *Medium-term storage (MTS)*: The active collection, which is generally larger than those meant for base collection, is conserved in the medium-term storage. The accessions are for regular distribution and therefore the time period for storage is not more than 5 yrs. The active collections are stored at temperatures ranging from 0 to 10 °C and relative humidity of 20 to 30%.
- *Long-term storage (LTS)*: This is the storage facility for base collection, where the seeds that meet all the above-mentioned criteria for NGB conservation are maintained at −18 to −20 °C.

Forage germplasm collected and conserved from Indian Himalayan region in the last 25 years by ICAR- IGFRI has been presented in Table 6.4.

6.3 Forage Genetic Resources and Varietal Improvement

Forage crop improvement is rather typical where individual plants have high genotypic and phenotypic heterogeneity. Many forage species are polyploid, a problem that is exacerbated by inbreeding across many grasses. Small flower morphology of forage crops is inconvenient to breeders for hybridization program. Even after successful hybridization, low seed formation is another limiting factor. Apart from these constraints, many types of forage are perennial in nature and flower once a year which discourage breeders to include forage crop in hybridization program.

Nutritional traits in forage crops are relatively complex. Many traits, such as cell wall constituents, are typically interrelated. Most of the traits are regulated by numerous enzymes that are involved in their biosynthesis or metabolism. Thus,

Table 6.4 Forage germplasm collection from Indian Himalayan region

Sr. no.	State	Area	Species	No. of accessions
1.	Himachal Pradesh	Hamirpur	*Zea mays* (1)	1
		Palampur	*Dactylis glomerata* (1), *Poa annua* (4)	5
		Chamba, Kangra, Mandi, Palampur	*Heteropogon contortus* (6) *Pennisetum pedicellatum* (2) *Sorghum bicolor* (9) *Zea mays* (6)	23
		Palampur, Mandi, Kullu	*Bothriochloa pertusa* (12), *Cenchrus ciliaris* (17), *Cenchrus setigerus* (3), *Chloris gayana (12), Chrysopogon fulvus* (36), *Dichanthium annulatum* (38), *Ermopogon sps (8), Heteropogon contortus* (27), *Iseilema laxum* (9), *Sehima nervosum* (22)	174
		Mandi	*Medicago sativa* (5), *M. falcata* (11), *Panicum* sp. (1), *Trifolium* (5), *Vigna unguiculata* (1)	23
		Lahaul	*Medicago* spp. (17)	17
2.	NE Region Assam Tripura Manipur Meghalaya	Silchar Barapani Agartala Imphal	*Panicum maximum* (6)	6
3.	Arunachal Pradesh Manipur Meghalaya	Barapani West Siang Basar	*Dichanthium annutulam* (15), *Hetropogon contortus* (27)	42
4.	Sikkim	Sandra East, Namli, Daramdin	*Zea mays* (20)	20
5.	Assam and Meghalaya	Kamrup, Nalbari, Barpeta, Bongaigaon, Chirang, Golaghat, Jorhat, Tinsukia, Sivasagar, Dibrugarh	*Avena sativa* (1) *Brachiaria mutica* (1) *Chrysopogon zizanioides* (1) *Coix lacryma-jobi* (1) *Eleusine indica* (2) *Miscanthus fuscus* (1) *Panicum antidotale* (1) *Panicum maximum* (5) *Pennisetum pedicellatum* (7) *Pennisetum purpureum* (1) *Saccharum ravennae* (1) *Setaria palmifolia* (1)	33

(continued)

Table 6.4 (continued)

Sr. no.	State	Area	Species	No. of accessions
			Setaria sphacelata (1) *Sorghum halepense* (1) *Themeda villosa* (1) *Tripsacum andersonii* (2) *Zea mays var. mexicana* (2) *Vigna umbellata* (1) *Vigna unguiculata* (2)	
6.	Jammu and Kashmir	Anantnag, Budgam, Ganderbal, Kulgam, Pulwama, Shopian, and Srinagar	*Andropogon polynurous* (1) *Andropogon viridis* (1) *Bothriochloa pertusa* (3) *Chrysopogon aciculatus* (3) *Cenchrus ciliaris* (22) *Cenchrus setigerus* (3) *Chloris gayana* (1) *Chrysopogon fulvus* (22) *Cymobopogon citrata* (1) *Desmodium gyrans* (5) *Dichanthium annulatum* (25) *Digitaria bentzii* (4) *Heteropogon contortus* (30) *Iseilema laxum* (9) *Lolium multiflorum* (1) *Lotus luniglosus* (1) *Neonotonia* sp. (1) *Panicum maximum* (5) *Panicum coloratum* (2) *Panicum* sp. (35) *Paspalum scrobiculatum* (2) *Pennisetum orientale* (1) *Pennisetum polystachyon* (2) *Pennisetum* sp. (21) *Sehima nervosum* (22) *Slephorum* (1) *Stylosanthes hamata* (2) *Tripogon bromoides* (2) *Vigna* sp. (4) and *Vigna trilobata* (2)	234
7.	Uttarakhand	Udham Nagar, Nainital, Almora Bageshwar, Pauri Garhwal, Dehradun, Haridwar, Rudraprayag & Karnprayag, Pithorgarh, Chamoli, Pithoragarh, Dehradun	*Aegilops taushii* (2) *Chloris gayana* (1) *Chrysopogon fulvus* (6) *Digitaria bentzii* (2) *Echinochloa frumentacea* (11) *Eleucine indica* (15) *Heteropogon contortus* (9) *Setaria sphacelata* (15)	124

(continued)

Table 6.4 (continued)

Sr. no.	State	Area	Species	No. of accessions
			Panicum maximum (14) *Paspalum scrobiculatum* (3) *Pennisetum glaucum* (6) *Pennisetum pedicellatum* (8) *Phaseolus vulgaris* (1) *Sehima nervosum* (1) *Sorghum bicolor* (8) *Sorghum halepense* (1) *Vigna unguiculata* (1) *Zea mays* (4) *Zea mays*. Var. *americanum* (1) and unidentified (15) *Paspalum scrobiculatum* (1)	

Source: Sahay et al. (2019)

selection for one trait often affects other traits as well and typically acts to modify the frequency of numerous functional genes within the population of plants. Thus, most forage nutritional value traits are considered to be under oligogenic or polygenic control and have been treated as quantitative genetic traits.

Conventional practices are adopted to improve forage quality using targeted phenotype which could have impact on nutritive value. Varietal improvement can be done following inter-varietal hybridization. This can help in introgression of traits like high dry matter and high yield followed by stringent selection procedures. Wide hybridization, like *Lolium-Festuca* hybrid has been adapted to introgress quality traits with drought tolerance and stability (Thomas and Humphreys 1991; Humphreys et al. 2003). These cultivars have the high-quality characteristics of *Lolium* combined with the stress tolerance and persistence found in *Festuca* (Ghesquière et al. 2010). Backcrossing of *Festulolium* has generated novel hybrids with more stable protein content when compared to parental lines (Humphreys et al. 2014). Distant hybridization also changes the ploidy level of the species altering the basic functioning and genetics of various traits. Hybridization can lead to heterosis in cross-pollinated species which provide a wide range of improvements in some crops like maize.

Evaluation of germplasm of different forages collected from diverse locations has provided variability which has been successfully utilized for the development of improved varieties through crop-breeding approaches. The germplasm superior in desirable traits like yield, tolerance to abiotic and biotic stress, and high nutritional value has been utilized in different breeding programs. The forage crop varieties developed in the past few years suitable for cultivation in different parts of the Himalayan region have been presented in Table 6.5.

Table 6.5 Forage crop varieties suitable for cultivation in different parts of Himalayan region

Crop	Variety	Yield and nutritional attributes	Institute involved
Oats	SKO-90	GFY (45 to 50 t/ha), CP (9.72%), ADF (40.20%), NDF (59.70%)	Srinagar, J & K
	SKO-96	GFY (47 t/ha), CP (8.94%), ADF (40.70%), NDF (57.90%)	Srinagar, J & K
	Palampur-1	GFY (50 t/ha)	CSKHPKV, HP
	UPO-94	–	Pantnagar, Uttarakhand
	UPO–212	–	Pantnagar, Uttarakhand
	Sabzaar (SKO-7)		Srinagar, J & K
	Bundel Jai-851	CP (10 to 12%), IVDMD (63.60%), NDF (57.50%), ADF (35.60%), CPY (8.10 to 12.30q/ha), Leaf/stem ratio (0.36 to 0.45)	IGFRI, Jhansi
	Bundel Jai 991 (JHO 99–1)	CP (9.50 to10.0%), IVDMD (57.40%), NDF (63.00%), ADF (47.00%)	IGFRI, Jhansi
	Bundel Jai-2004	CP (11%); IVDMD (57.60%)	IGFRI, Jhansi
Rye grass	Him Palam Rye grass-1	CP (12 to13%); GFY (35 to 40 t/ha)	CSKHPKV, HP
	PBRG-2	GFY (35 t /ha)	CSKHPKV, HP
Setaria	S-18	GFY: 50 to 60 t/ha; CP (10.80%); Cold and frost tolerant	CSKHPKV, HP
	Nandi	CP(7 to 8%) at 50% flowering, GFY (75 t/ha)	CSKHPKV, HP
	Setaria-92	GFY (30.0 t/ha), DMY (7.50 t/ha)	CSKHPKV, HP
	PSS-1 (Golden timothy)	GFY (57 t/ ha), low oxalates (2 to 3%) and CP (10%)	CSKHPKV, HP
	S-25	–	CSKHPKV, HP
Tall fescue	Hima-14	GFY: (20 to 35 t/ha); CP (13.60%); Drought and frost tolerant	CSKHPKV, HP
	Hima-4	GFY (45.0 t /ha) DMY (8.40 t/ha)	CSKHPKV, HP
	Hima-1	GFY (30 to 40 t/ha), DFY (5.0 to 5.20 t /ha)	CSKHPKV, HP
	EC-178182	GFY (25.0 to 30.0 t/ha)	CSKHPKV, HP
Cowpea	UPC-628	–	Pantnagar, Uttarakhand
	UPC-621	GFY (32.50 to 35.00 t/ha), DFY (5 to 5.50 t/ha) at 50% flowering, Seed yield (6 to 8 q/ha) in uncut crop. DM digestibility (60 to 65%) with 16 to 17% CP	Pantnagar, Uttarakhand
	UPO-06-2	–	Pantnagar, Uttarakhand

(continued)

Table 6.5 (continued)

Crop	Variety	Yield and nutritional attributes	Institute involved
Berseem	BL 22	GFY (70 t/ha)	PAU, Ludhiana
	Mescavi	GFY(65 t/ha)	HAU, Hisar
	UPB-10	GFY (70 to 75 t/ha)	Pantnagar, Uttarakhand
	BL-180	GFY (60 to 65 t/ha)	PAU, Ludhiana
	Wardhan	GFY (65 to 68 t/ha)	
Rice bean	VRB-3 (Him Shakti)	–	NBPGR, Shimla & VPKAS, Almora jointly
Maize	African Tall Composite	More dry matter and CP content, more number of leaves/plant, more leaf area, good grain and seed yield potential than other grain varieties. Resistant to foliar diseases and stem borer. GFY (60 to 70 t/ha)	MPKV, Kolhapur, Maharashtra
	J-1006	GFY (20.43 t/ha)	PAU, Ludhiana
Napier × Bajra Hybrid	NB-21	Low oxalic acid content	PAU, Ludhiana
	NB-37	CP (9 to 10%), low oxalates (2 to 3%), GFY (35 to 40 t/ha)	CSKHPKV, HP
Guinea Grass	Punjab Guinea Grass 1	–	PAU, Ludhiana
	PGG 9	GFY (45 to 50 t/ha) CP (8 to10%), Dry matter digestibility (55 to 70%)	PAU, Ludhiana
	Bundel Guinea-1 (JHGG-96-5)	CP (13.30%)	IGFRI, Jhansi
	Bundel Guinea-2 (JHGG 04–01)	GFY (50 to 60 t/ha), DMY (15 to 18 t/ha)	IGFRI, Jhansi
Alfalfa (Lucerne)	Anand-3	GFY (40 t/ha)	AAU, Anand
Cowpea	UPC-5286	GFY (35 t/ha)	Pantnagar, Uttarakhand
	UPC – 287	GFY (30 to 35 t/ ha)	Pantnagar, Uttarakhand
	Bundel Lobia-1 (IFC - 8401)	GFY (30 to 35 t/ha), DMY 3 to 5 t/ha	IGFRI, Jhansi
	UPC-622	–	Pantnagar, Uttarakhand
	UPC- 625	GFY (35 to 40 t/ha), Dry matter digestibility (65 to 70%), CP (15 to 17%)	Pantnagar, Uttarakhand

(continued)

Table 6.5 (continued)

Crop	Variety	Yield and nutritional attributes	Institute involved
Shaftal	SH-48		PAU Ludhiana
	Shaftal-48	GFY (55 t/ha)	CSKHPKV, HP
Red clover	PRC-3	GFY (40 to 60 t/ha)	CSKHPKV, HP
White clover	Palampur Composite 1	CP (23%), GFY (47 t/ha)	CSKHPKV, HP

Source: Ahmed et al. (2017)

Indian subcontinent is endowed with a rich genetic diversity of forage plant species which is largely due to diverse eco-climatic conditions and a variety of habitats and niches. The continent also has enormous diversity in native grasses, trees and shrubs having the potential to become important components of natural pastures and rangelands. In the past few years, there have been a lot of efforts for collection, evaluation, documentation, and conservation of germplasm of forage species from diverse locations. Different international and national institutions are actively involved in forage crops germplasm conservation and evaluation and conservation of rich diversity for forage crops improvement program.

References

Ahmed S, Sahay G, Sridhar K, Singh S, Saxena P, Roy AK (2017) Improved forage crop varieties of IGFRI. Technical Bulletin. Indian Grassland and Fodder Research Institute, Jhansi, p 64

Ghesquière M, Humphreys MW, Zwierzykowski Z (2010) Festulolium. In: Boller B, Posselt UK, Veronesi F (eds) Fodder crops and amenity grasses. New York, Springer, pp 288–311

Humphreys MW, Canter PJ, Thomas HM (2003) Advances in introgression technologies for precision breeding within the Lolium—Festuca complex. Ann Appl Biol 143:1–10

Humphreys MW, O'donovan SA, Farrell MS, Gay AP, Kingston-Smith AH (2014) The potential of novel Festulolium (2n = 4x = 28) hybrids as productive, nutrient-use-efficient fodder for ruminants. Food Energy Secur 3:98–110

Joshi V, Tyagi V, Kak A, Lal A (2009) Status of genetic resources of forage crops in India: a review. Indian J Plant Genet Resour 22(3):243–252

Kohli R, Batish K, Daizy R, Singh HP, Dogra KS (2006) Status, invasiveness and environmental threats of three tropical American invasive weeds (*Parthenium hysterophorous L, Ageratum conyzoides L., Lantana camara* L.) in India. Biol Invasions 8:1501–1510

Rana JC, Sharma Y, Singh A (2005) Invasive alien weeds encroaching forest-an issue need to be addressed for better forest management. In: Proceedings Regional Workshop on Forestry Extension Strategy Review. Organised by Himalayan Forest Research Institute (ICFR&E), Panthaghati, Shimla, pp 55–57. Accessed 27 Dec 2005

Rana JC, Dutta M, Rathi RS (2012) Plant genetic resources of the Indian Himalayan region—an overview. Indian J Genet 72:115–129

Roy AK, Malaviya DR, Kaushal P, Chandra A, Singh UP (2009) Descriptors for tropical forage legume Egyptian clover (*Trifolium alexandrinum* L.), Indian Grasses and Forest Research Station, Jhansi, India

Sahay G, Saxena P, Ahmed S, Singh T, Nayak S (2016) Role of genetic resources of forages in the present changing climatic scenario. In: Bisht J, Meena V, Mishra P, Pattanayak A (eds) Conservation agriculture. Springer, Singapore, pp 159–183

Sahay G, Sevanayak D, Tyagi VC, Bhardwaj N, Shmed S (2018) Tropical and subtropical forage genetic resources of India: their conservation and utilization. In: Forage for future, vol 6, p 3

Sahay G, Dikshit N, Singh T, Tyagi V, Nayak S, Shahid D, Roy A (2019) Forage genetic resources at ICAR-IGFRI Jhansi, accomplishments and future strategies. ICAR-IGFRI, Jhansi

Seghal JL, Mandal DK, Mandal C, Vadirelu S (1990) Agro-ecological regions of India. NBSS Publ. 24. National Bureau of Soil Survey & land Use Planning, (IACR), Nagpur

Singh RV, Chand D, Tyagi V, Singh AK, Singh SP, Singh S (2008) Important crop germplasm introduced into India during 2006. Indian J Plant Genet Resour 20:64–69

Singh AK, Varma N, Yadav SK, Mohanty A, Singh S, Singh S (2009) Indian forage genetic resources: perspectives and strategies. Progr Agric 9(2):250–256

Thomas H, Humphreys MO (1991) Progress and potential of interspecific hybrids of *Lolium* and *Festuca*. J Agric Sci 117:1–8

Tyagi V, Singh AK, Chand D, Singh RV, Dhillon BS (2008) Plant introduction in India during pre and post- CBD periods. An analysis. Indian J Plant Genet Resour 19:436–441

Approaches for Nutritional Quality Improvement in Forages

7

In the recent years, there are growing concerns for escalating demand for livestock products in food supply chain, however, to meet this demand, there is parallel requirement of a quality feed to the livestock. The development of improved forage crop varieties and their integration in farming system is essential to increase productivity of forages. Breeding forages is difficult as compared to food crops since different traits in forages related to yield and nutritional quality such as protein quality, total soluble carbohydrates content and cell wall digestibility, resistance to abiotic and biotic stresses and plant developmental processes are polygenic which means the most of the desirable traits are controlled by a number of genes. Moreover, difficulties are also associated with diversity in pollination behavior of forage crops, irregularities in fertilization behavior, seed setting and perennial nature of forage crops. There are certain unique problems associated with the range grasses and legumes. Most of the pasture crops still possess traits of wild plants. The major problems, which generally encountered in the forage breeding programs are:

(a) Overlapping of vegetative and reproductive growth phases
(b) Non-synchronous maturity and seed shattering in forages
(c) Inherent heterozygosity due to cross-pollinating behavior
(d) Self-incompatibility
(e) Difficulty in artificial hybridization due to small floral parts
(f) Poor seed producers
(g) Less seed viability
(h) Low seedling vigor and competitive ability
(i) Weak establishment of seedlings
(j) Polygenic control of desirable traits including forage quality
(k) Perennial nature of different forages

R. Katoch, *Nutritional Quality Management of Forages in the Himalayan Region*,
https://doi.org/10.1007/978-981-16-5437-4_7

(l) Fertility barriers in forages breeding due to the wild nature of species
(m) Inadequate knowledge of inter- or intraspecific variation in forages (Roy et al. 2015)

Genetic improvement of forage crops for increasing their nutritional quality is the cost-effective mechanism. It could be understood from the fact that even a small increase in digestibility results in appreciable improvement in livestock production and productivity. Many forage quality traits such as different cell wall constituents are typically intercorrelated and are regulated by different enzymes that are involved in their biosynthesis or metabolism. Therefore, selection for one quality trait often affects other traits. Moreover, traits interrelatedness also modifies the frequency of numerous functional genes within plant population. Therefore, most forage nutritional quality traits are considered to be under oligogenic or polygenic control (Quantitative genetic traits).

7.1 Forage Research: National and Regional Scenario

To meet the growing demand of quality feed for the livestock, there is a need for the development of forage varieties having high productivity and their adoption could gain higher monitory returns. The interlinking of livestock productivity and forage production is important for rural self-sustainability. Permanent pastures and grazing lands, fodder from forests, crop residues and to some extent cultivated forage crops are important forage resources in the Himalayan region. As the gap between demand and supply of feed available to livestock is continuously widening, there is a need for parallel efforts for improving the nutritional quality and availability of forages. This will help in economizing the feed cost which would result in better remuneration to farmers.

In India, collaborative and comprehensive efforts have been made for increasing genome resource base and for improving quality and productivity of forage crops. This has resulted in large collection of forage germplasm over the years through indigenous collection and introduction. The national agencies including NBPGR, (New Delhi), IGFRI (Jhansi, UP), AICRP (FCU) centers, state agriculture departments and state agriculture universities (SAU) are actively involved in the research and development of forage crop varieties.

AICRP (FCU) centers with the support of ICAR are carrying out germplasm evaluation and characterization activities for varietal development. Analysis of nutritional quality traits and yield attributes is carried out for developed germplasm in coordinated trials. Multilocation trials are also conducted to release promising forage crop varieties for different regions in the country.

Since the inception of IGFRI in the year 1962, the institute has made remarkable progress in the field of forage research, extension, and capacity building and has widened its visibility at national and international levels. The institute has achieved several milestones in generation of the various need-based technologies with ultimate aim of serve farming community with advanced technology. To address the continuously emerging challenges in providing feed and nutritional security to livestock, NBPGR, IGFRI, and AICRP (FCU) centers have widened their approach with focus on technological interventions and development of improved fodder varieties.

In the Indian Himalayan region, NBPGR and IGFRI institutes have established their regional stations in different states to carry out research and development programs on forages. These centers are actively involved in the collection of forage germplasm in temperate regions and morpho-physiological, biochemical, and molecular characterization. These centers conduct experiments on region-specific forages in collaboration with state universities and regional institutes for developing varieties excelling in production and quality.

7.2 Need for Nutritional Quality Improvement of Forages

Giving nutritiously wholesome feed to livestock is the major requirement to increase the economic viability of livestock production operation. Nutritional quality of forages depends upon their biochemical composition. Forage quality is often determined by measuring their intake by animals and digestibility. The composition and the level of nutritional and anti-nutritional constituents determine *D*-value of forages and amount of digestible energy available to the livestock. It has been estimated that even a small increase in digestibility would result in high livestock productivity (Reddy et al. 2005). The presence of anti-nutritive factors in forages affect the quality of forages that generally leads to reduced DMI, low DMD and nutritional imbalances in livestock and thereby have economic impact on livestock production system (Fribourg et al. 1991).

Improvement in the forage production is the main objective of the forage breeders. However, with realization of importance of forage quality in livestock production, breeding cultivars with improved quality traits has become increasingly important targets in forage breeding programs. In forages, digestibility and intake could be improved by decreasing lignin content or lignin cross-linkages between the cell wall carbohydrates and other cell wall components. High WSC (Water soluble carbohydrate) content and NSC (Nonstructural carbohydrates) content in forages can improve the production of protein from microbial origin. Forages having high protein content, the problem of rapid protein degradation is of major concern. Tannins like proanthocyanidins or CTs inhibit protein degradation by formation of complexes. The optimum tannins level in forages having higher protein content provides protection from the rapid protein degradation in rumen and prevent bloating. However, higher tannins content depresses the protein digestion and absorption (Piluzza et al. 2014). Therefore, lowering rapid degradation of protein in rumen is one of the important objectives in forage crops breeding.

Nutritional quality traits have now become important targets in forage breeding (Stewart and Hayes 2011). In recent years, significant progress has been made in the development of production of cultivars with improved nutritional quality characteristics aimed at improving nitrogen use efficiency in rumen.

Agronomic traits like days to maturity and other associated traits also influence forage quality. With advancing maturity, desirable components such as CP content and digestibility decrease, while negative components like NDF, ADF and CF increase. Keeping in view above issues, breeders often select forage species based on crop maturity. In smooth brome grass, the genetic variability that exists for forage nutritional attributes is entirely independent of maturity and little genetic variation exists for timing of reproductive maturity (Casler and Carlson 1995). Simultaneous selection for high nutritional quality and production has also been practiced in switch grass, alfalfa (Coors et al. 1986) and reed canary grass (Surprenant et al. 1990). In smooth brome grass, genotypes with high digestibility have more nondegradable tissues in both stem and leaf cross-sections than genotypes having low digestibility (Ehlke and Casler 1985). The diseased forages have the disadvantage of high lignin deposition and other cell wall constituents with reduction in the level of different nutritional constituents.

7.2.1 Targets for Nutritional Enhancement of Forages

Forage quality is defined as the extent of absorption and assimilation of nutrients from forages to produce desired results. The forages of good nutritional quality should have high palatability, DMI, digestibility, and low anti-nutritional content. The nutritional quality and digestibility of forage are determined by the level of protein content, WSC content, cellulose, hemicellulose, and lignin contents. Lignin and hemicellulose content in forages negatively affects their quality and digestibility. A number of plant-based and environmental factors affect the nutritional quality and digestibility of forages. Therefore, while considering for the improvement of forages, these factors should be considered during the selection of breeding method. The digestibility of forages may be increased through direct selection for high digestibility or indirect selection for high WSC content and protein content and low lignin and hemicellulose content. Different targets for nutritional enhancement of forages have been discussed under subsequent subheads:

(a) **Digestibility:** Forage digestibility has major influence on livestock performance. It is determined as efficiency of animal to metabolize nutrient components from forages. It also represents the amount of energy received from forages by animal for growth and development. If animals consume feed having low digestibility, its passage rate through alimentary canal is low which causes anorexia that ultimately leads to reduced feed intake, weight loss, nutrients deprivation, and decreased productivity from livestock. Feeding studies have established that a smaller improvement can significantly improve the energy intake and weight gain in animals. Therefore, improvement in digestibility of forage crops is a vital goal of forage breeders.

    Lignin is one of the important compounds for negative impact on digestibility of forages. Lignin increases with plant maturity and has negative relationship with digestibility. It cross-links with cellulosic polysaccharides and proteins of cell wall and blocks the access of these polymers to cell wall degrading enzymes. As a result of cross-linking, the amount of total digestible energy available to livestock is reduced and consequently results in incomplete utilization of nutrients. In grasses, cross-linking of lignin wall polysaccharides involves etherification of esterified ferulates of arabinoxylan. Any change in cell wall components content may enhance both intake and energy density in forage crops. Likewise, increasing digestibility of the cell wall polysaccharides would increase the amount of digestible energy to livestock. Cellular carbohydrate is an important contributor to nutritive value of grasses and may make up as much as 50% dry matter of some forage species. High water-soluble carbohydrates content in forages are positively related to digestibility and are thus important in breeding high-quality forage grasses for high livestock production. However, these components may be difficult to assess for breeding purposes since their level can vary widely within days as well as over longer periods. The inheritance of WSC content in *Lolium perenne* cultivars is complex and shows evidence of additive gene effects, although it is a reliable trait for improvement through breeding. Radojevic et al. (1994) observed that higher water-soluble carbohydrate content during summer in several perennial rye grass cultivars grown under Southern Australian conditions buffered the summer decline in digestibility caused by diminished neutral detergent fiber digestibility and resulted in the improvement in digestibility. In the study, WSC content was negatively correlated with nitrogen content in *Lolium perenne*. It has also been demonstrated that feeding rye grasses bred to express high WSC content improves the livestock growth and milk production (Miller et al. 2001). The provision of diet containing high diet WSC:N ratio increases nitrogen use efficiency by dairy cows, without affecting the milk production. Being an instant source of energy to livestock, improving WSC content in forages is one of the important objectives in forage breeding program.

(b) **Protein quality and utilization:** Provision of appropriate quantities of protein and energy-yielding substrates to livestock to the rumen microbial population improves absorption efficiency and nutrient utilization (Hall and Huntington 2008). The fast degradation of proteins by rumen microflora results in bloat, which is attributed to the formation of foam and trapping of gases in rumen, which traps gases and causes discomfort in animals. Reduction in protein degradability in the rumen is an important breeding objective in forages. This goal could be achieved by reducing the enzymatic degradation of proteins by rumen microflora.

Ruminal efficiency of the livestock could be increased by altering the degradability of forage proteins. A strategy for the improvement of microbial efficiency in livestock would be to protect proteins from ruminal degradation. Condensed tannins (CTs) could help in protecting forage proteins degradation and improve ruminal efficiency in livestock. The CTs bind with proteins under acidic to neutral conditions in the rumen, slow down the degradation rate, and increases the absorption of amino acids. The CTs content of 2 to 4% in forage dry matter can effectively reduce the ruminal protein degradation. Supplementation of forages of high protein content with forages having high CTs is an effective way to reduce the ruminal protein degradation. Forages containing high CTs content (*Lotus corniculatus*) could be useful in serving the purpose of protecting degradation of proteins in rumen. Marshall et al. (2008) reported considerable genetic variation in CTs content in birdsfoot trefoil that suggests importance of this forage in breeding improved varieties with optimum levels of CTs.

The aerobic conversion of monophenols to o-phenols and finally to *O*-quinones is catalyzed by polyphenol oxidase (PPO) enzymes. The activity of PPO enzymes results in the formation of protein–phenol complex (Parveen et al. 2010). Such complexes are resistant to enzymatic digestion by proteases and PPO mechanism may effectively reduce the protein breakdown in red clover silage (Sullivan and Hatfield 2006). Activity of PPO enzymes results in protection of over 80% protein content in red clover silage. This protected protein is true protein that provides an improved nitrogen use efficiency and product quality when fed to livestock (Winters and Minchin 2002). The inhibitory effects of PPO activity on proteolysis support the need of breeding improved forage varieties with increased PPO enzyme activity.

(c) **Anti-nutritional factors:** Many biochemical compounds can be detrimental to forage utilization by livestock, largely by the reducing palatability, digestibility, intake and/or health and livestock performance. These compounds are called anti-nutritional factors. By definition, anti-nutritional factors are substances that when present in animal feed they either by themselves or through their metabolic products reduce the availability of one or more nutrients. The presence of anti-nutritional factors in forages is the result of normal metabolism. Mostly, these anti-nutritional factors are disliked by animals and the toxicity of anti-nutrients depends upon their content and rate of degradation in gastrointestinal tract. Anti-nutrients include toxins, estrogenic compounds, alkaloids, saponins, tannins, estrogens, etc. Development of forage crop varieties with low anti-nutritional content is one of the important objectives of forage crop breeding program.

(d) **Feed use efficiency of livestock:** Improvement in feed conversion efficiency, which is defined as ratio of final product to the metabolized energy in livestock rumen, would help to improve animal gain and production. Feed energy conversion could be improved by improving dry matter digestibility and nutrients absorption efficiency, WSC content and amino acid and protein content and lowering down the lignin and cellulose content. Increasing the digestibility of feedstuffs is essential to increase energy use efficiency of livestock. Besides digestibility, other factors are increasingly important forage breeding targets (Casler and Van Santen 2010).

Feed dry matter digestibility is a major determinant of efficiency of feed use in rumen of livestock. Ruminal microflora uses liberated forage protein breakdown products to synthesize more microbial protein from which much of meat and milk protein is derived. For microbial synthesis of proteins, a source of energy is needed to drive microbial synthesis. However, when protein breakdown products are present in abundance, the colonizing microbial population is likely to be energy-limited due to much low degradation rate of their main energy-yielding substrates, the structural carbohydrates, cellulose, and hemicellulose content (Kingston-Smith et al. 2003). During bloat, ammonia accumulates and, if it is not incorporated into microbial proteins is absorbed from rumen and it is subsequently converted to urea in liver and is either recycled back to the rumen or excreted as waste nitrogen in urine (Moorby et al. 2006). There are two possible strategies for increasing the efficiency of conversion of forage nitrogen to microbial nitrogen. One strategy is to increase the amount of readily available energy accessible during the early part of fermentation. The second strategy is to provide a level of protection to forage proteins, thereby reducing rate at which their breakdown products are made available to colonizing rumen microflora.

Although, livestock are efficient at extracting energy from plant cell wall compared with mono-gastric animals, it has been known that methane production due to rumen fermentation is an indicator of feed use inefficiency. Increasing the efficiency of rumen function would have significant impact on livestock production. The amount of digestible energy from feed is a function of digestibility of feed which is a major target for forage breeding programs (Parsons et al. 2011).

Till 1980s the main aim of forage breeders was to increase animal production per unit area, which in present scenario is focused on reducing the economic and environmental costs of animal production. This has led to some changes in the relative importance of different traits. Due to the genetic linkage between desirable and undesirable genes, as well as pleiotropic effects of a minority of genes on more than one trait, it is usually necessary to assess all desirable traits in each generation of selection to produce improved varieties.

Improvement in nutritional quality of forages by conventional approaches cannot improve all nutritional quality attributes that are polygenic and very complex. Introduction and domestication are the oldest methods of crop improvement. Not only crop introduction, but introduction of superior forage varieties has also led to improvement directly or indirectly. Selection methods like mass selection, clonal selection, or recurrent selection following hybridization have led to the development of new forage varieties or improvement of existing forge varieties. Majority of the forage crops are cross-pollinated in behavior. Poly-crossing, composite breeding, or synthetic varieties are useful forage breeding approaches. In case of self-pollinated crops like rice bean, cowpea, etc., pure lines, mass selection, pedigree method of breeding, or backcross breeding could be of great significance.

7.3 Breeding for Improvement in Forage Quality

Breeding is the key for the future development of superior forages varieties. In all important forage crops, different varieties have been developed with conventional breeding for different growing conditions (Table 7.1). Studies have revealed that many quality and anti-quality traits are heritable and they can be stably modified by selection and breeding. For example, HCN content inheritance studies have revealed

Table 7.1 Improved forage crop varieties developed by different research organizations in India

Crop	Varieties	Institute involved in development	Breeding method used	Description
Maize	African tall composite	MPKV, Kolhapur	Composite breeding	High dry matter and CP content High number of leaves/plant Good fodder and seed production
	APFM-8	ANGRAU, Hyderabad	Mass selection	GFY (35 t/ha) DFY (7.50 t/ha)
	J-1006	PAU Ludhiana	Makki Safed 1-DR × Turpeno PB	–
Oats	HFO—114 (Haryana Javi—114)	CCS HAU, Hisar	Pure line selection	GFY (50 to 55 t/ha) DMY (13 t/ha)
	Palampur-1	CSK HPKV, Palampur	Selection	GFY (50 t/ha)
	OS-6	CCS HAU	Selection of the cross between HFO 10 × HFO 55 P2	Flag leaf remains erect at emergence of panicle, tolerant to diseases Suitable for temperate and sub-tropical regions GFY (54 t/ha)
	OL-9	PAU, Ludhiana	N.P. hybrid × Kent	GFY (40 to 50 t/ha)
	Bundel jai-822	IGFRI, Jhansi	Intervarietal hybridization and pedigree method of selection	Matures in 125 to 130 days GFY (50 t/ha) DFY (12 t/ha)
	UPO–212	GBPUA&T, Pantnagar	Intervarietal hybridization followed by pedigree breeding and selection	GFY (60 t/ha)
	OL-125	PAU, Ludhiana	Pedigree breeding and selection	GFY (58 t/ha)
	Bundel Jai-851	IGFRI, Jhansi	Selection from exotic Japanese germplasm	High regeneration potential Multicut nature, leafiness High tillering, and high CP content
	Bundel Jai 992 (JHO 99–2)	IGFRI, Jhansi	Pedigree method of selection	CP (10.70%), digestibility (58%) GFY (50 t/ha), DFY (10 t/ha)

(continued)

Table 7.1 (continued)

Crop	Varieties	Institute involved in development	Breeding method used	Description
	Bundel Jai 2004 (JHO 2000–4)	IGFRI, Jhansi	Induced polyploidy and pedigree method of selection	GFY (50 t/ha), DMY (10 t/ha), CP (11%), ADF (47.90%), NDF (62.45%) digestibility (57.60%) Tolerant to root rot, crown rust, leaf blight, and powdery mildew
	Bundel Jai 991 (JHO 99-1)	IGFRI, Jhansi	Pedigree method of selection	GFY (30 t/ha) DFY (7 t/ha), CP (9.50–10%), ADF (47%), NDF (63%) and digestibility (57.40%)
Guinea grass	Haritha	KAU, Kerala	Mutation breeding	GFY (49 t/ha) CP (7.60%)
	Marathakam	KAU, Kerala	Mutation breeding	GFY (55 t/ha) CP (8.20%)
Napier × Bajra hybrid	CO-1	TNAU, Coimbatore	Wide hybridization	GFY(300 t/ha)
	CO-2	TNAU, Coimbatore	Interspecific hybridization followed by clonal selection	GFY (350 t/ha)
	IGFRI-5	IGFRI, Jhansi	–	GFY (114 t/ha) DMY (37 t/ha) CP (6.35%) oxalate (2.8%)
	NB-37	CSK HPKV, Palampur	Interspecific hybridization	Low oxalate content (2 to 3%) and high CP (9 to 10%)
Tall fescue	HIMA-1	CSK HPKV, Palampur	Recurrent phenotypic selection	GFY (30 to 40 t/ha) DMY (5.00 to 5.20 t/ha)
	HIMA-4	CSK HPKV, Palampur	Recurrent phenotypic selection	
Golden timothy	Nandi	CSKHPKV, Palampur		GFY (75 t/ha)
	PSS 1 (Golden timothy)	CSKHPKV, Palampur	Selection	GFY (57 t/ha) oxalate content (2 to 3%) and CP (10%)
	Setaria 92	CSKHPKV, Palampur	–	GFY (30.0 t/ha), DMY (7.50 t/ha)

(continued)

Table 7.1 (continued)

Crop	Varieties	Institute involved in development	Breeding method used	Description
Berseem	Berseem Ludhiana-1 (BL-1)	PAU, Ludhiana	Selection from Mescavi	GFY(80–110 t/ha)
	Jawahar Berseem-1 (JB-1)	JNKVV, Jabalpur	Single plant selection from Chindwara followed by pedigree method of selection	GFY (70 to 75 t/ha), DFY (12 to 15 t/ha) dry fodder
	BL-22	PAU, Ludhiana	Irradiation (mutation breeding)	GFY (70 t/ha)
	UPB-10	GBPUA&T, Pantnagar	Composite of 5 lines followed by selection	GFY (70 to 75 t/ha)
	JB-5	JNKVV, Jabalpur	Recurrent selection	GFY (48 t/ha)
	BL-180	PAU, Ludhiana	Irradiation of variety BL 10 followed by selection	GFY (60 to 65 t/ha)
Lucerne	Type-9	Fodder Research Station, Sirsa	Mass selection	
	Co-1	TNAU, Coimbatore	Mass selection	CP (20%) GFY(80 to 90 t/ha)
	RL-88	MPKV, Rahuri	Selection from local collections	GFY (75 to 100 t/ha)
Rice bean	Bidhan-1 (BC 15/K 1)	BCKV, Kalyani	Selection from a local landrace	GFY (35 to 40 t/ha)
	Bidhan Rice bean 2 (KRB 4)	BCKV, Kalyani	Local landrace selection	GFY(35 to 40 t/ha)
Shaftal	Shaftal-48	CSK HPKV, Palampur	Selection in local material	GFY (55 t/ha)
Red clover	PRC-3	CSK HPKV Palampur		GFY (40–60 t/ha)
White clover	Palampur composite 1	CSKHPKV, Palampur	Composite breeding	High nitrogen-fixing capacity (200–400 kg N/ha), GFY (47 q/ha)

a single major gene intermediate in dominance, quantitative inheritance with additive genetic effects, and multiple loci affecting this trait. Sorghum forage with less than 750 ppm HCN on fresh weight basis is generally not detrimental to ruminant absorption *via* grazing.

7.3.1 Introduction and Selection

Introduction is one of the oldest methods of breeding. Alfalfa/lucerne (*Medicago sativa*) and berseem or Egyptian clover (*Trifolium alexandrinum*) are the oldest examples of global dispersion of forage crops. Both crops originated in Asia Minor and North-Western Persia (Iran) and were first introduced in Italy then in Spain. Berseem was of ancient cultivation in Egypt, from where it spread to India and Southern Europe and to USA and other countries. From Mediterranean basin, annual medics (*Medicago* spp.) and subterranean clover (*Trifolium subterraneum*) were introduced to Mediterranean-type climatic regions of Australia, Chile, California, and South Africa. Plant introduction has established berseem and lucerne in India as one of the major forage crops providing green and dry fodder to livestock. The varieties which have been developed by selection and introduction in different forage crops have been mentioned (Table 7.2).

To improve forages quality parameters, selection (plant selection and clonal selection) and various hybridization program have been used in forage breeding. Natural selection has a tremendous influence on phenotypic expression of forage crops. Despite the large knowledge base on the effects of natural selection pressures on adaptive traits of forage crops, little is known about the effects of natural selection pressures on traits related to quality of forages (Burton et al. 1993). Phenotypic recurrent selection has been the most commonly used procedure by breeders in perennial forages (Conaghan and Casler 2011). In contrast, genotypic selection involves selection of desirable parent genotypes followed by progeny tests to evaluate breeding value performance of parents and cultivar generation from few numbers of parents (Casler and Brummer 2008).

Recurrent selection is a selection process, which consists of repeated cycles of selection where each cycle involves the evaluation of a population of plants, identification of a subset of superior plants, and intercrossing of the selected plants. This process of breeding is used majorly in cross-pollinated and heterozygous population species, which increases the frequency of desirable and favorable alleles without restricting the genetic variability for other traits. Majority of the forage breeders use clonal propagation, but some researchers practice replication in time, repeated measures on unreplicated spaced plants to improve the precision of their selection criterion. Selection in the second or later stages is often based on the mean overall available harvests. Multistage selection for digestibility, nitrogen content, and forage yield in a switch grass population has been predicted to increase selection gain by 6 to 7% for each additional selection stage beyond the first (Godshalk et al. 1988). Recurrent selection method has been used in switch grass and smooth brome grass for high digestibility and low NDF content, respectively, where digestibility

Table 7.2 Introduction in major forage crop

Crop	Variety	Selection from (Parentage)	Cultivation area	Yield and nutritional quality description
Guinea grass	Bundel Guinea −2 (JHGG-04-1)	Selection from exotic germplasm, EC-400533 (IG 97–3) received from ILCA	Rainfed/irrigated conditions all over the country	GFY: (50 to 60 t/ha), DMY: (15 to 18 t/ha), CP % (7.80%), digestibility (69.40%), NDF (68.60%), ADF (52.60%), leaf/stem ratio (1.0)
	Bundel Guinea −4 (JHGG-08-1)	Selection from old accession 59,903 (IG-01-95)	Rainfed/irrigated conditions all over the country	GFY (80.0 t/ha), DMY (18.0 t/ha), CP (8.20)%
	Macueni	–	Kerala	
	Riverdale	–	Southern India	GFY (74 t/ha)
	Punjab Guinea grass 1	Australia (CPI 59985)	North- West hill zone	–
Setaria	PSS 1 (Golden timothy)	Kenyan introduction Narok-5	Sub-tropical regions of Himachal Pradesh between 1100 to 2100 amsl	GFY (57 t/ha), low oxalates (2 to 3%) CP (10%)
Cowpea	Kohinoor (S-450)	Selection from the material received from Iran (ACC no. IL-68-786)	–	GFY (40 to 45 t/ha), DMY (6.0 t/ha), CP (20 to 22%), digestibility (55 to 60%), NDF (43 to 46%), ADF (34 to 37%), cellulose (23 to 25%, hemi cellulose (5 to 6%)
Oats	Kent	Introduction from USA	Entire country	GFY (50 t/ha)
	Weston 11	–	Punjab	GFY (50 t/ha)
Berseem	Mescavi	Introduction from Egypt	Punjab, Haryana and Himachal Pradesh	GFY (65 t/ha)

increased after three cycles of selection, whereas after one cycle of selection, NDF content was reduced.

7.3.2 Hybridization

The biggest bottleneck in recombination breeding in grasses is apomictic reproduction behavior. The identification and generation of sexual lines in apomictic grasses may prove promising. On availability of appropriate sexual lines, the benefits of hybrid vigor can be fixed taking the advantage of apomixis. Hybrid breeding has been utilized in some forage grasses such as guinea grass, buffel grass, and *Bracharia* where either obligate or facultative sexual lines are available. In *Panicum maximum,* increased foliage percentage, seed production, rapid regrowth have been observed in F_1 hybrid in comparison to parents (Resende et al. 2004). Many wide adaptable varieties in guinea grass (PGG-9, PGG-14, PGG-19, PGG-518, and CO-2) have been developed and released from PAU (Punjab Agricultural University, Ludhiana) and TNAU (Tamil Nadu Agricultural University, Coimbatore). High forage quality and determined flowering cycle of *Bracharia ruziziensis*, the yield and resistance to spittlebug in *Bracharia brizantha*, and vigor and adaptation to acid, infertile soils in *Bracharia decumbens* grass have also been recorded (Peters and Lascano 2003).

7.3.3 Interspecific and Polyploidy Breeding

The role of interspecific hybridization in plant evolution is well known, and the origin of a number of domesticated crops is based on an interspecific hybrid combined with polyploidy. Attempts to produce synthetic species by artificially creating interspecific hybrids, followed by polyploidy induction, has limited success with regard to the production of commercial varieties. One of the most notable successes has been the development of tetraploid hybrids between Italian rye grass (*Lolium multiflorum*) and perennial (*L. perenne*) rye grass, which combine the complementary characters of rapid and early high-quality growth from Italian rye grass and good aftermath tillering and persistency from perennial rye grass.

Crossing between sexual and apomictic plants of *Brachiaria ruziziensis* directly or by chromosomal doubling with colchicine to make cross-compatible and to induce variability along with quality traits has been attempted in grasses (Lutts et al. 1994). The ploidy level resulted in the discovery of a sexual accession, which was artificially doubled to make crosses with tetraploid accessions feasible (Pinheiro et al. 2000). High seed-producing cultivars of *Pennisetum purpureum* have been developed by crossing *Pennisetum purpureum* and *Pennisetum glaucum* (L.). The progenies from this cross are triploid and sterile, but once doubled by colchicine become hexaploid and fertile (Pereira and Lédo 2008). IGFRI has developed world's largest ploidy series of *Panicum maximum* under the ploidy-regulated phenomenon at genetic level. Superior fodder-type genotypes of pearl millet have been generated through interspecific hybridization (*Pennisetum glaucum* × *Pennisetum squamulatum*; *Pennisetum glaucum* × *Pennisetum orientale)* at IGFRI (Jhansi). The resultant hybrid has shown higher biomass, better adaptability, better tolerance for moisture-deficit, and improved forage attributes compared to other tropical grasses.

In India and worldwide, Bajra × Napier hybrids have been recognized as a revolution in forage breeding program due to wide adaptation, high biomass yield, tolerance to abiotic stresses, and perennial nature of the hybrid. The interspecific cross between annual *Pennisetum glaucum* with perennial *Pennisetum purpureum* (presented below) was used to develop NB hybrid.

Pennisetum glaucum Diploid ($2n = 14$) × *Pennisetum purpureum*

Tetraploid ($2n = 28$)

↓

Triploid ($2n = 21$) Sterile.

Breeding in cool-season grasses has also focused on introgression of drought tolerance from fescue into rye grasses (Humphreys and Thomas 1993). Stay green gene has been successfully transferred from *Festuca pratensis* into *Lolium perenne*. Crown rust resistance has also been transferred from *Festuca Pratensis* to *Lolium multiflorum* (Oertel and Matzk 1999). The formation of allopolyploid hybrids between important forage grasses (*Lolium multiflorum* × *Dactilys glomerata*, *Festuca pratensis* × *Dactylis glomerata*, and *Festuca rubra* × *Lolium perenne*), hybrid rye grass from crosses between perennial × Italian rye grass and *Festucalolium* from crosses between *Lolium* spp. × *Festuca* spp. have also been attempted to generate variability and introgression of desirable traits. Some important examples of interspecific hybrids have been enlisted in Table 7.3.

The potential of rye grass× fescue amphiploids as commercial cultivars depends on control of chromosome pairing to maintain hybridity and to allow polyploid

Table 7.3 Interspecific hybrids in forages

Sr. no.	Interspecific hybrids	Trait	References
1.	*Lotus tenuis* × *Lotus corniculatus* hybrids	Lower proanthocyanidins (PAs) in edible tissues to prevent bloating in ruminants	Escaray et al. (2014)
2.	*Pennisetum americanum* × Napier grass (*Pennisetum purpreum* Schum.)	Hybrid Tift 23A × N23, yielded 34 and 27% more dry matter	Hanna and Monson (1980)
3.	Elephant grass (*Pennisetum purpureum* Schum.) × Pearl millet (*Pennisetum glaucum*)	Hybrids combined resistance to drought, tolerance to diseases, and seed size of pearl millet with the hardness, aggressiveness, and high dry matter production of elephant grass	Leão et al. (2011) and Obok et al. (2012)
4.	Buckwheat (*Agopyrum esculentum* Moench.) × *Fagopyrum tataricum* (L.) Gaertn	Trangressive segregants have improved self-pollination, fertility, frost tolerance, and higher content of beneficial compounds	Mendler-Drienyovszki et al. (2013)
5.	*Trifolium pratense* × *Trifolium medium*	Selection was practiced for forage yield and persistence	Řepková et al. (2013)
6.	*Trifolium uniflorum*, *Trifolium occidentale*, *Trifolium pallescens*, 2× *Trifolium ambiguum* and 6× *Trifolium ambiguum*	Interspecific hybridization to transfer gene for the induction of tolerance to drought, diseases, and salinity	Williams (2014)

Source: Rauf et al. (2016)

plants to remain stable and perform well agronomically. Identification of pairing genes is important in this context. Moreover, introgression between rye grass and fescue species depends on recombination between rye grass and fescue chromosomes and the ability to identify quickly the good agronomic/physiological characteristics. Linkage to genetic markers, such as isozymes and more genomically widespread RFLP systems (Restriction Fragment Length Polymorphism), is valuable in this context.

7.3.4 Mutation Breeding

Mutation breeding increases the frequency of natural mutation rates in nature and can be efficiently utilized to develop new breeding materials, which have not been identified in germplasm collections. ϒ-rays induce both a higher percentage of small deletions (1 to 6 bp) and a lower percentage of large deletions (10,000–100,000 bp) (Morita et al. 2009). ϒ -ray irradiation is one of the most powerful tools to induce mutations and favorable mutations are valuable for genome analysis, gene isolation, and development of the new varieties. Mutation breeding has been utilized in berseem to improve crop duration. The success has been achieved by the development of longer duration mutant in Mescavi variety through ϒ-ray treatment (Sohoo et al. 1985). BL-22 a variety with longer duration was released in the year 1988 for cultivation in temperate zone and North-West zone, BL-180 in the year 2006 for cultivation in North-West zone of India. Brown midrib mutants significantly reduce the level of enzyme-resistant lignin in plants and increase their palatability and digestibility (Cherney et al. 1991). Brown midrib silage with and without protein supplements significantly increased milk yield of lactating cows. Similarly, the extent of digestibility and cell wall degradation by rumen bacterium of leaf blades from bmr-12 sorghum was shown to be significantly higher than those from their respective wild-type isolines (Akin et al. 1986). A spontaneous, naturally occurring, recessive mutation called bloomless (bm: bloomless, no epicuticular wax on the stem surface) and brown midrib (bmr: brown mid-rib, low lignin content) have been successfully utilized in sorghum breeding for forage use in Japan. These characteristics are also suggested to improve sorghum bioethanol production through fermentation of cellulose of stems as well as the digestibility of the foliage. Vogler et al. (2009) reported that chemically induced mutants of sorghum significantly reduced lignin content in leaf, blade, sheath, stem, and panicle tissue. They also reported high cellulose and hemicellulose content in mutant sorghum plants.

7.3.5 Apomixis

Apomixis is a method of reproduction to generate clonal seeds and offers tremendous potential to fix heterozygosity and hybrid vigor (Kaushal et al. 2019). In other words, apomixis avoids meiosis and fertilization to produce embryos similar to maternal parent (Kaushal et al. 2005). Asexual propagation offers biological advantage of conserving even highly heterozygous genotypes indefinitely. Apomixis is a special type of asexual propagation *via* convenient vehicle of seed. It offers many advantages in a plant-breeding program. However, in the tropical grasses, only two types of apomixis have been reported. In both cases, a female gametophyte is formed (gametophytic apomixis), originating either from unreduced megaspore mother cell following meiosis failure (diplospory) or from a somatic, nucellar cell (apospory). In either case, the unreduced egg cell develops without fertilization (parthenogenesis) into an unreduced embryo, genetically identical to the mother plant. Cyto-embryological and molecular processes have been studied in model aposporous (*Brachiaria, Pennisetum, Cenchrus, Hieracium, Paspalum, Poa,* etc.) and diplosporous taxa (*Erigeron, Taraxacum, Tripsacum,* etc.) (Kaushal et al. 2019). The apomictic crops where the advantages of seed propagation are combined with the fidelity of asexual reproduction are several economically important forage grasses, generally recent domesticates or semidomesticates. For example, guinea grass, buffel grass, bahia grass, and several commercial species in the genus *Brachiaria*.

In buffel grass, an apomictic species, it has been shown that apomictic reproductive process is genetically controlled with progenies segregating plants that reproduce apomictically and plants that have normal sexual reproduction. With apomixis, genetic segregation is prevented and superior heterozygous plants may be uniformly propagated, as with vegetative propagation or cloning. Apomictic species, irregular chromosome numbers may be propagated in the progeny, giving rise to aneuploids and polyploids. Failure of apomictic plants to undergo gene recombination and segregation reduces the capability of anapomictic forage to adapt to changing environments and restricts ecotype formation. However, once a desirable apomictic genotype is produced, it may have greater persistence and vigor than sexual plants of same species.

Breeding systems for improving the apomictic forages are unique and are not useful for improving sexual species. Generally, most apomictic species also produce some sexual offspring (Facultative apomicts). In breeding programs, superior, naturally occurring apomictic ecotypes are identified using ecotype evaluation and selection procedure (Hanna and Bashaw 1987). The most vigorous and productive ecotypes are selected, increased, tested, and released as new cultivars. The success of ecotype selection for apomictic species is improved, if there is a large amount of genetic variation between and within ecotypes.

7.3.6 Wide Hybridization

As intensification of farming increases, plant breeders developed cultivars that are more productive, more nutritious, and better adapted to environments without the need for increased external inputs, such as the application of water and nutrients. Breeders can exploit the extended heritable variation that results from interspecific hybridization and improved rates of genetic gain can be achieved when new traits that are not known in either parent are expressed in hybrids. There are three gene pools from where breeders can obtain genetic diversity for cultivar development. The primary gene pool includes individuals of the same species, where mating occurs easily, and progeny is viable and fertile. When genes for traits of interest cannot be found within a species, breeders can look to related species for further variability. The secondary gene pool includes members of closely related species that will cross with the crop, and produce progeny, however they tend to be sterile, and often weak. The tertiary gene pool also contains related species; however, hybrid progeny tend to be abnormal, lethal, or completely sterile.

While wide hybridization offers many promising advances in plant breeding, problems including, but not limited to, the cross-incompatibility between the wild related species and cultivated crop species, F_1 hybrid sterility, reduced recombination between chromosomes of the two species, and the phenomenon of linkage drag. Failure of hybrids post-fertilization is commonly associated with endosperm abortion and the consequent starvation of the hybrid embryo. Hybrids have been produced by dissecting embryos from young ovules shortly after pollination then growing them on media containing nutrients and hormonal substances, a method known as embryo rescue. Endosperm failure because of crossing two species with different ploidy levels can be overcome by inducing polyploidy *via* colchicine treatment, or by using a bridge cross with a species at the lower ploidy level, followed by chromosome doubling, then subsequent hybridization with the cultivated species. Sterility can also be overcome by chromosome doubling using colchicine.

Interspecific and intergeneric crosses are mostly used to produce new allopolyploid forms or to increase the genetic variation within the species by introgression. Moreover, the production of haploids *via* chromosome elimination, resynthesis of allopolyploid species, production of alloplasmic lines, and analysis of genomes may be important objectives. Plant breeders are interested in combining characters from different species or transferring genes of one species into another. In several cases, interspecific and intergeneric hybrids have contributed to crop improvement. In wide hybridization, barriers prevent fertilization, disturb the development of embryos or cause hybrid lethality, weakness, sterility, or breakdown in F_2. Attempts to overcome pre- and post-zygotic barriers have been made extensively in wide crosses. Hybrid sterility was overcome in some cases by the production of amphidiploids. The grass species are the most important fodder grasses in temperate climates. Because of several outstanding features in *Lolium multifiorum*, *Lolium perenne* and *Dactylis glomerata*, the production of allopolyploids is of interest to breeders.

7.3.6.1 *Festuca-Lolium* Hybridization

Worldwide, perennial rye grass (*Lolium perenne*) and Italian rye grass (*L. multiflorum*) are important forage grasses of temperate grasslands. Both grasses are quick-growing and provide dense swards in short period of time. The fodder is of high nutritional quality and easily digestible forage however, there is lack of persistency during adverse climatic conditions. The close relative species from genus *Festuca* have high persistency to withstand abiotic and drought stresses but, they have low nutritional quality and low digestibility. *Lolium* and *Festuca* species

Table 7.4 CP content of *Festulolium* hybrid in comparison to *Lolium* and *Festuca* grass

Genotypes	CP (%)	CP yield (g/plant)
Elmet (*Lolium perenne* × *Festuca pratensis*)	12.45%	8.86
Prior (*Lolium multiflorum* × *Festuca pratensis*)	10.55%	7.19
Festuca pratensis	11.49%	7.10
Lolium multiflorum	9.78%	5.22

Source: Akgun et al. (2008)

hybridize naturally and exhibit high frequencies of gene exchange in the hybrid condition. To overcome the sterility of the *Lolium* × *Festuca* hybrids and the genetic instability of the Italian x perennial rye grass hybrid, the chromosome number of interspecific hybrids has been doubled using colchicine to produce fertile allopolyploids.

Festulolium hybrids, the intergeneric hybrids, between *Lolium* and *Festuca* species are being used to broaden the gene pool and to assist forage breeder with options to combine nutritional quality traits with ability to withstand a wide range and adverse climatic conditions. *Festulolium* varieties have high forage quality. Akgun et al. (2008) demonstrated that *Festulolium* cv. Elmet had higher CP content and CP yield than *Lolium* and *Festuca* individually (Table 7.4). Ostrem et al. (2015) analyzed two *Festulolium* hybrid Felopa and Hykor for nutritional quality and revealed that nutritional quality of *Festulolium* cv. Felopa was equal or better than that of the perennial rye grass. In addition to this, *Festulolium* hybrids have shown high degree of resistance to environmental stress, thus can be useful to improve grassland productivity, persistency, and benefit incomes. Humphreys et al. (2014) have reported that backcrossing of *Festulolium* has generated novel hybrids with more stable protein content in comparison to parents.

7.3.6.2 *Napier-Bajra* Hybrids

Napier-Bajra hybrid (NB hybrid) is the result of crossing common Napier grass with bajra (pearl millet). It is a triploid (AA′B genomes) interspecific hybrid ($2n = 3x = 21$). Hybrid Napier is a perennial, heavy tillering grass with leafiness of bajra with height of 3 to 4 m and leaves particularly near stem are profusely hairy and stiff. It provides nutritional fodder round the year and contains 8 to 10% CP content, 30 to 35% CF, 8 to 10% ash with calcium and phosphorus in optimum amount. It is useful in cut-and-carry system for stall-feeding systems and also good for hay and silage-making. NB hybrid performs well in areas having temperatures above 15 °C. In recent years, NB hybrids are becoming popular among farmers due to their wider adaptability under different agroclimatic conditions, low cost of cultivation, tolerance to grazing /damages by wild animals, propagation by vegetative materials, and high response toward manure/fertilizer applications along with timely irrigations/rainfall. Since the development of first NB hybrid, many BN hybrids [CO-1, CO-2, CO-3, IGFRI 5, IGFRI 7 and IGFRI 10, PBN-83, PGG-518, PGG-101, PG-616, PBN-233, Hybrid Napier-3 (Swetika), BNH-3, BNH-6, BNH-10, NB-21, NB-37, Suguna, Supriya, Sampoorna, KKM-1, APBN-

1, Yashwant (RBN-9)] have been developed consequential to research and development activities of ICAR institutes and state agricultural universities. These NB hybrids are superior in quality, more palatable, nutritious, high-yielding, quick-growing, less hairy, and with better regeneration capacity. Gate et al. (2018) carried out a study to compare nutritional quality of ten NB hybrids with their parents and checks. Hybrids were found statistically superior for various parameters studied. GB (var. giant bajra) × GBN-2001-4-2, GB × FD-444, GB × GBN-2001-8 and GB × GBN-2001-10 were superior over their parents. The highest dry matter was recorded in hybrids such as GB × FD-444 and GB × FD-436, whereas the digestibility was highest for GB × GBN-2001-4-2 hybrid. NB hybrids are being cultivated in different parts of the country by farmers on arable lands, field bunds, and common grazing lands for fodder production and conservation.

7.3.6.3 *Lolium-Dactylis* Hybridization

Lolium perenne and *Dactylis glomerata* are important forage species in temperate regions belonging to the grass family Poaceae. They are cross-pollinated species and present a broad genetic variance in varieties which are synthetics. These species are the excellent models to study mechanisms that lead to genetic changes in grassland populations.

Lolium-Dactylis hybrids are the interspecific hybrids developed to combine traits of both *Lolium* species and *Dactylis glomerata. Lolium multiflorum and Lolium perenne* can be hybridized with *Dactylis glomerata* (tetraploid) and vice versa with embryo rescue technique (Oertel et al. 1996). Oertel et al. (1996) hybridized *Lolium multiforum* and *Dactylis glomerata in vivo*. The viable plant was obtained from the cross *L. multiforum* cv. Malmi × *Dactylis glomerata*. The morphological characteristics of the hybrid were mostly intermediate when compared with their parents. In wide crosses, the endosperm is often lacking or disturbed and the immature embryos abort. Matzk (1991) showed the effect of different auxins in stimulating embryo development up to the stage of organ differentiation in the absence of endosperm. Thus, the regulation of embryogenesis by the endosperm can be simulated by exogenous auxin application and embryo rescue. Auxin treatment improves opportunities for overcoming interspecific and intergeneric cross-barriers. For embryo rescue, embryos containing caryopsis are dipped in 100 ppm aqueous solution of Dicamba (3,6-dichloro-2-methoxy benzoic acid) 2 days after pollination to stimulate embryo development (Matzk 1996). Eighteen days after pollination, the mature embryos were excised and placed on Kruse-medium (Kruse 1974). The production of amphidiploid hybrids using autotetraploid parents is a promising approach in wide crosses. In contrast to crosses between diploid parents, colchicine treatment of hybrids is not necessary to restore regularity in meiosis. The *Lolium multiforum* × *Dactylis glomerata* hybrid shows satisfying meiotic regularity but this situation does not result in high fertility.

The production of one hybrid and two BC plants between *Lolium multiforum* and *D. glomerata* could be the first step for transferring valuable characters from *D. glomerata* into *Lolium multiforum*. *Dactylis glomerata* is a high-yielding fodder grass with drought resistance, early spring growth, and low nutrient requirements.

Since *Lolium multiflorum* and *Dactylis glomerata* are related, more hybrids have to be produced for breeding purposes. The successful hybridization between *Festuca arundinacea* Schreb and *Dactylis glomerata* L. has been reported by Matzk (1981).

7.3.6.4 *Trifolium* Interspecific Hybridization

Wide hybridization has been practised in many crops of economic importance for their improvement. The hybridization of white clover with wild *Trifolium* relatives facilitates the introgression of genes for adaptive traits to enable improved performance in suboptimum environments.

The genus *Trifolium* is comprised of 250 to 300 species, more than 20 of which are cultivated as forages (Williams and Nichols 2011). The eight species and white clover within the white clover complex appear closely related and are therefore potential sources of genetic variation through interspecific hybridization (Nichols 2012). White clover (*Trifolium repens*) has been successfully hybridized with six *Trifolium* species such as *Trifolium nigrescens*, *Trifolium uniflorum*, *Trifolium argutum*, *Trifolium occidentale*, *Trifolium isthomocarpum*, and *Trifolium ambiguum*. In crosses with *Trifolium repens* × *Trifolium nigrescens* and *Trifolium repens* × *Trifolium uniflorum*, immature embryos are mainly excised 12 to 25 days after pollination and hybrid plants are raised. *Trifolium nigrescens* is part of the complex of species closely related to white clover and until recently was thought to be the source of one of its genomes (Badr et al. 2002). In recent years, crosses between *Trifolium repens* and *Trifolium nigrescens* have been practised to explore the evolutionary history of the genus and for the improvement of white clover. A ploidy series has been developed by backcrossing from a single F_1 hybrid (Hussain et al. 1997) and this approach has considerable potential for the introgression of desirable traits in white clover. The interspecific hybridaization of *Trifolium repens* with other *Trifolium* species have revealed increased pest and disease resistance, drought tolerance, and nutrient use efficiency of white clover.

7.4 Techniques for the Development of Improved Forage Crop Varieties

Interspecific and intergeneric hybridization program have been adopted for the improvement of forage crops for various desirable attributes including nutritive quality. Wide hybridization has also been practiced for the introgression of yield as well as forage quality attributes. However, wide hybridization has its own limits and needs to conquer post- and pre-fertilization barriers to develop hybrids. These barriers could be overcome using various biotechnological tools complementing conventional breeding. The embryo rescue, callus culture, somatic hybridization are important techniques in overcoming the fertility barriers in forage crops. Numerous wide hybridization programs have been practised to improve the nutritive value such as *Lolium-Festuca* hybrids, *Lolium-Dactylis*, *Trifolium* interspecific hybridization, etc.

7.4.1 Embryo Rescue

Tall fescue has a wider distribution than rye grass and generally greater adaptability to climatic and edaphic conditions. Although, superior to rye grass in persistency and ability to withstand extremes of temperature and water availability, tall fescue is characterized by poor establishment, low seedling vigor, and lower nutritive value than rye grass. It has long been a goal for forage breeders to combine the stress-tolerant characteristics of *Festuca* species with the earliness and high nutritional traits of *Lolium* species. Through breeding efforts, breeders have become successful in developing *Festulolium* hybrids with high palatability and forage quality with higher resistance to disease and abiotic stresses, such as cold and drought. The rate of success in *Lolium-Festuca* complex is low unless the embryo rescue technique is adopted. The reason for less success of hybrid fertility is attributed to the genomic imbalance or genomic incompatibility between the species or genera. Pasakinskine et al. (1997) produced amphiploids by crossing *Lolium multiflorum* (diploid) var. Prima Roskilde and var. S.22 with *Festuca arundinacea* var. Barundi. Embryo rescue was followed after 16 to 18 days of pollination. F_1 embryos were cultured on Lismaier-Skoog medium containing 30 g sucrose, 0.4 mg IAA, 0.25 mg kinetin, and 7 g agar per liter. The first successful *Trifolium* embryo culture was reported by Keim (1953) after obtaining normal mature plants by culturing immature embryos (8 days post-pollination) of red clover (*Trifolium pratense*).

7.4.2 Callus Culture

For the initiation of regenerable callus culture in forage grasses, different explants containing meristematic cells or showing a developmental proximity to the embryo have been used, for example, isolated mature and immature embryos (Boyd and Dale 1986), caryopses (Franklin et al. 1990), shoot apices (Osuna-Avila et al. 1995), young leaf bases (Mucciarelli et al. 1993), mesocotyls and leaf stem segments (Shatters et al. 1994). In *Lolium-Festuca* complex, callus culture of segments of sterile F_1 hybrids, *Lolium multiflorum* × *Festuca arundinacea* and *Lolium multiflorum* × *Festuca pratensis* crosses regenerate when cultured on Murashige and Skoog medium supplemented with growth regulators. Regenerated plants from former crosses had doubled chromosome number $2n = 56$ (Zwierzykowski et al. 1985). *In vitro* tissue cultures in forage grasses and legumes have shown generating adventitious shoots, and somatic embryos. In forage legumes, regeneration of alfalfa plants from *in vitro* cultures is achieved exclusively *via* somatic embryogenesis, while regeneration, in other forage legumes such as widely studied species of *Trifolium* and *Lotus*, has been achieved most often *via* shoot organogenesis (McKersie and Brown 1997).

7.4.3 Meristem Culture

Efficient methods for the *in vitro* culture of vegetative and floral apices have been developed for different forage grasses and legumes such as cocksfoot (*Dactylis glomerata*), fescues, rye grasses, timothy (*Phleum pratense*) (Dale and Dalton 1983), alfalfa (Cheyne and Dale 1980) and clovers (Barnett et al. 1975). Long-term storage of white clover meristems following cryogenic techniques has been developed. Meristem culture further enhances development of doubled haploids or haploid sterile plants with meiotic abnormalities following chromosome doubling. The virus elimination from infected plants, the limited scope for newly induced genetic variation, and an accelerated tiller production by the precocious development of shoots from axillary buds make the *in vitro* culture of vegetative and floral apices an attractive tool. Furthermore, *in vitro* cultured vegetative and floral apices may be used as targets for genetic transformation using micro-ballistic methods (Perez-Vicente et al. 1993).

Forage crops are crucial components of livestock-agriculture farming system. As primary producers, forage crops provide essential nutrients to livestock and convert into valuable product. In comparison to food crops, forage crops have been considered of low priority, which is mainly due to the fact of few research and development programs and weak policy support. Although, success has been achieved using traditional breeding approaches for improvement of forage crops, new tools have been developed which can boost the improvement programs on forage crops.

Forage improvement programs are difficult to conduct in comparison to food crops with lack of information on variability and useful genetic stocks for various traits. Therefore, concerted and precise efforts are required to collect, evaluate, cataloguing, and maintenance of forage germplasm. For regular supply of green fodder for a longer period, multicut varieties having profuse tillering, quick regeneration, faster growth, and capability of giving a minimum of four to five cuttings need to be developed. The high yield and multicut potential of forages could be further exploited in breeding programs to develop highly adapted and high-yielding stable multicut varieties of forages. Further, forage improvement programs should simultaneously focus on stability in production potential and nutrient content. The identification and development of male sterile lines and restorers suitable for hybrid production could be taken up on an intensive scale. The future strategies should be environmentally and nutritionally focused priorities for sustainable agriculture, prevailing challenge of climate change, and increased biodiversity. More efforts are required to conduct fundamental research to identify key traits and their improvement and development of improved varieties. There is an ample scope for application of molecular breeding and biotechnological approaches in forage breeding for faster and targeted development of elite forage varieties.

References

Akgun İ, Tosun M, Şengül S (2008) Comparison of agronomic characters of festulolium, festuca pratensis Huds. and *Lolium multiflorum* lam. Genotypes under high elevation conditions in Turkey. Bangladesh J Bot 37(1):1–6

Akin AE, Hanna WW, Rigsby LL (1986) Normal-12 and Brown midrib- 12 sorghum. I. Variations in tissue digestibility. Agron J 78:827–832

Badr A, Sayed-Ahmed H, El-Shanshouri A, Watson L (2002) Ancestors of white clover (*Trifolium repens* L.), as revealed by isozyme polymorphisms. Theor Appl Genet 106:143–148

Barnett W, Gibson PB, Sepo A (1975) A comparison of heat treatment and meristem-tip culture for obtaining virus-free plants of *Trifolium repens*. Plant Dis Rep 59:834–837

Boyd LA, Dale PJ (1986) Callus production and plant regeneration from mature embryos of *Poa pratensis* L. Plant Breed 97:246–254

Burton GW, Gates RN, Hill GM (1993) Registration of 'Tifton 85' bermudagrass. Crop Sci 33: 644–645

Casler MD, Brummer C (2008) Theoretical expected genetic gains for among-and-within-family selection methods in perennial forage crops. Crop Sci 48:890–902. https://doi.org/10.2135/cropsci2007.09.0499

Casler MD, Carlson IT (1995) Smooth bromegrass. In: Barnes RF, Miller DA, Nelson CJ (eds) Forages: an introduction to grassland agriculture. Iowa State University Press, Ames, IA, pp 313–324

Casler MD, Van Santen E (2010) Breeding objectives in forages. In: Boller B, Posselt UK, Veronesi F (eds) Handbook of plant breeding. Fodder crops and amenity grasses. Springer, New York, pp 115–136

Cherney JHD, Cherney JR, Akin DE, Axtell JD (1991) Potential of brown-midrib, low-lignin mutants for improving forage quality. Adv Agron 46:157–198

Cheyne VA, Dale PJ (1980) Shoot tip culture in forage legumes. Plant Sci Lett 19:303–309

Conaghan P, Casler M (2011) A theoretical and practical analysis of the optimum breeding system for perennial ryegrass. Irish J Agric Food Res 50:47–63

Coors JG, Lowe CC, Murphy RP (1986) Selection for improved nutritional quality of alfalfa forage. Crop Sci 26:843–848

Dale PJ, Dalton SJ (1983) Immature inflorescence culture in *Lolium, Festuca, Phleum and Dactylis*. Z Pflanzenphysiol 111:39–45

Ehlke NJ, Casler MD (1985) Anatomical characters of smooth bromegrass clones selected for *in vitro* dry matter digestibility. Crop Sci 25:513–517

Escaray FJ, Passeri V, Babuin FM, Marco F, Carrasco P, Damiani F, Pieckenstain FL, Paolocci F, Ruiz OA (2014) *Lotus tenuis* × *Lotus corniculatus* interspecific hybridization as a means to breed bloat-safe pastures and gain insight into the genetic control of proanthocyanidin biosynthesis in legumes. BMC Plant Biol 14:40. https://doi.org/10.1186/1471-2229-14-40

Franklin CI, Trieu TN, Gonzales RA (1990) Plant regeneration through somatic embryogenesis in the forage grass Caucasian bluestem (*Bothriochloa caucasica*). Plant Cell Rep 9:443–446

Fribourg HA, Hoveland CS, Gwinn KD (1991) Tall fescue and the fungal endophyte: a review of current knowledge. Tenn Farm Home Sci 160:30–37

Gate DV, Damame SV, Gore SB (2018) Assessment of forage nutritional quality of B × N hybrids between giant bajra and napier grass. Forage Res 43:279–282

Godshalk EB, Timothy DH, Burns JC (1988) Application of multistage selection indices to crop improvement. Crop Sci 28:23–26

Hall MB, Huntington GB (2008) Nutrient synchrony: sound in theory, elusive in practice. J Anim Sci 86:E287–E292

Hanna WW, Bashaw C (1987) Apomixis: its identification and use in plant breeding. Crop Sci 27: 1136–1139

Hanna WW, Monson WG (1980) Yield, quality, and breeding of pearl millet x napiergrass interspecific hybrids. Agron J 72:358–360

Humphreys MW, Thomas H (1993) Improved drought resistance in introgression lines from *Lolium multiflorum* x *Festuca arundinacea* hybrids. Plant Breed 111:155–161

Humphreys MW, O'Donovan SA, Farrell MS, Gay AP, Kingston-Smith AH (2014) The potential of novel Festulolium (2n = 4x = 28) hybrids as productive, nutrient-use-efficient fodder for ruminants. Food Energy Secur 3:98–110

Hussain SW, Williams WM, Mercer CF, White DWR (1997) Transfer of clover cyst nematode resistance from *Trifolium nigrescens* Viv. To *Trifolium repens* L. by interspecific hybridisation. Theor Appl Genet 95:1274–1281

Kaushal P, Zadoo SN, Malaviya D, Roy A (2005) Apomixis research in India: past efforts and future strategies. Curr Sci 89:1092–1096

Kaushal P, Dwivedi KK, Radhakrishna A, Srivastava MK, Kumar V, Roy AK, Malaviya DR (2019) Partitioning Apomixis components to understand and utilize gametophytic Apomixis. Front Plant Sci 10:256. https://doi.org/10.3389/fpls.2019.00256

Keim WF (1953) Interspecific hybridization in *Trifolium* utilizing embryo culture techniques. Agron J 45:601–606

Kingston-Smith AH, Bollard AL, Thomas BJ, Brooks AE, Theodorou MK (2003) Nutrient availability during the early stages of colonization of fresh forage by rumen micro-organisms. New Phytol 158:119–130

Kruse A (1974) An *in vivo/vitro* embryo culture technique. Hereditas 77:219–224

Leão FF, Davide LC, Campos JMSD, Pereira AV, Bustamante FO (2011) Genomic behavior of hybrid combinations between elephant grass and pearl millet. Pesq Agrop Brasileira 46:712–719

Lutts S, Ndikumana J, Louant BP (1994) Male and female sporogenesis and gametogenesis in apomictic *Brachiaria brizantha*, *Brachiaria decumbens* and F_1 hybrids with sexual colchicine induced tetraploid *Brachiaria ruziziensis*. Euphytica 78:19–25

Marshall AH, Bryant D, Latypova GA, Hauck B, Olyott P, Morris P, Robbins MP (2008) A high throughput method for the quantification of proanthocyanidins in forage crops and its application in assessing variation in condensed tannin content in breeding programmes for *Lotus corniculatus* and *Lotus uliginosus*. J Agric Food Chem 56:974–981

Matzk F (1981) Successful crosses between *Festuca arundinacea* Schreb. and *Dactylis glomerata* L. Theor Appl Genet 60:119–122

Matzk F (1991) A novel approach to differentiated embryos in the absence of endosperm. Sex Plant Reprod 4:88–94

Matzk F (1996) Hybrids of crosses between oat and Andropogoneae or Paniceae species. Crop Sci 36:17–21

McKersie BD, Brown DCW (1997) Somatic embryogenesis and artificial seeds. In: McKersie BD, Brown DCW (eds) Biotechnology and the improvement of forage legumes. CAB International, Wallingford, pp 111–143

Mendler-Drienyovszki N, Cal AJ, Dobránszki J (2013) Progress and prospects for interspecific hybridization in buckwheat and the genus *Fagopyrum*. Biotechnol Adv 31:1768–1775

Miller LA, Moorby JM, Davies DR, Humphreys MO, Scollan ND, MacRae JC, Theodorou MK (2001) Increased content of water-soluble carbohydrate in perennial ryegrass (*Lolium perenne* L.): milk production from late-lactation dairy cows. Grass Forage Sci 56:383–394

Moorby JM, Evans RT, Scollan ND, MacRae JC, Theodorou MK (2006) Increased content of water-soluble carbohydrate in perennial ryegrass (*Lolium perenne* L.). evaluation in dairy cows in early lactation. Grass Forage Sci 61:52–59

Morita R, Kusaba M, Iida S, Yamaguchi H, Nishio T, Nishimura M (2009) Molecular characterization of mutations induced by gamma irradiation in rice. Genes Genet Syst 84:361–370

Mucciarelli M, Gallino M, Scannerini S, Maffei M (1993) Callus induction and plant regeneration in *Vetiveria zizanioides*. Plant Cell Tiss Org Cult 35:267–271

Nichols SN (2012) Introgression of root and shoot characteristics in *Trifolium repens* x *Trifolium uniflorum* interspecific hybrids: a thesis submitted in partial fulfilment of the requirements for the degree of doctor of philosophy at Lincoln University. Lincoln University, Lincoln. https://hdl.handle.net/10182/5091

Obok EE, Aken'Ova ME, Iwo GA (2012) Forage potentials of interspecific hybrids between elephant grass selections and cultivated pearl millet genotypes of Nigerian origin. J Plant Breed Crop Sci 4:136–143

Oertel C, Matzk F (1999) Introgression of crown rust resistance from Festuca spp. into *Lolium multiflorum*. Plant Breed 118:491–496

Oertel C, Fuchs J, Matzk F (1996) Successful hybridization between *Lolium* and *Dactylis*. Plant Breed 115:101–105

Ostrem L, Volden B, Steinshamn H, Volden H (2015) *Festulolium* fibre characteristics and digestibility as affected by maturity. Grass Forage Sci 70:341–352

Osuna-Avila P, Nava-Cedillo A, Jofre-Garfias AE, Cabrera-Ponce JL (1995) Plant regeneration from shoot apex explant of foxtail millet. Plant Cell Tiss Org Cult 40:33–35

Parsons AJ, Rowarth JS, Rasmussen S (2011) High-sugar grasses. CAB Rev 6:1–12

Parveen I, Threadgill MD, Moorby JM, Winters A (2010) Oxidative phenols in forage crops containing polyphenol oxidase enzymes. J Agric Food Chem 58:1371–1382

Pasakinskine I, Anamthawat-JoÂnsson K, Humphreys MW, Jones RN (1997) Novel diploids following chromosome elimination and somatic recombination in *Lolium multiflorum* × Festuca arundinacea hybrids. Heredity 78:464–469

Pereira AV, Lédo FJS (2008) Melhoramento genetico de *Pennisetum purpureum*. In: Valle CB, Jank L (eds) Melhoramento de forrageiras tropicais Resende RMS. Embrapa, Campo Grande, pp 89–116

Perez-Vicente R, Wen XD, Wang ZY, Leduc N, Sautter C, Wehrli E, Potrykus I, Spangenberg G (1993) Culture of vegetative and floral meristems in ryegrasses: potential targets for microballistic transformation. J Plant Physiol 142:610–617

Peters M, Lascano C (2003) Forage technology adoption: linking on-station research with participatory methods. Trop Grassl 37:197–203

Piluzza G, Sulas L, Bullitta S (2014) Tannins in forage plants and their role in animal husbandry and environmental sustainability: a review. Grass Forage Sci 69:32–48

Pinheiro AA, Pozzobon MT, Valle CB, Penteado MIO, Carneiro VTC (2000) Duplication of the chromosome number of diploid *Brachiaria brizantha* plants using colchicine. Plant Cell Rep 19: 274–278

Radojevic I, Simpson RJ, St. John JA, Humphreys MO (1994) Chemical composition and in vitro digestibility of lines of *Lolium perenne* selected for high concentrations of water-soluble carbohydrate. Aust J Agr Res 45:901–912

Rauf S, Sienkiewicz-Paderewska D, Malinowski DP, Hussain MM, Niazi IAK, Kausar M (2016) Forages: ecology, breeding objectives and procedures. In: Al-Khayri J, Jain S, Johnson D (eds) Advances in plant breeding strategies: agronomic, abiotic and biotic stress traits. Springer, Cham, pp 149–201

Reddy MSS, Chen F, Shadle G, Jackson L, Aljoe H, Dixon RA (2005) Targeted down- regulation of cytochrome P_{450} enzymes for forage quality improvement in alfalfa (*Medicago sativa* L.). Proc Natl Acad Sci U S A 102:16573–16578

Řepková J, Nedělník J, Jakešová H et al (2013) Interspecific hybrids *Trifolium pratense* × *Trifolium medium* as the source of new diversity. In: Proceeding of 30th meeting of EUCARPIA fodder crops and amenity grasses section, 12–16 May 2013, Vrnjačka Banja Serbia, pp 25–30

Resende RMS, Jank L, Valle CB, Bonato ALV (2004) Biometrical analysis and selection of tetraploid progenies of *Panicum maximum* using mixed model method. Pesq Agrop Brasileira 39:335–341

Roy AK, Malaviya DR, Kaushal P (2015) Strategies to improve fodder legumes with special emphasis on clover and medics. In: Roy MM, Malaviya DR, Yadav VK, Singh T, Sah RP, Vijay D, Radhakrishna A (eds) Proceedings: Sustainable use of Grassland Resources for Forage Production, Biodiversity and Environmental Protection, The 23rd International Grassland

Congress New Delhi, India November 20–24, 2015. Range Management Society of India, pp 327–332
Shatters RG, Wheeler RA, West SH (1994) Somatic embryogenesis and plant regeneration from callus cultures of "Tifton 9 Bahiagrass". Crop Sci 34:1378–1384
Sohoo MS, Beri SM, Bhardwaj BL (1985) Radiation induced late flowering and high yielding mutant in berseem (*Trifolium alexandrinum* L.). J Nuclear Agric Biol 14:136–138
Stewart A, Hayes R (2011) Ryegrass breeding—balancing trait priorities. Irish J Agric Food Res 50: 31–46
Sullivan ML, Hatfield RD (2006) Polyphenol oxidase and o-diphenols inhibit postharvest proteolysis in red clover and alfalfa. Crop Sci 46:662–670
Surprenant J, Michaud R, Allard G (1990) Effect of one cycle of divergent phenotypic selection for crude protein, digestibility, and digestible yield in timothy. Can J Plant Sci 70:757–765
Vogler RK, Tesso TT, Johnson KD, Ejeta G (2009) The effect of allelic variation on forage quality of brown midrib sorghum mutants with reduced caffeic acid O-methyl transferase activity. Afr J Biochem Res 3:70–76
Williams WM (2014) *Trifolium* inter-specific hybridisation: widening the white clover gene pool. Crop Pasture Sci 65(11):1091–1110
Williams WM, Nichols SN (2011) *Trifolium*. In: Kole C (ed) Wild crop relatives: genomic and breeding resources, legume crops and forages. Springer-Verlag, Berlin, Heidelberg, pp 249–272
Winters A, Minchin F (2002) The effect of PPO on the protein content of ensiled red clover. In: Gechie LM, Thomas C (eds) Proceedings of 13th international silage conference SAC, 11--13 September 2002. Auchincruive, Ayr, Scotland, pp 84–85
Zwierzykowski Z, Slusarkiewicz-Jarzina A, Rybczynski JJ (1985) Regeneration of plants and chromosome doubling in callus cultures of *Lolium multiflorum* lam. (2n = 14) × *Festuca pratensis* Huds. (2n = 14) hybrid. Genet Pol 26:187–197

8 Nutritional and Anti-Nutritional Constituents in Forages

Nutritional quality of forages has direct influence on health and productivity of livestock. High crude protein (CP) in forages makes rumen microflora more active and increases the efficiency of livestock to digest and assimilate the nutrients. Feeding studies have revealed that the diets sufficient in protein quantity but deficient in protein quality have negative impact on animal growth and production. Anti-nutritive compounds often show ill effects on animal health by interfering directly or indirectly in utilization of available nutrients in forages. Leguminous forages have better nutritive profile than grasses, which is mainly attributed to high protein content, balanced amino acid profile and good proportion of minerals and vitamins. The knowledge about different nutritional as well as anti-nutritional constituents will provide better option for the selection of forages to prepare nutritious and wholesome diet for livestock.

8.1 Nutritional and Anti-Nutritional Constituents of Forages and Their Importance in Animal Nutrition

Feeding value and animal performance depend on three major factors which are intake, digestibility, and assimilation of nutrients. Forage intake has a positive relationship with digestibility and affects the intake potential of the animal. In forages, lignin, cellulose, and hemicellulose are the important cell wall component while NDF and ADF form the fiber component. More than 60% NDF content denotes low quality of forage as it interferes with digestion and assimilation of nutrients in animal. Different nutritional and anti-nutritional components in forages and their impact on production and productivity of livestock have been discussed below:

R. Katoch, *Nutritional Quality Management of Forages in the Himalayan Region*, https://doi.org/10.1007/978-981-16-5437-4_8

8.1.1 Nutritional Components

8.1.1.1 Dry Matter

Dry matter is not directly related to the nutritional component but refers to material remains after the removal of moisture from forages. In other words, dry matter (DM) is the total of all the solid components in a plant without moisture content. Therefore, dry matter provide estimates about the nutrients vital for animals, such as proteins, sugars, lipids, cellulose, starch, structural compounds, pigments, and minerals. Determination of dry matter content in forages is important to ensure that animals are receiving proper amount of nutrients through their diet. It is an indicator of the amount of nutrients that are available to animals in feed. Livestock need to consume a certain amount of dry matter daily to maintain health and production. Any possible increase in dry matter intake is associated with higher production. Dry matter is calculated as the difference between the total weight and the moisture content. In forages, feed nutrient composition is represented on DM basis. Most dry feeds often have DM content around 85 to 92% and the wet feeds have 25 to 35% DM content. Intermediate feeds generally have approximately 50 to 70% DM. Dry matter digestibility is estimated on the basis of animal feeding trials and forage ADF content. When ADF content in forages increases, the amount of digestible dry matter decreases. The optimum amount of feed must be calculated and adjusted on the basis of dry matter content and therefore specific amounts of nutrients provided to animal could be calculated. In forage crops the dry matter content is calculated as

$$\text{DM} = \text{Total weight} - \text{moisture content}$$

The amount of digestible dry matter of forages is calculated as:

$$\%\text{DDM} = 88.90 - (0.779 \times \%\text{ADF})$$

It is necessary to know both about the DM and moisture content of forage crops at different stages to determine appropriate harvest time, monitor quality and prevent spoilage of forages. Though dry matter increases with maturity but if crops are harvested too late, they become more fibrous and unpalatable. Moreover, there is a loss in nitrogen levels, which is an important nutrient. The estimation of dry matter in forages is also necessary for proper ensiling and baling of feeds.

8.1.1.2 Crude Protein

Proteins are important constructive and energetic compounds, which also play a bioactive role besides being the precursors of biologically active peptides with various physiological functions. Being rich in the protein content, leguminous forages are important component of livestock diet. Crude protein content in forages is measured by standard Kjeldahl procedure (Ruckle et al. 2017). This measures % nitrogen content which is multiplied by 6.25 to calculate crude protein content in forages.

Crude protein can be divided into ruminally undegraded protein (RUP) and ruminally degraded protein (RDP). True protein differs from nonprotein N, which does not have an amino acid profile but can be used as N source by ruminal microorganisms. Nonprotein N (NPN) is represented by the nitrogen present in nucleic acids and ammonia.

One of the well-known NPN compound used for livestock feeding is urea. These compounds are quickly degraded and converted into microbial protein by rumen microorganisms. RDP provides a mixture of peptides, free amino acids and ammonia that are used by microorganisms for growth and synthesis of microbial proteins, which is an important source of nutrition for the ruminant supplying majority of amino acids entering the small intestine. RUP is the second-most important source of absorbable amino acids to the animal. Feeds with high RUP content include heat-treated soybean meal and cottonseed meal. Endogenous CP comprises salivary and digestive secretions of ruminants, as well as sloughed epithelial cells. Metabolizable Protein (MP) represents the total amount of protein absorbed by the animal and it is supplied by the diet (RDP and RUP), microbial CP (MCP) and to a much less extent, endogenous CP.

More specifically for ruminants, adequate protein level ($>7\%$ CP) in the diet is required for growth and activity of ruminal microorganisms for producing desired microbial CP content and maximizing ruminal fermentation. In contrast, feeding diets with protein content $\leq 7\%$ CP may result in impaired growth of ruminal microorganisms. This is extremely important because MCP is the main protein source for ruminants and inadequate protein levels negatively impact ruminal function, performance and subsequent productivity of the livestock. Furthermore, not only protein amount, but type of protein in the diet is very important for the rumen microorganisms, given that RDP is used by rumen microorganisms for microbial CP synthesis. The balance between these two types of protein is critical, and in diets where protein is not limiting microbial growth and ruminal functions, feeding RDP in excess may harm the ruminal environment.

The upper limit of CP content in the forages may be up to 21% of DM and beyond this level N losses could be expected. Many types of forages especially legumes may exceed CP limits, therefore N loss is a feature of many grazed forages. Above the threshold limit, potentially large amounts of ingested protein is lost as ammonia (NH_3) across the rumen wall (Cruickshank et al. 1992). Cereals and legumes straw contain protein $<5.0\%$, while rye grass and clover hay contain about 10 to 12% crude protein. On an average, silages contain 7.50 to 9.00% crude protein content. Some micronutrients like proanthocyanidins or condensed tannins also change digestibility of protein, but they inhibit protein degradation in rumen. This may be sometimes advantageous as rapid protein degradation is causative of bloat, however too high tannin content in forages also results in low protein absorption.

8.1.1.3 Amino Acids Composition and Protein Quality

It is well accepted that amino acids are the building blocks of protein and are also the integral part of livestock feeding regimes. The quality of protein is determined by its potential to cover the physiological requirements in terms of amino acids for maintenance and performance. The quality of protein required is different depending on the animal species, age, genotype, and sex as well as on the performance levels. It therefore follows that there are two important factors with respect to protein quality, (a) the amino acid profile which is the ratio of essential amino acids in the protein and (b) their digestibility.

Legumes are rich source of protein and have excellent amino acid profile. For example, Sainfoin is a perennial leguminous forage crop with a high content of crude protein, high palatability and nutritive value. It is suitable for pastoral use due to the presence of condensed tannins and does not cause the animals to swell (bloat) while feeding on the green mass. It can also be used for hay and silage. In addition, sainfoin is also grown in mixtures with grasses. It also contains high lysine, histidine, proline, cystine, glycine, alanine, valine, leucine, tyrosine, phenylalanine, and tryptophan content whereas lucerne contains high aspartic acid, methionine and isoleucine content.

8.1.1.4 Crude Fiber

Crude fiber analysis involves alkali and acid treatments of sample to isolate indigestible portion of cell wall. It estimates the indigestible fraction of feed or those fractions of the feed that are fermented in the hindgut by microbes. Crude fiber includes different insoluble carbohydrates that are associated with plant cell wall and are resistant to action of digestive enzymes. Crude fiber is made up of plant cell structural components, including cellulose, hemicellulose, lignin, and pectin. These constituents appear in different ratios in various forages and their levels are affected by stage of maturity. For non-ruminant, crude fiber is of little value, however, it is important for maintaining hindgut health and microbial population. Crude fiber is important in the diets of ruminant animals, which could ferment its large portion. Long stem hay and pasture grasses contain over 20% crude fiber, whereas most grain mixes, even so-called "complete" feeds, contain less than 12% fiber.

To determine crude fiber in feed, a sample is dried, boiled in weak sulfuric acid solution (1.25% H_2SO_4), and filtered. The residue is boiled in a weak alkali solution (1.25% NaOH) and filtered and the remaining residue is dried and ashed. The difference between the filtered dried sample and ash is crude fiber. The two boiling processes simulate the pH conditions of the digestive tract, acidic in the stomach and alkaline in the small intestine. However, the enzymatic digestion in the digestive tract is not simulated in the procedure. A major problem with this procedure is that the acid and base solubilize some of the true fiber (particularly hemicellulose, pectin, and lignin) and some cellulose portion is partially lost in the process. Hence crude fiber underestimates true fiber in the test material. Most laboratories have phased out the crude fiber term and replaced it with the detergent fiber system (Table 8.1).

8.1.1.5 Ether Extract

Ether-soluble materials in feed include different organic compounds are soluble in organic solvents. In animal feeds, ether extract may include fats, fatty acid esters and fat-soluble vitamins, and hence are often referred to as crude fat. In addition to fats and oils, it includes plant pigments and fat-soluble vitamins. The primary objective

Table 8.1 Crude fiber content in different forages

Forages	CF (%) range
Berseem	12.00–25.50
Cowpea	14.50–26.50
Alfalfa	15.70–28.50
Albizzia	21.5–30.0
White clover	15.00–27.80
Grewia	20.95–24.50
Bauhinia	24.36–30.50
Birdsfoot trefoil	21.00–28.50
Rice bean	18.70–22.50
Red clover	10.00–25.00
Celtis	18.10–28.50
Sainfoin	18.50–30.0

of ether extracts is to isolate the fraction of the feedstuff that has a high caloric value. A portion of dried feed sample is boiled in ether for 4 h. Since fats are soluble in ether, ether extract is equivalent to fat. Provided the ether extract contains fats and fatty acid esters, this approach is valid. However, in samples that contain high level of other compounds soluble in organic solvents, such as plant waxes or resins, it may not give a true estimate of feed caloric value. However, this error is generally small in typical animal feedstuffs. Overall, this test does not indicate anything about the quality of the fat in the feed.

If a feed is high in fat, it may be susceptible to rancidity, which causes off-flavors, low palatability and potential toxic effects. Usually an antioxidant such as vitamin E is added to prevent a feed from becoming rancid. Because fat has more energy per unit of weight compared to carbohydrates, feeds with high fat content are more energy dense and are added to diets to increase calories, provide essential fatty acids and improve palatability in some cases. There is variation in fatty acid profile within forage species, with grasses tending to have more α-linolenic acid when compared to legumes and legumes in turn having high linolenic acid content (Boufaied et al. 2003).

8.1.1.6 Fatty Acids

Forages provide substantial lipids and fatty acids in diet. Lipids represent up to 8% of leaf DM in forages. They are often localized in leaf chloroplasts which contain 22 to 25% lipids on a DM basis. Complex lipids constitute most leaf tissues, mainly as glycolipids and phospholipids. The esterified lipids in forages represent two-thirds of the total lipids (5% of DM); their composition is 33% simple lipids, 50% galactolipids and 17% phospholipids. Fatty acid composition of lipid from forages is dominated by high proportions of polyunsaturated linolenic and linoleic acids (C18:2), but also small amounts of oleic acid (C18:1). In fresh grasses, C18:3 represents between 50 and 75% of total fatty acid content; C18:2 and palmitic acid (C16:0) are the next most abundant. In rye grass, C18:3 fatty acid represents 55 to 66% of total fatty acids content but only 40% in alfalfa.

Grasses have nearly uniform composition across species with a mean of 66% of total fatty acids provided by C18:3, 13% by C18:2, and 14% by C16:0. The fractional contribution of C18:3 fatty acids to total fatty acids is low and more variable in forbs than in grasses. Linoleic and α-linolenic acids are of primary importance in ruminant products where the diet is predominantly forages. Omega-3 fatty acids and conjugated linoleic acid have been shown to reduce cardiovascular disease, diabetes, cancer, and obesity. The primary conjugated linoleic acid found in ruminant products include the cis-9, trans-11 and trans-10, cis-12 isomers, with the former elevated when animals are fed forage-based diets. The diet composition of animals affects the fatty acids composition in the end products. Therefore, analysis of fatty acids in forages is imperative in order to increase the quality of livestock products. The fatty acid composition of important forages of Himalayan region is given in Table 8.2.

8.1.1.7 Ash

Inorganic residue left after ignition at high temperature 550–600 °C represents total ash. Ash content represents inorganic residues in forages and provides a rough estimate of total mineral content (Katoch et al. 2018) where more ash content can imply high level of minerals (Ncube et al. 2015). Mineral composition in forages depends on interaction of many factors including soil, plant species, maturity stage, yield and climatic conditions (McDowell 2003). Ash in forage comes from two sources: internal, e.g., minerals like calcium, magnesium, potassium and phosphorus, and external, e.g., dirt, bedding, sand, etc. The average internal ash content of alfalfa is about 8% and of grasses is about 6%. Additional ash in a hay or silage sample is contamination with dirt, sand, etc. The normal ash content of corn silage is approximately 5.0% of DM. On average cool-season grasses such as orchard grass or fescue harvested as hay or silage have about 7 to 9% ash content and legumes such as alfalfa harvested as hay or silage have average of 10 to 12% ash content. By nature ash or minerals are devoid of protein, calories, energy, or nutrients. However, we measure ash content of forages to calculate nonfiber carbohydrate content (NFC).

$$\text{Non-fiber carbohydrate content}\,[\text{NFC} = 100 - ((\text{NDF} - \text{NDFCP}) + \text{CP} + \text{Fat} + \text{Ash})$$

8.1.1.8 Nitrogen-Free Extract (NFE)

NFE represents nonstructural carbohydrates such as starches and sugars. The major roles of nonstructural carbohydrates in plants are intermediary metabolism, energy transport and energy storage. Nonstructural carbohydrate content varies greatly with the physiological status of the plant. Nonstructural carbohydrates are the primary substrates for microorganisms that grow on poorly preserved forages during storage. Monosaccharides, glucose and fructose, and disaccharides, sucrose and maltose, account for most of the soluble sugars in forages. Sucrose and other soluble sugars are generally present in higher content in legumes than grasses and are usually higher in cool-season grasses than warm-season grasses. Species that accumulate fructans in vegetative tissues generally store starch in their seeds. The degree of

Table 8.2 Fatty acid composition of important forages of the Himalayan region

Forage species	C12:0	C14:0	C16:0	C16:1	C18:0	C18:1	C18:2	C18:3	Total fatty acid content
	(mg g^{-1} of DM)								
Orchard grass	0.06 to 0.10	0.09 to 0.13	2.83 to 3.49	0.28 to 0.41	0.29 to 0.34	0.73 to 0.74	2.60 to 3.47	7.76 to 11.81	14.68 to 20.52
Timothy	0.04 to 0.07	0.10 to 0.13	2.83 to 3.62	0.27 to 0.43	0.28 to 0.41	1.12 to 1.13	3.15 to 4.63	6.76 to 11.16	14.59 to 21.63
Alfalfa	0.08 to 0.12	0.13 to 0.17	4.08 to 4.35	0.37 to 0.48	0.75 to 0.84	1.12 to 1.09	3.53 to 3.67	6.97 to 6.73	17.06 to 17.48
Red clover	0.06 to 0.12	0.12 to 0.17	3.73 to 4.63	0.46 to 0.65	0.74 to 0.88	1.56 to 1.54	3.82 to 4.52	8.61 to 9.39	19.13 to 21.94

Source: Boufaied et al. (2003)

polymerization of fructans varies with species. Species that accumulate long-chain fructans include members of the Aveneae tribe and some species of the Festuceae tribe including *Dactylis glomerata* and *Poa pratensis*. Other species of Festuceae tribe, including smooth *Bromus inermis* Leyss, *Festuca arundinacea*, and *Lolium perenne* as well as *Hordeum vulgare*, accumulate short-chain fructans.

$$\%NFE = \%DM - (\%EE + \%CP + \%Ash + \%CF)$$

8.1.1.9 Acid Detergent Insoluble Nitrogen (ADIN)

Some feeding systems for livestock use ADIN analysis as a measure of N availability. ADIN fraction is assumed to be both nondegradable in rumen and indigestible in small intestine and therefore amount of rumen undegraded protein in feed is predicted from ADIN content.

8.1.1.10 Water-Soluble Carbohydrates (WSC)

Carbohydrates content in forages include monosaccharides, oligosaccharides and fructosans. WSC are completely digestible and have an important role in animal nutrition, as they are a primary source of the readily available energy necessary for efficient microbial fermentation in the rumen. Fermentation of structural carbohydrates also provides energy directly to the microbial population and indirectly to the animal. High WSC content might indicate high fructan levels in grasses or high simple sugars in cereal grains. Carbohydrates are the most abundant class of compounds found in plants account for 50 to 80% of dry biomass of forages (Van Soest 1982). Carbohydrates play a key role in intermediary metabolism, energy transfer and storage. In forages, carbohydrates are important components of cell wall and play important role in maintaining the structural integrity of tissues. In animals, approximately 90% digestion of carbohydrates occurs within rumen, however, digestion of significant carbohydrates content can also occur in the small and large intestines under certain circumstances. The content and classes of carbohydrates isolated from forages vary greatly among genera, species, and to some extent within the species. Success in conserving legumes and grasses as silage often depends on the amount of readily available fermentable carbohydrates present in herbage. If their content is sufficiently high, conditions are favorable for establishment and growth of lactic acid bacteria, which leads to better fermentation and preservation. A minimum of 3.70% WSC in fresh grass is considered necessary to produce good quality silage without additives. WSC content in forages depends on a wide range of factors, including the plant part, ploidy level, plant maturity, diurnal and seasonal effects, temperature, light intensity, growth rate, endophyte, water status, as well as inherent differences between cultivars. WSC content in forage grasses varies considerably ranging from less than 5% to up to 50%. Humphreys (1989) mentioned considerable variation in WSC content of rye grass during both vegetative and reproductive growth phases. The range of variation for WSC content (13.90 to 28.80%) was greater than that variation found for CP or DMD and WSC content. Wilkins and Davies (1994) also reported significant differences for WSC content among four varieties of perennial rye grass under conservation management but the range was relatively small (30.50 to 33.60%). Significant variation for WSC in five phenological stages (vegetative, heading, pollination, milky and dough seed stage) has also been reported by Jafari et al. (2010) in tall wheatgrass (11.20 to 14.40%); Jafari and Rezaeifard (2010) in tall fescue; (8.38 to 12.69%) and by Rezaeifard et al. (2010) in cocksfoot (7.76 to 11.37%), respectively.

8.1.1.11 Total Digestible Nutrients (TDN)

TDN (%) is the sum of digestible crude fiber, digestible crude protein, digestible fat content multiplied by the factor 2.25, and digestible nitrogen-free extract (NFE). TDN is directly related to digestible energy and is often calculated based on ADF. Total digestible nutrients (TDN) and DMI (Dry matter intake) for grasses and legumes is calculated as (Table 8.3):

$$\begin{aligned}TDN_{(\text{legumes})}(\%\text{of DM}) = {} & (CP \times 0.93) + (FA \times 0.97 \times 2.25)\\ & + [NDFn \times (NDFD \div 100)] + (NFC \times 0.98) - 7\end{aligned}$$

$$\begin{aligned}TDN_{(\text{Grasses})}(\%\text{of DM}) = {} & (CP \times 0.87) + (FA \times 0.97 \times 2.25)\\ & + [NDFn \times (NDFDp \div 100)] + (NFC \times 0.98) - 10\end{aligned}$$

where,

NDFn = Nitrogen-free NDF = NDF – NDFCP, also estimated as NDFn = NDF × 0.93
NDFCP = Neutral detergent fiber crude protein
NDFD = 48-h *in vitro* NDF digestibility (% of NDF)
NFC = Nonfibrous carbohydrate (% of DM) = 100 – (NDFn + CP + EE + Ash)
EE = Ether extract (% of DM)
NDFDp = 22.70 + 0.664 × NDFD

Table 8.3 TDN (%) of some important forages of Himalayan region

Forage	Total digestible nutrients (TDN)%
Brachiaria brizantha	53.50
Chloris guyana	41.90
Cynodon dactylon	38.20
Panicum maximum	49.90
Pannisetum purpureum	50.50
Leucaena leucocephala	45.30
Macroptilium atropurpureum	63.40

8.1.1.12 Fat-Soluble Micronutrients

Carotenoids and vitamins A and E are fat-soluble micronutrients that play an important role in maintaining human as well as animal health. Carotenoids, which are present as pigments in plants and have protective effect in prevention of diseases. Moreover, carotenes, especially all β-carotene, possess a pro-vitamin A activity. Vitamin A is necessary for normal vision, cellular differentiation, growth and reproduction. Vitamin E is well known for its antioxidant properties and for promoting immunity. It also improve the nutritional quality of foods, notably of animal products such as milk. By varying the quality of forage fed to ruminants, it is possible to modulate carotenoids, vitamins A and E level in cow's milk.

8.1.2 Anti-Nutritional Constituents

Anti-quality components of forages are different factors inherent in forages that limits the ability of animal to reach their potential for growth, productivity and reproduction. These factors are synthesized during normal metabolism in plants and are useful in combating different biotic and abiotic stresses during growth and development of plants. These anti-nutritional factors are mostly disliked by animals due to bitter taste, poor palatability, and indigestibility. The reason for the presence of anti-nutritional factors in plants seems to be a way of storing nutrients or defending their structure and reproductive elements. Anti-nutritional factors have variable effects on animal productivity and may cause adverse health effects such as bloat, mineral disorders, nitrate toxicity, poisonings from plant toxins such as glycosides or alkaloids, neurological defects, and photosensitization when consumed in large quantities. Lignin and tannins are other important anti-quality attributes which reduce voluntary intake and digestibility of forages. The toxicity of anti-nutrients depends upon their rate of degradation by ruminal microbes. Based on their effects on nutritional value of the feedstuffs and biological responses in animals, anti-nutritional factors are classified into following subgroups (Huisman and Tolman 2001).

- Factors having reductive effect on protein digestion and mobilization, e.g., protease inhibitors, tannins, and saponins
- Factors depressing the mineral utilization, e.g., phytic acid
- Factors associated with immune-stimulatory effects, e.g., lectins, antigenic proteins
- Factors with negative effect on digestion and utilization of carbohydrates, e.g., polyphenolic compounds, α-amylase inhibitors

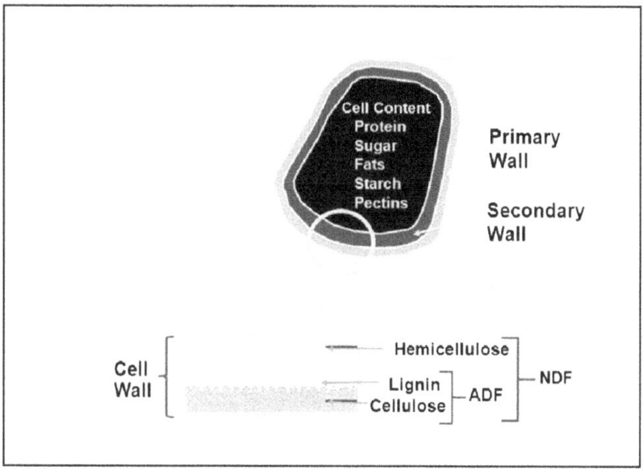

Fig. 8.1 Plant cell wall components

8.1.2.1 Neutral Digestible Fiber (NDF) and Acid Detergent Fiber (ADF)

Levels of NDF and ADF content are critical as their amounts determine the extent of forage digestion in livestock. NDF measures structural components in plant cells, i.e., lignin, hemicellulose, and cellulose (Fig. 8.1). Neutral detergent fiber (NDF; the residue after boiling in a neutral detergent) comprises hemicellulose, cellulose, lignin and ash, commonly referred to as the cell wall or fiber fraction of plants. NDF content in feeds is negatively correlated with their intake and amount of digestible energy available to livestock.

The stage of maturity at harvesting of forages has the greatest influence on NDF digestibility. As forage matures, digestibility of NDF content declines more than 40%. The decline in NDF digestibility in grasses is dramatic with advancing maturity. Alfalfa forage contains around 48% NDF during late bloom stage while during early pre-bloom contains less than 35% NDF. In timothy, at post-head stage, the NDF content is more than 65% and during pre-head stage it is less than 55%. Grass silage and hay have a wide range of NDF digestibility (22 to 75%). For legume forages, NDF content below 40% is desirable, while NDF content above 50% is considered poor. Grasses having NDF content less than 50% are considered of high quality and more than 60% are graded as low quality. ADF content also denotes the quality of forages, where higher values suggest their low quality. In forages, ADF content <35% is desirable. ADF is a measure of the plant components in forages that are the least digestible by livestock, including cellulose and lignin (Fig. 8.1). As ADF content increases in forages, their digestibility decreases, therefore forages with high ADF content are typically lower in energy. Alfalfa contains more than 41% ADF during late bloom stage, while less than 30% during pre-bloom stage.

8.1.2.2 Lignin

Forage quality represents intake of feed, digestibility of feed and efficiency of utilization of digested feed. Lignin, a constituent of secondary cell wall, has direct influence on the digestibility of forages and therefore considered as anti-quality component. As a forage crop matures, its lignin content increases whereas digestibility decreases; a close negative relationship exists between lignin content and digestibility. Thus, lignin counts among the most important chemical components related to forage digestibility and plays role as an indicator for evaluating the digestibility or nutritive value of forages.

Lignin is a complex heteropolymer resulting from oxidative coupling of three main monolignols precursors (*p*-coumaryl alcohol, coniferyl alcohol and sinapyl alcohol) (Rogers and Campbell 2004). The digestive system of ruminants and associated microbial populations are well adapted to obtain nutrients and energy from forages. Lignin content increases with advancing maturity stage of forages. ADL content of forage legumes and perennial grasses varies from 3.95 to 5.18 and 4.50 to 6.75%, respectively (Naydenova 2014). The lignin content in different feeds is given in Table 8.4.

The complex structure of lignin cross-links with cellulosic polysaccharides of cell wall and blocks the access of these polymers to ruminal microbes secreting cell wall degrading enzymes (Fig. 8.2). This makes lignin a major obstacle in forage digestibility. When animals consume less digestible feed, its passage rate through the alimentary canal is quite low as compared to high digestible feed. The slow passage of feed-in gut creates anorexic conditions, which are responsible for reduced feed

Table 8.4 Lignin content in different forages

	Lignin content (%DM)
Fresh grasses	
Lolium perenne (Perennial rye grass)	2.92
Perennial rye grass (Nil Endophyte)	3.38
Festuca arundinacea (Tall fescue)	3.78
Pennisetum clandestinum (Kikuyu)	3.82
Dactylis glomerata (Cocksfoot)	5.13
Paspalum dilatatum (Paspalum)	6.85
Legumes and herbs	
Trifolium repens (White clover)	5.87
Medicago sativa (Lucerne)	6.12
Trifolium pratense (Red Clover)	6.23
Cichorium intybus (Chicory)	6.97
Lotus corniculatus (Birdsfoot trefoil)	7.22
Conserved	
Zea maize (Maize silage)	4.36
Medicago sativa (Lucerne silage)	7.33
Medicago sativa (Lucerne hay)	8.04

Source: Chaves et al. (2002)

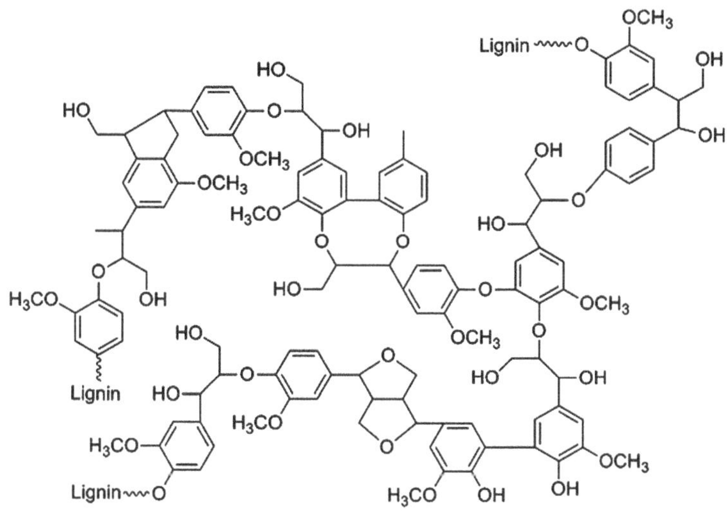

Fig. 8.2 The complexity of lignin structure

intake, weight loss, decreased nutrient utilization thereby decreases the livestock productivity.

Lignin is more prevalent in grass stem than leaf and absent from legume leaves. In contrast to grasses, there are minimal ferulate cross-linkages between lignin and hemicelluloses in legume stems, therefore, lignin is less detrimental to nutritive value of legumes as compared to grasses.

8.1.2.3 Polyphenols

Different phenolic compounds are known to inhibit the activity of digestive enzymes like α-amylase, trypsin, chymotrypsin, and lipase. Besides decreasing the digestibility of proteins, phenolic compounds also reduce the availability of carbohydrates, vitamins, and minerals. Tannins are complex polyphenolic substances which form strong chemical complexes with proteins, sugars and starches that are stable over a pH range from the neutral environment of the rumen (pH 7.0) to acidic conditions in the stomach (pH 3.40). In plants, the ability of tannins to form insoluble complexes with proteins and polysaccharides can effectively reduce herbivory from mammals, birds, reptiles and insects. When herbivores forage on tannin-rich plants, tannin-protein complexes reduces the digestion of forage protein. Tannins have direct influence on digestibility of plant cell walls in rumen. They also reduce the digestibility of cell wall carbohydrates by making indigestible complexes. The low digestibility of cell wall carbohydrates restricts the digestible energy available to livestock. The anti-nutritional effects of tannins have been given in Table 8.5.

Tannins have been classified into two classes: hydrolyzable and condensed tannins (Fig. 8.3) which differ in their nutritional and toxic effects.

Table 8.5 Tannins and their anti-metabolic effects

Browse species	Tannin type	Animal type	Effects
Abizzia chinensis (leaves)	CT	Goat	Reduced *in sacco* nitrogen digestibility
A. nilotica (leaves)	HT	Sheep	Precipitation of leaf proteins: Inhibition of rumen fermentation
Bauhinia variegata (leaves)	CT	Cattle	Reduced *in sacco* organic matter loss
Caliandra calothyrsus (leaves)	CT	Goat	Reduced *in sacco* nitrogen
Quercus gambelii (leaves)	CT	Goat	Reduced digestibility of cellular constituents and increased faecal nitrogen
Quercus grises (leaves)	CT	Goat	Reduction in feed intake digestibility and nitrogen retention
Quercus incana (leaves)	HT	Cattle	Reduced organic matter digestibility
Quercus calliprinos (leaves)	HT	Cattle	Anorexia, constipation, wasting, dullness, nephrosis
Quercus pyrenaica (leaves)	HT	Cattle	Reduction in rumen fermentation activity
	HT	Cattle	Proteinuria, elevated serum creatinine, and blood urea nitrogen
Quercus robur (leaves)	HT	Cattle	Inhibition of ruminal enzyme activity
Robinia pseudoacacia(leaves)	CT	Sheep	Reduced protein digestibility, reduced feed intake, and negative acid detergent fiber digestibility
		Rabbit	Decreased wool production and nitrogen retention
Leucaena leucocephala (leaves)	HT, CT	Goat	Inhibits digestibility
Salix tetrasperma (leaves)	CT	Cattle	Reduced organic matter loss in sacco
Lespedeza cuneata (leaves)	CT	Cattle	Inhibits rumen fermentation
Vicia faba (faba beans)	CT	Chickens	Decreased feed intake and weight gain; decreased retention of protein and calcium
Shorea robusta (salseed)	HT	Chickens	Depressed growth and feed conversion

Source: Bhat et al. (2013)

In livestock, higher condensed tannins content decreases the absorption of methionine and lysine amino acids. Decreased methionine availability could increase the toxicity of other plant compounds such as cyanogenic glycosides, as methionine is involved in the detoxification mechanisms. Tannins also have the property to precipitate proteins and make more complex bonds with starch, cellulose and

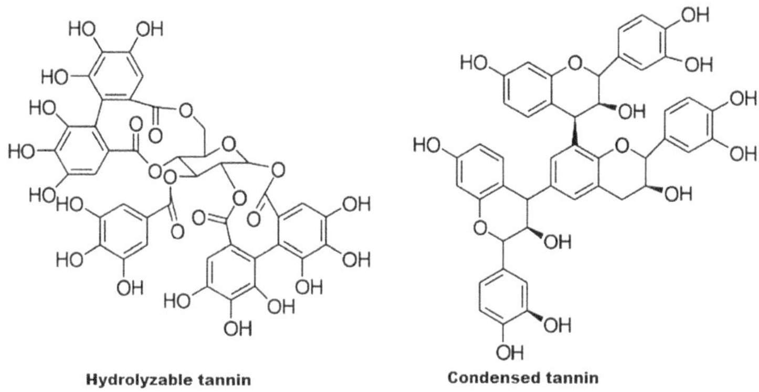

Fig. 8.3 Hydrolyzable tannins and condensed tannins

minerals affecting the digestibility. Tannins content more than 40 to 50 g kg^{-1} in forages may have significant influence on protein content and dry matter digestibility of forages. At low to moderate levels, condensed tannins increases the availability of protein to ruminal microflora.

Forages tannins have both negative as well as positive effects on nutritive value. In higher amounts, they may reduce digestibility of protein and carbohydrates, which often leads to reduced animal performance. Leaves of woody browse plants often contain enough tannins to reduce protein digestibility by 40%. At low to moderate levels, tannins could prevent bloat by binding with cellular complexes produced during mastication. In rumen, tannins eliminate foaming properties of legume forage proteins and reduce the rate of gas production during fermentation. Tannins also increase the flow of protein compounds through the rumen to the small intestine thereby escaping microbial fermentation (bypass protein). This is important because problems associated with extensive breakdown of proteins and amino acids in the rumen reduce protein quality and limits livestock production in modern feeding systems.

Tannins consumed by animals could otherwise protect them against diseases caused by parasitic worms. Tannins in oak such as tannic acid and gallic acid cause oak poisoning. Initial symptoms include anorexia, depression, clear watery nasal discharge, rumen stasis, excessive thirst and frequent urination. Constipation is followed by excretion of dark, thin, mucus-like, and often bloody feces. Ultimately, oak poisoning causes kidney, liver, and gastrointestinal lesions.

Condensed tannins (CTs) account for up to 20% of the dry matter in forage legumes used as ruminant feeds. Several forage legumes contain significant levels of tannins, including lespedeza (*Lespedeza* spp.), trefoil (*Lotus* spp.), sainfoin (*Onobrychis viciifolia*), crown vetch (*Coronilla coronata*) and *Hedysarum coronarium*. In rangelands, foraging livestock may encounter tannins in browse species. The content of condensed tannins in forages is affected by many factors including seasonal variations. *Albizzia chinensis* leaves have highest net tannins content (9.30%) during May, whereas *Grewia oppsitifolia* had least net tannin

Table 8.6 Condensed tannins content (CT) (g/100 g DM) of some important forage legumes of Himalayan region

Forage legume	CT content
Erect canary clover (*Dorycnium rectum*)	15 to 20 g/100 g
Sericea lespedeza (*Lespedeza cuneata*)	6 to 13 g/100 g
Panicledleaf ticktrefoil (*Desmodium paniculatu*)	20 to 22 g/100 g
Hedysarum coronarium	2 to 12 g/100 g
Big trefoil (*Lotus pedunculatus*)	5 to10 g/100 g
Sainfoin (*Onobrychis viciifolia*)	1 to 9 g/100 g
Birdsfoot trefoil (*Lotus corniculatus*)	0 to 5 g/100 g

content (1.40%) during the month of November in Himalayan region. *Grewia oppsitifolia* leaves have low level of condensed tannins (0.70%) in November, whereas *Albizzia chinensis* leaves retain high level of condensed tannins (4.40%) during the month of July. *Albizzia chinensis* leaves lopped during September have highest level of hydrolyzable tannins (5.10%), while significantly low content (0.70%) was recorded for *Grewia oppsitifolia* leaves in the month of November. The level of condensed tannins in some important forage legumes is presented in Table 8.6.

Various techniques have been practised for detannification of forages. Chopping and grinding increase the surface area that facilitates the interaction between tannin in leguminous forages and polyphenolic oxidases. The supplementation of tannin-rich leaves with polyethylene glycol-4000 (PEG-4000) is an effective method to reduce or neutralize tannin content (Ben Salem et al. 2000; Salem et al. 2007). PEG-4000 prevents the formation of tannin-protein complexes and assists the breakdown of already formed complexes (Reddy 2001). Treatment with potassium permanganate (0.03 M) and potassium dichromate (0.02 M) results in 95% reduction in tannin content in feeds (Makkar and Singh 1992). Treatment of forage with oxidizing chemicals convert tannins to quinones, which are not capable of forming complexes with proteins and wall carbohydrates. These oxidizing agents can be used for large-scale removal of tannins. Treatment of forages with alkalis like sodium hydroxide, sodium carbonate and sodium bicarbonate leads to oxidation of phenolics at higher pH (Makkar and Singh 1991, 1992). Among them, sodium hydroxide (0.05 M) is most effective in detannification of feeds, followed by sodium carbonate (0.05 M) and sodium bicarbonate (0.1 M). Ferrous sulfate is a tannin-complexing agent and at 0.015 M, it causes 85% detannification in tree foliage.

8.1.2.4 Alkaloids

Alkaloids are small organic molecules usually comprising several carbon rings with side chains and one or more of the carbon atoms being replaced by nitrogen (Fig. 8.4). They are synthesized by plants from amino acids. Decarboxylation of amino acids produces amines which react with amine oxides to form aldehydes. The characteristic heterocyclic ring in alkaloids is formed from Mannich-type condensation from aldehyde and amine groups. Alkaloids obstruct electrochemical

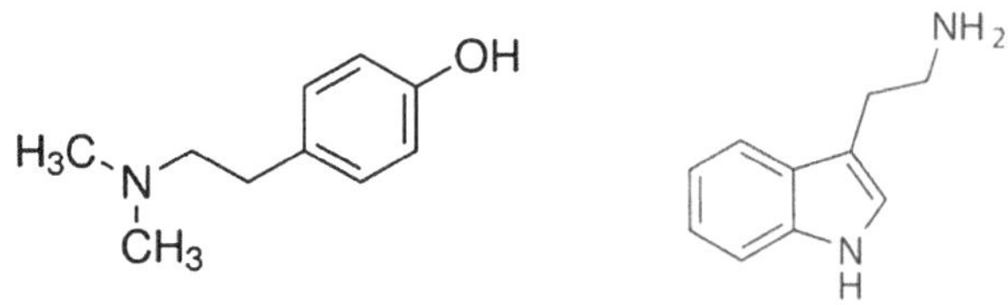

(a) (b)

Fig. 8.4 Structure of (**a**) Hordenine (**b**) Tryptamine

transmission of nerve impulse which results in several neurological dysfunctions. For example, alkaloid *N*-methyl-β-phenethylamine form Guajillo (*A. berlandieri*) causes limberleg, which is characterized by locomoter ataxia of hind quarters of sheep and goats. There are few grasses such as tall fescue, perennial rye grass, and reed canary grass (*Phalaris arundinaceae*) that produce alkaloids through a mutualistic relationship with an endophytic fungus. The negative effects of alkaloids produced by fungus include heat intolerance, fat necrosis, fescue foot, decreased weight gains, rye grass staggers, and death. Tryptamine and tyramine alkaloids produced by reed canary grass are responsible for toxicity in animals. Nine alkaloids have been isolated from reed canary grass and these include one phenol (hordenine), five indoles (graminae and four derivatives of tryptamine), and three derivatives of β-carboline. Total alkaloid concentration is negatively associated with palatability. Low palatability could lead to low intake by ruminants and may explain the poor performance sometimes reported for animals grazing reed canary grass.

Ergot alkaloids from *Claviceps* species and grasses such as rye grass, *Phalaris* spp., *Sorghum* spp., *Paspalum* spp, could cause reproductive problems in livestock. They have also been reported in bermudagrass. Ergot alkaloids mainly include ergovaline, ergotamine, ergovine, and lysergamine. All ergot alkaloids have similar effects on cattle but differ in their potency, with ergovaline being the most potent alkaloid. Ergot alkaloids have many adverse pharmacological effects on cattle, which are commonly termed fescue toxicosis. Mean total (μg/g DM) and proportional ergot alkaloids in three plant parts of tall fescue are given in Table 8.7. Signs of fescue toxicosis include heat intolerance, fat necrosis, lameness, and associated decreased performance. Tall fescue also produces alkaloids independent of endophytic fungus that include perloline and perlolidine. The alkaloids in perennial rye grass produce rye grass staggers. Alkaloids represent the main anti-nutritional factor in lupins. Wild lupin contains 0.40 to 3.00% of alkaloids in the dry weight of their seeds and 0.30 to 0.50% in the dry weight of green forage (Brummund and Święcicki 2011). In the narrow-leafed lupin, the major alkaloids are lupanine, 13-hydroxylupanine and angustifoline (Kurlovich et al. 2002). According to Petterson (1998), quinolizidine alkaloid profile is characteristic for sweet cultivars of narrow-leafed lupin: lupanine (42 to 59%), 13-hydroxylupanine (24 to 45%), angustifoline (7 to 15%) and isolupanine (1 to 1.50%).

Ammonization of fescue hay has resulted in consistent improvement in animal performance. Energy supplementation with concentrated feedstuffs like grains and molasses-based feeds may overcome some of the negative effects of alkaloid-containing grass forages. Mineral supplementation with selenium, copper, or cobalt

Table 8.7 Mean total (µg/g DM) and proportional ergot alkaloids in three plant parts of tall fescue

		Proportion of total ergot alkaloids				
Plant part	Total ergot alkaloids	Ergine	Ergovaline	Aci-ergovaline	Didehydroergovaline	Ergosine + ergonine
Seed	6.32	0.35	0.37	0.19	0.02	0.07
Culm	1.30	0.46	0.51	0	0	0.04
Plant	0.51	0.48	0.43	0	0	0.10

Source: Shelby et al. (1997)

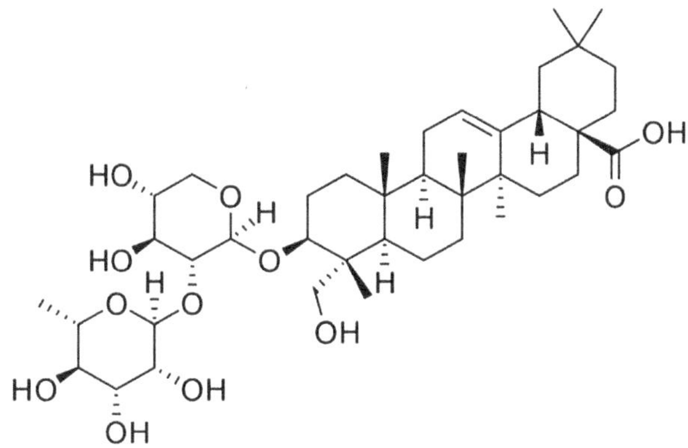

Fig. 8.5 Structure of saponins

has improved animal resistance to alkaloids in some cases. Ensiling has been proposed as a way to mitigate negative effects of fescue alkaloids to some extent.

8.1.2.5 Saponins

Saponins are complex glycosides that are widely distributed throughout the plant kingdom. Historically, alfalfa (*Medicago sativa*) and yams (*Dioscorea* spp.) have been recognized as rich in saponins which bind with nutrients and reduce the nutritional value of forages. Saponins contain a polycyclic aglycone either C_{27} steroid or C_{30} triterpenoid (Fig. 8.5). Saponins form stable and soap-like foams in aqueous solutions, containing a carbohydrate moiety (mono/oligosaccharide) linked to a hydrophobic aglycone (sapogenin). The physiological activity of saponins arises due to their powerful physical interactions, as surface-active agents with other components of the digestive system and their ability to interact with the membranes of mucosal cells. Saponins form insoluble complexes with 3-β-hydroxysteroids and are known to form large, mixed micelles with bile acids and cholesterol. Saponins are divided into two groups, including the steroidal saponins, which occur as glycosides in some pasture grasses, and the triterpenoid saponins that occur in many temperate legumes, particularly alfalfa. A total of some 24 saponins have been identified in alfalfa (Bialy et al. 1999), but the soyasapogenols, zanhic acid glycosides, and medicagenic acid are quantitatively the most important (Oleszek et al. 1992; Massiot et al. 1991). Symptoms of saponin toxicity, believed largely due to medicagenic and zanhic acid content in alfalfa include irritation to mouth and digestive tract, increased membrane permeability and, in acute cases, hemolysis. Zanhic acid glycosides may also cause production of intestinal gases. Ensiling of alfalfa can reduce total saponin and medicagenic acid content.

Saponins are bitter in taste and can affect livestock health and performance through erythrocyte hemolysis, depression of growth rate, inhibition of smooth muscle activity, bloat in ruminants, enzyme inhibition and reduction in nutrient absorption and assimilation. Fodder tree leaves rich in saponins give rise to toxic manifestations (listlessness, anorexia, weight loss, and gastroenteritis) in sheep. Saponin-rich diets also cause photosensitization and subsequently cause damage to liver and kidney in ruminants. The adverse effects of saponins can be reversed by the addition of dietary cholesterol, presumably because saponins form insoluble complexes with cholesterol. It has been reported that repeated soaking/washing effectively removes bitterness associated with saponins (Joshi et al. 1989).

8.1.2.6 Glycosides

Glycosides are naturally occurring compounds found in many rangeland plants and forages. Glycosides are a chemically diverse group of compounds that bear little resemblance to each other and they can form toxic compounds upon hydrolysis. The variations on this chemical theme yield a variety of powerful toxic effects when animals eat plants containing glycosides. These toxic effects can be observed as restlessness, uncontrolled bleeding, convulsions, or rapid death.

8.1.2.6.1 Nitro Group Containing Glycosides

Glycosides containing a toxic nitro-group are found in several species of legumes. Miserotoxin has been identified in milk vetch (*Astragalus* spp.) and is responsible for stock losses on western rangelands. Karakin, a glycoside-like compound is found in several legumes (*Astragalus*, *Coronilla*, *Indigofera*, and *Lotus*) and has been implicated in neural degeneration in mammals. Acute clinical symptom of toxicity caused by nitro-containing glycosides includes lack of coordination, distress, labored breathing, bluish skin or tongue, muscular weakness and collapse. Death may occur within a few hours because of acute poisoning. In chronic poisoning, animals lose weight and develop respiratory distress, a poor hair coat, hind limb paralysis and nasal discharge.

8.1.2.6.2 Cyanogenic Glycosides

Cyanogens are glycosides of a sugar or sugars and a cyanide containing aglycone. They are widely distributed in the plant kingdom but, within the temperate forage legumes, they are considered as a cause of concern only in some cultivars of white clover and *Lotus corniculatus*. The difference in cyanogenic glucoside content is caused by variation in two genes: *Ac* regulates the production of the cyanogenic glycosides linamarin and lotaustralin and *Li* regulates the production of hydrolyzing enzyme linamarase. Cyanogenesis is the ability of some plants to synthesize cyanogenic glycoside that releases cyanohydric acid (HCN), known as prussic acid when enzymatically hydrolyzed. In most cases, hydrolysis is accomplished by β-glucosidase producing sugars and a cyanohydrin that spontaneously decomposes to HCN and a ketone or aldehyde. Ruminant animals are more susceptible to HCN poisoning than non-ruminants due to fast microbial breakdown of cyanogenic glycosides. The cyanogenic glucosides have negative impacts on grazing animals as they suffer from direct effects of cyanide poisoning and/or indirect effects of cyanide and its derivative thiocynate on metabolism of iodine, selenium, and sulfur metabolism, including nutritional myopathy, reduced erythrocyte glutathione peroxidase activity and lower functionality of thyroid gland. The cyanide is extremely toxic because it blocks the vital cellular process of aerobic respiration. Clinical signs of subacute and acute poisoning in cattle include rapid heart rate, rapid breathing, increased pinkness of the mucous membranes around the eyes and mouth, and convulsive contractions. In liver, rhodanese enzyme converts cyanide to thiocyanate. The lethal dose of HCN for cattle and sheep is 2.0 to 4.0 mg per kg body weight. Administration of nitrite-thiosulfate supplemented with oxygen is the preferred treatment to alleviate cyanogens toxicity, especially. The major forage crop with considerable HCN content is sorghum fodder. The feeding of fresh sorghum fodder is safe when it has below 100 ppm HCN content while it is toxic to the cattle when it goes beyond 200 ppm.

8.1.2.6.3 Cardiac Glycosides

The consumption of cardiac glycosides in higher quantities causes toxicity in animals. Membrane-bound proteins in the heart are the major receptors for cardiac glycosides. The presence of these glycosides results in more forceful contractions of the heart. Medicinal doses of cardiac glycosides have been used for the treatment of congestive heart failure in humans. They are toxic to herbivores when consumed in higher quantities. Sub-acute to acute signs of poisoning in cattle and sheep include restlessness, labored breathing, frequent urination and defecation and irregular or rapid heartbeat. Treatments to alleviate cardiac glycosides toxicity include administering activated charcoal, potassium chloride, atropine, digitoxin-specific antibodies, β-adrenergic blocking agents, procainamide and phenytoin.

Bio-activation and toxicity of the glycosides mainly depend on the: (A) rate of digestion and hydrolysis by rumen microbes; (B) rate of detoxification and; (C) degree of absorption from the gastrointestinal tract. Livestock management to reduce the risk of glycoside toxicity is difficult because of the immense variety of glycoside compounds and the toxic effects they produce. However, wise grazing management of range and pastureland containing glycosidic plants should be based on:

(a) Knowledge of plants containing toxic glycosides.
(b) Knowing the life cycle of potentially toxic plants and adjusting the timing of grazing when the plants are least toxic.
(c) Providing the opportunity to the animals to select a diverse diet.
(d) Carefully selecting plants for pasture planting.

8.1.2.7 Oxalates

Many forage plants such as golden timothy (*Setaria* spp.), buffel grass (*Cenchrus ciliaris*), pangola grass (*Digitaria decumbens*), kikuyu grass (*Pennisetum clandestinum*), and Napier grass (*Pennisetum purpureum*) are high oxalate-containing forage grasses for livestock. However, these species can sometimes accumulate oxalate (as soluble and/or insoluble form) to potentially toxic level (Dhillon et al. 1971; Cheeke 1995; Rahman et al. 2006). When ruminant consumes an oxalate-containing plant, the oxalate is metabolized in four possible ways. First, soluble oxalate may be degraded by rumen microflora. Second, soluble oxalate may combine chemically with Ca to become insoluble oxalate and results in the reduction of absorption of Ca. Third, soluble oxalate may be absorbed from the rumen into the bloodstream where it can combine with serum Ca to form insoluble oxalate crystals (Blaney et al. 1982). The precipitation and accumulation of crystal in kidneys can cause kidney failure (Lincoln and Black 1980). Fourth, ingested insoluble oxalate from plants may pass through the digestive tract without a harmful effect on the body's metabolism. Soluble oxalate usually forms complexes with sodium (Na^+), potassium (K^+), and ammonium (NH_4^+) ions, and insoluble oxalate forms with Ca^{2+}, magnesium (Mg^{2+}), and iron (Fe^{2+}) ions (Savage et al. 2000). Insoluble calcium oxalate complex formation leads to disturbance in calcium to phosphorus ratio, which results in deficiency of calcium in body (Blaney et al. 1982). The excessive mobilization of calcium and phosphorus from bone results in secondary hyperparathyroidism or osteodystrophy fibrosa (Noonan and Savage 1999; Rahman and Kawamura 2011). The demineralized bones become fibrotic and misshapen, causing lameness and bighead in horses (McKenzie et al. 1981). The oxalate content in BN hybrid, guinea grass and fodder bajra range from 2.58 to 5.62%, 1.05 to 2.40%, and 0.63 to 1.98%, respectively.

Various strategies have been identified for the reduction of oxalate content in plants. Pretreatment by leaching soluble oxalate from the leaves or the conversion of soluble oxalates in the leaves to insoluble oxalates, are important approaches for the reduction of oxalate content. Studies have found that soluble oxalate levels decreased with high levels of calcium were consumed along with diet containing high levels of soluble oxalates. Savage et al. (2000) observed that soaking and cooking in water can reduce the soluble oxalate content of leafy foods as soluble oxalates easily leach into the water at room temperature. The addition of calcium compounds such as calcium sulfate, calcium chloride, calcium citrate, and calcium carbonate has also been observed effective in reducing the soluble oxalate content.

8.1.2.8 Protease Inhibitors and α-Amylase Inhibitors

Protease inhibitors inhibit proteolytic enzymes activity in gastrointestinal tract of animals. The presence of protease inhibitors in feed decreases the apparent nutritional quality of proteins in the diet by affecting the ability of digestive enzymes to degrade dietary proteins, thereby limiting the intake of essential amino acids. Protease inhibitors could also lead to pancreatitis and a reduction in the apparent availability of sulfur amino acids. Trypsin inhibitors are the most extensively studied protease inhibitors due to their importance in human and animal nutrition. Trypsin inhibitors irreversibly bind trypsin, making enzyme incapable of performing its role in the breakdown of proteins. Starches are complex carbohydrates that cannot be absorbed unless they are first broken down by the digestive enzyme amylase and other secondary enzymes. α-amylase inhibitors (starch blockers) bind with α-amylase from pancreas (which act on starch and hydrolyzed into a mixture of small oligosaccharides like maltose, maltotriose, and oligoglucans) and delay carbohydrate absorption and result in a reduction of peak postprandial plasma glucose levels in blood. Certain processing methods, such as cooking and soaking can lead to removal of protease inhibitors and α-amylase inhibitors from forages.

8.1.2.9 Nitrates

Excess nitrates accumulate in plants when they are stressed. Drought or hot dry winds put forage under water stress often resulting in nitrate accumulation. Nitrates are present in Sudan grass, pearl millets, and oats. Feeds can be safely fed to the animals only when the content of nitrate is 0 to 1000 ppm. Once the level of nitrate in forage reaches 1500 to 2000 ppm, it is recommended to limit the particular feed to 50% of ration's total dry matter. As the content of nitrate increases in the forage feed, the more of it is limited to be given to the animals. Feeds containing over 4000 ppm of nitrate are potentially toxic and never given as feed to the animals. When livestock consume forages with normal nitrate content, they convert nitrate to nitrite to ammonia to amino acid to protein in the rumen. However, the situation is reverse when forages having high nitrate content are fed to animals. As a result, nitrite accumulates which is then absorbed into the bloodstream and converts hemoglobin to methemoglobin, which cannot carry oxygen. Due to nitrate (nitrite) poisoning, animal dies from asphyxiation or lack of oxygen. Nitrate toxicity can be easily confused with prussic acid poisoning because the two afflictions share many of the same symptoms. However, the blood is bright red in prussic acid poisoning versus chocolate brown in nitrate toxicity. Proper diagnosis is very important because the treatment for nitrate toxicosis can cause death in animals suffering from prussic acid poisoning.

Intravenous administration of 2 mg methylene blue per 400 pounds of body weight of animal has been effective in countering nitrate toxicity in animals. Another treatment for excessive nitrate consumption is to flush the animal's rumen. This crude operation can be achieved with cold water delivered to the rumen with a flexible tube placed down the esophagus. To be effective the rumen contents should be sucked out through tube. Supplementing livestock with vitamin A and iodine is also recommended when animals are consuming feeds high in nitrates. Supplementing cattle with 2 g sodium tungstate per day also reduces the occurrence of nitrate toxicosis. Animal must be fed with sufficient minerals, such as molybdenum, copper, iron, magnesium, and manganese, because they are necessary for the conversion of nitrate to ammonia in the rumen. Altering management of pastures and crops can help minimize susceptibility of livestock to nitrate toxicity.

8.1.2.10 Nonprotein Free Amino Acids

The presence of nonprotein amino acids in forages obstructs amino acids biosynthesis or activation and incorporation of amino acids into growing polypeptide chain during protein synthesis. Nonprotein free amino acids can disrupt the proteins synthesis by competing with the protein amino acids on the messenger RNA, but some toxic amino acids such as mimosine have additional anti-nutritional effects (Guillaume et al. 1999). Mimosine and its metabolite, 3-hydroxy-4-(1H)-pyridone (DHP) are toxic to livestock. In ruminants, mimosine is a depilatory, whereas DHP is a goitrogen. Poisoning can be acute or chronic and is characterized by alopecia, poor growth, oral ulcerations and goiter. Toxicity of mimosine in livestock has been reported from the different parts of the world and is known by different vernacular names such as jumbey in the West Indies, lamtoro in Indonesia, and koa haole in

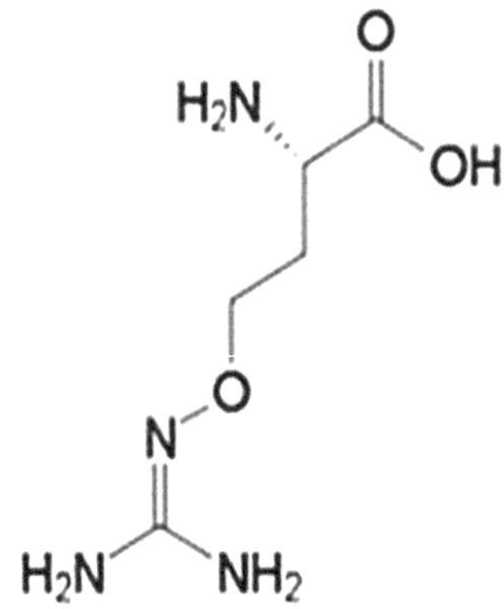

Fig. 8.6 Structure of L-canavanine

Hawaii. A variation of animal susceptibility to mimosine toxicity in different parts of the world is due to the geographic distribution of ruminal bacteria capable of degrading DHP.

CAN (canavanine) is produced in many legumes including jack bean and lucerne (alfalfa). This arginine structural analogue (Fig. 8.6) is typically incorporated into proteins by arginyl-tRNA synthetase, leading to inhibition of proteins. In alfalfa plants, canavanine content ranges from 1.30 to 2.40% of the dry matter. Alfalfa seeds have also been found rich in this toxic amino acid as the canavanine content varied from 1.40 to 1.80% of the dry matter. On average, the tested seeds contained 1.54% canavanine. The canavanine content has been found comparable to the levels found in the seeds of representative members of the genus *Canavalia*, which are amongst the more abundant sources of this antimetabolite. Canaline is a toxic metabolic product of canavanine. Toxicity of canavanine has been found in those animals having the capability to convert canavanine into caline.

8.1.2.11 Crop Specific Anti-Nutrients

Anti-nutrients are the components that decide the intake and feeding value of forage crops. Anti-nutrients like HCN are present in sorghum and phytoestrogens are predominant in white clover. Likewise *Leucaena* foliage contains mimosine whereas Robin, a lectin is specifically present in *Robinia pseudocaccia.*

8.1.2.11.1 Mimosine in *Leuceana leucocephala*

Mimosine, a nonprotein amino acid is structurally similar to tyrosine (Fig. 8.7). It occurs in a few species of mimosa and all species of genus *Leuceana*. The toxic effects of mimosine are due to formation of complex with pyridoxal phosphate and inhibition of catalytic activity of transaminases. In non-ruminants, it causes irregularities in cellular mitosis, alopecia, reduced appetite, reduced weight gain and often death (Kumar 2003). In case of ruminants, mimosine degradation in rumen produces toxic metabolite (3, 4-dihydroxypyridine) which is responsible for poor health and performance, loss of hair and wool in animals, swollen and raw coronets above the hooves, lameness, and lesions in gut.

Fig. 8.7 Structure of Mimosine and DHP

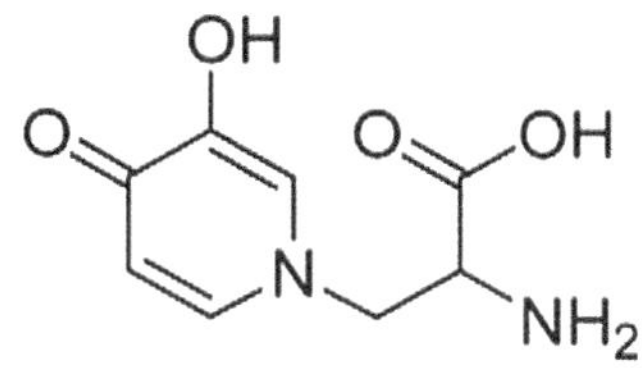

Mimosine

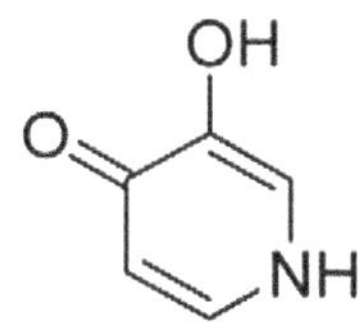

DHP

Different treatments have been developed to reduce the mimosine content in *Leucaena*. Simple heating or drying improves quality of proteins by inactivating heat-labile, anti-physiological factors and by unfolding the protein structure, which makes them more susceptible to attack by digestive enzymes. Vitti et al. (2005) reported drying of *Leucaena* leaves at 90 °C for 24 h reduces the tannin content. Akbar and Gupta (1985) observed that moist heat at 70 to 100 °C causes 50% reduction in mimosine content. Mali et al. (1990) reported 17 to 19% reduction by dry heating at 100 °C and 19 to 23% reduction by autoclaving in mimosine content. D'Mello and Acamovic (1982) observed ferrous sulfate at 0.5% level and aluminum at 1.17% level are also effective in increasing the mimosine excretion through feces. They also suggested that ferric sulfate is more effective than ferrous sulfate. Tsai and Ling (1973) revealed that efficacy of aluminum salts formed stronger complexes with mimosine than ferrous sulfate for inducing faecal excretion of mimosine. Sunaria and Vidya-Sagar (1989) observed application of ferrous sulfate at 1.26% optimum for reducing mimosine content in *Leucaena* leaf meal. Treatment with sodium acetate (0.05 N) has also been reported to cause 95% reduction in mimosine content in *Leucaena* leaves (Tawata et al. 1986). Sharma et al. (1990) observed similar effects with 0.20% ferrous sulfate and calcium sulfate in reducing the mimosine content in *Leucaena* leaves. The significant reduction in mimosine content with different chemical treatments may be the result of the chelation of metal ions with mimosine and/or leaching during sample preparation for mimosine estimation. Our group has also identified different treatments for reducing mimosine content in *Leucaena* foliage harvested at different maturity stages. Soaking of leaves, application of sodium acetate (0.05 N), ferrous sulpahte (0.50%), calcium chloride (2%), and calcium sulfate (0.20%) were effective in reducing mimosine content in tender, premature, and mature *Leucaena* leaves. Treatment of tender and mature *Leucaena* leaves with sodium acetate was highly effective for reducing mimosine content to the extent of 85.04% and 86.61%, respectively, while calcium chloride treatment reduced the mimosine content to the extent of 86.49% in premature leaves in comparison to control.

8.1.2.11.2 Lectin in *Robinia pseudocacia*

Hemagglutinins are glycoproteins which are also referred as phytoagglutinins or lectins. Lectins exhibit reversible binding to specific carbohydrate moieties on the surface of erythrocytes and agglutinate them, without altering the structure of surface carbohydrates (Fig. 8.8). Robin, a lectin from *Robinia pseudocacia* has been reported for anorexia, lassitude, weakness, and posterior paralysis in cattle (Emire et al. 2013). The binding of lectin to carbohydrate moieties expressed by intestinal epithelial cells alters the cellular morphology and metabolism of stomach and/or small intestine. These lectins induce changes in digestive, absorptive, protective, and

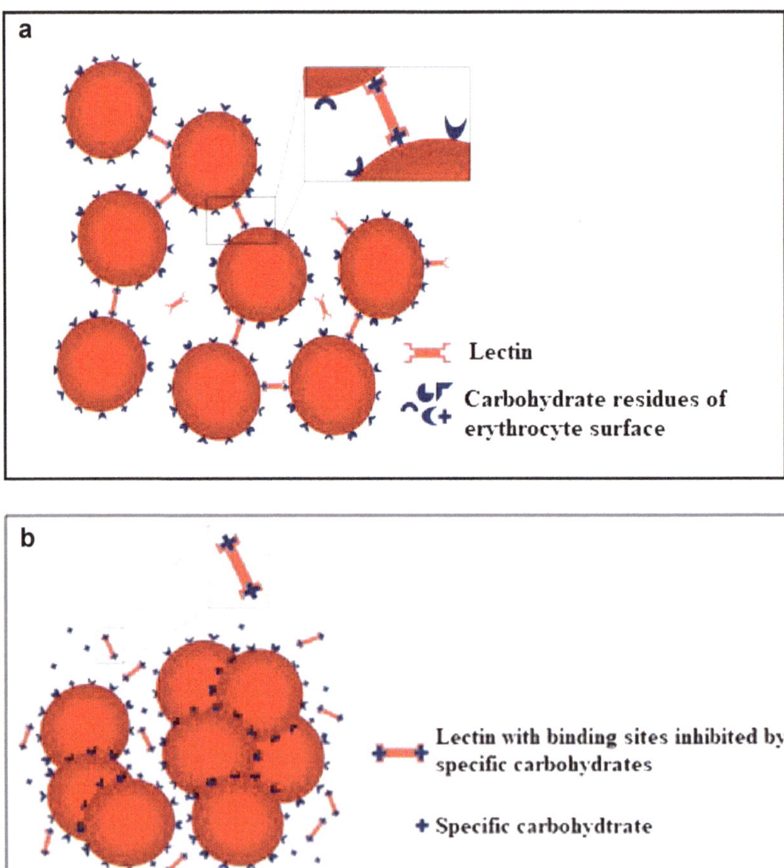

Fig. 8.8 (**a**) Agglutination and (**b**) agglutination inhibition of lectin. (Source: Santos et al. (2013))

secretary functions which ultimately affect cellular proliferation and turnover. Lectins also have immune-stimulatory responses.

8.1.2.11.3 Alkaloids in Reed Canary Grass

Toxicosis has been associated with reed canary grass (*Phalaris arundinacea* L.). Reed Canary grass contains alkaloids that could result in sudden death or neurological staggers in livestock. At least eight alkaloids are known to be present in canary grass including one phenol (hordenine), five indoles (gramine and 4 derivatives of tryptamine), and two derivatives of β-carboline. Total basic alkaloid content in genotypes of reed canary grass is negatively correlated with their palatability to ruminant animals. The tryptamine alkaloids present in some reed canary grass lines are in fact potentially toxic to sheep and cattle. Toxicity of prussic acid has also been reported in animals if they consume large amounts of reed canary grass. Leaves usually produce 2 to 25 times more prussic acid than stems in forage grasses. Young, upper leaves have more prussic acid than lower leaves. New grass shoots often contain high prussic acid content.

8.1.2.11.4 Hydrogen Cyanide (HCN) in Sorghum

Prussic acid is sometimes referred to as hydrogen cyanide. Hydrogen cyanide (HCN) is an anti-nutritional chemical that can cause mild to severe reactions in animals when grazing forage sorghum. When plants are damaged due to drought, freeze, herbicides or insects, or even mechanical injury, the cyanogenic glycoside, dhurrin, comes into contact with enzymes as the plant tissue degrades to form the toxic HCN compound. HCN content of less than 200 mg/g dry weight in sorghum is preferred for consumption by livestock. Once the forage containing HCN is consumed, the HCN is released and transported in the blood. The HCN then prevents oxygen absorption at the cellular level, leading to animal death due to asphyxiation.

8.1.2.11.5 Phytoestrogens in Clovers

A phytoestrogen is a plant-derived xenoestrogen not generated within the endocrine system, but consumed by eating phytoestrogenic plants. Also called a dietary estrogen, it is a diverse group of naturally occurring nonsteroidal plant compounds that, because of its structural similarity with estradiol (17-β-estradiol), have ability to cause estrogenic and/or anti-estrogenic effects. Two types of phytoestrogens (i) coumesterols (coumestrol) related to coumarins and quantitatively more important in alfalfa and (ii) isoflavonoids are widely distributed in *Trifolium* spp. Levels of coumestrol in alfalfa forage range from 2.99 to 104.37 ppm. The structures of the more important isoflavonoids recognized in red clover are shown in Table 8.8. Formononetin and biochainin A are the two isoflavones found in the greatest amounts in forage legumes and together can reach 15 g kg^{-1} dry matter in some red clover cultivars. Their content in white clover are usually substantially lower (0.5 g kg^{-1} dry matter). The major metabolic transformation of the isoflavones occurs in the rumen. Biochainin A is demethylated to genistein and *via* ring cleavage to 4-ethylphenol and organic acids with the loss of all estrogenic activity. Formononetin is mainly demethylated to daidzein and then to

Table 8.8 Structure of the isoflavonoids identified in red clover

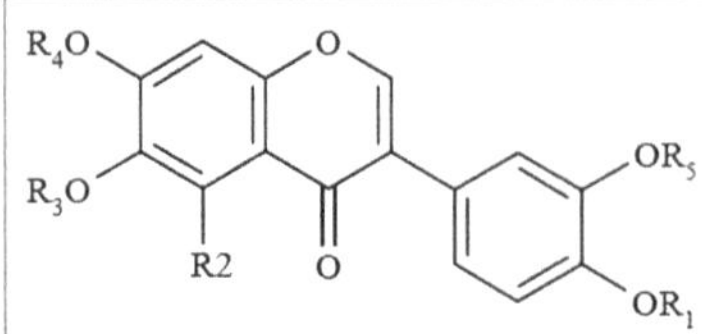

Compound	R1	R2	R3	R4	R5
Daidzein	H	H	H	H	H
Daidzin	H	H	H	Glucose	H
Genistein	H	H	OH	H	H
Genistin H	H	H	OH	Glucose	H
Formononetin	CH_3	H	H	H	H
Ononin	CH_3	H	H	Glucose	H
Biochainin A	CH_3	H	OH	H	H
Sissotrin	CH_3	H	OH	Glucose	H
Trifoside	Glucose	H	H	CH_3	H
Calycosin	CH_3	H	H	H	OH
Pectolinarigenin	CH_3	OH	OCH_3	H	H
Pratensein	CH_3	OH	H	H	OH
Pseudobaptigenin	$-CH_2$	H	H	H	$-O^-$

equol by hydrogenation and ring cleavage. Coumesterol, present as the major phytoestrogen in alfalfa and white clover, is not metabolized by the rumen flora and is absorbed in its original form. Coumesterol is known to be approximately 30-fold more effective than genistein in mice and to cause estrogen-related disorders in animals. Its content in healthy plant rarely exceed 30 to 60 mg kg^{-1} dry matter.

Animals-fed with estrogenic forage may suffer with impaired ovarian function, often accompanied by reduced conception rates and increased embryonic loss. Males are relatively unaffected, but the mammary glands in females and castrate males may undergo hypertrophy of the duct epithelium, accompanied by secretion of clear or milky fluid. In cows, clinical signs resemble with those associated with cystic ovaries.

8.1.2.11.6 Glucosinolates in *Brassica* Species

The glucosinolates, precursors of mustard oils, are mainly found in the mustard family (Brassicaceae), but they are also found in other families of tropical plants. Some glucosinolates are degraded into compounds that cause enlargement of the thyroid gland (i.e., goiters). Two types of goitrogens are derived from glucosinolates which affect the thyroid gland in different ways. One compound, the thiocyanate ion, inhibits uptake of inorganic iodine by the thyroid gland and the inhibition could be reversed with iodide supplements. The other type of goitrous compound, cyclic thiouracils, causes enlargement of the thyroid, which cannot be reversed. Several drugs are available to treat these thyroid effects including, methimazole, propylthiouracil, amphenone and chlorpromazine. Ingestion of mustard (*Brassica* spp.) seeds by cattle can result in lesions in the gastrointestinal tract including profuse edema of the rumen, death of mucous tissue and intestinal bleeding. The digestion of several *Brassica* species and other mustards can cause toxic effects in the pancreas and kidneys.

Knowledge about the nutrient composition of forage is essential for balancing diets in order to maintain animal health and productivity. Imbalance in dietary constituents can lead to variable effects on livestock productivity. Crude protein, fiber, minerals, carbohydrates etc., are important feed components. Fatty acids and crude fats components for energy production. Anti-nutritional factors can cause unpredictable effects on animals when fed in large quantities. They are associated with reduced feed intake, lower feed conversion, bind to protein and other important nutrients required by the animal and can also cause death in some cases. The application of modern biotechnological techniques could help in lowering of important anti-nutrients from important forage crops.

References

Akbar MA, Gupta PC (1985) Proximate composition, tannin and mineral contents of different cultivars and of various parts of subabul (*Leucaena leucocephala)*. Indian J Anim Sci 51:57–58

Ben Salem H, Nefzaoui A, Ben Salem L, Tisserand JL (2000) Deactivation of condensed tannins in *Acacia cyanophylla* Lindl. Foliage by polyethylene glycol in feed blocks. Effects on feed intake, diet digestibility, nitrogen balance, microbial synthesis and growth by sheep. Livestock Prod Sci 64:51–60

Bhat TK, Kannan A, Singh B, Sharma OP (2013) Value addition of feed and fodder by alleviating the antinutritional effects of tannins. Agric Res 2:189–206

Bialy Z, Jurzysta M, Oleszek W, Piacente S, Pizza C (1999) Saponins in alfalfa (*Medicago sativa* L.) root and their structural elucidation. J Agric Food Chem 47:3185–3192

Blaney BJ, Gartner RJW, Head TA (1982) The effects of oxalate in tropical grasses on calcium, phosphorus and magnesium availability to cattle. J Agric Sci (Camb) 99:533–539

Boufaied H, Chouinard PY, Tremblay GF, Petit HV, Michaud R, Bélanger G (2003) Fatty acids in forages. I. Factors affecting concentrations. Can J Anim Sci 83:501–511

Brummund M, Święcicki W (2011) The recent history of lupin in agriculture. In: Naganowska B, Kachlicki P, Wolko B (eds) Lupin crops: an apportunity for today a promise for the future. Proceedings of the 13th international Lupin conference, Poznan, pp 15–23

Chaves AV, Waghorn GC, Tavendale MH (2002) A simplified method for lignin measurement in a range of forage species. Proc N Z Grassl Assoc 64:129–133

Cheeke PR (1995) Endogenous toxins and mycotoxins in forage grasses and their effects on livestock. J Anim Sci 73:909–918

Cruickshank GJ, Poppit DP, Sykes AR (1992) The intake, digestion and protein degradation of grazed herbage by early-weaned lambs. Br J Nutr 68:349–364

D'Mello JPF, Acamovic T (1982) Apparent metabolizable energy value of dried *Leucaena* leaf meal for young chicks. Trop Agric 59:329–332

Dhillon KS, Paul BS, Bajwa RS, Singh J (1971) A preliminary report on a peculiar type of napiergrass (*Pennisetum purpureum*, 'Pusa giant') poisoning in buffalo calves. Indian J Anim Sci 41:1034–1036

Emire S, Jha Y, Mekam F (2013) Role of anti-nutritional factors in food industry. Beverage Food World 2:23–28

Guillaume JC, Kaushik S, Bergot P, Métailler R (1999) Nutrition et alimentation des poissons et des crustacés. Inra, Paris-Ifremer, Issy les Moulineaux (co Eds). pp. 489

Huisman J, Tolman GH (2001) Antinutritional factors in the plant proteins of diets for non-ruminants. In: Garnsworthy PC, Wiseman J (eds) Recent developments in pig nutrition. Nottinghum University Press, Nottingham, pp 261–322

Humphreys MO (1989) Water-soluble carbohydrates in perennial ryegrass breeding. III. Relationships with herbage production, digestibility and crude protein content. Grass Forage Sci 44:423–430

Jafari AA, Rezaeifard M (2010) Effects of maturity on yield and quality traits in tall fescue (*Festuca arundinacea Schreb*). Am Eur J Agric Environ Sci 9(1):98–104

Jafari AA, Anvari H, Nakhjavan S, Rahmani E (2010) Effects of phenological stages on yield and quality traits in 22 populations of tall wheatgrass grown in Lorestan, Iran. J Rangeland Sci 1(2): 13–20

Joshi DC, Katiyar RC, Khan MY, Banerji R, Misra G, Nigam SK (1989) Studies on Mahua (*Bassia latifolia*) seed cake saponin (*Mowrin*) in cattle. Indian J Anim Nutr 6:13–17

Katoch R, Tripathi A, Thakur N (2018) Comparative analysis of three *Rumex* species for livestock feeding prevailing under mid-Himalayan region. Indian J Agric Biochem 31(2):181–185

Kumar R (2003) Anti-nutritive factors, the potential risks of toxicity and methods to alleviate them. http://www.fao.org/DOCREP/003/T0632E/T0632E10.html

Kurlovich BS, Kartuzowa LT, Heinanen J, Benken II, Chmeleva ZV (2002) Biochemical composition. In: Kurlovich BS (ed) Lupins. Geography, classification, genetic resources and breeding. INTAN, St Petersburg, pp 241–267

Lincoln SD, Black B (1980) Halogeton poisoning in range cattle. J Am Vet Med Assoc 176:717–718

Makkar HPS, Singh B (1991) Effect of drying conditions on tannin, fiber and lignin levels in mature oak (*Quercus incana*) leaves. J Sci Food Agric 54:323–328

Makkar HPS, Singh B (1992) Effect of wood ash on tannin content of oak (*Quercus incana*) leaves. Bioresour Technol 41:85–86

Mali JM, Kute LS, Jambhale ND, Kadam SS (1990) Effect of leaf processing on antinutrients in leucaena seeds. Indian J Anim Sci 60:385–388

Massiot G, Lavaud C, Besson V, Le Men-Olivier L, Van Binst G (1991) Saponins from the aerial part of alfalfa (*Medicago sativa*). J Agric Food Chem 39:78–82

McDowell LR (2003) Minerals in animal and human nutrition, 2nd edn. Amesterdam, Elsevier Science, BV

McKenzie RA, Gartner RJW, Blaney BJ, Glanville RJ (1981) Control of nutritional secondary hyperparathyroidism in grazing horses with calcium plus phosphorus supplementation. Aust Vet J 57:554–557

Naydenova Y (2014) Forage quality analysis of perennial grass and legumes species in pure stands and mixtures. J Anim Sci 48:176–183

Ncube S, Saidi PT, Halimani TE (2015) Effect of stage of growth and processing methods on the nutritional content of *Acacia angustissima* leaf meal harvested for broiler feeding. Livestock Res Rural Dev 27:89

Noonan SC, Savage GP (1999) Oxalate content of foods and its effect on humans. Asia Pac J Clin Nutr 8:64–74

Oleszek W, Jurzystra M, Ploszynski M, Colquhoun IJ, Price KR, Fenwick GR (1992) Zanhic acid tridesmoside and other dominant saponins from alfalfa (*Medicago sativa* L) aerial parts. J Agric Food Chem 40:191–196

Petterson DS (1998) Composition and food uses of lupins. In: Gladstones JS, Atkins CA, Hamblin J (eds) Lupins as crop plants. Biology, production and utilization. CAB International, Oxon, pp 353–384

Rahman MM, Kawamura O (2011) Oxalate accumulation in forage plants: some agronomic, climatic and genetic aspects. Asian Aust J Anim Sci 3:439–448

Rahman MM, Niimi M, Ishii Y, Kawamura O (2006) Effects of seasons, variety and botanical fractions on oxalate content of napiergrass (*Pennisetum purpureum* Schumach). Grassl Sci 52: 161–166

Reddy DV (2001) Principles of animal nutrition and feed technology. Oxford IBM, New Delhi

Rezaeifard M, Jafari AA, Assareh MH (2010) Effects of phenological stages on forage yield quality traits in cocksfoot (*Dactylis glomerata*). J Food Agric Environ 8(2):365–369

Rogers LA, Campbell MM (2004) The genetic control of lignin deposition during plant growth and development. New Phytol 164:17–30

Ruckle M, Meier M, Frey L, Eicke S, Kölliker R, Zeeman S, Studer B (2017) Diurnal leaf starch content: an orphan trait in forage legumes. Agronomy 7:16. https://doi.org/10.3390/agronomy7010016

Salem AZM, Robinson PH, El-Adawya MM, Hassan AA (2007) In vitro fermentation and microbial protein synthesis of some browse tree leaves with or without addition of polyethylene glycol. Anim Feed Sci Technol 138:318–330

Santos AFS, Napoleão TH, Bezerra RF, Carvalho EVMM, Maria C, Paiva PMG, Coelho LC (2013) Strategies to obtain lectins from distinct sources. In: Berhardt LV (ed) Advances in medicine and biology, vol 63. Nova Science, New York, NY, pp 33–60

Savage GP, Vanhanen L, Mason SM, Ross AB (2000) Effect of cooking on the soluble and insoluble oxalate content of some New Zealand foods. J Food Compos Anal 13:201–206

Sharma T, Purohit GR, Bhatia JS, Sharma YK (1990) Effect of chemical treatment on subabul (*Leucaena leucocephala*) on the nutrient utilization in karakul x Malpura F2 rams. Indian J Anim Nutr 7:295–298

Shelby RA, Olsovska J, Havlicek V, Flieger M (1997) Analysis of ergot alkaloids in endophyte infected tall fescue by liquid chromatography/ electrospray ionisation mass spectrometry. J Agric Food Chem 45:4674–4679

Sunaria KR, Vidya-Sagar V (1989) Subabul (*Leucuenu leucocephala*) leaf meal its chemical composition, amino acid make up and detoxification of mimosine and tannins. Indian J Anim Nutr 6:223–226

Tawata S, Hongo F, Sungawa K, Kawastima Y, Yoga S (1986) A simple reduction method of mimosine in the tropical plant *Leucaena*. Sci Bull Coll Agric Univ Rykyus Okinawa 33:87–93

Tsai WE, Ling KH (1973) Study on the stability constant of some metal ion chelatea of mimosine and 3,4- tihydroxypyridine. J Chin Biochem Soc 2:70–86

Van Soest PJ (1982) Nutritional ecology of the ruminant livestock. Cornell University Press, Ithaca, NY

Vitti DMSS, Nozella EF, Abdalla AL, Bueno ICS, Silva Filho JC, Costa C, Long MS, Vieira MEQ, Cabral Filho SLS, Godoy PB, Mueller-Harvey I (2005) The effect of drying and urea treatment on nutritional and anti-nutritional components of browses collected during wet and dry seasons. Anim Feed Sci Technol 122:123–133

Wilkins PW, Davies RW (1994) Progress in combining high dry matter yield with reduced flowering intensity and improved digestibility in perennial ryegrass. In: Proceeding of the 19th EUCARPIA fodder crops section meeting, Brugge, Belgium, pp 247–250

9 Nutritional Quality Estimation of Forages

Forage quality encompasses their nutritive value and intake. In other words, forage quality is a function of intake, rate and extent of digestion and absorption of nutrients in animals (Barnes and Marten 1979). Forage quality should be expressed in terms of meeting the nutritional requirements of livestock, rather than solely as a descriptor of nutrient content and/or intake potential of forages. Forage quality has direct effect on animal performance and ultimately profitability of livestock operation. This can be understood from the fact that when livestock is fed with low-quality feed, its passage rate through alimentary canal is low which results in reduced feed intake, weight loss, nutrients deprivation and ultimately resulting in low livestock productivity. Forage quality mainly depends on the level of different nutritional constituents. The level of these components in forages should be estimated to prepare a nutritionally balanced diet. The methods for the estimation of different nutritional quality components in forages include wet laboratory methods and *in vivo* studies. Most important and earliest methods of quality estimation in forages are Weende system and Van Soest system, together with *in vitro* technique for the estimation of digestibility developed by Tilley and Terry (1963). Near-Infrared Reflectance Spectroscopy (NIRS) and Fourier-Transform Infrared Spectroscopy (FTIR) techniques have also been utilized for forage quality analysis.

9.1 Scheme for Forage Quality Analysis

The accurate forage quality estimation depends on sampling techniques, appropriate handling of samples after collection and use of reliable analytical procedures. During forage quality evaluation, sampling is a crucial step and the collected sample should represent the quality of the complete lot. The scheme for collection and preparation of samples is discussed below.

R. Katoch, *Nutritional Quality Management of Forages in the Himalayan Region*, https://doi.org/10.1007/978-981-16-5437-4_9

1. *Collection, drying and storage of sample*: Prior to the collection of samples for quality analysis, determination of the stage of maturity is imperative since forage maturity has significant effect on the quality of forages. Collection and transportation of samples in plastic/polythene bags should be avoided. Poor sampling technique and inadequate number of subsamples are sources of error in forage quality analysis.
2. *Grinding of sample*: The sample must be finely ground and approximately 100 g of sample should be passed through a 0.50 mm sieve or depending upon the parameter to be analysed. At any stage of grinding, the sample temperature should be below 40 °C.
3. *Quality analysis*: Nutritional quality estimation of forages is carried out using destructive and nondestructive methods. The destructive methods include Weende and Van Soest systems and the nondestructive methods include NIRS and FTIR techniques.

9.2 Destructive Methods for Forage Quality Estimation

9.2.1 The Weende Scheme

In the mid-nineteenth century, Henneberg and Stohmann developed Weende scheme for forage analysis. This method is also known as proximate analysis and has been used for more than one century. This system involves the destruction of forage samples and analysis of moisture, crude protein, crude fiber, ether extract and ash content. Nitrogen-free extract (NFE) is calculated as [100 − (% Moisture + % CP + EE + CF + Ash)]. This system is both comparative and predictive in nature. Crude fiber contains cellulose and some part of lignin. A negative relationship exists between crude fiber content and digestibility and energy value of forage. Figure 9.1 indicates different steps involved in proximate analysis of feeds and forages. This system enables us to use proximate principles to estimate components associated with animal performance, like digestibility and feed intake. Though proximate analysis system is a rather old scheme, it allows one to compare feeds on the basis of the level of nutritional constituents and to some extent for predicting the components of animal performance.

9.2.2 The Van Soest System

Another destructive method for forage quality estimation was developed by American scientist Van Soest in the year 1963 (Fig. 9.2). According to this system, the cell content of forage is practically 100% digestible and the cell wall is a measure of less digestible parts. The cell content contains proteins, total ash, fat, nonstructural carbohydrates, organic acids and enzymes. The NDF content in cell wall refers to hemicellulose, cellulose and lignin content of forage. In this technique, NDF estimation is based on lauryl sulfate, which is used to dissolve cell content. Acid detergent

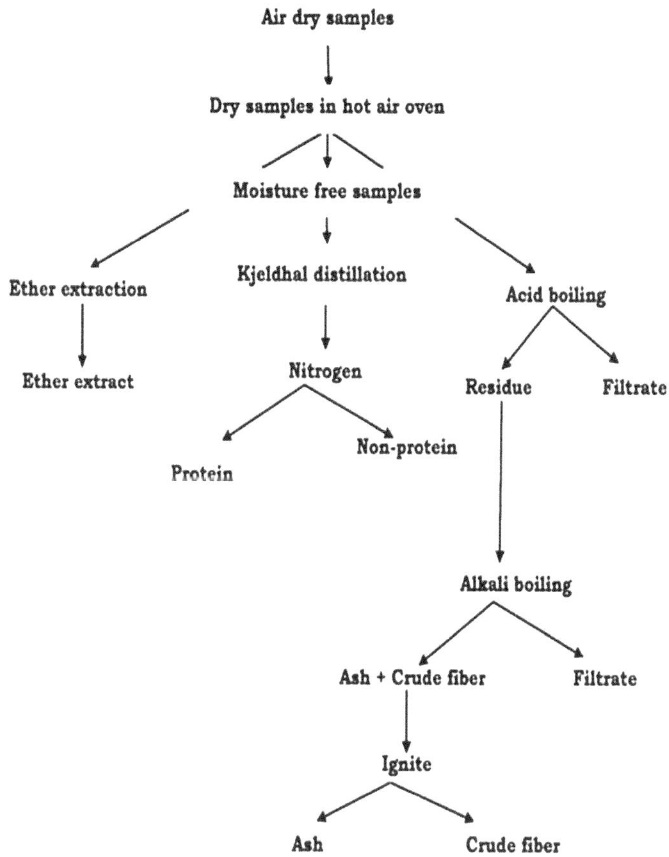

Fig. 9.1 Weende proximate analysis

soluble component is used to determine ADF content. Permanganate is used to determine lignin from ADF fiber. Lignin has very complex structure and composition. Lignification of cell wall in forages is a major deterrent to microbial digestion in animals. In the first step of determination of lignin, forage sample is treated with a neutral detergent solution and rinsed with amylase enzyme to make sugars, starches and pectins soluble. The remaining residue consists of less digestible cell wall substances, such as hemicellulose, cellulose and lignin. In the second step, hemicellulose is made soluble using an acid detergent solvent. The residue, consisting of cellulose and lignin is then treated with concentrated sulfuric acid, which dissolve cellulose and leave lignin in the residue. These steps can be performed consecutively or separately to determine NDF, ADF, and ADL.

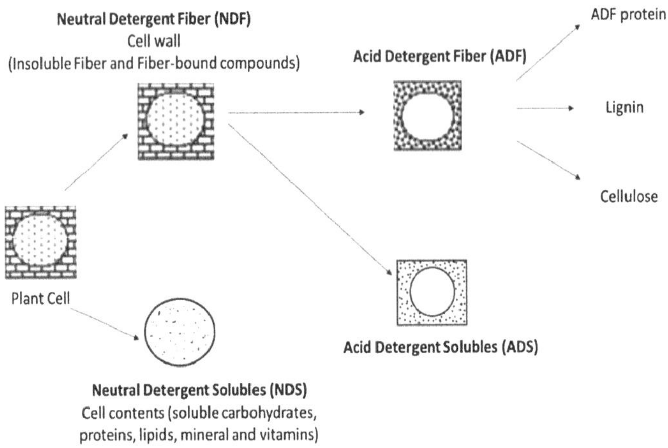

Fig. 9.2 Detergent partitioning of fiber fraction in forages

9.3 Nondestructive Methods for Forage Quality Estimation

9.3.1 Near-Infrared Reflectance Spectroscopy (NIRS) Technique

Near-infrared reflectance spectroscopy (NIRS) is a spectrographic analytical technique used for the estimation of forage quality. In this technique, near-infrared radiation is used to predict quality components in forages and was developed by Norris et al. (1976). The technique involves drying and grinding of feed samples, which are then exposed to infrared light in spectrophotometer. The reflected infrared radiation is converted to electrical energy and the signals are interpreted by the computer. Different organic components in forages absorb and reflect near-infrared light of different wavelengths. Quantification of forage quality is crucial for agricultural research and forage management programmes. The estimation of different quality parameters, such as CP, NDF, ADF and *in vitro* digestibility, by classical analytical techniques is laborious, time consuming and expensive. The evolution of computers and multivariate statistical techniques has enabled the use of NIRS in the estimation of forage quality. NIRS technique is based upon the interactions between light reflectance in wavelength ranging between 750 and 2500 nm and organic compounds in plant biomass. The prediction of forage quality with NIRS method involves analyzing particular forage with both traditional lab analysis and spectrophotometery and then developing a predictive equation by pairing information in the calibration dataset.

9.3.1.1 Principle

NIRS is a rapid and nondestructive technique in which organic compounds are characterized in terms of their absorption properties. All bonds, such as C-H, N-H, and O-H, have absorption bands in near-infrared region (NIR) and NIRS could detect these bonds in fractions of fats, proteins and carbohydrate component in forages. NIR region ranges between 700 and 2500 nm in the electromagnetic spectrum. Infrared (IR) light is emitted and absorbed by all biological compounds. When a sample is scanned, the spectrometer projects NIR light in the sample and the radiant energy is absorbed by the sample molecules according to the frequency of a specific vibration, which results in a unique spectrum. Spectral data in near-infrared region are collected and transformed by calibration models developed for representative feed samples.

9.3.1.2 Procedure

1. *Sampling*: It is important that the feed sample should be representative. All feed samples should usually be kept under conditions that do not change the composition of the sample from the time of sampling to the time of analysis.
2. *Preparation of feed sample*: Feed samples should be prepared in a similar manner as in the case of preparation of validation samples. Before the analysis, the sample should be taken in such a way that it should be the representative of the feed which is to be analyzed. Moist samples should be dried in a hot air oven prior to grinding for 48 h until a constant weight is obtained. After drying, the samples should be allowed to equilibrate with atmospheric temperature and humidity. Alternatively, microwave ovendrying can be used to increase the speed of analysis. Sample analysis should be ground with a cyclone mill fitted with a 1 mm screen. The sample is held in the holder with a disposable back made of rubber of foam core.
3. *Measurement*: NIRS measurement is very sensitive to the temperature of the instrument and sample. NIRS instrument work within normal room temperatures (25 ± 5 °C). The temperature must be equal to room temperature. A constant relative humidity of 60 ± 2% is recommended. Instrument must be properly warmed up. Two diagnostic tests should be performed before beginning each work session. First, noise of the instrument needs to be ascertained. Second, accuracy of wavelength needs to be verified. The development of equation is an essential step in the NIRS analysis.
4. Before using NIRS, calibrations are required. The calibration process establishes a relationship between a spectrum and a reference property through creation of a spectrochemical prediction model. This model aims at making the most accurate and precise prediction for a parameter of interest. Calibration equations can be calculated from the relationship between spectral properties of samples and the results obtained by a reference laboratory method. It is necessary that the laboratory values and techniques should be precise and accurate to develop suitable calibration equations; the lower the error in reference values, the better the precision of the model. NIRS has been successfully used by forage researchers engaged in experimental plant breeding, crop management, plant physiology, and animal nutrition. In all cases, successful use of NIRS technique has been accompanied by the following four important rules:
   (a) Calibration samples represent unknown samples.
   (b) Select correct wavelengths.

(c) Conduct accurate laboratory analyses of calibration samples for validation of results.
(d) Select correct data processing techniques to extract maximum information from spectral data.

The quality of the calibrations in NIRS can be evaluated in terms of the coefficient of determination (R^2), which represents the proportion of variability in the reference data that are accounted for by the regression equation. Other important variables are SEC (Standard Error of Calibration) or SECV (Standard Error of Cross-validation), which represents the variability in the difference between the predicted values and values obtained by the reference methods when the equation was developed from the calibration dataset and RMSEC (Root Mean Square Error of Calibration). When the spectra and the reference data are included and calibration is finalized with high R^2 and low SEC values, it is necessary to validate them, which will evaluate prediction and accuracy of the calibration process. After the first process, it is necessary to include new sample spectra and new reference data to evaluate the quality of the calibration equation, which will generate validation equation and parameters, such as R^2, SEP (Standard Error of Prediction) and RMSEP (Root Mean Square Error of Prediction). With good validation equation parameters and the inclusion of new spectra, it is then possible to make a prediction of selected parameters of new samples. To select NIRS calibration and validation equations, they must have the highest R^2 and lowest errors associated with measurements (Fig. 9.3).

9.3.2 Fourier-Transform Infrared Spectroscopy (FTIR)

Currently, forage quality analysis is not only focused on chemical characteristics and functions of feed ingredients, but is also aimed at understanding their combination in feed and real assessment of their nutritional impact on livestock performance. Infrared (IR) spectroscopy is a technique based on the principle that the type of atoms and atomic bonds within molecules determine the wave numbers in which the electromagnetic vibrations happen when molecules are excited with laser. Therefore, analysis of feeds by IR spectroscopy provides chemostructural information that can be applied by forage nutritionists using chemometric approaches as an alternative method to traditional wet chemistry analysis. These IR techniques require only relatively small amounts of material and generally allow rapid, low-cost, and robust predictions. To this end, NIRS has been successfully used to predict feed composition and kinetics of nutrient degradation in the rumen.

Mid-IR spectroscopy scans a wider spectral range than NIRS (600 to 4000 cm^{-1} v/s 800 to 2500 cm^{-1}, respectively). Like NIR, mid-IR spectroscopy radiation causes vibration of chemical bonds in a sample. However, mid-IR spectroscopy has several important advantages: it is more strongly absorbed by a sample, so it will detect chemical components at much lower levels in a complex mixture; it is absorbed strongly by a much greater range of chemical functional groups, allowing a greater range of chemical components to contribute to the signal and a single

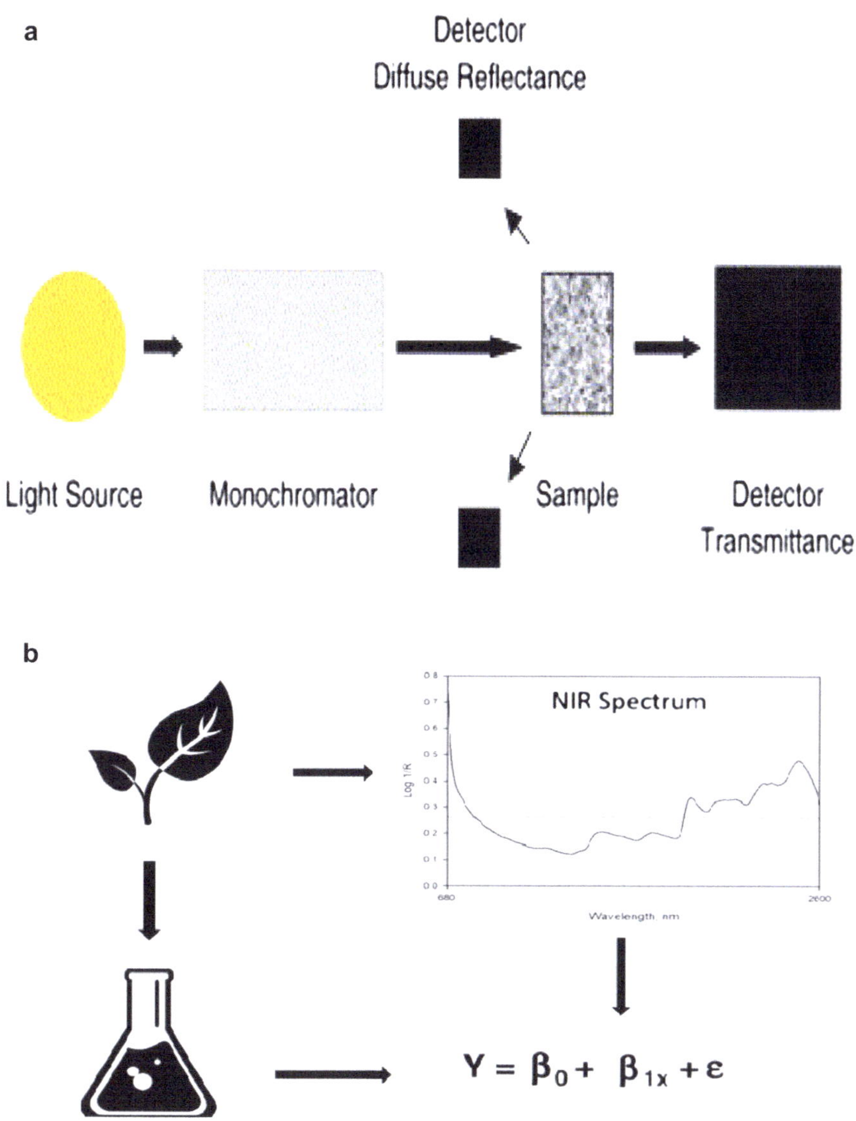

Fig. 9.3 (**a**) Basic principle of NIRS technology (**b**) Illustration of application of NIRS for forage quality predictions

chemical functional group absorbs a narrower range of MIR wavelengths, allowing chemical functional groups and the chemical components associated with those functional groups to be more easily distinguished using MIR radiation. Mid-IR spectroscopy generates information on fundamental molecular vibrations, rather than on harmonic and overtone absorptions and gives better insight into molecular bonds present in a sample. This means that the mid-IR spectrum of a sample should be directly related to its chemical composition. Over the recent years, mid-IR spectroscopy has been revolutionized by the development of Fourier-Transform Infrared (FTIR) Spectrometers. FTIR spectroscopy coupled with multivariate statistical data analysis has enormous potential to facilitate the measurement of forage quality attributes. Generally, the application of statistical methods in feed as well as science allows the highlighting of effective trends, to study and investigate relationships, and to draw conclusions from experimental data.

FTIR spectroscopy represents a rapid, less destructive and high-throughput method for feed analysis. It provides simplified handling and enables analysis of samples directly in their original state. FTIR is a method that predominantly measures the vibrations of chemical bonds within functional groups and generates a spectrum that is considered as a biochemical or metabolic fingerprint of the sample. The characteristic absorption bands are associated with these components, e.g., the amide signals for protein, the carbonyl ester and C-H signals associated with fat, the C-OH bands for carbohydrates and the O-H bending absorption of water. The main advantages of this technique are being simple, inexpensive and the spectrum could be obtained in a few seconds with minimal sample preparation. FTIR spectrometer simultaneously collects high spectral resolution data in a wide spectral range. This provides a significant advantage over a dispersive spectrometer, which measures intensity over a narrow range of wavelengths at a time. Fourier-Transform Infrared Spectroscopy (FTIRS) analysis is commonly used to estimate milk fat and protein concentrations, the ratio of which has been proposed as an indicator of energy balance. Milk FTIRS is also able to predict the concentration of various fatty acids in milk, the composition of which is known to change when body tissue is mobilized, that is, when the cow is in negative energy balance. Energy balance is mathematically very similar to residual energy intake (REI), a suggested measure of feed efficiency of animals (Fig. 9.4).

9.4 Forage Quality Indices

The ultimate measure of forage quality is the animal performance which is a function of forage nutritive value and intake. Forage quality indices are a valuable tool for matching forage quality with nutritional demands of livestock. The indices represent nutritive value metrics and animal intake predictions in a single value. The greater the value, the more likely it is that the forage will meet the nutrient demands of livestock. Various forage quality indices have been developed on the basis of animal requirements and forage intake and they have potential practical applications in livestock feeding as well as harvest and marketing of forages.

(i) *Nutritive value index (NVI)*: NVI, developed by Crampton et al. (1960) is based on the observation that voluntary dry matter intake (DMI) and digestibility of gross energy of forage are essential components of an overall index of forage quality.

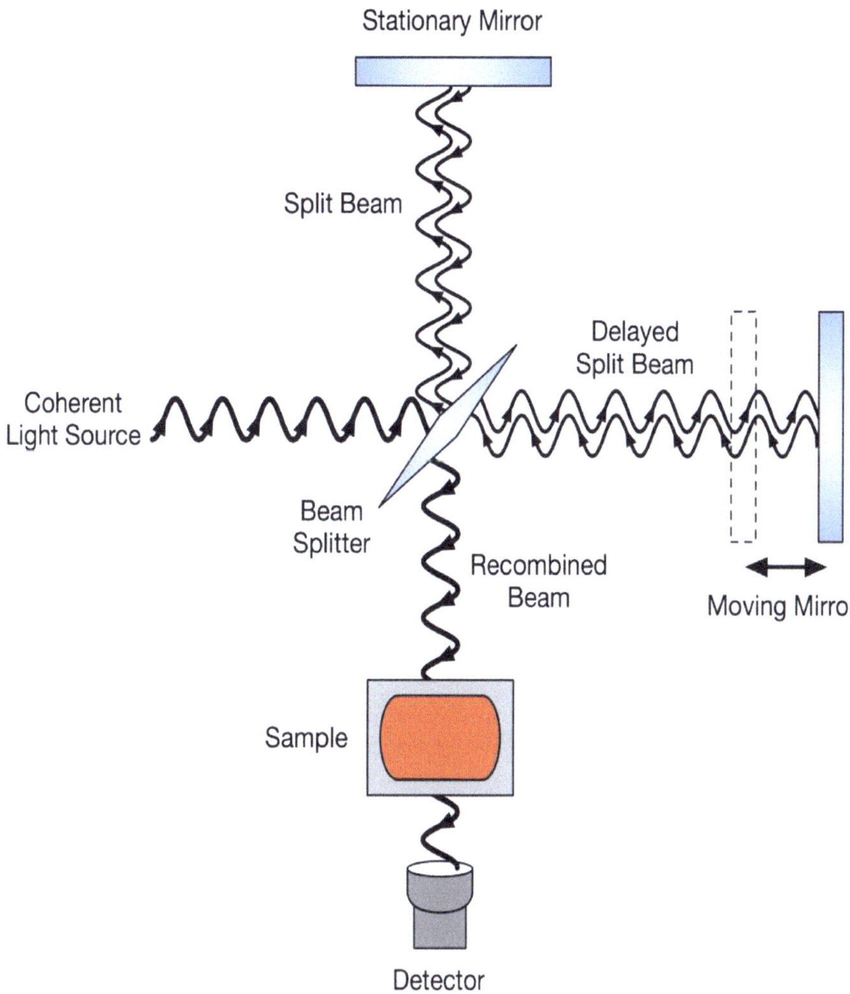

Fig. 9.4 Basic principle of FTIR

$$\text{NVI} = \text{Relative intake} \times (\%\text{Gross energy digestibility})/100$$

where

$$\text{Relative intake} = \frac{\text{Observed DM intake}}{\text{Standard DM intake}} \times 100$$

Dry matter intake can be expressed as grams per day per kg metabolic weight [MW = (live body weight in kg)$^{0.75}$]. The NVI equations may be rearranged as follows:

$$\text{NVI} = \frac{\text{Observed DM Intake} \times \text{Gross energy digestibility\%}}{\text{Standard DM Intake}}$$

(ii) *Digestible energy intake (DEI)*: Heaney et al. (1966) and Heaney (1970) suggested daily digestible energy intake (DEI, kcal/MW) as a forage quality index. The digestible energy intake can be calculated as

$$\text{DEI(Digestible energy intake)} = \text{Observed DM intake} \times \text{Digestible energy concentration}$$

Observed DM intake is expressed as g/MW and DE concentration as kcal/ gm DM.

(iii) *Fill unit and feed unit*: Fill unit relates to voluntary intake and feed unit is with nutritive value. Both fill units and feed units can be expressed in terms of reference feeds. The fill unit (per kg DM) can be calculated as

$$\text{Fill Unit} = \text{Reference DM intake}/\text{Observed DM intake}$$

Feed unit is the ratio of Net energy (NE) content of a feed to NE content of "standard" feed for a particular type of production. One feed unit corresponds to the amount of NE supplied by 1 kg of standard feed for maintenance. The feed unit can be calculated as follows:

$$\text{Feed unit} = \frac{\text{Metabolizable energy concentration} \times \text{Partial efficiency}}{\text{Net energy concentration}}$$

Feed unit index considers variation in both forage intake and nutritive value. Because feed unit is a measure of net energy, the index does consider the efficiency of utilization of digestion end products.

(iv) *Quality index (QI)*: QI is defined as voluntary TDN intake as a multiple of TDN requirement for maintenance. Moore et al. (1990) defined quality index as voluntary intake of forage TDN divided by TDN requirement for maintenance.

$$\text{QI} = \frac{\text{Observed DM intake} \times (\text{Digestible OM, \%})/100}{\text{TDN requirement for maintenance}}$$

If a forage has QI value = 1.0, the animal would just maintain itself if the forage is the only source of dietary energy. Forages with QI values >1.0 are expected to support animal performance.

(v) *Relative feed value (RFV)*: RFV has been used to compare the quality of legume and legume/grass hays and silages. It is calculated based on the two laboratory-determined parameters *viz.*, NDF and ADF levels in forage. NDF has been used as an indicator of forage intake because it takes into account all fiber components and ADF has been used as an indicator of digestibility since it includes cellulose and lignin. Altogether, ADF and NDF take into account the most important traits of forage, intake potential and digestibility and are used to calculate RFV.

$$\text{DDM} = \text{Digestible Dry Matter} = 88.9 - (0.779 \times \%ADF)$$

$$\text{DMI} = \text{Dry Matter Intake}\,(\%\text{of BW}) = 120/(\%\text{NDF})$$

$$\text{RFV} = (\text{DDM} \times \text{DMI})/1.29$$

Limitations of RFV:

- DDM and DMI are assumed constants for all forages.
- ADF and NDF contents are the only laboratory values used in the calculation.
- CP content of forage is not used in the calculation.
- RFV cannot be used in the formulation of ration.

(vi) Relative forage quality (RFQ): To overcome the limitations in using RFV index, RFQ index is developed. This index takes into consideration the differences in digestibility of fiber fractions and can be used to predict animal performance and match animal requirements. The RFQ uses NDF digestibility (NDFD) to account for DMI and true TDN and is calculated as

$$\text{RFQ} = (\text{DMI}, \%\text{of BW}) \times (\text{TDN}, \%\text{of DM})/1.23$$

The value 1.23 ensures that the equation has a mean and range similar to that of RFV. Total digestible nutrients (TDN) and Dry matter intake (DMI) for grasses and legumes can be calculated as

$$\begin{aligned}\text{TDN}_{(\text{legumes})}(\%\text{of DM}) = {} & (\text{CP} \times 0.93) + (\text{FA} \times 0.97 \times 2.25) \\ & + [\text{NDFn} \times (\text{NDFD} \div 100)] + (\text{NFC} \times 0.98) - 7\end{aligned}$$

$$\begin{aligned}\text{TDN}_{(\text{Grasses})}(\%\text{of DM}) = {} & (\text{CP} \times 0.87) + (\text{FA} \times 0.97 \times 2.25) \\ & + [\text{NDFn} \times (\text{NDFDp} \div 100)] + (\text{NFC} \times 0.98) - 10\end{aligned}$$

where

- NDFn = Nitrogen-free NDF = NDF − NDFCP, also estimated as NDFn = NDF × 0.93
- NDFCP = Neutral detergent fiber crude protein
- NDFD = 48 h *in vitro* NDF digestibility (% of NDF)
- NFC = Nonfibrous carbohydrate (% of DM) = 100 – (NDFn + CP + EE + Ash)

- EE = Ether extract (% of DM)
- NDFDp = 22.7 + 0.664 × NDFD

$$\mathrm{DMI}_{(\mathrm{Legumes})}(\%\mathrm{of\,BW}) = (120 \div \mathrm{NDF}) + \{(\mathrm{NDFD} - 45) \times 0.374 \div 1350 \times 100\}$$

where

- 120 is a base NDF intake of 1.2% of body weight (legumes only)
- 45 is an average value for fiber digestibility of alfalfa and alfalfa/grass mixtures, and 1350 is the body weight in pounds of a typical cow (used for legumes only)
- NDF (Neutral detergent fiber) as % of DM (Dry matter) and NDFD (48-h *in vitro* NDF digestibility) as % of NDF

$$\mathrm{DMI}_{(\mathrm{Grass})}(\%\mathrm{of\,BW}) = -2.318 + 0.442 \times \mathrm{CP} - 0.0100 \times \mathrm{CP}^2 - 0.0638 \times \mathrm{TDN} + 0.000922 \times \mathrm{TDN}^2 + 0.180 \times \mathrm{ADF} - 0.00196 \times \mathrm{ADF}^2 - 0.00529 \times \mathrm{CP} \times \mathrm{ADF}$$

9.5 Estimation of Forage Quality Components

9.5.1 Moisture

Animal feeds contain six classes of nutrients including water, carbohydrates, proteins, fats, vitamins and minerals. Among these water or moisture in feedstuffs is an important factor for transportation and economic values of feeds. Therefore, determination of moisture is the first step in the quality analysis of feeds. An accurate assessment of moisture content in feed ingredients is important because moisture variations influence the nutritional evaluation of feedstuffs. The amount of dry matter in a feed is inversely related to the amount of moisture it contains. The moisture determination ultimately influences the nutritional evaluation of ingredients when converting nutrient contents on a dry matter basis (Thiex and Richardson 2003).

Procedure: Moisture determination in a sample is carried out in an aluminum cup. The procedure involves weighing 5 g of ground sample in the aluminum cup and drying the aluminum cup in oven at high temperature for about 15 to 30 min. Aluminum cup is allowed to cool and is weighed. The process is repeated until the difference between two successive weighings is minimal (<1 mg). Toluene determination of moisture (AOCC 1990) is another procedure to determine moisture content in feeds having high volatile acids content. Basic principle of this procedure is to distill water from sample and trap it under a layer of toluene to determine moisture content.

9.5.2 Estimation of Dry Matter (DM)

DM is the percentage of the feed without moisture. Fresh grass has higher moisture content and lower DM content compared to hay. Expressing nutrients on a dry matter basis allows comparison among feeds by correcting "as-fed" moisture. Diets are formulated on DM basis, but ingredients are weighed out or mixed on as-fed basis. Dry matter (DM) is the total of all solid components in forages without the water content. The dry matter therefore estimates cellulose, starch, sugars, structural compounds, proteins, fats, pigments and minerals. Using dry matter, it is possible to calculate the optimum amount of nutrients in forages. If the feed amounts are not adjusted according to DM content and the same weight of feed is given to the animals, they will either get too much or too little proteins and energy. DM increases with plant age when harvested. The lower the DM, the more the moisture and the lower the nutrient density in the fresh feed. High moisture may decrease the keeping quality of a feed (through molding) unless it is made into silage. In forages, both DM content and moisture content are interrelated as

$$\text{DM}(\%) = \text{Total weight} - \text{Moisture content}$$

9.5.2.1 Dry Matter Estimation in Bulk Feed Samples

Procedure: Sample is weighed in required quantity in a dry, pre-weighed moisture box and put in a hot air oven at 100 °C ± 2 °C overnight. Quickly after cooling, moisture box is weighed at room temperature. Difference in weight is expressed as moisture percentage from which DM percent is calculated. Ovendried sample is stored in airtight containers for further estimation of various quality parameters.

Calculation

$$\text{Moisture}(\%) = \frac{(B - A) - (C - A)}{(B - A)} \times 100$$

$$\text{Dry matter}(\%) = 100 - \text{Moisture}(\%)$$

$$\text{Dry matter}(\%) = \frac{(C - A)}{(B - A)} \times 100$$

where

A = Empty weight (g) of moisture box
B = Weight (g) of moisture box with sample before ovendrying
C = Weight (g) of moisture box with sample after ovendrying

9.5.2.2 Dry Matter Estimation in Laboratory Samples

Procedure: Moisture box is dried in a hot air oven at $100 \pm 2\,^{\circ}\mathrm{C}$ for 10 to 15 min. The moisture box is allowed to cool and weigh. Ground sample is spread on a sheet of white paper to attain room temperature and then transferred into moisture box and weighed. Sample is dried in the hot air oven at $100 \pm 2\,^{\circ}\mathrm{C}$ for 8 to 12 h. The moisture box is cool in a desiccator with closed lid and weighed carefully. The drying, cooling and weighing process is repeated until constant weight of moisture box is achieved. The ovendried sample is stored in airtight containers for the estimation of other quality parameters.

Calculation

$$\mathrm{Moisture}(\%) = \frac{(B-A)-(C-A)}{(B-A)} \times 100$$

$$\mathrm{Dry\ matter}\,(\%) = 100 - \mathrm{Moisture}(\%)$$

$$\mathrm{Dry\,matter}(\%) = \frac{(C-A)}{(B-A)} \times 100$$

where

A = Empty weight (g) of moisture box
B = Weight (g) of moisture box and sample before ovendrying
C = Weight (g) of moisture box and sample after ovendrying

9.5.2.3 Dry Matter Estimation in Silage, Haylage and Molasses

Procedure: Add 400 mL toluene to distillation flask containing 70 to 80 g of sample. The content of flask is distilled for 6 to 8 h at the rate of 2 to 3 drops per second. After observing two equal consecutive volumes at 15 min interval, distillation is stopped. Two or three washings are given to the condenser with toluene to remove traces of moisture and allow distillation for another 15 min. Keep receiver flask in a water bath at 20 °C for 20 min and note the volume of water. Transfer aqueous layer to volumetric flask of 100 mL capacity and make up the volume with distilled water. About 40 mL of neutralized ethanol solution is added (80% alcoholic ammonia solution neutralized to phenolphthalein) to 10 mL diluted distillate solution and titrate against 0.10 N NaOH using phenolphthalein indicator.

Calculation

$$\mathrm{DM} = 100 - \frac{99.8\,B}{A} \times (t - f/100)$$

where

A = Weight (g) of sample
B = Volume (mL) of aqueous phase
t = Volume (mL) of 0.1 N NaOH utilized
f = Factor depending on the constituents of acids of sample (Normally 0.00555 is taken for all practical purposes)

9.5.2.4 Estimation of Dry Matter by Near-Infrared Reflectance Spectroscopy (Marten et al. 1989)

NIRS (Near-infrared Reflectance Spectroscopy) is a rapid and low-cost computerized method to analyze forage as well as food crops for their nutritive value. Instead of using chemicals, as in conventional methods, NIRS uses near-infrared light to determine protein, fiber, energy, and mineral contents. This analytical method involves drying and grinding of samples, which are then exposed to infrared light in spectrophotometer. The reflected infrared radiation is converted to electrical energy and fed to a computer for interpretation. Each major organic component of forages will absorb and reflect near-infrared light differently. By measuring these different reflectance characteristics, the NIRS unit and a computer determine the quantity of these components in the feed sample. The detection of specific nutrients is possible because reflectance spectra from forage samples of established nutrient values by wet chemistry procedures are programmed into a computer. When a similar feed sample is evaluated by NIRS, the computer compares wavelength reflections by feed sample and matches them to previously tested feed samples. NIRS method of determining forage quality is very rapid and less expensive than wet chemistry methods. Accuracy depends on good sample collection, storage, and consistent drying, grinding, and mixing of samples prior to analysis. The calibration set that is used must be developed from an adequate number of wet chemistry samples similar to those being analyzed. Without proper calibration, the NIRS analysis can have serious errors.

Dry matter estimation using NIRS is applicable for determining dry matter of ground, air-dried or partially dried forage in laboratory. Samples are finely ground through grinder; random portions of a sample are loaded into an NIR sample holder, and light reflected from the ground sample is measured in the infrared region. Equations derived from the spectral data are used to calculate dry matter content of feed and forage samples.

The NIRS technique has many advantages, which are as follows:

- It can be used anywhere: in pastures, farms, silos, or barns
- As it gives rapid results, a farmer can test several bales of hay or parts of a crop per day
- NIR can be used with all forage crops
- It is suitable for determining harvest time and quality during storage
- It is useful for quality estimation of silages and hay

9.5.3 Estimation of Crude Protein (AOAC 1970)

Proteins are the most important nutrients for livestock, as they support rumen microbes that consequently degrade forage. Crude protein is an estimate of the level of protein in the feed based on the amount of nitrogen present. Since some, but not all, of the nitrogen is in the form of true protein, it is termed "crude" protein. CP also does not give individual amino acid profiles. CP includes both true protein and nonprotein nitrogen. Because feeds high in protein are generally more expensive, therefore, comparison of feed costs is significant on the basis of cost per pound of CP rather than per pound of DM. For swines, poultries and other monogastrics, knowing the amount of each essential amino acid in the feed is critical. Some protein fractions are more digestible than others, but in general, the higher the protein level, the more digestible is the feed. CP content of forage crops is high at initial stages and begins to decline with the advancement in maturity. Legumes such as alfalfa and clovers are generally higher in protein than grasses and other forages. Seed oil meals have higher CP values than whole grains. The protein quality of grasses declines more rapidly than that of legumes with plant age, and protein quality of tropical forages decline more rapidly with plant age than temperate forages.

The Kjeldahl method is a commonly used procedure for the estimation of percent nitrogen content and in turn protein content in samples. During the procedure, the amino (-NH_2) nitrogen in samples is oxidized by sulfuric acid in the presence of a catalyst to $(NH_4)_2SO_4$. The ammonium ion is converted to ammonia by NaOH. The liberated ammonia is collected in a container by distillation and is quantitatively titrated against standard acid (HCl or H_2SO_4) of known strength and nitrogen in the sample is computed. The determination of crude protein involves multiplication of percent nitrogen content by a factor of 6.25. Multiplication by the factor 6.25 is based on the assumption that most of the feeds have 16% nitrogen. It provides an estimate of total protein without any consideration of quality. It also does not distinguish the nitrogen contributed from true protein and nonprotein nitrogenous substances.

Procedure: Approximately 0.25 g finely ground moisture-free feed sample is taken in digestion tubes and 15 mL conc. H_2SO_4 and 1 g of digestion mixture (mixture of potassium sulfate and copper sulfate in 9:1 ratio) are added. The feed sample is digested in Kjeldahl heating unit (400 to 500 °C) until the solution becomes colorless. After digestion, the final volume in tubes is made up to 50 mL with distilled water. Ten milliliters of 4% boric acid solution is taken in a conical flask of 100 mL capacity. The receiving flask is placed in such a way that condenser outlet completely dips into boric acid solution. Aliquot of digested sample (5 mL) is transferred to the Kjeldahl distillation apparatus and 5 to 8 mL of 40% NaOH solution is added to digested sample. Immediately close the stopcock and pass steam through a steam chamber in order to distill ammonia till about 30 to 40 mL of distillate is collected in receiving flask. Receiving flask is removed and condenser outlet tip in the receiving flask is rinsed with distilled water. The content in the flask is titrated against 0.01 N HCl until the bluish green color changes to light pink.

Calculation

$$1\,\text{mL of } 0.01\,\text{N HCl} = 0.00014\,\text{g nitrogen}$$

$$\text{Volume of } 0.01\,\text{N HCl used for blank solution} = a\,\text{mL}$$

$$\text{Volume of } 0.01\,\text{N HCl used for sample} = b\,\text{mL}$$

$$\text{Volume used in titration} = a - b\,\text{mL}$$

$$\text{Volume of sample for distillation} = 5\,\text{mL}$$

$$\text{Total volume made of the digested sample} = 50\,\text{mL}$$

$$\text{Nitrogen present in 5 mL of digested sample} = (a - b) \times 0.00014\,\text{g}$$

$$\text{Nitrogen present in 50 mL digested sample} = \frac{(a - b) \times 0.00014 \times 50\,\text{g}}{5}$$

$$\%\text{N in 100 g sample} = \frac{(a - b) \times 0.00014 \times 50\,\text{g} \times 100}{5}$$

$$\%\text{protein content in sample} = \frac{6.25 \times (a - b) \times 0.00014 \times 50\,\text{g} \times 100}{5}$$

9.5.4 Estimation of Nonprotein Nitrogen in Feed Samples

Nutrients such as protein, fat, calcium salt, etc. are equally important for health and productivity of livestock. Among these nutrients, protein plays a vital role in the energy and body growth of livestock. Normally nutritionist makes the formula of feed from the analysis record of feed raw materials. Protein-rich raw materials contain nitrogen and the protein is measured on the basis of nitrogen content in the raw materials but they also contain nonprotein nitrogen (NPN).

The ruminal microflora of cattle and sheep can utilize NPN to synthesize proteins that can replace a portion of their total dietary protein requirements. Given that feeding NPN sources is typically cheaper than feeding expensive true protein, the practice of adding NPN to ruminants' diets is relatively commonplace. Feed grade urea ($CO[NH_2]_2$) is the cheapest, most effective, and most available nitrogen source. Other NPN compounds that may be less readily available include ammonium acetate ($CH_3CO_2NH_4$), ammonium bicarbonate (NH_4HCO_3), ammonium carbamate ($NH_2CO_2NH_4$), ammonium lactate ($CH_3CHOHCO_2NH_4$), ammonium sulfate ($[NH_4]_2SO_4$), monoammonium phosphate ($NH_4H_2PO_4$), and biuret ($NH_2CONHCONH_2$). All these NPN sources consist of varying nitrogen content and protein equivalent; so the addition of these products into the diet requires stringent nutritional dietary examination. Many of these supplements are solids and can be directly mixed into the mineral or grain component of the diet. Some, such as urea, can be added to molasses-based liquid products or into solid mineral blocks. NPN in feed samples is calculated by the difference between total crude protein nitrogen and the value of precipitated true protein nitrogen.

Procedure: Moisture-free ground feed sample (0.50g) is weighed in 125 mL Erlenmeyer flask. 50 mL of distilled water is added to the sample and allowed to stand for 30 min. 10 mL of 10% tricholoroacetic acid solution is added and allowed to stand for 20 to 30 min. The content of flask is filtered with Whatman No. 54 filter paper and washed twice with TCA solution. Filter paper is transferred to digestion tube and the amount of residual nitrogen is determined by micro-Kjeldahl method. The amount of NPN is calculated by subtracting residual nitrogen from total nitrogen. The amount of NPN is expressed as crude protein ($N \times 6.25$) or as a percent of total feed nitrogen.

9.5.5 Estimation of Urea Nitrogen in Feed Samples

To determine the protein content of a feedstuff, it is usual to ascertain first the percentage of nitrogen by chemical analysis. This estimate is then multiplied by 6.25 as the average protein content of a feed is 16% nitrogen ($6.25 \times 16 = 100$). The resulting value is called crude protein, as distinguished from true protein, because some of the nitrogen analyzed is not derived from protein. In most grasses and other green feeds, only a part of the nitrogen comes from protein and the balance consists of inorganic nitrogen salts, amino nitrogen, amides, and other forms. This is of no importance, however for ruminants as they can utilize inorganic nitrogen as well as protein nitrogen through the microbial activity in rumen, where bacteria thrive on nonprotein nitrogen and incorporate it in their own proteins. The protein in the bodies of microorganisms is then digested in the intestinal tract of ruminant and absorbed. Hence, instead of feeding livestock expensive true protein, cheaper sources of nitrogen can be equally effective.

Urea contains 46% nitrogen, thus each kilogram of urea is equivalent to 2.88 kg of crude protein. Fertilizer urea is hygroscopic and cakes very easily, making it difficult to mix with solid feeds. To improve flow characteristics, urea is processed into feed grade urea (42% nitrogen), in which each grain of urea is covered with kaolin or some other nonhygroscopic substance. The cheaper fertilizer urea can be used, however, when mixed with liquid feeds or even in solid feeds if added in the form of a suspension or solution in molasses. Urea stops bacterial growth and fermentation in concentrations over 10% but it has a very bitter taste and limits intake if used at high levels. Urea is the most widely used nonprotein nitrogen. The seeds of some legumes, especially soybeans, contain enzyme urease, which breaks down urea and renders feed unpalatable. Urease is largely destroyed by heat during processing of beans so that oil meals can usually be mixed with urea. A simple test to ensure that no ingredients that are to be mixed with urea contain urease is to moisten a mixture of urea and the suspected ingredients and allow it to stand under cover for 1 h. Urea is not necessary in the diet of livestock; it is fed as a replacement for a part of the protein in a ration. Since urea is used as an additive for improving the quality of livestock diets, the estimation of urea nitrogen is essentially required. The procedure to estimate urea nitrogen in feed samples is described as follows:

Standardization of Urease Solution Dissolve 0.10 g commercial urease enzyme in 50 mL of distilled water. Determine the alkalinity by titrating against 0.10 N HCl solution using methyl red indicator. Prepare neutralized urease solution as standardized above by dissolving equal volumes of 0.10 N HCl and urease solution of 0.10 g in distilled water, therefore every 10 mL of neutralized solution may convert more than 0.10 g of urea into nitrogen.

Determination of Enzyme Activity Determine the urease enzyme activity by preparing 50 mL of 1% neutralized solution. Add different quantities of neutralized urease solution to each 0.10 g pure urea. Distill the content after digestion with enzyme. Calculate the activity of urease solution.

Procedure: The estimation of urea nitrogen in feed samples involves shaking 2 g of ground feed sample in Kjeldahl distillation flask with 250 mL of distilled water and 10 mL of urease solution. Thereafter, reaction is allowed for 1 h at room temperature. About 2 g carbonate-free magnesium oxide, 1 mL calcium chloride (0.25%), and 5 mL defoaming solution (Dissolve 50 g diglycol stearate in a mixture of 375 mL benzene, 75 mL alcohol and 250 mL dibutyl phthalate) are added to the content of flask. Add methyl red indicator by directly connecting Kjeldahl distillation flask to a condenser until 100 mL of filtrate is collected in a receiver flask containing standard sulfuric acid. The content of flask is titrated against standard sodium hydroxide solution.

Calculation

$$\text{Urea nitrogen}(\%)(\text{as total ammonical N}) = \frac{X(A - B)N}{W}$$

where

A = mL of standard sodium hydroxide required to neutralize excess acid in test
B = mL of standard sodium hydroxide required to neutralize excess acid in blank
N = Normality of standard sodium hydroxide
W = Weight of moisture-free sample
X = Factor for converting 1 mL of standard sulfuric acid into nitrogen

9.5.6 Determination of Fiber Content in Forages

9.5.6.1 Crude Fiber Estimation (AOAC 1965)

Estimation of crude fiber is one part of the original system of analyzing digestible fraction of feedstuffs. The analytical method uses sequential acid and alkali extraction. It was developed by Henneberg and Stohmann during the 1860s at the Weende

Experiment Station in Germany and is part of the Weende system of proximate analysis. Since crude fiber accounts for most of the cellulose but only a portion of the hemicellulose and lignin and no ash, it underestimates true fiber. Crude fiber values are less than ADF values. Thus crude fiber is not a perfect indicator of digestibility in ruminants.

Procedure: About 2 g moisture-free sample is weighed and transferred to 1 L beaker. In the next step, 200 mL of 1.25% H_2SO_4 solution is added and placed on a hot plate, covered with round bottom flask condenser, and allowed to reflux for 30 min. The content of beaker is shaken every 5 min and after 30 min of boiling, beaker is removed from the hot plate and filtered through a muslin cloth using vacuum suction. The residue is washed with hot water to remove acid. Material is transferred to the same beaker and 200 mL of 1.25% NaOH (*w/v*) solution is added and again the content is refluxed for 30 min. The content of beaker is filtered again through the muslin cloth with the help of vacuum pump and the residue is washed with hot water to remove alkali. Total residue is transferred to a crucible and allowed to dry to a constant weight at 80 to 110 °C and the weight is recorded. Residue is ignited in a muffle furnace at 550 to 600 °C for 3 h and weighed. The loss of weight due to ignition is the weight of crude fiber.

Calculation:

$$\text{Crude fiber}(\%) = \frac{\text{Weight of crude fiber}}{\text{Original weight of sample}(\text{g})} \times 100$$

9.5.6.2 Estimation of Detergent Fibers

Animal nutritionists have replaced crude fiber assay with Van Soest detergent fiber analysis system. The technique of using detergents to separate digestible and indigestible parts of plant tissues was originally proposed by Van Soest in 1963. Plant cell substances can be divided into less digestible cell wall constituents (hemicelluloses, cellulose, and lignin) and highly digestible cell contents (starch and sugars). The highly digestible cell contents are separated from cell wall using two different detergent systems: Neutral detergent fiber (NDF) and Acid detergent fiber (ADF).

Acid Detergent Fiber (ADF) This fibrous component represents the least digestible fiber portion of forage or other roughage. This highly indigestible part of forage includes lignin, cellulose, silica, and insoluble forms of nitrogen, but not hemicelluloses. Forages with high ADF values have low digestible energy than forages with low ADF values. During laboratory analysis, ADF is the residue remaining after boiling a test material in acid detergent solution. ADF is often used in nutritional equations to calculate digestibility, total digestible nutrients (TDN), and/or net energy for lactation (NEL).

Neutral Detergent Fiber (NDF) NDF is the residue or insoluble fraction left after boiling a feed material in neutral detergent solution. The NDF contains insoluble

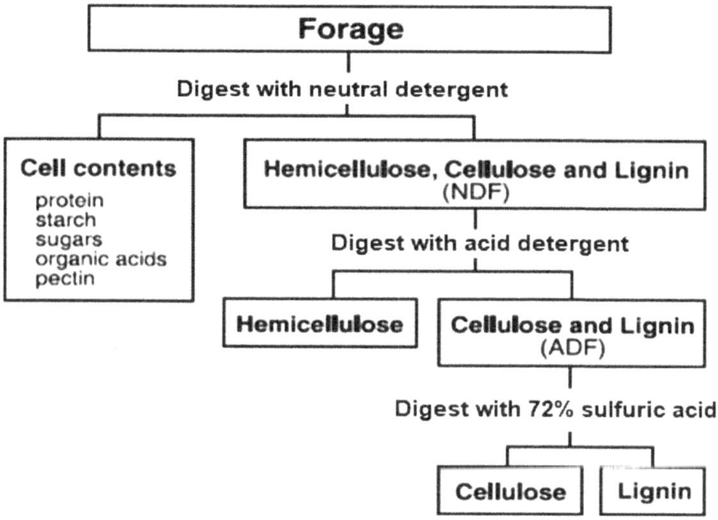

Fig. 9.5 Scheme for the determination of detergent fibers

plant cell wall components that include cellulose, hemicelluloses, lignin, silica, and cutins. The hemicelluloses, cellulose, and lignin represent fibrous content of forage (Fig. 9.5).

(a) *Determination of neutral detergent fiber (NDF)* (Van Soest and Wiens 1967)
Neutral detergent solution (1 L) contains sodium lauryl sulfate (30 g), EDTA disodium salt (18.61 g), sodium borate decahydrate ($Na_2B_4O_7.10\ H_2O$) (6.81 g), sodium phosphate (dibasic) (Na_2HPO_4) (4.56 g), and 2-ethoxy ethanol (10 mL). To prepare 1 L of neutral detergent solution, put EDTA and sodium borate in 1 L beaker and add some of the distilled water and heat until chemicals are completely dissolved. Take sodium phosphate (dibasic) in a separate beaker and add a few mL of distilled water and heat until it is completely dissolved. Mix both the solutions and maintain pH between 6.90 and 7.10.

Procedure: Weigh 0.5 to 1 g moisture-free sample in a clean beaker and add 100 mL neutral detergent solution, 2 mL decahydronapthlene, and 0.5 g sodium sulfite. Boil the content for 5 to 10 min and reflux the content for 60 min. Filter the content through weighed crucible and wash with hot distilled water. Repeat the washing procedure thrice. Wash the residue with acetone and break all the lumps so that the solvent may have contact with all particles. Wash the residue with hexane and finally dry at 100 °C for 8 h and weigh.

Calculation

$$\mathrm{NDF}(\%) = \frac{(\text{Weight of crucible} + \text{fiber}) - \text{Weight of crucible}}{\text{Weight of sample}} \times 100$$

$$\text{Cell contents}\,(\%) = 100 - \text{cell wall constituents}$$

$$\text{Insoluble ash in neutral detergent}(\%) = \frac{\text{Weight of crucible} + \text{Ash} - \text{Weight of crucible}}{\text{Weight of dry sample}} \times 100$$

(b) *Determination of acid detergent fiber (ADF)* (Van Soest and Wiens 1967)
For the preparation of neutral detergent solution (1 L), add 20 g of CTAB in 1 N H_2SO_4 and make the final volume to 1 L with 1 N H_2SO_4.
Procedure: Weigh 1 g moisture-free sample in a clean beaker and add 100 mL acid detergent solution and 2 mL decahydronapthlene. Boil the content for 5 to 10 min and avoid foaming during boiling. Reflux the content for 60 min. Filter the content through pre-weighed crucible and rinse the sample with hot distilled water. Repeat the washing procedure thrice. Again wash the residue with acetone and break all the lumps. Wash the residue with hexane and finally dry at 100 °C for 8 h and weigh.
Calculation

$$\text{ADF}(\%) = \frac{(\text{Weight of crucible+fiber}) - \text{Weight of crucible})}{\text{Weight of the sample}} \times 100$$

9.5.7 Estimation of Cell Wall Constituents

9.5.7.1 Estimation of Cellulose

The major component in rigid cell wall in plants is cellulose. Cellulose is a linear polysaccharide polymer with many glucose monosaccharide units. The acetal linkage is β, which makes it different from starch. This peculiar difference in acetal linkages results in a major difference in digestibility. Human beings are unable to digest cellulose because of β-linkage. Indigestible cellulose aids in smooth working of intestinal tract. Animals such as cows, horses, sheep, goats, and termites have symbiotic bacteria in the intestinal tract. These symbiotic bacteria secrete necessary enzymes to digest cellulose in gastrointestinal tract.

Analysis of cellulose content in forages involves hydrolysis of acetylated cellodextrins (acetylated cellulose with acetic acid and nitric acid) with 67% H_2SO_4. The hydrolysis releases glucose molecules which are further dehydrated to form hydroxymethyl furfural. The dehydrated product forms blue-green product with anthrone reagent and intensity of color is measured at 630 nm.

Procedure: To 0.50 g dried sample, 3 mL acetic/nitric reagent (mix 150 mL of 80% acetic acid and 15 mL of conc. nitric acid) is added in a test tube and stirred. The tube is placed in a water bath at 100 °C for 30 min. Cool and centrifuge the contents for 15 to 20 min. Supernatant is discarded and the residue is washed with distilled water. Add 10 mL of 67% sulfuric acid and allow to stand for 1 h. Take 1 mL of above solution and dilute it to 100 mL. To 1 mL of diluted solution, 10 mL of anthrone reagent is added (dissolve 0.20 g of anthrone in 100 mL of 95% sulfuric acid). The reagent is always prepared fresh. The content is heated in a boiling water bath for 10 min. Measure the intensity of blue-green color at 630 nm. Set a blank with anthrone reagent and distilled water. Similarly, run a blank by taking cellulose as a standard. Draw a standard curve and calculate the amount of cellulose in feed sample.

Preparation of standard curve: Aliquots of working standard of cellulose (prepare from standard cellulose solution containing 100 mg cellulose in 100 mL of distilled water) corresponding to 10, 15, 20, 25, 30, and 35 mg of cellulose. Add 3 mL acetic/nitric reagent (mix 150 mL of 80% acetic acid and 15 mL of conc. nitric acid) to a test tube and stir. Place the tube in a water bath at 100 °C for 30 min. Cool and centrifuge the contents for 15 to 20 min. Supernatant is discarded and the residue is washed with distilled water. Add 10 mL of 67% sulfuric acid and allow it to stand for 1 h. Take 1 mL of above solution and further dilute it to 100 mL. To 1 mL of diluted solution, 10 mL of anthrone reagent is added (dissolve 0.20 g of anthrone in 100 mL of 95% sulfuric acid). Prepare fresh, chill on ice and mix well. Heat the content of tubes in a boiling water bath for 10 min. Cool and measure the intensity of color at 630 nm. Plot absorbance (nm) values against concentration to prepare a standard curve.

Calculation

$$\text{Cellulose}(\%) = \frac{\text{Standard concentration} \times \text{Total volume} \times \text{OD of sample}}{\text{Standard OD} \times \text{Weight of sample} \times \text{Aliquot used}} \times 100$$

9.5.7.2 Estimation of Silica (Van Soest and Wine 1968)

Silica is deposited on the cell walls of plants as part of the cell maturity. In grasses, silica is considered an anti-nutritional component for its negative impact on the nutritional availability of plant fiber. It interferes with the digestion of cell wall polysaccharides by acting as a physical barrier to microbial enzymes. The procedure for the estimation of silica content in forages is described below.

Procedure: Prepare acid detergent fiber according to the procedure described previously. Take 1 gm sample (ADF) in pre-weighed crucibles. Saturated potassium permanganate is prepared by dissolving 50 g $KMnO_4$, 0.05 g Ag_2SO_4 in 1 L distilled water. Silver sulfate is added to dehalogenate the reagent. Lignin buffer solution is prepared by dissolving 6 g ferric nitrate nonahydrate and 0.15 g silver nitrate in distilled water. Combine this solution with 500 mL acetic acid and 5 g potassium acetate. Add 400 mL of tertiary butyl alcohol solution, mix, and raise the final volume to 1 L with distilled water. Permanganate solution is prepared by mixing saturated potassium permanganate and lignin buffer solution in the ratio of 2:1 by volume before use. About 25 mL of permanganate solution is added to the crucible in an enamel pan containing cold water. Level (2 to 3 cm) of water in the pan is adjusted to reduce the flow of solution out of crucible. Stir the contents to break lumps with a small glass rod and to draw permanganate solution upon the sides of crucible to wet all particles. Allow crucible to stand at 20–25 °C and add permanganate solution if necessary. Purple color must be present at all times. Crucibles from filtering apparatus are removed and placed in an enamel pan and crucible is filled half with demineralizing solution (dissolve 50 g of oxalic acid dehydrate in 700 mL of 95% ethanol). Add 50 mL of conc. HCl (12 N) and make the final volume to 1 L with distilled water. Demineralizing solution may be added directly to the crucible in case filtering is difficult. Care must be taken to avoid spillage by foaming. After 5 min, suck dry on filter and refill half with demineralizing solution. The process is repeated after the second interval if a solution is very brown. Rinse the sides of the crucible with solution from a wash bottle with a fine stream. Treat fiber until it is totally white. Fill and thoroughly wash the crucible and contents with 80% ethanol. The process is repeated twice and washed twice in a similar manner with acetone. The residue is ignited at 500 °C for 3 h, cooled, and weighed. Calculate residual ash as the difference between this weight and original weight of crucible.

The presumptive analysis of silica content in forages is carried out by treating residue left after the estimation of permanganate lignin with hydrobromic acid. Add enough drops of 48% HBr (4 mL) to residue and allow to stand for 1 to 2 h. Add more drops of HBr if much red color forms. Excess acid is removed and washed once with acetone. Ignite at 500 °C, cool, and weigh.

Calculation

$$\text{Silica}(\%)\text{on DM basis} = \frac{(\text{Weight of crucible} + \text{ash after HBr washing}) - \text{Weight of empty crucible}}{\text{Weight of oven dried sample taken for ADF analysis}} \times 100$$

9.5.7.3 Lignin Analysis

Forages are an important source of energy and a major portion of this energy comes from the plant cell wall. Lignin is a complex heteropolymer present in the secondary cell wall of plants. After complete deposition in the secondary cell wall, lignin provides a coating for cellulose–hemicellulose microfibrils thus ensuring strength and impermeability for lignified tissues. The complex structure of lignin includes ether and carbon–carbon bonds which cross-links with wall polysaccharides and make them unavailable to cell wall degrading enzymes. Though lignin ensures the proper functioning of plant systems, its presence at higher amounts in forages is not nutritionally desirable. In forages, the main anti-quality role of lignin is in limiting the digestion of structural polysaccharides that reduces the amount of total available digestible energy and dry matter intake. If animals consume less digestible forages, the slow feed passage rate through the alimentary canal reduces feed intake resulting in weight loss and nutrients deprivation in livestock.

The more commonly used methods of lignin analysis are classified into two categories: i) methods that remove the cell wall constituents except for lignin and ii) methods that break down the lignin polymer separating it from the cell wall structure. Treatment with conc. sulfuric acid (72%) results in an insoluble acid residue after hydrolysis of cell wall polysaccharides. Such is the case of acid detergent lignin (ADL) and Klason lignin. Lignin can also be determined by the difference in weight after lignin degradation with oxidizing reagents such as potassium permanganate.

The acetyl bromide and thioglycolic acid methods are based on solubilization of lignin and determination of absorbance at 280 nm. The acetyl bromide protocol is based on the formation of acetyl derivatives in non-substituted OH groups and bromide replacement of carbon OH groups for complete solubilization of lignin under acidic conditions. Thioglycolic acid method is based on the formation of thioethers of benzyl alcohol groups in lignin structure, resulting in solubilization of lignin under alkaline conditions. The Klason method is a gravimetric assay used for the direct quantification of lignin. In this method, an insoluble lignin fraction is extracted from plant tissues after digestion with 72% sulfuric acid followed by partial solubilization of cell wall polysaccharides.

Methods for the estimation of lignin content in forages are described as follows:

9.5.7.3.1 Determination of Acid Detergent Lignin (ADL) (Van Soest and Wiens 1967)

To determine ADL content in forage samples, preparation of ADF fiber is a preliminary step. The detergent removes protein and other acid-soluble materials which create interference in lignin determination. The ADF residue consists of cellulose, lignin, cutin, and acid insoluble ash which mainly include silica. Treatment with 72% H_2SO_4 solution dissolves cellulose. Ashing of the residue will determine crude lignin fraction including cutin.

Procedure: Prepare ADF (see Sect. 9.5.6.2), then place crucible in glass tray in such a way that one end of the higher to drain acid from crucibles. Cover the contents of the crucible with cold 72% H_2SO_4 solution (take 417 mL of distilled water in a volumetric flask and add 583 mL of H_2SO_4 gently) and stir with a glass rod to smooth paste, breaking all lumps. Fill crucible about half with acid and stir. Refill crucible with 72% H_2SO_4 solution and stir at 1 h interval. After 3 h, remove excess acid. Wash residues with distilled water and leave overnight at 100 °C for drying and then weigh. Ignite crucible in a muffle furnace at 500–550 °C for 3 h and then cool and weigh again.

Calculation

$$\mathrm{ADL}(\%) = \frac{(\text{Weight of crucible} + \text{lignin}) - (\text{Weight of crucible} + \text{ash})}{\text{Weight of sample}} \times 100$$

9.5.7.3.2 Estimation Of Permanganate Lignin

Prepare acid detergent fiber (see Sect. 9.5.6.2). Take 1 g sample (ADF) in pre-weighed crucibles and place them in a shallow enamel pan containing cold water to a depth of about 1 cm. Prepare saturated potassium permanganate by dissolving 50 g $KMnO_4$ and 0.05 gm Ag_2SO_4 in 1 L of distilled water. Add silver sulfate to dehalogenate the reagent. Prepare lignin buffer solution by dissolving 6 gm ferric nitrate nonahydrate and 0.15 g silver nitrate in 200 mL distilled water in a beaker. Add 500 mL acetic acid, 5 gm potassium acetate, and 400 mL tertiary butyl alcohol solution to the content of beaker. Mix and make the final volume up to 1 L with distilled water. Prepare combined permanganate solution by mixing saturated potassium permanganate solution and lignin buffer solution in 2:1 ratio by volume before use. Add about 25 mL of combined solution to crucible in an enamel pan containing cold water. Adjust level (2 to 3 cm) of water in the pan to reduce the flow of solution out of crucible. Allow the crucible to stand at 20–25 °C and add permanganate solution if necessary. Purple color must be present at all times. Remove crucible from filtering apparatus and place it in an enamel pan and fill crucible half with demineralizing solution [dissolve 50 gm of oxalic acid dehydrate in 700 mL of 95% ethanol]. Add 50 mL of conc. HCl (12 N) and make the final volume to 1 L with distilled water. Care should be taken to avoid spillage by foaming. After 5 min, suck dry on filter and refill half with demineralizing solution. Repeat the process. Rinse the sides of crucible and treat the fiber until it turns totally white. Fill and thoroughly wash the crucible and contents with 80% ethanol. Repeat the process twice and wash twice in a similar manner with acetone. Ignite the residue at 500 °C for 3 h, cool, and weigh. Calculate residual ash as the difference between weight of crucible after ignition and original weight of the crucible.

Calculation

$$\text{Lignin}(\%)\text{on DM basis} = \frac{\text{Weight of ADF} - \text{Weight of permanganate fiber residue}}{\text{Weight of oven dried sample taken for ADF analysis}} \times 100$$

9.5.7.3.3 Quantification of Lignin by Thioglycolic Acid, Acetyl Bromide, and Klason Methods (Moreira-Vilar et al. 2014)

Protein-free cell wall preparation: Homogenize 0.30 g dry sample in 50 mM potassium phosphate buffer (7 mL, pH 7.0) and transfer into a centrifuge tube. Centrifuge at 1400 × *g* for 5 min and wash the pellet twice with phosphate buffer (pH 7.0; 7 mL), thrice with 1% (*v/v*) Triton X − 100 in pH 7.0 buffer (7 mL), twice with 1 M NaCl in pH 7.0 buffer (7 mL), twice with distilled water (7 mL), and twice with acetone (5 mL). Dry the pellet at 60 °C for 24 h and cool.

1. *Thioglycolic acid method*: Take 0.10 gm sample into a screw cap centrifuge tube containing reaction mixture (1.20 mL of thioglycolic acid and 6 mL of 2 M HCl). Heat the reaction mixture at 95 °C for 4 h. After cooling at room temperature, centrifuge the sample at 1400 × *g* for 5 min and discard the supernatant. The resulting pellet contains lignin thioglycolic acid (LTGA). Wash the pellet thrice with distilled water (7 mL) and extract LTGA by shaking at 30 °C for 18 h in 0.5 M NaOH (6 mL). After centrifugation (1400 × *g*, 5 min), retain supernatant and wash pellet with 0.5 M NaOH (3 mL). Mix supernatant and acidify with conc. HCl (1.80 mL). After precipitation (4 °C, overnight), recover LTGA by

centrifugation (1400 × *g*, 5 min), and wash twice with distilled water (7 mL). Dry the pellet at 60 °C, then dissolve in 0.50 M NaOH and dilute. Measure the absorbance at 280 nm. Express lignin content as mg LTGA g^{-1} dry weight. Prepare a standard curve for lignin by taking alkali 2-hydroxy-propyl ether as a substitute for lignin (Absorptivity value = 18.31 g^{-1} L cm^{-1}).

2. *Acetyl bromide method*: Take 0.10 g sample into a centrifuge tube containing 0.5 mL of 25% acetyl bromide (*v/v* in glacial acetic acid) and incubate at 70 °C for 30 min. After complete digestion, cool the sample in an ice bath and then mix with 0.90 mL of 2 M NaOH, 0.10 mL of 5 M hydroxylamine-HCl, and a volume of glacial acetic acid sufficient for complete solubilization of lignin extract. After centrifugation (1400 × *g*, 5 min), measure the absorbance of supernatant at 280 nm. Prepare standard curve for alkali lignin (absorptive value = 22.90 g^{-1} L cm^{-1}). Express amount of lignin as mg lignin g^{-1} cell wall.
3. *Klason method*: Digest 1.0 g sample with 72% H_2SO_4 at 47 °C for 7 min by vigorous stirring. After complete digestion, filter to separate the soluble and insoluble fractions. Measure the absorbance of soluble fractions at 280 nm and 215 nm. Calculate the soluble lignin content using the following formula:

$$S = (4.53A_{215} - A_{280})/300$$

This formula is the simultaneous resolution of two equations: (a) $A_{280} = 0.68$ F + 18 S and (b) $A_{215} = 0.15$ F + 70 S, where A_{280} is the absorbance value at 280 nm, A_{215} is the absorbance value at 215 nm, F is the furfural concentration (g L^{-1}), and S is the soluble lignin content (g L^{-1}). The values 0.68 and 0.15 are molar absorptivities of furfurals at 280 and 215 nm, respectively. The values 18 and 70 are molar absorptivity coefficients for soluble lignin at 280 and 215 nm, respectively. The insoluble lignin content in forages is calculated by taking the difference in the weight of dry mass and total ash. The total ash content is estimated by igniting the insoluble fraction at 550 °C for 4 h. The total lignin content is estimated by adding insoluble and soluble lignin and expressed as mg g^{-1} cell wall.

9.5.8 Estimation of Carbohydrate Content in Forages

Carbohydrates account for 50–80% of the dry biomass of forages. They constitute most of the plant cell wall portion and therefore play an important role in the structural integrity of individual cells, tissues, and organs. Carbohydrates are extremely important from a nutritional perspective, providing the primary source of energy in ruminant diets. In ruminants, nearly all carbohydrate digestion occurs within the rumen, although under certain circumstances, such as high rates of passage, a significant amount of carbohydrate digestion can occur in the small and large intestines. Simple sugars are rapidly fermented within the rumen to yield volatile fatty acids which are absorbed into the blood through the rumen wall. Polysaccharides must be degraded to simple sugars before being utilized.

Nonstructural polysaccharides, such as starch and fructans, are rapidly and generally completely degraded within the rumen, while the degradability of structural polysaccharides varies considerably.

9.5.8.1 Estimation of Total Carbohydrates (Hedge and Hofreiter 1962)

Total carbohydrates estimation involves hydrolysis of complex polysaccharides in the forage sample into simple sugars with dilute hydrochloric acid. In a hot acidic medium, glucose is dehydrated to hydroxymethyl furfural and forms blue-green complex with anthrone reagent. The complex has absorption maximum at 630 nm.

Procedure: Weigh 0.10 g sample in a test tube and hydrolyze with 5 mL of 2.50 N HCl solution for 3 h in a boiling water bath and then allow to cool to room temperature. Sodium carbonate is added to neutralize the acid content until the effervescence ceases. Make up the final volume to 100 mL and centrifuge. Collect the supernatant and take 0.20 to 1 mL aliquots for analysis.

Working standard solution is prepared by diluting 10 mL of stock solution (100 mg glucose/100 mL) to 100 mL with distilled water. Subsequently, standard curve is prepared by taking aliquots of 0.20 to 1 mL working standard solution and the final volume is made up to 1 mL with distilled water. The tube having 1 mL distilled water serves as blank. In each tube, add 4 mL of anthrone reagent (Dissolve 0.2 g anthrone in 100 mL of concentrated sulfuric acid) and heat for 8 min in a boiling water bath. Cool the content in a tube and read the absorbance of green to dark green color spectrophotometrically at 630 nm.

$$\text{Total carbohydrates}(\%) = \frac{\text{Standard concentration} \times \text{Total volume} \times \text{OD of sample}}{\text{Standard OD} \times \text{Weight of sample} \times \text{Aliquot used}} \times 100$$

9.5.8.2 Estimation of Starch

Starch is the major nonstructural polysaccharide. It is an important energy reserve stored in various plant organs. Starch is the predominant storage polysaccharide in higher plants. Starch polymers are composed of α-D-glucose units and occur in two forms in plants: amylose and amylopectin. Amylose is a linear molecule of α-D-glucose units linked with α-1-4 glycosidic linkage. It has a relatively low molecular weight with typically less than 2000 glucose residues per molecule. Amylose is soluble in water and takes on a helical conformation in solution. Amylopectin is a highly branched molecule with 1–4 linked α-D-glucose chains joined 1–6 at intervals after 15–20 glucose residues. Amylopectin has a higher molecular weight than amylose and generally ranges from 2000 to 220,000 glucose residues per molecule.

The estimation of starch in forages involves the extraction of sugars by treating forage samples repeatedly with hot 80% (*v/v*) alcohol. The residue is then treated with cold perchloric acid to solubilize starch. After filtration, starch in perchloric acid extract is hydrolyzed to glucose in hot acidic medium which undergoes dehydration reaction and forms hydroxymethyl furfural. This derivative then condenses with anthrone reagent to give a blue-colored complex having absorption maxima at 630 nm.

Procedure: Take 0.20 g forage sample in a 50 mL centrifuge tube and add 20 mL of hot 80% ethyl alcohol. Shake tubes vigorously for 5 to 10 min. After centrifugation at 3000 rpm for 10 min, discard supernatant and repeat extraction process 5 to 6 times till supernatant is free of sugars. Add 6.50 mL of 52% perchloric acid (add 270 mL of 72% perchloric acid to 100 mL of water) while stirring with a glass rod. Allow the solution to stand for 15 min and then centrifuge at 3000 rpm. Collect the supernatant and repeat the extraction step 4 to 5 times. Combine all supernatants and make the final volume up to 100 mL with distilled water. Dilute the above extract so that it nearly contains 5 to 20 μg of glucose per mL. Take 5 mL aliquot of diluted extract in test tubes and place tubes in a cold water bath. Add 10 mL freshly prepared anthrone reagent (Dissolve 0.20 g of anthrone in 100 mL of cold 95% conc. H_2SO_4 and store at 4 °C), mix properly, and transfer tubes in a boiling water bath for 8 min. After cooling, note the absorbance of solution at 630 nm. Prepare a standard curve using 0 to 100 μg glucose and calculate the amount of glucose in the sample.

Calculation

1. From the standard curve, determine the amount of glucose present in 5 mL of diluted aliquot of sample extract.
2. Calculate the quantity of glucose in the original extract by multiplying the dilution factor.
3. Multiply obtained value with 0.90 for conversion of glucose value to starch. Calculate the amount of starch in 1 gm of the powdered sample.

9.5.8.3 Estimation of Non-Starch Polysaccharides

Nonstructural carbohydrate minus starch and sugars equals to non-starch polysaccharide content in forages. Starch and sugars are measured directly. The net fraction is calculated by the difference using one of the two formulas:

1. Nonstructural carbohydrate = 100 – (NDF + protein + fat + ash)
2. Nonstructural carbohydrate = 100 – (NDF - NDF protein) + protein + fat + ash

In the second formula, neutral detergent insoluble protein is less digestible and therefore excluded. The non-starch polysaccharides do not include hemicellulose and cellulose contents which are only a part of lignified cell wall matrix. The determination of organic matter, starch, protein, fat, NDF, and lignin is also useful in estimating non-starch polysaccharides indirectly or by directly estimating the different components of a non-starch polysaccharide such as cellulose, hemicellulose, β-glucans, pentosans, pectins, galactomannans, and L-galactosides.

9.5.8.4 Determination of Nitrogen-Free Extract (NFE)

Nitrogen-free extract (NFE) in forages is calculated by subtracting the total dry weight of the fed sample from the total sum of ash, crude fiber, crude fat, and crude protein on a dry weight basis.

Calculation

$$\text{NFE}(\%) = 100 - (\%\text{ash} + \%\text{crude fiber} + \%\text{crude fat} + \%\text{protein})$$

9.5.9 Determination of Digestibility of Forages

Forage digestibility represents the extent of availability of nutrients for absorption in livestock vis-à-vis the nutrients for maintaining the growth and productivity of livestock. Digestibility provides an index of nutritive value of feeds because more digestible feeds are normally consumed to a greater extent. Digestibility measurement has numerous advantages. One of the most significant is determining the nutritive value of a feed and predicting animal performance. Digestibility data offer an insight into the proper feeding and management of animals.

9.5.9.1 *In Sacco* Determination of Forage Digestibility

In sacco method (Nylon bag method) involves incubation of feed samples in the rumen of fistulated animals for different incubation periods and subsequent determination of the rate of disappearance of feed components. The method uses bags of 6.50 × 14 cm size made of nylon mesh (30 to 50 mm). Feed sample of known weight is tightly sealed in nylon bags and placed in the rumen of fistulated animal. After completion of incubation period, the sample is removed, washed, dried and weighed. Degradability of the feed sample is determined by weight loss during the incubation period. The major factors which affect nylon bag degradation include placement of nylon bags in rumen, the particle size of feed sample vis-à-vis pore size of nylon bag, loss of feed particles through the bag (a function of fineness of grinding, cloth pore size, and feed material), method of washing, incubation period, and rumen environment.

Procedure: Weigh 3 g feed sample and transfer it to nylon bag. Note the weight of nylon bag plus feed sample. Attach nylon bag to respective plastic tubes for incubation and place them with bags in rumen and tie the bags with cannula with a piece of nylon thread. Incubate nylon bags for different incubation periods such as 4, 8, 16, 24, 48, 72, and 96 h. Then remove nylon bags and immediately place them at low temperature to prevent further fermentation. Transfer nylon bags to the washing machine for 20 min for cold water washing. Detach nylon bags from tubes by cutting rubber bands. Dry nylon bags at 60 to 65 °C for 48 h and after drying weigh bags immediately.

Calculation

- Empty nylon bag weight: (A)
- Nylon bag + feed sample before incubation: (B)
- Nylon bag + residue after incubation: (C)
- Per cent dry matter in the feed sample: (DM %)

- Sample dry matter weight $(B - A) \times$ DM % = (D)
- Residue dry matter weight $(C - A) = (E)$

$$\text{DM disappearance}(\%) = \frac{(D - E)}{D} \times 100$$

Degradation kinetics:

The degradation kinetics of the fermented feedstuff can be calculated by curvilinear regression of DM and other degradable components of the feed determined by the nylon bag technique.

$$\text{Potential degradability}(P) = A + B(1 - e^{ct})$$

where

- A: y-axis intercept at time 0 that represents soluble and completely degradable substrate washed out of the bags (0 h disappearance)
- B: The difference between the intercept (a) and the asymptote that represents the insoluble but potentially degradable substrate
- t: Incubation time
- C: The rate of disappearance of component "b" per hour (rate constant)
- $A + B$: The asymptote of the disappearance curve

$$\text{Effective degradability}(E) = A + Bc/(c + k)$$

where,

k = rumen small particle outflow rate

The above equations assume that component "B" disappears at a constant fractional "c" per hour.

9.5.9.2 *In Vitro* Determination of Forage Digestibility

In vitro digestibility technique provides quick, inexpensive, and precise determination of digestibility of forages in livestock. Tilley and Terry's method is the most commonly followed procedure for the estimation of forage digestibility. The method includes two consecutive digestion phases. During the first digestion phase, the feed sample is incubated under anaerobic conditions with rumen microorganisms for 48 h at 39 °C followed by a 24-h acid-pepsin digestion phase at 39 °C, under anaerobic conditions.

Reagents

Solution A: Dissolve disodium hydrogen orthophosphate (anhydrous) (18.5 g) and sodium hydrogen carbonate (49.0 g) in 1 L of distilled water.

Solution B: Dissolve sodium chloride (23.50 g), potassium chloride (28.5 g), magnesium chloride (6.0 g), and calcium chloride (2.0 g) in 1 L of distilled water.

Stock solution: Prepare stock solution by mixing solutions A and B in the ratio 9:1. Add solution B only after complete dissolution of phosphate and bicarbonate in solution A.

Solution C: Dilute one part of above stock solution with four parts of water immediately before use and warm up to 39 °C ± 1 °C. To this cloudy solution, add CO_2 till the solution becomes clear.

Buffer nutrient solution: Dilute one part of strained rumen liquor to four parts of solution C immediately before use.

Procedure: Take 0.5 g of ovendried sample in a 250 mL centrifuge tube fitted with gas release cork. Add nutrient buffer solution (50 mL) to each tube. Pass carbon dioxide gas over the surface of tube contents for 10 s. Close the tube immediately and incubate at 39 °C. Shake gently during incubation to resuspend the substrate. After 48 h, filter the contents of the tube through the muslin cloth fitted on the top of the Buckner's funnel under suction. Resuspend the residue in 50 mL of 0.20% pepsin in 0.10 N HCl. Incubate tubes for 24 h at 39 °C and then filter through sintered glass crucibles. Dry residue at 100 °C overnight and weigh. Calculate the IVDMD (*In vitro* dry matter digestibility) using the following formula:

$$\text{Weight of empty crucible} = W_1$$

$$\text{Weight of (crucible + residual material)} = W_2$$

$$\text{Weight of sample} = W_3$$

$$\text{Weight of residual material} = W_2 - W_1 = W_4$$

$$\text{IVDMD}(\%) = \frac{W_3 - W_4}{W_3} \times 100$$

9.5.9.3 Calculation of Organic Matter Digestibility from Dry Matter Digestibility

Digestibility is the difference between the amount of a feed component eaten and the amount of that component excreted in feces. It accounts for a major proportion of metabolizable energy (ME) in a feed. The digestibility of a feed can be expressed as

$$\text{Dry matter digestibility (DMD)}\% = 100 \times (\text{feed DM} - \text{faecal DM})/\text{feed DM}$$

$$\text{Organic matter digestibility (OMD)\%} = 100 \times [(\text{feed DM} - \text{feed ash}) - (\text{faecal DM} - \text{faecal ash})]/(\text{feed DM} - \text{feed ash})$$

$$\text{Digestible organic matter in the DM (DOMD)\%} = 100 \times [(\text{feed DM} - \text{feed ash}) - (\text{faecal DM} - \text{faecal ash})]/\text{feed DM}$$

DOMD can also be calculated based on DMD values:

$$\text{DOMD\%} = 0.95\,\text{DMD(\%)} - 0.90$$

A major limitation with this equation is that it can only be applied to feeds with ash content in the range of 9–12%. As many feeds have ash content falling outside this range. The following equations can be used to calculate DOMD in a wide range of feed resources (Kaiser et al. 2005):

$$\text{DOMD\%} = 6.83 + 0.847\,\text{DMD(\%)}\,(\text{For forages/roughages})$$

$$\text{DOMD\%} = 2.11 + 0.961\,\text{DMD(\%)}\,(\text{For feed concentrates})$$

9.5.9.4 Calculation of Metabolizable Energy

Metabolizable energy (ME) represents the fraction of total energy in a feed that is available to the animal in order to maintain health and production. It is calculated as the difference between gross energy and sum of energy excreted in feces, in urine, and energy consumed in methane production during digestion. The calculated ME values are all expressed on a DM basis as megajoules (MJ) per kg DM. Equations have been given to calculate metabolizable energy from feed resources:

- For roughages other than silages:

$$\text{ME(MJ/kg DM)} = 0.203\,\text{DOMD(\%)} - 3.001$$

- Silages:

$$\text{ME(MJ/kg DM)} = 0.16\,\text{DOMD(\%)}$$

- Grains and concentrates:

$$\text{ME(MJ/kg DM)} = 0.858 + 0.138\,\text{DOMD(\%)} + 0.272\,\text{EE(\%)}$$

9.5.10 Mineral Analysis of Forages

Minerals are essential nutrients for all animals. In livestock, they are involved in growth, fertility, production and maintenance of health. Without mineral elements, the macronutrients of protein, carbohydrate and fat, which provide energy, would be worthless. Forages are an important source of minerals for ruminants. In some instances, forages may provide adequate quantities of all essential minerals required by ruminants. However, in other situations, forages are deficient in one or more minerals and supplementation is required for improved livestock performance. Severe mineral deficiencies still occur to some degree, but marginal mineral deficiencies are widespread. Marginal mineral deficiencies may result in no clinical deficiency signs and result only small decreases in metabolic functions, but the overall impact on growth, reproduction, or health of ruminants can be substantial. The mineral content in forages varies greatly depending on soil, plant, and management factors.

9.5.10.1 Determination of Ash Content (AOAC 2000)

Ash is a measure of the total mineral content of the feed, but it does not tell us how much of each mineral is present. Ash is not digestible by animals. High ash content of feeds may dilute the amount of nutrients available to the animal. If a plant feed is high in ash content, it may be due to soil contamination during the harvest of the plant material.

Ash is an inorganic material left after either ignition or complete oxidation of the feed sample. In other words, the total ash is the material left after combustion of feed sample in muffle furnace at 500 °C. The process of dry ashing involves the use of muffle furnace capable of maintaining temperatures of 500 to 600 °C. Wet ashing is a procedure for oxidizing organic substances using acids and oxidizing agents or their combinations. Wet ashing often is preferable over dry ashing for specific elemental analysis.

Procedure: Place crucible in a muffle furnace at 600 °C for 1 h and allow it to cool in desiccator and weigh. Weigh 2 g of sample into pre-weighed silica crucible. Place in the muffle furnace and adjust the temperature to 600 °C for 6 h. Transfer the crucible to a desiccator and cool down. Weigh the crucible quickly to prevent moisture absorption.

Calculation

$$\text{Ash}(\%) = \frac{\text{Weight of ash}}{\text{Weight of sample}} \times 100$$

9.5.10.2 Estimation of Acid Insoluble Ash

The residue left after dissolving the inorganic portion of total ash represents acid insoluble ash (AIA), chiefly comprising sand and silica.

Procedure: Calculate the amount of total ash in the feed sample by following the procedure described in Sect. 9.5.10.1. Add 25 mL of dilute (1:2) HCl to the total ash and quantitatively transfer to 250 mL beaker by giving repeated hot distilled water washings. Boil the contents for 5 to 10 min and filter in a volumetric flask. Retain filtrate in the volumetric flask containing dissolving inorganic salts for further estimation of minerals after making up the volume when cooled. Transfer filter paper with the retained residue to a pre-weighed silica basin. Dry in a hot air oven at 100 ± 2 °C to a constant weight. Transfer the silica basin to muffle furnace after decarbonization and ignite at 550–600 °C for 1 to 2 h. Cool in a desiccator and weigh the silica basin with left-back AIA.

Calculation

$$\text{Acid insoluble ash}(\%) = \frac{A - B}{W} \times 100$$

where

A = Weight (g) of silica basin with acid insoluble ash
B = Empty weight (g) of silica basin
W = Weight (g) of moisture-free sample taken for ashing

9.5.11 Estimation of Flavanols

Flavonoids as benzopyrone derivatives of a phenolic compound include a large family of thousands of hydroxylated polyphenolic compounds. They have the general structure of a 15-carbon skeleton, which consists of two phenyl rings and a heterocyclic ring. The majority of flavonoids generally remain conjugated with sugars as glycosides. Flavonoids have different roles in plants which include help in pollination, reduce environmental stress, disposal of microbial infection and regulating cell growth. They have positive effects on the productivity and health of animals as well as rumen fermentation and control of nutritional stress such as bloat and acidosis. Flavonoids as polyphenolic compounds have action similar to monensin and other types of antibiotics. Both rumen microbial ecosystem and ruminant's performance can be influenced by the use of flavonoids in the diets. Generally, flavonoids can modify the fermentation characteristics and can serve as antimicrobial agents to decrease pathogens and methane-producing bacteria as well as improve the antioxidant content to decrease lipid oxidation and increase the quality of the milk/meat production. Incorporation of moderate levels of flavonoids/polyphenols into milk/meat and dairy products through the feeding of animals with food sources rich in flavonoids and phytochemicals may provide a novel way to increase flavonoids/polyphenol content of daily food for consumers.

Flavanols, also known as flavan-3-ols, are important flavonoids. Most commonly they are present in glycosidic combinations. Due to their widespread nature and importance as an antioxidant, anti-inflammatory agent, and anticancer factor, they should be estimated in plants.

Procedure: Take 1 mL of the extract in a test tube and evaporate to dryness on a water bath maintained at 70 °C. Dissolve the residue in 10 mL of 0.1 M aluminum chloride solution (Dissolve 2.41 g of aluminum chloride hexahydrate in 1 L of distilled water) and thereafter, measure the absorbance of the solution at 420 nm. Prepare a blank simultaneously and calculate the concentration of flavanols by taking rutin as standard and express results as mg g^{-1} dry weight.

9.5.12 Estimation of Chlorophyll Content in Forages (Arnon 1949)

Chlorophyll is an important molecule for photosynthesis. The estimation of chlorophyll content in forages can be used to estimate dry matter yield and CP content and as a marker of the need for nitrogen fertilization.

Procedure: Weigh about 1 g forage leaves and grind to fine powder in 20 mL of prechilled 80% acetone. Centrifuge at 5000 rpm for 5 min and transfer supernatant to a 100 mL volumetric flask. Grind the resulting residue again with 20 mL of 80% acetone, centrifuge, and transfer the supernatant to a volumetric flask. Repeat the procedure until the residue is colorless. Wash the mortar and pestle thoroughly with 80% acetone and collect washings in volumetric flask. Make up the final volume to 100 mL with 80% acetone and measure the absorbance of solution spectrophotometrically at 645, 663, and 652 nm against blank.

Calculation

Calculate chlorophyll content using the following equations:

$$\text{mg chlorophyll a/g tissue} = 12.70(A_{663}) - 2.69(A_{645}) \times \frac{V}{1000 \times W}$$

$$\text{mg chlorophyll b/g tissue} = 22.90(A_{645}) - 4.68(A_{663}) \times \frac{V}{1000 \times W}$$

$$\text{mg total chlorophyll(per g tissue)} = 20.20(A_{645}) + 8.02(A_{663}) \times \frac{V}{1000 \times W}$$

where

A = Absorbance at specific wavelengths
V = Final volume of chlorophyll extract in 80% acetone
W = Fresh weight of the tissue extracted

9.5.13 Estimation of Antioxidant Activity

Antioxidants are free radical scavengers, singlet oxygen quenchers, inactivators of peroxides and other reactive oxygen species (ROS), metal ion chelators, quenchers of secondary oxidation products, and inhibitors of pro-oxidative enzymes. These substances exert their inhibitory effect against oxidation processes through different mechanisms and with varied activities. Antioxidants in forages play an important role in their preservation by inhibiting oxidation processes and contributing to health promotion rendered by many dietary supplements, nutraceuticals, and functional food ingredients.

Procedure: Prepare a working standard solution of catechin (0.10 mg/mL) by dissolving 10 mg of catechin in 100 mL of methanol. Simultaneously prepare a stock solution of DPPH (200 μM) by dissolving 7.80 mg of DPPH in a volumetric flask containing 100 mL of methanol. Wrap solution with aluminum foil and leave it for 1 h at room temperature. Extract 1 g feed sample with 10 mL of methanol for 12 h at 37 °C. Evaporate methanol from the sample mixture. Add 1 to 2 mL of methanol with sample and shake vigorously. To prepare a standard curve of catechin, take aliquots (20, 40, 60, 80, 100, 120, and 140 μL) of catechin in test tubes and make the final volume to 3 mL with methanol. Add 1 mL of 200 μM DPPH solution to the test tubes containing methanolic solution of catechin. Vortex the content and allow to incubate at 30°C for 30 min in dark. Record the absorbance at 517 nm after 30-min incubation in a hot air oven at 30 °C. Calculate the percentage of DPPH free radical scavenging activity (% inhibition) of sample and standard catechin using the following equation:

$$\%\text{Inhibition} = \frac{\text{Absorbance}\,(\text{Control}) - \text{Absorbance}\,(\textit{Test})}{\text{Absorbance}\,(\text{Control})} \times 100$$

Calculate IC_{50} value (the amount of antioxidant necessary to decrease the initial DPPH-free radical concentration by 50%) from the regression line obtained from the plot of % inhibition against the concentration of each solution using the following equation:

$$IC_{50}\,\text{value} = \frac{(50 - y\,\text{intercept})}{\text{Slope}}$$

9.6 Estimation of Anti-Nutritional Factors in Forages

Forages provide nutrients and energy to livestock, however, the presence of anti-nutritional compounds limits the amount of nutrients and energy availability to the livestock. These substances when consumed by animals in large quantities not only diminish animal productivity but also results in toxic effects. Feeding such toxin-containing feed and fodder above critical limit is fatal and their regular intake even at lower levels reduces the growth, production, reproduction and even quality of livestock products. Different roughages, legumes, shrubs, herbs, trees and other nontraditional feeds contain some anti-nutritional components.

9.6.1 Estimation of Phenols (Makkar 2003)

High phenolic content inhibits the growth and activity of ruminal microbiota in livestock thereby affecting digestibility and absorption of nutrients from forages. The estimation of phenolic content in forages involves the reaction of phenols with phosphomolybdate in Folin–Ciocalteu reagent under alkaline conditions and leads to the formation of blue-colored complex. The absorbance of blue-colored complex is spectrophotometrically measured at 725 nm.

Procedure: Weigh 0.20 g sample, mix it with 10 mL of 70% acetone and extract for 2 h in a shaking water bath at 37 °C. Centrifuge the homogenate at 10000 rpm for 20 min. Save the supernatant for the estimation of phenols and tannins. Pipette out different aliquots (0.2 to 1 mL) in test tubes. Make the volume of each tube to 1 mL with distilled water. Add 2 mL of 20% Na_2CO_3 in each tube and incubate for 3 min. Thereafter, add 0.50 mL FC reagent in each tube. Vortex the tubes and incubate for 40 min at room temperature. Take absorbance of the standard tubes at 725 nm. For simple phenols, add 1 mL of 1% PVP to 1 mL extract and place the solution in cold condition for 15 min, thereafter the procedure is same as described for the estimation of total phenols. Simultaneously run blank in a similar manner.

Preparation of standard curve: Prepare a working standard solution by diluting 10 mL of standard tannic acid solution (0.5 to 100 mL with distilled water). Prepare the standard curve by taking aliquots of 0.20 to 1 mL of working standard solution and make up the final volume to 1 mL with distilled water. Add 2 mL of 20% Na_2CO_3 in each tube and incubate for 3 min. Thereafter, add 0.50 mL FC reagent in each tube. Vortex the tubes and incubate for 40 min at room temperature. Take absorbance of the standard tubes at 725 nm. For simple phenolic content, add 1 mL of 1% PVP to 1 mL extract and place the solution in cold condition for 15 min, thereafter the procedure is same as described for the estimation of total phenols.

$$\text{Total phenols}(\%) = \frac{\text{Standard concentration} \times \text{Total volume} \times \text{OD of sample}}{\text{Standard OD} \times \text{Weight of sample} \times \text{Aliquot used}} \times 100$$

$$\text{Simple phenols } (\%) = \frac{\text{Standard concentration} \times \text{Total volume} \times \text{OD of sample}}{\text{Standard OD} \times \text{Weight of sample} \times \text{Aliquot used}} \times 100$$

9.6.2 Estimation of Tannin Content Using *FOLIN-DENIS* Reagent (AOAC 1995)

Tannins bind with proteins and form reversible and irreversible complexes due to the existence of a number of phenolic hydroxyl groups. Hydrolyzable tannins and condensed tannins are two different classes of tannins. Generally, tree and shrub leaves contain both types of tannins. The condensed tannins have a more profound digestibility-reducing effect than hydrolyzable tannins, whereas the latter may cause varied toxic manifestations due to hydrolysis in the rumen. Tannins form a less digestible complex with dietary proteins and may bind and inhibit endogenous proteins such as digestive enzymes. The tannin–protein complexes are astringent and adversely affect feed intake and all plants contain phenolic compounds but their type and amount may cause negative animal responses. When an animal consumes high amount of tannins, tannin–protein complexes reduce the digestion of forage proteins. Tannins directly affect the digestibility of the cell wall by binding with microbial enzymes in the rumen. The reduced digestibility of cell wall compounds restricts the digestible energy that animals gain from forage plants.

Estimation of tannin content in samples using *Folin–Denis* reagent involves reduction of phosphotungstomolybdic acid in alkaline solution and formation of blue-colored solution. The intensity of blue-colored solution which is measured in a spectrophotometer at 700 nm is proportional to the amount of tannins in the sample.

Procedure: Weigh 0.50 g in a 250 mL conical flask, add 75 mL of distilled water to it, and boil for 30 min. Centrifuge at 2000 rpm for 20 min and collect the supernatant. Transfer 1 mL sample extract to 100 mL volumetric flask containing 75 mL distilled water. Prepare Folin–Denis reagent by dissolving 100 g sodium tungstate and 20 g phosphomolybdic acid in 750 mL distilled water in a volumetric flask and thereafter add 50 mL of phosphoric acid. Reflux the mixture for 2 h and make up the final volume to 1 L with distilled water. Always prepare the reagent in dark. Add 5 mL Folin–Denis reagent, then add 10 mL of sodium carbonate solution and dilute to 100 mL with distilled water. Mix well and measure the absorbance at 700 nm after 30 min. Prepare a blank with distilled water instead of a sample. If the absorbance is greater than 0.70, make appropriate dilution. Prepare a working standard solution by diluting 10 mL of standard tannic acid solution (0.5 to 100 mL with distilled water). Prepare the standard curve by taking aliquots of 0.20 to 1 mL of working standard solution and make up the final volume to 1 mL with distilled water. The tube having 1 mL distilled water serves as blank. In each tube, add 5 mL of Folin–Denis reagent and 10 mL of sodium carbonate solution. Shake well and measure the absorbance at 700 nm after 30 min.

$$\text{Tannins content}(\%) = \frac{\text{Standard concentration} \times \text{Total volume} \times \text{OD of sample}}{\text{Standard OD} \times \text{Weight of sample} \times \text{Aliquot used}} \times 100$$

9.6.3 Determination of Condensed Tannins (Proanthocyanidins)

Procedure: Pipette out 0.50 mL of tannin extract in a test tube and dilute with 70% acetone. Add 3.0 mL butanol-HCl reagent (Mix 950 mL *N*-butanol with 50 mL conc. HCl (37%) and 0.10 mL of ferric reagent). Vortex the tubes and cover the mouth of each tube with a glass marble and put the tubes in a heating block adjusted at 97 to 100 °C (or in a boiling water bath) for 60 min. Cool the tubes and record absorbance at 550 nm. If the extract has flavan-4-ols, a pink color develops without heating. If this happens, use one heated blank for each sample, comprising 0.50 mL of the extract, 3 mL of butanol, and 0.10 mL of ferric reagent (2% ferric ammonium sulfate in 2 N HCl). The condensed tannins (% in dry matter) content as leucocyanidin equivalent is calculated as

$$(A_{550}\,\text{nm} \times 78 : 26 \times \text{Dilution factor})/(\%\text{dry matter})$$

The formula assumes that the effective $E^{1\%}$, $^{1\text{cm}}$, $^{550\text{ nm}}$ of leucocyanidin is 460. The dilution factor is equal to 1 if 70% acetone is not added and the extract is made from 200 mg sample in 10 mL solvent. When 70% acetone is added (for example, to prevent the absorbance from exceeding 0.6), the dilution factor is 0.50 mL/volume of extract taken. The presence of pigments may interfere in this method though can be removed by extracting dried leaves with petroleum ether containing 1% acetic acid. Ascorbic acid does not interfere with the determination of condensed tannins in forages using this method.

9.6.4 Estimation of Oxalate Content (Mahadevan 1979)

Oxalic acid is one of a number of anti-nutrients found in forage. It can bind with dietary calcium (Ca) or magnesium (Mg) to form insoluble Ca or Mg oxalate, which then may lead to low serum Ca or Mg levels as well as to renal failure because of precipitation of these salts in the kidneys. Therefore, the estimation of oxalate content in forages is essential.

The estimation of oxalic acid in forages involves defatting of feed sample with ether and then extracting with sodium hydroxide and water. The water layer is treated with calcium chloride buffer (pH, 4.5) and centrifuged. After centrifuge, the resulting pellet is treated with acetic acid saturated with calcium oxalate, centrifuged, and the residue is dissolved in sulfuric acid. Finally, the extract is titrated against standard 0.02 N potassium permanganate solution and the amount of oxalic acid is calculated.

Procedure

Sample extraction: Mix 500 mg moisture-free sample with 1 g asbestos and 1.50 mL of 4 N sulfuric acid. Fill the extraction thimble to a depth of 2 cm with ground glass. Place two circular pieces of cheesecloth over the grounded glass. Transfer sample mixture to thimble. Place thimble in Soxhlet apparatus and extract with

500 mL diethyl ether for 48 h. Add 5 mL solution of 1 N NaOH and 7 mL of distilled water to the extract and shake well. Evaporate ether layer in a rotary evaporator and transfer water phase to centrifuge tube, add 4 mL of calcium-chloride-acetate buffer [Dissolve 25 g of anhydrous calcium chloride in 500 mL of 50% acetic acid. Dissolve 330 g of sodium acetate trihydrate in distilled water and make up the volume to 550 mL. Mix two solutions and adjust the pH to 4.5] and allow to stand overnight.

Estimation: Centrifuge content at 3000 × *g* for 10 min and discard supernatant and wash pellet with 5 mL of 5% acetic acid saturated with calcium oxalate. Repeat centrifugation process and dilute residue in 5 mL of 4 N sulfuric acid and heat at 80 to 90 °C on a water bath. Filter the content while hot and titrate against standard 0.02 N potassium permanganate solution. Calculate the amount of oxalic acid (mg/100gm) present in the sample using the following relationship:

$$1\ \text{mL of } 0.02\,N\ \text{Potassium permanganate} = 1.2653\ \text{mg of oxalic acid}$$

9.6.5 Estimation of Saponins Content (Akhidue and Monago 2002)

Saponins reduce nutrients uptake in the gut through intra-lumenal physicochemical interaction. Symptoms of saponins toxicity include listlessness, anorexia, weight loss, and gastroenteritis. In nonruminants (chicks and pigs), retardation of growth rate, primarily due to reduction in feed intake, is a major concern due to presence of saponins.

Procedure: Put 3 g moisture-free feed sample in a thimble and extract in Soxhlet apparatus with methanol (110 mL) at 20 °C for 10 h. After 10 h, reduce the volume of extract to about 3 to 5 mL by gentle heating on a hot plate. Cool the extract at room temperature and add 15 to 20 mL of acetone in dropwise manner till white precipitate appears. Filter and ovendry the white precipitate, and finally take the weight of filter paper with precipitate.

Calculation

$$\text{Saponin content} = \frac{\text{Weight of dry filter paper with precipitate} - \text{Weight of empty filter paper}}{\text{Weight of sample}}$$

9.6.6 Estimation of Cyanogenic Glycosides (AOAC 1995)

Cyanogenic glycosides are present in various forage species including Sorghum spp., *Trifolium* spp., and *Lotus* spp. The release of toxic hydrogen cyanide (HCN) from endogenous cyanogenic glycosides leads to a serious problem (Cyanogenesis) in livestock.

9.6.6.1 Qualitative Test

Procedure: Dip filter paper strips in 1% picric acid solution and after drying, further dip into 10% Na_2CO_3 solution and dry again. Store the strips in a stoppered bottle. Place feed sample in test tube and insert a piece of moistened sodium picrate paper into the tube while taking care that it does not come in contact with the sample. Add few drops of chloroform and stopper tube hermetically. The sodium picrate paper gradually turns orange and then brick red if plant tissue contains cyanogenic glycosides. The rapidity of change in color depends upon the amount of free HCN present. This test works well with fresh plant materials, but relatively dry samples should be ground and moistened with H_2O and allowed to hydrolyze in a stoppered test tube containing sodium picrate paper.

9.6.6.2 Titrimetric Method for Quantitative Test of alkaloids

9.6.6.2.1 Acid Titration Method

Procedure: Place about 10 to 20 g finely ground feed sample (sieve no. 20) in 800 mL Kjeldahl flask. Add 100 mL H_2O and macerate at room temperature for 2 h. Further, add 100 mL H_2O and distill for collecting distillate in 20 mL 0.01 N $AgNO_3$ acidified with 1 mL HNO_3. Adjust the distillation apparatus so that the tip condenser dips below the liquid in receiver. Collect 150 mL distillate and then pass it through Gooch crucible. Wash the receiver and crucible with little amount of water and titrate excess $AgNO_3$ in the combined filtrate and give washings with 0.02 N KHCN, using ferric alum indicator.

Calculation

$$1\,\text{mL of } 0.02\,N\,\text{silver nitrate solution} = 1.08\,\text{mg of HCN}$$

9.6.6.2.2 Alkaline Titration Method

Procedure: Take 10 to 20 g powdered feed sample in 800 mL Kjeldahl flask, add 200 mL of distilled water, and allow to stand for 2 to 4 h. Distill the sample and collect 150 to 160 mL of distillate in NaOH solution (0.5 g NaOH in 20 mL distilled water) and dilute it to a known volume. Add 8 mL of 6 N ammonium hydroxide and 2 mL of 5% potassium iodide solution. Titrate with 0.02 N silver nitrate solution until faint but permanent turbidity is obtained.

Calculation

$$1\,\text{mL of } 0.02\,N\,\text{silver nitrate solution} = 1.08\,\text{mg of HCN}$$

9.6.7 Estimation of Nitrates and Nitrites (Wiseman and Jacobson 1965)

In animal tissues, nitrite is naturally present mainly as the result of endogenous nitrate conversion. Nitrate levels in forages are naturally high and the interconversion of nitrate to nitrite is the largest contributor to nitrite exposure in food-producing animals. Dietary intake of nitrite in livestock is due to feeding of vegetable feeds and forages, fertilizers/manures, or when used as a preservative in silages and fish meal. In monogastric animals, most nitrite is formed and absorbed in the upper digestive tract. In contrast in ruminants, nitrite and nitrate are metabolized by the rumen flora.

Nitrate is reduced to nitrite by zinc and manganese sulfate. The reaction proceeds with diazotization of sulfonic acid by nitrite ion and subsequent coupling with 1-naphthylamine to form a red dye. Copper aids the reaction at low ppm level (0.20), but over 1.20 ppm it interferes. Interfering iron is complexed with citrate. Reaction is optimum at pH range 1.70–3.00.

Reagents

- 20% acetic acid reagent: (200 mL acetic acid +5 mL copper sulfate, final volume 1 L with distilled water)
- Bray's indicator (1 L)
- 100 g barium sulfate ($BaSO_4$)
- 10 g manganese sulfate ($MnSO_4.H_2O$)
- 2 g zinc (metallic Zn)
- 75 g citric acid
- 4 g sulfanilic acid
- 2 g 1-naphthylamine

Zinc and citric acid are powdered; manganese sulfate, zinc, sulfanilic acid, and 1-naphthylamine are mixed separately with a part of barium sulfate and then mixed with citric acid and the remaining part of barium sulfate. The indicator is stored in a dark bottle and stored away from light.

- Hydrochloric acid (0.1 N): 8.18 mL. 37–38% HCl in 1 L distilled water
- Activated charcoal
- Standard nitrate solution
- Stock solution (1 mg/mL): 1.37 g sodium nitrate in 1 L distilled water and working solution (10 μg/mL); dilute 1 mL of stock solution to 100 mL with distilled water

Procedure: Take 1 g air-dry material or 3 to 5 g moist sample (nitrate content ranging between 1 and 8 mg) in a conical flask of 150 mL capacity. To above add 100 mL of 0.1 N HCl and place it for 1 h with occasional shaking. If the extract is strongly colored, decolorize with activated charcoal and then filter through Whatman No. 1 filter paper. The sample extract, blank, and standard are run simultaneously.

Shake the above solutions avoiding exposure to light. Centrifuge at 3000 rpm for 5 min. Remove any film on top and collect the clear red supernatant. Measure absorbance at 520 nm and find out the concentration of nitrate in the sample from the standard curve drawn taking OD against concentration.

Calculation

$$\text{Total nitrate}(\text{nitrate} + \text{nitrite}) = \frac{\text{Conc.of nitrate}(\mu\text{g/mL})\text{ in final solution from standard curve}}{g\text{ sample taken}(\text{DM basis}) \times 10}$$

$$\text{Nitrite concentration}(\mu\text{g/mL}) = \frac{\text{Conc.in final solution from standard curve}}{g\text{ sample taken}(\text{DM basis}) \times 10}$$

9.6.8 Estimation of Nitrate-N (AOAC 1970)

The complex formed by nitration of salicylic acid under highly acidic condition absorbs maximally at 410 nm in basic solution. Absorbance of chromosphere is directly proportional to amount of nitrate-N present in feed sample.

Procedure: Take 0.10 g of ground feed sample (in duplicate) into the tubes. Add about 10 mL of distilled water into the tubes. Place the tubes in a water bath for 1 h at 40 °C. After that, filter the samples and take 0.2 mL of filtrate for the estimation. Add 0.80 mL of 5% (*w/v*) salicylic acid - H_2SO_4 to filtrate. Mix the contents thoroughly and place them at room temperature. After 20 min, add 19 mL of 2 N NaOH to increase the pH more than 12. Bring the temperature of feed sample to room temperature and measure absorbance at 410 nm. Amount of nitrate-N in the feed sample can be calculated from a standard curve prepared simultaneously using potassium nitrate (0 to 50 μg) as standard.

9.6.9 Estimation of Trypsin Inhibitor (Roy and Rao 1971)

Trypsin is an endogenous enzyme involved in the breakdown of different proteins, including as part of digestion in humans and other animals. Trypsin inhibitors (TI) are serine protease inhibitors that inhibit the activities of endogenous trypsin thus having anti-nutritional effects. These inhibitors are widely seen across the plant kingdom and are proteinaceous in nature.

Procedure: Take 4 g finely ground feed sample in a stoppered 250 mL conical flask and treat with 40 mL of 0.10 M sodium phosphate buffer (pH 7.5) and 40 mL of distilled water. Shake suspension for 3 h on shaker and centrifuge at 700 × *g* for 30 min at 15 °C. Dilute the supernatant so that 40 to 60% activity of the control enzyme is inhibited. Prepare incubation mixture containing 0.50 mL of trypsin enzyme solution, 2.0 mL of casein solution (2%), 1.0 mL of 0.1 M sodium phosphate buffer, 0.4 mL 0.001 M of HCl, and 0.10 mL extract or supernatant. Make the volume of incubation mixture to 4.0 mL. Incubate the mixture at 37 °C for 20 min.

Add 6.0 mL of 5% TCA to stop the reaction. Run parallel blank. Record the absorbance on the spectrophotometer at 280 nm.

Calculation

One trypsin unit is defined as an increase of 0.01 absorbance unit at 280 nm in 20 min for 10 mL reaction mixture and the trypsin inhibitory activity as the number of trypsin unit inhibits (TUI) per mg of protein. To express the value of specific activities, the total TUI is expressed as per mg of protein. The value for trypsin inhibitory activity for 40 to 60% inhibition gives reproducible and accurate results.

9.6.10 Estimation of Mimosine Content in Forages (Mastumoto and Sherman 1951)

Mimosine, a nonprotein amino acid structurally similar to tyrosine, occurs in a few species of mimosa and all species of closely allied genus *Leucaena*. In *L. leucocephala*, the level of mimosine in the leaf is about 2 to 6% and varies with seasons and maturity. In nonruminant animals, mimosine causes poor growth, loss of hair and wool, swollen and raw coronets above the hooves, lameness, mouth and esophageal lesions, depressed serum thyroxine level, and goiter. Some of these indications may be due to mimosine and others due to 3, 4 dihydropyridine, a metabolite of mimosine in the rumen. Toxic signs like skin lesions also resemble Zn deficiency.

Procedure: Weigh 40 g moisture-free sample and place it in a beaker. Add 400 mL of 0.10 N HCl and boil the content at 85 °C on a hot plate for 1 h. Transfer the content into a 250 mL storage bottle and allow it to stand until most of the solids get settled to the bottom. Repeat the HCl digestion process twice.

Preparation of standard and calibration curve: Weigh 0.025 g of L-mimosine carefully and dissolve in 25 mL of 0.10 N HCl in a conical flask. Transfer aliquots of 1, 2, 3, 4, and 5 mL into different conical flasks and add 10 mL of 0.10 N HCl and 4 mL of 0.5% $FeCl_3$ into a conical flask and make the final volume of each aliquot up to 100 mL with distilled water. Determine the absorbance of each aliquot spectrophotometrically at 535 nm and plot it against concentration. Adjust the concentration of mimosine as 10, 20, 30, 40, and 50 μg/mL.

9.6.11 Estimation of Alkaloids

Alkaloids are naturally occurring compounds that contain nitrogen atoms in their structures. These nitrogen atoms are the cause of alkalinity in these compounds. The nitrogen atoms are usually situated in some ring (cyclic) system. In their pure form, alkaloids are colorless, nonvolatile, crystalline solids and have a bitter taste. Alkaloids have been reported to cause gastrointestinal and neurological disorders in livestock if fed in higher amounts.

Procedure: Homogenize 10 mg sample in a pestle and mortar and add 20 mL of methanol:ammonia (68:2) solution and after 24 h, decant ammonia solution. Add fresh methanolic ammonia and repeat the procedure thrice and pool the extracts. Evaporate the extracts using a flash evaporator, treat the residue with 1 N HCl, and keep it overnight. Extract acidic solution with 20 mL of $CHCl_3$ thrice, pool organic layers, and evaporate to dryness. Adjust the pH of acidic layer with sodium hydroxide to 12 and extract with $CHCl_3$ (20 mL) thrice. Pool the $CHCl_3$ layers, dry over absorbent cotton, and evaporate to dryness. Weigh the fraction and express it as mg/100 g.

9.6.12 Determination of Phytic Acid Content (Wheeler and Ferrel 1971)

Anti-nutritional factors cause depression in growth and performance and could negatively affect the health of the animals. Phytate impedes nutrient digestions by forming protein–phytate complexes and is resistant to normal digestions. Phytic acids have a negative impact on amino acid digestibility and have been shown to increase the endogenous nutrient losses in pigs and poultry.

Phytic acid is extracted with trichloroacetic acid and precipitated as ferric salt. The iron content of the precipitate is determined spectrophotometrically and phytate phosphorus content is calculated from this value assuming a constant 4Fe: 6P ratio in the precipitate.

Procedure: Weigh a finely ground sample and extract in 50 mL of 3% trichloroacetic acid solution for 30 min with mechanical shaking. Centrifuge suspension at 3000 rpm for 10 min and transfer 10 mL aliquot of supernatant to a conical centrifuge tube. Add 4 mL of $FeCl_3$ solution (Dissolve 583 mg $FeCl_3$ in 100 mL of 3% trichloroacetic acid solution) to aliquot and heat the contents on a boiling water bath for 45 min. (If the supernatant is not clear after 30 min, add one or two drops of 3% sodium sulfate and continue heating). Centrifuge at 3000 rpm for 10 min and carefully decant the supernatant. Wash precipitate twice by dispersing in 20 to 25 mL of 3% TCA solution, heat in boiling water bath for 5 to 10 min, and again centrifuge. Repeat the washing process with distilled water. Disperse the precipitate in a few mL of distilled water and add 3 mL of 1.50 N NaOH with mixing. Bring volume to approximately 30 mL with distilled water and heat in boiling water bath for 30 min. Filter the content through Whatman No. 2 filter paper. Wash the precipitate in 60 to 70 mL of hot water and discard the filtrate. Dissolve precipitate in 40 mL of hot 3.20 N HNO_3 in a volumetric flask of 100 mL capacity. Give several washings to paper with distilled water and collect washings in the same flask. Transfer 5 mL aliquot to another 100 mL volumetric flask and dilute to approximately 70 mL. Add 20 mL of 1.50 M KSCN (Dissolve 29.15 g KCSN in 200 mL of distilled water) and measure the intensity of color immediately (within 1 min) at 480 nm. Run a reagent blank with each set of samples.

Standard $Fe(NO_3)_3$ solution: Weigh 433 mg of $Fe(NO_3)_3$ and dissolve in 100 mL of distilled water in a volumetric flask. Dilute 2.5 mL of this stock standard and make the final volume to 250 mL in a volumetric flask. Pipette out 2.5, 5, 10, 15, and

20 mL aliquots of working standard solution into a series of 100 mL volumetric flasks and proceed in a similar manner.

Calculation

$$\text{Phytate}\,(\text{mg}/100\,\text{g sample}) = \frac{\mu\text{g Fe} \times 15}{\text{Weight of sample}(\text{g})}$$

9.7 Estimation of Quality of Grasslands and Pastures

Grasslands and pastures are the most important sustainable feed sources for livestock. Additionally, grasslands and rangelands provide numerous beneficial services to the human population. They also play a major role in maintaining the global plant biodiversity, carbon sequestration, water cycling, and other functions. Generally in grasslands and pastures, there is predominance of grasses (50 to 70%) followed by legumes (15 to 25%) and other shrubs and fodder trees. These range plants have been classified into three categories according to their forage value.

- High forage value designates plants that are nutritious and palatable and generate abundant forage.
- Medium forage value denotes plants that have the ability of providing adequate nutrients; however, these plants are not preferred by animals or do not produce abundant forage.
- Low or poor forage value describes plants that do not provide adequate nutrients and have higher amounts of anti-quality compounds.

Nutritional quality of grasslands and pastures is affected by abiotic and biotic environmental factors including soil type, climatic regime, botanical composition and management. Herbivores also influence botanical composition through selective grazing and alter the organic and mineral properties of grassland communities. The presence of a large number of weeds or inferior grass species also affects the quality of grasslands and pastures. The evaluation of quality of grasslands and pastures is an utmost requirement for providing nutritionally excellent feed to livestock. Using laboratory methods, it is difficult to generate a true picture of the quality of grasslands and pastures. Therefore, to determine the quality of grasslands and pastures, a combinational approach (Including chemical analysis and evaluation of quality based on floristic composition and forage values of individual plant species in the evaluated grassland) is required. There are also well-known scales for forage values of range plants. Vries et al. (1942) used a scale (from 0 to 10) to sort range plants according to their quality from toxic to nutritional ones. Stahlin (1971) sorted 1057 range plant species according to their quality into monocotyledonae (grasses and other monocotyledonous species), dicotyledonae (legumes and other dicotyledonous species), and plants producing spores (pteridophyte ferns and mosses). He calculated the percent distribution of range plant species in grassland and determined

Table 9.1 Classification of valuability of range plants in livestock feeding based on their forage value

Forage value	Plant species
7–8	Highly valuable to most valuable
6–7	Valuable to highly valuable
4–6	Less valuable to valuable
2–4	Least valuable to less valuable
1–2	Worthless to least valuable
0–1	Deleterious to worthless
(−1)	Deleterious to slightly toxic
(−1)–(−3)	Slightly toxic to highly toxic
(−3)–(−4)	Highly toxic to death causing

the nutritional quality of grassland. The quality of grasslands and pastures can be assessed by separating out range plant species on the basis of palatability scale. The scale of palatability values was created by Klapp et al. (1953) ($-1 =$ poisonous, $0 =$ no palatability value, $1 =$ very low palatability value, $2 =$ low palatability value, $4 =$ medium palatability value, $6 =$ high palatability value, $8 =$ very high palatability value). Usually, 9-stage scale, ranging from $1 =$ poisonous to $9 =$ very high palatability value is used for the determination of the grassland quality. All the methods for the evaluation of quality of grasslands and pastures which have been published are very complicated for application in practice. In the year 2004, Jan Novak gave a simpler method using 13-point scale system to determine the quality of grasslands and pastures. He gave the following equation to calculate the quality of grasslands and pastures:

$$E_{SQ} = \frac{(D \times FV)}{8}$$

$$E_{GQ} = \frac{\Sigma(D \times FV)}{8}$$

where

E_{SQ} = Forage quality of individual plant species
E_{GQ} = Grassland and pasture quality
D = Predominance of species (%)
FV = Forage value of species

This method utilizes 13-point scale system (from - 4 to 8) for the forage value of species (see Tables 9.1, 9.2, 9.3 and 9.4).

The determination of quality of grasslands and pastures using this method can be easily understood by the following example:

Table 9.2 Composition of typical grasslands and pastures

Plant species	*D* (%)	Forage value	E_{SQ}
Lolium perenne	22	8	22.00
Trifolium repens	19	8	19.00
Taraxacum officinale	7	5	4.37
Festuca pratensis	6	8	6.00
Poa pratensis	5	8	5.00
Cynosurus cristatus	5	6	3.75
Festuca rubra	4	5	2.50
Carum carvi	4	5	2.50
A. millefolium	3	5	1.87
Poa trivialis	3	4	1.50
Lentodon autumnalis	3	5	1.87
Tragopogon pratensis	2	4	1.00
Trifolium pratense	2	7	1.75
Phleum pratense	2	8	2.00
A. vulgaris	2	5	1.25
Ranunculus repens	1	-1	−0.12
Plantago lanceolata	1	6	0.75
Ranunculus acris	1	−3	−0.37
Prunella vulgaris	1	2	0.25
Mosses, empty places	8	–	–
E_{GQ}	100		76.87

Table 9.3 Quality of grasslands and pastures based on E_{GQ} value

E_{GQ}	Grasslands and pastures
90–100	Highly valuable - most valuable
70–90	Valuable - highly valuable
50–70	Less valuable - valuable
25–50	Least valuable - less valuable
15–25	Worthless - least valuable
0–15	Deleterious - worthless
< 0	Toxic

9.7.1 Determination of Quality of Grasslands and Pastures Using Remote Sensing Technology

Measurement of quality of grasslands' and pastures' attributes has been largely dependent on "wet lab analysis" and laboratory-based NIRS analysis. The Association of Official Agricultural Chemists (AOAC) has approved laboratory-based analytical methods as a standard approach for estimating the quality of grasslands and pastures. NIRS approach has also been accepted for quality evaluation of grasslands and pastures. However, these labor-intensive and time-consuming methods are constrained by destructive analysis, difficulties in getting "feed representative samples," delays between sampling time and the time of analysis. Moreover, these methods are difficult to implement on sites that are remote or difficult to

Table 9.4 Forage value of some prevalent plant species in grasslands and pastures

Plant species	Forage value
A. canina L	3
A. capillaris L.	5
A. gigantea Roth	7
Brachypodium pinnatum (L.) P. Beauv	2
Bromus inerinis Leyss	5
Festuca arundinacea Schreb.	4/2
Lolium perenne	8
Trifolium repens	8
Taraxacum officinale	5
Trifolium pratense	7
Festuca pratensis	8
Trifolium arvense	4
Lolium multiflorum	7
Lolium perenne	8
Phtagmite australis	2
Dactylis glomerata	7
Lotus corniculatus	7/5
Melilotus albus	3
Medicago lupulina	7
Astragulas glycyphyllos	7
Lathyrus tuberosus	7/5
Medicago falcnta	6
Trifolium dubium	6
Aconitum circum	−4
Datura stramonium	−4
Saxifraga granulata	2
Rumex obtusifolius	1
Rumex crispus	1
Rumex alpinus	2
Taraxacum offlcinale	5/3
Urtica dioica	1
Equisetum arvense	−2
Equisetum palustre	−1
Lysimnchia nummularia	1
Trifoliuin hybridum	6
Broums erectus	5
Poa pratensis	7
Lathyrus pratensis	7/5
Mentha arvensis	0
Trifolium dubium	6

access. To overcome these limitations, various remote sensing techniques have been developed and considered as potential tools for determining the quality of grasslands and pastures. The main advantages of remote sensing include: (1) the ability to obtain data from every location, (2) efficiency in data collection and processing, and (3) low cost.

(a) *Proximal remote sensing*: Extending laboratory NIRS techniques to the field, remote sensing technologies have been developed for real-time and nondestructive estimation of the quality of grasslands and pastures. Proximal remote sensing has great potential to predict pasture quality (Pullanagari et al. 2012; Kawamura et al. 2009; Biewer et al. 2009) because of high spectral resolution and negligible interference of atmospheric conditions in data collection. Compared to wet chemistry analysis, the relative cost involved in determining crude protein content using proximal remote sensing is low.
(b) *Aerial and space-based remote sensing*: With the development of aerial and satellite-based remote sensing, the ability to quantify characteristics of grasslands and pastures has expanded to regional and national levels. However, little research has been completed to quantify pasture quality parameters. In the early 1990s, NASA started examining foliar biochemistry in forest landscapes using the airborne instrument High-Resolution Imaging Spectrometer (HIRIS). Knox et al. (2011) mapped forage nutrients (nitrogen and phosphorus) in African savannas using an airborne sensor (CAO Alpha sensor). Similarly, Ramoelo et al. (2012) mapped canopy nitrogen at a regional level using the RapidEye sensor in the Kruger National Park (KNP), South Africa. Though the cost of determining the quality of grasslands and pastures is slightly higher than proximal remote sensing, a large area can be covered in a short period of time using aerial and space-based remote sensing.
(c) In comparison to air and space-based remote sensing, proximal sensors have been widely used in research to establish methods to quantify herbage quality, due to their wide availability and relative ease of operation (Thulin et al. 2012). Studies demonstrating the potential of remote sensing to estimate quality parameters of grasslands and pastures have been summarized in Table 9.5.

Forage quality is the expression of nutritional characteristics that affect the consumption of forages and ultimately the performance of livestock. How well animals consume forage and how efficiently the nutrients in the forage are assimilated in animals and converted into animal products, depend on the quality of forages. Therefore, knowledge of the forage quality is essential to develop the best feeding strategy for livestock. Forage quality is highly variable among and within forage types for nutrient composition as well as digestibility. Livestock are most productive when fed a ration that meets their nutritional needs. The laboratory analysis of feeds is the best indicator for predicting animal performance on a ration before it is fed. Livestock feed analyses provide producers with useful data that can help to determine the nutrient content of a mixed feed or a particular ration component decrease feed costs through lower cost rations and more efficient

Table 9.5 Use of remote sensing in the evaluation of quality of grasslands and pastures

Pasture quality parameters (% DM)	Range	Remote sensing type	Grassland type	Reference
Crude protein (%)	9.00–19.50	Proximal	Managed hill country pastures	Kawamura et al. (2008)
	9.38–28.41	Proximal	Managed pastures	Pullanagari et al. (2012)
	4.39–30.45	Proximal	Managed pastures	Thulin et al. (2012)
Nitrogen (%)	0.53–1.44	Spaceborne	Rangelands	Ramoelo et al. (2012)
	0.83–3.42	Spaceborne	Unmanaged rangelands	Mutanga and Skidmore (2004)
	0.31–0.91	Airborne	Unmanaged rangelands	Knox et al. (2011)
	0.83–3.42	Airborne	Unmanaged rangelands	Skidmore et al. (2010)
ADF (%)	25.20–41.00	Proximal	Managed pastures	Kawamura et al. (2008)
	19.99–38.19	Proximal	Managed pastures	Pullanagari et al. (2012)
NDF (%)	41.00–67.60	Proximal	Managed pastures	Kawamura et al. (2008)
	28.66–67.32	Proximal	Managed pastures	Pullanagari et al. (2012)
Digestibility (%)	44.99–85.56	Proximal	Managed pastures	Thulin et al. (2012)

utilization, evaluate forage production practices such as plant species selections, fertilizer schedules, and time of harvest. The results of the analysis could only be useful when the sample tested truly represents the lot of feed that the animals will consume. Poor sampling technique results in misleading feed analysis values, which may lead to increased feed costs, reduced animal performance, and economic loss.

References

Akhidue CV, Monago VV (2002) Estimation of tannin, saponin, oxlate, cyanogens and cardiac glycosides in Garsinia kola. J Appl Sci 6:22–25

AOAC (1965) Official methods of analysis, 10th edn. Association of Official Analytical Chemists, Washington, DC

AOAC (1970) Official methods of analysis, 11th edn. Association of Official Analytical Chemists, Washington, DC

AOAC (1995) Modification of tannin estimation by Folin-Denis reagent (Konard D, Christensen, IAEA). AOAC, Washington, DC

AOAC (2000) Official methods of analysis, 17th edn. Association of Official Analytical Chemists, Washington, DC

AOCC (1990) Official methods of analysis, 15th edn. Association of Official Analytical Chemists, Washington, DC

Arnon DI (1949) Copper enzymes in isolated chloroplasts. Polyphenoloxidase in *Beta vulgaris*. Plant Physiol 24:1–15

Barnes RF, Marten GC (1979) Recent developments in predicting forage quality. J Anim Sci 48: 1554–1561

Biewer S, Fricke T, Wachendorf M (2009) Development of canopy reflectance models to predict forage quality of legume-grass mixtures. Crop Sci 49:1917–1926

Crampton EW, Donefer E, Lloyd LE (1960) A nutritive value index for forages. J Anim Sci 19: 538–544

Heaney DP (1970) Voluntary intake as a component of an index to forage quality. p. C1-C10. In: Barnes RF, Clanton DC, Gordon CH, Klopfenstein TJ, Waldo DR (eds) Proc. Natl Conf forage Qual Eval Util, Lincoln, NE. 3–4 sept. 1969. Nebraska Center for Continuing Education, Lincoln

Heaney DP, Pigden WJ, Pritchard GJ (1966) Comparative energy availability for lambs of four timothy varieties at progressive growth stages. J Anim Sci 25:142–149

Hedge JE, Hofreiter BT (1962) In: Whistler RL, Be Miller JN (eds) Carbohydrate chemistry, vol 17. Academic, New York

Kaiser A, Freer M, Flinn P, Black J (2005) Prediction of the metabolisable energy content of forages and concentrates from measurements of digestibility. In: Paper presented to the AFIA quality evaluation committee meeting, Adelaide, 17 February 2005, 8 pp

Kawamura K, Watanabe N, Sakanoue S, Inoue Y (2008) Estimating forage biomass and quality in a mixed sown pasture based on partial least squares regression with waveband selection. Grassl Sci 54:131–145

Kawamura K, Betteridge K, Sanches ID, Tuohy MP, Costall D, Inoue Y (2009) Field radiometer with canopy pasture probe as a potential tool to estimate and map pasture biomass and mineral components: a case study in the Lake Taupo catchment, New Zealand. N Z J Agric Res 52:417–434

Klapp E, Boeker P, Konig F, Stahlin A (1953) Wertzahlen der Griinlandpflanzen. Das Grunland 2: 38–42

Knox NM, Skidmore AK, Prins HHT, Asner GP, van der Werff HMA, de Boer WF, van der Waal C, de Knegt HJ, Kohi EM, Slotow R, Grant RC (2011) Dry season mapping of savanna forage quality, using the hyperspectral Carnegie airborne observatory sensor. Remote Sens Environ 115:1478–1488

Mahadevan A (1979) Physiology of host–parasite interaction. Today and Tomorrow Publication, New Delhi, pp 167–171

Makkar HPS (2003) Effects and fate of tannins in ruminant animals, adaptation to tannins and strategies to overcame detrimental effects of feeding tannin-rich feeds. Small Rum Res 49:241–256

Marten GC, Shenk JS, Barton FE (1989) Near infrared reflectance spectroscopy (NIRS): analysis of forage quality (handbook 643), 2nd edn. United States Department of Agriculture, Washington

Mastumoto H, Sherman DG (1951) A rapid colorimetric method for the determination of mimosine. Arch Biochem Biophys 33:195–200

Moore JE, Kunkle WE, Denham SC, Allshouse RD (1990) Quality index: expression of forage quality and predictor of animal performance. J Anim Sci 68(Suppl. 1):572

Moreira-Vilar FC, Siqueira-Soares Rde C, Finger-Teixeira A, de Oliveira DM, Ferro AP, da Rocha GJ, Ferrarese L, dos Santos WD, Ferrarese-Filho O (2014) The acetyl bromide method is faster, simpler and presents best recovery of lignin in different herbaceous tissues than Klason and thioglycolic acid methods. PLoS One 9(10):e110000

Mutanga O, Skidmore AK (2004) Integrating imaging spectroscopy and neural networks to map grass quality in the Kruger National Park, South Africa. Remote Sens Environ 90:104–115

Norris KH, Barnes RF, Moore JE, Shenk JS (1976) Predicting forage quality by near infrared reflectance spectroscopy. J Anim Sci 43:889–897

Novak J (2004) Evaluation of grassland quality. Ekol Bratis 23(2):127–143

Pullanagari RR, Yule IJ, Tuohy MP, Hedley MJ, Dynes RA, King WM (2012) In-field hyperspectral proximal sensing for estimating quality parameters of mixed pasture. Precis Agric 13:351–369

Ramoelo A, Skidmore AK, Cho MA, Schlerf M, Mathieu R, Heitkönig IMA (2012) Regional estimation of savanna grass nitrogen using the red-edge band of the spaceborne RapidEye sensor. Int J Appl Earth Observ Geoinform 19:151–162

Roy DN, Rao PS (1971) Evidence, isolation, purification and some properties of a trypsin inhibitor in *Lathyrus sativus*. J Agric Food Chem 19:257–259

Skidmore AK, Ferwerda JG, Mutanga O, Van Wieren SE, Peel M, Grant RC, Prinsm HHT, Balcik FB, Venus V (2010) Forage quality of savannas- simultaneously mapping foliar protein and polyphenols for trees and grass using hyperspectral imagery. Remote Sens Environ 114:64–72

Stahlin A (1971) Gutezahlen von Pflanzenarten in frischen Grundfutter. In: Sonderheft 5, der Zeitschrift "Das wirtschafHeigene Futter". DLG - Verlag, Frankfurt (Main)

Thiex N, Richardson CR (2003) Challenges in measuring moisture content of feeds. J Anim Sci 81: 3255–3266

Thulin S, Hill MJ, Held A, Jones S, Woodgate P (2012) Hyperspectral determination of feed quality constituents in temperate pastures: effect of processing methods on predictive relationships from partial least squares regression. Int J Appl Earth Observ Geoinform 19:322–334

Tilley JMA, Terry RA (1963) A two-stage technique for the in vitro digestion of forage crops. J Br Grassl Soc 18:104–111

Van Soest PJ, Wiens RH (1967) Use of detergents in the analysis of fibrous feeds. Determination of plant cell wall constituents. J Assoc Off Anal Chem 50:50–55

Van Soest PJ, Wine R (1968) The determination of lignin and cellulose in acid-detergent fiber with permanganate. J Assoc Off Anal Chem 51:780

Vries DM, t'Hart ML, Kniijne AA (1942) Een waardeering van grasland op grond van de plantkundige samenstelling. *Landbouwkundig Tijschrift* 54:245–265

Wheeler EL, Ferrel RE (1971) A method for phytic acid determination in wheat and wheat fractions. Cereal Chem 48:312–320

Wiseman HG, Jacobson WC (1965) Nitrate in plant material, determination of nitrate in silages and forages. J Agric Food Chem 13:36–39

Nutritional Quality of Major Forage Grasses of Himalayan Region 10

Forages are an important feed resource providing essential nutrients to livestock for maintaining their health and productivity. The Himalayan region has diversity in climatic conditions, geographical distribution and species distribution including forages. Livestock rearing is an essential component of hill farming where they contribute to the livelihoods and food security of rural households. Higher productivity from livestock requires the provision of forage with good nutritional quality. For this purpose, it is essential to have knowledge of the nutritional composition of forages and feed ingredients for providing nutritionally sound diet which could maintain health and productivity of animal. The nutritional quality of different grass species of the Himalayan region has been discussed.

10.1 Range Forages

Rangelands are the beauty of the Himalayan region constituting different forage grasses and legumes. *Cynadon dactylon, Cyperus rotundus, Imperata cylindrica, Themeda* spp., *Paspalum dilatatum, Cassia mimisoides, Digitaria stricta, Pennisitum flaccidum, Setaria homonyma, Panicum maximum, Setaria italic, Arun dodonax, Apluda mutica, Cyrtococcum accresecns, Cyrtococcum longipes, Dichanthium annulatum, Panicum auritum, Paspalidium punctatum Pennisetum orientale, Saccharum longisetosum* and *Brachiaria mutica* are important range grasses of the Himalayan region.

10.1.1 Guinea Grass (*Panicum maximum* Jacq.)

Guinea grass is used for cut-and-carry systems and is also used for silage production and haymaking. It is well preferred by grazing livestock, with a particularly high intake of grass at early maturity stages (Cook et al. 2005). The CP content in young

R. Katoch, *Nutritional Quality Management of Forages in the Himalayan Region*, https://doi.org/10.1007/978-981-16-5437-4_10

grass, mature grass and straw range from 11 to 12.50%, 4.50 to 5.40% and 6.50 to 7.70%, respectively. CF content ranges from 27.00 to 46.90% of DM. NDF and ADF content ranges from 50 to 70% and 40 to 55%, respectively. Lignin content of fresh grass contains about 5–6% lignin. Ash and EE content range from 6.10 to 13.20% and 0.80 to 3.80%, respectively. The *in vitro* dry matter digestibility values range from 55 to 65%. The level of different pasture quality parameters decreases from the young to mature stages as a result of differences in plant composition between levels of maturity. The higher proportion of plant stems, typical of older plants, may restrict access to leafy parts and force animals to consume low-quality herbage. Advancing maturity has a negative effect on the nutritional value of *P. maximum* pasture, indicating that this forage could be best utilized at young stages of development. It is generally suggested to supplement guinea grass with concentrates or forage legumes (Aschfalk et al. 2002; Brown and Adjei 1995).

10.1.2 Bermuda Grass (*Cynadon dactylon* Pers.)

Bermuda grass is considered as medium-quality forage. Fresh bermuda grass has 9–16% CP content, 45–85% NDF and 20–45% ADF (DM basis). It is highly tolerant to drought and heavy grazing. It is also used for cut-and-carry, hay and deferred feed. Bermuda grass hay is usually of low quality than fresh forage it contains about 8–10% CP content, 70–75% NDF and 35–40% ADF content. Nutritional quality of bermuda grass hay is affected by environmental conditions and storage time period. Young bermuda grass or well N-fertilized grass has good nutritional quality. The grass contains high HCN content and low level of oxalates (1% of DM), however the content changes with maturity. With good management practices for the crop and favorable weather conditions, the CP content and TDN content could be reach up to the extent of 12–14% and 55%, respectively.

10.1.3 Cogon Grass [*Imperata cylindrica* (L.) Raeuschel]

Imperata cylindrica grows well from sea level up to 2000 m in the Himalayas, in sub-humid or humid grasslands or open woodlands. The grass is used as pasture at early growth stages. Like many other grasses, maturity is the important factor influencing the CP content in grass. The protein content of *Imperata* grass varies from 4 to 11.50%. *Imperata* grass contains varied amounts of NDF, ADF and lignin depending mainly on the stage of maturity. The NDF, ADF ADL, EE, and ash content in grass ranges from 70.10 to 77.20%, 41.50 to 46.80, 6.60 to 10.20%, 1.30 to 3.0% and 4.70 to 9.0%, respectively. Digestibility of *Imperata* grass ranges from 36.80 to 55.0%. *Imperata* grass is less digestible than *Setaria anceps*, *Cenchrus ciliaris*, and *Pennisetum purpureum* at the same maturity stage.

10.1.4 Dallis Grass (*Paspalum dilatatum* Poiret)

Paspalum dilatatum is valuable forage due to its vigor, persistence and ability to withstand heavy grazing and trampling (Cook et al. 2005). It has 10–15% CP content, 68–74% NDF, 37–40% ADF, 4.60–5.20% lignin, 5.40–11.60% ash content and 57–63% TDN. The grass is very palatable from early growth stage to the pre-flowering stage. However, palatability declines with advancement in maturity (Cook et al. 2005). It is more palatable than tall fescue and bahia grass (Nada and Sawamura 1985). Dry matter digestibility ranges between 45 and 60%. As *Paspalum dilatatum* does not have high nutritional value, supplementation of the grass with maize silage, cereal grains, feed concentrates, and forage legumes is must for livestock feeding.

10.1.5 Para Grass [*Brachiaria mutica* (Forssk.) Stapf]

Para grass is good quality forage particularly suited to poorly drained and swampy areas. Para grass can be grazed, used in cut and carry systems, or made into hay or silage. It is usually used as green chopped forage. The grass contains 3.22–5.48% CP content and 39–77% DM digestibility. CF ranges from 25.70 to 43.50% with mean CF content of 35.50%. The grass has considerable amount of ash content (9.70%). NDF and ADF contents are in the range of 65–70% and 40–50%, respectively. The grass contains 5–7% lignin. EE content in para grass varies from 2.44 to 5.22%. Its nutritive value decreases with plant age whereas the fiber content increases. Its CP and DM digestibility varies with growth stage and leaf: stem ratio and N fertilizer application. It is a moderate to poor source of macro-minerals. Its feeding requires supplemental minerals in the diet.

10.1.6 Guria Grass (*Chrysopogon fulvus* (Spreng.) Chiov.)

Guria grass grows throughout the hilly regions of India. It is easily grown in low to medium rainfall areas with light red gravelly soil. CP content varies between 5.50 and 15.0% depending on age and nitrogen application. Digestibility varies from 45 to 65% depending on maturity and leafiness. The CF content varies from 31.60 to 36.80%. The grass contains 55–70% NDF and 35–40% ADF content with 4–5% lignin content. The ash content in grass varies from 8.70 to 11%. This grass is considered a useful pasture grass and has been used to reseed poor natural pastures. It should not be grazed during the establishment year. This grass is relatively palatable even at a mature stage. It is used for haymaking and silage preparation and thus provides a good roughage source for livestock. The *in vitro* dry matter digestibility of the grass varies with maturity. The grass is well eaten by livestock with particularly high intake at young stages.

10.1.7 Marvel Grass [*Dichanthium annalatum* (Forssk.) Stapf]

Marvel grass is quite palatable and popular pasture grass despite its poor feeding value (Pacheco et al. 1983). It is commonly grazed by small ruminants in India. CP content and CF content vary between 6 to 15% and 35 to 45%, respectively, depending upon the stage of growth. The CP content of improperly managed grass hay may be as low as 4%. It is a good source of macro and trace minerals. Marvel grass is appreciated by all classes of ruminants. In mixed pastures, marvel grass is preferred to all other grasses (Cook et al. 2005). Marvel grass can be used in pastures, in cut-and-carry systems, or for hay and silage making if it is cut before flowering. The grass has 65–70% NDF and 40–45% ADF content, 6–7% lignin and 1–1.6% EE content. The grass contains an appreciable amount of ash content (7–12%). It can constitute the main forage component of the diet but with supplementation.

10.1.8 Creeping Bluegrass (*Bothriochloa insculpta* (Hochst. ex A. Rich.) A. Camus)

Creeping bluegrass is used as permanent pasture and haymaking. *Bothriochloa insculpta* can be grazed or cut to make good quality hay. Forage quality of the grass is relatively low and declines with maturity and more rapidly with the onset of flowering. The level of CP content can drop from 9% in young leaves to about 5% in mature growth. The NDF and ADF contents vary 60–75% and 40–50%, respectively. Grass contains 27–32% CF, 5–6.5% lignin and 9.90–15.0% ash content (Keba et al. 2013). Quality declines with age and more rapidly with the onset of flowering. At the early growth stage, IVDMD is higher than the mature stage and could vary between 40 and 60% depending on growth stages.

10.1.9 Dinanath Grass (*Pennesitum pedicellatum* Trin.)

This grass is widely used as green fodder for livestock but found poor to sorghum fodder mainly due to its low voluntary intake and digestibility. Supplementation of mature grass with protein sources increases its intake and digestibility. The grass has low CP content (9.60% DM at early stage and 1.60% at straw stage), digestibility and voluntary intake, which decrease with advancement in maturity (Kaboré-Zoungrana et al. 2008). Dinanath grass is palatable to goats but less than bermuda grass. The grass has high NDF (70–78%) and ADF (40–45%) contents while lignin content varies from 5.80 to 10%. Crude fiber content, ash content and ether extract content ranges from 33 to 47%, 6 to 13% and 0.5 to 2.5%, respectively.

10.1.10 Congo Grass (*Brachiaria ruziziensis* Germ. & Evrard)

Congo grass is a short-lived perennial grass of the Eastern Himalayan region. Congo grass is valuable nutritive forage for livestock (Schultze-Kraft and Teitzel 1992). It is mostly used for direct grazing of permanent pastures. CP and CF content in grass varies from 8 to 15% and 23 to 42%, respectively. The NDF, ADF, lignin, ash and ether extract contents in congo grass ranges from 60 to 75%, 31 to 45%, 3 to 11%, 4.9 to 14% and 0.70 to 3.50%, respectively. *In vitro* organic matter digestibility, dry matter digestibility and protein digestibility ranges between 40 to 50%, 50 to 55% and 60 to 65%, respectively (Herrero et al. 2001; Khanum et al. 2010; Lopes et al. 2010; Meale et al. 2012). Ground Congo grass hay can be used as a fiber source in balanced complete feeds at up to 22% of the diet (Bianospino et al. 2010).

10.1.11 Alpine Timothy (*Phleum alpinum* L.)

The grass occurs in higher elevations and higher latitudes worldwide. It is a cold-tolerant grass and is found in wet alpine or sub-alpine meadows, grassy slopes, marshes, and along stream banks. It naturally regrows in open or disturbed sites in western mountains. The forage value of alpine timothy is good, being palatable to all domestic livestock. It survives in many areas despite heavy spring grazing but produces much more herbage where it is moderately grazed and given occasional rest from heavy grazing. It produces good forage that remains green throughout the summer and is valuable as a late-season feed. *Phleum alpinum* grass has 5–13%, 25–40%, 60–70%, 35–40%, 2.5–6%, 4–8% and 2–2.5% DM, CP, CF, NDF, ADF, lignin, ash, and EE content, respectively. The OM digestibility ranges between 44 and 70%.

10.1.12 Kentucky Bluegrass (*Poa pratensis* L.)

Kentucky bluegrass is a common perennial grass native to Europe, Asia, North America, and northern Africa. The ecological variability of this grass allows it to adapt to cold winter climates. Quality and palatability of the grass decrease with the advancement in maturity. Kentucky bluegrass has CP content in the range of 7.60–26.60% and NDF content from 43.50 to 58.70%. The grass has about 3.40–4.50% EE with 5.0–6.50% ash content. CF content varies from 24 to 28%.

10.1.13 Bahia Grass [*Paspalum notatum* Flueggé]

Paspalum notatum is nutritious when young or at an early stage of regrowth. It is widely used as a component of intensively grazed pastures. Its palatability is good at early growth stages but decreases with the advancement in maturity. Addition of N fertilizer may improve intake since it has a positive impact on protein content in grasses. Feeding value varies greatly with age of regrowth, genotype, and soil fertility. CP level is high (14%) at an early stage of regrowth which declines as the maturity advances (6%). Palatability varies with age, genotype, and soil fertility. It is essential to maintain grazing pressure to avoid decline in palatability. NDF, ADF, and lignin content vary from 60 to 75%, 25 to 40%, and 4.5 to 6.4%, respectively, depending on the stage of maturity. CF, ash, EE content and OM digestibility ranges from 28 to 34%, 6.5 to 13.5%, 1.2 to 4.0% and 58 to 67%, respectively.

10.1.14 Mission Grass [*Pennisetum polystachion* (L.) Schultes]

The grass is a valuable fodder, either grazed or cut to be used as hay by cattle. Its nutritive value and palatability are high before seedling and drops dramatically after it; it is thus recommended to prevent flowering by 6-week cutting intervals. It cannot stand heavy grazing, which results in pastures being invaded by other species. The grass contains 6.50–12.30% CP content. CF content varies from 27 to 41%. NDF and ADF contents are in the range of 72–78% and 30–50%, respectively. The grass has 2.2–11.0% ADL with 1.1–2.00% EE and 8.60–15.90% ash content (Senanayake 1995).

10.1.15 Big-Leaf Bristle Grass or Buffalo Grass [*Setaria palmifolia* (J. Koenig) Stapf]

Setaria palmifolia grass is highly palatable at younger stages and is useful in the cut-and-carry system (Melletti and Burton 2014). As a pasture grass, it is good for grazing at early growth stages (Hankey and Mashinini 2002). The grass is a moderately nutritional with 4.60–11.50% CP content, 33.50–42.50% CF, 0.50–1.20% EE and 5.80–12.0% ash content. The ether extract and ash content of the grass is around 1.20% and 8.40% DM. The grass contains high ADL content (5.0–6.60%) with 70–74% and 40–45.00% NDF and ADF content, respectively.

10.1.16 Spear Grass [*Heteropogon contortus* (L.)]

The grass has a low nutritional value with high cell wall constituents (70–75% NDF, 40–42.00% ADF, 3–7.5% lignin), it is low in protein content (3–9%) and deficient in minerals. Native spear grass pastures can be oversown with legumes to increase the nitrogen content of the sward (Orr et al. 2010). Spear grass has low digestibility, therefore, supplementation is necessary to sustain ruminant requirements and to avoid impairing fertility and growth. This grass is usually used as fodder in its young stage. It can be grazed or cut for haymaking. After flowering, its nutritive value declines, and the sharp-pointed seeds and tangled awns may injure animals and human beings too. At mature stage, it is less palatable and not liked by grazing animals. It is a good source of Ca and K and a moderate source of other macro and trace elements. Spear grass has moderate palatability compared to many other tropical grasses. It is moderately or little grazed. Its palatability depends on the season and growing conditions. The low palatability of spear grass compared to other grasses such as doob grass can be explained by a higher tensile resistance of the stems of spear grass (Benvenutti et al. 2009).

10.1.17 Bristle Grass, Rough Bristle Grass and Bristly Foxtail [*Setaria verticillata* (L.) P. Beauv]

Setaria verticillata is a species of grass known by the common names hooked bristle grass, rough bristle grass, and bristly foxtail. It is native to Europe, but it is known on most continents as an introduced species and often a noxious weed. The bristle grass is palatable, and is, therefore, used as pasture, hay, and forage. CP content varies between 6 and 20% depending on the age of the material and nitrogen fertilization, with CP digestibility ranging from 44 to 77%. Dry matter digestibility of about 70% has been recorded in 3-week-old regrowth in young leaves, falling to 50–55% at 6–8 weeks. Palatability is high in the young stage and becomes stemmy and unacceptable with the advancement in maturity. The grass has high NDF (60–75%) and ADF contents (50–60%) (Brown et al. 2016).

10.1.18 Blue Panic (*Panicum antidotale* Retz)

Panicum antidotale provides a palatable pasture and is valuable as a fodder crop for light sandy soils in dry areas (Sarwar et al. 2006). It is readily grazed by livestock before flowering. At flowering, the spikes have hard and woody stalks that decrease palatability rapidly. *Panicum antidotale* has a large range of CP content (4.0–15.60% DM). Crude fiber content ranges between 31 and 44% DM. At the late flowering stage, *Panicum antidotale* can acquire a bitter taste and accumulate large amounts of oxalic acid that may cause kidney disorders (Göhl 1982). It is rich in cell walls. The NDF and ADF content of the grass varies between 55 to 70% and 40 to 50%, respectively, depending on the growth stage. *Panicum antidotale* provides a palatable pasture and is valuable as a fodder crop for light sandy soils in dry areas. Livestock grazes it readily up to pre-flowering and after flowering spikes have hard and woody stalks that decrease its palatability.

10.1.19 Rhode Grass (*Chloris gayana* Kunth.)

This grass makes good quality forage for ruminants when grazed or harvested at an early stage of maturity. However, its nutritive value strongly decreases with maturity, especially after the first cut. Hay harvested at a later stage of maturity has low protein content and high fiber content, particularly in the stems, and it should be supplemented when fed to ruminants with nutritional requirements higher than those necessary for maintenance. IVDMD varies from 40 to 80%. Rhodes grass is generally harvested for hay at a later stage of maturity when the protein content is low (5–8%). Time of harvest and the sensitivity of variety to leaf shattering have an effect on the nutritional composition of hay. Rhodes grass is known to be a selenium accumulator and when grown on selenium-rich soils, its selenium content can cause mortality or morbidity in livestock. Acute toxicity occurs at 3 mg of selenium per kg body weight in cattle and causes death within a few days after intoxication. Another limitation of mature *Chloris gayana* hay is its low intake by the animal.

10.1.20 Buffel Grass or Anjan Grass (*Cenchrus ciliaris* L.)

Cenchrus ciliaris is considered to be highly nutritive and excellent for pasture in hot, dry areas. It is valued for its production of palatable forage and intermittent grazing during dry periods. CP values are mostly in the range of 5–15% but maybe as low as 4% in mature grass and IVDMD and CP digestibility vary from 50 to 60%, depending on the age of growth, cultivar, and soil fertility. Grass has variable amounts of oxalate and with soluble oxalate levels of 1–2% in the DM, there is rarely a problem in adult ruminants. Ruminants are less susceptible to oxalate poisoning but may suffer from hypocalcemia (milk fever in dairy cows) and develop renal problems. Calcium supplementation by feeding limestone or dolomite may help controlling hypocalcemia (Fig. 10.1).

10.2 Cultivated Forage Grasses of the Himalayan Region

Natural pastures and grassland are important feed resources to livestock in the Himalayan region. Forages including grasses and legumes are important sources of feed to livestock. Though forage grasses and legumes are important in the Himalayan region, their cultivation for livestock feeding is limited. Only small part of the area in the Himalayan region is utilized for forage cultivation. The nutritional attributes of some of the important cultivated forage grasses has been discussed.

10.2.1 Tall Fescue (*Festuca arundinacea* Schreb)

Tall fescue is the predominant cool-season perennial grass. It is a bunchgrass grown for pasture, hay and silage. It also provides stockpiled forage for late autumn and winter grazing due to its autumn growth and ability to maintain quality and palatability in adverse climatic conditions. It is one of the best temperate perennial grasses for supplying nutritious feed in late spring and early summer. Tall fescue has 22–27.10% dry matter, 9.33–12.54% CP, 62.40–71.40% NDF, 46.06–53.06%

Fig. 10.1 The portfolio of some important range forages in the Himalayan grasslands and rangelands. In reading order: (**a**) Guinea grass, (**b**) Bermuda grass, (**c**) Cogon grass, (**d**) Dallis grass, (**e**) Para grass, (**f**) Guria grass, (**g**) Marvel grass, (**h**) Creeping bluegrass, (**i**) Dinanath grass, (**j**) Congo grass, (**k**) Alpine Timothy, (**l**) Kentucky bluegrass, (**m**) Bahia grass, (**n**) Mission grass, (**o**) Big-leaf bristle grass or buffalo grass, (**p**) Spear grass, (**q**) Bristle grass, rough bristle grass, and bristly foxtail, (**r**) Blue panic, (**s**) Rhode grass, (**t**) Buffel grass Anjan Gras

ADF, and 11.23–19.73% hemicellulose content. Lignin content in tall fescue varies from 2.40 to 3.0% based on the time of harvest. Variation in hemicellulose and cellulose contents in tall fescue ranges from 28.80 to 34.00% and 25.00 to 28.10%, respectively. Crozier et al. (1997) reported 8.00% NSC (Nonstructural carbohydrates) in tall fescue. *In vitro* dry matter digestibility (IVDMD) of tall fescue ranges from 47 to 56.00%. Bagley et al. (1983) recorded average ash content of 8.90% in tall fescue harvested at different intervals.

10.2.2 Perennial Rye Grass (*Lolium perenne* L.)

Perennial rye grass is one of the most valuable pasture grasses in temperate regions of the world. Perennial rye grass can grow throughout the year if there is sufficient soil moisture. During the autumn break or when there are significant summer rains an established perennial rye grass sward can respond almost immediately. Perennial rye grass has the potential to produce good quantity of quality forage with 70% digestible dry matter and 10–15% CP content. The grass could be used for pasture, hay, silage and conservation purposes. It contains 70–80% TDN in early vegetative growth and 55–64% in sundried hay. The NDF, ADF, and lignin fractions increase from the vegetative to milk stages and remain unchanged through the hard dough stage.

10.2.3 Golden Timothy (*Setaria* spp.)

Golden timothy grass is used as green cut fodder, silage and hay (Katoch et al. 2012). The grass gives satisfactory silage with molasses. CP and CF content of grass range from 7.0 to 12.0 and 24 to 34%, respectively. It has good digestibility and cattle accept its stubble grazing. It has a good fodder yield, extended green period and quality as compared with other conventional grasses. *Setaria* has low dry matter yield in comparison to *Brachiaria* and *Panicum maximum* grass. The digestibility of CP ranges from 44 to 77%. Dry matter digestibility of 70% has been recorded in 3-week regrowth, falling to 50–55% at 6–8 weeks. With a supplement of 0.25 kg feed per liter of milk produced, golden timothy produces the same nutritive value as elephant grass. Varieties PSS-1, S-92, and S-20 are good performers in the Himalayan region. *Setaria* grass is also known to have variable levels of oxalates. Highest levels of oxalate has been observed during night and the lowest in mid-afternoon. Golden timothy forage contains moderate to high levels of oxalates, depending on variety, stage of growth, fertilizer level, the part of the plant used. The grass contains 70–72% NDF and about 40–44% ADF content. Cattle should be introduced gradually to the *Setaria*, before more regular exposure. This allows them to develop rumen flora that can detoxify oxalate. Grass staggers can also occur. The disease is characterized by low magnesium in the blood system. The combined use of calcium borogluconate and magnesium hypophosphite is used for the treatment of grass staggers disease in cows.

10.2.4 Orchard Grass/Cock's Foot (*Dactylis glomerata* L.)

Dactylis glomerata is also known as cock's-foot or orchard grass and cat grass. It is a cool-season perennial C_3 bunchgrass native throughout most of Europe, temperate Asia, and northern Africa. The main advantage of cocksfoot is greater forage production during summer compared to other forage grasses: it stays green after most prairie grasses have dried. Dry matter ranges from 15.30 to 27.00%, while CP varies from 9.80 to 15.00%. The protein content declines with maturity and can be lower than 10% DM at the end of flowering. It contains 60–65% NDF and 30–35% ADF. It provides energy of 15–18 MJ/kg DM. The grass is free from alkaloids and mycotoxins (Barnes et al. 2007). Organic matter digestibility of fresh cock's foot is 52–78% and produces 7–11.50 MJ/kg DM of energy.

10.2.5 Elephant Grass (*Pennisetum purpureum* Schumach.)

Elephant grass is much used in paddocks for cutting and is the important forage for silage making because of high production per area and good nutritional value. However, its high cell wall content and low water-soluble carbohydrates (WSC) content impair the ensiling process. When used for pasture, elephant grass should be heavily grazed so that most of the young leaves and shoots, which have the highest nutritive value, are available to ruminants. Elephant grass is extremely palatable when young and leafy (Cook et al. 2005). However, it becomes coarse and unpalatable with the advancement in maturity. NDF content in elephant grass varies from 55 to 75% DM (Moran 2011). The grass contains 85–88.50% OM, while CP may vary from 5 to 10%. Elephant grass contains nitrate which causes poisoning in cattle when used as the sole component of the diet (Cook et al. 2005). Elephant grass is low in energy and protein content (Artus-Poliakoff et al. 1991). *In vitro* DM digestibility of the grass is similar to signal grass (*Brachiaria decumbens*), but higher than guinea grass (*Panicum maximum*) (Benedetti et al. 2008).

10.2.6 Bajra × Napier Hybrids (BN Hybrids)

Bajra × Napier hybrids have been developed by crossing bajra with elephant grass. This hybrid has a higher number of tillers. The BN hybrids contain 6.40–8.70%, 0.80–1.10%, 31.10–34.10%, 6.00–7.20%, and 2.30–2.80% of CP, crude fat, crude fiber, silica and oxalic acid contents, respectively. Calcium, phosphorus, potassium, magnesium and sulphur contents range from 0.31 to 0.49%, 0.32 to 0.44%, 1.59 to 2.06%, 0.63 to 0.89% and 0.14 to 0.20%. The BN hybrids are popular due to their higher yields. The grass is cultivated largely on bunds and wastelands due to its capacity to resist drought stresses with poorly fertile soils.

10.3 Cereal Fodders

Cereal crops belong to grass family Poaceae and are grown for their edible starch seeds. Cereal fodders are determinate in growth habits and their nutritional quality starts declining after flowering. Different cereal fodders like maize, sorghum, barley, and oats provide energy-rich herbage to livestock. These cereal forages have wider adaptability and variability in terms of growth, regeneration potential, yield, and nutritional quality of fodder.

10.3.1 Fodder Maize (*Zea mays* L.)

Maize is an important fodder crop of Himalayan region that gives highly succulent and nutritionally rich fodder. Its green forage has relatively low protein content (6–13% of DM) and high crude fiber content (19–36% of DM). Depends on the harvesting stage, it can be rich in starch (up to 30% or more) and therefore, serve as a good energy source for animals. Digestibility and nutritive value of the whole maize plant, fresh or ensiled, have been widely studied. Feed value of maize is much greater, up to 35% more when the whole plant is harvested.

Maize is traditionally cultivated as staple food, primarily for household consumption. It is often grown in mixed farming systems where maize stover makes a significant contribution to livestock feeding during the scarcity of green fodder. Increasing demand for fodder, shortage of arable land and water together with shrinking and deteriorating food resources is further increasing the demand for maize as a food-feed crop (Katoch and Kumar 2014). The cultivation of dual-purpose maize genotypes is preferred that provide green fodder besides grains. Maize silage is also used for winter animal feed in cold temperate regions. Maize crop has great potential in improving smallholder food security and benefits from livestock through superior dual-purpose maize cultivars providing both food and feed. The higher economic returns from the cultivation of dual-purpose maize cultivars could help in doubling the income of farmers.

10.3.2 Oats (*Avena sativa* L.)

Oats are a major cereal grain crop worldwide and the sixth grain cereal after maize, rice, wheat, barley and sorghum. Oats are a valuable feed resource to livestock in the Himalayan region. Oat forage is mainly used for livestock feeding. It may be used for grazing, cut-and-carry system could be dried to hay and preservation as silage or balage. Oat forage contains approximately 7–10% CP with about 50–58% NDF, 40–45% ADF, 3–4.50% lignin and 25–30% crude fiber. The oat crop is used as a winter cover to protect soil from erosion and to trap N that would otherwise be lixiviated during winter. Using oat forage prior to grain production is a valuable way of managing oats. This method may avoid crop lodging and decrease foliage fungal diseases. Oat forage sown during spring out yields barley DM production but has a lower nutritive value (Suttie and Reynolds 2004). However, grazing oats in autumn provides a system whereby livestock could substantially improve their body condition when perennial forages are of poorer quality and limited supply.

10.3.3 Sorghum [*Sorghum Bicolour* (L.) Moench]

Sorghum is a food-feed crop. Some sorghum varieties have been developed solely for grain production, others have been developed for forage production. Some sorghum varieties are dual-purpose (Harada et al. 2000). Commercial hybrid fodder sorghums are usually based on Sudan grass and grain sorghum. Fodder sorghum retains the multicut qualities of Sudan grass but has a much higher yield potential. CP content varies from 7 to 15%, whereas NDF and ADF content of 50 to 58% and 35 to 45%, respectively, have been reported. It contains around 3.30–5.00% of lignin and 25–35% of crude fiber. Organic matter digestibility of sorghum has been observed and reported to be 60–65% in livestock.

Forage sorghum can withstand heavy grazing pressure. Its green chop can help to provide summer feed supplies. Dhurrin, a cyanoglycoside, is present in the aerial parts of sorghum. It is present in higher concentration in young leaves and tillers and in plants that are suffering from drought. Freshly cut sorghum should be consumed rapidly to avoid the formation of HCN (Undersander et al. 2003). Forage sorghum could contain HCN content between 100 and 800 mg/kg plant. It is lethal to ruminants if consumed at 2 mg/kg body weight. Sorghum is a nitrate accumulator and while nitrates are generally nontoxic to ruminants, their transformation into ammonia within the rumen leads to the formation of toxic nitrites.

10.3.4 Barley (*Hordeum vulgare* L.)

Barley is a major cereal crop primarily grown for its grain, but it also yields valuable forage that can be grazed, cut for hay or silage while still green or cut after grain harvest as straw (Göhl 1982). The widespread use of barley forage for feeding purposes is relatively new. Using barley forage prior to grain production may be a valuable way of managing barley and it may avoid crop lodging and decrease foliage fungal diseases while feeding livestock. Because of its good carbohydrate content, barley is easy to ensile and with a rapid fall in pH produces good quality silage. The protein and NDF content of barley forage vary only slightly between flowering and the dough stage, decreasing from 12 to 9% DM for protein, 63 to 56% DM for NDF. The OM digestibility of barley forage varies slightly according to the stage of maturity and increases only from 61 to 65% between flowering and the dough stage. Though it has less feed value than oat straw, barley straw is an inexpensive way to feed cattle. It can be offered as the sole roughage during winter provided it is adequately supplemented.

10.3.5 Pearl Millet (*Pennisetum glaucum* (L.)

Pearl millet is an erect annual grass with a profuse root system. It can grow up to 3 m in height. It has the ability to thrive well even in soils with low nutrient and moisture contents. The CP and NDF content of pearl millet forage harvested at 50% heading range from 7 to 11%, 52 to 65%,respectively (Ward et al. 2001; Hancock 2009). The nutrient profile of grain is also high as it contains about 12% protein, 70% carbohydrate (starch), 5% lipid, and ~5% fiber and ash (Hadimani et al. 2001). The grain is also rich a rich source of Ca, Fe, and lysine. Lysine is an important amino acid in dairy nutrition, as it promotes body growth and contributes to Ca absorption, which supports high milk production in dairy cattle (Fayed et al. 2016). Pearl millet forage produces a greater leaf-to-stem ratio, thus resulting in lower lignin and greater CP content, and higher digestibility than sorghum and corn (Kumar 1989). Jahansouz et al. (2014) observed a leaf-to-stem ratio of 1.50 in pearl millet, contrary to 0.92 in sorghum and 0.72 in corn when harvested for forage at 100 days after planting. Machicek (2018) also observed a 35% greater leaf-to-stem ratio in pearl millet than in sorghum when harvested as forage at 90 days after sowing.

10.3.6 Indian Barnyard Millet [*Echinochloa frumentacea* (Link) Ridl.]

Barnyard millet has a wide adaptation capacity and can grow up to an altitude of 2000 m above mean sea level during summer (Gupta et al. 2009). Barnyard millet grown under natural precipitation is a highly valued food-feed crop. It grows at a faster pace and produces voluminous fodder. The fodder is highly palatable and can be used for making hay or silage. It is considered superior to rice and oat straw because of its high protein and calcium content. Barnyard millet straw contains up to 61% total digestible nutrients and a fairly good amount of protein and digestible fiber. Banyard millet fodder contains 7–10% protein, 20–25% digestible fiber, 10–12% ash and 1–2.0% ether extract content. Because of its superior feed quality, higher digestibility and nitrogen content, barnyard millet is used as a potential livestock feed crop in the dry areas of the Deccan plateau to the extreme hills of the temperate sub-Himalayan region (Singh and Singh 2005; Bandyopadhyay 2009; Yadav and Yadav 2013; Sood et al. 2015). In Uttarakhand, barnyard millet fodder constitutes contributes about 10% of the total fodder consumption of the State (Singh and Singh 2005). The grains of Indian barnyard millet are also fed to animals. It has superior feed quality when fed young to the animals; however, protein content declines up to 6% at maturity.

10.3.7 Finger Millet [*Eleusine coracana* (L.) Gaertn]

Finger millet is cultivated as fodder grass in different regions of the world. It provides excellent hay and is used as green forage for cattle, sheep, and goats. The straw resulting from the grain harvest is valuable and can be grazed directly by the animals or used in cut-and-carry feeding systems. Finger millet forage has low nutritive value. The fresh forage contains about 5–8% protein (Bakhashwain et al. 2010). Like other cereal straws, finger millet straw is highly fibrous (NDF > 70% DM) and poorer in protein than the fresh forage (about 5% DM). Finger millet straw can be used as forage in diets for crossbred dairy cows supplemented with a balanced concentrate mixture (energy and protein).

10.3.8 Foxtail Millet [*Setaria italica* (L.) Beauv.]

Foxtail millet is a grain and forage crop well adapted to nutrient-poor soils. Foxtail millet (*Setaria italica*) is grown for green grass, hay, and silage and is good forage for livestock. Foxtail millet forage has 9–12.5% protein in fresh forage and less than 8% in the hay. It contains 48–72% NDF and 33–44% ADF content. The OMD and ME range from 55 to 59% and 6 to 8.3 MJ/kg DM, respectively. Foxtail millet hay is of lower quality than alfalfa hay. However, *in situ* NDF rumen degradability is similar for foxtail millet and alfalfa hay (Castillo et al. 1992) (Fig. 10.2).

10.4 Studies on Forage Grasses of the Himalayan Region

10.4.1 Comparative Biochemical Profile of Tall Fescue (*Festuca arundinacea* Schreb.) Varieties Grown in the Himalayan Region

Tall fescue owing to its ability to substantially tolerate adverse conditions, better nutritional profile, suitability and flexibility into animal production system is viewed as a potential livestock feed for the Himalayan livestock. Since the information on the systemic characterization of nutritional attributes of tall fescue for the Himalayan region is meager, therefore, the study was conducted to evaluate the nutritional potential and digestibility of four tall fescue varieties, Hima-1, Hima-4, Hima-5 and EC-178182 grown under mid-Himalayan conditions (Sood et al. 2019). The results of the study in tall fescue varieties have been discussed under the following subheads:

(a) **CP and carbohydrate content**: The CP content in tall fescue varieties ranged from 8.82 to 12.81%, of which, varieties EC-178182 and Hima-5 were observed with significantly higher (12.81%) and lower (8.82%) CP content, respectively (Table 10.1). Varieties Hima–4 and Hima-5 revealed statistically at par values of CP content (8.96% and 8.82%, respectively). The level of the total carbohydrate content in tall fescue varieties ranged from 9.74% (Hima-5) to 13.34% (EC-178182) (Table 10.1). The reducing sugar content in tall fescue varieties ranged from 1.26% (Hima-4) to 1.83% (EC-178182) (Table 10.1).
(b) **Fiber content and cell wall constituents**: The fiber content (ADF and NDF) has a direct influence on the digestibility of forages in animals. High fiber

Fig. 10.2 The portfolio of some important cultivated forages; a) cereal forages of the Himalayan region. In reading order: (**a**) Tall fescue, (**b**) Perennial rye grass, (**c**) Golden timothy, (**d**) Orchard grass/Cock's foot, (**e**) Elephant grass, (**f**) Bajra × Napier hybrids, (**g**) Fodder maize, (**h**) Oat, (**i**) Sorghum, (**j**) Barley, (**k**) Pearl millet, (**l**) Indian barnyard millet, (**m**) Finger millet, (**n**) Foxtail millet

Table 10.1 Biochemical composition of tall fescue varieties

Variety	CP (% DM)	TC (% DM)	TS (% DM)	RS (% DM)	NDF (% DM)	ADF (% DM)	HC (% DM)	ADL (% DM)	CL (% ADF)	CC (% DM)	Ash (% DM)	IVDMD (%)
Hima-1	9.45^{b}	11.41^{b}	4.89^{b}	1.47^{c}	69.36^{cd}	50.28^{d}	19.08	7.30^{bc}	43.02^{d}	30.64	6.68^{bcd}	47.14^{bc}
Hima-4	8.96^{cd}	10.30^{c}	4.69^{cd}	1.26^{d}	69.52^{cd}	51.28^{c}	18.24	7.40^{abc}	43.94^{c}	30.48	6.42^{bcd}	46.93^{bc}
Hima-5	8.82^{cd}	9.74^{d}	4.71^{cd}	1.69^{b}	71.12^{a}	52.04^{ab}	18.40	6.50^{d}	45.54^{ab}	28.88	6.51^{bcd}	49.45^{a}
EC-178182	12.81^{a}	13.34^{a}	5.09^{a}	1.83^{a}	70.40^{b}	52.72^{ab}	18.36	7.50^{ab}	45.22^{ab}	29.60	8.01^{a}	46.55^{d}
CD (5%)	0.461	0.187	0.089	0.048	0.546	0.691	NS	0.153	0.783	NS	0.327	0.245

Table 10.2 Anti-quality components in tall fescue varieties

Variety	TP (%DM)	SP (%DM)	TT (%DM)	Ox (%DM)
Hima-1	1.40[c]	0.47[c]	0.93[c]	2.52[c]
Hima-4	2.14[a]	0.77[a]	1.37[a]	2.87[ab]
Hima-5	1.62[bc]	0.54[bc]	1.08[b]	2.86[ab]
EC-178182	0.78[d]	0.39[d]	0.39[d]	1.26[d]
CD (5%)	0.095	0.028	0.101	0.042

Table 10.3 Chlorophyll and carotenoids content in tall fescue varieties

Variety	Chl a (mg/g)	Chl b (mg/g)	Total Chl (mg/g)	Carotenoids (μg/g)	TF (mg/100 g)
Hima-1	0.67[c]	0.98[a]	1.65[a]	25.44[d]	1.21[cd]
Hima-4	0.58[d]	0.38[c]	0.96[d]	39.95[c]	1.16[cd]
Hima-5	0.84[ab]	0.33[d]	1.17[c]	47.94[a]	2.54[a]
EC-178182	0.83[ab]	0.52[b]	1.36[b]	41.86[b]	1.48[b]
CD (5%)	0.016	0.013	0.044	0.169	0.099

content in feeds reduces the voluntary intake and digestible energy. The NDF content in tall fescue varieties ranged from 69.36% (Hima-1) to 71.12% (Hima-5) (Table 10.1). Cellulose and hemicellulose content in tall fescue varieties varied from 43.02% to 45.54%, 18.24 to 19.08%, and 28.88 to 30.64%, respectively. Significantly high and low level of ADL content was observed for varieties EC-178182 (7.50%) and Hima-5 (6.50%), respectively.

(c) **Phenolic and anti-nutrient content**: In tall fescue varieties, phenolic compounds *viz.*, total phenols, simple phenols, and total tannins varied from 0.78 to 2.14%, 0.39 to 0.77%, and 0.39 to 1.37%, respectively (Table 10.2). The oxalate content in tall fescue varieties varied from 1.26% (EC-178182) to 2.87% (Hima-4) (Table 10.2).

(d) **IVDMD and phytochemical constituents**: The *in vitro* dry matter digestibility of tall fescue varieties ranged from 46.55% (EC-178182) to 49.45% (Hima-5) (Table 10.1). Different phytochemical constituents in tall fescue varieties have been presented in Table 10.3. Chlorophyll "a" content was significantly higher in variety Hima-5 (0.836 mg/g) which was statistically at par with variety EC-178182 (0.826 mg/g). The chlorophyll "b" content was significantly high in Hima-1 (0.976 mg/g) whereas, a significantly low value was observed for variety Hima-5 (0.334 mg/g). The total chlorophyll content of tall fescue varieties ranged from 0.962 mg/g (Hima-4) to 1.648 mg/g (Hima-1). Variety Hima-5 had high carotenoid content (47.94 μg/g), whereas, significantly lower carotenoid content (25.44 μg/g) was observed in variety Hima-1. Varieties *viz.*, Hima-5 and Hima-4 exhibited significantly high (2.54 mg/100 g) and low levels (1.16 mg/100 g) of total flavonoid content, respectively.

(e) **Antioxidant activity and mineral content**: The antioxidant activity in terms of inhibition constant (IC50) among different tall fescue varieties has been represented in Table 10.2. The significantly high and low inhibition constant (IC50) was observed in EC-178182 (116.08 μg/ml) and Hima-5 (65.53 μg/ml), respectively. The ash content was significantly higher in the variety EC-178182 (8.01%). The other varieties, Hima-1 (6.68%), Hima- 4 (6.42%) and Hima-5 (6.51%) revealed statistically at par values of ash content (Table 10.1). The level of macrominerals *viz.*, Ca, K, Na, P and Mg varied from 0.36% (Hima-4) to 0.69% (EC-178182), 2.43% (Hima-1) to 2.62% (EC-178182), 0.016% (Hima-1) to 0.030% (EC-178182), 0.26% (Hima-4) to 0.36% (EC-178182) and 0.18 (Hima-4) to 0.24% (EC-178182), respectively. Micro-minerals *viz.*, Cu, Fe, Mn, Zn and B in tall fescue varieties ranged from 0.54 mg/100 g (Hima-1) to 0.67 mg/100 g (EC-178182), 10.98 mg/100 g (Hima-1) to 12.78 mg/100 g (EC-178182), 2.22 mg/100 g (Hima-1) to 4.46 mg/100 g (EC-178182), 1.56 mg/100 g (Hima-4) to 1.76 mg/100 g (EC-178182) and 0.85 mg/100 g (Hima-4) to 1.06 mg/ 100 g (EC-178182), respectively. Tall fescue variety EC-178182 was observed superior with appreciable levels of macro as well as micro-minerals over other varieties (Table 10.4).

10.4.2 Productivity and Biochemical Attributes of Different Maize Varieties

Maize (*Zea mays* L.) is one of the most important coarse grains and is an important staple food in many countries of the world. Among cereal crops, maize has the highest average yield per hectare and remains third after wheat and rice in total area and production in the world. India is the fourth largest producer of maize in the world, contributing about 2% of total global production. The production and productivity of maize are continuously increasing in India. In the 11th 5-year plan, maize achieved the highest growth rate (6.7%) among the cereals against the required growth rate of 4.7 % set by the planning commission of India. Besides, being intensively used as human food and animal feed, it is also a source of a large number of diversified industrial products including corn oil, corn starch, corn oil, baby corns, popcorns, alcoholic beverages, food sweeteners, etc., which is of great economic importance.

Different maize varieties are grown for grain maize, starch maize, sweet corn, baby corn, popcorn, quality protein maize, high oil maize and fodder maize. Stover from maize provides dry season feed for livestock. Its silage is also used for winter animal feed in cold temperate regions. In the areas where production is high, maize stalks are used as soil mulch. Maize is a crop of great significance in the countries like India where the increasing human population jeopardizes food supplies. It is often grown in crop–livestock farming systems where stover makes a significant contribution in livestock feeding during the scarcity of green fodder. The rapid increase of both human and livestock populations, along with a restricted land

Table 10.4 Macro and micronutrient composition in tall fescue varieties

	Macronutrients (%)				Micro-nutrients (mg/100 g)				
Variety	Ca	K	Na	P	Mg	Cu	Fe	Mn	Zn
Hima-1	0.40^{bcd}	2.43^{bcd}	0.016^{d}	0.29^{bcd}	0.20^{bcd}	0.54^{bcd}	10.98^{cd}	2.22^{d}	1.60^{bcd}
Hima-4	0.36^{bcd}	2.48^{bcd}	0.026^{abc}	0.26^{d}	0.18^{bcd}	0.55^{bcd}	11.14^{bcd}	3.48^{bc}	1.56^{cd}
Hima-5	0.40^{bcd}	2.46^{bcd}	0.024^{bc}	0.30^{bc}	0.20^{bcd}	0.56^{bcd}	11.50^{bcd}	3.55^{bc}	1.66^{bc}
EC-178182	0.69^{a}	2.62^{a}	0.030^{a}	0.36^{a}	0.24^{a}	0.67^{a}	12.78^{a}	4.46^{a}	1.76^{a}
CD (5%)	0.056	0.050	0.003	0.025	0.026	0.039	0.525	0.186	0.041

In Tables 10.1, 10.2, 10.3, and 10.4, the significant differences according to CRD in the same column are indicated by different alphabets. Superscripts "a," "b," "c," and "d" represent significant values at ($p \leq 0.05$). Same alphabets represent that their values are statistically at par ($p \leq 0.05$); *CP; CP; TC* Total carbohydrates, *TS* Total sugars, *RS* Reducing sugars, *NDF* Neutral detergent fiber, *ADF* Acid detergent fiber, *HC* Hemicellulose, *ADL* Acid detergent lignin, *CC* Cell content, *IVDMD In Vitro* dry matter digestibility, *TP* Total phenols, *SP* Simple phenols, *TT* Total tannins, *Ox* Oxalate, *TF* Total flavonoids, *Chl* Chlorophyll

Table 10.5 Morpho-physiological and yield parameters of different maize varieties

Sr. No.	Name	Morphological traits: Plant height (cm)	Dry matter (%)	Dry matter yield (q/ha)	Yield traits: Green fodder yield (q/ha)	Baby corn yield (q/ha)	Maize green fodder equivalent yield (q/ha)
1.	Girija	217.00	15.30	61.06	426.60	12.20	549.00
2.	KH-581	219.50	15.36	78.77	512.50	07.40	587.00
3.	Early Composite	202.00	16.90	86.95	515.20	19.70	712.00
4.	VL-78	232.40	14.93	70.99	464.00	14.90	613.00
5.	HIM-129	212.90	17.73	66.60	379.70	33.60	715.00
6.	VL-Makka-42	186.80	14.10	42.33	298.60	29.80	664.00
7.	VL-Baby Corn-1	204.20	14.03	50.47	357.30	30.90	666.00
8.	Kesri	126.00	17.56	48.46	282.60	29.80	581.00
9.	Seetal	210.70	16.93	66.82	389.80	27.70	667.00
10.	Parkashi	211.00	16.36	42.96	262.40	19.70	363.00
11.	JH-3459	214.00	12.53	42.98	339.70	22.90	569.00
12.	PMH-2	194.00	14.23	55.51	384.00	23.40	618.00
13.	1006	202.90	15.20	57.01	374.90	16.00	534.00
14.	African Tall	218.00	17.26	76.47	454.90	13.30	588.00
	G.M	**207.61**	**15.60**	**60.53**	**388.76**	**21.60**	**604.33**
	CD 5%	**NS**	**NS**	**NS**	**NS**	**3.48**	**NS**
	CV%	**9.26**	**17.96**	**29.42**	**24.94**	**9.58**	**16.06**
	SE (m)±	**11.09**	**1.618**	**10.28**	**55.98**	**0.39**	**18.67**

base and an increased demand for livestock products has further increased the demand for maize as a food-feed crop.

The study was conducted on different maize germplasm lines for studying their morpho-physiological, economic and biochemical parameters (Katoch and Kumar 2014). Among 14 different maize germplasms, variety VL-78 revealed maximum plant height (232.40 cm). The dry matter (%) was highest in variety HIM-129 (17.73%) and lowest for JH-3459 (12.53%). Variety KH-581 exhibited highest dry matter yield of 78.77 q/ha and VL-Makka-42 revealed dry matter yield of 42.33 q/ha. Early Composite was superior in green fodder yield (515.2 q/ha). The green fodder yield from different varieties varied from 262.40 to 515.20 q/ha. Varieties Early Composite, KH-581, VL-78, and African Tall were promising in green fodder yield. Baby corn yield and maize green fodder equivalent yield were superior in HIM-129 with 33.60 q/ha and 715.0 q/ha, respectively. Morpho-physiological and yield parameters of different maize varieties have been presented in Table 10.5. After the first harvest, the CP content varied from 5.54% in Parkashi to 8.16% in the variety Girija. Variety JH-3459 has the highest crude fiber, NDF, and ADF content

(31.20%, 65.13%, and 57.0%, respectively). After the second harvest, the CP content reduced from 5.25% (Parkashi) to 8.16% (VL-78). The crude fiber content was highest in variety Early Composite, while NDF and ADF were with higher value in VL-Baby Corn-1 (64.73%) and VL-Makka-42 (57.73%), respectively. The cost of cultivation of HIM-129 was 23,032 Rs/ha with Net returns of Rs. 48,534. Variety KH-581 had least cost of cultivation (Rs. 19,896/ha). Among all varieties, the Early Composite resulted highest gross returns of Rs. 71,233 and net returns of Rs. 49,865 with a cost of cultivation of Rs. 21,360/ha while, variety Parkashi has least gross and net returns. Varieties, HIM-129, Early Composite, Seetal, KH-581, and VL-Baby Corn-1 resulted in 73.27%, 70.00%, 66.52%, 66.10%, and 65.93% economic gains, respectively. The cultivation of dual/multipurpose maize cultivars for procuring baby corn and green fodder was economically viable rather than the cultivation of only fodder-type varieties. These varieties also increases the green fodder yield, green fodder equivalent yield and net returns. The economic returns were higher in dual/multipurpose maize cultivars as compared to those grown for single purpose. The nutritional equivalence of dual-purpose maize varieties to African Tall grown solely for fodder purposes suggests that the harvesting of baby corn does not have any influence on the quality of harvested fodder.

10.4.3 Studies on Quality Parameters and Production Economics of Maize Under Different Use Patterns

Maize is traditionally cultivated as a staple food, primarily for household consumption, but its demand for feed and industrial uses has increased rapidly in the recent past. Nowadays, maize is often grown in mixed farming systems where maize stover makes a significant contribution to livestock feeding during the scarcity of green fodder. Increasing demand for fodder, shortage of arable land and water together with shrinking and deteriorating food resources are further increasing the demand for maize as a food-feed crop. The cultivation of dual-purpose maize genotypes is preferred which also provide green fodder besides grains. Maize silage is also used for winter animal feed in cold temperate regions.

Maize crop has great potential in improving smallholder food security and fodder for the livestock by cultivation of dual-purpose maize cultivars providing both food and feed. The higher economic returns from the cultivation of dual-purpose maize cultivars could help improving the income of farmers. Due to immense maize growing potential of the state of Himachal Pradesh, 2-year experiment comprising 12 maize varieties under different use patterns was conducted to evaluate their growth, yield, biochemical attributes and economic viability. The stover yield among the varieties varied from 34.81 q/ha in Girija variety used mainly for grain purpose to 429.62 q/ha in African Tall and J-1006, during first year, whereas the stover yield varied from 51.84 to 411.04 q/ha in Early Composite and J-1006, respectively during second year (Fig. 10.3). Variety VL-Baby Corn-1 revealed highest baby corn yield (24.81 q/ha), whereas, variety VL-Makka-42 reveled higher seed yield (26.29 q/ha). When varieties are specifically for green cob purposes,

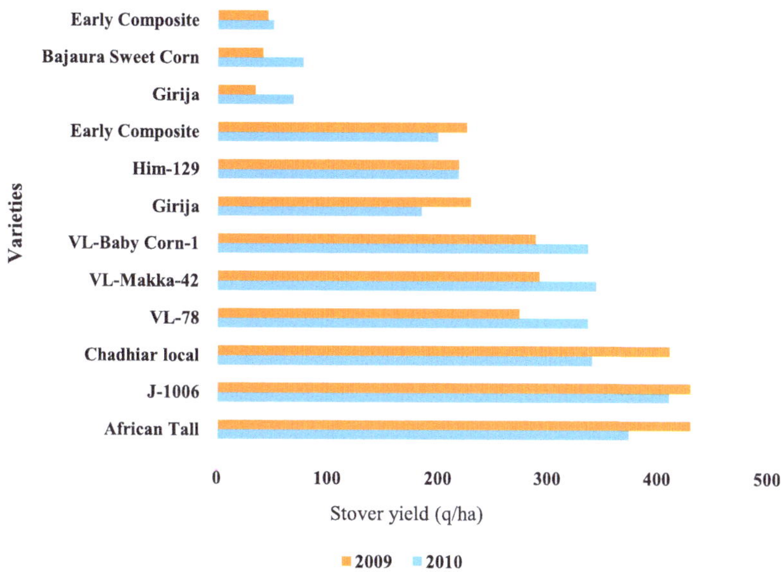

Fig. 10.3 Fodder/ stover yield in maize varieties

Him-129 resulted highest cob yield of 58.89 and 52.58 q/ha, 2009 and 2010, respectively. Maize variety Girija revealed highest grain yield of 28.51 q/ha in 2009, while variety Early Composite had maximum grain yield of 19.99 q/ha in 2010 (Tables 10.6 and 10.7).

Dry matter per cent of maize varieties varied from 15.36 to 76.23% in year 2009. The fodder dry matter per cent of grain maize varieties namely, Girija, Bajaura Sweet Corn, and Early Composite was significantly higher than the varieties of other types of use pattern. Dry matter yield (q/ha) was highest for the fodder type maize varieties *viz.,* African Tall (65.84) and Chadhiar local (69.28) followed by VL-Makka-42 (60.72) and VL-Baby Corn-1 (62.24) varieties of baby corn type. The lower dry matter yield was observed in the varieties for grain purpose. During the year 2010, the dry matter content among various maize varieties grown under different use patterns varied from 15.50 to 75.80%. Dry matter yield (q/ha) of baby corn maize varieties *viz.,* VL-Baby Corn-1 (84.29), VL-Makka-42 (74.69), and VL-78 (61.36) was on higher side followed by fodder type maize varieties *viz.,* J-1006 (74.60) and African Tall (61.05). The lowest dry matter yield (q/ha) was exhibited by the Early Composite variety (39.35). During both years, the CP content of maize varieties of fodder type namely, J-1006, Chadhiar Local, and African Tall and two varieties of baby corn type, namely VL-78 and VL-Makka-42 was significantly higher than other varieties. The crude fiber content varied from 24.80 to 34.50%. The crude fiber content was significantly higher in maize varieties of grain maize type *viz.,* Girija, Early Composite, and Bajaura Sweet Corn followed by Him-129 and VL-Makka-42 grown for green cob. Ash content, which represents the total mineral content varied significantly from 4.63 to 7.77%. Ash content was

Table 10.6 Biochemical parameters of different maize varieties

Sr. No.	Name	1st harvest				2nd harvest				
		CP (%)	Crude fiber (%)	NDF (%)	ADF (%)	CP (%)	Crude fiber (%)	NDF (%)	ADF (%)	IVDMD (%)
1.	Girija	8.16	28.60	61.00	46.59	7.58	29.73	61.20	48.66	52.00
2.	KH-581	5.83	30.00	61.53	49.60	5.83	29.23	63.00	50.46	58.50
3.	Early Composite	6.41	30.63	58.13	49.93	6.12	31.13	61.66	53.60	56.40
4.	VL-78	7.99	27.36	61.20	51.46	8.16	28.80	62.06	52.66	55.60
5.	HIM-129	6.12	30.60	62.26	53.46	5.54	30.83	63.60	53.60	55.65
6.	VL-Makka-42	6.41	30.20	62.80	55.73	6.70	29.83	63.86	57.73	58.30
7.	VL-Baby Corn-1	6.70	29.53	63.26	55.06	6.12	30.63	64.73	55.46	61.50
8.	Kesri	7.58	28.50	59.66	50.86	7.58	29.80	62.40	52.06	55.40
9.	Seetal	6.12	28.80	62.13	53.00	6.41	28.00	61.80	54.53	59.70
10.	Parkashi	5.54	31.06	63.00	54.60	5.25	29.26	63.66	57.26	52.80
11.	JH-3459	7.58	31.20	65.13	57.00	6.70	29.96	64.33	55.33	52.70
12.	PMH-2	7.00	28.93	62.33	53.80	7.87	29.00	63.93	54.93	55.20
13.	1006	6.70	30.76	62.53	53.00	7.58	30.60	64.33	54.46	56.90
14.	African Tall	7.87	29.26	62.76	55.26	7.58	29.53	64.40	54.93	60.30
	G.M.	**6.87**	**29.67**	**61.98**	**52.81**	**6.78**	**29.75**	**63.20**	**53.98**	**56.49**
	CD at 5%	**0.91**	**1.014**	**2.98**	**1.35**	**0.72**	**NS**	**1.66**	**2.08**	**NS**
	CV%	**7.94**	**2.04**	**2.87**	**1.52**	**6.36**	**3.52**	**1.57**	**2.30**	**5.99**
	SE (m)±	**0.31**	**0.35**	**1.03**	**0.46**	**0.25**	**0.60**	**0.57**	**0.71**	**2.39**

Table 10.7 Baby corn, green cob, and grain yield of different maize varieties

Sr. No.	Varieties	Yield (q/ha)					
		Baby corn		Green cob		Grain	
		2009	2010	2009	2010	2009	2010
1.	VL-78	22.32	25.17	–	–	–	–
2.	VL-Makka-42	22.59	26.29	–	–	–	–
3.	VL-Baby Corn-1	24.81	24.44	–	–	–	–
4.	Girija	–	–	49.26	39.99	–	–
5.	Him-129	–	–	58.89	52.58	–	–
6.	Early Composite	–	–	54.44	43.33	–	–
7.	Girija	–	–	–	–	28.51	15.92
8.	Bajaura sweet corn	–	–	–	–	19.25	18.51
9.	Early Composite	–	–	–	–	20.00	19.99
	G.M	23.24	25.30	54.19	45.3	22.50	18.14

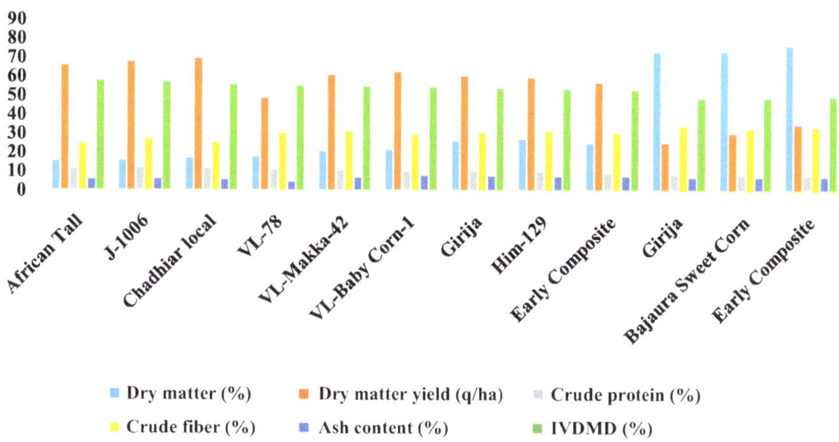

Fig. 10.4 Quality parameters of maize varieties under different use patterns (2009)

higher in Early Composite (Grain maize and Green cob type) followed by VL-Baby Corn-1 and Girija. *In vitro* dry matter digestibility (IVDMD) among the maize varieties ranged from 49 to 57.80% and 51.80 to 56.20% during 2009 and 2010, respectively. The results revealed that the cultivation of varieties for the baby corn is more beneficial as in addition to green fodder, the baby corn produce also has high economic value. The data on different parameters has been presented in Figs. 10.4 and 10.5.

The production economics of maize varieties have been presented in Table 10.8. The gross returns from VL-Baby Corn-1 cultivation were higher during year 2009

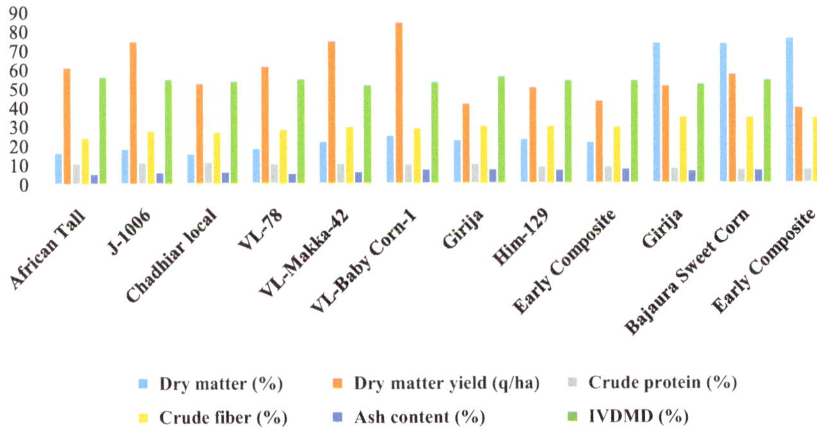

Fig. 10.5 Quality parameters of maize varieties under different use patterns (2010)

(Rs. 128,128.30 ha) and 2010 (Rs. 113,660.33/ha). Maize varieties for baby corn and green cob types accounted for significantly higher gross returns than those of fodder type which could be attributed to the dual yield (cobs and fodder). The highest net returns were observed from the varieties of baby corn type followed by those of green cob type. Though the cost of cultivation was more for the baby corn type varieties, yet they resulted higher net returns. The mean values of both years revealed that the highest monetary return and benefit: cost ratio was observed for maize varieties of baby corn type followed by green cob and fodder type. The maize varieties grown solely for grain purposes resulted least monetary return and benefit–cost ratio.

The cultivation of multipurpose maize varieties is economically more viable and profitable. The nutritive quality of fodder obtained from baby corn and varieties grown for green cobs is comparable to the fodder type varieties; therefore the farmer could grow maize for baby corn and green cobs and could harvest the fodder with the nutritive value equivalent to maize grown solely for fodder purpose.

Forage grasses are the dominating plant species in the natural grasslands and pastures and even in the forests of the Himalayan region. Other than range forages, some of the cultivated forages such as Napier grass and BN hybrids have revealed high yielding potential. Keeping in view the forage quality, Cock's foot and perennial rye have now been considered as quality forage grasses with high CP and digestibility. Barley and oats are important cereal forages in the Himalayan region. Maize and sorghum are the important dual-purpose cereal forages, where farmers could gain benefit from both from the fodder and the grain.

Table 10.8 Production economics of maize varieties

Sr. No.	Varieties	Maize green fodder equivalent yield (q/ha		Gross returns (Rs/ha)		Cost of cultivation (Rs/ha)		Net returns (Rs/ha)		Benefit: cost ratio	
		2009	2010	2009	2010	2009	2010	2009	2010	2009	2010
1	African Tall	429.62	373.94	42,962.33	37,393.67	18,463.00	18,463.00	24,499.00	18,930.67	1.33	1.03
2	J-1006	429.62	411.07	42,962.33	41,107.00	18,463.00	18,463.00	24,499.00	22,644.00	1.33	1.23
3	Chadhiar local	411.11	340.71	41,110.67	34,070.67	18,463.00	18,463 00	22,648.00	15,607.67	1.23	0.85
4	VL-78	1166.73	973.80	116,006.66	97,380.33	24,461.00	24,461 00	91,545.00	72,919.33	3.73	2.98
5	VL-Makka-42	1196.18	1084.81	119,618.66	108,481.00	24,461.00	24,461.00	95,157.00	84,020.00	3.88	3.43
6	VL-Baby Corn-1	1281.28	1136.60	128,128.30	113,660.33	24,461.00	24,461.00	103,667.30	89,199.33	4.24	3.65
7	Girija	722.19	585.10	72,219.33	58,510.00	21,301.00	21,301.00	50,918.00	37,209.00	2.39	1.75
8	Him-129	807.38	744.33	80,738.33	74,433.00	21,301.00	21,301.00	59,437.00	53,132.00	2.79	2.49
9	Early Composite	770.39	633.25	77,039.00	63,324.67	21,301.00	21,301.00	55,738.00	42,023.67	2.61	1.97
10	Girija	181.81	228.45	18,180.67	22,845.00	23,006.00	23,006.00	-4825.00	−161.00	−0.21	−0.01
11	Bajaura sweet corn	196.99	263.24	19,698.67	26,323.67	23,006.00	23,006.00	−3307.00	3317.67	−0.14	0.14
12	Early Composite	208.85	25–1.74	20,885.33	25,174.33	23,006.00	23,006.00	−2120.00	2168.33	−0.09	0.09
	G.M	**650.18**	**585.59**	**64,962.53**	**58,558.64**	**21,807.75**	**21,807.75**	**43,154.80**	**36,750.89**	**1.93**	**1.63**
	CD at 5%	**104.91**	**77.23**	**10,543.40**	**7722.33**	–	–	**10,543.10**	**7722.51**	**0.50**	**0.33**
	CV %	**9.53**	**7.79**	**9.59**	**7.79**	–	–	**14.43**	**12.41**	**15.23**	**11.85**
	SE (m) ±	**35.77**	**26.33**	**3595.26**	**2633.29**	–	–	**3595.18**	**2633.36**	**0.17**	**0.11**

References

Artus-Poliakoff F, Champannet F, Gayalin M (1991) Production fourragère et élevage ovin à la Martinique. Rev Elev Méd Vét Pays Trop 44:91–98

Aschfalk A, Steingass H, Muller W, Drochner W (2002) *Merremia tridentata* as a supplementary feed to the grass *Panicum maximum* for young West African Dwarf sheep. Trop Anim Health Prod 34(2):145–150

Bagley CP, Fontenot JP, Blaser RE, Webb JKE (1983) Nutritional value and voluntary intake of tall fescue (*Festuca arundinacea* Schreb.) fed to sheep. J Anim Sci 57:1383–1391

Bakhashwain A, Sallam SMA, Allam AM (2010) Nutritive value assessment of some Saudi Arabian foliages by gas production technique *in vitro*. JKAU: Met Env Arid Land Agric Sci 21:65–80

Bandyopadhyay BB (2009) Evaluation of barnyard millet cultivars for fodder yield under single and double cut treatments at higher elevation of hills. Agric Sci Dig 29:66–68

Barnes RF, Nelson CJ, Moore KJ, Collins M (2007) Forages: the science of grassland agriculture, vol II, 6th edn. Wiley-Blackwell

Benedetti E, Rodriguez NM, Campos WE, Goncalves LC, Borges I (2008) *In vitro* and *in situ* digestibility of three tropical forages collected manually and by oesophageal fistulated cows. Acta Sci Anim Sci 30:203–210

Benvenutti MA, Gordon IJ, Poppi DP, Crowther R, Spinks W, Moreno FC (2009) The horizontal barrier effect of stems on the foraging behaviour of cattle grazing five tropical grasses. Livest Sci 126:229–238

Bianospino E, Wechsler FS, Fernandes S, Roça RO, Moura ASAMT (2010) Growth, carcass and meat quality traits of straight bred and crossbred Botucatu rabbits. World Rabbit Sci 14:237–246

Brown WF, Adjei MB (1995) Urea ammoniation effects on the feeding value of guineagrass (*Panicum maximum*) hay. J Anim Sci 73:3085–3093

Brown D, Ng'ambi JW, Norris D (2016) Voluntary intake and palatability indices of Pedi goats fed different levels of *Acacia karroo* leaf meal by cafeteria method. Indian J Anim Res 50:41–47

Castillo AR, Gallardo MR, Gaggiotti MC, Quaino OR (1992) Administration of hay to grazing dairy cows. 2. Ruminal environment and digestion *in situ*. Rev Arg Prod Anim 12:373–382

Cook BG, Pengelly BC, Brown SD, Donnelly JL, Eagles DA, Franco MA, Hanson J, Mullen BF, Partridge IJ, Peters M, Schultze-Kraft R (2005) Trop forages. CSIRO, DPI&F(Qld), CIAT and ILRI, Brisbane

Crozier JA, Allen VG, Jack NE, Fontenot JP, Cochran MA (1997) Digestibility, apparent mineral absorption and voluntary intake by horses fed alfalfa, tall fescue and Caucasian bluestem. J Anim Sci 75:1651–1658

Fayed MH, Salem MSA, Abd El-Kader OMAM (2016) Pearl millet (*Pennisetum glaucum* L.) as affected by some agricultural treatments. J Plant Prod 7:393–400

Göhl B (1982) Les aliments du bétail sous les tropiques. FAO, Division de Production et Santé Animale, Rome

Gupta A, Mahajan V, Kumar M, Gupta HS (2009) Biodiversity in the barnyard millet (*Echinochloa frumentacea* Link, Poaceae) germplasm in India. Genet Resour Crop Evol 56:883–889

Hadimani NA, Muralikrishna G, Tharanathan RN, Malleshi NG (2001) Nature of carbohydrates and proteins in three pearl millet varieties varying in processing characteristics and kernel texture. J Cereal Sci 33:17–25

Hancock D (2009) Georgia forages: pearl millet. Univ. of Georgia. http://georgiaforages.caes.uga.edu/species-and-varieties/warm-season/pearl-millet.html

Hankey A, Mashinini L (2002) *Setaria megaphylla* (Steud) Dur. & Schinz. South African National Biodiversity Institute, South Africa

Harada H, Yoshimura Y, Sunaga Y, Hatanaka T (2000) Variations in nitrogen uptake and nitrate-nitrogen concentration among sorghum groups. Soil Sci Plant Nutr 46(1):97–104

Herrero M, do Valle CB, Hughes NRG, Jessop NS (2001) Measurements of physical strength and their relationship to the chemical composition of four species of *Brachiaria*. Anim Feed Sci Technol 92(3–4):149–158

Jahansouz MR, Afshar RK, Heidari H, Hashemi M (2014) Evaluation of yield and quality of sorghum and millet as alternative forage crops to corn under normal and deficit irrigation regimes. Jordan J Agric Sci 10:699–715

Kaboré-Zoungrana C, Diarra B, Adandedjan C, Savadogo S (2008) Nutritive value of *Balanites aegyptiaca* as feed for ruminants. Livest Res Rural Dev 20(4):56

Katoch R, Kumar N (2014) Productivity and quality attributes of maize varieties (*Zea mays* L.). Range Manage Agroforest 35(1):32–35

Katoch R, Thakur M, Kumar N, Bandhari JC (2012) Golden Timothy—present status and future perspectives in North-West Himalayas. Range Manage Agroforest 33(1):1–7

Keba HT, Madakadze IC, Angassa A, Hassen A (2013) Nutritive value of grasses in semi-arid rangelands of Ethiopia: local experience based herbage preference evaluation versus laboratory analysis. Asian Aust J Anim Sci 26(3):366–377

Khanum SA, Hussain HN, Hussain M, Ishaq M (2010) Digestibility studies in sheep fed sorghum, sesbania and various grasses grown on medium saline lands. Small Rumin Res 91:63–68

Kumar KA (1989) Pearl millet: current status and future potential. Outlook Agric 18:46–53

Lopes FCF, Paciullo DSC, Mota EF, Pereira JC, Azambuja AA, Motta ACS, Rodrigues GS, Duque ACA (2010) Chemical composition and *in situ* ruminal degradability of four *Brachiaria* species. Arq Bras Med Vet Zootec 62(4):883–888

Machicek J (2018) Evaluating forage sorghum and pearl millet for forage production and quality in the Texas high plains. M.S. thesis. West Texas A & M Univ., Department of Agricultural Sciences, Canyon, TX

Meale SJ, Chaves AV, Baah J, McAllister TA (2012) Methane production of different forages in *in vitro* ruminal fermentation. Asian Aust J Anim Sci 25(1):86–91

Melletti M, Burton J (2014) Ecology, evolution and behaviour of wild cattle: implications for conservation. Cambridge University Press, Cambridge

Moran J (2011) Improving the utilisation of Napier grass by dairy cows through fractionating the stems into juice and fibrous residue. In: Successes and failures with animal nutrition practices and technologies in developing countries, vol 11. FAO Anim Prod Health Proc, pp 97–100

Nada Y, Sawamura H (1985) Seasonal productivity of tall fescue, Dallis grass and Bahia grass grown in pure pastures, and rotational grazing combining these pastures in Kyushu district. J Japan Soc Grassl Sci 30:428–433

Orr DM, Burrows WH, Hendricksen RE, Clem RL, Back PV, Rutherford MT, Myles DJ, Conway MJ (2010) Impacts of grazing management options on pasture and animal productivity in a *Heteropogon contortus* (black speargrass) pasture in central Queensland. 1. Pasture yield and composition. Crop Pasture Sci 61:170–181

Pacheco ME, Brown RD, Bingham RL (1983) Nutritive value and intake of Kleberg Bluestem by beef cattle. J Range Manage 36:222–224

Sarwar M, Khan MMA, Mushtaque M (2006) Chemical composition, herbage yield and nutritive value of *Panicum antidotale* and *Pennisetum orientale* for Nili buffaloes at different clipping intervals. Asian Aust J Anim Sci 19:176–180

Schultze-Kraft R, Teitzel JK (1992) In: Mannetje L't, Jones RM (eds) *Brachiaria decumbens* Stapf. Record from Proseabase. PROSEA (Plant Resources of South-East Asia) Foundation, Bogor

Senanayake SGJN (1995) The effects of different light levels on the nutritive quality of four natural tropical grasses. Trop Grassl 29:111–114

Singh HS, Singh K (2005) Status and needs of pasture and fodder management in Uttaranchal. In: Bisht JK, Srivastava AK (eds) Road map for pasture and fodder development in NWHR for livestock sustenance. Vivekananda Parvatiya Krishi Anusandhan Sansthan (Indian Council of Agriculture Research), Almora, pp 39–64

Sood S, Khulbe R, Kumar RA, Agrawal PK, Upadhyaya H (2015) Barnyard millet global core collection evaluation in the sub mountain Himalayan region of India using multivariate analysis. Crop J 3:517–525

Sood S, Tripathi A, Katoch R (2019) Comparative biochemical profile of tall fescue (*Festuca arundinacea* Schreb.) varieties grown in Himalayan region. Indian J Agric Biochem 32:166–171

Suttie JM, Reynolds SG (2004) Fodder oats: a world overview. Plant Production and Protection Series No. 33. FAO, Rome

Undersander DJ, Smith LH, Kaminski AR, Kelling KA, Doll JD (2003) Sorghum forage. In: Alternative field crop manual. University of Wisconsin-Extension, Co-operative Extension

Ward JD, Readfern DD, McCormick ME, Cuomo GJ (2001) Chemical composition, ensiling characteristics, and apparent digestibility of summer annual forages in a subtropical double-cropping system with annual ryegrass. J Dairy Sci 84:177–182

Yadav R, Yadav V (2013) Comparative performance of Indian and Japanese barnyard millet cultivars under varied fertility conditions for dual use in Indian Central Himalaya. Range Manage Agroforest 34:175–178

Forage Legumes in Himalayan Region 11

Since the dawn of civilization and the domestication of animals and plants, forages have a special position due to the well-defined association between agriculture and animal production that sustains the livelihoods of rural households. Forages are indispensable component in the livestock production system being the cheapest and reliable source of nutrients to livestock. They also have significant value in sustaining the environment. Leguminous crops are an important proteinaceous and overall nutritional dietary source for both humans and animal consumption. They can be grouped under four categories, which include (1) dual-purpose grain legumes used as feed and grain; (2) dual-purpose legumes of which aerial parts (leaf and stem) are used as fodder while grains are used as feed or grain; (3) pasture legumes, which are part of grasslands/rangelands and (4) annual or perennial cultivated fodder. The legumes provide a better level of nutrition than grasses at a similar stage of growth.

Leguminous forages generally have 16–20% CP and less than 50% NDF and 40% ADF content. Nutritional quality of leguminous forages varies from excellent to poor due to the presence of anti-nutrients. Besides yielding highly nutritious fodder, forage legumes also offer several advantages to farming systems. The appreciable amount of atmospheric nitrogen is fixed by forage legumes in association with *rhizobium* bacteria (Peoples et al. 1995). Many forage legumes have soil covering growth habits and deep root systems that can contribute to maintain soil structural and textural properties. When used as a cover crop, forage legumes could also be useful in controlling the growth of weeds.

In the Himalayan region, a major problem of livestock farming is the unavailability of nutritionally rich feed resources. One of the effective strategies to overcome this problem is the integration of forage legumes in livestock feeding. Legumes are an excellent source of proteins, minerals, and other essential nutrients as compared to crop residues and grasses. They could be easily grazed, harvested and fed fresh or stored as hay or silage in comparison to forage grasses (Amole et al. 2013). Forage legumes also provide a higher quality protein that may be of equal or greater importance in the case of nonruminant livestock species. Legumes also contain

R. Katoch, *Nutritional Quality Management of Forages in the Himalayan Region*, https://doi.org/10.1007/978-981-16-5437-4_11

higher digestible energy than grass/cereal forages due to the structure and development of the legume cell wall. Indeed, the cell wall of legume plants contains lesser hemicelluloses and more pectin compared to that of cereals, thus increasing their digestibility.

A sustainable strategy of improving the nutritional quality of low-quality forages such as grasses and crop residues is through supplementation with forage legumes. Forage legumes are also a component of nutritional security for livestock as they are an integral component for the increased availability of livestock products. Nutritional quality of forage legumes is affected by developmental morphology, which in turn is greatly affected by harvesting management. Alfalfa (*Medicago sativa* L.), white clover (*Trifolium repens* L.), Red clover (*Trifolium pretense* L.), Berseem (*Trifolium alexandrium* L.), birdsfoot trefoil (*Lotus corniculatus* L), sainfoin (*Onobrychis vicifolia* L.), *Cliotoria ternatea* and *Medicago tranculata* are important range legumes of the Himalayan region. Ricebean (*Vigna umbellata* L.), cowpea (*Vigna unguiculata* L.), grass pea (*Lathyrus sativa*) and garden pea (*Pisum sativa*) are important dual-purpose forage legumes. Berseem is one of the most important forage legumes of the Himalayan region and has been described as *King of Fodders*. It provides highly palatable fodder having 17–24% CP content, 25–28% CF, and 60–65% TDN. *Lucerne* is another valuable forage legume with 18–22% CP and 25–35% CF content. Cowpea is an quick-growing dual-purpose legume, which has significant value in both humans as well as animal nutrition. Cowpea forage contains 18–22% CP, 3–5% EE, 26–30% CF and 55–60% TDN. Clovers also produce highly palatable fodder with 17–25% CP and 25–29% CF content. However, clovers also contain cyanogenic glycosides, which could be harmful to livestock, if fed in large quantities.

The nutritional quality of important forage legumes of the Himalayan region has been presented in Table 11.1.

11.1 Cultivated Forage Legumes of the Himalayan Region

Major cultivated forage legumes and their nutritional characteristics have been discussed below.

11.1.1 Alfalfa (*Medicago sativa* L.)

Alfalfa is one of the important forages because of its high protein content, dietary intake and dry matter digestibility. The fodder obtained from alfalfa is energy-rich which supports lactation and its rapid degradability increases voluntary intake. It has good nutritive profile with 12.50–25.0% CP, 28.57–34.0% CF, 9.91–12.51% ash, 34.50–49.0% NDF, 29.40–41.67% ADF, 6.30–11.30% ADL content, 13.70–57.70% hemicellulose and 22.90–30.20% cellulose content (Mirzaei-Aghsaghali et al. 2008; Millic et al. 2011). Alfalfa is used for livestock nutrition in different forms, most frequently as hay, but also dried/dehydrated in the form of

Table 11.1 Nutritive quality of common leguminous forages of Himalayan region

	Nutritional composition (%)										
Forage crop	DM	CP	CF	EE	NDF	ADF	HC	ADL	Cellulose	Ash	References
Berseem (*Trifolium alexandrium* L.)	9.74–20.00	15.80–22.96	19.70–35.10	1.20–3.28	33.63–55.97	24.43–43.68	12.30–19.20	3.23–10.96	17.80–29.10	12.70–17.00	Chaturvedi et al. (2000); Fernandes and Waditake (2006); Berimavandi et al. (2010); Heuzé et al. (2016b)
Cowpea (*Vigna unguiculata* L.)	11.10–26.40	13.50–24.30	11.50–35.90	1.30–4.10	33.00–60.70	26.58–58.10	5.10–6.50	1.60–14.10	39.90	3.90–15.10	Karachi and Matata (1996); Rao et al. (2003); Mislevy et al. (2005) Mirzaei-Aghsaghali et al. (2008); Agza et al. (2012); Ayana et al. (2013); Solomon and Kibrom (2014); Heuzé and Tran (2015)
Alfalfa (*Medicago sativa* L.)	84.90–96.00	12.50–27.90	15.70–38.50	1.50–4.40	32.30–54.70	22.00–41.50	3.37–30.50	6.30–10.10	23.10–30.20	8.10–16.80	Peterson et al. (1992); Katic et al. (2001); Millic et al. (2011);

(continued)

Table 11.1 (continued)

Forage crop	Nutritional composition (%) DM	CP	CF	EE	NDF	ADF	HC	ADL	Cellulose	Ash	References
											Homolka et al. (2012) Salehi et al. (2014); Heuzé et al. (2016c)
Rice bean (*Vigna umbellata* (Thunb.) Ohwi and Ohashi)	13.80–32.00	13.59–26.83	18.70–38.00	1.40–1.80	52.67–59.67	38.90–47.00	12.67–13.77	–	–	4.27–13.00	Iqbal et al. (1998); Katoch et al. (2010) Katoch et al. (2007b)
White clover (*Trifolium repens* L.	9.80–30.50	19.50–29.60	15.00–27.80	10.40–21.80	18.50–57.40	15.50–40.00	3.40–17.40	3.90–8.20	18.20–20.90	9.5–13.00	Rattray and Joyce (1974) Peterson et al. (1992); Ayres et al. (1998); Peiretti et al. (2016)
Red clover (*Trifolium pretense* L.)	12.70–34.70	15.52–27.70	10.00–36.50	1.80–5.30	36.40–41.40	16.20–34.50	9.80–14.80	4.10	22.50	7.70–13.30	Homolka et al. (2012) Ricardo et al. (2013)
Birdsfoot trefoil (*Lotus corniculatus* L.)	12.10–32.00	13.71–20.90	21.00–31.00	10.60–22.90	42.80–53.50	35.90–45.50	5.10–12.0	–	–	7.20–10.20	Vuckovic et al. (2007); Kaplan et al. (2009); Peiretti et al. (2016); Heuzé et al. (2016d)
Sainfoin (*Onobrychis vicifolia* L.)	16.00–29.80	12.40–21.30	18.50–39.00	3.80–4.50	25.60–45.40	20.20–39.10	3.30–6.30	–	–	3.80–9.00	Scharenberg et al. (2009); Heuzé et al. (2015)

briquettes, silage, and haylage. Alfalfa has a direct influence on the performance of beef and dairy animals as well as the cost of production. The decline in nutritional quality with increasing harvest interval is the consequence of progressing maturity, along with the associated effects of increasing stem growth and decreasing leaf proportion and decreasing stem nutritive value. The digestibility of alfalfa depends on the cellulose and lignin content. As lignin is indigestible, intensive lignification of cell walls in later stages of alfalfa development tends to reduce digestibility. Since alfalfa foliage is preferably eaten by animals and has better nutritional quality than stems, the estimation of the appropriate stage of maturity for harvesting quality fodder is important for livestock feeding. Nutritive quality of alfalfa fodder is positively co-related with the protein content, which primarily depends on the proportion of leaves in dry matter yield, which may vary depending on maturity at harvest, handling, and rain damage. Grazed or fresh alfalfa at vegetative or mid-bud stage may cause bloat in sheep and cattle. Drying of forage cut at flowering time in the field reduces the bloating effect. Alfalfa contains phytoestrogens that are reported to reduce conception rates in cattle and sheep fed with alfalfa forage prior to mating. Saponin content in alfalfa forage may have adverse hemolytic effects on livestock. However, it also gives the plant resistance to pests.

11.1.2 Berseem (*Trifolium alexandrium* L.)

Berseem is a valuable, high-yielding annual forage legume which is well adapted to a wide range of environments. It is widely grown in pure stands or in mixture. Berseem provides quality fodder to livestock and is ranked second after alfalfa. It saves concentrate mixture while making a balance livestock diet for animals. The fodder contains 15–25% CP content, 40–50% NDF, 30–35% ADF, 19.70–25.0% CF, 10–12% ash, 60–65% TDN and 1.20–3% EE (Mandal and Banerjee 2009). Chauhan et al. (1992) reported CP content in berseem fodder ranging from 11.40 to 20.50%. Ensiling is one of the best methods to preserve the nutritional quality of berseem fodder (Berimavandi et al. 2010). Berseem could be mixed with maize for generating silage of excellent nutritional quality (Shaug et al. 2000). The hay has nutritional quality comparable to alfalfa hay harvested at comparable maturity. Dry matter intake (DMI) ranges between 2.40 and 3.0% for growing heifers, buffalo, sheep and goats. Organic matter digestibility (OMD) ranges from 63 to 71% in growing cows and CP digestibility ranges from 63 to 75% in growing goats. Berseem can also be used as a straw, after threshing and seed collection. However, nutritional quality is low than reported for hays but remains higher than cereal straws with a protein content of about 8–8.50%, dry matter digestibility of about 45–50%. Like other leguminous fodders, berseem has a good buffering capacity and high moisture content that could delay a drop in pH during ensiling. Berseem is often mixed before ensiling with wheat straw and molasses.

11.1.3 White Clover (*Trifolium repens* L.)

White clover is considered as a high-quality forage legume. The feeding value of white clover is attributed to greater intake, rapid particle breakdown, digestion, absorption, and metabolizable energy content. Nutrients arising from digestion and absorption of white clover result in high efficiency of metabolizable energy utilization for live weight gain. White clover is an excellent forage legume for grazing, cutting and carrying, ensiling and hay preparation. It is recommended to offer between 20 and 50% white clover in a grass-based diet or with a small amount of concentrate in the case of a straw-based diet.

White clover contains 22–28% CP, 2.70–3.30% crude fat, 9.40–11.90% ash, 20.93% cellulose, 9.70% hemicellulose, and 15.70–21.01% CF (Rattray and Joyce 1974; Hamdy et al. 2009). Nutritional quality of white clover fodder changes with the advancement in maturity. Neutral detergent fiber (NDF) increases from 184 to 301 g kg^{-1} dry matter, nitrogen declines from 36 to 20 g kg^{-1} dry matter, *in vitro* digestibility from 0.74 to 0.65 and metabolizable protein content from 144 to 67 g kg^{-1} dry matter from early flowering to ripe seed stage (Ayres et al. 1998). Sanderson and Wedin (1989) reported N content in ADF in clover twice than the protein content in forage grass. This is one of the reasons for relatively higher N digestion in sheep fed with white clover than with perennial rye grass. Apart from its usage as a fodder crop, white clover is also valuable in rejuvenating permanent pasture that lacking forage legumes in temperate regions and cold subtropical areas. White clover has low level of phytoestrogens (0.02–0.06%) as compared to red clover (*Trifolium pratense*) (1–2.50%) (Saloniemi et al. 1995). White clover can cause bloat in ruminants if fed in large quantities. However, it could be managed by gradually introducing livestock to pasture and by supplementing with hay or straw. White clover shows polymorphism for cyanogenesis that means cyanogenic and acyanogenic plants can exist in same population. The cyanide content in clover could change during growing season,

11.1.4 Red Clover (*Trifolium pratense* L.)

Red clover is an important forage legume of temperate regions and grows in high-altitude areas in subtropical regions. This forage legume is relished by livestock and is used for pasture cut-and-carry systems, hay, and silage. Red clover has high seedling vigor, rapid growth and tolerance to acidic and humid conditions (Katoch et al. 2007a). Red clover can also be used for silage preparation but has low dry matter content, low WSC content and high buffering capacity. Before ensiling, it is advisable to wilt red clover for 24–48 h so that it reaches 25% DM, concentrates WSC and reduces water losses during fermentation. Nutritional quality of red clover is mainly attributed to 7.74–26.50% CP, 15.50–27.40% CF, 37.70–44.50% NDF, 25.80–35.22% ADF, 8.16% of mineral matter, 2.50% EE, 4.95–9.87% ADL and 6.80% ash content (Homolka et al. 2012; Ricardo et al. 2013). Red clover silage is superior to lucerne silage for generation of high milk production and improvement of live weight gain in lambs (Broderick et al. 2001; Dewhurst et al. 2003; Bertilsson and Murphy 2003; Fraser et al. 2004). Additionally, replacing grass silage with red clover silage can also lead to desirable changes in the fatty acid composition of milk (Dewhurst et al. 2006; Moorby et al. 2009).

Red clover has more cell wall constituents than white clover and therefore has low *in sacco* degradability and *in vivo* digestibility. Like other forage legumes such as alfalfa or other clover species, red clover may cause bloat in livestock. It may be preferable to avoid grazing in pure stands of red clover and prefer mixed stands of red clover with grasses or legume species that are not prone to produce bloat. Red clover may cause an increase in urinary calculi in sheep due to the presence of benzocoumarins (phytoestrogens) (Hall 2007). Red clover also contains isoflavones compounds, including daidzein, genistein, formononetin and biochanin A.

11.1.5 Birdsfoot Trefoil (*Lotus corniculatus* L.)

Birdsfoot trefoil is a widely distributed perennial forage legume in wild and naturalized populations in temperate regions of the world including the Himalayan region. It is a perennial self-generating legume grown in wild and naturalized populations of temperate regions, where it is primarily used as a fodder crop. In some parts of the world, birdsfoot trefoil was introduced for improving pasture for increasing animal production since it is non-bloating and highly palatable in nature. Nutritional quality of birdsfoot trefoil is comparable to other legumes (Waghorn and Shelton 1992). It contains 137–174 g/kg CP content, 10.60–22.90 g/kg lipid content, 428–535 g/kg NDF content, 359–450 g/kg ADF content, 75–98 g/kg lignin content (Peiretti et al. 2016). Birdsfoot trefoil fodder also has a significant amount of minerals particularly calcium, phosphorus, and potassium (Donovan 2006).

Birdsfoot trefoil fodder has high tannins content mainly procyanidins, which reduce the numbers of proteolytic rumen bacteria and protect protein from rumen degradation (Min et al. 2005). These tannins are also responsible for reduction in urinary N excretion and prevent bloat (by reducing methane and ammonia emissions) together with anthelminthic activity. Condensed tannins content in birdsfoot trefoil fodder is variable with values ranging from 2 to 55 g/kg DM. The condensed tannins of birdsfoot trefoil do not affect OM digestibility in sheep. In mixed diets, the nutritive value of birdsfoot trefoil forage for sheep is not affected when condensed tannins do not exceed 1% of total dietary DM. However, the high content of condensed tannins in birdsfoot trefoil reduces the voluntary feed intake and digestibility and depresses weight gain and wool growth in sheep.

11.1.6 Caribbean Stylo [*Stylosanthes hamata* (L.) Taub.]

Caribbean stylo legume is mainly used as permanent pasture but is also cut to be fed fresh in cut-and-carry systems. *Stylosanthes hamata* contains about 15–20% CP content while fresh stylo forage harvested at 50% flowering stage has about 50–65%

dry matter digestibility (Nandanwar et al. 1991). CF content of stylo ranges from 20 to 34% DM. It has been observed that stylo has great potential for increasing the feeding quality of rice straw. Being nutritive and palatable, it is used as feed for all types of animals in the form of hay and silage. Among *Stylosanthes* species, *Stylosanthes hamata* is rich in CP and low in fiber content as compared to *Stylosanthes scabra* and *Stylosanthes sebrana*. It is also a good source of Ca, P, K, Cu, and Mn. Average digestibility of *S. hamata* forage ranges from 50 to 65%. Despite of its moderate protein value, *Stylosanthes hamata*, like other legumes is a valuable fodder for providing protein to livestock fed with grasses or low protein feeds.

11.1.7 Sainfoin (*Onobrychis vicifolia* L.)

The Onobrychis genus comprises a few agronomically important forage legume species. Sainfoin (*Onobrychis viciifolia*) is also known as St. Foin, holy clover or holy grass. *Onobrychis viciifolia* has a long history of traditional culture worldwide. It thrives well in warm-temperate conditions in regions where annual rainfall is above 330 mm. It has low productivity and is more difficult to maintain than other legumes. It has low protein content and digestibility than other leguminous fodder but has excellent nutritive value as compared to conventional grasses (Karnezos et al. 1994). Sanfoin has 12.40–21.30% CP, 3.80–4.50% EE, 6.50–10.60% lignin, 25.60–45.40% NDF, 22.30–39.10% ADF and 7.0–10.90% ash content (Scharenberg et al. 2009). Sainfoin silage has a beneficial effect on dry matter intake, dry matter digestibility and cellulose digestibility in sheep. It has several other beneficial properties due to its tannin and polyphenol composition. Condensed tannins present in *Onobrychis* species confer anthelmintic properties, increase protein utilization and prevent bloating.

11.1.8 Sweet Clover (*Melitotus officinalis* L.)

Sweet clover, known as field melilot, ribbed melilot and yellow melilot, is a dicotyledonous summer annual forage legume. It is considered a highly palatable livestock forage. The legume contains 12–23% CP, from 24 to 38% ADF, 28–40% NDF and from 60 to 70% TDN. Bloat caused by sweet clover is less of a problem than with alfalfa and clover species. Forage quality declines with maturity. It is less palatable than many other legumes because of its bitter taste due to the presence of coumarin. It has good palatability for cattle, sheep and horses.

11.1.9 Shaftal or Persian Clover (*Trifolium resupinatum* L.)

Persian clover is an annual legume, up to 20–60 cm high, similar to berseem but shorter in length. It grows up to an altitude of 2500 m in the Afghan uplands and

2750 m in the Himalayan region. Persian clover has rapid regrowth after grazing, high tolerance to water-logging and moderate tolerance to salinity. It is mostly used for supplying highly palatable and nutritive fodder and hay. Its high protein and moisture content may make it unsuitable for ensiling. Persian clover forage is rich in protein and minerals. In comparison to other temperate forages, it has relatively low fiber and lignin content. Persian clover fodder contains 14–27% protein (about 21% on average). NDF, ADF, and lignin contents ranges from 20 to 35%, 15 to 27% and 2 to 5%, respectively. Mineral content is about 12–14% DM. Cases of bloat have been reported in cattle, but not in sheep. The risk of bloat may be reduced by sowing Persian clover with grass to reduce the amount of legume in the grazed diet, by feeding cattle with dry hay before they enter the sward, by turning cattle out of stands and by using anti-foaming agents such as poloxalene or molasses/urea blocks. *Trifolium resupinatum* contains saponins (Simonet et al. 1999) and phenolic compounds (Janda et al. 2009) which can be deleterious for livestock. Nevertheless, their levels are generally low and should not be a problem, especially when Persian clover hay is mixed with other feeds.

11.1.10 Moth Bean (*Vigna aconitifolia* L.)

Moth bean is a multipurpose legume that provides both palatable pasture and hay for livestock and seeds/pods for human consumption. The pasture is relished by livestock and it can make valuable hay for livestock. After seed harvest, the straw remains nutritious and palatable. The stems are small and the leaves do not easily fall off when the plant is dried (Sherasia et al. 2017). Moth bean seeds contain a significant amount of nonprotein nitrogen. Globulin is the major protein fraction (Adsule 1997). Vitamin levels in seeds are relatively lower than those in soybean. Moth bean forage contains moderate protein content (9–17% DM), which is slightly lower than other legume forages. ADF content ranges from 21 to 30% DM. Ash content ranges from 13 to 18% DM. Moth bean forage has a feeding value almost equal to that of alfalfa hay. *In vitro* dry matter digestibility of moth bean forage is slightly lower than alfalfa fodder.

11.1.11 Butterfly Pea (*Clitoria ternatea* L.)

Clitoria ternatea is a good protein-rich forage that is used either alone or as a protein supplement for animals. Its CP content is comparable to alfalfa with values for fresh forage typically higher than 18%. Forage quality persists even when maturity is advanced, without affecting digestibility or feed intake. The lignin content is about 10% and in some cases it may be more than 15%. Protein content in *Clitoria ternatea* varies from 15 to 23% in the vegetative state (42 days regrowth) to 19% at seeding (82 days). ADF content ranges from 38 to 47%. The lignin content is high, usually about 10% and higher than 15% in some cases. Unlike other legumes, *Clitoria ternatea* seems to be relatively free of toxic compounds and could be fed to ruminants and monogastric species, though its higher fiber content may be limiting for pigs and poultry (Cook et al. 2005). *In vivo* organic matter and dry matter digestibility values for hay range from 50 to 60% which reflect high fiber content in forage (de Bustamante et al. 2002).

11.1.12 Phasey Bean [*Macroptilium lathyroides* (L.) Urban.]

Phasey bean is generally considered to be an annual forage plant. It does not usually regenerate in pastures, but it can show reasonable persistence. It grows sea level up to an altitude of 1800–2000 m. It mainly thrive well in wet places along roadsides, on waste lands, in open fields, pastures and in open situations along streams and rivers. Phasey bean forage is rich in protein (may be up to 27%). Fiber content is also relatively high (average ADF content of 25%). With the advancement in maturity, the protein content of leaves decreases from 30 to 20% while NDF increases from 50 to 60%. ADF content of leaves increases from 35 to 45% with maturity (Nakanishi et al. 1993). Longer cutting intervals also have depressive effect on leaf protein content. Ensiling rice forage and phasey bean forage together improves fermentation quality and palatability of rice forage (Tobisa et al. 2005). Ensiling phasey bean forage with Napier grass (*Pennisetum purpureum*) improves the nutritive value of mixed feed compared to Napier grass alone (Yunus et al. 2001). Phasey bean makes good quality hay provided that it is cut and handled early enough to preserve a maximum of leaf material.

11.2 Dual-Purpose Legumes

Dual-purpose legumes are also known as food-feed crops. Although cowpea, rice bean, garden pea and grass pea majorly fall in the category of grain legumes, these legumes also have significant value in livestock feeding as a forage crop. The nutritive quality of some important dual-purpose legumes of the Himalayan region has been discussed under the following subheads.

11.2.1 Cowpea (*Vigna unguiculata* L.)

Cowpea is highly valued for contributing both toward human food and livestock feed. Its dual-purpose character makes it a very attractive crop for increasing the productivity and profitability of mixed farming systems. The vines and leaves and sometimes pods from cowpea are often used as a feed source. It has great potential to be grown as a mixed crop with maize, sorghum and millets to produce an ideal cereal-legume fodder mixture. Cowpea fodder is good for green feeding, haymaking and ensiling with other crops. It generally has high content of nutritional attributes, which make cowpea quite comparable to lucerne and soybean in fodder quality (Bressani 2002). Cowpea provides high-quality forage, rich in protein (14–24%).

Leaves and shoots usually contain more than 20% protein, depending on the stage of maturity and seasonal variation. Cowpea pod husks are characterized by a high CF content (about 31%) and a relatively low level of CP content (12–13%). It also contains 26.58–53.0% ADF, 32.10–58.10% NDF, 13.10% ADL, 2.90–14.20% ash, 2.30% fat, and 56.24–63.10% carbohydrate content (Ayan et al. 2012; Sreerama et al. 2012; Lauriault et al. 2012; Gwanzura et al. 2012; Owolabi et al. 2012). Cowpea residue contains 6–8% CP content. The IVDMD content of fresh foliage is more than 80% whereas digestibility of crop residues is low (55–65%).

11.2.2 Garden Pea (*Pisum sativum* L.)

Garden pea is an important legume after soybean, groundnut and beans in India. They are an important food-feed crop in temperate regions. Garden pea can be fed fresh and as hay or ensiled. Pea forages are generally rich in protein. However, protein content is lower than other temperate forage legumes such as alfalfa and clover. Garden pea foliage contains 15–20% CP, 38–56% NDF, 32–39% ADF, about 4–6% ADL and 60–65% digestibility. Nutritional quality changes with the advancement in maturity (Hayashi et al. 2007). Pea straw has 5–10% CP content, 1.50–2% calcium and 7–12% mineral content. In temperate regions, pea straw is suitable for livestock with low nutritional requirements. Starch content is extremely variable (5–25%), depending on seed maturity (Borreani et al. 2007). Protein degradability decreases with the advancement in maturity, for example, protein degradability at the bud stage is 96% and pod formation stage is 88% (Iliev et al. 2002). The whole crop pea silage is an excellent feed for livestock. Pea oat silage is liked by cattle over corn silage and is easy to incorporate in growing and finishing rations.

11.2.3 Grass Pea (*Lathyrus sativus* L.)

Grasspea is an important food-feed crop. The crop is grown primarily as a winter pulse crop for animal feed or sometimes as human food. Grass pea is a dual-purpose annual legume grown for its seeds for human consumption and fodder for livestock feeding. Grass pea has considerable protein, carbohydrates and mineral content, although deficient in some essential sulfur-containing amino acids. It contains a toxic component that may cause paralysis in humans and livestock if consumed in excessive amounts. It can be grown as a sole crop in intercropping systems or in mixtures. Grass pea foliage contains 17–22% CP, 40–48% NDF, 25–36% ADF, 26–30% CF, 2.50–3.30% EE, 8.30–11.10% ash content. *Lathyrus sativus* seeds and vegetative parts contain a neurotoxic nonprotein amino acid, β-oxalyl-L-α, β-diaminopropionic acid (ODAP or BOAA). Ingestion of ODAP causes neurolathyrism, a neurodegenerative disease that damages motor neurons causing irreversible paralysis of the lower limbs and sometimes death in humans and animals (Spencer and Schaumburg 1983). The plant also contains trypsin and chimotrypsin inhibitors, amylase inhibitors, lectins, tannins, phytates and oligosaccharides. Except for ODAP, most of these anti-nutritional factors are in lower amounts in grass pea than in other legumes. Various processing techniques have been developed which could be used for the management of anti-nutrients from forages.

11.2.4 Common Bean (*Phaseolus vulgaris* L.)

The common bean is a major grain legume consumed worldwide for its edible seeds and pods. It is a highly polymorphic, warm-season and herbaceous annual plant. The legume also produces a good amount of fodder. Common bean is majorly used for cut-and-carry systems. They are generally mixed with small grains to increase silage protein content (Linn et al. 2002). Ensiling of common bean fodder with maize gives silage of good quality (Dawo et al. 2008). The whole plant is of minor importance as fodder in livestock feeding, but beans, straw and other crop residues and processing by-products are also used occasionally. Leaves have 21–28% CP content at different growth states. Digestibility ranges from 55 to 76%. Nutritional composition of bean crop residues depends on the proportions of stems, pod husks and leaves. The stem portions and pod husks have 8 and 4%, respectively (Pieltain et al. 1998). Common bean straw (haulms) contains 5–11% protein and 38–45% crude fiber. However, like other legume straws, it has better nutritive value than cereal straws due to a higher protein and low fiber content (Lopez et al. 2005).

11.2.5 Rice Bean (*Vigna umbellata* (Thunb.) Ohwi and Ohashi)

Rice bean, a potential legume crop with yield potential, ability to grow in aberrant growing conditions and a good nutritional profile. This crop has multipurpose usage as potential livestock fodder, green manure and cover crop. Our group has conducted elaborate studies on different aspects of rice bean (Katoch 2011).

11.3 Studies on Forage Legumes in the Himalayan Region

11.3.1 Rice Bean: A Potential Forage Legume

Rice bean is a commercially unexploited crop, with origin from South and South-East Asia (Katoch 2011). This crop is mainly grown in remote areas where semi-subsistence consumption prevails. Among Asian countries, rice bean has a predominant occurrence in the mountainous regions of India, Burma, China, Malaysia, Indonesia, Korea, and the Philippines. Sporadic cultivation of rice bean has also been reported from South-Western Japan, where this crop is associated with slash and burn type of agriculture system. In India, this crop is gaining interest particularly in the North-Eastern region, Eastern and Western Ghats, Punjab, Uttar Pradesh, Rajasthan, and Himachal Pradesh (Arora et al. 1980). Rice bean, also known as the

Climbing bean, Rajmoong, Red bean, Oriental bean, Haricot bean, is a highly photosensitive, annual herbaceous plant of either an erect, semi-erect (trailing) or climbing growth habit. This crop has an average height of 0.12–0.15 m and could be grown up to 2.5 m in height (Katoch 2011, 2013). The leaves are trifoliate having 6–9 cm long leaflets with hairy pubescence. Rice bean reveals high heterogeneity in different phenotypic attributes such as pod shape, pod color, pod length, seed shape and seed color. Apart from being an untraditional pulse, rice bean has dual-purpose uses, which make it a very attractive crop in the areas where arable land is becoming scarce. Owing to luxuriant growth, high yield potential and fodder quality, this legume is gaining popularity among farmers (Katoch et al. 2007b). Like other leguminous crops, rice beans fix nitrogen in symbiotic association with *rhizobium* and improve the fertility of the soil. Despite of having high yield potential and exceptional nutritional attributes, rice bean so far has not been included in the category of major legumes. However, due to high fodder production potential, rice bean is now receiving attention as an explicit source of fodder for grazing animals (Katoch 2010). The palatability and nutritional quality of rice bean fodder are richer than other legume fodders like cowpeas (*Vigna unguiculata*) and moth (*Vigna acontifolia*) particularly being rich in protein calcium and phosphorus (Chatterjee and Das 1989; Ahuja et al. 2001). It remains lush green for a long period and can supply green fodder during lean periods when usually other leguminous fodder resources are scarce (Chaudhury and Prasad 1972; Mukherjee et al. 1980). In the areas, where acute competition is present for the land to be utilized for food and fodder production, it is ideally suited as a short duration crop (Katoch et al. 2007b). Rice bean varieties with indeterminate growth habits and late-flowering are excellent for producing a high quantity of fresh fodder yield in comparison to early flowering and determinate genotypes (Katoch et al. 2007c) (Fig. 11.1).

Rice bean is a short-duration forage crop that produces a considerable quantity of dry matter. In Bengal (India), rice bean forage yield has been reported from 5 to 7 t DM/ha from the crop harvested from May to June and 8 to 9 t DM/ha in November–December under favorable conditions (Chatterjee and Dana 1977). However,

Fig. 11.1 Luxuriant vegetative growth of rice bean

variations have been observed in different growing conditions. In a study conducted on 15 rice bean genotypes, fresh fodder yield of more than 15 t/ha has been recorded by Mandal and Mukherjee (1999). Although, a high forage yield (37.50 t/ha) has also been recorded by Bhattacharya and Mukherjee (1998) from rice bean cultivar K-1 after 100 days of sowing. Rice bean hay increases dry matter intake, digestible protein and N retention along with a positive effect on rumen microflora resulting in increased VFA production and lower CH_4 emissions (Chanthakhoun and Wanapat 2010; Chanthakhoun et al. 2011). CP content of rice bean cultivars varies from 15 to 27%. It is rich in limiting amino acids methionine and tryptophan and also valine, tyrosine, and lysine. In rice bean fodder varieties, methionine content varies from 0.74 to 0.99% and starch content from 58.20 (RBL-1) to 71.10% (RBL-50). Fresh rice bean forage contains 17–23% protein on DM basis and 2% calcium. It is relatively rich in lysine but poor in sulfur-containing amino acids. It has high starch content, (52–57% of DM), with variable amylase content. Rice bean forage contains variable amounts of condensed tannins, trypsin inhibitors and low phenols and phytic phosphorus content.

We have collected and characterized rice bean germplasm from diverse geographical locations of the country including North-East India, Northern hilly region and other parts of the country. The germplasm from the diverse locations has also been obtained from NBPGR, New Delhi. The under-mentioned table presents the information on the collections of the rice bean germplasm and their growing locations which were utilized in our research programme for the assessment of different traits of the crop (Table 11.2).

11.3.1.1 Fodder Production Potential of Rice Bean

The study on the fodder production potential and nutritional quality of forage from different rice bean genotypes categorized in Group-I, Group-II, and Group-III was conducted to assess the potential of the crop in feeding livestock. The results of the study have been discussed under the following subheads:

(a) **Fodder yield of rice bean genotypes of group-I**

Green fodder yield, being a complex trait, is the interactive effect of various yield-attributing traits. The green fodder yield from the first group (G-1) of rice bean genotypes after 115 days of sowing has been presented in Table 11.3. Among 30 genotypes, the highest fodder yield was recorded in genotype IC-137200 (356.66 q/ha) while least fodder yield was recorded in genotypes *viz.*, Panchrukhi (151.10 q/ha) Baroi (160.98 q/ha) and Dhagwar (163.60 q/ha). The lower forage yield from local germplasm may be attributed to their low height and early flowering. It was observed that the genotype flowering early produced less fodder. The plant height after 50 days after sowing also had a positive correlation with fodder yield (0.537) along with 25 and 50% flowering. The co-relational analysis also revealed a significant positive correlation between plant height after 115 days of sowing and fodder yield (0.629).

Table 11.2 Rice bean germplasm characterized for forage yield and quality traits

Sr. no.	Rice bean germplasm (Group-I)	Collection site	Sr. no.	Rice bean germplasm (Group-II)	Collection site	Sr. no.	Rice bean germplasm (Group-III)	Collection site
1.	IC-137186	Haryana	1.	LRB-1	Punjab	1.	JCR-12	Assam
2.	IC-137187	Sikkim	2.	LRB-40-1	Punjab	2.	JCR-20	Assam
3.	IC-137188	Sikkim	3.	LRB-40-2	Punjab	3.	JCR-32	Assam
4.	IC-137189	Sikkim	4.	LRB-45	Punjab	4.	JCR-52	Assam
5.	IC-137190	Sikkim	5.	LRB-126	Punjab	5.	JCR-54	Assam
6.	IC-137191	Sikkim	6.	LRB-127	Punjab	6.	JCR-76	Assam
7.	IC-137194	Uttar Pradesh	7.	LRB-128	Punjab	7.	JCR-79	Assam
8.	IC-137195	Uttar Pradesh	8.	LRB-134	Punjab	8.	JCR-81	Assam
9.	IC-137199	Uttar Pradesh	9.	LRB-135	Punjab	9.	JCR-93	Assam
10.	IC-137200	Uttar Pradesh	10.	LRB-140	Punjab	10.	JCR-107	Assam
11.	IC-140795	Punjab	11.	LRB-141	Punjab	11.	JCR-149	Assam
12.	IC-140796	Punjab	12.	LRB-156	Punjab	12.	JCR-152	Assam
13.	IC-140798	Punjab	13.	LRB-158	Punjab	13.	JCR-162	Assam
14.	IC-140802	Punjab	14.	LRB-164	Punjab	14.	JCR-163	Assam
15.	IC-140803	Punjab	15.	LRB-168	Punjab	15.	JCR-171	Assam
16.	IC-140804	Punjab	16.	LRB-176	Punjab	16.	JCR-178	Assam
17.	IC-140805	Punjab				17.	BRS-2	Uttarakhand
18.	IC-140808	Punjab						
19.	IC-016771	Manipur						
20.	IC-016789	Manipur						
21.	IC-016801	Manipur						

(continued)

Table 11.2 (continued)

Sr. no.	Rice bean germplasm (Group-I)	Collection site	Sr. no.	Rice bean germplasm (Group-II)	Collection site	Sr. no.	Rice bean germplasm (Group-III)	Collection site
22.	IC-019352	Orissa						
23.	EC-48223-B	China						
24.	Check	Baroi						
25.	Check	Dhagwar						
26.	Check	Panchrukhi						

Table 11.3 Fodder yield from different rice bean genotypes (Group-1)

Genotype	Fodder yield (q/ha)	Genotype	Fodder yield (q/ha)
IC-137186	248.49	IC-140805	319.54
IC-137187	228.76	IC-140808	282.88
IC-137188	251.74	IC-016771	236.16
IC-137189	226.25	IC-016789	300.43
IC-137190	311.33	IC-016801	200.45
IC-137191	284.74	IC-019352	282.38
IC-137194	253.93	EC-48223-B	259.30
IC-137195	312.92	JCR-12	242.33
IC-137199	244.63	JCR-32	241.70
IC-137200	356.66	JCR-52	190.56
IC-140795	258.76	JCR-76	214.20
IC-140796	291.39	Dhagwar	163.60
IC-140798	301.59	Panchrukhi	151.10
IC-140802	266.90	Baroi	160.98
IC-140803	290.83	CD (5%)	3.042
IC-140804	284.20	SE (d)	1.517
		CV	0.728

(b) **Fodder yield from rice bean genotypes of Group-II**
Green fodder yield from the different rice bean genotypes of LRB series (G-2) has been presented in Table 11.4. The genotypes LRB-176 (402.67 q/ha) followed by LRB-127 (393.06 q/ha), and LRB-156 (349.33 q/ha) were high yielding. The low fodder yield was observed for LRB-40-1 (240.67 q/ha) followed by LRB-128 (240.67 q/ha), LRB-134 (244.00 q/ha), LRB-135 (256.00 q/ha) and LRB-158 (268.67 q/ha). Although, genotype Baroi was of low height it produced considerably higher fresh fodder due to exuberant lateral growth. It was observed that the plant height is not always a single criterion for higher green fodder yield in rice bean but branching and lateral growth pattern are also contributing factors for yield.

(c) **Fodder yield from rice bean genotypes of Group-III**
Variation in fodder yield of rice bean genotypes of group-III (JCR series) has been presented in Table 11.5. The higher fodder yield was recorded in genotype JCR-93 (384.67 q/ha), JCR-152 (333.33 q/ha), JCR-162 (303.33 q/ha) and JCR-20 (296.67 q/ha). The genotype with lower fodder yield were JCR-81 (191.33 q/ha) followed by JCR-171 (198.00 q/ha), JCR-54 (204.67 q/ha) and JCR-149 (240.67 q/ha). Green fodder yield is directly related to plant height but other factors such as number of branches and leaf area also have a significant effect on fodder yield.

The rating of rice bean genotypes based on production of green biomass suggests that the genotypes IC-137200, IC-137190, IC-137195, IC-140796, IC-140798, IC-140805, IC-016789 and IC-140804 from group-I, while LRB-176,

Table 11.4 Fodder yield from different rice bean genotypes (Group-II)

Genotype(s)	Green fodder yield (q/ha)
LRB-1	291.33
LRB-40-1	240.67
LRB-40-2	253.73
LRB-45	293.87
LRB-126	252.40
LRB-127	393.06
LRB-128	240.67
LRB-134	244.00
LRB-135	256.00
LRB-140	333.33
LRB-141	310.00
LRB-156	349.33
LRB-158	268.67
LRB-164	270.67
LRB-168	292.67
LRB-176	402.67
Local Baroi	290.00
GM	**293.12**
CV	**32.94**
CD (%)	**NS**
SE(m)	**–**

Table 11.5 Fodder yield from rice bean genotypes of Group-III

Genotype(s)	Green fodder yield (q/ha)
JCR-20	296.67
JCR-54	240.67
JCR-79	266.00
JCR-81	191.33
JCR-93	384.67
JCR-107	302.00
JCR-149	204.67
JCR-152	333.33
JCR-162	303.33
JCR-163	228.67
JCR-171	198.00
JCR-178	278.67
Local	266.67
GM	**284.51**
CV	**40.91**
CD (%)	**NS**
SE(m)	**–**

Table 11.6 Promising ricebean genotypes for fodder yield

Group	Genotype
I	IC-137200, IC-137190, IC-137195, IC-140796, IC-140798, IC-140805, IC-016789, IC-140804
II	LRB-176, LRB-127, LRB-156, LRB-140, LRB-141
III	JCR-93, JCR-152, JCR-107, JCR-162, JCR-20

LRB-127, LRB-156, LRB-140, LRB-141 from group-II were promising in fodder yield. In group-III, the genotypes JCR-93, JCR-152, JCR-107, JCR-162, and JCR-20 were high fodder yielding genotypes (Table 11.6).

11.3.1.2 Biochemical Composition of Rice Bean Forage

Good nutritive quality forages are the main requisites of any livestock operation. The quality of forages defines as the optimum level of essential macro and micronutrients that can be derived from it by animals and the level of any toxic compounds, which could reduce animal performance or threaten animal health. The nutritional quality of forages also defines a function of nutrient content in the herbage, intake potential and nutrient availability and partitioning of metabolized products in the animals.

In most situations, intake of energy and protein determine the animal performance and the forage quality. Intake of available energy is primarily a function of plant cell wall constituents. Available energy and protein in forages are often low relative to animal requirements. The forage maturity is the most important factor determining the amount of available energy and protein in forages. Additionally, environment also has a significant impact on the quality of forages. Livestock productivity is the ultimate test of forage quality. If the animal has the genetic potential, animal production of forage-based diets depends on the nutritive value of forage.

Importantly, the performance of livestock depends on the intake of the forage. Rice bean fodder is highly nutritious fodder for livestock with appreciable nutritional content. The foliage from rice bean is more palatable and with protein content compared with other fodder legumes like cowpea (*Vigna unguiculata*) and moth bean (*Vigna aconitifolia*) (Heuzé et al. 2016a).

Dry matter (DM) is an important component for the determination of chemical composition and nutritive value. It varies greatly with different types of feed and fodders. The dry matter in rice bean increases with the advancement of maturity. There is a positive relationship between the dry matter accumulation and plant maturity. The highest yield of digestible dry matter is at the first flowering stage in legumes. The dry matter content in rice bean fodder varies from 14.74 to 20.81% (Katoch et al. 2010) (Table 11.7).

11.3.1.2.1 Nutritional Constituents in Rice Bean Forage

(a) **Crude protein**

Livestock need substantial quantities of protein in their feed. In animals, protein helps in the repair of body tissue, synthesis of enzymes and hormones and also in the energy supply. The level of CP content is commonly assess forage quality.

Table 11.7 Comparative nutritional profile of different leguminous forages

Nutritional constituents	Rice bean	Cowpea	Mung bean	Alfalfa	Berseem	White clover
Dry matter (%)	21.40	20.90	26.90	19.90	12.50	16.80
CP (%)	19.00	18.10	17.10	20.60	19.90	24.90
CF (%)	30.80	24.10	22.50	26.70	22.30	19.60
NDF (%)	59.70	38.60	28.40	39.30	44.80	27.50
ADF (%)	38.90	27.10	–	30.90	27.60	22.10
EE (%)	1.80	2.80	3.00	2.90	3.20	2.70
Ash (%)	10.40	11.30	11.40	11.50	15.40	11.30

Source: Heuzé et al. (2016a)

Higher the CP content in forages the more will be the nutritional quality of forage. Rice bean forage is good source of protein for livestock. The protein content in rice bean fodder is comparable or higher than other traditional leguminous forages. The biochemical evaluation of rice bean genotypes of Group-I revealed CP content ranging from 13.59% (IC-137188) to 18.92% (IC-137190) (Table 11.8) (Katoch et al. 2007b). The genotypes IC-137188, IC-137195, IC-137200, IC-140802 and IC-140803 have low CP content. The genotype IC-137190 was recorded with higher protein content in the group.

In LRB series (Group-II) of rice bean genotypes, the CP content varied from 13.12% (LRB-164) to 16.92% (LRB-156) (Table 11.9). Among these genotypes, LRB-164, LRB-40-1, LRB-45, LRB-127 and LRB-168 were with lower CP content.

In rice bean genotypes of JCR series, the protein content was within the range of 13.12% (JCR-54) to 17.20% (JCR-79) (Table 11.10). Genotypes JCR-54, JCR-93, JCR-20, JCR-107, and JCR-152 were with lower protein content. The rice bean fodder enhances the intake of other roughages with comparatively lower nutritional value, when fed in combination.

(b) **Crude fibre**

CF also called roughage, consists of complex structural carbohydrates. The main constituents of CF are NDF, ADF, cellulose, hemicellulose, lignin, pectin and cutin substances. The high fiber content in forages is postulated to have positive physiological effects in animals like reducing the transit time of food material in the gastrointestinal tract. This helps in relieving the problem of constipation and diverticular disease.

The CF content in rice bean genotypes of group-I was in the range of 18.70% (Dhagwar and Panchrukhi) to 25.87% (IC-140805) (Table 11.8). The mean CF content in rice bean genotypes of LRB series (Group-II) was from 21.10 to 35.70% (Table 11.9). The genotypes with higher CF content were LRB-40-1 (35.70%) followed by LRB-135 (34.70%), LRB-164 (34.20%) and LRB-134 (33.70%). The CF in fodder of group-III genotypes (JCR series) was in the range from 27.13 to 32.30% (Table 11.10). Highest CF was recorded in the

Table 11.8 Biochemical components in foliage of rice bean genotypes (G-I)

Genotype(s)	CP (%)	CF (%)	Carbohydrate (%)	Total ash (%)
IC-137186	16.07	19.93	10.44	1.80
IC-137187	17.34	19.97	9.82	1.33
IC-137188	13.59	22.53	12.69	1.27
IC-137189	15.99	20.43	11.72	1.53
IC-137190	18.92	19.73	9.99	1.27
IC-137191	17.15	22.87	8.29	1.67
IC-137194	15.20	23.37	10.60	1.33
IC-137195	14.07	24.37	10.06	1.33
IC-137199	15.39	22.47	10.00	1.60
IC-137200	14.17	24.70	7.78	1.13
IC-140795	15.09	23.73	9.90	1.00
IC-140796	15.11	23.33	10.05	1.53
IC-140798	15.86	24.93	10.61	1.60
IC-140802	14.35	25.60	10.06	1.20
IC-140803	14.25	22.83	9.48	1.60
IC-140804	17.81	20.37	8.83	1.33
IC-140805	16.62	25.87	8.95	2.33
IC-140808	16.60	22.63	9.20	2.13
IC-016771	16.93	22.97	10.99	1.83
IC-016789	15.33	23.80	11.28	1.93
IC-016801	17.10	21.30	7.79	1.33
IC-019352	15.29	20.83	8.70	2.33
EC-48223-B	18.77	19.40	7.33	1.06
JCR-12	15.44	19.03	9.02	1.60
JCR-32	16.13	19.80	8.81	1.27
JCR-52	16.8	19.93	9.12	1.40
JCR-76	17.20	20.37	9.29	1.27
Dhagwar	15.95	18.70	12.21	2.50
Panchrukhi	15.11	18.70	11.90	0.80
Local (C)	16.79	17.80	12.20	1.47
CD (5%)	2.28	3.77	N.S.	0.66
SE(m)	0.81	1.33	1.21	0.28

genotypes, JCR-54 (32.30%), JCR-163 (31.67%), JCR-93 (31.13%) and JCR-20 (31.06%).

(c) **Total soluble carbohydrates**

Carbohydrates form one of the important nutritional attributes as these are considered readily available sources of energy. The level of total available carbohydrates in plants is helpful in indicating the period of storage and also helpful in evaluating the potential for regrowth and persistence in adverse climatic conditions. In plants, there are two major types of carbohydrates, referred to as structural and storage carbohydrates. The structural carbohydrates make up the fiber portion of the plants, whereas storage carbohydrates make up

Table 11.9 Biochemical components in foliage of rice bean genotypes (G-II)

Genotype(s)	CP (%)	CF (%)	Carbohydrates (%)	Total ash (%)
LRB-1	14.87	30.30	11.44	0.73
LRB-40-1	13.41	35.70	12.19	1.07
LRB-40-2	15.16	27.30	15.96	0.40
LRB-45	13.41	21.10	16.85	1.00
LRB-126	16.33	26.00	17.46	1.60
LRB-127	14.29	28.07	14.65	0.27
LRB-128	16.04	32.67	13.92	0.60
LRB-134	15.16	34.70	14.00	0.60
LRB-135	14.58	33.77	12.11	0.27
LRB-140	16.33	27.53	12.38	0.33
LRB-141	16.04	30.07	12.48	0.33
LRB-156	16.92	29.53	12.02	0.73
LRB-158	16.92	29.20	13.25	0.73
LRB-164	13.12	34.20	15.79	0.27
LRB-168	14.87	33.17	13.43	0.60
LRB-176	16.62	29.70	12.34	1.07
Local Baroi	16.33	30.73	13.56	0.40
GM	15.30	30.57	13.92	0.65
CV	12.80	13.21	2.06	18.80
CD (5%)	NS	NS	8.93	0.20
SE(m)	1.06	2.33	0.72	0.068

Table 11.10 Biochemical components in foliage of rice bean genotypes (G-III)

Genotype(s)	CP (%)	CF (%)	Carbohydrates (%)	Total ash (%)
JCR-20	14.29	31.06	11.74	1.53
JCR-54	13.12	32.30	11.41	0.60
JCR-79	17.20	28.26	8.86	0.47
JCR-81	15.45	30.66	7.44	1.13
JCR-93	13.70	31.13	10.35	1.06
JCR-107	14.29	30.97	10.21	0.93
JCR-149	16.91	28.40	8.82	0.67
JCR-152	14.87	27.13	10.34	0.93
JCR-162	15.16	30.43	7.14	0.93
JCR-163	16.62	31.67	8.31	0.87
JCR-171	16.33	28.83	9.01	1.00
JCR-178	16.04	30.37	9.71	1.27
Local	14.58	30.90	8.62	0.93
GM	15.27	30.16	9.38	0.95
CV	8.00	12.95	0.89	47.45
CD (5%)	2.05	NS	5.69	NS
SE(m)	0.70	2.25	0.14	0.258

the cell content. The performance and productivity of livestock are highly associated with the level of nonstructural carbohydrates in forages. High level of nonstructural carbohydrates allows more efficient utilization of nitrogen in the rumen. Amount of total carbohydrates in fodder also determines the nutritive value of forage.

Total carbohydrates in forage of rice bean genotypes of group-I were in a range from 7.33% (EC-48223-B) to 12.69% (IC-137188) (Table 11.8). The level of total carbohydrate in forage from the rice bean genotypes of group-II was from 11.44 to 17.46%. The genotypes LRB-126 (17.46%), LRB-145 (16.85%), and LRB-164 (15.79%) had the higher level of total carbohydrates (Table 11.9). In the JCR series of rice bean, the total carbohydrates content was in the range of 7.14 to 11.74%. Total carbohydrates were high in genotype JCR-20 (11.74%) followed by JCR-54 (11.41%), JCR-93 (11.35%) and JCR-152 (11.34%) (Table 11.10).

(d) **Total ash**

Ash is the inorganic residue remaining after the water and organic matter is removed by heating in the presence of oxidizing agents and represents the total of mineral elements as well as nutritionally undesirable silica content. Minerals are important factors in nutrition of animals both as a whole and as individual ingredients. Farm animals, especially those in active stage of growth, in advanced state of pregnancy, and producing liberal quantities of milk are prone to suffer from lack of calcium and phosphorus. The feeding stuffs vary markedly in their calcium and phosphorus contents and their absorbability in the animal system depends on the combination with other constituents of feeds. Apart from sodium chlorine (common salt), calcium and phosphorus, microelements such as iron, copper, manganese, iodine and cobalt are required in small quantities by farm animals and when these elements are insufficient or absent in feeding stuffs cause a considerable setback in animal proformance.

In forage from 30 rice bean genotypes (group-I), the ash content was in range of 1.06–2.50%. The highest ash content was observed in genotype Dhagwar (2.50%) followed by IC-140805 (2.33%), IC-019352 (2.33%) and IC-140808 (2.13%). The genotypes *viz.*, IC-140795 (1.00%), followed by EC-48223-B (1.06%), IC-137200 (1.13%), IC-140802 (1.20%) and IC-137188 (1.27%) had low ash content (Table 11.8). In group-II (LRB series), the high ash content was recorded in genotype LRB-126 (1.60%) followed by LRB-40-1 (1.07%), LRB-176 (1.07%) and LRB-45 (1.00%), while the ash content was low in LRB-127 (0.27%), LRB-164 (0.27%), LRB-140 (0.33%), LRB-141 (0.33%) and LRB-127 (0.027%) (Table 11.9). The ash content in JCR series of rice bean genotypes (Group-III) ranged from 0.47 to 1.53%. The genotypes JCR-20 (1.53%), JCR-178 (1.27%), JCR-81 (1.13%) and JCR-93 (1.06%) were with higher ash content (Table 11.10).

(e) **Cell wall constituents**

Hemicellulose and cellulose are two important constituents of plant cell walls. Cellulose is the single most abundant component in cell walls and is composed of linear glucose chains linked by β-1, 4 glycosidic linkages. Cellulose chains are bundled together and make cellulose fibrils whereas hemicellulose consists of polymers of pentoses, hexoses and sugar acids. Hemicelluose serves as a cementing material between lignin and cellulose fibrils and makes cell walls more rigid. Cellulose and hemicellulose could be broken down by the enzymatic action of bacteria and other microbes in the animal's digestive tract, though their digestion is markedly slower than the digestion of sugars, starches and other freely available nonstructural carbohydrates. In contrast, lignin is not carbohydrate-based but is a phenolic compound. As such, lignin is not digestible. The presence of lignin acts as a physical barrier to the microbial enzymes that break down cellulose and hemicellulose.

High neutral detergent fiber (NDF) content indicates high fiber content in forage, therefore lower NDF content in fodder is desirable. ADF fraction of forages is moderately indigestible. Feeding trials and laboratory testing have shown that high ADF values are associated with decreased digestibility; therefore, a low ADF is better for the good digestibility of forage. NDF content has traditionally been used as a predictor of forage intake, while ADF has been used as a predictor of forage digestibility. These relationships are used to calculate relative feed value (RFV). The neutral detergent fiber (NDF) content represents the total fiber fraction (cellulose, hemicellulose, and lignin) that make up the cell wall. The NDF content in forages varies from 10 to 80%. High level of NDF content in the cell wall reduces the digestibility of any forage crop, hence, considered as an important indicator of forage digestibility. The ADF content has also been used as a predictor of forage digestibility. The content of ADF increases with the advancement in the maturity of the forage crops.

Lignin is considered an anti-quality component in forages because of its negative impact on the availability of nutrients. Lignin is an integral component of plant cell walls, which form the structural framework of plant architecture. The most important function of lignin is to provide strength and rigidity to plants. It is also important in limiting the water loss by rendering the permeability of the cell wall. All of these attributes are desirable from the perspective of plant function and survival, but it limits the nutritive value of forage crop for livestock. The main anti-quality role of lignin in forages is in limiting digestion of the structural polysaccharides. Lignification controls the amount of fiber that can be digested and therefore, has a direct and often important impact on digestible energy and dry matter intake. The undigested portion of the forage passes slowly through the gastrointestinal tract and contributes to the fill effect of the diet. Greater the content of undigested fiber in the diet, lesser it will be consumed by livestock. Therefore, lignification reduces nutritive value of forages by decreasing digestible energy and limiting dry matter intake.

(f) **Neutral detergent fiber (NDF)**

NDF content in foliage of rice bean genotypes (Group-I) varied from 45.73 to 52.97%. The highest NDF content was observed in the genotypes IC-019352 (52.67%) followed by JCR-52 (52.20%), IC-140802 (51.60%) and Dhagwar (51.40%), whereas the lowest values were recorded for genotypes IC-016771 (45.27%), IC-137199 (45.53%), IC-137186 (45.73%), IC-137188 (46.00%) and

Table 11.11 Cell wall constituents in foliage of rice bean genotypes (G-I)

Genotype(s)	NDF (%)	ADF (%)	Hemi-cellulose (%)	Lignin (%)
IC-137186	45.73	40.07	5.53	12.00
IC-137187	51.13	39.33	11.80	11.73
IC-137188	46.00	40.00	6.00	10.73
IC-137189	47.00	40.06	6.90	12.93
IC-137190	46.47	39.67	6.80	14.27
IC-137191	49.20	42.33	6.90	13.60
IC-137194	45.53	40.33	5.20	15.00
IC-137195	46.13	39.73	7.00	12.20
IC-137199	46.40	38.60	7.10	13.87
IC-137200	50.47	41.33	8.30	15.40
IC-140795	50.00	43.27	7.50	12.50
IC-140796	48.93	40.67	8.50	12.20
IC-140798	48.33	39.33	8.40	10.70
IC-140802	51.60	41.20	10.80	13.87
IC-140803	50.53	40.60	9.40	15.73
IC-140804	48.33	39.80	7.40	15.80
IC-140805	50.33	42.20	9.10	15.20
IC-140808	50.20	39.20	10.80	18.08
IC-016771	45.27	41.00	4.30	14.33
IC-016789	47.00	41.50	5.60	12.27
IC-016801	48.33	40.40	7.10	14.73
IC-019352	52.67	42.93	10.50	15.27
EC-48223-B	51.50	40.47	10.50	13.07
JCR-12	48.60	40.67	7.50	14.27
JCR-32	48.27	40.07	5.50	12.33
JCR-52	52.20	43.73	9.50	15.53
JCR-76	47.13	41.40	5.40	16.67
Dhagwar	51.40	40.93	11.10	17.40
Panchrukhi	49.00	40.67	8.10	15.27
Local (C)	45.93	40.07	5.90	17.47
CD (5%)	4.36	1.51	NS	1.59
SE(m)	1.54	0.53	–	0.56

IC-137195 (46.13%) (Table 11.11). The NDF content in rice bean forage (Group-II) varied from 45.80 to 52.27%. The highest value for NDF content was observed in LRB-164 (52.27%) and the lowest value was observed for LRB-128 (45.80%) (Table 11.12). In JCR series, NDF% content varied from 48% (JCR-79) to 52.93% (Panchrukhi). The other genotypes JCR-20 (52.53%), JCR-152 (52.40%) and JCR-162 (51.30%) had higher NDF content, whereas JCR-149 (48.06%) and JCR-171 (48.47%) had lower NDF content (Table 11.13).

Table 11.12 Cell wall constituents in foliage of rice bean genotypes (G-II)

Genotype(s)	NDF (%)	ADF (%)	Hemicellulose (%)	Lignin (%)
LRB-1	49.80	41.67	8.13	15.60
LRB-40-1	51.67	40.60	11.07	15.87
LRB-40-2	50.06	42.07	8.00	16.80
LRB-45	49.67	39.00	10.67	14.20
LRB-126	48.87	40.20	8.67	14.87
LRB-127	48.60	41.67	6.93	12.40
LRB-128	45.80	40.60	5.20	12.67
LRB-134	47.93	40.67	7.26	13.70
LRB-135	49.80	42.87	6.93	10.80
LRB-140	49.60	41.80	7.80	12.20
LRB-141	50.27	43.33	6.93	11.87
LRB-156	49.13	42.00	7.13	13.73
LRB-158	51.27	38.93	12.33	10.53
LRB-164	52.27	47.00	5.27	10.00
LRB-168	46.53	39.93	6.60	12.93
LRB-176	51.40	43.73	7.67	13.60
Local Baroi	48.00	37.67	10.33	8.87
GM	49.45	41.40	8.05	12.97
CV	NS	4.12	27.04	19.94
CD (5%)	5.16	6.00	3.62	4.30
SE(m)	1.47	2.05	1.25	1.49

Table 11.13 Cell wall constituents in foliage of rice bean genotypes (G-III)

Genotype(s)	NDF (%)	ADF (%)	Hemicellulose (%)	Lignin (%)
JCR-20	52.53	43.13	9.40	9.93
JCR-54	50.33	42.33	8.00	9.73
JCR-79	48.00	39.86	8.14	9.40
JCR-81	48.00	41.60	6.40	11.47
JCR-93	50.07	43.60	6.47	10.73
JCR-107	49.33	44.86	4.47	11.87
JCR-149	48.06	42.20	5.87	10.27
JCR-152	52.40	42.00	10.40	11.47
JCR-162	51.33	45.46	5.87	11.00
JCR-163	49.40	43.00	6.40	14.07
JCR-171	48.47	41.80	6.67	13.00
JCR-178	49.93	45.40	4.53	10.53
Local	52.93	43.87	9.06	13.20
GM	50.06	43.01	7.05	11.28
CV	6.65	6.58	47.40	NS
CD (5%)	NS	NS	NS	20.76
SE(m)	2.18	1.63	19.29	1.35

(g) **Acid detergent fiber (ADF)**
ADF content in foliage from rice bean genotypes of group-I ranged from 39.20 to 43.27%. The genotypes IC-140808 (39.20%), IC-137199 (38.60%), IC-140798 (39.33%), IC-137187 (39.33%) and IC-137190 (39.67%) had the lowest value of ADF content. These genotypes with low ADF are suitable for fodder purpose. The higher values of ADF content observed for genotypes IC-140795 (43.27%), IC-019352 (42.93%), IC-137191 (42.33%) and IC-140805 (42.20%) (Table 11.11). In LRB series (Group-2), the ADF content varied from 38.93 to 42.87%. The lowest ADF content recorded in the genotype Baroi (37.67%), LRB-158 (38.93%), LRB-45 (39.00%) and LRB-168 (39.93%) (Table 11.12). The ADF content in JCR series (group-3) of rice bean genotypes (group-III) varied from 41.60 to 45.46%. The lowest values of ADF content was observed in the genotype JCR-79 (39.86%), JCR-81 (41.60%), JCR-171 (41.80%) and JCR-152 (42.00%) (Table 11.13).

(h) **Hemicellulose content**
Hemicellulose content in foliage of rice bean genotypes (Group-I) varied from 4.30 to 11.80%. Genotypes IC-137187, Dhagwar, IC-140802, IC-140808, IC-019352 and EC-48223B have high levels of hemicellulose. The lowest hemicellulose was observed for the genotypes IC-016771 (4.30%), IC-137194 (5.20%), JCR-76 (5.40%), IC-137186 (5.53%), and IC-016789 (5.60%) (Table 11.11). The hemicellulose content in forage from rice bean genotypes (Group-II) varied from 5.20 (LRB-128) to 12.33% (LRB-158). All genotypes except LRB-45 (10.67%), LRB-40-1 (11.07%) and LRB-158 (12.33%) had lower hemicellulose content (Table 11.12). In JCR series, the genotypes JCR-152 (10.40%), JCR-20 (9.40%), Panchrukhi (9.06), JCR-79 (8.14%) and JCR-54 (8.00%) had high hemicellulose, whereas, genotype JCR-178 (4.53%), JCR-162 (5.87%) and JCR-149 (5.87%) were with low hemicellulose content (Table 11.13).

(i) **Lignin content**
Lignin content in the foliage of rice bean genotypes of group-I ranged from 10.70 to 17.47% (Katoch et al. 2007b). The genotypes Baroi (17.47%), Dhagwar (17.40%), IC-140804 (15.80%), IC-140803 (15.73%), JCR-52 (15.53%) and IC-137200 (15.40%) had high lignin content in their foliage. The low level of lignin content was recorded in genotype IC-140798 (10.70%) followed by IC-137188 (10.73%), IC-137187 (11.73%) and IC-137186 (12.00%) (Table 11.11). The lignin content in the foliage of rice bean genotypes of group-II varied from 8.87% in Baroi to LRB-40-2 (16.80%). The genotypes with high lignin content were LRB-40-1 (15.87%), LRB-1 (15.60%), LRB-126 (14.87%) and LRB-45 (14.20%) (Table 11.12). In the foliage of rice bean genotypes of JCR series, the lignin content varied from 9.40 to 14.07%. The genotypes with higher lignin content were JCR-163 (14.07%), Panchrukhi (13.20%), JCR-171 (13.00%) and JCR-107 (11.87%). The low lignin content was observed in genotypes JCR-79 (9.40%) followed by JCR-54 (9.73%) and JCR-20 (9.93%) (Table 11.13).

Nutritionally superior genotypes from Group-I

Forage nutritional factors	Genotype
CP	IC-137190, IC-140804, IC-137187, EC-48223B
CF	IC-140798, IC-140802, IC-140808, IC-137195
Total soluble carbohydrates	IC-137188, Dhagwar, Baroi, Panchrukhi, IC-137189
Ash content	Dhagwar, IC-019352, IC-140808, IC-140805,
NDF	IC-016771, IC-137194, IC-137186, Baroi
ADF	IC-137199, IC-140808, IC-137187, IC-140808, IC-137190
Hemicellulose content	IC-016771, IC-137194, JCR-76, IC-137186
Lignin content	IC-140798, IC-137188, IC-137187, IC-137186

Nutritionally superior genotypes from Group-II

Forage nutritional factors	Genotype
CP	LRB-140, LRB-128, LRB-135, LRB-40-1
CF	LRB-40-1, LRB-134, LRB-164, LRB-168
Ash content	LRB-40-1, LRB-176, LRB-45, LRB-1
Lignin content	LRB-40-1, LRB-40-2, LRB-1, LRB-126
NDF	LRB-168, LRB-128, LRB-134, LRB-127
ADF	LRB-158, LRB-45, LRB-168, LRB-126
Hemicellulose content	LRB-164, LRB-168, LRB-156, LRB-176
Lignin content	LRB-164, LRB-158, LRB-135, LRB-141

Nutritionally superior genotypes from Group-III

Forage nutritional factors	Genotype
CP	JCR-149, JCR-163, JCR-171, JCR-178
CF	JCR-54, JCR-163, JCR-93, JCR-20
Ash content	JCR-20, JCR-178, JCR-81, JCR-93
Lignin content	JCR-20, JCR-54, JCR-79, JCR-149
NDF	JCR-79, JCR-81, JCR-149, JCR-171
ADF	JCR-79, JCR-81, JCR-171, JCR-152
Hemicellulose content	JCR-107, JCR-149, JCR-162, JCR-81
Lignin content	JCR-79, JCR-54, JCR-20, JCR-149

11.3.1.2.2 Anti-nutrients in Rice Bean Forage

Rice bean foliage has the prospects for green feeding, haymaking, grazing and also for ensiling in mixtures with low-quality feeds. Though, rice bean forage is a potential feed for the livestock, the varying level of different anti-nutrients have been recorded. Normally harvesting rice bean at 25% flowering is beneficial for obtaining good quality fodder with minimal level of anti-nutrients. Tannins are known to form complexes with proteins under certain pH conditions. Tannins-protein complexes are reported to be responsible for low protein digestibility, decreased amino acid availability and increased fecal nitrogen. Tannins are categorized into two subgroups, condensed tannins and hydrolyzable tannins. They have negative effects on animal production include inhibition of microbial and animal enzymes, prevention of nutrient absorption and limited forage intake resulting from reduced palatability of feed.

11.3.1.2.2.1 Phenolic Content in Group-I Ricebean Genotypes

The forage from different rice bean genotypes of group-I revealed variation in total phenolic content from 0.83% (IC-137194) to 1.21% (IC-137195), of which genotypes IC-137195 (1.21%) and IC-019352 (1.18%) had higher phenolic content (Table 11.14). Genotypes IC-137194 (0.83%), IC-140805 (0.83%), Dhagwar (0.86%), Panchrukhi (0.87%), JCR-76 (0.93%), IC-137186 (0.93%) and JCR-52 (0.94%) had comparatively low level of total phenols.

In the case of simple phenols, all the genotypes have content ranging from 0.29 to 0.41% (Table 11.14). The total tannins content ranged from 0.48% (IC-137194) to 0.86% (IC-137195). The lowest tannin levels were observed in IC-137194 followed by IC-140805 (0.50%), Panchrukhi (0.52%), Dhagwar (0.54%) and JCR-76 (0.56%).

The condensed tannins in foliage from group-I rice bean genotypes varied from 0.046 to 0.066% (Table 11.14). The genotypes IC-137194 and IC-137195 had lowest condensed tannins (0.036%) followed by IC-137188 (0.037%), IC-140803 (0.039%) and IC-137200 (0.040%).

The hydrolyzable tannins in rice bean fodder (group-I) exhibited a range of variation from 0.44 to 0.82% (Table 11.14). Among the genotypes, IC-137194 had the lower level of hydrolyzable tannins followed by IC-140805 (0.46%), Panchrukhi (0.48%) and Dhagwar (0.50%).

11.3.1.2.2.2 Phenolic Content in Group-II Ricebean Genotypes

The rice bean genotypes belong to group-II revealed the total phenolic content ranging from 2.57% (LRB-45) to 4.13% (LRB-135) (Table 11.15). The genotypes Local Baroi (4.12%), LRB-164 (4.10%), and LRB-176 (4.08%) were with higher levels of total phenolic content whereas genotype LRB-140 (2.76%), and LRB-1 (3.14%) were with low level of total phenols.

The simple phenols among the LRB genotypes varied from 0.72% (LRB-176) to 2.43% (LRB-135). Higher level of simple phenols was observed in LRB-135 (2.43%) followed by LRB-128 (2.20%) and LRB-45 (1.69%), whereas the genotypes LRB-140 (0.75%), and LRB-176 (0.72%) had low simple phenolic content. The condensed tannins varied from 0.029% (Local Baroi) to 0.063% (LRB-141) (Table 11.15). Genotypes LRB-134, LRB-45 and LRB-140 had least level of condensed tannins. The total tannin content in 13 rice bean genotypes of group-II ranged from 1.51% (LRB-128) to LRB-176 (3.20%) while the level of hydrolyzable tannins ranged from LRB-45 (0.94%) to LRB-176 (3.15%) (Table 11.15).

Table 11.14 Anti-nutrients in forage from group-I rice bean genotypes

Genotype (s)	Total phenols (%)	Simple phenols (%)	Total tannins (%)	Condensed tannins (%)	Hydrolyzable tannins (%)
IC-137186	0.93	0.35	0.58	0.044	0.53
IC-137187	1.14	0.35	0.79	0.047	0.74
IC-137188	1.03	0.33	0.70	0.037	0.66
IC-137189	0.96	0.29	0.66	0.050	0.60
IC-137190	1.15	0.39	0.75	0.056	0.69
IC-137191	1.11	0.33	0.78	0.041	0.74
IC-137194	0.83	0.35	0.48	0.036	0.44
IC-137195	1.21	0.35	0.86	0.036	0.82
IC-137199	1.05	0.38	0.68	0.041	0.64
IC-137200	1.03	0.35	0.67	0.040	0.63
IC-140795	0.99	0.34	0.65	0.042	0.60
IC-140796	1.13	0.35	0.78	0.042	0.74
IC-140798	1.06	0.33	0.73	0.043	0.68
IC-140802	0.95	0.31	0.61	0.044	0.57
IC-140803	1.06	0.35	0.71	0.039	0.67
IC-140804	0.97	0.31	0.66	0.042	0.62
IC-140805	0.83	0.32	0.50	0.043	0.46
IC-140808	1.08	0.32	0.76	0.041	0.72
IC-016771	1.04	0.29	0.74	0.045	0.69
IC-016789	0.98	0.29	0.68	0.050	0.63
IC-016801	1.06	0.32	0.75	0.043	0.70
IC-019352	1.18	0.33	0.85	0.054	0.80
EC-48223-B	1.08	0.33	0.75	0.053	0.69
JCR-12	0.95	0.32	0.62	0.043	0.58
JCR-32	0.98	0.41	0.57	0.042	0.53
JCR-52	0.94	0.35	0.59	0.042	0.55
JCR-76	0.93	0.38	0.56	0.043	0.51
Dhagwar	0.86	0.31	0.54	0.041	0.50
Panchrukhi	0.87	0.35	0.52	0.043	0.48
Local (C)	1.08	0.36	0.73	0.041	0.68
GM	1.01	0.34	0.66	0.044	0.63
CV	13.23	11.27	21.44	18.63	22.70
CD (5%)	0.21	NS	NS	NS	NS
SE(m)	–	–	–	–	–

11.3.1.2.2.3 Phenolic Content in Group-III Ricebean Genotypes

The forage from the rice bean genotypes belong to JCR series revealed variation in the total phenolic content from 2.54% (JCR-163) to 3.44% (JCR-81) whereas, the variation in simple phenolic content ranged from 0.61 to (JCR-163) to 1.25% (JCR-93) (Table 11.16).

Table 11.15 Anti-nutrients in the forage from Group-II rice bean genotypes

Genotype (s)	Total phenols (%)	Simple phenols (%)	Total tannins (%)	Condensed tannins (%)	Hydrolyzable tannins (%)
LRB-1	3.14	0.96	2.99	0.052	2.29
LRB-40-1	3.65	0.94	2.14	0.057	2.08
LRB-40-2	3.70	1.56	2.80	0.048	2.09
LRB-45	2.57	1.64	0.98	0.040	0.94
LRB-126	3.18	1.37	1.83	0.047	1.79
LRB-127	3.82	1.63	2.56	0.051	2.14
LRB-128	3.93	2.20	1.51	0.054	1.59
LRB-134	4.13	2.43	1.68	0.049	1.64
LRB-135	3.69	1.07	2.19	0.037	2.15
LRB-140	2.76	0.75	2.01	0.040	1.97
LRB-141	3.71	1.14	2.56	0.063	2.50
LRB 156	3.85	0.82	2.85	0.051	2.80
LRB-158	3.88	0.93	2.43	0.052	2.88
LRB-164	4.10	1.09	3.00	0.040	2.96
LRB-168	4.05	1.32	2.70	0.049	2.65
LRB-176	4.08	0.72	3.20	0.052	3.15
Baroi	4.12	1.21	2.91	0.029	2.88
GM	3.67	1.29	2.40	0.04	2.26
CV	0.60	0.30	0.47	NS	0.66
CD (5%)	4.85	14.31	11.99	26.0	17.60
SE(m)	0.20	0.10	0.16	0.07	0.23

Genotypes JCR-152 (0.51%) and JCR-81 (0.45%) were with low level of simple phenols. The level of condensed tannins varies from 0.034 (JCR-171) to 0.065% (JCR-20) (Table 11.15).

Genotypes, JCR-79, JCR-162, and JCR-152 have the highest level of condensed tannins. The total tannins among 13 JCR genotypes ranged from 1.67% (JCR-20, JCR-54) to 2.98% (JCR-81). The genotypes, JCR-20 (1.67%), JCR-79(1.89%), and JCR-149 (1.86%) had low level of total tannins.

Variations from 1.61% (JCR-20) to 2.92% (JCR-81) were observed for hydrolyzable tannins. JCR-107 had highest level of hydrolyzable tannins, whereas the genotypes JCR-93 (2.03%), JCR-171(1.98%), JCR-163 (1.88%), and JCR-178 (1.85%) were observed with low level of hydrolyzable tannins (Table 11.16).

11.3.1.3 Rice Bean a Potential Dual-Purpose Crop

The global food and feed demands have been projected to double in the twenty-first century, which will further increase the pressure on the pattern of using existing farmlands and other resources.

Rice bean is a potential underutilized crop, which has the potential to mitigate the problem of food and nutritional security. The forage from this crop has an

Table 11.16 Anti-nutrients in the forage from Group-III rice bean genotypes

Genotype (s)	Total phenols (%)	Simple phenols (%)	Total tannins (%)	Condensed tannins (%)	Hydrolyzable tannins (%)
JCR-20	2.82	1.14	1.67	0.065	1.61
JCR-54	2.72	1.05	1.67	0.041	1.62
JCR-79	3.00	1.10	1.89	0.059	1.83
JCR-81	3.44	0.45	2.98	0.060	2.92
JCR-93	3.39	1.25	2.13	0.048	2.08
JCR-107	3.34	0.88	2.45	0.051	2.39
JCR-149	2.63	0.76	1.86	0.037	1.82
JCR-152	2.90	0.53	2.36	0.055	2.31
JCR-162	2.95	0.75	2.19	0.057	2.14
JCR-163	2.54	0.61	1.92	0.037	1.88
JCR-171	2.72	0.71	2.01	0.034	1.98
JCR-178	2.74	0.84	1.90	0.049	1.85
Local	3.00	0.73	2.29	0.042	2.24
GM	2.94	0.83	2.10	0.048	2.05
CV	0.12	0.07	0.13	0.008	0.13
CD (5%)	2.38	5.37	3.92	9.79	4.01
SE(m)	0.04	0.025	0.04	0.002	0.047

good nutritional profile and high production potential. Harvest of rice bean foliage as well as grain from the same plant (dual-purpose) is a primary goal for farmers, which allows them to exploit the nutritional benefits of both fodder and seed.

The dual-purpose characteristics of rice bean resulted in the identification of the genotypes having potential usage as food-feed crops. The list of potential dual-purpose genotypes has been given in Table 11.17.

11.3.1.3.1 Surplus Rice Bean Seeds in Animal Feeding

Rice bean seeds have an excellent nutritional profile owing to high protein content and other essential nutrients. Utilization of surplus rice bean seeds in animal feeding can replace 50% of commercial feed concentrates in the preparation of livestock feed (Ahuja et al. 2001). Feeding sheep with rice bean seeds can replace 50% of metabolizable energy from oat hay and have no deleterious effect on nitrogen balance in animals (Krishna et al. 1989). Processed rice bean seeds could be included in the poultry diets (Gupta et al. 1992). Feeding rice bean fodder ad libitum and sorghum sudangrass mixture could support growth of 456 g/day in crossbred calves (age 22–23 months) (Singh et al. 2000). Rice bean hay is a good protein source to supplement poor-quality roughage-based diets (Gupta et al. 1981). Wanapat et al. (2012) reported that Holstein × Friesian crossbred dairy cows receiving rice bean hay mixed with Ruzi grass tended to have higher digestibility of dry matter. Thang et al. (2008) observed that higher daily weight gain (609 g/day), feed efficiency and reduced feeding cost in growing crossbred heifers fed on a mixture of cassava hay

Table 11.17 Dual-purpose rice bean genotypes with quality traits

Genotype	Seed yield (g)	Green fodder yield (q/ha)	Nutritional characteristics (Seeds) CP (%)	CF (%)	EE (%)	Vit. C (mg/100 g)	CHO (%)	Tryptophan (g/16 g N)	Methionine (g/16 g N)	Nutritional characteristics (Forage) CP (%)	CF (%)	NDF (%)	ADF (%)	CHO (%)	Ash (%)	Lignin (%)
JCR-178	286.50	227.00	22.45	4.67	0.70	82.67	49.88	1.15	3.05	24.50	27.76	43.33	33.07	16.15	0.46	6.67
JCR-20	387.43	267.00	19.24	5.10	1.06	103.32	51.44	1.36	2.95	22.16	29.76	48.13	34.07	15.17	0.73	7.93
JCR-152	308.67	233.00	22.45	5.17	0.76	82.67	49.94	0.78	2.36	23.91	28.06	44.00	35.00	11.49	0.86	7.20
LRB-168	369.00	220.00	21.58	5.30	0.90	92.99	50.49	1.27	3.14	22.75	24.30	41.33	30.26	13.30	0.80	7.19
LRB-141	301.20	267.00	21.29	4.93	1.03	82.67	52.76	1.06	2.41	19.24	26.93	31.36	34.06	17.32	1.06	11.13
IC-140808	217.60	237.00	21.87	5.07	1.30	77.51	51.56	1.02	2.50	25.29	25.36	42.80	30.60	14.90	0.60	7.00
IC-019352	276.73	253.00	23.62	5.13	1.50	67.16	52.53	1.06	2.26	23.91	27.10	42.33	30.27	14.95	0.93	8.33
IC-140798	256.70	260.00	23.03	5.30	1.40	82.67	49.88	1.08	2.32	20.99	21.30	49.40	29.40	12.24	0.80	8.40
IC-016789	310.37	247.00	22.45	5.10	0.90	77.51	50.30	1.06	2.45	18.66	33.90	47.66	35.46	14.48	0.66	7.40
IC-140796	392.37	243.00	21.29	5.10	1.20	87.82	52.00	1.25	2.42	21.58	31.20	45.60	36.40	15.68	0.46	10.47
IC-140802	256.17	187.00	21.87	5.13	1.03	87.82	49.69	1.17	2.06	20.12	29.76	44.33	35.46	11.83	0.66	12.26
IC-137195	323.63	243.00	21.87	5.00	1.10	67.16	48.35	1.15	2.16	23.62	29.53	41.40	32.46	12.66	1.06	8.47
LRB-176	312.03	175.00	22.45	5.30	1.10	67.16	49.66	1.16	2.18	18.95	28.56	45.59	34.46	13.33	1.06	9.06
IC-137200	273.27	173.00	23.04	5.00	1.03	77.51	47.98	1.26	2.71	19.24	29.53	44.98	33.73	14.37	0.46	9.20
BRS-2	266.27	227.00	22.16	5.30	1.06	87.82	48.04	1.25	2.40	21.29	27.36	46.46	35.13	14.54	1.00	7.40
GM	302.53	230.60	22.05	5.11	1.07	81.63	49.25	1.14	2.49	21.75	28.70	43.89	33.98	14.16	0.77	8.54
CD (5%)	NS	–	NS	NS	0.36	12.39	NS	NS	0.56	NS	NS	NS	NS	1.20	0.20	NS

(continued)

Table 11.17 (continued)

Genotype	Seed yield (g)	Green fodder yield (q/ha)	Nutritional characteristics (Seeds)							Nutritional characteristics (Forage)						
			CP (%)	CF (%)	EE (%)	Vit. C (mg/100 g)	CHO (%)	Tryptophan (g/16 g N)	Methionine (g/16 g N)	CP (%)	CF (%)	NDF (%)	ADF (%)	CHO (%)	Ash (%)	Lignin (%)
CV	24.14	–	7.64	6.81	20.4	9.07	14.59	18.17	13.61	13.77	15.02	16.57	15.55	5.07	15.64	36.87
SE (m)	–	–	0.97	0.20	0.13	4.27	4.41	0.119	0.19	1.72	2.48	4.19	3.05	0.414	0.22	1.81

CP: CP; CF: CF; EE: EE; CHO Total carbohydrates, *NDF* Neutral detergent fiber, *ADF* Acid detergent fiber and rice bean hay (3:1 ratio) replaced with 60% of feed concentrate in forage-based diet (*Pennisetum purpureum* + Urea treated rice straw). Das (2002) reported that local goats fed with grass and rice bean hay (15% of the total diet) did not increase grass intake, while total dry matter intake and nutrient digestibility were increased. Increasing the level of rice bean hay above 15% had no further effect on digestibility. Chanthakhoun and Wanapat (2010) also reported that supplementation of rice bean hay in the diet of buffalo increases cellulolytic rumen bacteria, thus improving the utilization of feeds having high fiber content. Supplementation of rice bean hay at 600 g dry matter/head/day is beneficial for swamp buffaloes fed with rice straw as a basal roughage, as it results in increased dry matter intake, reduced protozoal and methane gas production in the rumen, increased nitrogen retention as well as better efficiency of ruminal microflora (Chanthakhoun et al. 2011).

11.3.2 Effect of Altitude on Nutritional Profile of White Clover

White clover (*Trifolium repens* L.) is a well-recognized forage legume, has good nutrient profile as compared to conventional grasses in temperate regions where it contributes to increasing sustainability and reduces the input cost for forage production. Various environmental as well as plant factors are well known to have a significant impact on the nutritional quality of forage plants. In one of the studies, we have evaluated the nutritional profile of white clover plants collected from different altitudes to establish a relationship between altitude and nutritive value of white clover (Katoch and Tripathi 2020). The study revealed influence of altitude on the nutritional composition of white clover collected from different locations in the mid-hill region of Himachal Pradesh (Table 11.18). The lowest protein content was observed in white clover plants from low elevations. White clover plants from lower altitudes had high ADF and NDF content (49.73% and 67.53%, respectively) in comparison to higher altitudes (46.60% and 59.06%, respectively). At lower altitudes, high temperature increases the synthesis of different cell wall constituents which ultimately affects forage digestibility. ADL content increased from 1.30 to 4.73%, while hemicellulose content increased from 14.57 to 17.80% in white clover collected from lower altitudes. Cell content, total ash, and reducing sugars were in the range of 32.47–40.37%, 8.23–10.46%, and 89.10–53.32 mg/g, respectively whereas total phenols, simple phenols, and total tannins reduced from 1.84 to 0.82%, 0.81 to 0.31%, and 1.03 to 0.51%, respectively, with increase in altitude. Hydrocyanic acid content increased in white clover plants proportionally with the increase in altitude from 0.360 to 1.044 mg/g. The study revealed that white clover in higher altitudes have better nutritional proportions as compared to lower altitudes; however, elevated levels of HCN content could affect the feeding value of white clover.

Table 11.18 Effect of altitude on nutritional composition of white clover

Location	Elevation (m)	CP (%)	ADF (%)	NDF (%)	ADL (%)	HC (%)	CC (%)	Cellulose (%)
Kandi	1581	21.23 ± 0.89	44.60 ± 0.40	59.06 ± 0.60	1.30 ± 0.26	14.57 ± 0.60	40.37 ± 0.61	43.30 ± 0.17
Kandbari	1471	20.18 ± 0.92	44.86 ± 0.50	59.74 ± 0.60	1.93 ± 0.23	14.87 ± 0.11	40.26 ± 0.60	42.93 ± 0.70
Bandla	1392	20.41 ± 1.28	45.20 ± 0.0.93	60.13 ± 0.13	2.20 ± 0.35	14.93 ± 2.60	39.87 ± 0.13	43.00 ± 2.50
Uttrala	1379	19.83 ± 0.56	47.13 ± 0.23	62.13 ± 0.03	2.30 ± 0.26	15.00 ± 0.21	37.87 ± 0.03	44.83 ± 0.21
Jiya	1340	20.53 ± 0.64	48.06 ± 0.31	63.19 ± 0.19	2.40 ± 0.20	15.12 ± 0.50	36.81 ± 0.19	45.66 ± 0.50
Chauntra	1305	19.13 ± 0.88	48.13 ± 1.00	63.39 ± 0.30	2.53 ± 0.31	15.26 ± 1.19	36.61 ± 0.26	45.60 ± 1.06
Univ. Campus, Palampur	1231	24.38 ± 1.06	48.73 ± 0.93	64.06 ± 0.2	2.46 ± 0.31	15.33 ± 1.40	35.94 ± 0.20	46.26 ± 1.79
Jogindernagar	1198	22.86 ± 0.26	48.80 ± 0.87	64.46 ± 0.20	3.40 ± 0.20	15.66 ± 1.80	35.54 ± 0.20	45.40 ± 1.40
Menjha	1093	23.45 ± 0.77	49.06 ± 0.31	64.79 ± 0.50	3.56 ± 0.21	15.72 ± 0.40	35.21 ± 0.26	45.50 ± 0.36
Paprola	975	20.30 ± 0.95	49.13 ± 0.64	64.99 ± 0.50	4.36 ± 0.38	15.86 ± 0.90	35.01 ± 0.50	44.76 ± 0.55
Malan	951	19.01 ± 0.18	49.20 ± 0.40	65.66 ± 0.90	4.46 ± 0.23	16.46 ± 1.00	34.34 ± 0.50	44.73 ± 0.61
Mahakal	904	21.11 ± 0.88	49.26 ± 0.31	66.72 ± 0.32	4.60 ± 0.20	17.45 ± 0.70	33.28 ± 0.32	44.66 ± 0.31
Nagrota	857	18.66 ± 0.89	49.73 ± 0.12	67.53 ± 0.53	4.73 ± 0.12	17.80 ± 0.54	32.47 ± 0.53	45.00 ± 0.36

Location	Elevation (m)	RS (mg/g)	Ash (%)	TP (%)	SP (%)	TT (%)	HCN (mg/g)
Kandi	1581	89.08 ± 3.28	10.46 ± 0.31	0.82 ± 0.03	0.31 ± 0.01	0.51 ± 0.03	1.04 ± 0.06
Kandbari	1471	84.39 ± 1.46	10.20 ± 0.20	0.93 ± 0.01	0.33 ± 0.01	0.60 ± 0.03	0.86 ± 0.11
Bandla	1392	80.75 ± 1.14	9.63 ± 0.15	0.99 ± 0.01	0.34 ± 0.01	0.65 ± 0.01	0.82 ± 0.16
Uttrala	1379	74.38 ± 2.15	9.53 ± 0.15	1.02 ± 0.01	0.35 ± 0.02	0.67 ± 0.03	0.82 ± 0.06
Jiya	1340	73.78 ± 1.15	9.46 ± 0.31	1.06 ± 0.08	0.36 ± 0.04	0.70 ± 0.08	0.79 ± 0.06
Chauntra	1305	70.14 ± 3.70	9.36 ± 0.15	1.16 ± 0.01	0.51 ± 0.09	0.65 ± 0.02	0.76 ± 0.11
Univ. campus, Palampur	1231	64.69 ± 1.39	9.33 ± 0.31	1.27 ± 0.01	0.53 ± 0.05	0.74 ± 0.02	0.76 ± 0.19
Jogindernagar	1198	59.39 ± 2.92	8.66 ± 0.06	1.32 ± 0.02	0.54 ± 0.01	0.78 ± 0.02	0.65 ± 0.19
Menjha	1093	56.15 ± 3.87	8.53 ± 0.31	1.47 ± 0.01	0.59 ± 0.03	0.88 ± 0.01	0.61 ± 0.06
Paprola	975	55.90 ± 1.64	8.46 ± 0.35	1.50 ± 0.01	0.65 ± 0.04	0.85 ± 0.01	0.58 ± 0.12
Malan	951	54.23 ± 5.69	8.36 ± 0.15	1.49 ± 0.02	0.69 ± 0.01	0.80 ± 0.03	0.54 ± 0.11
Mahakal	904	53.63 ± 1.37	8.33 ± 0.15	1.62 ± 0.04	0.71 ± 0.01	0.91 ± 0.05	0.50 ± 0.06
Nagrota	857	53.32 ± 2.68	8.23 ± 0.25	1.84 ± 0.03	0.81 ± 0.03	1.04 ± 0.01	0.36 ± 0.12

Data presented in Mean ± SD

Table 11.19 Effect of diseases and insect pests on fodder yield of cowpea

Treatment	Year	Incidence (%)			No. sucking pests/leaf	Yield (q/ha)
		Wilt/root rot	Leaf blight	YMV		
Unprotected crop						
After 30 days of sowing	Ist year	7.7	12.0	1.0	0.5	
	IInd year	21.4	19.6	2.8	0.7	
	Mean	14.5	15.8	1.9	0.6	
After 60 days of sowing	Ist year	32.7	43.3	10.7	5.7	53.4
	IInd year	41.7	46.0	9.8	4.6	49.9
	Mean	37.2	44.7	10.3	5.1	51.6
Protected crop						
After 30 days of sowing	Ist year	6.0	5.0	0.2	0.2	
	IInd year	10.6	6.9	1.1	0.3	
	Mean	8.3	6.0	0.6	0.2	
After 60 days of sowing	Ist year	26.7	19.7	2.9	3.6	60.1
	IInd year	16.8	20.1	3.0	2.1	60.3
	Mean	21.8	19.9	3.0	2.9	60.2

11.3.3 Effects of Foliar Diseases and Insect Pests on Quality Parameters of Legumes

The incidence of diseases and insect pests can significantly alter both the yield and quality of forages. Insect feeding generally lowers forage quantity but has variable effects on forage quality. Foliar diseases induce the production of undesirable constituents while in another case they cause defoliation in plants reducing their quality. To investigate the relationship between the incidence of different diseases and insect pests on the nutritional quality of cowpea fodder, a 2-year experiment with two sets of cowpea plants was conducted. One set of cowpea was protected with seed treatment with imidacloprid 70 Ws @ 5 g/kg seed + carbendazim @ 2 g/kg seed followed by foliar sprays of imidacloprid 17.80 SL @ 0.30 mL/L at 15 days interval to manage insects and alternate foliar spray of mancozeb and metalaxyl + mancozeb @ 2.5 g/L at 10 and 15 days interval for the management of foliar diseases whereas one set remained as control.

The effect of foliar diseases and insect pests on the quality and yield attributes of fodder cowpea have been presented in Tables 11.19 and 11.20. Unprotected cowpea plants showed high wilt and root rot incidence (14.50–37.20%), leaf blight severity (15.80–44.70%) and YMV incidence (1.90–10.30%). Under protected conditions, the incidence of these diseases varied from 8.30 to 21.80%, 6.00 to 19.90%, and 0.60 to 3.00%, respectively. Number of sucking pests per leaf was also high (0.60–5.20

Table 11.20 Effect of foliar diseases and insect pests on quality of forage cowpea

Treatment	Quality characters													
	DM	CP %	NDF %	ADF %	Hemicellulose %	Chl a (mg/g)	Chl b (mg/g)	Total Chl (mg/g)	Ash %	K (ppm)	Ca (ppm)	IVDMD %	ADF-N (CP %)	CHO (g/100 g)	Phenolics (g/100 g)
Unprotected crop															
After 30 days of sowing	31.1	19.5	60.5	51.6	8.92	0.40	0.35	0.73	4.47	158.5	318.3	64.23	5.23	4.13	0.47
After 60 days of sowing	37.8	15.3	63.2	55.3	7.90	0.43	0.28	0.71	3.97	116.7	252.8	60.98	5.96	3.37	0.59
Protected crop															
After 30 days of sowing	31.7	20.4	58.7	50.4	8.42	0.49	0.36	0.86	4.61	162.4	290.9	66.58	4.34	4.40	0.40
After 60 days of sowing	38.5	18.6	61.7	53.5	8.20	0.44	0.35	0.79	4.02	132.1	269.4	62.36	5.67	3.97	0.50

insects/leaf) in unprotected plants (0.20–2.90 insects per leaf). The DM%, CP%, chl a, chl b, and total chl (mg/g), ash, K (ppm) and Ca (ppm), IVDMD and total carbohydrate were higher in protected crop as compared to unprotected one. NDF, ADF, hemicellulose, ADF-N (CP %) and phenolics (g/100g) were lower in chemically protected crop. In 60-day-old crops, DM, NDF, ADF, and phenolics were higher as compared to 30-day-old crops in both protected as well as unprotected crops (Table 11.20). In protected cowpea plants, DM and CP varied from 37.8 to 38.5 and 15.3 to 15.66% and NDF and ADF from 63.21 to 61.78% and 55.32 to 53.53% as compared to unprotected crops after 60 days of sowing, respectively. Chlorophyll contents (a, b and total) were high in protected crop (0.44, 0.35 and 0.79 mg/g), after 60 days of sowing. In 60-day-old crop, the ash (4.02%), K (132.1 ppm), Ca (269.2 ppm), IVDMD (62.32%), and CHO (3.97 g/100 g) were higher in the protected crop in comparison to 3.97%, 116.70 ppm, 258.80 ppm, 60.98% and 3.37 g/100 g in 30-day-old crop, respectively. The results of the study revealed that the incidence of diseases and insect pests has a direct influence on the forage quality of cowpea.

- **Protected**: Seed treatment with imidacloprid 70 Ws @ 5 g/kg seed + carbendazim @ 2 g/kg seed followed by foliar sprays of imidacloprid 17.8 SL @ 0.3 mL/L at 15 days interval for the management of insects and alternate foliar spray of mancozeb and metalaxyl +mancozeb @ 2.5 g/L at 10 and 15 days interval for the management of diseases.
- Unprotected

For improved animal productivity, sufficient quantity of nutritive fodder need to be provided to livestock for meeting the energy requirements. Forage legumes are explicit sources of protein and other essential nutrients and are considered as the best source of feed over conventional grasses. Certain anti-nutritional factors reduce the nutritive quality of leguminous forages and their content changes considerably with the advancement in growth season affecting the nutritive quality of fodder. The integration, cultivation and intensification of leguminous forages in livestock feeding is essential for achieving higher animal productivity in sustainable manner (Fig. 11.2).

Fig. 11.2 The portfolio of some important forage legumes of Himalayan region; In reading order (**a**) Berseem; (**b**) White clover; (**c**) Alfalfa; (**d**) Cowpea; (**e**) Rice bean; (**f**) Bird's foot trefoil; (**g**) Red clover; (**h**) *Lathyrus*; (**i**) Siratro; (**j**) Perennial soybean; (**k**) Centro; (**l**) Sweet clover; (**m**) Persian clover; (**n**) Moth bean; (**o**) Butterfly pea; (**p**) Perennial horse gram; (**q**) Faba bean; (**r**) Guar bean; (**s**) *Macroptilium lathyroides*; (**t**) Cicer *microphyllum*; (**u**) *Macrotyloma uniflorum*; (**v**) Caragana versicolor; (**w**) *Astragulus alpines*; (**x**) *Desmodium heterocarpon*

References

Adsule RN (1997) Moth bean. In: Nwokolo E, Smartt J (eds) Food and feed from legumes and oilseeds. Springer, Berlin, p 419

Agza B, Binyam K, Solomon Z, Eskinde A, Ferede A (2012) Animal feed potential and adaptability of some cowpea (Vigna unguiculata) varieties in northwest lowlands of Ethiopia. J Agric Res 1: 478–483

Ahuja AK, Kakkar VK, Gupta BK (2001) Nutritional evaluation of ricebean in buffalo calves. Indian J Anim Nutr 18:172–175

Amole TA, Odugawa BO, Shittu O, Famakinde A, Okwelum N, Ojo VOA, Dele PA, Idowu OJ, Ogunlolu B, Adebiyi AO (2013) Herbage yield and quality of *Lablab purpureus* during the late dry season of western Nigeria. Slovak J Anim Sci 46:22–30

Arora RK, Chandel KPS, Joshi BS, Pant KC (1980) Ricebean-tribal pulse of eastern-India. Econ Bot 34:260–263

Ayan I, Mut H, Ugur B, Acar Z, Onal O (2012) Forage potential of cowpea (Vigna unguicullata L. Walp). Turk J Field Crops 17:135–138

Ayana E, Tadesse E, Mengistu A, Hassen A (2013) Advanced evaluation of cowpea (Vigna unguicullata) accessions for fodder production in the central rift valley of Ethiopia. J Agric Ext Rural Dev 5:55–61

Ayres JF, Nandra KS, Turner AD (1998) A study of the nutritive value of white clover (Trifolium repens L.) in relation to different stages of phenological maturity in the spring primary growth phase. Grass Forage Sci 53:250–259

Berimavandi AR, Akhgari H, Kaviani B (2010) Determination of the best method for silage of berseem clover (Trifolium alexandrium L.) in humid weather conditions. Int J Agric Biol 12: 141–144

Bertilsson J, Murphy M (2003) Effects of feeding clover silages on feed intake, milk production and digestion in dairy cows. Grass Forage Sci 58:309–322

Bhattacharya N, Mukherjee AK (1998) Herbage growth of rice bean under different dates of sowing and levels of phosphorus. Forage Res 24:121–123

Borreani G, Peiretti PG, Tabacco E (2007) Effect of harvest time on yield and pre-harvest quality of semi-leafless grain peas (Pisum sativum L.) as whole-crop forage. Field Crops Res 100:1–9

Bressani R (2002) Factors influencing nutritive value in food grain legumes: Mucuna in comparison to other grain legumes. In: Flores M, Eilitta M, Myhrman ML, Carew R (eds) Mucuna as a food and feed: current uses and the way forward. Proceedings of an international workshop held April 26–29, 2000 in Tegucigalpa, Honduras, pp 164–188

Broderick GA, Walgenbach RP, Maignan S (2001) Production of lactating dairy cows fed alfalfa or red clover silage at equal dry matter or CP contents in the diet. J Dairy Sci 84:1728–1737

Chanthakhoun V, Wanapat M (2010) Effect of legume (Phaseolus calcaratus) hay supplementation on rumen cellulolytic bacterial populations in swamp buffaloes investigated by the real-time PCR technique. J Anim Vet Adv 9:1654–1659

Chanthakhoun V, Wanapat M, Wachirapakorn C, Wanapat S (2011) Effect of legume (*Phaseolus calcaratus*) hay supplementation on rumen micro-organisms, fermentation and nutrient digestibility in swamp buffalo. Livest Sci 140(1/3):17–23

Chatterjee BN, Dana S (1977) Rice bean (Vigna umbellata (Thunb) Ohwi and Ohashi). Trop Grain Legume Bull 10:22–25

Chatterjee BN, Das PK (1989) Forage crop production-principle and practices. Oxford and IBH, New Delhi, p 277

Chaturvedi VB, Singh P, Jain VK (2000) Chemical composition, nitrate and β-carotene contents in berseem fodder at different cuts. Indian J Anim Nutr 17:347–349

Chaudhury AP, Prasad B (1972) Flowering behavior and yield of rice bean (*Phaseolus calcaratus* Roxb.) in relation to dates of sowing. Indian J Agric Sci 24:627–630

Chauhan TR, Gupta R, Chopra AK (1992) Comparative nutritive values of legume hays fed to buffaloes. Buffalo J 3:243–246

Cook BG, Pengelly BC, Brown SD, Donnelly JL, Eagles DA, Franco MA, Hanson J, Mullen BF, Partridge IJ, Peters M, Schultze-Kraft R (2005) Tropical forages. CSIRO, DPI&F(Qld), CIAT and ILRI, Brisbane

Das A (2002) Effect of rice bean (*Vigna umbellata* thumb) supplementation on utilization of mixed jungle grass in goats. Indian J Anim Nutr 19(1):47–50

Dawo MI, Wilkinson JM, Pilbeam DJ (2008) Interactions between plants in intercropped maize and common bean. J Sci Food Agric 89:41–48

de Bustamante GJ, Villanueva AJF, Bonilla CJA, Rubio CJV, de Bustamante Guerrero JJ (2002) Use of clitoria (*Clitoria ternatea* L.) hay in the feeding of lactating Brown Swiss cows. Tecnica Pecuaria en Mexico 42(3):477–487

Dewhurst RJ, Fisher WJ, Tweed JKS, Wilkins RJ (2003) Comparison of grass and legume silages for milk production. 2. In vivo and in sacco evaluation of rumen function. J Dairy Sci 86:2612–2621

Dewhurst RJ, Shingfield KJ, Lee MRF, Scollan ND (2006) Increasing the content of beneficial polyunsaturated fatty acids in milk produced by dairy cows in high-forage systems. Anim Feed Sci Technol 131:168–206

Donovan M (2006) Conservation evaluation of the bog Bird's-foot trefoil, *Lotus pinnatus*, in Canada. Can Field Nat 120:157–162

Fernandes AP, Waditake SK (2006) Comparative analysis of berseem (*Trifolium alexandrium*) varieties for yield and fodder quality. Anim Nutr Feed Technol 6:301–306

Fraser MD, Speijers MHM, Theobald VJ, Fychan R, Jones R (2004) Production performance and meat quality of grazing lambs finished on red clover, lucerne or perennial ryegrass swards. Grass Forage Sci 59:345–356

Gupta BN, Singh RB, Chatterjee D (1981) Chemical-composition and nutritive-value of Rice bean (*Phaseolus calcaratus* Roxb.) hay. Indian Vet J 58:727–730

Gupta JJ, Yadav BPS, Gupta HK (1992) Rice bean (*Vigna umbellata*) as poultry feed. Indian J Anim Nutr 9(1):59–62

Gwanzura T, Ng'ambi JW, Norris D (2012) Nutrient composition and tannin contents of forage sorghum, cowpea, lablab and mucuna hays grown in Limpopo province of South Africa. Asian Australas J Anim Sci 6:256–262

Hall E (2007) Red clover. Pastures Australia, a collaboration between AWI, GRDC, MLA, RIRDC and Dairy Australia

Hamdy AA, Ismail HM, AbdEl-Moneim A, Gomma NF (2009) Determination of flavonoid and phenolic acid contents of clover, cotton and citrus floral honeys. J Egypt Pub Health Assoc 84: 245–258

Hayashi Y, Devkota NR, Kumagai H (2007) Effects of field pea (*Pisum sativum* L.) hay feeding on dry matter intake and milk production of Murrah buffaloes (*Bubalus bubalis*) fed rice straw *ad libitum*. Anim Sci J 78:151–158

Heuzé V, Tran G (2015) Cowpea (*Vigna unguiculata*) seeds. Feedipedia

Heuzé V, Tran G, Lebas F (2015) Sainfoin (*Onobrychis viciifolia*). Feedipedia

Heuzé V, Tran G, Boval M (2016a) Rice bean (*Vigna umbellata*). Feedipedia, a programme by INRA, CIRAD, AFZ and FAO

Heuzé V, Tran G, Boudon A, Bastianelli D, Lebas F (2016b) Berseem (*Trifolium alexandrinum*). Feedipedia

Heuzé V, Tran G, Boval M, Noblet J, Renaudeau D, Lessire M, Lebas F (2016c) Alfalfa (*Medicago sativa*). Feedipedia

Heuzé V, Tran G, Nozière P, Lebas F (2016d) Birdsfoot trefoil (*Lotus corniculatus*). Feedipedia

Homolka P, Koukolova V, Podsendnicek M, Hlavackova A (2012) Nutritive value of red clover and lucerne forages for ruminants by *in-vitro* and *in-vivo* digestibility methods. Czech J Anim Sci 57:454–568

Iliev F, Kozelov L, Chindarska Z, Ignatova M (2002) Determination of energy and protein value of pea hay and dehydrated alfalfa. Zhivotnov'dni Nauki 39(3):37–39

Iqbal K, Tanveer A, Ali A, Ayub M, Tahir M (1998) Growth and yield response of rice bean fodder to different levels of N and P. Pak J Biol Sci 1:212–214

Janda B, Stochmal A, Montoro P, Piacente S, Oleszek W (2009) Phenolics in aerial parts of Persian clover *Trifolium resupinatum*. Nat Prod Commun 4:1661–1664

Kaplan M, Atalay AI, Medjekal S (2009) Potential nutritive value of wild birdsfoot trefoil (*Lotus corniculatus*) plants grown in different sites. Livestock Res Rural Dev 21:1–14

Karachi M, Matata Z (1996) Preliminary evaluation of pigeon-pea accessions for forage production and nutritive value and seed yield in Western Tanzania. Trop Agric 73:253–258

Karnezos TP, Matches AG, Brown CP (1994) Spring lamb production on alfalfa, sainfoin, and wheatgrass pastures. Agron J 86(3):497–502

Katic S, Mihailovic V, Pataki I, Karagic Đ, Vasiljević S (2001) Produktivnost i hemijski sastav suve materije lucerke. Arhiv za poljoprivredne nauke 62:83–91

Katoch R (2010) Effect of different fertilizer levels on root nodulation and fodder quality in ricebean (*Vigna umbellata*) genotypes. Range Manag Agrofor 31:41–47

Katoch R (2011) Morpho-physiological and nutritional characterization of ricebean (*Vigna umbellata*). Acta Agron Hung 59:125–136

Katoch R (2013) Nutritional potential of ricebean (*Vigna umbellata*): an underutilized legume. J Food Sci 78:8–16

Katoch R, Tripathi A (2020) Effect of altitude on nutritional profile of white clover (Trifolium repens L.). Range Manag Agrofor 41(1):173–177

Katoch R, Chand U, Bandana (2007a) Red clover: present status and future needs. Forage Res 33: 1–8

Katoch R, David B, Priya KN, Bandhari JC (2007b) Fodder yield and nutritional attributes of ricebean (*Vigna umbellata*). Range Manag Agrofor 28:260–262

Katoch R, Sood S, Kumar N, Bandhari JC (2007c) Fodder production and nodulation efficiency of different rice bean genotypes. Forage Res 33:73–77

Katoch R, Sood S, Kumar N, Bhandari JC (2010) Response of NPK on fodder production potential of ricebean genotypes under rainfed mid hill conditions of Himachal Pradesh. Range Manag Agrofor 3:94–95

Krishna G, Mandal AB, Paliwal VK, Yadav KR (1989) Rice bean (*Vigna umbellata*) as a feed for adult sheep. Indian J Anim Nutr 6:365–368

Lauriault LM, Angadi SV, Marsalis MA, Vanleeuwen DM (2012) A comparison of pigeon pea and cowpea forage yield and nutritive value in the southern high plains of the USA. Bulletin 802. N M State University, Las Cruces, NM, pp 1–6

Linn JG, Hutjens MF, Shaver R, Otterby DE, Howard WT, Kilmer LH (2002). Feeding the dairy herd. University of Minnesota Extension, pub. 346

Lopez S, Davies DR, Giraldez FJ, Dhanoa MS, Dijkstra J, France J (2005) Assessment of nutritive value of cereal and legume straws based on chemical composition and *in vitro* digestibility. J Sci Food Agric 85:1550–1557

Mandal L, Banerjee S (2009) Evaluation of nutritive value of berseem (Trifolium alrexandrium) on Garole sheep. J Crop Weed 5:43–44

Mandal SR, Mukherjee AK (1999) Forage yield and qualities of some ricebean (Vigna umbellata) germplasms. Environ Ecol 17(3):749–751

Millic D, Karagic D, Vasiljevic S, Mikic A, Mijic B, Katic S (2011) Leaf and stem chemical composition of divergent alfalfa cultivars. Biotechnol Anim Husb 27:1505–1511

Min BR, Attwood GT, McNabb WC, Molan AL, Bary TN (2005) The effect of condensed tannins from *Lotus corniculatus*on the proteolytic activities and growth of rumen bacteria. Anim Feed Sci Technol 121(1–2):45–58

Mirzaei-Aghsaghali A, Maheri-Sis N, Mirza-Aghazadeh A, Safaei AR, Aghajanzadeh-Golshani A (2008) Nutritional value of alfalfa varieties for ruminants with emphasis on different measuring methods: a review. Res J Biol Sci 3:1227–1241

Mislevy P, Blount A, Martin F, Scully B (2005) Soybean and Clay Cowpea grown for forage production in the subtropics. Crop Manage 4(1). https://doi.org/10.1094/CM-2005-0926-01-RS

Moorby JM, Lee MRF, Davies DR, Kim EJ, Nute GR, Eills NM, Scollan ND (2009) Assessment of dietary ratios of red clover and grass silages on milk production and milk quality in dairy cows. J Dairy Sci 92:1148–1160

Mukherjee AK, Roquib MA, Chatterjee BN (1980) Ricebean for the scarcity period. Indian Farm 30:26–28

Nakanishi Y, Tsuru K, Bungo T, Shimojo M, Masuda Y, Goto I (1993) Effects of growth stage and sward structure of *Macroptilium lathyroides* and *Macroptilium* atropurpureum on selective grazing and bite size in goats. Trop Grassl 27:108–113

Nandanwar RS, Deshmukh AM, Patil AN (1991) Evaluation of promising stylo species and its animal productivity in western Vidarbha (Maharashtra). Range Manag Agrofor 12(2):139–145

Owolabi AO, Nadidi US, James BD, Amune FA (2012) Proximate, antinutrient and mineral composition of five varieties (improved and local) of cowpea (Vigna unguicullata) commonly consumed in Sumaru community, Zaria-Nigeria. Asian J Food Sci Technol 4:70–72

Peiretti PG, Gai F, Alonzi S, Tassone S (2016) Nutritive value and fatty acid profile of birdsfoot trefoil (*Lotus corniculatus*) and white clover (*Trifolium repens*) in alpine pastures. Livestock Res Rural Dev 28:218

Peoples MB, Ladha JK, Herridge DF (1995) Enhancing legume N2 fixation through plant and soil management. Plant and Soil 174:83–101

Peterson PR, Sheaffer CC, Hall MH (1992) Drought effects on perennial forage legume yield and quality. Agron J 84:774–779

Pieltain MC, Castañón AAR, Flores MP, Castañón JIR (1998) Nutritive value of postharvest bean by-products (*Phaseolus vulgaris* L.) for ruminants. Anim Feed Sci Technol 62(4):271–275

Rao SC, Phillips WA, Mayeux HS, Phatak SC (2003) Potential grain and forage production of early maturing pigeonpea in the southern Great Plains. Crop Sci 43:2212–2217

Rattray PV, Joyce JP (1974) Nutritive value of white clover and perennial ryegrass. N Z J Agric Res 17:401–406

Ricardo LAJ, Olivo CJ, Bern CM, Anguiree PF, Quartin MP, Avilla SC (2013) Nutritional value and chemical composition of pastures of peanut forage or red clover. Anim Sci 35:57–62

Salehi S, Lashkari S, Rahman AE, Kamangar H (2014) Nutrient digestibility and chemical composition of potato (*Solanum tuberosum* L.) vine as alternative forage in ruminant diets. Agric Comm 2:63–66

Saloniemi H, Wähälä K, Nykänen-Kurki P, Kallela K, Saastamoinen I (1995) Phytoestrogen content and estrogenic effect of legume fodder. Proc Soc Exp Biol Med 208:13–17

Sanderson MA, Wedin WF (1989) Nitrogen in the detergent fiber fractions of temperate legumes and grasses. Grass Forage Sci 44:159–168

Scharenberg A, Kreuzer M, Dohme F (2009) Suitability of Sanfoin (*Onobrychis vicifolia*) hay as a supplement to fresh grass in dairy cows. Asian Aust J Anim Sci 22:1005–1015

Shaug S, Lu C, King W, Buu R, Lin J (2000) Forage production and silage making for berseem clover. J Taiwan Livest Res 33(1):105–110

Sherasia PL, Garg MR, Bhanderi BM (2017) In: Calles T, Makkar HPS (eds) Pulses and their by-products as animal feed. FAO, Food and Agriculture Organization of the United Nations, Rome

Simonet AM, Stochmal A, Oleszek W, Macias FA (1999) Saponins and polar compounds from *Trifolium resupinatum*. Phytochemistry 51:1065–1067

Singh RB, Saha RC, Singh S (2000) Effect of feeding ricebean and sorghum Sudan mixed fodder on growth and nutrient utilization in crossbred calves. Indian J Anim Nutr 17(2):160–161

Solomon G, Kibrom G (2014) Forage production potential and nutritive value of cowpea (Vigna unguicullata) genotypes in the northern lowlands of Ethiopia. J Agric Res Dev 5:66–71

Spencer PS, Schaumburg HH (1983) Lathyrism: a neurotoxic disease. Neurobehav Toxicol Teratol 5(6):625–629

Sreerama YN, Sashikala VB, Pratape VM, Singh V (2012) Nutrients and anti-nutrients in cowpea and horsegram flours in comparison to chickpea flour: evaluation of their flour functionality. Food Chem 131:462–468

Thang CM, Sanh MV, Wiktorsson H (2008) Effects of supplementation of mixed cassava (*Manihot esculenta*) and legume (*Phaseolus calcaratus*) fodder on the rumen degradability and performance of growing cattle. Asian Aust J Anim Sci 21:66–74

Tobisa M, Nakano Y, Paek J, Mochizuki T, Shimojo M, Masuda Y (2005) Fermentation quality and palatability of rice plant silages as affected by mixing with phasey bean (*Macroptilium lathyroides* (L.) Urb.). J Japan Soc Grassl Sci 51:274–280

Vuckovic S, Stojanovic I, Prodanovic S, Cupina B, Zivanovic T, Vojin S, Jelacic S (2007) Morphological and nutritional properties of birdsfoot trefoil (*Lotus corniculatus* L.) autochthonus populations in Serbia and Bosnia and Herzegovina. Genet Resour Crop Evol 54:421–428

Waghorn GC, Shelton ID (1992) The nutritive value of lotus for sheep. Proc N Z Soc Anim Prod 52: 89–92

Wanapat M, Wongnen N, Sangkloy W, Pilajun R, Kanpukdee S (2012) On-farm use of legume (Phaseolus calcaratus) and Ruzi grass on rumen fermentation and milk production in lactating dairy cows. Agric Sci 3:355–360

Yunus M, Ohba N, Tobisa M, Nakano Y, Shimojo M, Furuse M, Masuda Y (2001) Improving fermentation and nutritive quality of napiergrass silage by mixing with phasey bean. Asian Aust J Anim Sci 14:947–950

12 Fodder Trees in Mitigating Forage Demand

Fodder trees play a significant role in mixed farming systems. The importance of fodder trees as a potential livestock feed increases with aridity. They are the essential components in the dry seasons when most other feed resources depreciate in quality and quantity. Fodder trees remain lush green for longer periods and are less affected by seasonal dry conditions due to extensive root systems enabling them to extract water and nutrients from deep soil profiles. Fodder shortage has been a serious problem for livestock holders in the Himalayan region, especially during dry periods and winter. Generally, trees occupy a significant place in farming systems and the overall way of life in animal production in the region. Fodder trees and shrubs are often higher in crude protein and other nutrients and play a vital role as dietary supplements of low-quality grasses in dry seasons. The intake of browse forage increases the total dry matter and crude protein intake, and also improves the digestibility of low-quality forages if mixed with browse forage. The utilization of foliage from fodder trees is helpful in overcoming feed scarcity during dry periods. The integration of fodder trees into local farming systems through agroforestry is imperative in order to increase livestock productivity and also to support rural livelihoods. It is also important to steer further research support for cultivation, management, processing and use of fodder shrubs and trees for improved livestock production and productivity in the Himalayan region.

12.1 Fodder Trees: An Important Forage Resource in the Himalayan Region

Livestock are key components of hill farming. They play a crucial role in the sustenance of the rural livelihoods in the Himalayan region. Many small-scale farmers in the Himalayan region keep livestock, particularly cattle, goats and sheep for different products like milk, meat, leather, wool, field ploughing and manure. Fodder trees are an important source of energy and protein to keep animals

R. Katoch, *Nutritional Quality Management of Forages in the Himalayan Region*,
https://doi.org/10.1007/978-981-16-5437-4_12

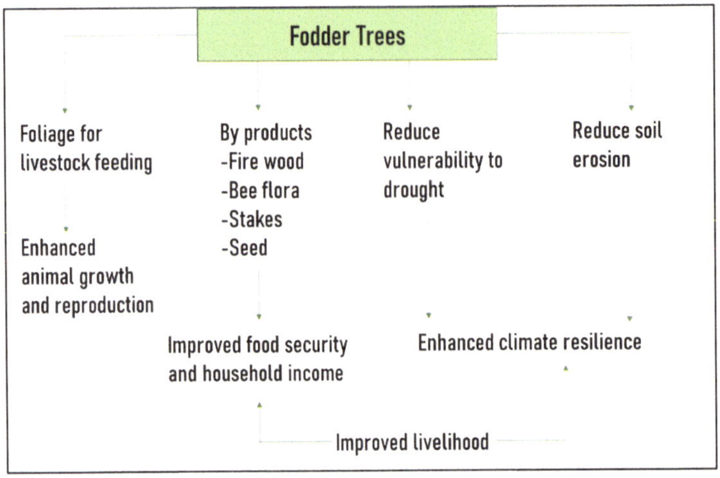

Fig. 12.1 Fodder trees in mixed farming system in Himalayan region

healthy, improve growth rates and even increase milk and wool production. They are recognized as important feed components, particularly as suppliers of protein and other essential nutrients, especially during harsh environmental conditions. Fodder trees are effective insurance against seasonal feed shortages, supplementing the quality and quantity of pasture components. They are less affected by dry conditions because of their more extensive root systems and longer life span.

Utilization of foliage from fodder trees in livestock feeding has been accepted as one of the critical strategies for round the year fodder availability. Besides the importance of fodder trees in livestock feeding, they also provide shade, wind shelter, live fences, improved fallow, bee forage, human food, fuelwood, timber, fiber, resins, dyes, medicines, wildlife habitat, nutrient cycling, and farm diversity. These advantages justify the more intensive and wider utilization of fodder trees in feeding systems. Figure 12.1 presents the potential benefits of fodder trees in mixed farming, which could lead to the sustained livelihoods of rural farmers in the Himalayan region.

12.1.1 Fodder Trees as Nutritional Feeding Resource

Inclusion of protein-rich feeds is essential for nutritionally rich diet for livestock. For example, *Leucaena* provides a valuable source of protein, energy and sulphur for the rumen bacteria. The livestock is often fed with diet having large proportions of low-quality feed components, which have low protein content, sugar and high fiber content. Therefore, it is essential to supplement the basic diet with feed resources of good nutritional quality. Fodder tree leaves contain high protein content ranging from 10 to 30% and appropriate amount of minerals and other essential nutrients with high digestibility (Makkar and Becker 1998). Fodder tree foliage could substitute feed concentrates to some extent without adverse effects on health and productivity of livestock.

12.1.2 Fodder Trees as Dry Season Supplement

During dry period, trees remain green for longer period than grasses. When the availability of green fodder decreases during dry period, the fodder trees provide the required energy and proteins to livestock. Thus, they reduce the risk of drought and provide continuous supply of fodder to animals during adverse seasonal conditions.

12.1.3 Multiple Usages of Fodder Trees

Fodder trees besides providing foliage to livestock; also provide shade, shelter, timber, fiber, resins, dyes and wildlife habitat. They also improve the structural and textural properties of soil. To promote use of fodder trees on smallholder farms, farmers must be provided with information on good quality fodder trees and the practices for their efficient utilization. They should also be encouraged to start planting fodder trees in their farming systems or establishing fodder gardens on fallow lands.

12.2 Tree Fodder: Alternative Source of Quality Fodder

Mixed farming-based agriculture production system involves complementary interactions between crop and livestock. This system is most commonly followed in Himalayan region. The fodder scarcity during lean periods has often been associated with serious economic losses to small-hold farmers. Therefore, identification and introduction of alternative feed resources which do not compete with human food and have superior nutritional quality require utmost attention. The propagation of such feed resources could be utilized to combat desertification, drought, providing economic benefits and bridge the gap between supply and demand of animal feeds. Fodder trees have better nutritive profile as compared to grasses and capacity to provide adequate amount of protein and minerals when other quality fodder sources are unavailable (Katoch 2009; Rana et al. 1999; Bhardwaj et al. 2018; Sahoo et al. 2016). Moreover, fodder trees are less susceptible to sudden climatic changes and have longer life spans than conventional forages which aid their importance as a potential alternative feed resource. To promote the utilization of fodder trees as a component in livestock feeding and for maintaining the sustainability of agroforestry systems, one must consider:

- Easily establishing species
- Growth pattern and adaptability in local environment conditions
- Tolerance to various management practices
- Effect of different environmental conditions
- Voluntary intake of tree species by the livestock
- Nutritive value of tree foliage
- Ability to regenerate foliage

The under-given points should be considered for efficient management and utilization of fodder trees

- Fodder trees could be grown in backyards, farm roads, farm bunds, common property resources and marginal and sub-marginal poor fertile soils.
- Package of practices particularly seed /seedling requirement, spacing, fertilizer schedule, harvesting schedule should be adopted.
- Quality saplings need to be procured or produced on the farm for better establishment. Nursery bags should be used for growing saplings in a shaded area on the farm.
- The trees meant for fodder purposes should be allowed to grow for about 6 months without cutting the main trunk, but the side branches should be pruned until then. After 6 months, the main trunk should be cut at a height of 6 ft from the ground level.

Grewia oppositifolia, Morus serrata, Bauhinia variegata, Quercus leucotrichophora, Quercus floribunda, Ilex dipyrena, Oogeinia oojeinensis, Boehmeria rugulosa, Morus alba, and *Celtis australis* are important fodder trees in the Himalayan region due to:

(a) Availability of high quality and digestible forage resource, available in and around the farm;
(b) Supplement to nutrient-deficient locally available feeds resulting in enhancement of the microbial growth and digestion of cellulosic biomass in the rumen;
(c) Source of minerals to complement deficiencies in the basal fed resources and
(d) Decrease dependency on feed concentrates.

The important fodder trees of the Himalayan region are listed in Table 12.1.

Table 12.1 Dominating fodder trees in the Himalayan region

Species name	Family	Species name	Family
Abies pindrow	Pinaceae	*Machilus duthei*	Lauraceae
Acer acuminatum	Aceraceae	*Melia azedarach*	Meliaceae
Acer caesium	Aceraceae	*Meliosma dilleniifolia*	Sabiaceae
Acacia nilotica	Fabaceae	*Morus alba*	Moraceae
Acacia catechu	Fabaceae	*Morus serrata*	Moraceae
Aesculus indica	Hippocastanaceae	*Morus multicaulis*	Moraceae
Alnus nepalensis	Betulaceae	*Neolitsea cuipala*	Lauraceae
Albizzia lebbek	Fabaceae	*Oogeinia oojeinensis*	Fabaceae
Albizzia stipulata	Fabaceae	*Phoebe laceolata*	Lauraceae
Albizzia chinensis	Fabaceae	*Populus ciliata*	Salicaceae
Albizia procera	Fabaceae	*Prunus armeniaca*	Rosaceae
Bauhinia purpurea	Caesalpiniaceae	*Prunus cerasoides*	Rosaceae
Bauhinia variegata	Caesalpiniaceae	*Prunus cornuta*	Rosaceae
Betula alnoides	Betulaceae	*Pyrus pashia*	Rosaceae
Boehmeria rugulosa	Urticaceae	*Quercus leucotrichiphora*	Fagaceae
Bridelia retusa	Euphorbiaceae	*Quercus semecarpifolia*	Fagaceae
Broussonetia papyrifera	Moraceae	*Quercus floribunda*	Fagaceae
Bambusa arundinacea	Poaceae	*Quercus glauca*	Fagaceae
Celtis australis	Ulmaceae	*Rhamnus persica*	Rhamnaceae
Debregeasia longifolia	Urticaceae	*Robinia pseudoacacia*	Fabaceae
Dodecadenia grandiflora	Lauraceae	*Salix alba*	Salicaceae
Engelhardtia spicata	Juglandaceae	*Salix viminalis*	Salicaceae
Eranthemum palchellum	Acanthaceae	*Salix tetrasperma*	Salicaceae
Euonymus pendulus	Celastraceae	*Schleichera oleosa*	Sapindaceae
Ficus hispida	Moraceae	*Sterculia villosa*	Sterculiaceae
Ficus palmata	Moraceae	*Swida oblonga*	Cornaceae
Ficus racemosa	Moraceae	*Toona cilliata*	Meliaceae
Ficus religiosa	Moraceae	*Terminalia bellirica*	Combretaceae
Ficus semicordata	Moraceae	*Terminalia chebula*	Combretaceae
Grewia optiva	Tiliaceae	*Toona Hexandra*	Meliaceae
Grewia oppositifolia	Tiliaceae	*Ulmus wallichiana*	Ulmaceae
Leucaena leucocephala	Mimosaceae	*Salix alba*	Salicaceae

12.3 Nutritional Composition of Fodder Trees of Himalayan Region

Fodder trees constitute an integral part of the diet of livestock in Himalayan region due to their attractive nutrient profile feeding value, tolerance of a wide range of management practices, longevity and capacity to provide fodder when other fodder species are scare. In general, foliage from fodder tree have high protein and mineral content particularly calcium and phosphorus as compared to conventional grasses.

Tree foliage also have a high level of microminerals well above critical levels. The nutrients from fodder trees initiate the microbial activity in digestive tract of ruminants, which increases their ability to digest fiber and enable livestock for better utilization of dry season pastures. However, their nutritive value is significantly affected by various plant and environment-based factors such as plant maturity, plant parts, season, altitude and harvesting time.

CP in tree leaves from mid-Himalayan region generally ranges from 9.13 to 22.08% while calcium and phosphorus content ranged from 0.50 to 6.31 mg/100 g and 0.12–0.27 mg/100 g of dry matter, respectively (Sharma et al. 1966). Pal et al. (1979) observed variation in CP from 10.29% in *Ficus benghalensis* to 20.99% in *Albizzia stipulata*, CF from 14.38% in *Morus alba* to 33.74% in *Atriplex stipulata*, NFE from 35.41% in *Bambusa nutans* to 60.41% in *Syzygium cumini*, total ash from 7.40% in *Bauhinia variegata* to 17.41% in *Cordia dichotoma*, insoluble ash from 0.35% in *Syzygium cumini* to 8.05% in *Banksia nutans*, calcium from 0.76% in *Dendrocalamus hamiltonii* to 4.79% in *A. marmelos* and phosphorus from 0.11% in *Syzygium cumini* and *Quercus incana* to 0.25% in *Callicarpa dichotoma*. *Bauhinia variegata* has comparatively better nutritive profile and collated well with green leguminous fodder in its nutritive value (Fig. 12.2).

Sahoo et al. (2016) reported calcium content from 1.20 to 2.70% in fodder trees of the sub-temperate Himalayan region. High calcium content in *Ficus palmata* (2.70%) is useful in the early stage of lactation in animals. Khosla et al. (1992) observed high protein content (32%) in immature leaves of *Grewia optiva*, *Celtis australis* and *Robinia pseudoacacia*. Protein digestibility of *Grewia oppositifolia* and *Bambusa arundinacea* was reported to resemble leguminous forages (Sharma et al. 1966). The comparative nutritional analysis of different fodder trees was also carried out by Khatta and Katoch (1983) and they reported that *Bambusa arundinacea* is a nutritionally superior, ecologically adaptable, fast-growing plant with more palatable foliage than other fodder trees (Fig. 12.3).

Mulberry (*Morus alba* var. *multicolis*) is considered as high quality fodder tree due to its nutritional components. *Morus alba* foliage contains 15.00–27.60% CP, 2.30–8.0% EE, 9.10–15.30% CF, 48.00–49.70% NFE, 63.30% TC, 14.30–22.90% of ash, 2.42–4.71% Ca, 0.23–0.97% P, 30–50 ppm Fe, 0.50–1.00% K, 33–46% NDF, 28 to 35% ADF, 5–10% HC, 19–25% cellulose and around 11% lignin content on dry matter basis (Singh et al. 1989). Makkar and Becker (1998) reported protein content in fodder trees prevailing in foothills of the Himalayas ranging from 8.00 to 25.90%. Dhungana et al. (2012) carried out a nutritional analysis of important locally preferred fodder trees of mid-Hills of Nepal which included *Artocarpus lakoocha, Ficus subinisa, Ficus roxburghii, Ficus semicordata, Litsea monopetala, Ficus lacor, Monardella odoratissima, Ficus glaberrima, Brassaiopsis hainla,* and *Ficus hispida*. They reported CP content variation in these species from 15 to 29%. *Artocarpus lakoocha* had the highest CP content (28.64%) followed by *Litsea monopetala* (26.70%) and *Brassaiopsis hainla* (25.16%). Among the fodder trees, *Monardella odoratissima* had the lowest CP content (15.54%). Crude fat content was highest in *Brassaiopsis hanila* (3.22%) and lowest in *Ficus semicordata* (1%). CF content is highest in *Ficus lacor* (42.07%) and lowest in *Ficus hispida* (23.03%).

Fig. 12.2 Tree foliage for livestock feeding

Fig. 12.3 Portfolio of important fodder trees of the Himalayan region. In reading order: (**a**) *Bambusa arundinacea*; (**b**) *Bauhinia variegata*; (**c**) *Celtis australis*; (**d**) *Ficus palmata*; (**e**) *Ficus hispida*; (**f**) *Morus alba*; (**g**) *Grewia optiva*; (**h**) *Ficus religiosa*; (**i**) *Quercus leucotricophora*; (**j**) *Robinia pseudoacacia*; (**k**) *Leuceana leucocephala*; (**l**) *Toona ciliate*; (**m**) *Prunus padus*; (**n**) *Albizia chinensis*; (**o**) *Salix alba*; (**p**) *Quercus semicarpifolia*

Machilus odoratissima had highest total carbohydrates content (21.92%) and lowest in *Ficus hispida* (0.34%). In case of ash content, *Ficus hispida* (17.50%) had highest and *Litsea monopetala* (5%) had lowest ash content. Sahoo et al. (2016) also carried out nutritional analysis of tree foliage (*Grewia oppositifolia, Ficus nemoralis, Ficus palmate, Ficus roxburghii, Celtis australis, Bauhinia variegate, Quercus*

Table 12.2 Organic matter, crude protein, ether extract and digestibility of fodder trees

Fodder tree	OM	CP	EE	IVDMD
	(% DM)			
Grewia oppositifolia	92.60	21.10	3.90	77.20
Ficus nemoralis	93.30	14.50	5.70	72.80
Ficus palmata	91.50	13.80	4.90	68.50
Ficus roxburghii	91.20	13.30	5.90	66.50
Celtis australis	91.10	16.50	5.80	63.90
Bauhinia variegata	93.20	15.10	4.90	62.30
Quercus semicarpifolia	97.40	9.50	5.30	65.40
Quercus leucotricophora	96.70	10.50	5.60	64.70
Quercus floribunda	96.40	9.70	4.40	62.30
Quercus glauca	90.60	11.00	5.20	52.60
Alnus nepalensis	92.80	10.20	5.50	56.40

Source: Sahoo et al. (2016)

semicarpifolia, Quercus leucotricophora, Quercus floribunda, Quercus glauca, and *Alnus nepalensis*) in the sub-temperate Himalayan region. The results of their study revealed that OM, CP EE contents, and digestibility in fodder trees varied from 90.60 to 97.40%, 9.50 to 21.10%, 3.90 to 5.90%, and 52.60 to 77.20%, respectively (Table 12.2). *Grewia oppositifolia* had highest CP content (21.10%) followed by *Celtis australis* (16.50%), *Bauhinia variegata* (15.10%) and *Ficus nemoralis* (14.50%). Among the fodder trees, *Quercus semicarpifolia* had lowest CP content (9.50%). EE content was highest in *Ficus roxburghii* (5.90%) and lowest in *Grewia oppositifolia* (3.90%). The digestibility of foliage of *Grewia oppositifolia* (77.20%) followed by *Ficus nemoralis* (72.80%), *Ficus palmate* (68.50%), *Ficus roxburghii* (66.50%), *and Quercus semicarpifolia* (65.40%). Tree foliage of sub-temperate Himalayan region has also been found a rich source of Fe and Mn with a variation from 121 (*Quercus leucotricophora*) to 177 ppm (*Bauhinia variegata*) and 42 ppm (*Ficus palmata*) to 402 ppm (*Quercus leucotricophora*), respectively. The zinc content in tree leaves varies from 32 ppm (*Ficus roxburghii*) to 83 ppm (*Ficus nemoralis*). Se and Co content are highest in *Ficus roxburghii* and lowest in *Celtis australis* (Sahoo et al. 2016) (Table 12.3). Table 12.4 presents nutritional composition of some important fodder trees of the Himalayan region where CP, CF, EE, Ash, NFE, NDF and ADF contents varied from 8.09 to 26.40%, 9.94 to 41.11%, 1.00 to 12.92%, 3.87 to 17.15%, 36.12 to 63.25%, 21.66 to 70.90% and 12.89 to 59.63%, respectively. Nutritional profile of *Bauhinia variegata, Arratocarpus lakoocha, Albizia chinensis, Albizia lebbek, Morus alba, Grewia oppositifolia, Celtis australis* and *Bambusa arundinacea* and *Leucaena leucocephala* (Unpublished data) have also been investigated (Katoch et al. 2017) and the results of their study have been discussed and depicted in Figs. 12.4a, b and 12.5a, b:

Table 12.3 Mineral composition of tree foliage

	Ca	P	Mg	Fe	Mn	Zn	Mo	Cu	I	Se	Co
Fodder tree	% DM basis			(ppm)							
Grewia oppositifolia	2.10	0.38	0.34	155	57	47	0.69	6.30	0.07	0.25	0.39
Ficus nemoralis	2.30	0.35	0.34	153	107	83	0.71	5.80	0.06	0.32	0.38
Ficus palmata	2.70	0.38	0.35	163	42	52	0.68	5.20	0.07	0.27	0.37
Ficus roxburghii	2.60	0.50	0.34	139	123	32	0.58	7.90	0.09	0.33	0.49
Celtis australis	1.70	0.27	0.23	173	68	33	0.79	7.00	0.07	0.21	0.24
Bauhinia variegata	1.30	0.21	0.21	177	149	50	0.63	6.90	0.08	0.25	0.25
Quercus semicarpifolia	1.30	0.23	0.22	134	242	48	0.39	6.40	0.06	0.25	0.41
Quercus leucotricophora	1.70	0.27	0.25	121	402	37	0.49	8.00	0.09	0.29	0.38
Quercus floribunda	1.20	0.24	0.22	143	116	51	0.51	7.00	0.09	0.32	0.47
Quercus glauca	1.40	0.25	0.18	174	146	37	0.54	6.20	0.05	0.22	0.35
Alnus nepalensis	1.60	0.25	0.27	155	127	49	0.59	6.70	0.07	0.23	0.35

Source: Sahoo et al. (2016)

Table 12.4 Biochemical composition of important fodder trees of Himalayan region

	(% DM basis)						
Fodder trees	CP	CF	EE	Ash	NFE	NDF	ADF
Albizzia lebbek	17.59	33.10	3.88	8.83	36.62	52.68	46.28
Bauhinia variegata	15.25	24.36	3.42	6.72	50.25	59.81	46.59
Grewia oppositifolia	18.32	20.95	3.39	10.63	46.71	45.56	29.67
Celtis australis	18.08	18.10	2.19	16.43	45.20	37.21	27.12
Albizzia stipulata	10.35	32.75	3.86	7.71	45.33	68.37	33.29
Bambusa arundinacea	18.02	29.25	5.05	11.54	36.14	70.05	30.06
Morus alba	26.40	11.63	5.06	16.28	40.63	41.20	23.77
Zizyphus nummularia	14.99	16.48	2.20	10.70	55.63	70.90	59.63
Artocarpus lakoocha	14.45	21.10	2.40	12.13	49.92	54.48	38.72
Robinia pseudoacacia	17.55	18.02	4.28	4.09	43.94	51.68	34.20
Albizzia chinensis	13.41	25.16	2.70	4.50	–	60.20	37.80
Ficus lacor	17.61	42.07	1.91	12.30	–	60.40	53.70
Ficus glaberrima	17.11	27.40	1.24	9.50	–	45.70	37.90
Ficus roxburghii	10.90	26.31	2.34	14.70	–	42.10	34.60
Ficus semicordata	19.70	24.77	1.00	12.50	–	31.20	26.20
Leucaena leucocephala	22.20	16.60	4.88	8.40	50.45	53.00	37.90
Melia azedrach	21.60	13.91	3.82	9.84	51.26	45.80	32.80
Bauhinia purpurea	18.10	–	–	7.10	–	58.60	47.30
Quercus incana	16.47	18.29	5.71	3.87	55.66	40.70	23.35
Morus multicaulis	21.25	12.60	7.60	8.46	40.29	33.10	23.04
Salix viminalis	17.44	14.02	2.85	10.44	55.25	55.30	35.89
Ficus religiosa	13.48	23.18	3.47	15.32	44.66	43.53	33.73
Grewia optiva	25.56	13.05	3.26	8.40	49.69	21.66	13.70
Acacia nilotica	15.53	18.03	6.30	5.80	55.14	41.04	34.87
Salix alba	18.85	11.82	3.76	6.06	59.51	29.25	16.16
Albizia procera	13.10	25.23	3.29	6.43	51.95	54.08	38.69
Quercus leucotricophora	10.80	39.40	5.30	4.10	–	64.40	49.30
Ficus nemoralis	14.50	–	5.70	9.50	–	44.60	44.20
Ficus palmata	17.45	12.45	4.66	17.15	48.29	38.40	40.30
Quercus semicarpifolia	9.50	26.70	4.80	5.30	54.60	64.50	60.50
Salix tetrasperma	8.09	9.94	12.92	5.80	63.25	49.90	40.10
Terminalia bellerica	8.60	18.60	8.00	4.70	60.10	27.40	19.50
Toona cilliata	15.11	22.00	3.33	10.34	49.22	35.30	29.60
Acacia catechu	8.49	21.95	6.55	9.56	53.45	45.02	28.87
Melia composite	16.04	13.43	2.32	12.52	55.58	21.86	12.89
Morus serrata	14.97	13.58	4.37	15.35	51.77	27.59	17.77
Olea glanduliflora	10.50	19.02	2.87	8.15	59.46	60.59	29.29
Quercus glauca	10.60	41.11	5.25	7.46	36.12	61.13	43.79
Ouigenia oojeinensis	12.64	29.74	3.17	9.95	44.50	38.15	25.15

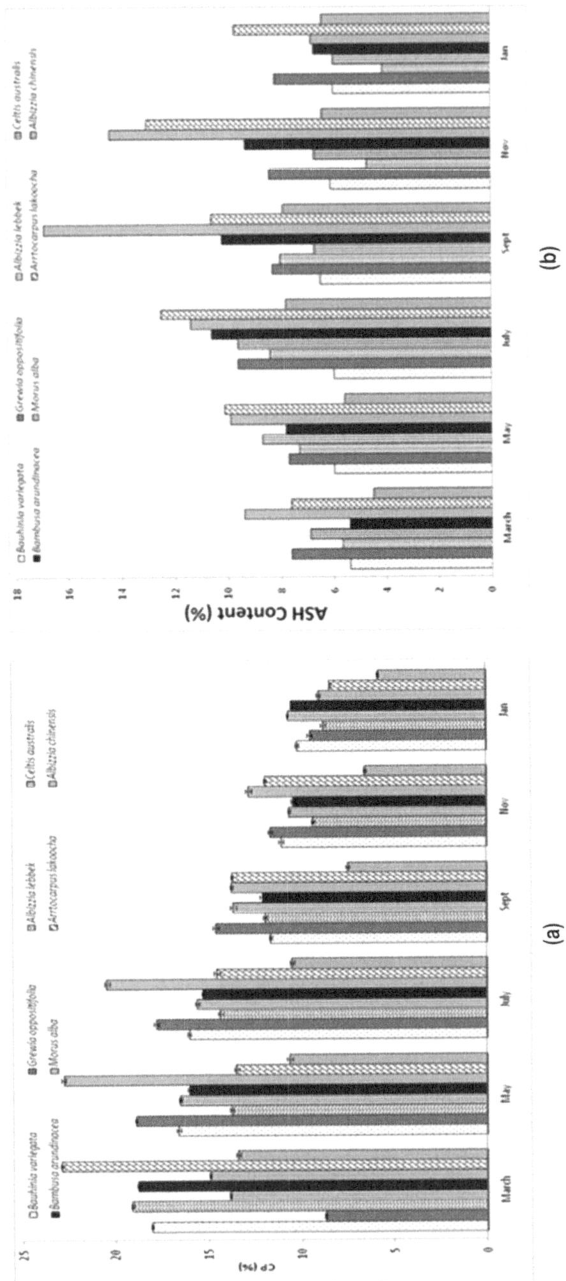

Fig. 12.4 (**a** and **b**) CP and Ash content in fodder trees of mid-Himalayan region (Katoch et al. 2017)

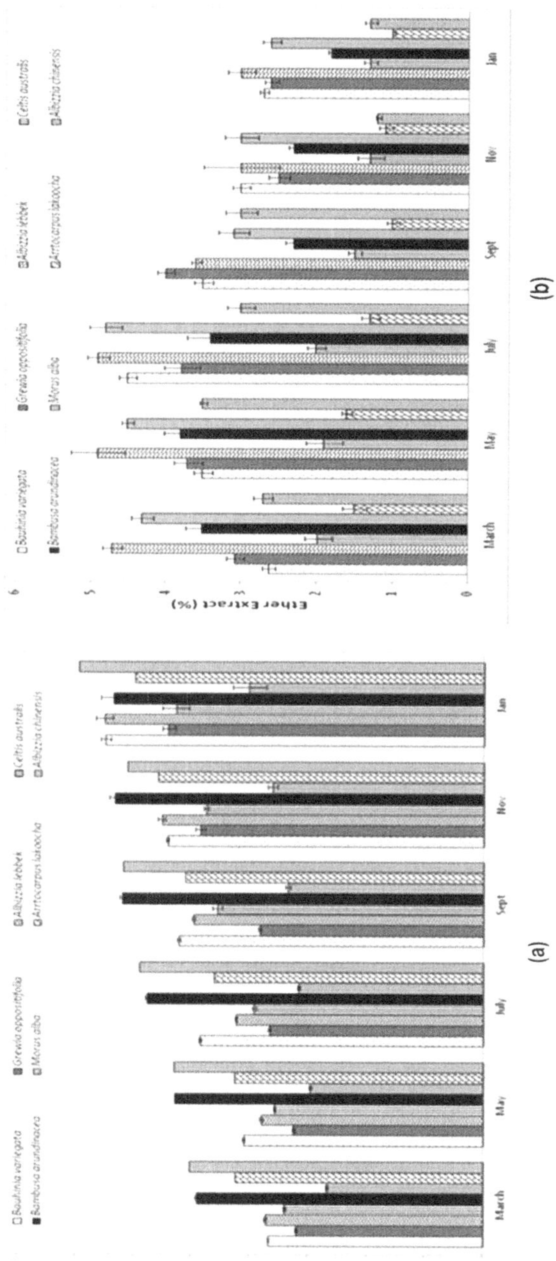

Fig. 12.5 (**a** and **b**) CF and Ash content in fodder trees of mid-Himalayan region (Katoch et al. 2017)

- CP content in *Bauhinia variegate* leaves varies from 10.20 to 18.08%. The foliage also has appreciable level of EE content (2.62–4.50%), ash content (5.40–6.50% and CF content (18.33–32.38%).
- *Arratocarpus lakoocha* foliage has CP content from 8.45 to 22.88%. The variation in EE, ash, and CF contents ranged from 1.0 to 1.60%, 7.60 to 13.00% and 21.16 to 29.83%, respectively.
- The foliage of *Albizzia chinensis* has 5.83–13.41% CP, 1.20–3.50% EE, 4.50–7.90% ash and 25.16–34.73% CF content. Variation in CP, EE, ash, and CF content in *Albizia lebbek* leaves ranged from 8.80 to 19.13%, 3.00 to 4.90%, 4.10 to 8.40% and 18.58 to 32.50%, respectively.
- *Morus alba* foliage has 9.04–22.77% CP and 2.60–4.30% EE, 13.30–20.03% CF content and 6.80–16.90% ash content.
- *Grewia* foliage has variation in CP and EE content from 9.50 to 18.92% and 2.50 to 4.00%, respectively. It also has 7.60–9.60% ash and 15.96–27.00% CF content. *Bambusa arundinacea* foliage revealed variation in CP and EE content from 10.40 to 18.81% and 1.80 to 3.80%, respectively. It also had an appreciable amount of ash and CF content 5.40–10.60% and 24.47–31.73%, respectively.
- Variation in EE, CP, ash, and CF content of *Celtis australis* foliage ranged from 1.30% to 2.00%, 10.64 to 16.53%, 6.00 to 6.90%, and 16.90 to 26.36%, respectively.
- CP content in tender, premature and mature *Leucaena* foliage varied from 25.70 to 34.85%, 25.33 to 32.64% and 21.84 to 28.65%, respectively. Total soluble carbohydrates content also revealed significant variations (45.24–48.51%, 39.83–41.72%, and 35.88–8.59%, respectively). Ash content varied seasonally from 6.45 to 7.94%, 7.39 to 9.14%, and 8.77 to 10.94%, in tender, premature and mature leaves, respectively.

12.4 Anti-nutritional Components of Fodder Trees

Although the nutritional composition of tree forage is of high quality, sometimes the presence of anti-nutrients in tree species affects their consumption and utilization in livestock feeding. Anti-nutritional factors are mainly those natural or synthetic compounds, which interfere with nutrients absorption and act to reduce nutrient intake, digestion and utilization and produce other adverse effects. They are also called toxic factors, aversive factors or incriminating factors. The presence of anti-nutrients in forages is the result of normal metabolism. At early growth stages, most of the anti-nutrients are present in higher concentrations in forages that may be due to the susceptibility of plant to different stresses (Pandey et al. 2011). Among the different categories of anti-nutrients, polyphenolic compounds particularly tannins are important anti-nutritional constituents in fodder trees (Davis 1981). Tannins reduce palatability due to astringency. They are of two types, hydrolyzable and condensed tannins. Hydrolyzable tannins are present in higher amounts particularly in leaves, pods, fruits, and galls of trees such as oak, chestnut and other species, whereas condensed tannins are more widespread in trees. A major characteristic of

Table 12.5 Anti-nutritional compounds in tree forage (Kumar 2003)

Sr. No	Anti-nutritional compounds	Tree foliage
1	**Polyphenolic compounds**	All vascular plants
	(A) Tannins	All vascular plants
	(B) Lignins	
2	**Nonprotein amino acids**	
	(A) Mimosine	*Leucaena leucocephala*
	(B) Cannavaine	*Canavalia ensiformis*
3	**Glycosides**	
	(A) Cyanogens	*Acacia giraffae, A. cunninghamii, A. sieberiana*
		Bambusa bambos,
	(B) Saponins	*Albizia stipulate, Bassia latifolia, Sesbania sesban, Agave sisalana*
4	**Phytohemagglutinins**	
	(A) Robin	*Bauhinia purpurea, Robinia pseudoacacia*
5	**Alkaloids**	
	(A) *N*-methyl-β-phenethylamine	*Acacia berlandieri*
	(B) Sesabine	*Sesbania vesicaria*
		Sesbania drummondii
		Sesbania punicea
6	**Oxalates**	*Acacia aneura*

tannins' is their ability to form complexes not only with proteins but also with other compounds. Tannins, because of their protein-binding properties are known to be highly astringent. Their content in plants is affected by species, genotype and growth stage and may vary with plant part, growth season and other specific environmental factors (Bhat et al. 2013). *Quercus semicarpifolia* foliage was observed with 10.71%, *Ficus palmata* with 0.92% while, *Bauhinia variegata* with 6.82% total tannins content. Apart from tannins, there are many other anti-nutrients, which play a crucial role in decreasing the feeding value of tree foliages. Hydrogen cyanide is possibly the most serious toxin occurring in *Acacia* spp. and this toxin is released when cyanogenic glycoside and hydrolyzing enzyme react in an animal's digestive tract (Goodchild and McMeniman 1986). *Albizzia stipulata* foliage has high saponins content (Kumar 2003) where *ad libitium* feeding of *Albizia* for 4–7 weeks results in toxic manifestation in sheep. The symptoms of saponins toxicity include listlessness, anorexia, weight loss and gastroenteristis while Robin is a lectin present in trees of *Robinia pseudoacacia* reported to cause poisoning in livestock. Major anti-nutritional components of fodder trees have been mentioned in Table 12.5.

Leucaena leucocephala has good nutritional profile but the presence of mimosine limits its use in livestock feeding in large proportions (Kumar 2003). In nonruminant animals, mimosine cause poor growth, loss of hair and wool, swollen and raw coronets above the hooves, lameness, mouth and esophegal lesions and goiter (Kumar 2003). Ramli et al. (2017) reported 2.03–4.89% mimosine in the dry matter

Table 12.6 Common anti-quality components of fodder trees

Scientific name	NDF (%)	ADF (%)	TT (%)	CT (%)	HT (%)
Grewia oppositifolia	52.30	44.40	1.80	0.05	1.75
Ficus nemoralis	43.00	48.70	3.33	2.83	0.50
Ficus palmata	38.40	40.30	0.92	0.02	0.90
Ficus roxburghii	47.50	53.50	3.88	2.87	1.01
Celtis australis	48.50	40.10	1.94	0.08	1.86
Bauhinia variegata	62.10	59.30	6.82	3.35	3.47
Quercus semicarpifolia	64.50	60.50	10.71	1.51	9.20
Quercus leucotricophora	66.20	60.10	8.08	3.05	5.03
Quercus floribunda	58.80	58.20	7.86	5.82	2.04
Quercus glauca	69.40	70.50	9.93	5.55	4.38
Alnus nepalensis	63.50	65.70	7.83	3.22	4.61

Source: Sahoo et al. (2016)

of *Leucaena* leaves, 0.68% in bark, 0.11% in xylem, 6–12% in the growing tips, 3–5% in young pods, 3.90–5% in seeds and 2% in green stems. Seasonal variations also affect the levels of anti-nutrients in foliage. Rana et al. (2006) and Amaral et al. (2010) observed that total phenols increase during summer and decrease during the winter season. Onyeonagu and Asiegbu (2013) also reported that tree foliage has high levels of alkaloids, phytate, saponins, and tannin content from July to October. The condensation of tannins during the winter season contributes to the plant defense system in order to avoid any damage from stress full environment (Salaj and Karmutak 1995).

Various researchers have conducted studies to investigate the anti-nutritional profile of fodder trees of the Himalayan region. Sahoo et al. (2016) reported that NDF and ADF content in fodder trees of temperate sub-Himalayan region from 38.40 to 69.40% and 40.10 to 70.50%, respectively. *Quercus glauca* has the highest NDF content (69.40%) followed by *Quercus leucotricophora (66.20%)* and *Quercus semicarpifolia* (64.50%). Among the fodder trees, *Ficus palmata* had the lowest NDF content (38.40%). The ADF content was also lowest in *Ficus palmata* foliage (40.30%) and highest (70.50%) in *Quercus glauca* foliage, respectively. Total tannins content was highest in *Quercus semicarpifolia* (10.71%) followed by *Quercus glauca* (9.93%) and *Quercus leucotricophora* (8.08%) and lowest in *Ficus palmata* (0.92%) (Table 12.6). Condensed tannins content varied from 0.02% in *Ficus palmate* foliage to 5.82% in *Quercus floribunda* foliage. Hydrolyzable tannins content was lowest in *Ficus nemoralis* (0.50%), followed by *Ficus roxburghii* (1.01%), *Grewia oppositifolia* (1.75%), *Celtis australis* (1.86%), and *Quercus floribunda* (2.04%). The tannins content was highest in *Quercus semicarpifolia* foliage (9.20%). The study revealed that tannins content of tree foliage were low to moderate as compared to arid and semi-arid region of India where the tree leaves were reported to have a high level of tannin than the temperate Himalayan region.

Our group investigated the variation in anti-nutritional profile of *Bauhinia variegate, Arratocarpus lakoocha, Albizia chinensis, Albizia lebbek, Morus alba, Grewia oppositifolia, Celtis australis,* and *Bambusa arundinacea* round the year for two consecutive seasons and inferences for individual fodder trees have been presented (Tables 12.7, 12.8, 12.9, 12.10 and 12.11) (Katoch 2019a, b).

- ***Bauhinia variegata***: The tree foliage has high total phenolic content (7.80%) during May, thereafter in subsequent months total phenolic content declined from 7.00 to 4.00%. Simple phenolic content ranged from 2.50 to 1.60% while condensed tannins content from 0.80 to 1.40%. High level of hydrolyzable tannins (4.10% and 3.60%, respectively) was observed during the months of May and July, respectively. In the case of NDF and ADF content, the variation was observed from 58.00 to 67.00% and 47.20 to 56.00%, respectively.
- ***Arratocarpus lakoocha:*** The foliage has a low levels of total phenols (7.30%) during the month of November, which subsequently increased in the month of January to March (9.20–9.30%). Net tannins were lowest during the months of July and November. The simple phenolic content ranged from 2.50 to 3.40% to while condensed tannins content and hydrolyzable tannins content were 1.90 to 2.60% and 3.90 to 4.10%, respectively.
- ***Albizia chinensis***: The variation in simple phenolic content in the foliage ranged from 2.50 to 4.70%. High level of total phenols was observed during the period between the months of May and July. Net tannins content ranged from 5.30 to 9.30%. Foliage of *Albizia* tree has high condensed tannins content during the months of May and July (4.20 and 4.40%, respectively). The fiber content (NDF and ADF) varied from 60.20 to 69.60% and 37.80 to 51.00%, respectively. Level of hydrolyzable tannins ranged from 2.20 to 5.10%.
- ***Albizia lebbek***: The foliage has higher levels of net tannins during the month of May (7.50%), thereafter a reduction in the net tannin content was observed during the month of November (5.00%). From the month of May to July, total phenolic content was high. The decrease in total phenolic content was observed during November (8.60%). High levels of simple phenolics were observed in the months of May and July. The foliage had high condensed tannins content (2.80 and 2.50%) during the months of May and July, respectively, however low level of condensed tannins were observed in the month of March (1.30%). During the months of May and March, the level of hydrolyzable tannins was high as compared with other months (4.70 and 4.40%, respectively).
- ***Morus alba***: The foliage has higher levels of total phenolic content in the months from May to July. The low levels of total phenolics were observed during the month of January (6.30%). The foliage has high level of simple phenols (3.50%) during the month of July. However, reduction in simple phenolic content was observed during the month of January (1.70%). Higher net tannins content was observed during the months of May to July. However, a consistent reduction in tannin content was observed up to the month of January. During the months of May and July (3.20 and 3.50%) the condensed tannins were higher; however, the content decreased during the month of November and January. Higher

Table 12.7 Variation in total phenols in the foliage from common fodder trees of mid Himalayas

Sr. No.	Name	May	July	September	November	January	March	Mean	SE
1.	*Bauhinia variegate*	7.8^{a}	7.0^{d}	4.3^{bd}	4.0^{bd}	5.3	6.9	5.88	0.64
2.	*Grewia oppositifolia*	**4.8^{ad}**	**5.6^{ad}**	**3.2^{bd}**	**3.1^{bd}**	**3.7^{d}**	**4.1^{d}**	4.08	0.40
3.	*Albizzia lebbek*	10.3^{a}	9.2^{a}	7.3	7.0^{b}	7.5	8.2	8.25	0.52
4.	*Celtis australis*	8.4^{a}	9.6^{a}	5.3	5.3	5.2^{b}	6.3	6.68	0.77
5.	*Bambusa arundinacea*	6.4^{d}	7.1^{ad}	5.0^{b}	5.1^{b}	5.7	7.0^{a}	6.05	0.38
6.	*Morus alba*	9.7^{a}	10.3^{a}	7.2	7.0	6.3^{b}	7.9	8.07	0.65
7.	*Arrtocarpus lakoocha*	8.9	8.0	8.1^{c}	7.3^{bc}	**9.2^{ac}**	9.3^{ac}	8.47	0.32
8.	*Albizzia chinensis*	**13.6^{ac}**	**13.9^{ac}**	**9.2^{c}**	**8.6^{bc}**	9.1^{c}	**10.3^{c}**	10.78	0.97
	Mean	8.74	8.84	6.20	5.93	6.50	7.50		
	SE (m) ±	0.94	0.91	0.73	0.66	0.69	0.67		

*Superscripts "a" and "b" represent significantly higher and lower levels of total phenols during different months, whereas, "c" and "d" represent significantly higher and lower levels of total phenols among trees during a particular month

Table 12.8 Variation in simple phenols in the foliage from common fodder trees of mid Himalayas

Sr. No.	Name	May	July	September	November	January	March	Mean	**SE**
1.	*Bauhinia variegate*	2.5^{a}	2.3^{d}	1.9	1.6^{bd}	2.3	2.3	2.15	0.14
2.	*Grewia oppositifolia*	1.9^{d}	2.8^{a}	1.6^{d}	1.7	1.5^{bd}	2.0	1.92	0.19
3.	*Albizzia lebbek*	2.8^{a}	2.6^{a}	2.0^{b}	2.0^{b}	2.0^{b}	2.5	2.32	0.15
4.	*Celtis australis*	2.3^{a}	2.5^{a}	1.8	1.9	1.5^{bd}	1.9^{d}	1.98	0.15
5.	*Bambusa arundinacea*	**1.8^{ad}**	**1.9^{ad}**	**1.1^{bd}**	**1.2^{bd}**	**1.5^{d}**	**1.7^{d}**	1.53	0.13
6.	*Morus alba*	2.3	3.5^{ac}	2.6	2.5^{c}	1.7^{b}	2.1	2.45	0.25
7.	*Arrtocarpus lakoocha*	3.4^{ac}	3.1	2.9^{c}	2.5^{bc}	3.0^{c}	3.0^{c}	2.98	0.12
8.	*Albizzia chinensis*	**4.3^{c}**	**4.7^{ac}**	**3.9^{c}**	**2.7^{bc}**	**2.5^{bc}**	**3.9^{c}**	3.67	0.36
	Mean	2.66	2.93	2.23	2.01	2.00	2.43		
	SE (m) ±	0.29	0.31	0.31	0.18	0.20	0.25		

*Superscripts "a" and "b" represent significantly higher and lower levels of simple phenols during different months, whereas, "c" and "d" represent significantly higher and lower levels of simple phenols among trees during the particular month

Table 12.9 Variation in net tannins in the foliage from common fodder trees of mid Himalayas

Sr. No.	Name	May	July	September	November	January	March	Mean	SE
1.	*Bauhinia variegate*	5.3^{a}	4.7^{a}	2.4^{bd}	2.4^{bd}	3.0^{d}	4.6	3.73	0.52
2.	*Grewia oppositifolia*	$\mathbf{2.9^{ad}}$	$\mathbf{2.8^{ad}}$	$\mathbf{1.6^{bd}}$	$\mathbf{1.4^{bd}}$	$\mathbf{2.2^{d}}$	$\mathbf{2.1^{d}}$	2.17	0.25
3.	*Albizzia lebbek*	7.5^{ac}	6.6	5.3^{c}	5.0^{bc}	5.5^{c}	5.7	5.93	0.38
4.	*Celtis australis*	6.1^{a}	7.1^{a}	3.5^{b}	3.4^{bc}	3.7	4.4	4.70	0.63
5.	*Bambusa arundinacea*	4.6^{d}	5.2^{a}	3.9^{b}	3.9^{b}	4.2	5.3	4.52	0.25
6.	*Morus alba*	7.4^{ac}	6.8^{a}	4.6^{b}	4.5^{b}	4.6^{b}	5.8	5.62	0.51
7.	*Arrtocarpus lakoocha*	5.5	4.9^{b}	5.2^{c}	4.8^{b}	6.2^{ac}	6.3^{ac}	5.48	0.26
8.	*Albizzia chinensis*	$\mathbf{9.3^{ac}}$	$\mathbf{9.2^{ac}}$	$\mathbf{5.3^{bc}}$	$\mathbf{5.9^{c}}$	$\mathbf{6.6^{c}}$	$\mathbf{6.4^{c}}$	7.12	0.70
	Mean	6.08	5.91	3.98	3.91	4.50	5.08		
	SE (m) ±	0.70	0.68	0.50	0.52	0.54	0.50		

*Superscripts "a" and "b" represent significantly higher and lower levels of condensed tannin during different months, whereas, "c" and "d" represent significantly higher and lower levels of condensed tannin among trees during a particular month

Table 12.10 Variation in condensed tannins in the foliage from common fodder trees of mid Himalayas

Sr. No.	Name	May	July	September	November	January	March	Mean	SE
1.	*Bauhinia variegate*	**1.2^{d}**	**1.1^{d}**	**1.0^{d}**	**1.0^{d}**	**0.8^{bd}**	**1.4^{ad}**	1.08	0.08
2.	*Grewia oppositifolia*	1.7^{ad}	1.5^{ad}	0.9^{d}	**0.7^{bd}**	0.9^{d}	1.0^{d}	1.12	0.16
3.	*Albizzia lebbek*	2.8^{a}	2.5^{a}	2.0^{c}	1.9c	2.0^{c}	1.3^{bd}	2.08	0.21
4.	*Celtis australis*	3.6^{ac}	3.7^{ac}	2.3^{c}	2.1^{c}	1.6^{b}	1.9	2.53	0.37
5.	*Bambusa arundinacea*	1.8^{d}	1.9	1.7^{b}	1.7^{b}	1.9	2.1^{a}	1.85	0.06
6.	*Morus alba*	3.2^{a}	3.5^{ac}	1.3	1.2^{b}	1.1^{bd}	1.7	2.00	0.44
7.	*Arrtocarpus lakoocha*	2.6^{a}	2.4	1.9^{b}	1.9^{bc}	**2.3^{c}**	2.2^{c}	2.22	0.11
8.	*Albizzia chinensis*	**4.2^{ac}**	**4.4^{ac}**	**2.1^{c}**	**1.9^{bc}**	**1.9^{b}**	**2.6^{c}**	2.85	0.47
	Mean	2.64	2.63	1.65	1.55	1.56	1.78		
	SE (m) ±	0.36	0.41	0.19	0.18	0.20	0.19		

*Superscripts "a" and "b" represent significantly higher and lower levels of condensed tannin during different months, whereas, "c" and "d" represent significantly higher and lower levels of condensed tannin among trees during particular month

Table 12.11 Variation in hydrolyzable tannin in the foliage from common fodder trees of mid Himalayas

Sr. No.	Name	May	July	September	November	January	March	Mean	SE
1.	*Bauhinia variegate*	4.1^{a}	3.6^{a}	1.4^{bd}	1.4^{bd}	2.2	3.2	2.65	0.47
2.	*Grewia oppositifolia*	$\mathbf{1.2^{d}}$	$\mathbf{1.3^{ad}}$	$\mathbf{0.7^{bd}}$	$\mathbf{0.7^{bd}}$	$\mathbf{1.3^{ad}}$	$\mathbf{1.1^{d}}$	1.05	0.11
3.	*Albizzia lebbek*	4.7^{ac}	4.1^{c}	$\mathbf{3.3^{bc}}$	3.1^{bc}	3.5	4.4^{ac}	3.85	0.26
4.	*Celtis australis*	2.5	3.4^{a}	1.2^{bd}	1.3^{bd}	2.1^{d}	2.5	2.17	0.34
5.	*Bambusa arundinacea*	2.8	3.3^{a}	2.2^{b}	2.2^{b}	2.3	3.2^{a}	2.67	0.21
6.	*Morus alba*	4.2^{a}	3.3^{b}	3.3^{bc}	3.3^{bc}	3.5	4.1^{ac}	3.62	0.17
7.	*Arrtocarpus lakoocha*	2.9	2.5^{b}	3.3^{c}	2.9	3.9^{ac}	4.1^{ac}	3.27	0.26
8.	*Albizzia chinensis*	$\mathbf{5.1^{ac}}$	$\mathbf{4.8^{c}}$	$\mathbf{2.2^{b}}$	$\mathbf{4.0^{c}}$	$\mathbf{4.7^{c}}$	**3.8**	4.10	0.43
	Mean	3.44	3.29	2.20	2.36	2.94	3.30		
	SE (m) ±	0.46	0.37	0.37	0.41	0.40	0.38		

*Superscripts "a" and "b" represent significantly higher and lower levels of hydrolyzable tannin during different months, whereas, "c" and "d" represent significantly higher and lower levels of hydrolyzable tannins among trees during particular month

hydrolyzable tannins were observed during May and March (4.20 and 4.10%, respectively), while low levels were observed from the month of July to November.

- ***Grewia oppositifolia***: The foliage exhibited a range of variations in simple phenolic content from 1.50 to 2.80%. Higher net tannins content was observed during the months of May to July whereas their level reduced in the month of September to November. The foliage harvested during May and July has higher condensed tannins content (1.70 and 1.50%, respectively). Subsequently, there was a reduction in the content during month of November (0.70%). The foliage revealed variation in hydrolyzable tannins content ranging from 0.70 to 1.30%. Fiber content (NDF and ADF) varied from 55.00 to 62.60% and 38.00 to 51.20%, respectively.
- ***Celtis australis***: The total phenolic content was 5.20% during the month of January. Higher levels of net tannins were observed from May to July (3.60 and 3.70%, respectively). The foliage also has high simple phenolic content during May and July. However, reduction in simple phenolic content was observed during the month of January (1.50%). The foliage exhibited high amount of hydrolyzable tannins (3.40%) during the month of July, however a reduction in hydrolyzable tannins content was recorded in September and November (1.20 and 1.40%, respectively).
- ***Bambusa arundinacea***: High levels of total phenolic content were observed during the months of March and July (7.0 and 7.10%, respectively). However, there was a reduction in content during the months of September and November (5.0 and 5.10%, respectively). Tree foliage has high simple phenolic content during the months of May and July. Thereafter, reduction in simple phenolic content was observed from September to November. During the month of July, the foliage has high net tannins content of 5.20% which reduced to 3.90% during the months of September and November. Highest condensed tannins levels were observed during the month of March (2.10%), however, during September and November, there was reduction in their content. Foliage has high hydrolyzable tannins content during July and March (3.30 and 3.20%, respectively), whereas low level (2.20%) of hydrolyzable tannins were observed during the months of September and November.
- ***Leucaena leucocephala***: The NDF content of tender, premature and mature *Leucaena* leaves ranged from 28.75 to 33.89%, 32.47 to 37.36%, and 36.91 to 41.51%, respectively, whereas ADF content ranged from 19.90 to 23.73%, 23.59 to 26.11% and 28.46 to 34.67%, respectively. In tender and premature leaves, the highest NDF and ADF contents were recorded during the month of June whereas the lower values were there in the month of January. In mature leaves, the NDF and ADF content were lower during the month of February and highest during the month of June. In tender and premature leaves, the ADL content ranged from 3.58 to 6.75% and 4.21 to 8.23%, respectively. Highest ADL content was observed during January and the lowest during the month of June. In mature leaves, the ADL content varied from 8.30 to 11.20%, where highest value was observed during December and lowest during the month of June. Hemicellulose and

cellulose content revealed a range of variation from 8.18 to 10.21%, 15.71 to 17.38%, respectively, in tender leaves, 8.74 to 11.97%, 16.04 to 17.88%, respectively, in premature leaves, and 6.38 to 8.45%, 18.81 to 23.51%, respectively, in mature leaves. There was a direct relationship between lignin, cellulose, and hemicellulose content and the maturity of *Leucaena* foliage.

- Variation in total phenolic content of tender, premature and mature leaves of *Leucaena* ranged from 12.15 to 13.60%, 10.35 to 12.69%, and 10.19 to 12.45%. In tender and premature leaves, total and simple phenolic contents were recorded highest during November while in mature leaves; the contents were high during June and November, respectively. Total tannins content in tender, premature and mature *Leucaena* leaves varied from 7.33 to 8.41%, 4.85 to 7.37%, 5.23 to 7.21%, respectively. The lowest tannins content in tender leaves was observed during the month of December whereas in premature and mature leaves during August and September, respectively. High polyphenolic content in tender leaves could be explained by the fact that any damage at this stage is more destructive than the later stages of growth, therefore the high amount of polyphenolic compounds in tender leaves seems to be an adaptive process to avoid any damage by the harsh winter season. The mimosine content in tender, premature and mature leaves varied from 1.87 to 3.61%, 1.53 to 3.48%, and 1.47 to 2.84%, respectively. In *Leaucaena* foliage, mimosine content was highest during the winter season (November to February) while minimum during the rainy season (June to September). Tender leaves had the highest mimosine content followed by premature and mature leaves.

12.5 Forages Availability During Lean Period

Green forages are backbone of livestock farming as they play a vital role in transforming livestock production into a profitable business. In Himalayan region, livestock farming is an important subsidiary occupation of farmers. It contributes to the economy of the region by providing milk, meat, wool, etc. Comparatively, the low productivity of the animals is ascribed chiefly to inadequate supply of feeding resources. Poor supply of nutrients to livestock during scarcity period remains a matter of concern. Both quantitatively and qualitatively, there exists a huge gap in the availability and supply of nutrients which is further compounded during the lean period. Inadequate supply of quality fodder has been identified as one of the reasons for poor livestock productivity in the Himalayan region.

Forages are a principal component in the diet of livestock and are also considered a cheap source of nutrient to livestock. During lean period of green fodder availability mainly from the month of October to March, farmer faces acute shortage of green forages for livestock feeding. In order to meet the feed as well as nutritional requirements of livestock, farmers depend on lean season forages. Dried roughages are an important feed resource to livestock during the period of scarcity of forages. These lean season feed resources can provide bulk to the belly and meet 70–80% supply of dry matter and total digestible nutrients (TDN) required by cattle, sheep, and goats. However, these resources are not nutritious and are often deficient in vital nutrients such as protein and total digestible nutrients and unable to meet even the maintenance requirement of livestock and thus lead to higher rates of nutrition-related health problems.

Tree foliage is an important lean season feed resource, which not only meets the nutritional requirement but also fulfills the feed requirement of the livestock. Tree foliage is rich source of protein and other nutrients which are vital from the perspective of livestock health and production. *Grewia oppositifolia, Leucaena leucocephala*, *Bambusa* spp., *Albizzia* spp., and *Robinia pseudoacacia* are important fodder trees in the Himalayan region. Bijalwan (2017) conducted a survey in Western Himalayan region to investigate practices of livestock feeding during lean period and observed that *Grewia optiva* and *Celtis australis* are important fodder trees. The crops residue mainly of *Eleusine coracana* (finger millet) and *Echinochloa frumentacea* (barnyard millet) are also used to feed livestock during lean period.

Apart from tree fodder, other fodder varieties have also been released which are tolerant to frost and drought and could mitigate to some extent the scarcity of green fodder during lean period of the year. Varieties S-92 (*Setaria*) PSS-1 (*Setaria*) and EC-178182 (Tall fescue) have been released for cultivation in Himachal, Uttrakhand, and Jammu Kashmir. These varieties have good nutritional profile, production potential and ability to resist drought and frost. These varieties are cultivated by the farmers for meeting green forage demand during feed scarcity in lean season.

The conservation of forages in the form of silage and hay also provides a mean of preserving forage. This also provides a mean of meeting the nutritional as well feed requirements of livestock during lean season. Silage consists of chopping green forage and storing them in a specially constructed underground or above-ground structure, ensuring that no air or moisture enters into it. Silage particularly corn silages is a major forage source for dairy animals. Maize silage contains about 25–35% dry matter, 5–10% protein content, and 15–27% fiber with highly variable starch content (18–37%).

Haymaking, the other method of forage preservation, consists of drying the harvested forage crops at an appropriate stage of growth to a safe level of around 15% moisture content without much bleaching, wetting, or shattering of leaves. The forage preserved in this manner is called "hay." Haylage sometimes also called low-moisture silage, hay crop silage, or drylage, can be defined as the hay put up at 40–60% moisture as compared with less than 15% for dried hay and 60–75% for silage. There is high energy output in high-temperature dehydration and therefore cannot be termed as economically feasible. Forage can be profitably preserved either as silage or hay for providing nutrients during scarcity. Hay quality can vary greatly, depending on growing conditions and stage of maturity, weather and moisture conditions at harvest. Factors that can affect nutritional value include plant species in the hay preparation, fertility of soil, harvesting methods (whether the hay was conditioned or crimped to dry faster and lose less leaves and nutrients during drying) and curing time. The nutritional value of hay is also related to the leaf mass. The

Fig. 12.6 Temporary structures for hay and straw storage in Himalayan region

grass hay has more nutrients and are more digestible when the plant was immature and growing, and more fiber when the plant has reached full growth. Legume leaves, by contrast, do not have the same structural function and do not change much as the plant grows, but the stems become coarser and more fibrous. Alfalfa stems, for example, are woody, serving as structural support for the plant. Leaf-to stem ratio is the most important criterion in judging nutrient quality in an alfalfa plant. The digestibility, palatability and nutrient values are highest when the plant is young with more leaves and less stems. Early bloom alfalfa (cut before the blossoms open) has about 18% crude protein, compared with 9.80% for early bloom timothy (before seed heads fill), 11.40% for early bloom orchard grass and lower levels for most other grasses. Alfalfa cut at full bloom has 15.50% crude protein, compared to 6.90% for late bloom timothy and 7.60% for late bloom orchard grass. Thus, legume hay when harvested at an early stage is more appropriate to meet the protein and mineral needs of a young growing, pregnant or lactating animals (Fig. 12.6).

Fodder trees provide an alternative feed resource to overcome fodder scarcity due to their less susceptibility to sudden climatic changes and longer life spans than conventional forages. These trees develop new sprouts and leaves, flowers, fruits and pods which are more nutritious than the foliage. In addition to fodder, these trees are also important for the sustainability of different agroforestry systems. Fodder trees have a better nutritive profile and the potential to provide quality feed to the livestock particularly during off-seasons. Tree fodders species also have varying levels of anti-nutritional compounds which may affect livestock health. *Morus alba* superior nutritive value with less anti-nutritional compounds and is considered a good quality tree fodder. *Grewia* spp. also complement the nutritional requirements of the livestock during lean periods. Growing fodder trees with good management practices could help farmers in providing quality feedstuff to their livestock particularly during lean season of forage availability. However, in existing farming systems, additional interventions like the introduction of promising local fodder tree species, supply of best planting material suitable to particular niche areas and facilitation of knowledge-intensive practices are required for more sustainability.

References

Amaral JV, Valentao P, Andrade P, Martins R (2010) Phenolic composition of hazelnut leaves: influence of cultivar, geographical origin and ripening stage. Sci Hortic 126:306–313

Bhardwaj DR, Sharma P, Bishist R, Navale MR, Kaushal R (2018) Nutritive value of introduced bamboo species in the northwestern Himalayas, India. J Forest Res 30:1–2. https://doi.org/10.1007/s11676-018-0750-2

Bhat TK, Kannan A, Singh B, Sharma OP (2013) Value addition of feed and fodder by alleviating the antinutritional effects of tannins. Agric Res 2:189–206

Bijalwan A (2017) Indigenous agroforestry as a source of fodder in the Western Himalaya, India. Int J Curr Res Biosci Plant Biol 4:118–122

Davis AM (1981) The oxalate, tannin, crude fiber and crude protein composition of young plants of some *Atriplex* species. J Range Manag 34:329–331

Dhungana S, Tripathee HP, Puri L, Timilsina YP, Devkota KP (2012) Nutritional analysis of locally preferred fodder trees of middle hills of Nepal: a case study from Hemja VDC, Kaski District. Nepal J Sci Technol 13:39–44

Goodchild AV, McMeniman NP (1986) Nutritive value of *Acacia* foliage and pods for animal feeding. In: Turnbull JW (ed) Australian acacias in developing countries. ACIAR, Canberra, pp 1–196

Katoch R (2009) Tree fodder: an alternative source of quality fodder in Himachal Pradesh. Range Manage Agroforest 30:16–24

Katoch R (2019a) Tree fodder for mitigating forage requirement in Himalayan region. Range Manage Agroforest 40(2):173–180

Katoch R (2019b) Seasonal alterations in anti-nutritional factors of common fodder trees of mid Himalayas. Range Manage Agroforest 40:143–149

Katoch R, Singh SK, Tripathi A, Kumar N (2017) Effect of seasonal variation in biochemical composition of leaves of fodder trees prevalent in the mid-hill region of Himachal Pradesh. Range Manage Agroforest 38:234–240

Khatta V, Katoch BS (1983) Nutrients composition of some fodder tree leaves available in submountainous region of Himachal Pradesh. Indian Forest 109:17–24

Khosla PK, Toky OP, Bisht RP, Hamidullah S (1992) Leaf dynamics and protein content of six important fodder trees of the western Himalaya. Agrofor Syst 19:109–118

Kumar R (2003) Anti-nutritive factors, the potential risks of toxicity and methods to alleviate them. http://www.fao.org/DOCREP/003/T0632E/T0632E10.html

Makkar HPS, Becker K (1998) Do tannins in leaves of trees and shrubs from African and Himalayan regions differ in the level and activity? Agrofor Syst 40:59–68

Onyeonagu CC, Asiegbu JE (2013) Harvest frequency effect on plant height, grass tiller production, plant cover and percentage dry matter production of some forage grasses and legumes in the derived savannah. Nigeria Afric J Agric Res 8:608–617

Pal RN, Dogra KK, Singh LN, Negi SS (1979) Chemical composition of some fodder trees in Himachal Pradesh. Forage Res 5:109–115

Pandey S, Meena BS, Sharma P, Dwivedi RN (2011) Gender involvement in decision making of on farm and off farm activities. J Comm Mobil Sustain Dev 6:42–45

Ramli N, Jamaludin, Ilham AA (2017) Mimosine toxicity in leucaena biomass: a hurdle impeding maximum use for bioproducts and bioenergy. Int J Environ Sci Nat Resour 6:555700. https://doi.org/10.19080/IJESNR.2017.06.555700

Rana RS, Yano F, Khanal SK, Pandey SB (1999) Crude protein and mineral content of some major fodder trees of Nepal. Lumle Seminar paper No. 99/13, Lumle Agriculture Research Center, Nepal

Rana KK, Wadhwa M, Bakshi MPS (2006) Seasonal variations in tannin profile of tree leaves. Asian-Austral J Anim Sci 19:1134–1138

Sahoo B, Garg AK, Mohanta RK, Bhar R, Thirumurgan P, Sharma AK, Pandey AB (2016) Nutritional value and tannin profile of forest foliages in temperate sub-Himalayas. Range Manage Agroforest 37:228–232

Salaj J, Karmutak A (1995) Structural change in mesophyll cells of *Abies alba* Mill. during the autumn spring period. Biol Bratislava 50:93–98

Sharma DD, Gill RS, Chander S, Negi SS (1966) Chemical composition of some fodder tree leaves in Kangra district. J Res 3:438–442

Singh B, Makkar HPS, Negi SS (1989) Rate and extent of digestion and potentially digestible dry matter and cell walls of various tree leaves. J Dairy Sci 72:3233–3239

Prospects of Nonconventional Feed Resources of Himalayan Region 13

Livestock rearing plays a significant role in agricultural sector. Both livestock and agricultural production are interdependent components of the farming system in the country. In the Himalayan region, the farmers are still conserving feed resources and continuing with conventional practices of livestock feeding. However, the ample availability of feed resources has become challenging in the present phase of urbanization and increased anthropogenic activities in the region. Therefore, the limited availabilty of green fodder, continuous rise in prices of feeds available in markets and overdependence on low-quality feeds are emerging as major concerns in achieving sustainable livestock production not only in the Indian Himalayan region but also in other countries under similar topographies. A viable strategy for sustainable livestock productivity is the efficient use of available feed resources including enlargement of the feed resource base through a quest for novel feed resources, particularly those not competing with human food (Katoch et al. 2018a). Therefore, it is high time to look for alternative feeding resources for supplementation of available resources for meeting the nutritional as well as feed requirements of livestock. Under these circumstances, the nonconventional feed resources (NCFR) is a viable option for filling the gap between the supply and demand of animal feeds and nutritional sufficiency.

13.1 Nonconventional Feed Resources in Livestock Feeding

Since the gap between the demand and supply of conventional feed resources for livestock feeding is continuously widening, the utilization of nonconventional feed resources could play a crucial role in supplementing the available forage resources. The nonconventional feed resources (NCFR) refer to all those feeds that have not been traditionally used in animal feeding and or are not normally used in commercially produced feeds for livestock (Amata 2014). Therefore, NCFR could be more appropriately termed as 'new feeds'. The feasibility of using alternative feeds for

R. Katoch, *Nutritional Quality Management of Forages in the Himalayan Region*, https://doi.org/10.1007/978-981-16-5437-4_13

ruminants depends on the feed value of novel feeds, animal production responses and feed costs compared with conventional feeds. In addition, the environmental footprint of feed and animal production and the economic value of novel feeds in alternative uses such as energy production are of great importance. Various feed resources have been included in the category of NCFR and their numbers are increasing gradually. Even though many resources under the category of NCFR are available, the farmer perceptions on NCFR vary from country to country and region to region. Sometimes, it is also difficult to distinguish between traditional feeds and NCFR. Therefore, identification of available NCFR is necessary to exploit their potential in order to bring changes in livestock feed programs.

Features of NCFR

- Have low economic value
- Noncompetitive in terms of human consumption
- Good source of fermentable nutrients
- Biodiversity conservator
- Compatible with farming system
- End products of various production processes and consumption

13.2 Sources of NCFR

13.2.1 Fruit and Vegetable Waste

Annually, huge amount of fruit and vegetable waste and by-products from processing units are generated throughout the world. In India, fruit and vegetable processing, packing, distribution and consumption in the organized sector generate approximately 1.81 tonnes of fruit and vegetable wastes. For example, every year there are about 9% losses in total cabbage and cauliflower production whereas, in the state of Himachal Pradesh, India, approximately 18.50% losses are there in cabbage and cauliflower production in form of vegetable waste. Fruit and vegetable waste (FVW) is either composted or dumped in landfills or rivers causing environmental hazards. Alternatives to such disposal methods could be recycling through livestock by using them as feed resources. Cauliflower (*Brassica oleracea var. botrytis*) stem, leaves, and discarded curds, cabbage (*Brassica oleracea var. capitata*) leaves, sugar beet leaves (*Beta vulgaris*), black chickpea (*Cicer arietinum*), pea (*Pisum sativum* var. *arvense*) vines, radish leaves (*Raphanus sativus*), summer squash vines (*Citrullus vulgaris*), carrot leaves, carrot waste, potato haulms, groundnut haulms, fruit and vegetable peas are important by-products, which can be used as a feed resource in livestock feeding or they can be used as a potential supplement of mix ration feed.

13.2.2 Agro-Industrial by-Products (AIBP)

The squeezing of natural pastures and grasslands has resulted in partial dependence on crop residues. The increasing human demands for different types of foods has led to a increase in area for cultivation of respective crops. By-products derived in the industry after processing the main products are referred to as 'agro-industrial by-products'. In comparison to crop residues, agro-industrial by-products are less fibrous and more concentrated and have a higher nutrient content. The use of agro-industrial by-products relies on their chemical and physical properties, which influence production system outputs. In developing countries, grain, which forms the bulk of concentrate feeds is in short supply and expensive due to direct competition with human food uses. Bagasse that refers to the remaining residue after the juice extraction from sugarcane constitutes approximately 15–20% of sugarcane tops. Bagasse has principle use as a fuel but its bulky and fibrous nature makes it a valuable feed source. It has been estimated that digestibility of bagasse-based diets supplemented with molasses and copra meal has high dry matter intake by large ruminants (Roxas et al. 1969).

13.2.3 Wood-Derived Products

Wood is the most abundant source of carbohydrates where cellulose (400–450 g/kg DM) and hemicellulose (200–300 g/kg DM) are principle components of wood carbohydrates (Halmemies-Beauchet-Filleau et al. 2018). Technologies have been developed to break the link between the lignin and cell wall carbohydrates and improve the digestibility of wood-derived ligno-cellulose materials. The residue generated from pulping and papermaking may contain different proportions of hemicellulose and/or cellulose with or without lignin. Wood-derived cellulose can be used as a feed for livestock. Hemicelluloses in a liquid form are often called wood molasses or wood sugar concentrates. They have successfully been used as diet components for ruminants at up to 10% of DM intake (Herrick et al. 2012). Wood-derived feeds may also provide a valuable feed source during scarcity of other feeds. They may fit best into the diet of animals having low-energy requirements (Halmemies-Beauchet-Filleau et al. 2018).

13.2.4 Weeds

Weeds are ubiquitous in the world and invade pastures and annual or perennial crops grown for livestock feed. They are also the source of fodder but may reduce the quantity and quality of harvested hay or grazed forage and could be toxic or poisonous to livestock or cause mouth injury to grazing animals. Many grass weed species are readily eaten and provide quality nutrition to grazing livestock. The grazing of weed species could be an important biological control strategy as part of an integrated weed management plan for some invasive plants. It is therefore necessary to know the potential quality of individual weed species for making their management decisions. Some of the species around the bunds, in the fields, home gardens, orchards, etc. may be utilized in feeding livestock with due knowledge and precautions. Some weeds are toxic or poisonous to livestock and other may be unpalatable. Several weed species have thorns or spines that could injure the grazing animal's mouth and/or irritate their eyes. Weeds like *Rumex nepalensis*, *R. hastitus*, *R. opptusifolius*, *Chenopodium* spp., *Amaranthus* spp., Chicory, *Taraxacum officinale* (dandelion), *Echinochloa* spp., *Setaria glauca,* etc. has the potential for livestock feeding involving a few management practices.

13.3 Nutritive Value of NCFR

13.3.1 Fruits and Vegetable Wastes

Fruits and vegetable wastes are the most common wet waste. In villages and even in urban areas, wet wastes are generally used in animal feeding. Most of the fruit and vegetable wastes contain more than 7.00% protein which is adequate for the maintenance requirement of animals. Vegetable waste contains more protein than fruit waste. Invariably, the fat contents are more in fruit and vegetable wastes than the cultivated cereal fodders. Fiber contents (NDF, ADF and cellulose) of fruit and vegetable wastes are low than the straws, stovers and mature grasses.

Negesse et al. (2009) studied the nutritive value of various NCFR. The results of their study revealed that papaya and onion peels has high CP (24.10% and 15.90%, respectively) and cabbage leftover had medium levels of CP (13.70%). Except for avocado peel, all the kitchen waste had lower than 50 g/kg DM of EE. With the exception of pineapple peel, all others have lower than 300 g/kg DM of NDF. Most vegetable waste, except potato, pineapple and banana have almost similar NDF and ADF values. The kitchen wastes has high ME (9.30–13.60 MJ/kg DM) and OMD (683–810 g/kg DM) with only very low value of OMD in avocado peel (365 g/kg DM). Kitchen wastes contain a large amount of anti-nutritional compounds like tannins, phytic acid, saponins, and total phenols. Avocado peel contains a high amount of lignin (12.50%), followed by banana (7.70%) and mango peels (4.70%). Mango peel contains 15.10% whereas cabbage leftover contains 12.30% of total phenolics among the kitchen wastes. Orange, avocado and onion peels contain low levels of phenolic content (5.10%, 4.90%, and 4.50%, respectively). Banana peel and onion peel contain 2.40% and 2.00% phytic acid, respectively, whereas, potato and avocado peels contain their lower levels (0.60% and 0.90%, respectively) (Negesse et al. 2009). Tannins are the main anti-nutritional factors of bananas which are mostly present in the peels. Ripening causes those tannins to migrate to the pulp or for degradation by polyphenol oxidases and peroxidases. On the other hand, the dry matter contents of fruit and vegetable waste are quite low and similar to the early growth stage of cultivated fodder crops.

13.3.2 Crop by-Products as NCFR

India has a wide diversity in terms of crop species, where a large number of crops are cultivated for grain purposes. Among them, major contribution is from cereals and legumes. Cereal crops form the bulk of crop residues after harvesting the grains. Some part of crop residue is burnt in the fields, whereas other is processed to produce husks, brans and straws. Cereal husks and brans are the major NCFRs used in the Himalayan region, whereas, leguminous husks are used to a negligible extent. Paddy husk can be mixed with high-energy feed ingredients. Rice bran, an oil-rich by-product of rice, is considered a good source of energy. It also contains a considerable amount of CP (9.60%) and 65% of TDN. Besides providing quality proteins and energy, rice bran also has considerable amounts of vitamins B, E, and antioxidants. Rice husks contain 8–11% moisture, 15.60–22.60% ash, 14.50–17.50% acid insoluble ash, 2.90–3.60% CP, 39–42% CF, and 25–29% nitrogen-free extract. Sole feeding of rice bran may result in colic pain. Hence, rice bran should always be mixed with other components.

Wheat bran is a by-product of the dry milling of common wheat (*Triticum aestivum* L.) into flour and is one of the major agro-industrial by-products used in animal feeding. It consists of the outer layers (cuticle, pericarp, and seed coat) combined with small amounts of starchy endosperm of the wheat kernel. Wheat bran is relatively rich in protein (14–19% DM, sometimes higher) and minerals (4–7% DM), notably calcium (0.07–0.2% DM) and phosphorus (0.90–1.30% DM). It contains 45.20% NDF and 13.40% ADF, in which lignin content is around 3.80%. The energy that could be obtained from wheat bran is 18.90 MJ/kg/DM. Maize husk is also a by-product obtained after the milling of maize. It contains about 7.44% of crude protein with 88% of dry matter, 71.14% NDF and 49.15% ADF and 14.87% ADL content.

Pomace is the major by-product of apple processing units. Pomace consists of crushed flesh, stalks, peels, seeds, etc. (Rahmat et al. 1995). Apple pectin pulp is the residue after pectin extraction from apple pomace (Ajila et al. 2007).

Apple pomace is generally poor in protein (3–11% of DM) and rich in fiber and sugars. Crude fiber ranges from 18 to 50% NDF, 48 to 75% ADF, 36 to 67% and lignin from 16 to 35%. Apple pomace is high in fructose (14–35% of DM) (Kołodziejczyk et al. 2007). Apple pomace and pectin pulp (wet, dried or ensiled), are suitable feeds for animals. It is palatable to cattle and sheep; pectin pulp is less palatable to dairy cows than apple pomace. The addition of molasses increase the palatability of pectin pulps. Apple pomace is often mixed with alfalfa or corn for ensiling.

Pear pulp and pear molasses are also palatable to animals. Pear molasses has a lower nitrogen and ash content and higher total digestible nutrients than cane molasses. Pear molasses are more palatable than beet molasses. Pear pulp and pear molasses have 60 and 86% total digestible nutrients, respectively.

Tomato processing wastes can be divided into three categories according to the type of by-product that could be recovered. Cull tomato feed is made from cull tomatoes processed into tomato pulp or tomato pomace. A second category is peel residue, a waste product from canning whole tomatoes; about 12% of the original tomato is removed as peel and adhering pulp.

The third category is tomato pomace, the residue from the manufacture of juice, paste, puree and sauce. Pomace contains the skin, seeds, fiber and some adhering pulp. Tomato pomace contains 14.30% digestible CP and 65% TDN and thereby the dried and ground tomato pomace could be used in feed. Tomato pomace is relatively rich in protein (17–22% DM) and fat (10–15% DM). Fat content can even exceed 20% if the seed proportion is high (Battaglini and Costantini 1978). Fiber content is high: crude fiber is in the 33–57% DM range. NDF (50–72% DM) consists largely of ADF (39–60%). It is also a good source of vitamin A and B.

Citrus fruits are consumed as fresh fruit or processed juice. After juice extraction, remains mainly comprise residue peel (flavedo and albedo), pulp, rag, and seeds. These components, either individually or in various combinations are the material from which citrus feedstuffs are produced. Nutritional composition of citrus by-products is influenced by factors that include the source of the fruit and type of processing. Total ash, crude fat, crude fiber, crude protein, NFE, total sugar, lignin, and pectin ranges from 1.70 to 7.0%, 1.20 to 2.10%, 5.70 to 8.60%, 2.20 to 4.20%, 65 to 75%, 10.20 to 16.50%, 1.30 to 6.50% and 4.40 to 12.80%, respectively, for different types of citrus wastes (Ali et al. 2010).

Potato waste is a good source of energy for livestock having energy value similar to corn and barley while being low in protein and calcium. Potato waste meal contains 7.60% CP, 7.0% EE, 4% CF, and 4% ash (Hulfan et al. 1982).

Bagasse is highly fibrous by-product after sugar cane is crushed to remove sucrose. Many sugar milling factories around the world release large quantities of bagasse as a part of their by-products, some even dispose of it as waste. Table 13.1 presents the nutritional composition of bagasse. Molasses contains 20.60% moisture, 3.20% CP, 60.80% total sugar, 2.20% soluble gums, 8.20% ash and 5.00% free acids. The spent grain is the most abundant brewery by-product corresponding to approximately 85% of the total by-products generated. Brewer's spent grain is a lignocellulosic material containing about 17 cellulose, 28% noncellulosic polysaccharides, mainly arabinoxylans and 28% lignin.

13.3.3 Microalgae

Microalgae could be an alternative feed source. Microalgae are unicellular or simple multicellular microorganisms having good nutritive profile. Microalgae have high CP content, lipid and carbohydrates content. Studies have revealed that feeding

Table 13.1 Nutritional composition of bagasse

Parameters	
Dry matter	31.30–62.90%
CP	1.40–2.40%
CF	35.80–50.30%
NDF	72.60–91.90%
ADF	55.10–62.20%
Lignin	11.00–13.60%
EE	0.40–0.70%
Ash	2.70–10.60%
Minerals (g/kg DM)	
Calcium	0.30–7.10%
Phosphorus	3.30–0.20%
Potassium	0.60–3.30%
Sodium	0.10%
Magnesium	0.40–1.40%

Sources: (Basile and Machado 1990; Keir et al. 1997; Onwuka et al. 1997; Gowda et al. 2004; Rivero et al. 2004; Suksombat 2004)

microalgae could increase trans-11 18:1, cis-9, trans-11 18:2 and n-3 content in milk (Boeckaert et al. 2008). Microalgae or defatted microalgae residues high in CP (*Spirulina platensis* and *Chlorella vulgaris*) could substitute conventional protein (Lamminen et al. 2017) and energy feeds (Van Emon et al. 2015). Amino acid composition of microalgae is nearly comparable to soybean and rapeseed meal. Sometimes, the protein degradability of many microalgae species is higher than that of rapeseed (Lamminen et al. 2017), soya bean and cottonseed meals (Costa et al. 2016) and influenced by growing and harvesting conditions (Lodge-Ivey et al. 2014).

13.3.4 Weeds

Weeds generally cause nuisance due to their large-scale dissemination and fast growth rate. Studies have revealed that weeds with good nutritional quality have prospects of their use in livestock feeding. *Rumex hastatus*, *Rumex obtusifolius,* and *Rumex nepalensis* have considerable merits in forage quality parameters. *Berteroa incana, Helianthus tuberosus Taraxacum officinale*, *Silene alba* and *Sonchus arvensis* at their vegetative and bud stages have high/similar digestibility than alfalfa. *Amaranthus retroflexus* and *Ambrosia artemisiifolia* weeds have higher digestibility than alfalfa. *Chenopodium album*, *Setaria glauca* and *Echinochloa crusgalli* weeds have similar digestibility to alfalfa (Marten and Anderson 1975). The study conducted by Marten and Anderson (1975) revealed higher digestibility of common ragweed, velvetleaf (*Abutilon theophrasti*), redroot pigweed and barnyard grass in comparison to oats (*Avena sativa*). Common lambs quarters, Pennsylvania smartweed (*Polygonum pensylvanicum*), redroot pigweed, and common ragweed have similar or slightly lower NDF proportions than alfalfa (Temme et al. 1979). *Chenopodium album* grows naturally as a weed in fields of wheat, barley, mustard, gram and other crops. Chenopod leaves retain 26–64 g/kg protein, 78–190 mg/kg β-carotene, 0.50–2.40 g/kg vitamin C, 2.60–5.0 g/kg nitrate and 9–39 g/kg oxalate content. The protein quality of grains is similar to that of milk and contains high lysine (6 g/100g protein), methionine (2.30 g/100g protein) and cysteine (1.20 g/100g). Its green leaves are comparable to any leaf vegetable being rich in iron, β-carotene and low in oxalate content. Guerrero and Isasa (1997) reported high vitamin C content (155 mg/100 g) and carotenoids (12.50 mg/100 g) in the leaves of *Chenopodium album*. Digestibility of some weeds decreases rapidly than cultivated forages as the plant matures. However, winter annual henbit (*Lamium amplexicaule*) and winter or summer annual *Carolina geranium* (*Geranium carolinianum*) have high IVDMD at later maturity stages. The summer annual/ perennial weeds like coffee senna (*Cassia occidentalis*), prickly sida (*Sida spinosa*), crowfoot grass (*Dactyloctenium aegyptium*), Texas panicum (*Panicum texanum*), yellow foxtail, bermuda grass (*Cynodon* spp.) and pearl millet (*Pennisetum glaucum*) have around 16–18% CP (Temme et al. 1979) (Table 13.2). Marten et al. (1987) observed that smooth brome grass (*Bromus inermis*) and quack grass (*Elytrigia repens*) consistently has higher NDF content. Marten and Anderson (1975) reported that the grassy weeds including yellow foxtail, green foxtail (*Setaria viridis*), giant foxtail (*Setaria faberi*) and barnyard grass, along with oats have more cell wall constituents as compared to herbaceous weeds. Common lambs quarters, Pennsylvania smartweed

Table 13.2 Nutritional composition of *Rumex* species in mid-Himalayas

Content (%)	*Rumex hastatus*	*Rumex obtusifolious*	*Rumex nepalensis*
Crude protein	22.05 ± 1.21	31.05 ± 0.88	31.15 ± 0.35
NDF	43.53 ± 0.55	42.26 ± 0.23	41.00 ± 0.20
ADF	33.00 ± 0.60	29.20 ± 0.20	27.73 ± 0.31
Hemicellulose	10.53 ± 1.41	13.06 ± 0.12	13.26 ± 0.50
Cell content	56.46 ± 0.55	57.73 ± 6.38	59.00 ± 0.20
ADL	9.33 ± 0.42	11.40 ± 0.40	10.33 ± 0.31
Cellulose	23.66 ± 0.23	16.80 ± 2.12	17.40 ± 0.53
Total carbohydrates	24.32 ± 1.66	24.24 ± 0.07	39.75 ± 1.61
Ash	18.13 ± 0.47	10.26 ± 0.04	7.40 ± 0.10
Total phenols	7.46 ± 0.96	4.60 ± 0.75	3.43 ± 0.03
Simple phenols	3.35 ± 0.36	2.01 ± 0.28	1.40 ± 0.04
Total tannins	4.11 ± 0.62	2.59 ± 0.48	2.03 ± 0.06
Oxalates	5.50 ± 0.47	5.03 ± 0.57	3.60 ± 0.22
Organic matter digestibility	59.80 ± 0.40	55.04 ± 0.83	55.94 ± 0.40
Dry matter digestibility	63.25 ± 0.86	66.15 ± 0.15	66.96 ± 0.77
Metabolizable energy (MJ/kg DM)	8.78 ± 0.10	9.24 ± 0.025	9.38 ± 0.13
Dry matter intake (% b. wt)	2.83 ± 0.02	2.75 ± 0.03	2.92 ± 0.01

Data presented in Mean ± S.D.

(*Polygonum pensylvanicum*), redroot pigweed and common ragweed were reported to have NDF content similar to or slightly lower than alfalfa (Temme et al. 1979). ADF content of common lambs quarters, Pennsylvania smartweed, redroot pigweed and common ragweed have similar or lower ADF content than alfalfa (Table 13.2). Marten and Anderson (1975) observed that common ragweed has higher ADF content than alfalfa, but similar digestibility.

Redroot pigweed, common lambs quarters and Pennsylvania smartweed have lesser ADF content than alfalfa (Marten and Anderson 1975). Many species of genus *Rumex* are considered nuisance weeds, but some species have edible quality and are commonly used in soups and salads (*R. acetosa* L., *R. sanguineus* L., *R. patientia* L., and *R. scutatus* L. *Rumex acetosa*) (Tuazon-Nartea and Savage 2013). *Rumex hastatus, Rumex obtusifolius, and Rumex nepalensis* are important weeds of this genus which have the prospects for use a fodder. These *Rumax* species possess certain characteristics that are important from forage perspectives (Hameed and Dastagir 2009; Humaira et al. 2014; Abbasi et al. 2015). Makkar et al. (2016) reviewed the nutritive value of seaweed indicating that some species have the potential to contribute towards protein and energy needs of livestock (e.g., *Macrocystis pyrifera, Palmaria palmatata, Laminaria digitata, Ulva lactuca*), while others contain a number of bioactive compounds, which could be used as prebiotics for enhancing production and health status of animals (e.g. *Ascophyllum nodosum*). Some seaweed species have shown potential to mitigate ruminal *in vitro* methane production depending on the basal diet (Maia et al. 2016).

Katoch et al. (2018b) investigated the prospects of *Rumex hastatus, Rumex obtusifolius and Rumex nepalensis* weeds in livestock feeding. The weed species displayed marked variations in their crude protein content (22.05–31.15%). The highest level of crude protein content was observed for *Rumex nepalensis*. Highest ADF (33.0%) and NDF (43.53%) content were observed in leaves of *R. hastatus,* while the lowest ADF (27.73%) and NDF (41.0%) contents were in *R. nepal.* Comparitevely lower levels of fibre content (27.73 and 41.00%) was observed in *Rumex nepalensis*. Low levels of hemicellulose and ADL content were observed in *Rumex hastatus* (10.53 and 9.33%) whereas, higher levels were observed in the leaves of *Rumex nepalensis* (13.26 and 10.33%) *Rumex obtusifolius* (13.06 and 11.40%), respectively (Table 13.2). The leaves of *Rumex hastatus* had comparatively higher levels of cellulose content whereas lower level was observed in *Rumex obtusifolius* leaves. Total ash content in *Rumex* species varied from 7.40 to 18.13% of which highest ash content was observed in *Rumex hastatus* and lowest in *R. nepalensis* (7.40%). The level of total carbohydrates in *Rumex* species ranged from 39.75 to 24.24%. Highest value of total carbohydrates content was observed in *R. neplensis* while lowest in *R. obtusifolious.*

Variable levels of anti-nutrients like total phenols, simple phenols, total tannins and oxalates were recorded in all three *Rumex* species. Total phenolic content was highest in *R. hastatus* and lowest in *R. nepalensis* (7.46% and 3.43%, respectively). The simple phenolic content varied from 1.40% in *R. nepalensis* to 3.35% in *R. hastatus*. The level of total tannins was highest in *R hastatus* (4.11%) followed by *R. obtusifolius* (2.59%) and *R. nepalensis* (2.03%).

Table 13.3 Crude protein, IVDMD, NDF and ADF content of pasture weeds

Weeds	Growth stage	CP%	IVDMD%	NDF%	ADF%
Common lambs quarters	Bud	22	73	22	17
Common lambs quarters	Flower	18	67	27	19
Shepherds purse	Green seed	19	55	37	29
Shepherds purse	Seed	16	53	41	34
Pennsylvania smartweed	Flower	18	47	24	19
Pennsylvania smartweed	Late flower	15	44	32	19
Redroot pigweed	Flower	18	74	22	16
Redroot pigweed	Early seed	15	73	27	20
Yellow foxtail	Early seed	17	63	52	27
Yellow foxtail	Seed	14	60	54	30
Common ragweed	Vegetative	26	77	21	17
Common ragweed	Vegetative	21	70	26	21

Oxalic acid content was low (3.60–5.50%), far below the value of potentially hazardous plants (up to 10%). OMD (Organic matter digestibility), DMD (Dry matter digestibility), ME (Metabolizable energy), and DMI (Dry matter intake) are important forage quality indices that are used to allocate forages to the particular livestock class with a given level of expected performance. Digestible dry matter is an estimate of the total digestibility of the feed and is calculated from the acid detergent fiber.

Metabolizable energy is also an important indicator of forage quality that makes up the diets of animals. The values for OMD, DMD, ME, and DMI in *Rumex* species varied from 55.04 to 59.80% 63.25 to 66.96% and 8.78 to 9.38 MJ/kg DM, respectively. The study revealed that *Rumex* species (*R. hastatus, R. obtusifolious, and R. nepalensis*) have considerable merits in the nutritive traits with remarkably high value of crude protein. Although the anti-nutrients were comparatively higher than the conventional forages, still nonconventional forage resources including *Rumex* have the potential to be utilized as a feed source and their integration in livestock feeding could be a possible avenue for addressing feed security for expanding livestock population. These nutritive nonconventional forage resources have the prospects of blending in conventional feeds. Although high in nutritive quality, further studies are required to assess the effect of feeding of *Rumex* species on animal health and production (Table 13.3 and Fig. 13.1).

Aquatic plants also have the potential to be used as feed sources due to their high protein content and low lignin content. Immature plants of aquatic fern contain a considerable amount of CP (10.45%) but low dry matter (20.35%), EE (1.37%), CF (26.96%) and ash (6.96%) than matured aquatic fern. Duckweed (*Lemna minor*) is a potent weed used as livestock feed. High level of CP in duckweed makes its use for supplementing low-quality forages. It is also rich in carbohydrate content (14.10–43.60%) and has a high level of essential amino acids, traces minerals and pigments such as xanthophylls and carotenes (Fasakin et al. 1999; Negesse et al.

Fig. 13.1 Portfolio of some weeds which can be used as NCFR. In reading order: (**a**) *Rumex nepalensis*; (**b**) *R. obtusifolius*; (**c**) *R. hastatus*; (**d**) *Amaranthus* spp.; (**e**) *Chenopodium album*; (**f**) *Setaria glauca*; (**g**) *Echinochloa* spp.; (**h**) *Mimosa pudica*; (**i**) *Shepherds spurse*

2009). Duckweed has higher levels of the essential amino acids lysine and methionine than most plant proteins and it has good proportions of trace minerals and pigments such as xanthophylls and carotene that make duckweed meal an especially valuable supplement for poultry, fish and animal diets. Water hyacinth leaves have a higher amount of CP (25.40%) when compared to water hyacinth petiole (17.01%).

Petiole and leaves contain 2.10% and 1.90% EE, respectively, whereas CF content has been reported to be 16.32% and 19.90% in petiole and leaves of water hyacinth, respectively. NDF of water hyacinth petiole is 55.90% while water hyacinth leaves contain higher NDF (64.70%). ADF content of water hyacinth petiole and leaves has been observed to be 35.91% and 36.50%, respectively.

Azolla also provides a sustainable feed for livestock feeding. Gouri et al. (2012) reported that floating fern contain 90.10% dry matter; 79.70% organic matter; 15.40% CP; 14.10% CF; 2.70% EE; 47.40% NFE; 20.40% ash content; 15.60% hemicelluloses; 6.80% celluloses and 17.50% lignin content which are required for all classes of livestock.

13.4 Constraints in Using Nonconventional Feeds

Nonconventional feeds derived from agro by-products, crop residues, weeds from pastures and grasslands, from browse plant species, and from other sources may vary in their nutritive value depending upon the processing technologies and storage methodologies. Nutrient levels of these feedstuff also vary depending upon the factors like soil, climate, temperature, rain, variety, harvest methods, and processing. The loss of nutrients during harvesting, processing and storage has been explained in other chapters. The major constraints in the availability and utilization of NCFR are:

- Production of NCFR is scattered and in some cases, the quality produced is low, especially for processing.
- High cost of collection of some of the NCFR.
- Noncompetitive costs and un-remunerative prices.
- Processing is difficult and problematic.
- Lack of managerial and technical skills to utilize them as feed.
- Limitations in the uses of products.
- Uncertainty about the marketability of end products.
- Lack of managerial skills and capital resources for the purchase and operation of suitable technology.
- Small farmers have neither the resources and know-how nor the quantities of residue to take individual action.
- Availability in terms of location, seasonality and storage facilities.
- Difficulty in handling, separation, transportation and physical processing of the residues.
- Limited knowledge about the composition of the residues, such as proximate components (e.g., crude protein, crude fiber, and minerals) intake and nutritive value (e.g., digestible energy and proteins).
- Inadequate demonstration of their potential in feeding systems due to low priority research.

13.5 Improving Quality and Digestibility of NCFR

Different technologies for improving the nutritive value of low-quality feed sources are:

13.5.1 Supplementation

The supplementation of low-quality feed with grains and concentrate feeds is the common strategy to provide a balanced feed. However, in drought conditions, the use of feed concentrates is usually high and this could lead to an increase in total feeding cost. The supplementation of NCFR with browse foliage could partially or totally replace the concentrate feeds without causing any negative effects on livestock performance.

13.5.2 Chemical Treatment

An alternative strategy is to treat NCFR with chemicals to improve its quality. Among chemical pretreatments, alkali pretreatment has been reported to have a significant impact on the nutritive value of NCFR. However, the use of sodium hydroxide on small farms is uneconomical and sometimes dangerous. The use of ammonium and calcium hydroxide has attracted the researchers because it breaks the bond between cell wall constituents resulting in swelling and flexibility of fiber content that ultimately increases the crude protein content, feed intake and digestibility. Improving the feeding quality of crop residues *via* increasing nonprotein nitrogen content with feed grade urea has been widely practised. The positive effects of urea application on the nutritive value of NCFR are because of the conversion of urea into ammonia by urease and the effect of ammonia on complex structures in cell walls.

13.5.3 Ensiling

Although a number of agro-industrial by-products are available, most of them are not widely used for livestock feeding. For example, due to the high moisture content of olive cake and tomato pulp, they tends to become more rancid and moldy. Ensiling techniques can be safely used to extend the storage period of these by-products separately or combined with other by-products such as molasses or wheat bran.

13.5.4 Feed Block Technology

Agro-industrial by-products having high moisture content can be efficiently used through feed block technology (Salem and Nefzaoui 2003). The ingredients of feed block can be divided into major and minor components. Major components are feed source and minor components include micronutrients and feed additives. This technology provides flexibility to add cheap and easily available ingredients. The inclusion of agro-industrial by-products and tree leaves could partially replace expensive concentrate feeds. Feed blocks may also be supplemented with anti-helminthic medicines to control gastrointestinal parasites in browsing animals (Anindo et al. 1998).

13.5.5 Removal of Anti-nutritional Factors

The presence of anti-nutrients in NCFR reduces their potential utilization in livestock feeding because consumption of anti-nutrients is often associated with decreased bioavailability of nutrients. Therefore, it important to look for methodologies for lowering the level of anti-nutrients and enhancing the availability of macro as well as micro-nutrients from NCFR. Various methodologies have been developed for the removal of anti-nutrients from forages. These methodologies can also improve the nutritional quality of NCFR by reducing the level of different anti-nutrients and by inhibiting the growth of microorganisms causing spoilage. The effectiveness of different processing methods depends not only on the type of NCFR but also on the properties of the anti-nutritive compounds and selected processing techniques.

13.6 Using NCFR in Livestock Feeding

For improving productivity, a sufficient quantity of nutritive fodder is required by the livestock for fulfilling their energy demand. A number of alternative feed options including NCFR are available for feeding, however, the presence of some anti-nutritional factors hinders their utilization in feed chain. Different pretreatments like chemical, biological and physical treatments provide an option for managing the anti-nutritional factors present in NCFR. Lignocellulolytic microorganisms have also been deployed to reduce lignin and other anti-nutritional factors in these feed resources. The urea, ammonia and alkali treatments are now well known to suppress the effects of different anti-nutritional factors. For future utilization of NCFR in livestock feeding the efforts must be focused on:

- Assessment of availability, nutritive value and utilization of NCFR.
- On-farm trials for assessing the potential of NCFR in commercial production.
- Integration of NCFR in the existing farming system.
- Use of biotechnological approaches for producing high-quality NCFR.

References

Abbasi AM, Shah MH, Khan MA (2015) Wild edible vegetables of lesser Himalayas. Ethanobotan Nutraceut Aspect 1:1–360

Ajila CM, Bhat SG, Prasada Rao UJS (2007) Valuable components of raw andripe peels from two Indian mango varieties. Food Chem 102:1006–1011

Ali J, Hamida A, Hussain A (2010) Study on some micronutrient composition in peels of different citrus fruits grown in NWFP. J Chem Soc Pak 32:83–86

Amata IA (2014) The use of non-conventional feed resources (NCFR) for livestock feeding in the tropics: a review. J Glob Biosci 3:604–613

Anindo D, Toé F, Tembely S, Mukasa-Mugerwa E, Lahlou-Kassi A, Sovani S (1998) Effect of molasses-urea-block (MUB) on dry matter intake, growth, reproductive performance and control of gastrointestinal nematode infection of grazing Menz ram lambs. Small Rumin Res 27:63–71

Basile F, Machado PF (1990) Feeding value of steam treated sugar cane bagasse in ruminant rations. Livest Res Rural Dev 2(1):1–6

Battaglini M, Costantini F (1978) Byproducts from the tomato industry in diets for growing rabbits. Coniglicoltura 15(10):19–22

Boeckaert C, Vlaeminck B, Dijkstra J, Issa-Zacharia A, Van Nespen T, Van Straalen W, Fievez V (2008) Effect of dietary starch or micro algae supplementation on rumen fermentation and milk fatty acid composition of dairy cows. J Dairy Sci 91:4714–4727

Costa DFA, Quigley SP, Isherwood P, McLennan SR, Poppi D (2016) Supplementation of cattle fed tropical grasses with microalgae increases microbial protein production and average daily gain. J Anim Sci 94:2047–2058

Fasakin EA, Balogun AM, Fasuru BE (1999) Use of Duckweed, *Spirodella polyrrhiza* L. Schleiden, as a protein feedstuff in practical diets for Tilapia, *Oreochromis niloticus* L. Aquacult Res 30:313–318

Gouri MD, Jagadheesh SS, Gopinath CR, Kalibavi CM (2012) Importance of Azolla as a sustainable feed for livestock and poultry. Agric Rev 33:93–103

Gowda NKS, Ramana JV, Prasad CS, Singh S (2004) Micronutrient content of certain tropical conventional and unconventional feed resources of Southern India. Trop Anim Health Prod 36(1):77–94

Guerrero JLG, Isasa MET (1997) Nutritional composition of leaves of *Chenopodium* species (*C. album L., C. murale L. and C. opulifolium*). Int J Food Sci Nutr 48:321–327

Halmemies-Beauchet-Filleau A, Rinne M, Lamminen M, Mapato C, Ampapon T, Wanapat M, Vanhatalo A (2018) Review: alternative and novel feeds for ruminants: nutritive value, product quality and environmental aspects. Animal 12:S295–S309

Hameed I, Dastagir G (2009) Nutritional analysis of *Rumex hastatus* D. Don, *Rumex dentatus* Linn and *Rumex nepalensis* Sreng. Afr J Biotech 8:4131–4133

Herrick KJ, Hippen AR, Kalscheur KF, Anderson JL, Ranatunga SD, Patton RS, Abdullah M (2012) Lactation performance and digestibility of forages and diets in dairy cows fed a hemicellulose extract. J Dairy Sci 95:3342–3353

Hulfan HW, Proudfoot FG, Zarkadas CG (1982) Potato waste meal. Compositional analysis. Can J Anim Sci 62:1161–1169

Humaira S, Rahmatullah Q, Shahid I, Mirza FQ (2014) Seasonal availability and palatability of native flora of Santh Saroola Kotli Sattian, Rawalpindi, Pakistan. Afr J Plant Sci 8:92–102

Katoch R, Tripathi A, Sood S (2018a) Possibilities of non-conventional feed resources in livestock feeding—a review. Forage Res 44:141–151

Katoch R, Tripathi A, Thakur N (2018b) Comparative analysis of three *Rumex* species for livestock feeding prevailing under mid-Himalayan region. Indian J Agric Biochem 31:181–185

Keir B, Nugyen VL, Preston TR, Ørskov ER (1997) Nutritive value of leaves from tropical trees and shrubs: *in vitro* gas production and *in sacco* rumen degradability. Livest Res Rural Dev 9(4) http://www.lrrd.org/lrrd9/4/bren941.html
Kołodziejczyk K, Markowski J, Kosmala M, Król B, Płocharski W (2007) Apple pomace as a potential source of nutraceutical products. Pol J Food Nutr Sci 57(4 B):291–295
Lamminen M, Halmemies-Beauchet-Filleau A, Kokkonen T, Simpura I, Jaakkola S, Vanhatalo A (2017) Comparison of microalgae and rapeseed meal as supplementary protein in the grass silage based nutrition of dairy cows. Anim Feed Sci Technol 234:295–311
Lodge-Ivey SL, Tracey LN, Salazar A (2014) Ruminant nutrition symposium: the utility of lipid extracted algae as a protein source in forage or starch-based ruminant diets. J Anim Sci 92: 1331–1342
Maia MRG, Fonseca AJM, Oliveira HM, Mendonça, Cabrita ARJ (2016) The potential role of seaweeds in the natural manipulation of rumen fermentation and methane production. Sci Rep 6: 32321
Makkar HP, Tran G, Heuzé V, Giger-Reverdin S, Lessire M, Lebas F, Ankers P (2016) Seaweeds for livestock diets: a review. Anim Feed Sci Technol 212:1–17
Marten GC, Anderson RN (1975) Forage nutritive value and palatability of 12 common annual weeds. Crop Sci 15:821–827
Marten GC, Sheaffer CC, Wyse DL (1987) Forage nutritive value and palatability of perennial weeds. Agron J 79:980–986
Negesse T, Makkar HPS, Becker K (2009) Nutritive value of some non-conventional feed resources of Ethiopia determined by chemical analysis and *in vitro* gas method. Anim Feed Sci Tech 154: 204–217
Onwuka CFI, Adetiloye PO, Afolami CA (1997) Use of household wastes and crop residues in small ruminant feeding in Nigeria. Small Rumin Res 24:233–237
Rahmat H, Hodge RA, Manderson GJ, Yu PL (1995) Solid sub-strate fermentation of Kloeckera apiculata and Candida utilis on apple pomace to produce an improved stock feed. World J Microbiol Biotechnol 11:168–170
Rivero ML, Nouel Borges G, Sanchez Blanco R, Rojas Castellanos J, Jimenez JL (2004) Poultry litter and rice polishings as supplements to sugar cane bagasse for lamb fattening. Livest Res Rural Dev 16(10):1–5
Roxas DB, Perez CB Jr, Trinidad EE (1969) The feeding value of sugarcane bagasse based rations for ruminants. Digestibility and nitrogen balance by cattle, carabaos and sheep. Phil J Anim Sci 6:33–42
Salem BH, Nefzaoui A (2003) Feed blocks as alternative supplements for sheep and goats. Small Ruminant Res 49:275–288
Suksombat W (2004) Comparison of different alkali treatment of bagasse and rice straw. Asian Aust J Anim Sci 17(10):1430–1433
Temme DG, Harvey RG, Fawcett RS, Young AW (1979) Effect of annual weed control on alfalfa quality. Agron J 71:51–54
Tuazon-Nartea J, Savage GP (2013) Investigation of oxalate levels in sorrel plant parts and sorrel-based products. Food Nutr Sci 4:838–843
Van Emon ML, Loy DD, Hansen SL (2015) Determining the preference, *in vitro* digestibility, in situ disappearance, and grower period performance of steers fed a novel algae meal derived from heterotrophic microalgae. J Anim Sci 93:3121–3129

14 Quality and Digestibility of Crop Residues

Feed unavailability has become a major concern in the sustenance of livestock rearing adopted as an occupation. To meet the increasing demand for livestock products, higher quantities of feed resources are required in the coming years. In this context, there is a need to look for alternate feed resources, which are available in sufficient quantity but have not been given much attention for their efficient utilization in livestock feeding. Among these, crop residues have enormous potential for their utilization as a potential livestock feed.

Crop residues are by-products of crop harvest and the quantity of their harvest is directly linked to all the factors that normally affect the crop yield. With increasing production of food grains and other agricultural produce, there is a corresponding increase in the quantity of crop residues. India being an agriculture-dominant country produces more than 500 million tons of crop residues annually, of which major share is contributed by wheat (110 million tons), rice (122 million tons), maize (71 million tons), millets (26 million tons), sugarcane (141 million tons) and pulses (28 million tons) (Devi et al. 2017). These crop residues are used for animal feeding, soil mulching, and manure making, thatching for rural homes, and fuel for domestic and industrial use. Thus, crop residues are of tremendous value to the farmers. Apart from being a valuable feed source during the scarcity of green fodder, the livestock productivity based on crop residue feeding is constrained by their low voluntary intake, low protein content and poor digestibility. This is because most of the crop residues have poor digestibility with high fiber content, lignin and silica content.

14.1 Crop Residues in Livestock Feeding

Crop residues form an abundantly available feeding resource for livestock (Mahesh and Mohini 2014). These are particularly valuable during dry periods when green fodder and other feeds are inadequate. The feeding of crop residues is constrained by

R. Katoch, *Nutritional Quality Management of Forages in the Himalayan Region*,
https://doi.org/10.1007/978-981-16-5437-4_14

their low voluntary intake and digestibility. Crop residues have low protein content (3–6%), high NDF (>70%) and ADF (>45%) content, high lignin content (6–15%) and silica content (1–13%). Most of the residues are also deficient in fermentable energy which is reflected by low organic matter digestibility and low availability of minerals, less protein content, and poor digestibility. These residues have a low rate of disappearance in the rumen and low passage rate through the gastrointestinal tract which leads to reduced feed intake, weight loss and nutrients deprivation in livestock. The physical properties of crop residues are to a larger extent determined by cell wall characteristics. Cell wall polysaccharides (cellulose and hemicellulose) account for more than 70% of dry matter in crop residues and are a major energy source for ruminal microflora. However, the cell wall content is negatively correlated with the voluntary intake of crop residues. High cell wall content increases rumination time and decreases the efficiency of livestock in converting metabolizable energy into net energy. The ability of rumen microorganisms to digest cell wall polysaccharides is limited by the presence of polyphenolics particularly lignin. Unlike cellulose and hemicellulose, lignin cannot be easily digested and is considered an impediment in the breakdown of cell wall polysaccharides by rumen microbes, affecting digestibility of crop residues.

Straws form the largest category of crop residues which comprise of stems and leaves of cereal crops left after the crop harvest and provide alternative feed source to livestock during lean periods of fodder availability. Wheat and rice are the major crops contributing to straw, however, oat, barley, millet and other cereal crops also produce a good amount of straws. Stovers are the field residues of cereal crops, such as maize and sorghum. Stubble consists of stumps of cereal and legume crops, left in the field after harvest. Vegetable waste is also an important feed source. One of the large sources of vegetable residue comes from cabbage harvest comprised of discarded leaves and heads contributing up to 6 tons of edible dry matter per hectare. Pulses residues or straw are the leftover parts of the plant after harvesting seeds or grains. Common bean, rice bean, cowpea, soybean, broad bean, chickpea, and adzuki bean are the important food legumes of the Himalayan region that provide fodder of good nutritional quality.

14.2 Nutritive Value of Common Crop Residues

In the Himalayan region, rice, wheat, barley, and oat straw along with the maize stover are the common crop residues used for livestock feeding. The nutritive quality and other characteristics of these crop residues have been discussed:

14.2.1 Rice Straw

Rice straw is generally used as lean period feed. It can also be fed to livestock in combination with other feed components. It has high dry matter (DM) content ranging from 92 to 96% but with low CP content ranging from 3 to 7% (Shen et al. 1998). Rice straw has a high percentage of cell wall constituents, neutral detergent fiber (ADF) and acid detergent fiber (ADF). It contains 25 to 45% cellulose, 25–30% hemicelluloses and 10–15% lignin (Yoswathana and Phuriphipat 2010) with low nitrogen, vitamins, minerals and high silica content (5–15%). High lignin and silica content hinder the nutrient availability to rumen microbes and eventually limits the necessary nutrient uptake (Sheikh et al. 2018). The utilization of rice straw as a sole feed without any supplementation results in poor livestock performance (Doyle et al. 1986) due to low intake, degradability, nitrogen content and unbalanced mineral composition. The voluntary intake and digestibility of rice straw depend on the proportion of different plant parts, their nutritional composition, distribution of nutrients in tissues and physical and chemical nature of cell walls (Devendra 1997). Rice straw has a high proportion of leaf (~60%) in comparison to other cereal straws such as barley (~35%), oats (~40%), and wheat (~20 to ~40%) (Theander and Aman 1984). Due to low NDF content, rice leaves contain less NDF, thus have high digestibility in comparison to stems (Vadiveloo 2000). Treatment with 4% urea solution for 21 days has been reported to increase the digestibility of rice straw (Vadiveloo 2000).

14.2.2 Wheat Straw

The main fractions of wheat straw are nodes, internodes and leaves. Wheat straw is composed of cellulose, hemicellulose, lignin, minerals (calcium and phosphorous), silica, acid detergent fiber, and ash. The macro and micronutrients vary from cultivar to cultivar, stages of plant growth, nature of soil, fertilizer and climatic conditions. Wheat straw contains 2.60–6.00% crude protein, 0.70–2.80% crude fat, 36–46% crude fiber, 4–10% ash, about 65–85% NDF and 45–55% ADF. Wheat straw is not much preferred by livestock due to the high level of cell wall constituents (cellulose: 35–40%; hemicelluloses: 25–30% and 5–9% ADL content) and fiber. High lignin, low digestibility, less protein content and poor palatability discourage the use of wheat straw as a sole animal feed. The feed value of wheat straw is lower than barley and oat straw.

14.2.3 Barley and Oat Straw

Untreated barley straw has poor digestibility and requires progressive adaptation by nitrogen supplementation. Oat straw has the least application as livestock feed. The IVDMD of oat straw is lower than that of barley straw (Lopez et al. 2005) and its rumen degradability is lower than that of wheat straw. Barley straw contains 2–3% ether extract, 7–10% ash content, 65–80% NDF and 40–50% ADF content whereas oat straw contains 70–75% NDF and 40–45% ADF. Oat straw has higher CP content (4–8%) in comparison to other cereal straws.

14.2.4 Maize Stover

Maize (*Zea mays* L.), also known as "Queen" of cereals crops is the third largest produced and consumed food crop after rice and wheat. The rapid increase in human and livestock population, continuously shrinking area under forages and mounting demands for livestock products have encouraged the use of maize as a dual-purpose (food-feed) crop. The maize stover left after harvesting of grain includes stalks, leaves, husks and cobs. It generally contains 3.50–8.70%, CP, 31.30–41.0% cellulose, 20.0–34.40% hemicellulose, 3.10–5.0% ADL and 4.20–7.50% ash (Chaudhary et al. 2012). The structural carbohydrates, (cellulose and hemicellulose) of maize stover are often present in lignocellulosic form. Unlike cellulose and hemicellulose, lignin cannot be easily digested and is considered as impediment in the degradation of cell wall polysaccharides by microflora in the gastrointestinal tract of the animals. Maize stover is stored in large heaps for use during lean periods and is also suitable for silage preparation and haymaking. It serves as a potential winter feed in cold temperate regions.

14.2.5 Sorghum Stover

Immediately after crop harvest, the stover is harvested preferably by cutting as green as possible which is then dried for animal feeding. Sorghum stovers has poor nutritive value with OM digestibility lower than 50% and need supplimentation with protein-rich concentrates for livestock feeding. It contains around 93–95% dry matter with low CP content (3.70–4.50%). It has a high NDF (70–74.80%) and ADF (43-46.60%) and around 6.00% ADL. In comparison to maize stover, it has low levels of EE content (1.30%).

14.2.6 Legume Straw

Legume straw has comparatively higher CP and low fiber content than cereal straws. It has a higher pectin content than grasses and these carbohydrates are important components of the intercellular spaces and are degraded promptly by rumen microorganisms. Chickpea straw differs from typical cereal straw in having more protein and metabolizable energy (ME) and low NDF content (Makkar et al. 1996). Makkar and coworkers (1996) reported low tannins and phenolic content in legumes straws. Chickpea straw is more palatable than wheat straw but livestock should be allowed to habitualize to the taste before offering large quantities. Pod husk and stem of common bean (*Phaseolus vulgaris*) contain 4–8% protein, while leaves contain up to 20% protein (Pieltain et al. 1996) with high fiber content (38–45%). Chickpea straw is slightly higher in nutritive value than wheat straw (44–46% TDN; 4.50–6.50% CP). The residual vines, leaves and stems from field peas are palatable and often resemble medium-quality grass hay in nutritional composition. Peas are harvested in the early autumn, so grazing or baling are options for utilizing pea

Table 14.1 Nutritional composition of major crop residues

		Chemical composition (% DM)						
Crop residues	DM (%)	CP (%)	EE (%)	CF (%)	NDF (%)	ADF (%)	ADL (%)	Ash (%)
Wheat straw	91.60	3.10	1.30	44.70	73.00	43.90	6.08	9.70
Maize stover	91.80	6.50	2.70	26.20	70.40	37.00	18.30	7.59
Rice straw	83.30	3.70	1.60	31.00	64.40	42.40	4.80	18.10
Sorghum stover	95.20	3.90	1.30	35.60	74.80	46.60	6.00	8.60
Barley straw	88.40	5.50	3.20	38.20	80.10	46.70	6.78	13.02
Oat straw	93.00	7.00	2.40	28.40	72.30	38.10	4.20	8.30
Millet straw	90.70	5.00	1.30	35.90	74.80	53.20	10.40	12.20
Chickpea straw	90.40	5.40	1.00	41.00	65.60	46.90	11.90	7.40
Pea straw	88.80	8.20	2.10	36.30	54.90	38.70	7.20	9.80
Common bean straw	88.00	7.10	1.10	41.00	69.70	48.50	8.00	8.90
Soybean straw	89.10	6.90	3.50	44.20	79.70	59.90	16.40	7.90
Cowpea haulms	95.00	13.70	2.20	29.90	49.00	35.40	8.50	11.00

Source: Haile et al. (2017)

straw. Field pea residue decomposes more rapidly than cereal crop residues therefore, it should be collected or grazed shortly after harvest to minimize loss of leaves and nutritional quality (Table 14.1).

14.3 Constraints in Utilization of Crop Residues as Sole Livestock Feed

Crop residues form an integral part of animal feeding in hilly areas. Dependency on crop residues is increasing due to the limited availability of green fodder in the Himalayan region, especially during the lean season. However, low voluntary intake and digestibility limit the utilization of crop residues as a maintenance type feed.

14.3.1 Lignin Content

Lignin represents the undesirable or anti-nutrient component of livestock feed. Lignin content of more than 5% has been reported in majority of crop residues. Lignin is not readily fermented by rumen bacteria but partly degraded by rumen anaerobic fungi (Krause et al. 2003). The fact that lignin has about 30% higher gross energy content than cellulose (Novaes et al. 2010), but this energy is barely accessible to livestock. Secondly, the low feed intake due to the association with polysaccharide constituents, lignin forms a physical barrier and thus hinders access of rumen microbes to fermentable cell wall components. Consequently, the passage rate of feeds through the rumen is reduced which influence the feed intake capacity.

14.3.2 Silica Content

Apart from lignin, the presence of silica also decreases the feeding value of crop residues. Rice straw has high silica content in comparison to other crop residues. It has about 13% silica in comparison to wheat straw which contains about 5% silica. Higher proportions of silica and lignin in crop residues is responsible for their difficult composting, digestion and hydrolysis. Apart from being a physical barrier, silica inhibits the action of hydrolytic enzymes in the rumen. Silica has more profound effects in reducing nutrient digestibility as compared to lignin content. Chemical and physical bonding of lignin and silica with hemicellulose and cellulose prevents swelling of fibrous feed resources thereby inhibiting degradation by digestive enzymes in rumen.

14.3.3 Fiber Content

Amount of fiber in livestock diet is considered as an important factor affecting the palatability and intake of feed. High fiber levels in diets could decrease the voluntary feed intake as a consequence of gut fill. Soluble fiber is usually susceptible to microbial degradation, increasing bacteria growth in the lower gut. The greater the amount of fiber in the diet, the higher is the disappearance of fiber (g of fiber disappeared/kg of feed DM) after fermentation in ruminants. Crop residues generally contain 30–40% CF.

14.3.4 Protein Content and Digestibility

The quality of various crop residues is determined by the protein and energy or DDM content of a particular residue. When the percentage of crude protein is low, the bacteria responsible for digestion cannot sustain adequate levels to process forage. High lignin content in crop residues restricts the intake potential of the animal, while low crude protein reduces the functioning of ruminal microflora, thereby affecting digestibility. Cell wall constituents of different crop residues have been presented in Table 14.2.

14.4 Improving Quality and Digestibility of Crop Residues

Crop residues constitute a bulk of feeds fed to the livestock in rural areas because of their large-scale availability; however, their utilization as a potential livestock feed is constrained by low nutritive value and digestibility due to high lignin, fiber and silica content. These issues are manageable to some extent with different treatments, processing techniques as well as feeding practices.

Different methodologies have been developed for treating crop residues before feeding for improving their nutritive value. The classification of methods for improving the quality of crop residues has been given in Fig. 14.1. These methods

Table 14.2 Cell wall constituents affecting nutritional value of crop residues

Crop residues	Hemicellulose (%)	Cellulose(%)	Lignin(%)	Silica(%)
Barley straw	27	44	7	3
Oat straw	16	41	11	3
Paddy straw	26	55	7	13
Wheat straw	36	39	10	6
Sorghum stover	30	31	11	3
Chickpea straw	20	30	10	2
Lucerne straw	19	38	11	1
Sugarcane bagasse	29	40	13	2
Sugarcane trash	26	36	10	2
Paddy hulls	14	39	11	22

Source: Jackson (1977)

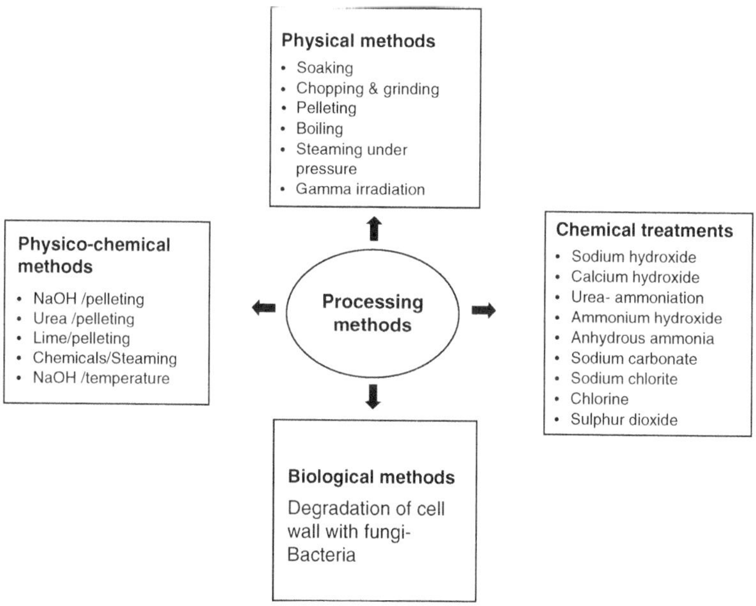

Fig. 14.1 Methods for improving nutritional value of crop residues

include physical treatments, chemical treatments and biological treatments which primarily affect the lignocellulose complex present in the cell wall. This complex is responsible for the indigestibility and low intake potential of crop residues in ruminants. Physical treatments like crushing, grinding, etc. increase the digestibility of crop residues by reducing the crystallinity of lignocelluloses that increase the accessibility of ruminal microbes for enzymatic hydrolysis (Katoch et al. 2017).

Chemical treatments solubilize hemicellulose and lignin fractions in the cell wall thereby decrease the total fiber content in crop residues. In biological treatments, inoculation with different microorganisms affects lignocellulosic polymer and degrade lignin. The degradation of lignin increases the access of cellulose for enzymatic hydrolysis by the enzymes secreted by ruminal microbiota.

14.4.1 Physical Treatment

Physical treatments help in improving the feeding value of the crop residues by soaking, boiling, pelleting and steaming under pressure and gamma radiation. The main purpose of this process is to reduce the crystallinity of lignocellulose in order to increase surface area and reduce the degree of polymerization and ultimately increase the accessibility of ruminal microbes for enzymatic hydrolysis of lignocelluloses (Saritha et al. 2012).

14.4.1.1 Chopping and Grinding

Chopping and grinding are the main physical processes to improve intake. Grinding reduces the particle size of crop residues. These physical processes reduce the particle size of the straw, providing easy access to the rumen microorganisms for degradation. The use of these techniques should consider the balance between the particle size and the retention time or passage rate of the ingested crop residues. The reduction in particles due to grinding or chopping promotes feed intake and increases the passage rate of the feed.

14.4.1.2 Pelleting and Cubing

The feeds containing high levels of cereal grain is finely ground and treated with steam for gelatinization of starch. Reducing feed particle size increases the surface area for microbial fermentation of fibrous components of diet in the rumen or hindgut. The process of cubing is mainly used for hay and crop residues for transport (Fig. 14.2).

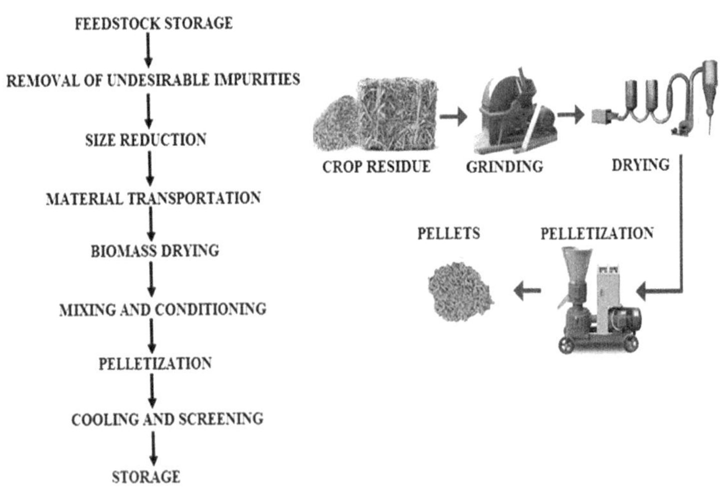

Fig. 14.2 Process of pelletization of crop residues

14.4.1.3 High Pressure/High-Temperature Steam Treatment

Soaking along with steaming technique has a direct effect on cell wall constituents, which removes a significant amount of lignin from crop residues. In this method, steam or exposure of lignocellulosic contents of crop residues provides a good environment for microbial enzymes for faster fermentation of nutrients. Steam treatment of crop residue is accomplished in a high-pressure vessel, i.e., a vertically mounted cylinder narrowing down to the shape of a cone bottom. Steam is released from a boiler beside the vessel containing substrate. For each treatment, the vessel is preheated and filled with wet samples. The vessel is then closed and charged with the desired steam pressure. The steam-treated straw is then collected from the vessel. Samples are analyzed for their properties. Mosier et al. (2005) reported that corn stover treated with water at 190 °C for 5 min, convert 90% of cellulose into glucose. Guggolz et al. (1971) reported marked increase in *in vitro* digestion for a variety of grass straws treated for 3 min at 232 °C and 28 kg/cm^2 in specially constructed pressure vessels. Considerable increases in both *in vitro* and *in vivo* organic matter digestibility were observed when sugarcane bagasse was treated for 45 s. Chemical studies have indicated extensive degradation of cellulose and hemicellulose and the production of undesirable phenolic compounds. Steam treatment of rice straw resulted increase in soluble carbohydrate content (Liu et al. 1999). Treatment of rice straw under high pressure of 15 bar for 5 min released hemicellulose, cellulose, lignin and sugars from rice straw (Rangnekar et al. 1982; Liu et al. 1999).

14.4.1.4 Ensiling

Ensiling is a fodder preservation method based on spontaneous lactic acid fermentation under anaerobic conditions. By this process, crop residues could be stored for longer periods with improved nutritional composition (Katoch et al. 2017). For example, ensiling of stover with molasses result in 7.29, 7.15, and 28.57% increase in CP, EE, and ash content, respectively (Wattanaklang et al. 2016). The ensiling of crop residues is carried out in six following phases:

1. **Phase I:** During this phase, the inoculation of aerobic microorganisms with the crop residues converts water-soluble carbohydrates to carbon dioxide, water and heat. In addition to this, other gases are produced due to the enzymatic proteolysis of protein. Phase I fermentation continues until either oxygen is depleted or total water-soluble carbohydrates are exhausted.
2. **Phase II**: Depletion of oxygen during the initial aerobic phase starts the anaerobic phase of fermentation and produces several fermentation end products SCFA's (acetate, lactate, propionate, and butyrate), ethanol and carbon dioxide. In addition to this, more nitrogenous end products can also be generated because of continued enzymatic proteolysis of protein. The second-phase bacteria create an environment for fermentation with lactic acid bacteria.
3. **Phases III–IV**: During these phases, fermentation occurs when lactic acid bacteria convert water-soluble carbohydrates to lactic acid. As the heat generated in phase 1 starts to dissipate and pH becomes lower, other LAB become active and produces lactic acid. This acid is most efficient for rapidly reducing the pH.

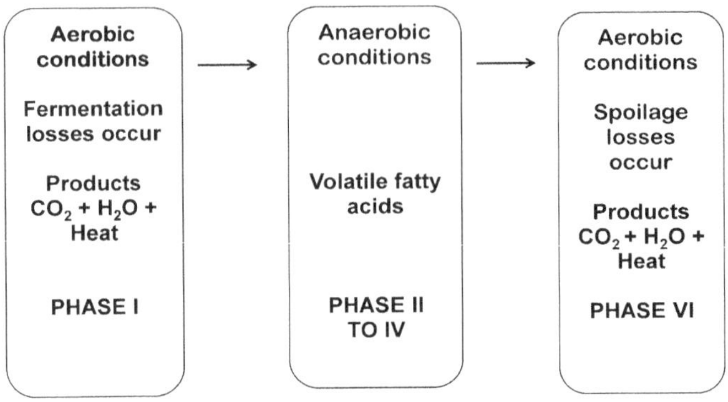

Fig. 14.3 Phases of ensiling process

A predominance of LAB relative to other silage acid-producing bacteria creates faster fermentation, conserve more nutrients in the form of water-soluble carbohydrates, peptides and amino acids. The time span of fermentation depends on the buffering capacity, moisture and maturity of the crop to be ensiled. The rate and end-point of the final pH drop in the ensiled crop depend largely on the type and moisture of forage being ensiled.

4. **Phase V**: The fifth stage of fermentation, the stable phase, lasts throughout storage. This phase is not static because other anaerobic bacteria become active during this period producing end products.
5. **Phase VI**: This is the final fermentation phase, which occurs when silage is fed from the storage structure. This phase is as important as the others but is often neglected. Up to 50% of dry matter losses result from secondary aerobic spoilage on the surface of the silage in storage. Aerobic microbial activity is stimulated because oxygen is introduced into the silo (Fig. 14.3).

14.4.2 Chemical Treatment

Chemical treatment improves digestibility and thereby voluntary intake of low-quality roughages by solubilizing lignin and hemicellulose fractions in cell walls (Saenger et al. 1982). Among different chemicals, sodium hydroxide, calcium hydroxide, urea, ammonia and dilute acid have been commonly used. After acid treatment, the solubilized lignin gets quickly condensed and precipitated. Improving the feeding quality of different crop residues by increasing the nonprotein nitrogen content with addition of feed grade urea has been widely practised. Ammonia treatment is also known to increase crude protein content, feed intake and digestibility of treated crop residues.

14.4.2.1 Ammoniation of Crop Residues

Ammoniation is an efficient method for improving the quality of crop residues. The main component of straw is fiber, including cellulose and hemicellulose that can be digested by ruminants. Some cellulose and hemicellulose are bound to lignin and resistant to microbial attack. Ammoniation usually increases digestibility by 20% and CP content up to 1–2 times. It can also improve palatability and consumption rate. The total nutritional value can be doubled, reaching 0.40–0.50 feed units for each kilogram of ammoniated straw. In addition, ammoniation reduces mold development, destroys weed seeds, parasite eggs, and bacteria.

(a) **NH_3 treatment**: Treatment of crop residues with anhydrous and aqueous ammonia, urea or ammonia releasing compounds improves their digestibility (Abou-EL-Enin et al. 1999; Selim et al. 2004). Ammonia treatment increases crop residues degradability and also adds nitrogen (Abou-EL-Enin et al. 1999) and preserves crop residues by inhibiting mold growth (Calzado and Rolz 1990). Selim et al. (2004) used gaseous ammonia (3 g NH_3 per 100 g dry matter) to treat rice straw packed in polyethylene bags for 4 weeks. The treatment increased the N content in the rice straw from 8.16 to 18.4 g/kg (CP content increased from 51 to 115 g/kg). Saenger et al. (1982) demonstrated that NH_3 treatment of straw along with corn supplements increased dry matter intake (1.71% of live weight) of livestock. The effect of ammonia treatment on protein, fiber, and digestibility of crop residues has been presented in Table 14.3.
(b) **Urea treatment**: Urea treatment is safer than anhydrous or aqueous ammonia. In addition it provides a source of nitrogen (crude protein) in which straw is deficient (Schiere and Ibrahim 1989). The application of urea has a positive effect on the nutritive value of crop residues which is the result of two processes: (a) conversion of urea into ammonia by urease and (b) effect of ammonia on the cell wall of residues (Katoch et al. 2017). These treatments increase the crude protein content of low-quality roughages by the addition of nonprotein nitrogen content and also break the bond between cell wall constituents resulting in swelling and flexibility of fiber content. These treatments solubilize the hemicellulose and lignin content which is responsible for the indigestibility of crop residues, thus enhancing the feeding value. Rice straw treated with urea

Table 14.3 Effect of ammonia treatment on protein, fiber and digestibility of crop residues

	Parameters (%)			
Feed	DM	CP	CF	Dry matter digestibility
Chinese wild-rye hay	90	5.90	32.00	52.00
Winter wheat straw	90	2.20	41.00	39.70
Ammoniated winter wheat straw	70	7.60	39.00	50.00
Rice straw	93	3.90	33.10	24.00
Ammoniated rice straw	90	7.80	32.50	48.00
Maize stover	90	3.70	30.50	42.00
Ammoniated maize stover	90	8.70	30.50	60.00

increases *in vitro* dry matter degradability. Treatment of rice straw with a combination of 3% urea plus 4% lime at 50% moisture for 3 weeks was the most effective treatment for improving its degradability (Sirohi and Rai 1995). Treatment of roughages with urea results in the production of ammonia by urease that affects the cell wall constituents of residues. Urea most effectively cleaves the ester bonds of uronic acids in hemicellulose leading to the reduction in NDF content in forages. Different studies have reported the positive effect of urea treatment on nutritional quality of stover, straw and bagasse (Woyengo et al. 2004; Aregheore 2005; Oji et al. 2007; Ramirez et al. 2007; Elias and Fulpagare 2015; Katoch et al. 2017).

14.4.2.2 Alkali Treatment

Treatment of crop residues with alkali involves solvation and saponification processes and makes crop residues more digestible to livestock (Hendriks and Zeeman 2009). Sodium hydroxide primarily affects the fiber content of crop residues by decreasing the hemicellulose and lignin content. Ololade et al. (1970) reported that the application of sodium hydroxide (8%) on maize stover results 4.88, 6.25 and 7.25% increase in ADF, lignin and cellulose content, respectively. Klopfenstein et al. (1972) reported 47.43% increase in organic matter digestibility and 7.30, 6.30, and 24.27% decrease in ADF, ADL and cell wall constituents, respectively, in maize stover treated with a 5% solution of sodium hydroxide.

Sodium hydroxide breaks the complex structure of lignocellulose and releases carbohydrate moieties resulting in increased carbohydrate content in stover (Theander 1981). The straw treated with NaOH treated is reported more digestible than pure straw up to 28% (Table 14.4). The Beckman method involves 1.50% NaOH for treatment of crop residues for 1 h before rinsing with tap water. NaOH reduces proteolysis and increases the rate of delignification by unlocking the linkage between lignin and cellulosic contents of straw. Treatment of rice straw and other crop residues using NaOH has also been reviewed by Jackson (1977), Berger et al. (1994), Arieli (1997), and Wang et al. (2004). These chemical reactions with NaOH on cell wall of rice straw are advantageous for the breakdown of esterified bonds between phenols group and cellulosic components of straw thus favoring enzymatic

Table 14.4 Effect of NaOH treatment on cell wall components of crop residues

Crop residue	Treatment	Hemicellulose (% DM)	Cellulose	Lignin	Silica
Paddy straw	Untreated	26.00	33.00	6.50	13.0
	Treated	11.00	32.00	4.90	7.00
Wheat straw	Untreated	36.00	39.00	9.60	5.50
	Treated	13.00	40.00	4.80	3.30
Oat straw	Untreated	16.00	41.00	10.50	2.70
	Treated	11.00	43.00	7.20	0. 20
Sugarcane bagasse	Untreated	29.00	40.00	12.50	1.50
	Treated	23.00	37.00	9.20	0.50

hydrolysis. However, there are some limitations of using NaOH for treatment of crop residues. There are reports of kidney damage in cattle fed with high levels of sodium over a long period. The Beckman method, which involves the treatment of crop residues with 1.5% NaOH for 18–20 h has been recommended by FAO (2012).

Treatment of crop residues with lime solution [$CaO/Ca(OH)_2$] can be carried out in two ways (1) by soaking and (2) ensiling. Lime treatment provides complementary effects with urea. The combination of lime and urea has been shown to increase digestibility and incrementing both calcium and nitrogen content of treated crop residue (Nguyen 2000). Pradhan et al. (1997) reported that ensiling rice straw with 4 or 6% $Ca(OH)_2$ had higher IVDMD than using 4 and 6% urea. Kim and Holtzapple (2005) reported that corn stover treated with an excess of calcium hydroxide in nonoxidative and oxidative conditions at different temperatures (25–55 °C) resulted 87.50% reduction of lignin content. Chen et al. (2009) reported that treatment of lime hydrolyzed 14.80% of hemicellulose to soluble sugar and reduced 34.80% lignin from corn stover. The enzymatic hydrolysis of alkali treated corn stover resulted in the conversion of glucan (91.30%) and xylan (51.80%) into glucose and xylose, respectively (Kim and Holtzapple 2005). The combination of lime and urea treatment could be useful in increasing degradability and also increasing both calcium and nitrogen content in crop residues.

14.4.2.3 Acid Hydrolysis

Dilute acid hydrolysis is also common among chemical hydrolysis of crop residues. It could be used as a pretreatment preceding enzymatic hydrolysis or as the actual method of hydrolyzing lignocellulose to the sugars.

1. **Dilute acid treatment**: In this treatment, low concentration of acid (0.2–2.5%), high temperatures (120–210 °C) and pressures are used for quick reaction (Menon and Rao 2012). The dilute acid pretreatment could significantly improve cellulose hydrolysis. The main advantage of this method is the possibility to recover a high portion of hemicellulosic sugars. Acid mainly affects the polysaccharides present in the cell wall and hydrolyze them into monomeric sugars. A variation of the process involves two stages of pretreatment: in the first stage, most of the hemicelluloses content in crop residues are solubilized in the presence of dilute acid, while the second stage higher acid concentration are used to hydrolyze cellulose and remaining hemicelluloses (Talebnia et al. 2010). Chen et al. (2009) observed that when corn stover was treated with dilute sulfuric acid (1.50%), it converted 76.60% hemicellulose to soluble sugar and reduced 12.20% lignin from raw material. Besides mineral acids (dilute acid H_2SO_4, HNO_3, and HCl), organic acids (Maleic, Fumaric, and Oxalic acids) have also been useful in the treatment of crop residues (Lee and Jeffries 2011; Li et al. 2011). Oxalic acid and maleic acid degrade hemicelluloses more efficiently than sulphuric acid. Oxalic acid treatment to corn stover at 200mM concentration gave best results at 160 °C for 10 min (Mtui 2012). Though dilute acid pretreatment has received wide attention from researchers yet it faces some drawbacks. A major demerit of this process is its requirement of special corrosion-resistant reactors which are usually expensive both in investment and operation.

14.4.2.4 Wet Oxidation Method

In this method, crop residue undergoes oxidation in an aqueous (acidic, neutral or alkaline conditions) solution on reaction with oxygen (air) at elevated temperatures (125–315 °C) and pressures (0.5–5 MPa) (Schutt and Abraham 2004). The pretreated suspension is filtered to separate the cellulose-rich solid from the hemicelluloses rich filtrate and the solid component is washed with deionized water before undergoing enzymatic hydrolysis. Wet oxidation pretreatment oxidizes the hemicellulose fractions into intermediates such as carboxylic acids- *via* peeling reactions and chain cleavage (Palonen et al. 2004; Rovio et al. 2012). The degree of fractionation, in most cases, is influenced by the reaction temperature more than the time and oxygen dose. High temperatures, pressures, pH and catalysts favor rapid oxidation; the catalysts further cause increases the acid concentration regardless of reaction temperature.

14.4.2.5 Use of Oxidizing Agents

Pretreatment involving the application of oxidizing agents has received less attention partly due to the high cost of the oxidants such as ozone and hydrogen peroxide.

Ozonolysis: In ozonolysis, ozone is introduced into a mixture of biomass and water at room temperature and specific time periods leading to the solubilization of lignin and hemicelluloses. The targets are the aromatic ring of lignin and the process is influenced by ozone concentration, biomass type, moisture content and air/ozone flow rate (Garcia-Cubero et al. 2009). Though the process is relatively expensive due to the large requirements of ozone, but with some benefits including effective removal of lignin, very low production of inhibitory products and reactions performed at atmospheric conditions (Schultz-Jensen et al. 2011; Travaini et al. 2013).

Lignin degradation products such as carboxylic acids could be eliminated by washing with water at room temperature. Other oxidants such as hydrogen peroxide, peracetic acid, sulphur trioxide and chlorine dioxide have also been used for solubilizing hemicellulose and lignin under mild alkaline or neutral conditions. Among the oxidants, hydrogen peroxide is the most used compound that finally degrades into hydrogen and oxygen and does not leave any residues in the biomass (Uppal et al. 2011). Sulphur trioxide in a process called sulphur trioxide microthermal explosion (STEX) has also been used to pretreat rice straw (Yao et al. 2011) (Table 14.5).

14.4.2.6 Studies on Chemical Treatment of Crop Residues

Crop residues are important feed resources in the Himalayan region particularly during the lean periods. However, due to low nutritional quality and digestibility, they cannot be used as a sole feed. Different treatments were investigated by our group for improving the quality and digestibility of crop residues.

Table 14.5 Effect of different chemical treatments on major crop residues

Treatment	Crop residues	Biochemical changes	References
Ammonia treatment	Rice straw	24% increase in NDF digestion	Abou-EL-Enin et al. (1999)
Ammonia treatment	Rice straw	120% increase in N content 125% increase in CP content	Selim et al. (2004)
Ammonia treatment	Maize stover	15% increase in IVDMD	Oji et al. (2007)
Ammonia treatment	Maize stover	Increase in dry matter intake	Saenger et al. (1982)
Urea + Calcium hydroxide	Rice straw	12.50% increase in IVDMD	Pradhan et al. (1997)
Urea	Maize stover	150% increase in CP content	Fall (1987)
Urea	Barley straw	10% increase in IVDOMD	Aregheore (2005)
Sodium hydroxide	Maize stover	4.88% ADF content 6.25% % decrease in lignin 7.25% % decrease in cellulose	Ololade et al. (1970)
Sodium hydroxide	Maize stover	47.43% increase in OMD 7.30% decrease in ADF content 6.30% decrease in ADL content 24.27% decrease in cell wall constituents	Klopfenstein et al. (1972)
Lime	Rice straw	15% increase in IVDMD	Pradhan et al. (1997)
Dilute sulfuric acid	Maize stover	12.20% decrease in ADL content	Chen et al. (2009)

14.4.2.6.1 Effect of Chemical Treatments on Nutritional Quality of Maize Stover

Maize stover is one of the abundant crop residues. It has high lignocelluloses content with low protein, fat and mineral content. Though maize stover has low nutritional quality yet it has been widely used for meeting the energy demand of livestock in the region. The effect of different chemical treatments on the quality of maize stover was investigated (Katoch et al. 2021). The maize stover was treated with: Urea (4.50%), NH_4OH (10%), NaOH (2%), $Ca(OH)_2$ (2%), and Dil. H_2SO_4 (1.50%). The effect of different chemical treatments on nutritional constituents of maize stover has been discussed:

(a) **Crude protein and total carbohydrates**: Treatment with different chemicals revealed variable effects on CP content of stover (Table 14.6). The CP content in untreated stover in different maize genotypes ranged from 7.00 (PMG-96) to 9.80% (PMG-93-2). Ammonium hydroxide treatment resulted increase in the CP content of stover up to 35.0% with the highest increase in genotype PMG-96. Treatment with sodium hydroxide and calcium hydroxide was comparatively not effective in improving CP content of stover as compared with

Table 14.6 Crude protein content (%) in raw and treated maize stover

Genotypes	Untreated	Urea (4.50%)	Aq. ammonia (10%)	NaOH (2%)	$Ca(OH)_2$ (2%)	NaCl (2%)	Dil. H_2SO_4 (1.50%)	SE	CD (5%)
PMG-3	7.70	9.27[a]	10.15[a]	5.60[b]	5.60[b]	8.92[a]	8.92[a]	0.46	0.99
SIRI-5455	8.22	8.75[*]	10.32[a]	5.42[b]	7.00[b]	9.10[*]	8.22[*]	0.54	1.17
PMG-9	8.05	10.50[a]	9.97[a]	4.55[b]	5.60[b]	8.40[*]	9.10[a]	0.46	1.00
PMG-14	7.87	11.20[a]	8.92[*]	4.90[b]	4.90[b]	8.57[*]	10.85[a]	0.66	1.42
PMG-18	8.40	9.80[a]	9.80[a]	5.77[b]	4.90[b]	9.27[*]	9.62[a]	0.42	0.90
Vivek	9.10	9.80[a]	9.62[*]	6.30[b]	4.72[b]	9.27[*]	9.10[*]	0.30	0.65
PMG-33	9.10	10.85[a]	9.80[*]	5.42[b]	6.30[b]	9.27[*]	8.75[*]	0.45	0.98
PMG-36	8.92	9.62[a]	9.80[a]	4.20[b]	5.77[b]	9.10[*]	8.75[*]	0.30	0.65
PMG-40	7.70	9.97[a]	10.15[a]	5.07[b]	7.35[*]	8.92[a]	8.92[a]	0.52	1.12
PMG-41	9.10	8.92[*]	10.50[a]	5.07[b]	7.00[b]	8.92[*]	10.32[a]	0.48	1.05
PMG-49	7.35	10.85[a]	8.75[*]	4.90[b]	5.95[*]	9.27[*]	10.15[a]	1.04	2.25
PMZ-4	9.27	10.32[a]	10.15[*]	4.90[b]	5.07[b]	9.45[*]	10.50[a]	0.46	0.99
PMG-52	7.70	8.40[a]	8.92[a]	4.20[b]	5.07[b]	8.75[a]	8.40[a]	0.25	0.55
PMG-52-1	8.92	9.97[*]	9.62[*]	5.42[b]	5.07[b]	9.27[*]	9.62[*]	0.63	1.36
PMG-62	8.57	10.50[a]	11.20[a]	5.42[b]	6.47[b]	8.92[*]	9.97[*]	0.72	1.56
PMG-93	9.10	9.80[*]	9.80[*]	5.77[b]	5.25[b]	9.27[*]	11.20[a]	0.44	0.95
PMG-93-1	8.57	10.15[*]	8.92[*]	6.12[b]	6.12[b]	8.75[*]	10.15[*]	0.86	1.85
PMG-93-2	9.80	11.55[a]	9.97[*]	4.90[b]	4.90[b]	9.62[*]	10.85[*]	0.76	1.65
PMG-95	7.70	9.45[*]	8.57[*]	5.07[b]	5.25[b]	9.45[*]	9.27[*]	0.84	1.82
PMG-96	7.00	9.27[a]	9.45[a]	4.02[b]	6.47[*]	9.45[a]	9.62[a]	0.48	1.05
Palam Shankar Makka	9.45	10.50[a]	9.45[*]	3.67[b]	5.95[b]	9.62[*]	9.45[*]	0.47	1.03

PMG-97	8.92	9.62^{*}	9.27^{*}	7.17^{b}	4.90^{b}	9.62^{*}	8.57^{*}	0.49	1.06
Local Chadhiar	9.10	9.80^{*}	9.62^{*}	7.00^{b}	4.72^{b}	9.27^{*}	9.97^{*}	0.76	1.65
African Tall	8.57	9.45^{*}	9.10^{*}	7.35^{b}	4.55^{b}	10.15^{a}	9.97^{a}	0.51	1.11

Superscripts "a," "b," and "*" represent significant increase, significant decrease, and statistically at par values as compared to untreated, respectively, at $p \leq 0.05$; *NS* nonsignificant, *CD* critical difference, *SE* standard error

other chemical treatments. Acid hydrolysis resulted in 38.09 and 37.42% increase in CP content of stover from genotype PMG-49 and PMG-96, respectively. In urea treated maize stover, CP content varied from 8.40% (PMG-52) to 11.55% (PMG-93-2) and the highest increase in the CP content (47.61%) was observed in stover from genotype PMG- 49 followed by 32.42% increase in stover from genotype PMG-96. Variation in total carbohydrates content of untreated stover ranged from 9.09 to 18.02% (Table 14.7). Treatment with sodium hydroxide and calcium hydroxide resulted increase in the carbohydrate content. Sodium chloride treatment resulted 39.27% increase in total carbohydrates content in stover (SIRI-5455) (Table 14.7). Urea and aqueous ammonia treatments revealed variable effects on total carbohydrates content of stover. The total carbohydrates content of stover treated with urea and aqueous ammonia varied from 8.01 to 13.94% and 6.12 to 19.54%, respectively. The highest increase in total carbohydrates content after the treatment was observed in stover from genotype SIRI-5455 and PMG-95.

(b) **Fiber and lignin content:** The variation in NDF content of stover from different maize genotypes was observed from 67.20 (Palam Shankar Makka) to 74.80% (PMZ-4). Urea treatment of maize stover (PMG-18) revealed a 14.76% reduction in NDF content as compared to untreated stover. Treatment with alkali resulted reduction in NDF content of stover to the extent of 13.77% (Table 14.8). Variation in ADF content of maize stover ranged from 40.50% (PMG-3) to 49.00% (PMG-62). Highest reduction (8.82%) in ADF content of maize stover (PMG-93-1) was observed after urea treatment. Among alkali treatments, calcium hydroxide treatment resulted 8.54% reduction (PMG-52), whereas sodium hydroxide treatment resulted in 8.02% reduction in ADF content of stover (PMG-40) (Table 14.9). Treatment with aqueous ammonia resulted in 8.0% reduction in ADF content of stover from genotype Palam Shankar Makka. Treatment of maize stover (PMG-93-1) with sodium chloride revealed 7.50% reduction in ADF content. Acid hydrolysis of stover (PMG-93-1) revealed 5.46% reduction in ADF content in comparison to untreated stover. ADL content in untreated stover of maize genotypes varied from 3.70 (PMG-41) to 7.20% (PMG-52-1).

Treatment with both sodium hydroxide and calcium hydroxide resulted in 52.08 and 56.25% reduction in ADL content of maize stover from genotypes Palam Shankar Makka and PMG-9, respectively. Acid hydrolysis also reduced the ADL content (41.61%) of stover (PMG-96) (Table 14.10).

(c) **Polyphenolic and silica content:** Total phenolic content in untreated maize stover was observed in the range of 9.12 (SIRI-5455) to 11.78 mg/g (PMG-93). Urea treatment reduced phenolic content up to 44.01% in stover from genotype PMG-49. Aqueous ammonia treatment resulted in the highest reduction in total phenolic content of stover from genotype African Tall. Stover from genotypes African Tall and PMG-52-1 revealed highest decrease in total phenolic content with sodium chloride and acid treatments, respectively. Among all chemical treatments, sodium hydroxide treatment resulted highest decrease in total phenolic content (11.43–3.17 mg/g) of stover from genotype PMZ-4. Calcium

Table 14.7 Total carbohydrates content (%) in raw and treated maize stover

Genotypes	Untreated	Urea (4.50%)	Aq. ammonia (10%)	NaOH (2%)	$Ca(OH)_2$ (2%)	NaCl (2%)	Dil. H_2SO_4 (1.50%)	SE	CD (5%)
PMG-3	14.06	11.60[b]	9.64[b]	13.85[*]	17.57[a]	13.23[*]	12.05[*]	1.09	2.35
SIRI-5455	9.09	11.64[a]	9.90[*]	16.21[a]	16.36[a]	12.66[a]	12.55[a]	1.09	2.35
PMG-9	9.64	9.45[*]	6.95[b]	14.94[a]	15.94[a]	11.11[a]	13.72[a]	0.61	1.31
PMG-14	14.62	13.94[*]	10.89[b]	20.55[a]	18.06[a]	16.59[a]	16.25[*]	0.80	1.74
PMG-18	11.03	11.72[*]	8.84[b]	14.74[a]	17.12[a]	13.94[a]	14.17[a]	0.58	1.26
Vivek	12.73	12.20[*]	12.47[*]	15.95[a]	18.82[a]	11.90[*]	12.96[*]	0.76	1.64
PMG-33	12.51	11.07[b]	12.81[*]	16.75[a]	18.93[a]	12.62[*]	12.32[*]	0.19	0.41
PMG-36	15.53	11.45[b]	7.86[b]	17.93[a]	18.25[a]	16.28[*]	11.68[b]	1.02	2.22
PMG-40	11.33	11.34[*]	10.39[*]	18.48[a]	17.53[a]	13.15[a]	13.91[a]	0.73	1.57
PMG-41	10.77	12.62[a]	10.84[*]	15.51[a]	18.25[a]	13.76[a]	12.85[a]	0.61	1.31
PMG-49	13.91	12.51[*]	12.09[*]	17.97[a]	25.20[a]	16.36[*]	11.90[*]	1.56	3.37
PMZ-4	11.86	11.03[*]	6.12[b]	16.32[a]	20.65[a]	13.49[*]	12.74[*]	0.78	1.69
PMG-52	13.45	13.72[*]	14.06[*]	16.76[a]	20.90[a]	14.32[*]	14.10[*]	0.46	1.01
PMG-52-1	12.92	12.74[*]	11.15[b]	15.32[a]	26.56[a]	12.43[*]	11.49[b]	0.57	1.24
PMG-62	14.06	11.98[*]	12.43[*]	13.99[*]	18.99[a]	13.15[*]	16.29[a]	1.02	2.21
PMG-93	12.89	11.00[*]	8.58[b]	11.92[*]	22.52[a]	9.11[b]	13.72[*]	1.02	2.21
PMG-93-1	13.30	13.26[*]	16.47[a]	20.22[a]	21.20[a]	17.04[a]	20.93[a]	1.39	3.00
PMG-93-2	13.00	12.51[*]	11.26[b]	14.51[a]	22.22[a]	14.17[*]	12.85[*]	0.64	1.38
PMG-95	11.03	13.04[a]	17.65[a]	17.72[a]	22.37[a]	14.78[a]	13.19[a]	0.20	0.44
PMG-96	18.02	10.70[b]	19.54[a]	17.15[b]	20.10[a]	14.81[b]	19.69[a]	0.26	0.57
Palam Shankar Makka	14.40	9.67[b]	12.40[b]	17.25[a]	20.33[a]	14.06[*]	12.58[b]	0.52	1.12
PMG-97	13.00	10.77[b]	7.86[b]	14.01[*]	19.42[a]	13.00[*]	16.48[a]	0.51	1.11

(continued)

Table 14.7 (continued)

Genotypes	Untreated	Urea (4.50%)	Aq. ammonia (10%)	NaOH (2%)	$Ca(OH)_2$ (2%)	NaCl (2%)	Dil. H_2SO_4 (1.50%)	SE	CD (5%)
Local Chadhiar	13.94	10.32^b	9.23^b	19.40^a	16.32^a	13.38^*	13.94^*	0.78	1.69
African Tall	13.04	8.01^b	8.05^b	13.81^*	20.59^a	11.79^*	13.53^*	0.85	1.84

Superscripts "a," "b," and "*" represent significant increase, significant decrease, and statistically at par values as compared to untreated, respectively, at $p \leq 0.05$; *NS* nonsignificant, *CD* critical difference, *SE* standard error

Table 14.8 NDF (%) in raw and treated maize stover

Genotypes	Untreated	Urea (4.50%)	Aq. ammonia (10%)	NaOH (2%)	$Ca(OH)_2$ (2%)	NaCl (2%)	Dil. H_2SO_4 (1.50%)	SE	CD (5%)
PMG-3	71.50	69.10	67.70	64.60	66.10	70.10	69.30	2.03	NS
SIRI-5455	70.00	66.30^{b}	68.40^{b}	67.50^{b}	68.60^{b}	69.00^{*}	69.00^{*}	0.66	1.42
PMG-9	68.30	62.40^{b}	66.20^{b}	67.30^{*}	66.70^{b}	66.20^{b}	67.00^{b}	0.52	1.12
PMG-14	72.40	67.60^{b}	70.10^{*}	66.20^{b}	64.60^{b}	70.40^{*}	70.50^{*}	1.48	3.20
PMG-18	71.80	61.20^{b}	69.70^{*}	71.00^{*}	69.80^{*}	68.50^{*}	69.90^{*}	1.78	3.86
Vivek	72.40	64.20^{b}	68.40^{b}	68.70^{b}	69.60^{b}	70.20^{*}	70.20^{*}	1.27	2.75
PMG-33	71.70	69.00	69.10	68.20	66.80	68.50	68.90	1.71	NS
PMG-36	69.10	68.00^{*}	66.00^{b}	66.10^{b}	66.80^{b}	66.70^{b}	66.90^{b}	0.75	1.62
PMG-40	71.90	67.40	67.80	66.90	65.40	68.00	69.80	1.80	NS
PMG-41	71.90	65.80	65.60	69.70	69.50	69.30	70.10	2.08	NS
PMG-49	71.20	68.40	69.40	69.50	67.30	68.70	71.90	1.64	NS
PMZ-4	74.80	67.70^{b}	70.80^{b}	67.50^{b}	64.50^{b}	72.40^{b}	72.10^{b}	0.78	1.69
PMG-52	70.70	68.30	67.70	65.50	65.70	69.10	67.50	2.27	NS
PMG-52-1	72.70	65.50^{b}	67.90^{b}	67.50^{b}	69.50^{b}	70.60^{*}	70.90^{*}	1.24	2.70
PMG-62	72.80	70.40	68.30	67.50	67.60	70.20	70.70	1.85	NS
PMG-93	73.20	70.70	67.50	67.70	67.90	70.20	70.90	2.13	NS
PMG-93-1	70.90	63.50^{b}	70.20^{*}	63.90^{b}	64.70^{b}	67.50^{*}	68.30^{*}	1.91	4.14
PMG-93-2	71.20	69.40^{*}	66.50^{b}	66.10^{b}	65.10^{b}	69.10^{*}	68.90^{*}	1.14	2.46
PMG-95	71.60	68.60	68.70	66.70	66.90	68.00	68.60	2.37	NS
PMG-96	72.20	62.60^{b}	68.30^{*}	64.60^{b}	66.40^{b}	67.70^{*}	68.80^{*}	2.22	4.82
Palam Shankar Makka	67.20	64.60^{b}	66.80^{*}	63.70^{b}	66.00^{*}	65.80^{*}	65.80^{*}	0.99	2.14

(continued)

Table 14.8 (continued)

Genotypes	Untreated	Urea (4.50%)	Aq. ammonia (10%)	NaOH (2%)	$Ca(OH)_2$ (2%)	NaCl (2%)	Dil. H_2SO_4 (1.50%)	SE	CD (5%)
PMG-97	72.30	62.70[b]	70.80*	66.80[b]	66.10[b]	68.90[b]	69.90*	1.48	3.21
Local Chadhiar	74.70	68.20[b]	68.80[b]	67.00[b]	65.80[b]	70.20[b]	73.00*	2.01	4.35
African Tall	69.00	67.00*	66.70*	63.30[b]	62.90[b]	65.20[b]	67.40*	1.54	3.34

Superscripts "a," "b," and "*" represent significant increase, significant decrease, and statistically at par values as compared to untreated, respectively, at $p \leq 0.05$; *NS* nonsignificant, *CD* critical difference, *SE* standard error

Table 14.9 ADF (%) in raw and treated maize stover

Genotypes	Untreated	Urea (4.50%)	Aq. ammonia (10%)	NaOH (2%)	$Ca(OH)_2$ (2%)	NaCl (2%)	Dil. H_2SO_4 (1.50%)	SE	CD (5%)
PMG-3	40.50	39.40^{b}	39.10^{b}	38.60^{b}	38.90^{b}	40.13^{*}	41.47^{*}	0.49	1.06
SIRI-5455	46.40	43.80^{b}	44.00^{b}	44.40^{b}	43.20^{b}	44.90^{b}	46.70^{*}	0.38	0.82
PMG-9	48.60	45.20^{b}	45.90^{b}	46.13^{b}	45.60^{b}	45.30^{b}	48.20^{*}	0.44	0.96
PMG-14	43.40	40.70	41.10	40.10	40.00	42.20	44.50	1.76	NS
PMG-18	46.70	43.30^{b}	43.80^{*}	43.90^{*}	43.40^{b}	44.40^{*}	47.90^{*}	1.42	3.05
Vivek	46.90	44.90^{b}	44.50^{b}	44.60^{b}	43.30^{b}	45.00^{b}	46.53^{*}	0.48	1.02
PMG-33	48.20	45.90^{b}	45.80^{b}	45.50^{b}	44.90^{b}	46.40^{b}	47.90^{*}	0.63	1.35
PMG-36	44.40	41.80^{b}	42.20^{b}	42.20^{b}	43.20^{*}	42.47^{b}	44.00^{*}	0.68	1.46
PMG-40	44.90	41.70^{b}	43.20^{*}	41.30^{b}	41.90^{b}	44.00^{*}	44.30^{*}	0.84	1.80
PMG-41	46.00	43.40	42.00	43.10	41.40	45.40	45.53	1.57	NS
PMG-49	47.50	45.40^{b}	44.00^{b}	44.50^{b}	44.50^{b}	46.67^{*}	46.10^{b}	0.58	1.25
PMZ-4	44.20	41.40^{b}	42.40^{*}	41.70^{b}	43.40^{*}	43.00^{*}	43.80^{*}	0.87	1.88
PMG-52	48.00	46.30^{b}	45.70^{b}	44.50^{b}	43.90^{b}	45.27^{b}	46.50^{*}	0.71	1.52
PMG-52-1	44.40	43.60	42.20	42.30	42.40	44.00	43.60	1.90	NS
PMG-62	49.00	46.20^{b}	45.70^{b}	45.80^{b}	45.50^{b}	46.27^{b}	47.00^{b}	0.38	0.82
PMG-93	46.40	43.30	44.10	43.70	45.00	43.60	46.00	1.27	NS
PMG-93-1	47.60	43.40^{b}	45.00^{b}	45.00^{b}	43.60^{b}	44.00^{b}	45.00^{b}	0.93	2.01
PMG-93-2	45.60	42.30	42.80	43.20	42.70	45.00	44.20	1.91	NS
PMG-95	46.20	42.70^{b}	43.50^{b}	42.90^{b}	43.20^{b}	45.00^{*}	44.70^{*}	0.95	2.05
PMG-96	43.00	41.40^{*}	43.50^{*}	39.70^{b}	40.30^{b}	42.00^{*}	42.80^{*}	0.94	2.02
Palam Shankar Makka	47.50	44.20^{b}	43.70^{b}	44.90^{b}	44.70^{b}	44.80^{b}	45.18^{b}	0.69	1.48

(continued)

Table 14.9 (continued)

Genotypes	Untreated	Urea (4.50%)	Aq. ammonia (10%)	NaOH (2%)	$Ca(OH)_2$ (2%)	NaCl (2%)	Dil. H_2SO_4 (1.50%)	SE	CD (5%)
PMG-97	45.60	44.10	43.70	43.20	42.70	44.90	44.40	1.10	NS
Local Chadhiar	47.80	46.00	44.70	44.50	45.60	45.60	46.00	0.91	NS
African Tall	46.20	43.10	43.50	42.80	43.30	44.50	46.00	1.58	NS

Superscripts "a," "b," and "*" represent a significant increase, significant decrease, and statistically at par values as compared to untreated, respectively, at $p \leq 0.05$; *NS* nonsignificant, *CD* critical difference, *SE* standard error

Table 14.10 ADL content (%) of raw and treated maize stover

Genotypes	Untreated	Urea (4.50%)	Aq. ammonia (10%)	NaOH (2%)	$Ca(OH)_2$ (2%)	NaCl (2%)	Dil. H_2SO_4 (1.50%)	SE	CD (5%)
PMG-3	5.30	4.50^{b}	3.90^{b}	3.50^{b}	2.60^{b}	4.33^{b}	5.20*	0.31	0.67
SIRI-5455	5.10	3.60^{b}	3.70^{b}	3.00^{b}	2.80^{b}	4.22*	4.80*	0.49	1.05
PMG-9	4.80	3.90^{b}	3.00^{b}	3.10^{b}	2.10^{b}	4.00^{b}	4.10^{b}	0.16	0.35
PMG-14	6.30	4.80^{b}	4.20^{b}	4.40^{b}	4.00^{b}	5.50^{b}	5.13^{b}	0.31	0.66
PMG-18	5.20	3.60^{b}	3.50^{b}	3.20^{b}	3.00^{b}	4.00^{b}	3.67^{b}	0.31	0.67
Vivek	5.40	4.10	3.70	3.70	3.10	3.40	4.80	1.25	NS
PMG-33	4.80	3.20^{b}	3.30^{b}	2.80^{b}	2.70^{b}	4.00^{b}	3.27^{b}	0.15	0.32
PMG-36	5.70	4.60	4.00	3.60	3.20	5.00	4.80	0.78	NS
PMG-40	6.50	4.70^{b}	4.70^{b}	4.60^{b}	3.90^{b}	4.50^{b}	5.50^{b}	0.41	0.89
PMG-41	3.70	2.60	2.40	3.00	2.10	2.80	3.00	0.54	NS
PMG-49	5.50	4.00^{b}	3.90^{b}	2.90^{b}	3.50^{b}	4.30^{b}	3.50^{b}	0.31	0.67
PMZ-4	5.90	4.40^{b}	4.60^{b}	3.50^{b}	3.40^{b}	4.50^{b}	4.13^{b}	0.25	0.53
PMG-52	5.70	4.10^{b}	3.80^{b}	3.20^{b}	3.30^{b}	4.60^{b}	5.20^{b}	0.23	0.49
PMG-52-1	7.20	5.00	5.30	4.80	4.80	5.60	6.20	0.88	NS
PMG-62	6.10	3.90	4.40	4.80	4.00	5.80	4.60	0.72	NS
PMG-93	4.40	2.90^{b}	3.40^{b}	2.40^{b}	2.70^{b}	3.47^{b}	3.00^{b}	0.29	0.62
PMG-93-1	4.70	3.00^{b}	4.10*	2.80^{b}	2.70^{b}	2.50^{b}	4.40*	0.36	0.77
PMG-93-2	5.50	4.00^{b}	4.00^{b}	3.20^{b}	3.30^{b}	4.30^{b}	5.00^{b}	0.24	0.53
PMG-95	6.30	4.50	4.80	5.00	4.30	5.00	4.53	0.58	NS
PMG-96	5.60	4.60*	4.40^{b}	3.80^{b}	3.90^{b}	4.40^{b}	3.27^{b}	0.53	1.13
Palam Shankar Makka	4.80	3.20^{b}	4.30^{b}	2.30^{b}	2.70^{b}	2.00^{b}	3.20^{b}	0.21	0.45

(continued)

Table 14.10 (continued)

Genotypes	Untreated	Urea (4.50%)	Aq. ammonia (10%)	NaOH (2%)	$Ca(OH)_2$ (2%)	NaCl (2%)	Dil. H_2SO_4 (1.50%)	SE	CD (5%)
PMG-97	6.20	4.60^b	4.90^b	3.90^b	3.80^b	5.10^b	5.40^b	0.10	0.22
Local Chadhiar	4.40	2.70^b	3.20^b	3.10^b	2.40^b	2.90^b	3.33^b	0.33	0.70
African Tall	4.70	2.50^b	2.90^b	2.70^b	2.10^b	3.40^b	4.00^b	0.11	0.24

Superscripts "a," "b," and "*" represent a significant increase, significant decrease, and statistically at par values as compared to untreated, respectively, at $p \leq 0.05$, *NS* nonsignificant, *CD* critical difference, *SE* standard error hydroxide treatment also revealed reduction in total phenolic content of stover from the genotype Vivek (Table 14.11). Total tannins content in untreated stover varied from 2.48 (Palam Shankar Makka) to 7.06 mg/g (PMG-52-1). The treatment with urea, aqueous ammonia and calcium hydroxide resulted in decrease in total tannin content and stover from genotype PMG-14. Sodium hydroxide treatment resulted in the reduction in total tannin content from 6.67 to 1.93 mg/g from genotype PMZ-4. Stover from genotype African Tall exhibited a significant reduction in total tannins from 6.17 to 2.30 mg/g after treatment with sodium chloride. Stover from genotypes PMG-52-1 and African Tall revealed reduction (32.72 and 11.67%, respectively) in total tannins content, after treatment with dilute acid (Table 14.12). After aqueous ammonia treatment, the highest reduction in silica content (2.60–1.20%) was observed in stover from the genotype PMG-9. Sodium hydroxide and calcium hydroxide treatments also reduced the silica content of stover from genotype PMG-93-1 and PMG-41, respectively. Acid hydrolysis revealed highest reduction in silica content from 2.60 to 1.07% in stover from the genotype Palam Shankar Makka (Table 14.13).

(d) ***In vitro* dry matter digestibility (IVDMD)**: *In vitro* dry matter digestibility of untreated maize stover ranged from 52.10 (PMG-3) to 59.40% (Palam Shankar Makka). Urea and aqueous ammonia treatments improved the digestibility of stover by 33.21 (PMG-3) and 29.17% (PMG-52-1), respectively. Treatment with sodium hydroxide and calcium hydroxide also increased the digestibility of stover to the extent of 29.51 (PMG-18) and22.37% (PMG-18), respectively. Sodium chloride treatment resulted in the improvement in the digestibility of stover to the extent of 17.47% (PMG-3) (Table 14.14). Like other treatments, acid hydrolysis also increased the digestibility of maize stover from 5.16 (PMG-62) to 22.56% (PMG-18).

14.4.2.6.2 Effect of Chemical Treatments on Quality of Hay, Wheat Straw and Rice Straw

Wheat (*Triticum aestivum*) and rice (*Oryza sativa*) are the important cereal crops in Himalayan region. Since wheat and rice are cultivated traditionally and commercially in this region, the availability of biomass production is also high. Farmers utilize biomass from these crops as livestock feed when green fodder is unavailable. The high fiber content, lignocellulosic complex, tannin–protein complex and silica in the cell wall are collectively responsible for low quality of these crop residues. SSG (Sorghum Sudan grass) and Guria grass (*Chrysopogon fulvus*) are two important grasses for haymaking in Himalayan region. The rapid establishment and ease in handling increase the importance of these grasses as a potential livestock feed. Despite of their availability, these grasses are not used as potential livestock feed on a large scale. We have investigated different chemical treatments for improving the quality of hay (SSG; Sorghum sudan grass; Guria grass; *Chrysopogon fulvus*) and crop residues (Rice straw and Wheat straw) (Katoch and Tripathi 2020).

Table 14.11 Total phenolic content (mg/g) of raw and treated maize stover

Genotypes	Untreated	Urea (4.50%)	Aq. ammonia (10%)	NaOH (2%)	$Ca(OH)_2$ (2%)	NaCl (2%)	Dil. H_2SO_4 (1.50%)	SE	CD (5%)
PMG-3	9.29	9.39*	6.28^b	4.38^b	6.68^b	9.47*	9.10*	1.04	2.24
SIRI-5455	9.12	8.30*	5.11^b	5.79^b	5.67^b	9.07*	9.28*	0.75	1.62
PMG-9	10.01	10.16*	6.58^b	5.96^b	7.99^b	9.07*	10.36*	0.72	1.56
PMG-14	11.58	7.37^b	4.94^b	5.24^b	3.53^b	11.15*	11.35*	0.34	0.73
PMG-18	9.87	9.03*	5.65^b	9.38*	4.27^b	9.69*	9.37*	0.70	1.51
Vivek	10.24	10.07*	6.15^b	6.96^b	3.08^b	7.92^b	9.93*	0.89	1.93
PMG-33	9.79	9.37*	5.85^b	6.24^b	3.72^b	8.30*	9.60*	0.75	1.61
PMG-36	10.02	8.98*	7.21^b	5.67^b	6.41^b	7.53^b	9.44*	0.72	1.56
PMG-40	11.43	8.96^b	5.60^b	4.68^b	7.52^b	10.39*	11.40*	0.99	2.13
PMG-41	11.22	9.97*	6.03^b	6.06^b	8.36^b	10.54*	10.92*	0.78	1.69
PMG-49	11.02	6.17^b	5.47^b	3.83^b	6.41^b	10.54*	10.37*	0.45	0.96
PMZ-4	11.43	8.88^b	4.59^b	3.17^b	6.43^b	9.53*	10.98*	0.97	2.09
PMG-52	11.08	10.33*	6.52^b	4.72^b	7.35^b	8.61^b	10.80*	0.82	1.76
PMG-52-1	11.53	10.69*	6.18^b	5.59^b	8.53^b	10.78*	9.42^b	0.80	1.72
PMG-62	11.83	10.00*	4.69^b	5.96^b	6.78^b	9.30*	11.37*	1.26	2.70
PMG-93	11.78	9.46^b	4.84^b	7.09^b	6.69^b	10.52*	10.41*	0.96	2.06
PMG-93-1	11.14	6.66^b	4.09^b	6.89^b	4.92^b	9.11*	10.45*	1.00	2.16
PMG-93-2	11.66	9.78^b	5.89^b	7.75^b	6.21^b	10.77*	10.27*	0.75	1.62
PMG-95	10.41	9.50*	4.45^b	7.28^b	7.87^b	9.14*	9.63*	0.85	1.84
PMG-96	10.26	8.42^b	3.75^b	6.68^b	5.45^b	9.15*	10.08*	0.54	1.16
Palam Shankar Makka	10.71	9.56^b	4.50^b	5.48^b	6.66^b	9.62*	10.05*	0.48	1.04

PMG-97	11.12	8.28^b	4.89^b	7.33^b	5.52^b	6.46^b	10.60^*	0.66	1.43
Local Chadhiar	9.20	9.17^*	4.71^b	8.72^*	6.12^b	8.76^*	8.47^*	0.66	1.43
African Tall	12.58	8.00^b	4.06^b	6.06^b	5.61^b	6.96^b	11.27^*	0.70	1.51

Superscripts "a," "b," and "*" represent significant increase, significant decrease, and statistically at par values as compared to untreated, respectively, at $p \leq 0.05$; *NS* nonsignificant, *CD* critical difference, *SE* standard error

Table 14.12 Total tannins content (mg/g) of raw and treated maize stover

Genotypes	Untreated	Urea (4.50%)	Aq. ammonia (10%)	NaOH (2%)	$Ca(OH)_2$ (2%)	NaCl (2%)	Dil. H_2SO_4 (1.50%)	SE	CD (5%)
PMG-3	3.96	3.93	2.85	2.77	2.93	3.53	3.50	0.70	NS
SIRI-5455	3.04	2.37	2.27	2.68	2.86	2.60	2.79	0.72	NS
PMG-9	4.86	4.54	4.45	3.27	4.43	3.65	4.68	0.90	NS
PMG-14	6.45	4.40^b	1.53^b	3.14^b	1.72^b	6.39^*	6.11^*	0.44	0.94
PMG-18	4.78	4.43	4.10	4.64	2.54	3.85	3.48	0.73	NS
Vivek	3.97	3.00^*	3.47^*	3.83^*	1.40^b	2.55^b	3.31^*	0.58	1.24
PMG-33	4.90	4.55^*	3.79^*	4.12^*	1.66^b	3.65^*	4.66^*	0.70	1.50
PMG-36	5.34	4.97	5.09	4.82	4.11	4.51	4.93	0.90	NS
PMG-40	5.13	4.87^*	3.52^b	2.67^b	4.85^*	4.78^*	4.80^*	0.34	0.73
PMG-41	5.99	5.76	4.21	4.25	4.62	5.05	5.13	0.62	NS
PMG-49	5.26	4.09^b	3.45^b	2.45^b	3.68^b	4.79^*	4.64^*	0.51	1.10
PMZ-4	6.15	5.78^*	2.76^b	1.93^b	4.47^*	5.26^*	5.58^*	0.81	1.75
PMG-52	6.67	5.65^*	3.92^b	3.08^b	4.75^b	5.15^b	6.33^*	0.64	1.37
PMG-52-1	7.06	6.47^*	4.34^b	3.93^b	5.88^*	6.71^*	4.75^b	0.75	1.61
PMG-62	6.94	6.84^*	2.14^b	4.24^b	4.31^b	4.15^b	6.51^*	0.99	2.12
PMG-93	4.00	3.46^*	2.45^b	3.91^*	3.61^*	2.63^b	3.97^*	0.51	1.10
PMG-93-1	4.87	4.09^*	2.30^b	4.44^*	3.19^b	2.97^b	4.29^*	0.62	1.34
PMG-93-2	6.16	5.72	3.75	4.40	3.21	5.79	6.00	1.07	NS
PMG-95	3.48	2.76	2.19	2.42	2.55	2.67	2.84	0.59	NS
PMG-96	3.41	2.75^*	1.68^b	3.05^*	3.29^*	2.43^b	3.33^*	0.45	0.96
Palam Shankar Makka	2.48	2.08	2.22	2.22	2.29	1.89	2.22	0.29	NS

PMG-97	4.43	4.37	3.08	4.23	3.70	3.02	4.36	0.83	NS
Local Chadhiar	3.18	2.67	3.13	2.97	3.02	2.70	2.68	0.44	NS
African Tall	6.17	4.90^{b}	1.94^{b}	3.78^{b}	3.40^{b}	2.30^{b}	5.45^{b}	0.31	0.67

Superscripts "a," "b," and "*" represent significant increase, significant decrease, and statistically at par values as compared to untreated, respectively, at $p \leq 0.05$; *NS* nonsignificant, *CD* critical difference, *SE* standard error

Table 14.13 Silica content (%) of raw and treated maize stover

Genotypes	Untreated	Urea (4.50%)	Aq. ammonia (10%)	NaOH (2%)	$Ca(OH)_2$ (2%)	NaCl (2%)	Dil. H_2SO_4 (1.50%)	SE	CD (5%)
PMG-3	2.10	3.70^{a}	2.10*	1.70^{b}	0.90^{b}	2.60^{a}	2.00*	0.09	0.20
SIRI-5455	1.90	2.30^{a}	2.00*	1.80*	1.20^{b}	2.13^{a}	1.80*	0.08	0.18
PMG-9	1.60	2.60^{a}	1.70*	0.90^{b}	0.70^{b}	1.87^{a}	1.43*	0.09	0.18
PMG-14	3.40	3.53*	2.10^{b}	2.60^{b}	1.90^{b}	3.53*	3.27*	0.16	0.34
PMG-18	2.60	2.93^{a}	1.20^{b}	1.70^{b}	1.70^{b}	2.93^{a}	2.47*	0.10	0.21
Vivek	2.50	3.30^{a}	1.40^{b}	1.90^{b}	1.40^{b}	2.80^{a}	2.20^{b}	0.12	0.26
PMG-33	1.90	2.27^{a}	1.50^{b}	0.70^{b}	1.10^{b}	2.27^{a}	1.40^{b}	0.15	0.32
PMG-36	2.50	2.80*	2.40*	1.30^{b}	1.60^{b}	2.67*	2.27*	0.17	0.35
PMG-40	2.50	2.67*	1.90^{b}	1.70^{b}	1.20^{b}	4.60^{a}	2.33*	0.16	0.34
PMG-41	2.10	2.40*	1.30^{b}	0.80^{b}	0.80^{b}	2.20*	1.67^{b}	0.15	0.32
PMG-49	1.30	2.60^{a}	1.10*	1.50*	1.60*	1.60*	1.00*	0.16	0.34
PMZ-4	2.70	2.80*	2.70*	1.50^{b}	1.80^{b}	2.87*	2.27^{b}	0.11	0.25
PMG-52	2.90	3.07^{a}	1.80^{b}	1.10^{b}	1.40^{b}	3.07^{a}	2.33^{b}	0.08	0.16
PMG 52-1	2.70	2.73*	2.40*	2.10^{b}	2.00^{b}	3.00*	2.47*	0.26	0.55
PMG-62	3.10	3.60^{a}	2.40^{b}	2.30^{b}	2.00^{b}	3.47^{a}	2.60^{b}	0.14	0.29
PMG-93	2.10	2.30*	1.80^{b}	1.00^{b}	1.50^{b}	2.27*	1.73^{b}	0.11	0.25
PMG-93-1	2.30	2.47*	2.07*	0.70^{b}	2.40*	2.27*	2.00*	0.13	0.28
PMG-93-2	1.60	2.40^{a}	1.90*	1.30^{b}	1.13^{b}	2.60^{a}	1.30^{b}	0.10	0.23
PMG-95	1.80	2.20*	1.67*	1.47*	1.70*	2.20*	1.33^{b}	0.20	0.42
PMG-96	2.50	2.60*	2.30*	1.80^{b}	1.70^{b}	3.20^{a}	2.27*	0.16	0.35
Palam Shankar Makka	2.60	3.00^{a}	1.70^{b}	1.10^{b}	1.20^{b}	2.90*	1.07^{b}	0.18	0.38

PMG-97	2.60	2.87*	1.40[b]	1.70[b]	2.00[b]	2.87*	2.13[b]	0.20	0.42
Local Chadhiar	1.50	1.60*	0.90[b]	1.07[b]	1.20[b]	1.70*	1.20[b]	0.12	0.27
African Tall	1.90	2.00*	0.90[b]	1.70*	1.70*	2.27[a]	1.50[b]	0.13	0.28

Superscripts "a," "b," and "*" represent a significant increase, significant decrease, and statistically at par values as compared to untreated, respectively, at $p \leq 0.05$; *NS* nonsignificant, *CD* critical difference, *SE* standard error

Table 14.14 *In vitro* dry matter digestibility (%) of raw and treated maize stover

Genotypes	Untreated	Urea (4.5%)	NH_4OH (10%)	NaOH (2%)	$Ca(OH)_2$ (2%)	NaCl (2%)	Dil. H_2SO_4 (1.5%)	SE	CD (5%)
PMG-3	52.10	69.40[a]	64.40 [a]	60.10 [a]	62.40 [a]	61.20 [a]	62.40 [a]	0.20	0.43
SIRI-5455	55.40	70.20 [a]	63.40 [a]	65.20 [a]	63.30 [a]	59.20 [a]	64.20 [a]	0.15	0.33
PMG-9	56.20	71.40 [a]	65.20 [a]	64.20 [a]	60.20 [a]	59.90 [a]	65.30 [a]	0.35	0.76
PMG-14	55.20	69.30 [a]	65.40 [a]	67.40 [a]	61.30 [a]	59.20 [a]	67.40 [a]	0.46	1.01
PMG-18	53.20	68.20 [a]	66.00 [a]	68.90 [a]	65.10 [a]	58.40 [a]	65.20 [a]	0.52	1.13
Vivek	56.20	67.20 [a]	68.20 [a]	61.20 [a]	65.20 [a]	60.20 [a]	61.00 [a]	0.33	0.71
PMG-33	57.60	69.30 [a]	66.40 [a]	63.40 [a]	64.20 [a]	61.40 [a]	63.20 [a]	0.22	0.47
PMG-36	57.10	69.60 [a]	68.20 [a]	65.90 [a]	63.90 [a]	60.20 [a]	62.20 [a]	0.37	0.80
PMG-40	58.60	70.10 [a]	61.20 [a]	67.20 [a]	66.30 [a]	62.40 [a]	64.30 [a]	0.38	0.83
PMG-41	59.20	68.60 [a]	63.20 [a]	66.40 [a]	67.40 [a]	61.00 [a]	69.40 [a]	0.49	1.07
PMG-49	58.30	69.70 [a]	64.20 [a]	67.20 [a]	68.40 [a]	61.20 [a]	65.20 [a]	0.12	0.25
PMZ-4	56.40	70.80 [a]	65.00 [a]	61.40 [a]	61.30 [a]	59.40 [a]	61.40 [a]	0.34	0.73
PMG-52	57.40	65.40 [a]	67.00 [a]	65.30 [a]	62.40 [a]	60.20 [a]	64.20 [a]	0.43	0.93
PMG-52-1	52.80	66.30 [a]	68.20 [a]	66.40 [a]	60.30 [a]	56.40 [a]	61.20 [a]	0.34	0.74
PMG-62	59.90	67.60 [a]	69.40 [a]	68.10 [a]	65.90 [a]	59.40*	63.00 [a]	0.26	0.56
PMG-93	57.80	69.10 [a]	61.20 [a]	63.40 [a]	61.20 [a]	60.20 [a]	64.20 [a]	0.14	0.31
PMG-93-1	56.40	68.90 [a]	63.40 [a]	65.30 [a]	63.40 [a]	59.40 [a]	62.90 [a]	0.36	0.79
PMG-93-2	56.30	69.00 [a]	65.90 [a]	66.90 [a]	66.70 [a]	59.00 [a]	66.00 [a]	0.32	0.69
PMG-95	53.40	67.00 [a]	66.00 [a]	62.90 [a]	62.40 [a]	56.00 [a]	61.20 [a]	0.41	0.88
PMG-96	54.80	70.10 [a]	65.00 [a]	63.80 [a]	62.10 [a]	59.20 [a]	63.00 [a]	0.37	0.80
Palam Shankar Makka	59.40	72.20 [a]	64.20 [a]	66.70 [a]	66.30 [a]	62.20 [a]	65.40 [a]	0.13	0.29

PMG-97	57.60	69.80 ^{a}	64.40 ^{a}	71.00 ^{a}	66.20 ^{a}	59.20 ^{a}	63.20 ^{a}	0.32	0.68
Local Chadhiar	58.60	69.40 ^{a}	67.00 ^{a}	67.20 ^{a}	69.30 ^{a}	61.20 ^{a}	64.20 ^{a}	0.46	1.00
African Tall	59.40	68.20 ^{a}	69.00 ^{a}	69.40 ^{a}	62.40 ^{a}	62.20 ^{a}	65.00 ^{a}	0.20	0.42

Superscripts "a," "b," and "*" represent a significant increase, significant decrease, and statistically at par values as compared to untreated, respectively, at $p \leq 0.05$; *NS* nonsignificant, *CD* critical difference, *SE* standard error

(a) **Crude protein (CP) and total carbohydrate content:** After different chemical treatments, CP content in wheat and rice straws improved from 3.85 to 9.57% and 5.48 to 9.10% (Tables 14.15 and 14.16). A significant increase in CP content of wheat straw was observed after treatment with sodium chloride, whereas aqueous ammonia and urea treatments resulted in 66.06 and 63.86% increase in CP content of rice straw. Application of urea, aqueous ammonia and dilute sulphuric acid resulted significant increase, whereas alkali and salt treatments reduced CP content in SSG and Guria grass. The highest increase in CP content of SSG and Guria grass was observed with urea (from 15.52 to 18.20%) and acid treatments (from 12.60 to 16.41%), respectively. Treatment with dilute sulphuric acid significantly increased (79.77%) the level of total carbohydrates in rice straw. The chemical treatments of grasses revealed variable effects on total carbohydrate content. In SSG, the maximum increase (64.73%) in the level of total carbohydrates was observed with sodium hydroxide treatment, whereas the aqueous ammonia treatment revealed 56.51% increase in total carbohydrates of Guria grass.

(b) **Fiber and lignin content:** Highest reduction in NDF content was 20.35% in wheat straw and 9.08% in rice straw treated with calcium hydroxide and sodium hydroxide, respectively (Tables 14.15 and 14.16). The ADF content reduced up to 10.34% in wheat straw and 6.73% in rice straw after treatment with dilute sulphuric acid and sodium chloride, respectively (Tables 14.15 and 14.16). Aqueous ammonia treatment resulted significant reduction in NDF content from 53.93 to 42.66% in SSG and from 50.93 to 39.60% in Guria grass. The dilute acid treatment resulted significant reduction in ADF content from 34.93 to 31.40% and from 34.73 to 30.47%, in SSG and Guria grass, respectively. Treatment of wheat and rice straw with dilute sulfuric acid resulted in reduction in ADL content up to 27.31 and 37.50%, respectively. With all chemical treatments, the ADL content reduced significantly in SSG and Guria grass. The highest reduction in ADL content in both grasses was with dilute sulfuric acid treatment (Tables 14.17 and 14.18).

(c) **Hemicellulose and cellulose content**: The degradation of hemicellulose content in wheat and rice straws treated with calcium hydroxide and sodium hydroxide was significantly higher than that of other treatments, whereas treatment with urea resulted significant increase in cellulose content. Except for sodium chloride and urea treatments, the other chemical treatments resulted reduction in hemicellulose and cellulose content of SSG whereas, in Guria grass, the hemicellulose and cellulose content reduced significantly with all chemical treatments. The highest reduction in hemicellulose and cellulose content of Guria grass was observed with aqueous ammonia and acid treatments, respectively (Tables 14.17 and 14.18).

(d) **Ash content**: Among different chemical treatments, the highest increase in ash content of wheat and rice straw was observed with dilute sulfuric acid and sodium chloride treatment, respectively (Tables 14.15 and 14.16). The increase in ash content of SSG and Guria grass was exhibited by sodium chloride (from 3.30 to 5.20%) and calcium hydroxide treatment (from 4.70 to 5.56%) (Tables 14.17 and 14.18).

Table 14.15 Biochemical composition of raw and treated wheat straw

Parameters	Untreated	Urea (4.50%)	Aq. ammonia (10%)	NaOH (2%)	Ca $(OH)_2$ (2%)	NaCl (2%)	Dil. H_2SO_4 (1.50%)
CP (%)	3.85^{e}	6.18^{cd}	6.07^{d}	6.77^{bc}	6.65^{cd}	9.57^{a}	7.35^{b}
TC (%)	7.14^{d}	4.87^{f}	18.35^{a}	11.33^{c}	6.30^{e}	13.86^{b}	13.78^{b}
NDF (%)	82.53^{a}	79.66^{b}	79.73^{b}	67.00^{d}	65.73^{d}	82.06^{a}	73.93^{c}
ADF (%)	51.53^{a}	51.73^{a}	49.33^{b}	49.26^{b}	48.80^{b}	47.27^{c}	46.20^{d}
IVDMD (%)	49.33^{e}	58.27^{a}	56.67^{b}	55.80^{bc}	54.80^{c}	52.67^{d}	54.80^{c}
CC (%)	17.47^{d}	20.34^{c}	20.27^{c}	33.00^{a}	34.27^{a}	17.94^{d}	26.07^{b}
ADL (%)	7.80^{a}	7.06^{b}	6.80^{b}	6.80^{b}	6.73^{b}	6.73^{b}	5.67^{c}
HC (%)	31.00^{b}	27.93^{c}	30.40^{b}	17.74^{d}	16.93^{d}	34.80^{a}	27.73^{c}
CL (%)	43.73^{b}	44.67^{a}	42.53^{c}	42.47^{c}	42.07^{c}	40.53^{d}	40.53^{d}
Ash (%)	2.30^{c}	2.70^{bc}	2.80^{bc}	3.00^{b}	2.67^{bc}	4.13^{a}	4.20^{a}
TP (mg/g)	8.71^{a}	6.99^{cd}	7.33^{bc}	6.96^{cd}	6.72^{cd}	6.35^{d}	7.93^{b}
SP (mg/g)	3.33^{a}	2.94^{b}	2.60^{c}	2.70^{c}	2.26^{d}	1.90^{e}	2.33^{d}
TT (mg/g)	5.38^{ab}	4.05^{c}	4.73^{bc}	4.26^{c}	4.46^{c}	4.45^{c}	5.60^{a}

Table 14.16 Biochemical composition of raw and treated rice straw

Parameters	Untreated	Urea (4.50%)	Aq. ammonia (10%)	NaOH (2%)	Ca $(OH)_2$ (2%)	NaCl (2%)	Dil. H_2SO_4 (1.50%)
CP (%)	5.48^{c}	8.98^{a}	9.10^{a}	7.35^{b}	5.72^{c}	7.35b	6.07^{c}
TC (%)	6.87^{d}	8.57^{c}	6.11^{de}	6.08^{de}	4.72^{e}	10.80^{b}	12.35^{a}
NDF (%)	76.40^{a}	77.93^{a}	77.20^{a}	69.46^{d}	71.00^{c}	70.26^{cd}	72.86^{b}
ADF (%)	48.60^{b}	49.40^{a}	46.53^{d}	46.46^{d}	47.53^{c}	45.33^{e}	45.60^{e}
IVDMD (%)	45.67^{d}	56.20^{a}	56.07^{a}	54.13^{b}	55.47^{ab}	52.07^{c}	53.80^{bc}
CC (%)	23.60^{d}	22.07^{d}	22.80^{d}	30.54^{a}	29.00^{b}	29.74^{ab}	27.14^{c}
ADL (%)	4.80^{a}	4.20^{b}	4.20^{b}	4.00^{b}	3.47^{c}	3.47^{c}	3.00^{d}
HC (%)	27.80^{b}	28.53^{b}	30.67^{a}	23.00^{d}	23.47^{cd}	24.93^{c}	27.26^{b}
CL (%)	43.80^{b}	45.20^{a}	42.33^{c}	42.46^{c}	44.06^{b}	41.86^{c}	42.60^{c}
Ash (%)	7.40^{d}	8.60^{abc}	8.80^{ab}	8.30^{bc}	8.00^{cd}	9.00^{a}	6.40^{e}
TP (mg/g)	6.63^{a}	6.02^{b}	6.27^{ab}	5.13^{c}	3.57^{e}	4.15^{d}	5.46^{ac}
SP (mg/g)	2.54^{b}	2.90^{a}	2.02^{c}	1.95^{c}	1.35^{d}	2.00^{c}	1.90^{c}
TT (mg/g)	4.08^{ab}	3.12^{c}	4.25^{a}	3.17^{c}	2.22^{d}	2.15^{d}	3.56^{bc}

Table 14.17 Biochemical composition of raw and treated sorghum sudan grass (SSG)

Parameters	Untreated	Urea (4.50%)	Aq. ammonia (10%)	NaOH (2%)	Ca $(OH)_2$ (2%)	NaCl (2%)	Dil. H_2SO_4 (1.50%)
CP (%)	15.52^{b}	18.20^{a}	17.77^{a}	11.55^{c}	9.80^{d}	10.15^{d}	18.08^{a}
TC (%)	9.10^{b}	3.66^{d}	8.12^{bc}	14.99^{a}	1.96^{e}	6.72^{c}	7.70^{bc}
NDF (%)	53.93^{a}	44.80^{e}	42.66^{f}	45.86^{d}	48.80^{c}	51.87^{b}	42.73^{f}
ADF (%)	34.93^{a}	34.93^{a}	33.06^{b}	32.60^{bc}	32.47^{bc}	32.20^{c}	31.40^{d}
IVDMD (%)	54.67^{b}	58.07^{a}	56.53^{ab}	58.13^{a}	59.07^{a}	58.07^{a}	57.80^{a}
CC (%)	46.07^{f}	55.20^{b}	57.34^{a}	54.13^{c}	51.20^{d}	48.13^{e}	57.27^{a}
ADL (%)	5.80^{a}	5.27^{b}	5.00^{b}	4.87^{bc}	5.20^{b}	5.00^{d}	4.53^{c}
HC (%)	19.00^{a}	9.87^{e}	9.60^{e}	13.26^{c}	16.33^{b}	19.67^{a}	11.33^{d}
CL (%)	29.13^{a}	29.66^{a}	28.06^{b}	27.73^{bc}	27.27^{bc}	27.20^{bc}	26.87^{c}
Ash (%)	3.30^{e}	4.50^{b}	3.73^{cd}	4.80^{b}	4.00^{c}	5.20^{a}	3.40^{de}
TP (mg/g)	15.13^{a}	7.04^{e}	10.19^{c}	4.18^{f}	6.77^{e}	12.01^{b}	7.79^{d}
SP (mg/g)	7.32^{a}	2.26^{f}	2.50^{e}	2.53^{e}	3.98^{d}	6.09^{b}	5.65^{c}
TT (mg/g)	7.81^{a}	4.78^{c}	7.69^{a}	1.65^{e}	2.78^{d}	5.92^{b}	2.14^{e}

Table 14.18 Biochemical composition of raw and pretreated Guria grass (*Chrysopogon fulvus*)

Parameters	Untreated	Urea (4.50%)	Aq. ammonia (10%)	NaOH (2%)	Ca $(OH)_2$ (2%)	NaCl (2%)	Dil. H_2SO_4 (1.50%)
CP (%)	12.60^{c}	14.58^{b}	15.37^{b}	9.45^{d}	9.10^{d}	7.12^{e}	16.41^{a}
TC (%)	10.76^{d}	2.41^{g}	16.84^{a}	13.97^{b}	6.23^{f}	7.49^{e}	13.29^{c}
NDF (%)	50.93^{a}	46.33^{cd}	39.60^{f}	42.66^{e}	46.73^{c}	48.80^{b}	45.73^{d}
ADF (%)	34.73^{a}	34.20^{ab}	32.73^{b}	33.33^{ab}	33.40^{ab}	32.87^{ab}	30.47^{c}
IVDMD (%)	50.53^{c}	56.53^{ab}	58.80^{a}	58.07^{ab}	56.40^{ab}	55.87^{b}	57.33^{ab}
CC (%)	49.06^{f}	53.67^{cd}	60.40^{a}	57.33^{b}	53.27^{d}	51.20^{e}	54.27^{c}
ADL (%)	6.53^{a}	6.06^{b}	5.73^{bc}	5.87^{bc}	5.53^{c}	5.67^{bc}	5.47^{c}
HC (%)	16.20^{a}	12.13^{c}	6.87^{e}	9.33^{d}	13.20^{bc}	15.93^{a}	15.26^{ab}
CL (%)	28.20^{a}	28.14^{a}	27.00^{a}	27.46^{a}	27.87^{a}	27.20^{a}	25.00^{b}
Ash (%)	4.70	5.00	5.04	5.07	5.56	5.00	5.50
TP (mg/g)	5.56^{a}	4.64^{b}	4.35^{bc}	2.88^{d}	4.13^{c}	4.36^{bc}	5.39^{a}
SP (mg/g)	4.21^{b}	2.49^{d}	2.17^{e}	2.73^{c}	2.78^{c}	2.24^{e}	4.70^{a}
TT (mg/g)	1.35^{b}	2.15^{a}	2.18^{a}	0.15^{d}	1.35^{b}	2.12^{a}	0.69^{c}

In Tables 14.15, 14.16, 14.17, and 14.18, each value represents mean of three replications. In the same row, significant differences according to CRD are indicated by different alphabets. Superscripts represent significant values at ($p \leq 0.05$). Same alphabets represent that their values are statistically at par ($p \leq 0.05$)

(e) **Phenolic and tannin content**: Total phenolics content reduced significantly to 27.10% in wheat straw and 46.15% in rice straw treated with sodium chloride and calcium hydroxide, respectively (Tables 14.15 and 14.16). In case of simple phenols, sodium chloride treatment reduced up to 42.94% in wheat straw, whereas 46.85% reduction was observed with calcium hydroxide treatment in rice straw (Tables 14.15 and 14.16). Wheat straw treated with urea, and rice straw with sodium chloride revealed maximum reduction in total tannin content from 5.38 to 4.05% and from 4.08 to 2.15%, respectively. Treatment of grasses with sodium hydroxide was promising in reducing the phenolic and tannin content as compared to other treatments (Tables 14.17 and 14.18).

(f) ***In vitro* dry matter digestibility *(IVDMD)***: Among the treatments, the highest increase in digestibility of wheat straw (from 49.33 to 58.27%) and rice straw (from 45.67 to 56.20%) was observed with urea treatment (Tables 14.15 and 14.16). The dry matter digestibility of grasses improved with all chemical treatments. The highest increase in digestibility of SSG (54.67–59.07%) and Guria grass (50.53–58.80%) was observed with calcium hydroxide and aqueous ammonia treatments, respectively (Tables 14.17 and 14.18).

14.4.3 Biological Treatment

Biological treatments are also effective in improving feeding value of crop residues. The chemical and physical treatments are constrained by their high cost, safety and potentially their negative effect on environment and animal health (Chen et al. 1995). Treatment of crop residues with different microorganisms (fungi, bacteria) is an alternative to physical and chemical treatments. Different category of microorganisms has been used for improving nutritional status of lignocelluloses biomass. Biological treatments increase the accessibility of hydrolytic enzymes on cellulosic polysaccharides by disrupting the lignocelluloses complex in crop residues (Ramirez-Bribiesca et al. 2010).

14.4.3.1 Lignocellulolytic Fungi

Lignocellulolytic fungi belonging to wood-decaying basidiomycetes are able to decompose and metabolize cell wall constituents by enzymes secreted by them (Eriksson et al. 1990). The ideal microorganisms should have high ligninolytic capability with low degradation of cellulose and hemicelluloses. The three major aerobic ligninolytic fungi are known to play a major role in lignin degradation of straw (Belewu and Okhawere 1998).

(a) **Brown-rot fungi**

Brown-rot fungi attack cellulose and hemicellulose, leaving lignin intact, and the decaying residue turns brown. This group of fungi is mainly humidifiers causing only limited changes in lignin. They do not cleave lignin's aromatic ring efficiently or if they open the rings, they are unable to make significant decomposition resulting in lignin fragments. This results in low digestibility compared to untreated substrate (Zadrazil et al. 1999). Examples are *Agrocybe aegerita* and *Flammulina velutipes*.

(b) **White-rot fungi**
White-rot fungi are capable of degrading lignin without much affecting cellulose and hemicelluloses, and the decayed residue turn white. White-rot fungi attack unaltered lignin polymers causing cleavage of interlignol bonds and aromatic ring cleavage which ultimately results increase in digestibility (Zadrazil et al. 1999). They mainly degrade polysaccharides with the action of cellulases and xylanases and lignin by oxidative ligninolytic enzymes such as lignin peroxidase, manganese peroxidase and laccase. The important white-rot fungi are *Abortiporus biennis, Agaricus bisporus, Dichomitus squalens, Pleurotus eryngii, Pleurotus sajor caju, Pleurotus ostreatus, Pleurotus flabellatus, Pleurotjus floridanus, Phanerochaete chrysosporium, Ganoderma* sp. rckk02, *Crinipellis* sp., *Pycnoporous sangeus, Coriolus versicolor, Lenzites striata, Poria plascenta,* etc.

(c) **Soft-rot fungi**
Soft-rot fungi leave the lignocellulosic material watery-soft and break down cellulose and hemicelluloses (Chahal et al. 1981). Example: *Chaetomium cellulolyticum*

The extracellular lignin-modifying enzymes secreted by lignocellulolytic fungi include lignin-peroxidase, peroxidase laccase and H_2O_2-producing oxidase. Some white-rot fungi are able to decompose free phenolic monomers and break the bonds with which lignin is cross-linked to polysaccharides in straw thus enhance digestibility. White-rot fungi are capable to colonize cereal straws and liberate water-soluble substrates from polymers fermentation and thus improve their digestibility. The potential of *Pleurotus* fungi such as *P. ostreatus* and *P. eryngii* to reduce indigestible cell wall components and increase the dry matter digestibility (DMD) of straw has been well documented. Treatment with *Pleurotus sajor caju* and *Pleurotus pulonarius* has been effective in improving the CP content and reducing the crude fiber, NDF, ADF and ADL content (Akinfemi et al. 2009). The degradation of lignin by two fungi may be due to their ability to secrete lignin-degrading enzymes such as lignin peroxidase and manganese peroxidase (Nerude and Misurcova 1995). Treatment of maize stover with oyster mushroom *(Pleurotus ostreatus)* also resulted in increase in the CP content (Ramirez-Bribiesca et al. 2010; Adenipekun and Okunlade 2012). The increase in CP may be due to the addition of protein-rich fungal biomass to the feeds (Ragunathan et al. 1996). Fazaeli et al. (2006) reported an 11.74% and 12.42% reduction in NDF and ADF content of rice straw treated with *Pleurotus* fungi, respectively. Fungal secretion of tannase enzyme also attribute to increased CP content in feeds. Treatment of rice straw with *Cyathus stercoreus* for 30 days resulted in degradation of almost half of lignin content. Rice straw contains a considerable amount of ash and silica. *Cyarhus srercoreus* growth was reported to have no effect on the solubility (in neutral detergent solution) of silica, but it increased the TNC content threefold. During fungal growth, part of the cell wall is converted into soluble sugars to provide energy. *Arachniotus* sp. is a white-rot fungus and has been used for economic utilization of many waste products (Shaukat et al. 2006). Higher production of microbial biomass protein by fermentation of maize stover with *Arachniotus* sp. has been reported by Ahmed et al. (2010). The microbial biomass protein increases by utilizing the cellulose and hemicellulose of the substrate as energy source for the synthesis of protein. Studies have shown that the increasing level of fungal-treated wheat straw has linear effects on DM intakes in livestock.

Treatment of crop residues with *Trichoderma viride* produces large amount of cellulolytic enzymes namely exo, endo-gluconases and β-glycosidase (Fadel 1983). El-Menniawy (2008) reported that treatment of sugarcane bagasse with *T. viride* decreases the NFE content. Yalchi and Hajieghrari (2010) observed a significant reduction in NDF and ADF content of wheat straw inoculated with isolates of *T. harzianum*. The treatment increased the digestibility may be due to a decrease in fiber and cell wall constituents. Inoculation of maize stover with *Trichoderma sp.* resulted in improved fiber quality. The lignin content is reduced to an apprcciable amount when treated with *Trichoderma* sp. in maize stover (Islamiyati et al. 2013). The decrease in ADL content may be due to the secretion of lignin peroxidase and manganese peroxidase by the microorganisms which aid enzymatic digestion of lignin (Nerude and Misurcova 1995). *Aspergillus niger* treatment on grass hay decreases cellulose content in crop residues. *Chaetomium cellulolyticum* transforms the hard residue into watery-soft material and breaks down cellulose and hemicelluloses (Chahal et al. 1981). Treatment of rice straw by *P. pulmonarius* increases the CP content from 4.50% to 4.60%, 4.78%, and 9.36%, at 10, 20, and 40 days after fermentation (Jonathan et al. 2012). Biological treatment of a mixture of straws (wheat and cotton) by *Pleorotus ostreatus*, *Pleorotus corniucopia* and *Pleorotus salmoneos* reduced 52.10, 59.30 and 39.40% lignin and improved organic matter digestibility.

Corn stover inoculated with 15% *Trichoderma viride* and incubated for 21 days resulted enhanced protein content from 6.52 to 10.28%, and decreased NDF from 64.27 to 55.39%, ADF from 44.49 to 37.77%, hemicellulose from 19.78 to 18.02% (Islamiyati et al. 2013). Fungal treatment of rice straw decreased crude fibre from 32.89% in control to 19.96% (Akinfemi and Ogunwole 2012). Bagasse inoculated with *Trichoderma viride* resulted increase in protein, cellulose, hemicellulose and ash content and decrease in the ADF, NDF and ADL content (Abdel-Azim et al. 2011). Reduction in NDF, ADF and ADL contents in rice straw was due to the ability of fungi to solubilize and utilize cell walls as carbon sources and changing the ratio of insoluble to soluble carbohydrates in straw. Shrivastava et al. (2012) also reported significant decrease in ADF, NDF, hemicellulose, lignin, and cellulose to 35.00, 38.88, 45.00, 37.48 and 37.86%, respectively, in *P. ostreatus* fermented straw, while 30.04, 33.85, 39.90, 31.29 and 34%, respectively, in *T. ersicolor* fermented straw. Shrivastava et al. (2012) also reported improvement in organic matter digestibility (OMD) of wheat straw (32.22%) with *Ganoderma* sp. rckk02. As a result of fungal growth, a significant decrease in ADF (24.77%), NDF (31.03%), hemicellulose (42.51%), lignin (34.95%) and cellulose (34.33%) contents were observed.

White-rot fungi *Phanerochaete chrysosporium*, *Dichomitus squalens,* and *Cyathus stercoreus* improve the degradability of cell wall, structural polysaccharides, lignin, and phenols. *Phanerochaete chrysosporium* degrade hemicellulose and cellulose affecting the digestion of cell wall components. *Dichomitus squalens* and *Cyathus stercoreus* degrade hemicellulose and lignin in wheat straw. *Cyathus stercoreus* increased digestion of cellulose by 50% in wheat straw after treatment (Agosin et al. 1986).

14.4.3.2 Bacterial Treatment

Treatment of bean straw with *Ruminococcus albus* increased CP content from 7.33 to 15.22% (Abd El-Galil et al. 2011). Abd El-Galil and Ebtehag (2008) reported that bacterial treatments (*Ruminococcus albus* and *Clostridium cellulovorans*) resulted an increase in CP content from 1.45 to 15.16% and a decrease in CF content from 44.08 to 28.44% of rice straw. Gado et al. (2009) reported that *Cellulomonas* and *Bacillus* isolated from camel converted cellulose content of bagasse to the extent of 44.17 and 30.50%, respectively. The reduction in phenolic content in forages treated with *Pseudomonas fluorescens* may be due to the utilization of phenol as a sole source of carbon and energy (Mahiudddin and Fakhruddin 2012). Treatment of rice straw with *Azospirillum brasilense* and *Saccharomyces cerevisiae* reduced 27.54% of CF content whereas *Bacillus megatarium* and *B. circulans* reduced the CF content from 40.33 to 34.38 and 34.89%, respectively (Zayed 2018).

14.4.3.3 Studies on Biological Treatment of Crop Residues

14.4.3.3.1 Effect of Biological Treatments on Quality and Digestibility of Maize Stover

The effects of microorganisms *viz., Trichoderma harzianum, T. viride, Saccharomyces cerevisiae,* and *Pseudomonas fluorescens* on nutritional quality and digestibility of maize stover was investigated by our group (Katoch et al. 2017). The results of the study have been discussed:

(a) **Crude protein (CP) and total carbohydrate content***:* Crude protein content in untreated stover from different maize genotypes were observed from 7.00% (PMG-96) to 9.80% (PMG-93-2). Treatment of stover with *P. fluorescens* revealed highest increase in CP content of stover from genotype PMG-14 followed by 78.64, 72.57, and 66.67% increase in stover from genotypes PMG-49, PMG-96 and PMG-49 treated with *T. harzianum*, *T. viride,* and *S. cerevisiae*, respectively (Table 14.19). Variation in total carbohydrates content in untreated stover ranged from 9.09 to 18.02% from different maize genotypes. Treatment with *T. viride* revealed an increase in carbohydrate content of stover from genotype Vivek to the extent of 83.61% whereas in the case of other fungal treatments the increase was up to 71.32%. The inoculation of stover with *P. fluorescens* was ineffective in improving carbohydrate content (Table 14.20).

Table 14.19 Crude protein content (%) in raw and inoculated maize stover

Genotypes	Untreated	*T. harzianum*	*T. viride*	*S. cerevisiae*	*P. fluorescens*	SE	CD (5%)
PMG-3	7.70	9.28*	10.85[a]	9.10*	13.13[a]	0.73	1.63
SIRI-5455	8.22	8.25*	9.28[a]	11.03[a]	10.33[a]	0.23	0.52
PMG-9	8.05	11.38[a]	11.88[a]	9.10[a]	12.08[a]	0.30	0.68
PMG-14	7.87	12.08[a]	12.05[a]	10.15[a]	15.75[a]	0.46	1.03
PMG-18	8.40	9.98[a]	13.83[a]	9.63[a]	8.58*	0.31	0.70
Vivek	9.10	8.75*	10.15*	11.73[a]	11.90[a]	0.73	1.62
PMG-33	9.10	10.83[a]	9.10*	8.23*	10.33[a]	0.55	1.22
PMG-36	8.92	9.28*	9.80*	11.38[a]	11.55[a]	0.66	1.47
PMG-40	7.70	8.25*	8.93*	11.03[a]	10.85[a]	0.94	2.09
PMG-41	9.10	10.85[a]	11.20[a]	9.63*	11.38[a]	0.50	1.12
PMG-49	7.35	13.13[a]	11.20[a]	12.25[a]	11.55[a]	0.45	1.00
PMZ-4	9.27	11.90[a]	13.13[a]	11.20[a]	11.20[a]	0.66	1.47
PMG-52	7.70	11.18[a]	9.28[a]	8.75[a]	11.20[a]	0.29	0.64
PMG-52-1	8.92	10.85[a]	8.40*	9.28*	13.65[a]	0.71	1.57
PMG-62	8.57	8.75*	9.28[a]	11.73[a]	13.48[a]	0.18	0.40
PMG-93	9.10	12.78[a]	10.83[a]	8.93*	13.83[a]	0.27	0.60
PMG-93-1	8.57	11.18[a]	9.28[a]	13.30[a]	9.98[a]	0.30	0.68
PMG-93-2	9.80	13.13[a]	11.55*	11.20*	15.05[a]	1.01	2.36
PMG-95	7.70	7.53*	8.93[a]	10.50[a]	9.63[a]	0.31	0.68
PMG-96	7.00	8.75[a]	12.08[a]	11.38[a]	12.25[a]	0.30	0.67
Palam Shankar Makka	9.45	9.28*	8.75[b]	12.40[a]	11.55[a]	0.16	0.36
PMG-97	8.92	8.75*	9.80[a]	9.63[a]	12.08[a]	0.36	0.79
Local Chadhiar	9.10	11.90[a]	10.33*	9.98*	11.03[a]	0.58	1.30
African Tall	8.57	12.25[a]	9.98[a]	10.50[a]	12.78[a]	0.42	0.93

Table 14.20 Total carbohydrates content (%) in raw and inoculated maize stover

Genotypes	Untreated	*T. harzianum*	*T. viride*	*S. cerevisiae*	*P. fluorescens*	SE	CD (5%)
PMG-3	14.06	5.37^{b}	14.97*	10.20^{b}	8.09^{b}	0.95	2.11
SIRI-5455	9.09	7.82*	16.69^{a}	9.26*	8.30*	0.78	1.73
PMG-9	9.64	6.92^{b}	15.44^{a}	15.19^{a}	8.88*	0.76	1.69
PMG-14	14.62	6.50^{b}	18.31^{a}	11.56^{b}	10.94^{b}	0.42	0.94
PMG-18	11.03	6.28^{b}	16.72^{a}	10.77*	8.73^{b}	0.64	1.42
Vivek	12.73	14.28^{a}	21.81^{a}	14.59^{a}	9.34^{b}	0.42	0.94
PMG-33	12.51	9.82^{b}	18.29^{a}	15.23^{a}	10.34*	1.16	2.58
PMG-36	15.53	9.71^{b}	15.38*	12.62^{b}	14.25*	0.98	2.19
PMG-40	11.33	8.01^{b}	15.70^{a}	12.77*	10.26*	0.82	1.83
PMG-41	10.77	12.55*	16.97^{a}	11.37*	10.55*	1.02	2.27
PMG-49	13.91	7.59^{b}	14.44*	15.98^{a}	8.69^{b}	0.28	0.62
PMZ-4	11.86	10.47^{b}	14.09^{a}	8.16^{b}	9.92^{b}	0.40	0.90
PMG-52	13.45	8.96^{b}	14.36*	13.04*	12.71*	0.83	1.85
PMG-52-1	12.92	10.70^{b}	17.10^{a}	13.15*	11.80^{b}	0.44	0.99
PMG-62	14.06	7.82^{b}	16.84*	9.71^{b}	11.57*	1.30	2.91
PMG-93	12.89	6.27^{b}	15.61^{a}	11.11^{b}	11.94^{b}	0.40	0.88
PMG-93-1	13.30	8.47^{b}	17.42^{a}	11.94^{b}	8.05^{b}	0.35	0.79
PMG-93-2	13.00	8.14^{b}	13.95*	9.87^{b}	13.13*	0.88	1.96
PMG-95	11.03	11.41*	17.44^{a}	11.79*	10.60*	0.50	1.11
PMG-96	18.02	6.43^{b}	19.58^{a}	13.79^{b}	9.38^{b}	0.52	1.16
Palam Shankar Makka	14.40	9.37^{b}	17.55^{a}	14.77*	13.58^{b}	0.25	0.56
PMG-97	13.00	13.64*	15.49^{a}	8.43^{b}	8.30^{b}	0.66	1.47
Local Chadhiar	13.94	7.56^{b}	12.57^{b}	8.73^{b}	9.16^{b}	0.31	0.69
African Tall	13.04	8.69^{b}	16.65^{a}	8.58^{b}	7.92^{b}	0.40	0.89

(b) **Fiber and lignin content:** The NDF content in stover from 24 maize genotypes were recorded from 67.20 (Palam Shankar Makka) to 74.80% (PMZ-4). Treatment with *T. harzianum*, resulted decrease (7.70%) in NDF content in stover from genotype PMG-95 followed by a 5.66% reduction with *T. viride* (Table 14.21). The variations in ADF content ranged from 40.50 (PMG-3) to 49.00% (PMG-62). The highest reduction in ADF content was observed in stover from genotype PMG-18 (27.34%) with *P. fluorescens* treatment followed by 16.91% reduction (PMG-18) with *T. viride*, 16.59% (PMG-33) with *T. harzianum*, and 15.91% (PMG-18) with *S. cerevisiae* treatments (Table 14.22). ADL content in untreated stover varied from 3.70 (PMG-41) to 7.20% (PMG-52-1). Treatment with *P. fluorescens* and *T. harzianum* resulted significant reduction in ADL content of stover from genotypes Local Chadhiar (38.64%) and PMG-18 (48.08%), respectively. Among all treatments, *T. viride* followed by *S. cerevisiae* treatment revealed maximum reduction in ADL content of stover from genotypes PMG-52-1 (58.33%) and PMG-96 (55.89%), respectively (Table 14.23).

(c) **Phenolic, tannin, and silica content:** The total phenolic content in untreated maize stover was observed in the range of 9.12 (SIRI-5455) to 11.78 mg/g (PMG-93). With *T. harzianum* and *P. fluorescens* treatments, the highest reduction in total phenolic content was observed in stover from genotype African Tall (34.42 and 34.34%, respectively). Treatment with *T. viride* also resulted reduction (32.64%) in total phenolic content. Maize stover treated with *S. cerevisiae* revealed variation in total phenolic content from 6.06 (SIRI-5455) to 10.91 mg/g (PMG-52-1) where the highest decrease in total phenolic content was observed in stover from genotype SIRI-5455 (33.55%) (Table 14.24). Total tannin content in untreated maize stover revealed variation from 2.48 (Palam Shankar Makka) to 7.06 mg/g (PMG-52-1). Treatment of maize stover with *T. viride* revealed significant decrease in tannin content. The highest reduction in total tannins content (33.08%) was observed in stover from genotype PMG-49. Treatment with *S. cerevisiae* resulted significant decrease in total tannin content ranging from 1.92 to 6.87 mg/g of stover from different maize genotypes where the highest significant decrease was for genotype PMG-9 (31.48%). Treatment with *S. cerevisiae* and *T. harzianum* resulted 31.48 and 18.96% reduction in total tannin content of stover from genotypes PMG-9 and African Tall, respectively (Table 14.25). The stover from genotypes PMG-14 (3.40%) and PMG-49 (1.30%) had higher and lower silica content, respectively. Treatment with *T. harzianum* and *T. viride* resulted in decrease in silica content to the extent of 41.18 and 46.15%, respectively. With *S. cerevisiae* treatment, the significant reduction in silica content was observed for stover from genotype Palam Shankar Makka (53.84%) (Table 14.26). Among all treatments, inoculation of stover with *P. fluorescens* was highly effective in reducing silica content to the extent of 66.67%.

Table 14.21 NDF content (%) in raw and inoculated maize stover

Genotypes	Untreated	*T. harzianum*	*T. viride*	*S. cerevisiae*	*P. fluorescens*	SE	CD (5%)
PMG-3	71.50	67.50	68.00	69.00	69.50	2.24	NS
SIRI-5455	70.00	65.70[b]	67.10[b]	67.60[b]	68.70[b]	0.52	1.16
PMG-9	68.30	65.60[b]	65.40[b]	65.80[b]	66.90[b]	0.31	0.70
PMG-14	72.40	69.20[b]	69.60[b]	69.90[b]	70.40[b]	0.35	0.79
PMG-18	71.80	67.80	67.60	68.30	70.20	1.79	NS
Vivek	72.40	68.90[b]	68.30[b]	69.70[b]	70.30[b]	0.78	1.74
PMG-33	71.70	67.90	67.60	68.70	69.30	1.50	NS
PMG-36	69.10	66.30[b]	65.80[b]	66.30[b]	66.70[b]	0.50	1.12
PMG-40	71.90	68.10	68.80	69.00	69.60	2.32	NS
PMG-41	71.90	68.90	68.00	69.40	68.70	2.68	NS
PMG-49	71.20	67.90	67.90	68.70	68.70	1.54	NS
PMZ-4	74.80	70.30[b]	70.90[b]	71.90[b]	72.40[b]	0.33	0.72
PMG-52	70.70	66.90	67.70	69.00	68.40	3.29	NS
PMG-52-1	72.70	68.20	68.60	70.20	70.60	1.96	NS
PMG-62	72.80	68.80[b]	68.90[b]	69.80[b]	70.40[b]	0.43	0.95
PMG-93	73.20	67.60	68.90	69.80	69.70	2.52	NS
PMG-93-1	70.90	66.80	68.20	68.40	68.20	1.76	NS
PMG-93-2	71.20	66.60[b]	67.90[b]	68.70[b]	69.20[b]	0.68	1.52
PMG-95	71.60	66.10	68.00	69.50	68.90	2.08	NS
PMG-96	72.20	66.80	68.30	69.70	69.80	3.21	NS
Palam Shankar Makka	67.20	66.30	64.80	65.30	65.80	1.01	NS
PMG-97	72.30	68.40	68.90	69.20	69.40	1.39	NS
Local Chadhiar	74.70	69.00	70.70	72.20	72.70	2.70	NS
African Tall	69.00	65.00	67.00	67.10	67.00	1.14	NS

Table 14.22 ADF content (%) in raw and inoculated maize stover

Genotypes	Untreated	*T. harzianum*	*T. viride*	*S. cerevisiae*	*P. fluorescens*	SE	CD (5%)
PMG-3	40.50	39.80^{b}	39.80^{b}	39.73^{b}	38.00^{b}	0.25	0.55
SIRI-5455	46.40	45.40^{b}	44.80^{b}	45.20^{b}	43.20^{b}	0.26	0.59
PMG-9	48.60	47.13^{b}	47.67^{b}	46.20^{b}	44.20^{b}	0.29	0.64
PMG-14	43.40	40.07^{b}	42.73*	40.00^{b}	35.40^{b}	1.17	2.62
PMG-18	46.70	40.80^{b}	38.80^{b}	39.27^{b}	33.93^{b}	0.86	1.91
Vivek	46.90	46.13*	44.00^{b}	42.13^{b}	35.20^{b}	0.39	0.88
PMG-33	48.20	40.20^{b}	38.93^{b}	42.40^{b}	36.40^{b}	0.17	0.38
PMG-36	44.40	43.33*	41.67^{b}	40.07^{b}	34.47^{b}	0.58	1.30
PMG-40	44.90	44.27*	42.53^{b}	44.20*	41.67^{b}	0.95	2.11
PMG-41	46.00	45.33*	44.20*	44.47*	41.53^{b}	1.04	2.33
PMG-49	47.50	45.87^{b}	40.07^{b}	45.80^{b}	42.27^{b}	0.42	0.94
PMZ-4	44.20	41.80^{b}	42.93^{b}	41.47^{b}	38.93^{b}	0.46	1.03
PMG-52	48.00	46.33^{b}	43.53^{b}	45.60^{b}	44.73^{b}	0.29	0.65
PMG-52-1	44.40	42.80	41.87	41.60	40.60	3.62	NS
PMG-62	49.00	41.70^{b}	44.80^{b}	46.60^{b}	43.47^{b}	0.39	0.87
PMG-93	46.40	42.93^{b}	45.60*	44.73*	40.73^{b}	1.14	2.54
PMG-93-1	47.60	43.07^{b}	45.40^{b}	47.00*	41.60^{b}	0.59	1.31
PMG-93-2	45.60	44.87*	43.80*	44.60*	41.50^{b}	1.17	2.61
PMG-95	46.20	45.40	45.27	44.87	44.73	0.85	NS
PMG-96	43.00	43.33*	41.40*	43.67*	39.87^{b}	0.92	2.05
Palam Shankar Makka	47.50	45.20^{b}	44.47^{b}	45.33^{b}	44.10^{b}	0.40	0.89
PMG-97	45.60	45.67*	45.13*	41.13^{b}	40.47^{b}	0.65	1.45
Local Chadhiar	47.80	46.93*	42.33^{b}	41.47^{b}	39.73^{b}	0.65	1.44
African Tall	46.20	45.97	44.03	46.00	43.70	2.16	NS

Table 14.23 ADL content (%) in raw and inoculated maize stover

Genotypes	Untreated	*T. harzianum*	*T. viride*	*S. cerevisiae*	*P. fluorescens*	SE	CD (5%)
PMG-3	5.30	4.10^{b}	3.20^{b}	3.47^{b}	4.40^{b}	0.19	0.43
SIRI-5455	5.10	4.70^{b}	3.60^{b}	3.93^{b}	4.30^{b}	0.18	0.40
PMG-9	4.80	4.33^{*}	3.80^{b}	4.47^{*}	4.00^{b}	0.23	0.51
PMG-14	6.30	5.50^{b}	5.57^{b}	4.47^{b}	5.00^{b}	0.26	0.59
PMG-18	5.20	2.70^{b}	4.47^{b}	3.53^{b}	4.40^{b}	0.30	0.67
Vivek	5.40	5.13	5.00	4.73	3.80	0.95	NS
PMG-33	4.80	4.40^{b}	3.70^{b}	3.30^{b}	3.60^{b}	0.15	0.35
PMG-36	5.70	4.13	4.00	3.53	4.90	0.73	NS
PMG-40	6.50	5.47^{b}	4.20^{b}	4.53^{b}	5.10^{b}	0.17	0.37
PMG-41	3.70	3.20	3.00	2.93	3.00	0.42	NS
PMG-49	5.50	4.20^{b}	5.07^{b}	4.70^{b}	4.30^{b}	0.15	0.33
PMZ-4	5.90	5.53^{b}	4.53^{b}	3.73^{b}	4.10^{b}	0.14	0.30
PMG-52	5.70	4.27^{b}	4.67^{b}	4.20^{b}	4.90^{b}	0.13	0.29
PMG-52-1	7.20	4.87^{b}	3.00^{b}	4.60^{b}	5.70^{b}	0.52	1.17
PMG-62	6.10	5.60^{*}	3.40^{b}	3.27^{b}	5.00^{*}	0.58	1.29
PMG-93	4.40	3.60^{b}	3.13^{b}	2.00^{b}	3.40^{b}	0.30	0.68
PMG-93-1	4.70	3.30^{b}	4.00^{*}	2.47^{b}	3.90^{*}	0.47	1.04
PMG-93-2	5.50	3.60^{b}	3.40^{b}	2.60^{b}	4.40^{b}	0.17	0.37
PMG-95	6.30	5.27^{*}	4.80^{b}	2.80^{b}	5.10^{b}	0.54	1.19
PMG-96	5.60	4.73^{*}	4.00^{b}	2.47^{b}	4.30^{*}	0.61	1.35
Palam Shankar Makka	4.80	2.87^{b}	2.47^{b}	3.00^{b}	3.60^{b}	0.24	0.53
PMG-97	6.20	4.73^{b}	4.33^{b}	3.47^{b}	5.00^{b}	0.28	0.63
Local Chadhiar	4.40	3.93^{*}	3.67^{b}	3.00^{b}	2.70^{b}	0.26	0.59
African Tall	4.70	3.60^{b}	2.00^{b}	4.20^{b}	2.90^{b}	0.15	0.35

Table 14.24 Total phenolic content (mg/g) in raw and inoculated maize stover

Genotypes	Untreated	*T. harzianum*	*T. viride*	*S. cerevisiae*	*P. fluorescens*	SE	CD (5%)
PMG-3	9.29	8.65^{b}	7.30^{b}	8.87^{b}	8.88^{b}	0.08	018
SIRI-5455	9.12	8.78*	8.41*	6.06^{b}	7.68^{b}	0.37	0.83
PMG-9	10.01	9.75*	7.40^{b}	8.46^{b}	9.43*	0.36	0.81
PMG-14	11.58	8.51^{b}	10.16^{b}	10.57^{b}	10.40^{b}	0.31	0.68
PMG-18	9.87	8.51	8.98	8.64	7.23	0.77	NS
Vivek	10.24	8.59^{b}	7.02^{b}	9.53^{b}	7.54^{b}	0.23	0.51
PMG-33	9.79	9.13	7.92	8.56	8.54	0.80	NS
PMG-36	10.02	9.23*	9.31*	8.61^{b}	7.25^{b}	0.61	1.37
PMG-40	11.43	9.42	10.27	9.92	8.87	1.18	NS
PMG-41	11.22	10.08	8.31	9.40	9.68	0.94	NS
PMG-49	11.02	10.16	9.98	10.57	10.70	0.33	NS
PMZ-4	11.43	11.28*	8.88^{b}	10.06*	9.22^{b}	0.72	1.61
PMG-52	11.08	9.23	10.30	8.72	10.53	1.62	NS
PMG-52-1	11.53	11.22*	8.21^{b}	10.91*	9.49^{b}	0.72	1.60
PMG-62	11.83	10.47	8.65	9.74	8.78	1.07	NS
PMG-93	11.78	11.35*	9.58^{b}	9.89^{b}	9.09^{b}	0.83	1.84
PMG-93-1	11.14	10.48*	9.83^{b}	8.27^{b}	9.37^{b}	0.37	0.82
PMG-93-2	11.66	9.98	10.11	10.84	10.71	1.55	NS
PMG-95	10.41	8.93	7.05	9.42	9.96	1.66	NS
PMG-96	10.26	9.93*	8.26^{b}	9.17^{b}	8.13^{b}	0.44	0.97
Palam Shankar Makka	10.71	10.37*	9.46^{b}	9.42^{b}	7.19^{b}	0.25	0.56
PMG-97	11.12	8.02^{b}	7.49^{b}	10.50^{b}	7.57^{b}	0.13	0.29
Local Chadhiar	9.20	7.29^{b}	8.54*	8.56*	7.50^{b}	0.40	0.90
African Tall	12.58	8.25^{b}	8.69^{b}	9.96^{b}	8.26^{b}	0.39	0.87

Table 14.25 Total tannins content (mg/g) in raw and inoculated maize stover

Genotypes	Untreated	*T. harzianum*	*T. viride*	*S. cerevisiae*	*P. fluorescens*	SE	CD (5%)
PMG-3	3.96	3.61*	2.93^{b}	3.88*	3.23^{b}	0.31	0.70
SIRI-5455	3.04	2.79	2.69	2.10	2.06	0.47	NS
PMG-9	4.86	4.14^{b}	3.36^{b}	3.33^{b}	3.72^{b}	0.28	0.63
PMG-14	6.45	5.27^{b}	4.50^{b}	5.86*	5.53*	0.47	1.06
PMG-18	4.78	3.76	4.34	4.39	3.79	0.44	NS
Vivek	3.97	3.27^{b}	3.13^{b}	2.98^{b}	3.52*	0.29	0.65
PMG-33	4.90	4.40	3.35	4.49	4.08	0.46	NS
PMG-36	5.34	4.58*	4.88*	4.02^{b}	3.20^{b}	0.38	0.84
PMG-40	5.13	4.89	4.23	4.93	3.93	0.68	NS
PMG-41	5.99	5.61*	5.54*	4.42^{b}	4.94^{b}	0.44	0.99
PMG-49	5.26	4.82*	3.52^{b}	4.76*	4.85*	0.49	1.09
PMZ-4	6.15	6.03*	5.37^{b}	5.21^{b}	4.99^{b}	0.34	0.77
PMG-52	6.67	5.70	6.33	5.65	6.00	1.71	NS
PMG-52-1	7.06	6.74	6.10	6.87	5.25	0.86	NS
PMG-62	6.94	6.09*	4.74^{b}	5.60^{b}	5.87*	0.54	1.20
PMG-93	4.00	3.59	3.79	3.47	3.48	0.59	NS
PMG-93-1	4.87	3.84^{b}	4.67*	4.22*	2.83^{b}	0.30	0.67
PMG-93-2	6.16	5.08	5.81	5.38	5.41	0.83	NS
PMG-95	3.48	3.17	3.11	2.81	2.91	0.35	NS
PMG-96	3.41	3.17	2.48	2.75	2.92	0.33	NS
Palam Shankar Makka	2.48	1.45	1.68	1.92	2.43	0.36	NS
PMG-97	4.43	3.95	3.55	3.92	3.70	0.39	NS
Local Chadhiar	3.18	2.87	3.06	2.44	2.35	0.35	NS
African Tall	6.17	5.00^{b}	5.05^{b}	5.86*	5.90*	0.40	0.89

Table 14.26 Silica content (%) in raw and inoculated maize stover

Genotypes	Untreated	*T. harzianum*	*T. viride*	*S. cerevisiae*	*P. fluorescens*	SE	CD (5%)
PMG-3	2.10	1.60^{b}	1.80^{b}	1.73^{b}	1.33^{b}	0.10	0.23
SIRI-5455	1.90	1.40^{b}	1.27^{b}	1.20^{b}	1.20^{b}	0.16	0.35
PMG-9	1.60	1.33^{*}	1.13^{b}	0.93^{b}	0.87^{b}	0.18	0.41
PMG-14	3.40	2.00^{b}	2.47^{b}	2.67^{b}	2.30^{b}	0.25	0.56
PMG-18	2.60	2.13^{b}	1.93^{b}	2.00^{b}	1.80^{b}	0.09	0.21
Vivek	2.50	2.07^{b}	1.67^{b}	1.60^{b}	1.20^{b}	0.12	0.28
PMG-33	1.90	1.60^{b}	1.20^{b}	1.00^{b}	1.10^{b}	0.12	0.26
PMG-36	2.50	2.13^{b}	1.87^{b}	1.73^{b}	2.10^{b}	0.14	0.30
PMG-40	2.50	1.90^{b}	1.80^{b}	2.07^{b}	1.47^{b}	0.13	0.29
PMG-41	2.10	1.73^{b}	1.40^{b}	1.80^{*}	1.70^{b}	0.14	0.32
PMG-49	1.30	1.00	0.73	1.07	0.73	0.20	NS
PMZ-4	2.70	2.20^{b}	1.73^{b}	1.87^{b}	1.70^{b}	0.18	0.40
PMG-52	2.90	2.47^{b}	1.93^{b}	2.00^{b}	1.53^{b}	0.17	0.37
PMG-52-1	2.70	2.33^{*}	2.33^{*}	1.47^{b}	1.73^{b}	0.25	0.55
PMG-62	3.10	2.27^{b}	2.53^{b}	2.73^{b}	2.10^{b}	0.12	0.26
PMG-93	2.10	1.73^{*}	1.40^{b}	1.20^{b}	1.00^{b}	0.19	0.42
PMG-93-1	2.30	1.60^{b}	2.00^{b}	2.00^{b}	1.40^{b}	0.13	0.29
PMG-93-2	1.60	1.33^{*}	1.40^{*}	1.00^{b}	0.73^{b}	0.14	0.31
PMG-95	1.80	1.20^{b}	1.13^{b}	0.87^{b}	0.60^{b}	0.17	0.39
PMG-96	2.50	2.00^{b}	2.13^{b}	1.67^{b}	1.60^{b}	0.16	0.36
Palam Shankar Makka	2.60	1.93^{b}	1.40^{b}	1.20^{b}	1.70^{b}	0.17	0.38
PMG-97	2.60	2.20^{*}	2.33^{*}	1.80^{b}	1.30^{b}	0.19	0.42
Local Chadhiar	1.50	1.40^{*}	1.47^{*}	1.07^{b}	1.00^{b}	0.12	0.28
African Tall	1.90	1.60^{b}	1.20^{b}	1.00^{b}	1.30^{b}	0.09	0.20

(d) ***In vitro* dry matter digestibility (IVDMD):** *In vitro* dry matter digestibility of untreated maize stover varied from 52.10% (PMG-3) to 59.40% (Palam Shankar Makka). Treatment with *T. harzianum* followed by *S. cerevisiae* resulted in significant increase (32.82 and 31.09%, respectively) in digestibility of stover from genotype PMG-3. Treatment with *T. viride* and *P. fluorescens* also increased the digestibility of stover to 29.10 (PMG-52-1) and 27.15% (PMG-95), respectively (Table 14.27).

14.4.3.3.2 Effect of Biological Treatments on Quality of Hay, Wheat and Rice Straw

Different biological treatments were also investigated by our group for improving quality and digestibility of quality hay (SSG; Sorghum sudan grass; Guria grass; *Chrysopogon fulvus*) and crop residues (Rice straw and Wheat straw). The observations after the treatment have been discussed:

(a) **Crude protein (CP) and total carbohydrate content**: After biological treatments, CP content of wheat and rice straw increased from 3.85 to 9.10% and 5.48 to 11.20%, respectively, with significant increase after treatment with *P. fluorescens* (104.38%) and *T. viride* (136.36%), respectively (Tables 14.28 and 14.29). The highest increase in CP content of SSG (33.05%) and Guria grass (13.89%) was recorded with *T. harzianum* and *P. fluorescens* treatments, respectively (Tables 14.30 and 14.31). After biological treatments, there was an increase in the CP content of hay from SSG and Guria grasses. Highest increase (33.05%) in CP content of SSG was recorded with *T. harzianum* treatment, whereas inoculation of *P. fluorescens* resulted 13.89% increase in CP content of Guria grass hay.

Inoculation with *T. viride* and *P. fluorescens* revealed an increase in total carbohydrates content of wheat straw and sorghum sudan grass hay, respectively (Tables 14.28 and 14.30). Treatment with *T. harzianum* resulted increase in total carbohydrates content of rice straw and Guria grass hay, respectively (Tables 14.29 and 14.31). Improvement in carbohydrate content of feeds may be due to the addition of cellular content of microorganisms and degradation of cell wall phenolics associated with carbohydrate moieties.

(b) **Fiber and cell wall constituents**: Fiber content (NDF and ADF) was reduced with the biological treatments and the highest reduction was observed after *P. fluorescens* treatment (Tables 14.28, 14.29, 14.30, and 14.31). Cell wall constituents (hemicellulose, cellulose, and ADL) of treated and untreated wheat straw and rice straw have been presented in Figs. 14.4 and 14.5, respectively. Inoculation with *P. fluorescens* and *T. harzianum* revealed a 33.33 and 38.96% reduction in ADL content of wheat and rice straw, respectively. Wheat and rice straw treated with *P. fluorescens* revealed lower hemicellulose content. Inoculation with *T. harzianum* and *P. fluorescens* resulted reduction in cellulose content of wheat straw and rice straw. All biological treatments were effective in delignification of SSG and Guria grass and the highest reduction in ADL content was observed after *T. viride* and *P. fluorescens* treatments. Hemicellulose content in SSG hay decreased significantly (11.58%) with *T. viride*

Table 14.27 IVDMD (%) of raw and inoculated maize stover

Genotypes	Untreated	*T. harzianum*	*T. viride*	*S. cerevisiae*	*P. fluorescens*	SE	CD (%)
PMG-3	52.10	69.20[a]	65.20 [a]	68.30 [a]	62.40 [a]	0.19	0.42
SIRI-5455	55.40	67.40 [a]	63.20 [a]	65.40 [a]	63.20 [a]	0.19	0.43
PMG-9	56.20	65.30 [a]	69.00 [a]	62.40 [a]	60.40 [a]	0.28	0.62
PMG-14	55.20	69.20 [a]	67.00 [a]	68.10 [a]	65.40 [a]	0.18	0.41
PMG-18	53.20	65.40 [a]	67.00 [a]	67.90 [a]	63.20 [a]	0.41	0.91
Vivek	56.20	67.20 [a]	68.20 [a]	67.20 [a]	68.40 [a]	0.17	0.40
PMG-33	57.60	69.00 [a]	67.20 [a]	68.00 [a]	62.20 [a]	0.43	0.97
PMG-36	57.10	68.20 [a]	65.40 [a]	67.20 [a]	69.40 [a]	0.40	0.90
PMG-40	58.60	69.40 [a]	67.20 [a]	65.40 [a]	68.00 [a]	0.31	0.69
PMG-41	59.20	70.00 [a]	69.40 [a]	66.20 [a]	65.40 [a]	0.16	0.36
PMG-49	58.30	72.20 [a]	70.00 [a]	65.30 [a]	67.30 [a]	0.10	0.22
PMZ-4	56.40	69.20 [a]	70.20 [a]	67.40 [a]	69.40 [a]	0.18	0.40
PMG-52	57.40	68.40 [a]	69.80 [a]	69.00 [a]	62.40 [a]	0.16	0.36
PMG-52-1	52.80	65.90 [a]	68.20 [a]	66.20 [a]	65.30 [a]	0.12	0.26
PMG-62	56.90	67.80 [a]	65.00 [a]	66.00 [a]	61.40 [a]	0.17	0.37
PMG-93	57.80	68.00 [a]	69.20 [a]	67.20 [a]	65.90 [a]	0.18	0.41
PMG-93-1	56.40	69.20 [a]	70.00 [a]	68.40 [a]	66.30 [a]	0.18	0.41
PMG-93-2	56.30	67.00 [a]	65.40 [a]	69.40 [a]	69.80 [a]	0.25	0.55
PMG-95	53.40	66.00 [a]	65.00 [a]	66.00 [a]	67.90 [a]	0.58	1.30
PMG-96	54.80	65.20 [a]	66.20 [a]	67.60 [a]	66.80 [a]	0.28	0.63
Palam Shankar Makka	59.40	69.20 [a]	69.20 [a]	68.20 [a]	67.40 [a]	0.19	0.43
PMG-97	57.60	67.40 [a]	68.00 [a]	64.30 [a]	67.80 [a]	0.15	0.34
Local Chadhiar	58.60	65.40 [a]	67.20 [a]	64.90 [a]	69.40 [a]	0.23	0.52
African Tall	59.40	70.00 [a]	69.40 [a]	65.40 [a]	66.60 [a]	0.41	0.91

In Tables 14.19, 14.20, 14.21, 14.22, 14.23, 14.24, 14.25, 14.26, and 14.27, superscripts “a,” “b,” and “*” represent a significant increase, significant decrease, and statistically at par values as compared to untreated, respectively, at $p \leq 0.05$. *NS* nonsignificant, *CD* critical difference, *SE* standard error

Table 14.28 Biochemical composition of raw and biologically treated wheat straw

Parameters	Untreated	*T. harzianum*	*T. viride*	*S. cerevisiae*	*P. fluorescens*
CP (%)	3.85^{d}	7.00^{b}	9.10^{a}	5.60^{c}	7.35^{b}
TC (%)	7.14^{d}	8.31^{c}	15.18^{a}	13.03^{b}	9.18^{c}
NDF (%)	82.53^{a}	77.40^{c}	77.60^{c}	79.60^{b}	75.33^{d}
ADF (%)	51.53^{a}	46.73^{c}	48.53^{b}	47.87^{b}	46.40^{c}
IVDMD (%)	49.33^{b}	54.20^{a}	53.00^{a}	53.67^{a}	54.93^{a}
CC (%)	17.47^{d}	22.60^{b}	22.40^{b}	20.40^{b}	24.67^{a}
Ash (%)	2.30^{c}	3.40^{b}	3.67^{b}	2.80^{bc}	4.60^{a}

Table 14.29 Biochemical composition of raw and biologically treated rice straw

Parameters	Untreated	*T. harzianum*	*T. viride*	*S. cerevisiae*	*P. fluorescens*
CP (%)	5.48^{d}	6.18^{cd}	8.28^{b}	6.65^{c}	11.20^{a}
TC (%)	6.87^{d}	17.10^{a}	8.68^{c}	13.52^{b}	8.99^{c}
NDF (%)	76.40^{a}	71.20^{c}	70.73^{c}	74.20^{b}	69.00^{d}
ADF (%)	48.60^{a}	42.33^{d}	43.27^{c}	46.73^{b}	42.27^{d}
IVDMD (%)	45.67^{b}	52.67^{a}	53.47^{a}	53.67^{a}	51.67^{a}
CC (%)	23.60^{d}	28.80^{b}	29.27^{b}	25.80^{c}	31.00^{a}
Ash (%)	7.40^{d}	8.80^{b}	9.20^{a}	9.20^{a}	8.20^{c}

Table 14.30 Biochemical composition of raw and biologically treated SSG hay

Parameters	Untreated	*T. harzianum*	*T. viride*	*S. cerevisiae*	*P. fluorescens*
CP (%)	15.52^{c}	20.65^{a}	19.83^{b}	19.95^{ab}	20.30^{ab}
TC (%)	9.10^{d}	10.40^{c}	10.94^{b}	9.17^{d}	12.01^{a}
NDF (%)	53.93^{a}	49.60^{bc}	48.80^{c}	49.80^{b}	47.27^{d}
ADF (%)	34.93^{a}	30.27^{d}	32.00^{c}	32.67^{b}	30.20^{d}
IVDMD (%)	54.67^{b}	58.00^{a}	56.47^{ab}	57.33^{a}	58.00^{a}
CC (%)	46.07^{d}	50.40^{bc}	51.20^{b}	50.20^{c}	52.73^{a}
Ash (%)	3.30^{c}	3.50^{c}	4.60^{a}	4.20^{b}	4.10^{b}

Table 14.31 Biochemical composition of raw and biologically treated Guria grass hay

Parameters	Untreated	*T. harzianum*	*T. viride*	*S. cerevisiae*	*P. fluorescens*
CP (%)	12.60	12.95	12.83	13.45	14.35
TC (%)	10.76^{c}	15.96^{a}	12.32^{b}	11.14^{c}	11.01^{c}
NDF (%)	50.93^{a}	46.60^{c}	47.07^{c}	48.93^{b}	45.46^{d}
ADF (%)	34.73^{a}	31.47^{b}	31.60^{b}	33.40^{ab}	29.40^{c}
IVDMD (%)	50.53^{b}	54.67^{a}	56.00^{a}	56.07^{a}	56.13^{a}
CC (%)	49.06^{d}	53.40^{b}	52.93^{b}	51.07^{c}	54.54^{a}
Ash (%)	4.70^{c}	5.50^{b}	5.33^{b}	6.30^{a}	5.50^{b}

In Tables 14.28, 14.29, 14.30, and 14.31, each value represents mean of three replications. In the same row, significant differences according to CRD are indicated by different alphabets. Superscripts represent significant values at ($p \leq 0.05$). Same alphabets represent that their values are statistically at par ($p \leq 0.05$)

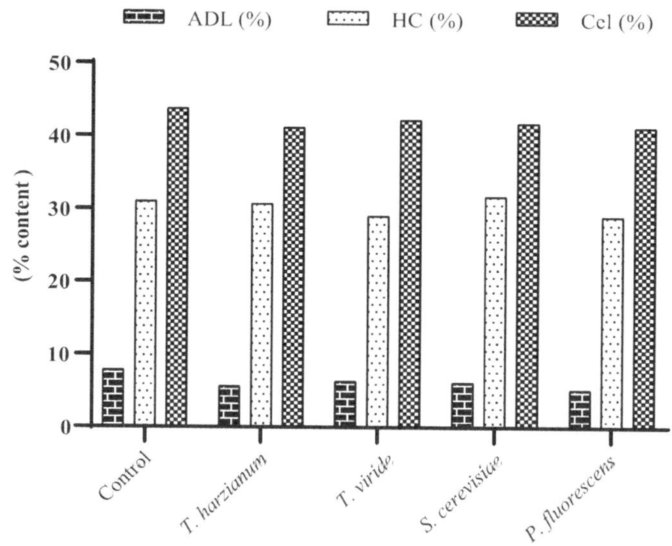

Fig. 14.4 Cell wall constituents of treated and untreated wheat straw

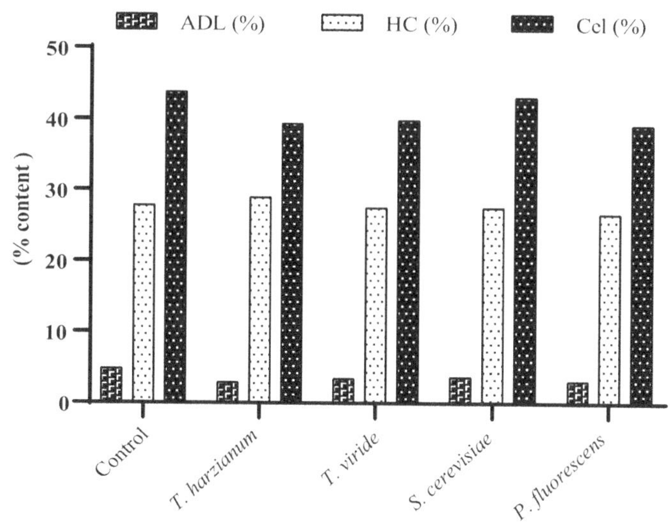

Fig. 14.5 Cell wall constituents of treated and untreated rice straw

treatment (Fig. 14.5), whereas reduction in hemicellulose content of Guria grass hay was not much significant with all biological treatments (Fig. 14.6). Wheat and rice straw inoculated with *P. fluorescens* resulted highest increment of 41.21 and 31.36% in cell content, respectively. Cell content increased significantly with all biological treatments in both SSG and Guria grass hay. The most significant increase was recorded with *P. fluorescens* treatment. The cellulose content in SSG hay reduced significantly with all biological treatments

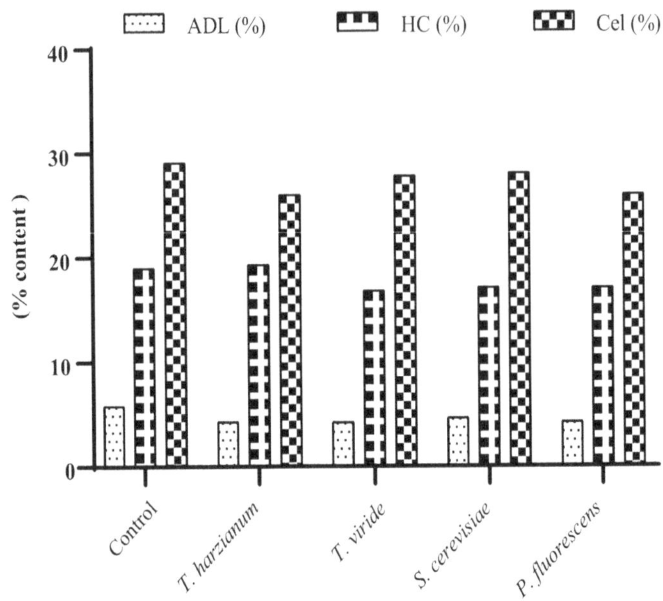

Fig. 14.6 Cell wall constituents of treated and untreated SSG hay

where the highest reduction (10.74%) was observed with *T. harzianum* and *P. fluorescens*. Inoculation with *P. fluorescens* resulted reduction in cellulose content of Guria grass hay to the extent of 14.43%.

(c) **Ash content:** Treatment with *P. fluorescens* resulted in increase (2.30 to 4.60%) in the ash content of wheat straw. Treatment of rice straw with *T. viride* and *S. cerevisiae* resulted in 24.32% increase in ash content. Treatment with *T. viride* exhibited 39.39% increase in ash content of SSG, whereas inoculation with *S. cerevisiae* revealed 34.04% increase in ash content of Guria grass.

(d) **Phenolic and tannin content:** Treatment with *S. cerevisiae* resulted reduction in total phenolic and tannin content of wheat straw (Fig. 14.3). Inoculation with *P. fluorescens* was the most effective treatment for minimizing the phenolic and tannin content of rice straw (Fig. 14.4). Reduction in total phenolic content of SSG and Guria grass was recorded with *S. cerevisiae* treatment (Figs. 14.7 and 14.8). The total tannin content of SSG hay also reduced significantly after treatment with *Trichoderma viride* (Fig. 14.7). The simple phenolic content of SSG and Guria grass hay reduced with all treatments, and the highest reduction was observed with *S. cerevisiae* treatment. Highest reduction in total tannin content of Guria grass hay was observed with *P. fluorescens* treatment (Fig. 14.8).

(e) **Dry matter digestibility:** Increase in dry matter digestibility of wheat and rice straw was recorded with *P. fluorescens* and *S. cerevisiae* treatments, respectively. Significant increase in the digestibility of SSG and Guria grass was observed with *P. fluorescens* treatment.

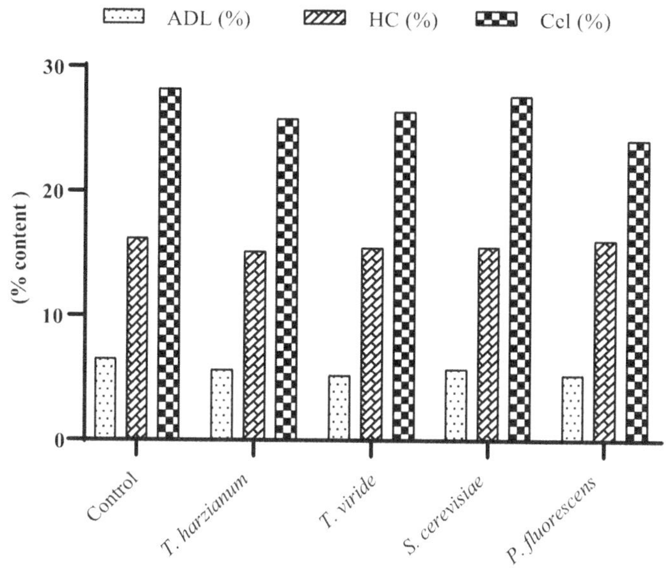

Fig. 14.7 Cell wall constituents of treated and untreated Guria grass hay

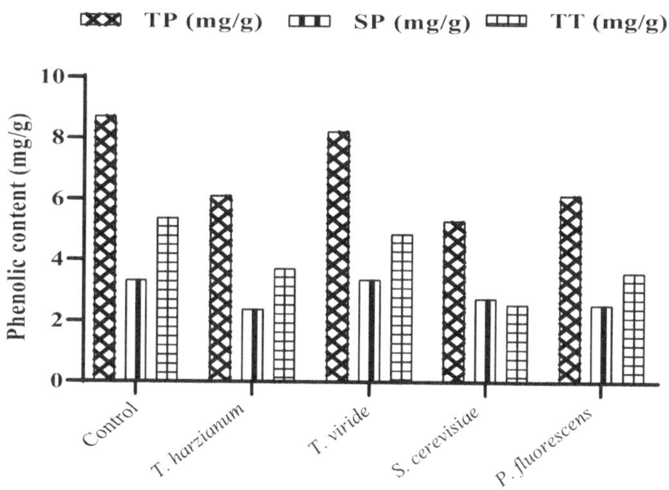

Fig. 14.8 Phenolic and tannin content of treated and untreated wheat straw

The study revealed a significant effect of different biological treatments on nutritive value and digestibility. The treatments with *T. viride* and *P. fluorescens* could be investigated further for the bioconversion of low-quality feeds into enriched feeds (Figs. 14.9, 14.10, and 14.11).

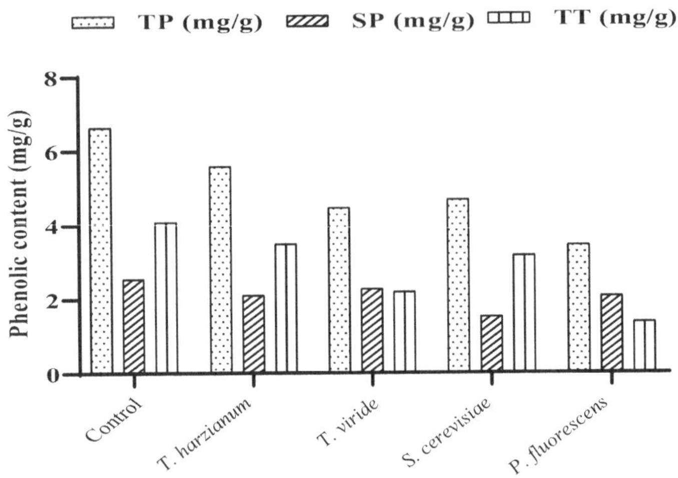

Fig. 14.9 Phenolic and tannin content of treated and untreated rice straw

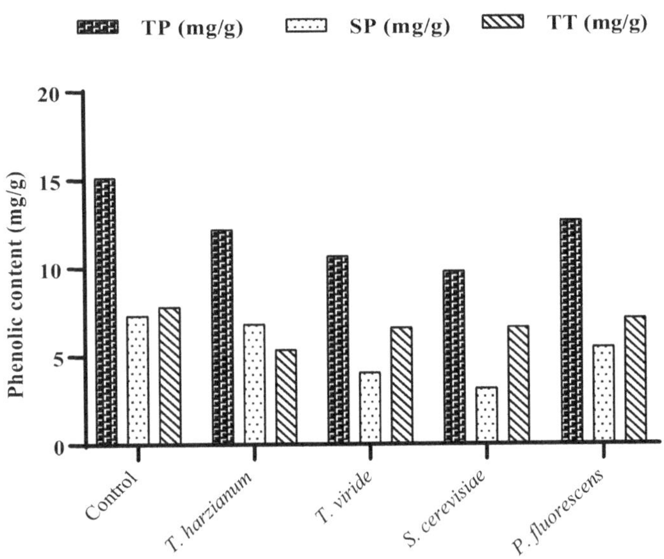

Fig. 14.10 Phenolic and tannin content of treated and untreated SSG hay

14.4.4 Exogenous Enzymatic Treatment

The exogenous enzymatic treatment for the crop residues includes cellulases and xylans. Commercial enzymes used in the livestock feed industry are generally of the fungal or bacterial origin. A number of studies have been carried out to demonstrate the effect of exogenous enzymes for improvement in fiber quality and digestibility of

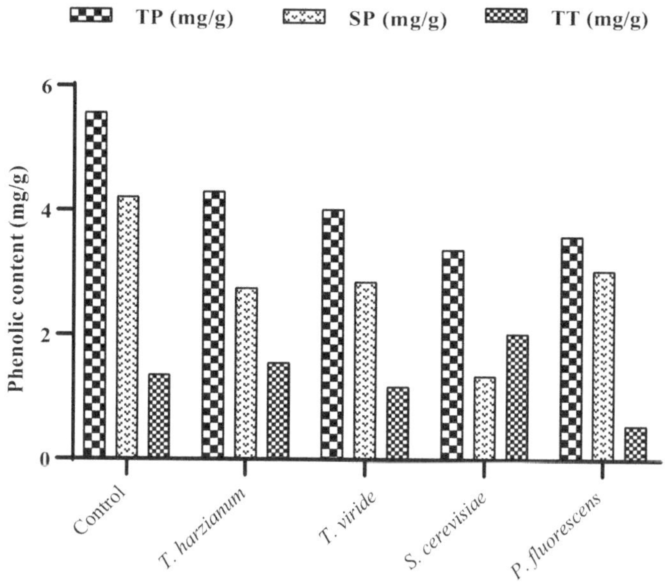

Fig. 14.11 Phenolic and tannin content of treated and untreated Guria grass hay

low-quality roughages. Utilization of fibrolytic enzymes in livestock feed have improved the average daily gain (Beauchemin et al. 1995), fleece weight, wool production of lambs (Jafari et al. 2005) and in milk yield of dairy animals (Yang et al. 2000). Eun et al. (2006) studied the effect of cellulose, xylanase, a combination of cellulose, hemicellulose and proteases on rice straw. These treatments increased digestibility and decreased hemicellulose content of rice straw. Fazaeli et al. (2006) used *Pleurotus* fungi to treat wheat straw and reported increased IVDMD by the action of enzymatic activity. In some studies, the utilization of fibrolytic enzymes alone were not effective in improving the digestibility of rice straw due to the inability of these enzymes to break esterified bonds within lignin carbohydrate complexes. The combination of pretreatments was effective in increasing the digestibility and fermentation characteristics of rice straw (Eun et al. 2006). Wang et al. (2004) treated rice straw with different enzymes such as xylanase, β-glucanase, carboxymethylcellulase and amylase in combination with NaOH. Zhu et al. (2005) treated rice straw with 1% NaOH and observed increment in cellulose and decline in lignin and hemicellulose content. Liu and Ørskov (2000) reported steam pressure in combination with cellulose enzyme is the best treatment for enhancing the nutritional quality of rice straw.

14.4.5 Supplementation

Supplimentation of nutrients also improves the nutritive value of low-quality feeds. The supplementation with grains and concentrates is one of the best option to improve the nutritional quality. Alternative feed sources available in the local area may be used for feeding small ruminants. Different by-products and browse foliage could partially or totally replace concentrate feeds without causing any negative effects on livestock performance (Amata 2014).

Livestock fed with green berseem or lucerne and cowpea during winter and summer/rainy seasons, respectively, along with 1–2 kg dry roughages sustained milk production up to 10 kg/day. But above 10 kg/day milk production, mere fodder does not balance the diet ratio. The supplementation of 1 kg for every 2.50–3.0 kg of extra milk production has to be made in addition to green fodder and dry roughages. Mixture of leguminous and cereal fodder at the ratio of 2:1 or 1:1 was effective in maintaining animal productivity.

Catalytic supplementation is a procedure in which straws are supplied with deficient nutrients like nitrogen, sulphur and phosphorous. This is performed for improved rumen function which results in increased intake and digestibility of straws. The supplement is known to have positive effect in digestion of crop residues and straw.

Another approach is to use the supplements only for the economically important or physiologically sensitive animals. This is called strategic supplementation. It is applied where better feeds are given to reproductive and young valuable animals. A typical, but a somewhat disguised form of strategic supplementation is "selective consumption", where the better animals are allowed to select the better feed parts, and the feed refusals are fed to less productive animals.

Main reasons for low livestock productivity are round the year unavailability of quality feed and high volatile prices of commercial feed concentrate leading to nutritional insecurity to the livestock. Every year considerable amount of crop residues such as straws, molasses and other agro-industrial wastes are produced but because of low nutritive value and uneconomical handling, their utilization remains limited in livestock feeding. Different strategies have been developed for significant improvement in nutritive value and digestibility of crop residues. However, these approaches need further refinement to simple and economically viable methods which could be practiced by small and marginal farmers. The combination of pretreatment of lignin-rich crop residues and other by-products with value addition will help in utilization of these resources in livestock feeding and to some extent addressing the demand for fodder.

References

Abd El-Galil EE, Ebtehag IMA (2008) Effect of bacterial treatments on chemical composition, cell wall constituents and digestibility of rice straw. Egyptian J Nutr Feeds 11(3):497–510

Abd El-Galil ER, Ebtehag IMA, Dokki, Egypt G (2011) Role of bacterial treatments for upgrading nutritive value of bean straw and native goats performance. J Am Sci 7:502–510

Abdel-Azim SN, Ahmed MA, Abo-Donia F, Soliman H (2011) Evaluation of fungal treatment of some agricultural residues. Egypt J Sheep Goat Sci 6:1–13

Abou-EL-Enin OH, Fadel JG, Mackill DJ (1999) Differences in chemical composition and fibre digestion of rice straw with, without, anhydrous ammonia from 53 rice varieties. Anim Feed Sci Technol 79:129–136

Adenipekun O, Okunlade OA (2012) Biodegradation of rattan wood and maize stovers by *Pleurotus ostreatus*. Nat Sci 10:49–57

Agosin E, Tollier MT, Brillouet JM, Thivend DP, Odier E (1986) Fungal pre-treatment of wheat straw: effects on biodegradability of cell walls, structural polysaccharides, lignin and phenolic acids by rumen microorganisms. J Sci Food Agric 37:97

Ahmed S, Ahmad F, Hashmi AS (2010) Production of microbial biomass protein by sequential culture fermentation of *Arachniotus* sp. and *Candida utilis*. Pak J Bot 42:1225–1234

Akinfemi A, Ogunwole OA (2012) Chemical composition and in vitro digestibility of rice straw treated with *Pleurotus ostreatus*, *Pleurotus pulmonarius* and *Pleurotus tuber-regium*. Slovak J Anim Sci 45:14–20

Akinfemi A, Adu OA, Doherty F (2009) Assessment of the nutritive value of fungi treated maize cob using in vitro gas production technique. Lives Res Rural Dev 21(11):0121–3784

Amata IA (2014) The use of non-conventional feed resources (NCFR) for livestock feeding in the tropics: a review. J Glob Biosci 3:604–613

Aregheore EM (2005) Effect of *Yucca schidigera* saponin on the nutritive value of urea-ammoniated maize stover and its feeding value when supplemented with forage legume (*Calliandra calothyrsus*) for goats. Small Rumin Res 56:95–102

Arieli A (1997) Whole cottonseed in dairy cattle feeding: a review. Anim Feed Sci Technol 72:97–110

Beauchemin KA, Rode LM, Sewalt VJH (1995) Fibrolytic enzymes increase fiber digestibility and growth rate of steers fed dry forages. Can J Anim Sci 75:641–644

Belewu MA, Okhawere OC (1998) Evaluation of feeding value of fungi treated Rice husk to Ram. In: Proc. WASAP Inaugural Conference, pp 320–321

Berger LL, Fahey GC, Bourquin LD, Tilgeyer EC (1994) Modification of forage quality after harvest. In: Fahey C (ed) Forage quality, evaluation, and utilisation. American Society of Agronomy, Inc, Madison, pp 922–966

Calzado JF, Rolz C (1990) Estimation of the growth rate of *Pleurotus* on stocked straw. J Ferment Bioeng 69:70–71

Chahal DS, Moo-Young M, Vlach D (1981) Effect of physical and physicochemical pretreatments of wood for SCP-production with *Chaetomium cellulolyticum*. Biotechnol Bioeng 23:2417–2420

Chaudhary DP, Kumar A, Mandhania SS, Srivastava P, Kumar RS (2012) Maize as fodder ? An alternative approach, Directorate of Maize Research, Pusa Campus, New Delhi. Tech Bull, p 32

Chen J, Fales SL, Varga G, Royse DJ (1995) Biodegradation of cell wall component of maize stover colonized by white-rot fungi and resulting impacts on *in vitro* digestibility. J Sci Food Agric 68: 91–98

Chen M, Zhao J, Xia L (2009) Comparison of four different chemical pretreatments of corn stover for enhancing enzymatic digestibility. Biomass Bioenergy 33:1381–1385

Devendra C (1997) Crop residues for feeding animals in Asia: technology development and adoption in crop/livestock systems. In: Renard C (ed) Crop residuals in sustainable mixed crop/ livestock farming system. CAB International, Wallingford, pp 241–267

Devi S, Gupta C, Jat SL, Parmar MS (2017) Crop residue recycling for economic and environmental sustainability. The case of India. Open Agric 2:486–494

Doyle PT, Devendra C, Pearce GR (1986) Rice straw as a feed for ruminants. International development program. Australia Universities and Colleges, Canberra. https://idl-bnc-idrc.dspacedirect.org/bitstream/handle/10625/8776/70328.pdf?sequence=3

Elias ST, Fulpagare YG (2015) Effects of urea treated maize stover silage on growth performance of crossbred heifers. J Agric Vet Sci 8:58–62

El-Menniawy AYM (2008) Studies on biological treatments in animal nutrition. Ph.D. thesis. Fac. Agri. Al-Azhar University

Eriksson K-E, Blanchette RA, Ander P (1990) Microbial and enzymatic degradation of wood and wood components. Springer series in wood science. Springer, Berlin
Eun JS, Beauchemin KA, Hong SH, Bauer MW (2006) Exogenous enzymes added to untreated or ammoniated rice straw: effects on in vitro fermentation characteristics and degradability. Anim Feed Sci Technol 131:86–101
Fadel M (1983) Nidobial protein production using agric. industrial solid wasted. M.Sc thesis. Faculty of Agriculture Cairo University, p 25
Fall S, Guerin H, Sall C, M'baye ND (1987) Les pailles de cereales dans le systeme d'alimentation des ruminants au Senegal, Rapport ISRA-LNERV No. 70, Septembre 1987, Dakar, Senegal, p 68
FAO (Food and Agriculture Organization) (2012). http://Teca.Fao.Org/Technology/Treating-Straw-Animal-Feeding-Beckmann-Method. Accessed 30 Mar 2012
Fazaeli H, Azizi A, Amile M (2006) Nutritive value index of treated wheat straw with *Pleurotus* fungi fed to sheep. Pak J Biol Sci 9:2444–2449
Gado HM, Rania AA, Younis Etab RI, Abd El-Galil EE (2009) Molecular identification of some cellulolytic bacteria from different species of ruminant animals and *in vitro* fermentation of cellulose of sugarcane bagasse. Egypt J Genet Cyto 38:165–185
Garcia-Cubero MT, Gonzalez-Benito G, Indacoechea I, Coca M, Bolado S (2009) Effect of ozonolysis pretreatment on enzymatic digestibility of wheat and rye straw. Bioresour Technol 100:1608–1613
Guggolz J, Kohler G, Klopfenstein T (1971) Composition and improvement of grass straw for ruminant nutrition. J Anim Sci 33:151–156
Haile E, Gicheha M, Njonge FK, Asgedom G (2017) Determining nutritive value of cereal crop residues and lentil (*Lens esculanta*) straw for ruminants. Open J Anim Sci 7:19–29
Hendriks A, Zeeman G (2009) Pretreatments to enhance the digestibility of lignocellulosic biomass. Bioresour Technol 100:10–18
Islamiyati R, Rasjid S, Natsir A, Ismartoyo (2013) Crude protein and fiber fraction of corn stover inoculated by fungi *Trichoderma* sp. and *Phanerochaete chrysosporium*. Int J Sci Tech Res 2: 149–152
Jackson MG (1977) Review article: the alkali treatment of straws. Anim Feed Sci Technol 2:105–130
Jafari A, Edriss MA, Alikhani M, Emtiazi G (2005) Effects of treated wheat straw with exogenous fibre-degrading enzymes on wool characteristics of ewe lambs. Pak J Nutr 4:321–326
Jonathan SG, Okorie AN, Babayemi OJ, Oyelakin AO, Akinfemi A (2012) Biodegradation of agricultural wastes (rice straw and sorghum stalk) into substrates of utilizable products using white rot fungus (*Pleurotus florida*). Nat Sci 10:131–137
Katoch R, Apoorva, Tripathi A, Sood S (2017) Improving nutritive value and digestibility of maize stover—a review. Forage Res 43(3):174–180
Katoch R, Tripathi A (2020) Improving nutritive value and digestibility of hay and crop residues. Forage Res 46(2):157–162
Katoch R, Apoorva TA, Manoj NV (2021) Effect of pre-treatments on quality of maize (zea mays l.) Stover. Forage Res 46(4):356–362
Kim S, Holtzapple TM (2005) Lime pre-treatment and enzymatic hydrolysis of corn stover. Bioresour Technol 96:1994–2006
Klopfenstein TJ, Krause VE, Jones MJ, Woods W (1972) Chemical treatment of low quality roughages. J Anim Sci 35:418–422
Krause DO, Denman SE, Mackie RI, Morrison M, Rae AL, Attwood GT, McSweeney CS (2003) Opportunities to improve fibre degradation in the rumen: microbiology, ecology, and genomics. FEMS Microbiol Rev 797:1–31
Lee JW, Jeffries TW (2011) Efficiencies of acid catalysts in the hydrolysis of lignocellulosic biomass over a range of combined severity factors. Bioresour Technol 102:5884–5890
Li X, Cai Z, Horn E, Winandy JE (2011) Oxalic acid pretreatment of rice straw particles and loblolly pine chips: release of hemicellulosic carbohydrates. TAPPI J 10:41–45

Liu JX, Ørskov ER (2000) Cellulase treatment of untreated and steam pre-treated rice straw-effect on in vitro fermentation characteristics. Anim Feed Sci Technol 88:189–200

Liu JX, Orskov ER, Chen XB (1999) Optimization of steam treatment as a method for upgrading rice straw as feeds. Anim Feed Sci Technol 76:345–357

Lopez S, Davies DR, Giraldez FJ, Dhanoa MS, Dijkstra J, France J (2005) Assessment of nutritive value of cereal and legume straws based on chemical composition and in vitro digestibility. J Sci Food Agric 85(9):1550–1557

Mahesh MS, Mohini M (2014) Crop residues for sustainable livestock production. J Adv Dairy Res 2:e108. https://doi.org/10.4172/2329-888X.1000e108

Mahiudddin M, Fakhruddin ANM, Abdullah-Al-Mahin (2012) Degradation of phenol via meta cleavage pathway by Pseudomonas fluorescens PU1. ISRN Microbiol 2: 212–218

Makkar HPS, Goodchild AV, Abd El-Moneim AM, Becker K (1996) Cell-constituents, tannin levelsby chemical and biological assays and nutritional valueof some legume foliage and straws. J Sci Food Agric 71:129–136

Menon V, Rao M (2012) Trends in bioconversion of lignocellulose: biofuels, platform chemicals & biorefinery concept. Prog Ener Combust Sci 38:522–550

Mosier N, Hendrickson R, Ho N, Sedlak M, Ladisch MR (2005) Optimization of pH controlled liquid hot water pretreatment of corn stover. Bioresour Technol 96:1986–1993

Mtui G (2012) Oxalic acid pretreatment, fungal enzymatic saccharification and fermentation of maize residues to ethanol. Afr J Biotechnol. https://doi.org/10.5897/AJB11.3032

Nerude F, Misurcova A (1995) Distribution of lignolytic enzyme in selected white rot fungi. Folia Microbiol 4(3):264–266

Nguyen XT (2000) Treatment and supplenetation of rice straw for ruminant feeding in Vietnam. Doctor Scientiarum thesis, Agricultural University of Norway. ISSN 0802-3220. ISBS 82-575-0440-8

Novaes E, Kirst M, Chiang V, Winter-Sederoff H, Sederoff R (2010) Lignin and biomass: a negative correlation for wood formation and lignin content in trees. Plant Physiol 154:555–561

Oji UI, Etim HE, Okoye FC (2007) Effects of urea and aqueous ammonia treatment on the composition and nutritive value of maize residues. Small Rumin Res 69:232–236

Ololade BG, Mowat DN, Winch JE (1970) Effect of processing method on the *in-vitro* digestibility of sodium hydroxide treated roughages. Can J Anim Sci 50:657–662

Palonen H, Thomsen AB, Tenkanen M, Schmidt AS, Viikari L (2004) Evaluation of wet oxidation pretreatment for enzymatic hydrolysis of softwood. Appl Biochem Biotechnol 117:1–17

Pieltain MC, Castañón AAR, Flores MP, Castañón JIR (1996) Nutritive value of postharvest bean by-products (*Phaseolus vulgaris* L.) for ruminants. Anim Feed Sci Technol 62(2–4):271–275

Pradhan R, Tobioka H, Tasaki I (1997) Effect of moisture content and different levels of additives on chemical composition and in vitro dry matter digestibility of rice straw. Anim Feed Sci Technol 68:273–284

Ragunathan R, Gurusamy R, Palaniswamy M, Swaminathan K (1996) Cultivation of Pleurotus spp. on various agro-residues. Food Chem 55:139–144

Ramirez GR, Aguilera-Gonzalez JC, Garcia-Diaz G, Nunez-Gonzalez AM (2007) Effect of urea treatment on chemical composition and digestion of *Cenchrus ciliaris* and *Cynodon dactylon* Hays and *Zea mays* residues. J Anim Vet Adv 6:1036–1041

Ramirez-Bribiesca JE, Soto-Sanchez A, Hernandez-Calva LM, Salinas-Chavira J, Galaviz-Rodriguez JR, Cruz-Monterrosa RG, Vargas-Lopez S (2010) Influence of *Pleurotus ostreatus* spent corn straw on performance and carcass characteristics of feedlot Pelibuey lambs. Int J Anim Sci 80:754–757

Rangnekar DV, Badve VC, Kharat ST, Sobale DN, Joshi AL (1982) Effect of high pressure steam treatment on chemical composition and digestibility in vitro of roughages. Anim Feed Sci Technol 7:61–70

Rovio S, Kallioinen A, Tamminen T, Hakola M, Leskela M, Siika-ahoa M (2012) Catalysed alkaline oxidation as a wood fractionation technique. Bioresources 7:756–776

Saenger PF, Lemenager RP, Hendrix KS (1982) Anhydrous ammonia treatment of corn stover and its effects on digestibility, intake and performance of beef cattle. J Anim Sci 54:419–425
Saritha M, Arora A, Nain L (2012) Biological pretreatment of lignocellulosic substrates for enhanced delignification and enzymatic digestibility. Indian J Microbiol 52:122–130
Schiere JB, Ibrahim MNM (1989) Feeding of urea ammonia treated rice straw: a compilation of miscellaneous reports produced by the straw utilization project (Sri Lanka). Pudoc, Wageningen
Schultz-Jensen N, Kadar Z, Thomsen AB, Bindslev H, Leipold F (2011) Plasma-assisted pretreatment of wheat straw for ethanol production. Appl Biochem Biotechnol 165:1010–1023
Schutt BD, Abraham MA (2004) Evaluation of a monolithreactor for the catalytic wet oxidation of cellulose. Chem Eng J 103:77–88
Selim ASM, Pan J, Takano T, Suzuki T, Koike S, Kobayashi Y, Tanaka K (2004) Effect of ammonia treatment on physical strength of rice straw, distribution of straw particles and particle-associated bacteria in sheep rumen. Anim Feed Sci Technol 115:117–128
Shaukat S, Bajwa MA, Waqar MA, Sohail T (2006) Production, optimization, purification and characterization of glucomalyase from *Arachniotus* sp. J Chem Soc Pak 28:368–373
Sheikh GG, Ganai AM, Reshi PA, Bilal S, Mir S (2018) Improved paddy straw as ruminant feed: a review. JOJ Scin 1(1): JOJS.MS.ID.555555
Shen HS, Ni DB, Sundstøl F (1998) Studies on untreated and urea-treated rice straw from three cultivation seasons: 1. Physical and chemical measurements in straw and straw fractions. Anim Feed Sci Technol 73:243–261
Shrivastava B, Nandal P, Sharma A, Jain KK, Khasa YP, Das TK, Mani V, Kewalramni NJ, Kundu SS, Kuhad RC (2012) Solid state bioconversion of wheat straw into digestible and nutritive ruminant feed by Ganoderma sp. rckk02. Bioresour Technol 107:347–351
Sirohi SK, Rai SN (1995) Associative effect of lime plus urea treatment of paddy straw on chemical composition and *in vitro* digestibility. Indian J Anim Sci 65:1346–1351
Talebnia F, Karakashev D, Angelidaki I (2010) Production of bioethanol from wheat straw: an overview on pre-treatment, hydrolysis and fermentation. Bioresour Technol 101:4744–4753
Theander O (1981) Chemical composition of low quality roughages as related to alkali treatment. In: Kategile JA, Said AN, Sundstol F (eds) Utilization of low quality roughages in Africa. Workshop Arusha, Tanzania. AUN-Agricultural Development Report 1, Aas, pp 1–15
Theander O, Aman P (1984) Anatomical and chemical characteristics. In: Sundstøl F, Owen E (eds) Straw and other fibrous byproducts as feed: developments in animal veterinary sciences. Elsevier, Amsterdam, pp 45–78
Travaini R, Otero MD, Coca M, Da-Silva R, Bolado S (2013) Sugarcane bagasse ozonolysis pre-treatment: effect on enzymatic digestibility and inhibitory compound formation. Bioresour Technol 133:332–339
Uppal SK, Kaur R, Sharma P (2011) Optimization of chemical pretreatment and acid saccharification for conversion of sugarcane bagasse to ethanol. Sugar Tech 13:214–219
Vadiveloo J (2000) Nutritional properties of the leaf and stem of rice straw. Anim Feed Sci Technol 83:57–65
Wang Y, Spratling BM, ZoBell DR, Wiedmeier RD, McAllister TA (2004) Effect of alkali pretreatment of wheat straw on the efficacy of exogenous fibrolytic enzymes. J Anim Sci 82: 198–208
Wattanaklang B, Abrar A, Cherdthong A (2016) Nutritional value of fermented maize stover as feed for ruminant. J Peternakan Sriwijaya 5(1):44–51
Woyengo TA, Gachuiri RG, Wahome RG, Mbugua PN (2004) Effect of protein supplementation and urea treatment on utilization of maize stover by red Maasai sheep. South Afr J Anim Sci 34: 23–30
Yalchi T, Hajieghrari B (2010) Effect of *Trichoderma* spp. inoculation on the chemical composition and *in vitro* digestibility of wheat straw. Afr J Biotechnol 9:4132–4137
Yang WZ, Beauchemin KA, Rode LM (2000) A comparison of methods of adding fibrolytic enzymes to lactating cow diets. J Dairy Sci 83:2512–2520

Yao R-S, Hu H-J, Deng S-S, Wang H, Zhu H-X (2011) Structure and saccharification of rice straw pre-treated with sulfur trioxide micro-thermal explosion collaborative dilutes alkali. Bioresour Technol 102:6340–6343

Yoswathana N, Phuriphipat P (2010) Bioethanol production from rice straw. Energy Res J 1:26–31

Zadrazil F, Gonser A, Lang E (1999) Influence of incubation temperature on the secretion of extracellular ligninolytic enzymes of *Pleurotus* sp. and *Dichomitus squalens* into soil. In: Proceedings of the conference on enzymes in the environment. Activity, Ecol. Applications Granada, Spain, p 526

Zayed MS (2018) Enhancement the feeding value of rice straw as animal fodder through microbial inoculants and physical treatments. Int J Recycling Organic Waste Agric 7:117–124

Zhu S, Wu Y, Yu Z, Liao J, Zhang Y (2005) Pre-treatment by microwave/alkali of rice straw and its enzymatic hydrolysis. Process Biochem 40:3082–3086

Biotechnological Techniques for Nutritional Quality Improvement in Forages 15

Comparative to food crops, forage crops have been given the least attention, which could be seen from the small number of research and development programs for their improvement. Though, success has been achieved using traditional breeding approaches for the improvement of forage crops, conventional breeding is a time requiring process and forage breeders have to overcome barriers in forage crop breeding such as genotypic and phenotypic heterogeneity, polyploid nature and polygenic control of agronomic and nutritional attributes (Vogel and Pedersen 1993). Due to these reasons, a substantial gain in the development of improved forage varieties through conventional breeding has not been achieved. Considering the genetic complexity of forage crops and associated difficulties in employing conventional breeding methods, the potential of biotechnological techniques in the development of improved forage cultivars has been recognized which further need exploitation on large scale for accelerating forage improvement programs (Bingham and Conger 1995).

Plant biotechnology is a branch of traditional plant breeding. The difference between these two approaches of crop improvement lies in that modern plant biotechnology allows the transfer of genetic information in a highly precise and controlled manner. Unlike traditional plant breeding, which involves the crossing and transfer of hundreds or thousands of genes, plant biotechnology offers opportunities and possibilities to generate variation in forage crops (Spangenberg 2005). Biotechnological tools such as genetic transfer techniques for direct gen introduction, somatic hybridization for combining germplasm from sexually incompatible species and development of molecular markers for cultivar identification and marker-assisted selection are promising in forage breeding programs. These approaches are expected to complement or accelerate conventional breeding, as they extend the range of sources from where genetic information may be obtained for improvement and thus offer new opportunities for molecular breeding. New biotechnological tools permit unlimited gene modifications in gene pool, therefore, holds potential for dramatic improvements in forage crops. Success has been

R. Katoch, *Nutritional Quality Management of Forages in the Himalayan Region*,
https://doi.org/10.1007/978-981-16-5437-4_15

achieved in improving nutritional quality and other desirable traits in some forage crops through genetic engineering and gene silencing technology; however, these approaches have some bottlenecks which need to be addressed (Jaganathan et al. 2018).

New biotechnological tools have equipped scientists with enormous options to tailor forage crops for any desirable attributes. These tools mainly include genome editing approaches like Clustered Regularly Interspaced Short Palindromic Repeats/ CRISPR associated 9 (CRISPR/Cas9), Transcription Activator-Like Effector Nucleases (TALENs), and Zinc Finger Nucleases (ZFNs) (Jinek et al. 2012; Schaart et al. 2016; Weeks et al. 2016). Although developed recently, CRISPR/Cas system has been vigorously used for the manipulation of desirable traits in different economic crops. Though CRISPR/Cas system techniques have shown success in improving the quality of major food crops, the potential of this technique is yet to be explored in forage breeding program.

15.1 Molecular Approaches for Forage Nutritional Quality Improvement

Conventional plant breeding techniques are the classical crop improvement approach, which includes pre-breeding, hybridization, selection and analysis. This approach of improvement has been utilized in the breeding of improved forage cultivars, but little emphasis has been given to forage quality. The conventional approaches have limitations being slow and laborious processes, moreover, forage quality is a complex trait governed by quantitative genes. Efforts for improving forage crops for different traits through conventional breeding are limited by the allogamous, self-incompatibility and polyploid nature of forage crops. Transfer of these quantitative genes is difficult through conventional plant breeding approaches. To overcome these limitations, genetic engineering approaches have edge in crop improvement programs.

15.1.1 Genetic Engineering

Genetic engineering together with plant breeding is an invaluable tool for crop improvement. Transgenics is the process of introducing foreign DNA into the host genome. The foreign DNA or transgene that is transferred to recipient could be from other individuals of the same species or even from unrelated species. *Agrobacterium*-mediated transformation and biolistics (particle bombardment) are the most commonly used transformation methods for forage grasses and legumes, respectively (Spangenberg et al. 2000). Genetic engineering has gained the attention of agriculturists for the efficient improvement of desired traits in plants. The genetic transformation provides unique methods to advance and supplement conventional breeding.

Genetic engineering offers the opportunity to generate unique genetic variation in forage crops by the introduction of foreign genes and/or by up- or downregulation of genes within the plant system (Wang and Brummer 2012). Using this technology, significant progress in forage breeding has been made (Wang and Ge 2006). Despite of different merits, certain negative issues are associated with this technology including generation of random phenotypes through off targeting, transgene carry-over, pleiotropic effects and regulatory hurdles on deregulation of transgenic forage crops. To address these issues, new alternatives and updates are coming up regularly.

15.1.2 Cisgenesis/Intragenesis

This technique involves the transfer of genetic material between the related species by avoiding any possibility of linkage drag (Holme et al. 2012). Since it do not involve transfer of foreign DNA, the bred cultivar is just similar to conventionally bred cultivars. However, this approach also has some limitations (Haverkort et al. 2008; Russell and Sparrow 2008).

15.1.3 RNA Interference (RNAi) and Antisense RNA Technologies

These technologies have also been used for improving quality of forage crops. The term RNA interference (RNAi) is a cellular mechanism that uses the gene's own DNA sequence to turn it off. In a wide variety of organisms, including animals, plants and fungi, RNAi is triggered by double-stranded RNA (dsRNA). Antisense technology is a tool that is used for the inhibition of gene expression. The principle behind it is that an antisense nucleic acid sequence base pairs with its complementary sense RNA strand and prevents it from being translated into a protein. RNAi and antisense RNA technology serve as a potential tool to manipulate a large number of pathways to either upregulate or downregulate the enzymes involved. These techniques have been mainly used for the manipulation of biosynthetic compounds related to forage quality. For example, Cinnamyl alcohol dehydrogenase (CAD) to downregulate lignin deposition in the plant cell wall, Levan sucrase (SacB) for increased sucrose production, Hydroxy cinnamoyl-CoA Hydratase/lyase (HCHL) and Cinnamyl alcohol dehydrogenase (CAD) for altering digestibility and lignin content in tall fescue and rye grass, while alteration of S-adenosylhomocysteine (SAH), alteration in lignin monomer ratio and alteration in Caffeic acid O-methyltransferase (COMT) to reduce lignin component in forages has been successfully achieved. RNAi and antisense RNA technologies also suffer from off-targeting, pleiotropic effects and regulatory barriers in the commercialization of genetically modified crops.

15.1.4 Genome Editing Technologies

The genome-editing technologies based on engineered nucleases have dominated the field of genome engineering. Various successful reports have been published on implication of gene-editing techniques for desirable changes in the genome of different organisms. Zinc finger nucleases (ZFNs), Transcription Activator-Like Effector Nucleases (TALENs) and CRISPR–Cas system have been implicated widely for genome engineering. Zinc finger nucleases (ZFNs) need substantial protein engineering to each DNA target, while the clustered regularly interspersed short palindromic repeats (CRISPR)/Cas9 system only requires a change in a 20-nt targeting sequence within the single-guide RNA (sgRNA). The advantage in the simplicity of cloning strategy and less limitations on potential target sites make the CRISPR/Cas9 system the standout choice for targeted genome editing in understanding gene function and developing valuable traits in model plants and food crops (Li et al. 2013; Shan et al. 2013).

15.1.5 QTLs and Identification of Candidate Genes

Candidate genes have known biological functions and they are directly or indirectly regulating the developmental processes in organisms. The identification of these genes is the must-follow step in the development of improved crops. A number of strategies have been developed for candidate genes identification. The position-dependent strategy requires genome-wide scanning in which candidate genes are identified on the basis of physical linkage information in a QTL-identified chromosomal segment. This approach primarily aims at the vicinity of known QTLs and candidate genes are identified from other genes present in the targeted chromosomal region. However, using this approach, it is difficult to prioritize candidate genes in the targeted region. At present comparative genomic analysis is widely used for the identification of candidate genes. This approach relies on the information on previously characterized candidate genes that is already available in public genome databases. In forage crops, many candidate genes have been identified using QTL (Quantitative Trait Loci) analysis or MAS (Marker-Assisted Selection) and studied in detail in relation to nutritional quality, biomass, and growth traits. Breeding programs for forage crops are accompanied by different hurdles due to high genotypic and phenotypic heterogeneity since many species are polyploid, a problem that is exacerbated by inbreeding in many grasses and few agronomic traits being linked to the distinct genes (Poehlman 1987; Vogel and Pedersen 1993).

Protein digestion and nutrient uptake and utilization are directly related to energy availability in livestock (Nocek and Russell 1988; McCarthy et al. 1989), it is important to increase the level of water-soluble carbohydrates in many forage crops, as they provide an instant source of energy for livestock. QTL for water-soluble carbohydrates have been identified on chromosomes 1, 2, 3, 6, and 7 in perennial rye grass (Humphreys and Turner 2003; Cogan et al. 2005). QTLs for various direct measures of digestibility such as DMD, OMD, and CWD have also been identified in *Lolium* on chromosome numbers 1, 2, 5, 6, 7 (Van Loo et al. 2003; Cogan et al. 2005).

Fructans, polymer of fructose are the main constituent of WSC in forage grasses (Ritsema and Smeekens 2003). Studies have revealed that high fructans content in temperate forage grasses increases their digestibility, stability of protein in rumen and live weight gain in livestock (Spangenberg 2005). The structural diversity of fructans is mainly controlled by fructosyltransferases. A number of FTs involved in fructan biosynthesis such as 1-SST (sucrose:sucrose 1-fructosyltransferase), 1-FFT (fructan:fructan 1-fructosyltransferase), sucrose:fructan 6-fructosyltransferase, and 6G-FFT (fructan:fructan 6G-fructosyltransferase), have been considered a potential candidate gene and are characterized from different plant species. In the past, efforts have been made to increase fructan biosynthesis in pasture plants by targeting these enzymes (Gao et al. 2010). Prolonged green period is a highly desirable trait in perennial grasses breeding as it is associated with forage quality and biomass accumulation (Xu et al. 2018). Leaf senescence occurs naturally in plants with aging or can be induced by biotic and abiotic stresses. Chlorophyll breakdown is a hallmark of leaf senescence. Protein degradation is tightly associated with accelerated chlorophyll catabolism during leaf senescence. Therefore, blocking or reducing chlorophyll breakdown and thereby improving chlorophyll and leaf protein contents is desirable for agronomic improvement in perennial forage grasses. Chlorophyll degradation is regulated by multiple Chl catabolic enzymes (CCEs) and associated genes, such as pheophytinase enzyme encoding gene and Stay Green [SGR, or NON-YELLOING 1 (NYE1)] gene (Schelbert et al. 2009). Wei et al. (2011) identified *FaNYE1*, an orthologue of *AtNYE1* from major tall fescue (*Festuca arundinacea* Schreb.) by RACE-PCR. This gene can be successfully used to induce stay green characters in other grasses too. An *Arabidopsis* mutant deficient in pheophytinase (PPH) (pph-1) has been reported unable to degrade chlorophyll during senescence and therefore exhibits a stay green phenotype. Furthermore, pph-1 accumulates phein during senescence. Therefore, PPH is an important component of the chlorophyll breakdown machinery of senescent leaves. These genes have been suggested as a potential target for improving green period and quality of grasses through genome engineering (Wei et al. 2011; Xu et al. 2018).

High lignin content in cell wall limits the amount of digestible energy to livestock. The targets for lignin manipulation in forages include gene(s) encoding the regulatory enzyme(s) catalyzing the synthesis of monolignols (Verma and Dwivedi 2014) and their polymerization (Wang et al. 2015). Moreover, transcription factor genes such as *MYB* genes regulating the lignification process have also been considered as a potential target for the engineering of lignin metabolic pathways in plants. Candidate genes involved in feruloylation of arabinoxylan, an important process involved in lignification of cell wall in grasses have also been identified (Mitchell et al. 2007). Condensed tannins (CT) are secondary metabolites that complex with proteins and reduce their ruminal degradation. They are considered as anti-nutrients at higher their amount but moderate content provide protection to proteins which are rapidly digested in rumen and responsible for bloat. Improvement in CT content in tannins negative forages such as *Medicago sativa* and *Trifolium repens* would be useful in enhancing their nutritional quality (Salunkhe et al. 2017). Since the role of transcriptional regulators R2R3 MYB, bHLH, and WD40 protein in CT biosynthesis has been well characterized, they can be used to manipulate CT concentration in forage crops (Lepiniec et al. 2006). Lorenzo et al. (2020) suggested *MsFTa1* gene (Member of FlOWERING LOCUST gene family) as a potential candidate for improving the digestibility of alfalfa. Amount of resistant starch (RS) is important for determining the digestibility and nutritional content of forage crops. In general, lower RS in forages denotes their high digestibility. QTLs for resistant starch have been found on chromosome 3 in *Lolium* independent of QTLs of WSC. Studies have also reported that sulphur-containing amino acids, methionine and cysteine play important role in the improvement of dairy, wool, and meat production (Pickering and Reis 1993). Different proteins rich in sulfur amino acids, resistant to degradation by rumen fluid degradation, but susceptible to intestinal proteases have been identified, and genes encoding these proteins have been used for enhancing the nutritional quality of forage crops (Wang et al. 2001; Tabe and Droux 2002). Advances in sequencing methods could provide more transcriptomic information including SNPs which can be useful to improve forage grasses. Drafting of genomes of forage crops is becoming increasingly common as well as more evidence that model species like *Brachypodium* can direct research for the improvement of other forage crops.

15.2 Nutritional Quality Enhancement of Forage Crops with New Biotechnological Techniques

Plant biotechnology employs the use of molecular techniques to improve the beneficial traits of plants. It involves introduction of new genes and up and downregulation of target genes to achieve desired traits. The primary focus of early biotechnology was the improvement of agronomic traits or input traits, such as yield, herbicide and pesticide tolerance and disease resistance producing the so-called first generation of transgenic crops. Today, there is increasing interest in improving output traits, which have a direct impact and are more favorably perceived by consumers. The genetic engineering of biosynthetic pathways in plants has led to the development of varieties with excellent nutritional profiles. Forages are the ultimate source of nutrition for livestock, however, there has been limited biotechnological intervention in forages as compared to food crops. In the recent years efforts have been directed towards improving the nutritional quality of forages by introducing potential genes and through the engineering of metabolic pathways.

15.2.1 Transfer of Nutritionally Potential Genes

15.2.1.1 Gene Introgression for Higher Amino Acid Content

While forage crops are not specifically deficient in sulphur-containing amino acids, dietary supplementation with methionine or cysteine may increase live weight gain, milk production and wool growth (Tabe and Droux 2002). In addition to nitrogen losses during rumen fermentation, ruminant production may suffer from a loss of sulfur during the conversion of ingested plant protein to microbial protein, thus limiting sulfur utilization in the abomasum and small intestine. Different proteins are known to be relatively resistant to degradation in the rumen and are rich in sulphur-containing amino acids, which can be easily assimilated. Sunflower seed albumin (SSA), rich in both methionine and cysteine, is highly resistant to degradation during rumen fermentation (Tabe et al. 1993). Both alfalfa and subterranean clover were successfully transformed with a chimeric SSA gene which in the case of subterranean clover stably expressed through one generation of sexual reproduction (Khan et al. 1996). Transgenic plants with highest levels of SSA expression had 0.10 and 0.75% of soluble leaf protein in the form SSA in alfalfa and subterranean clover, respectively (Khan et al. 1996; Tabe et al. 1995). For subterranean clover, this level meets the lower end of the range at which a wool-growth response would be expected, based on dietary supplementation research (Khan et al. 1996). The gene encoding β-phaseolin protein under the control of 35S promoter was also introduced in alfalfa (Bagga et al. 1992). Bagga et al. (2004) co-expressed zein gene from maize into alfalfa and reported stable accumulation of δ-zein in endoplasmic reticulum-derived protein bodies formed by β-zein. The gene encoding δ zein has been targeted for expression in some important leguminous forages (Bellucci et al. 1997, 2002; Bagga et al. 2004). Transformation of alfalfa with cystathionine γ-synthase gene from *Arabidopsis* driven by *RuBisCo* small subunit promoter increased the accumulation of sulphur-containing amino acids and water-soluble proteins (Avraham et al. 2005). Transgenic subterranean clover plants (*Trifolium subterraneum* L.) expressing gene encoding sulphur amino acid-rich protein from sunflower has also been generated (Rafiqul et al. 1996). The transgenic alfalfa expressing the *Arabidopsis* cystathionine γ-synthase (*AtCGS*) gene has reported up to 32-fold, 19-fold and 2.20-fold increase in total methionine content, S-methylmethionine (SMM) content and water-soluble proteins content than wild-type (Avraham et al. 2005). Cho et al. (2000) reported an increase in tryptophan synthesis in a forage legume *Astragalus sinicus* by expressing tobacco feedback-insensitive anthranilate synthase (ASA2) gene. Gebril et al. (2015) constitutively expressed sucrose phosphate synthase (SPS) gene from maize in alfalfa and reported an increase in growth and crude protein content in leaves. Tong et al. (2014) co-expressed two bacterial genes, aspartate kinase and adenylylsulfate reductase to increase the level of sulfur amino acids in alfalfa.

Sulphur-containing amino acids, (methionine and cysteine) are the most limiting essential amino acids in livestock nutrition (Orskov and Chen 1990). In particular, wool growth in sheep is frequently limited by the supply of sulphur amino acids under normal conditions (Reis 1979; Higgins et al. 1988). Rumen fermentation contributes partly to the S amino acid deficiency since rumen microflora degrades feed protein and in some conditions resynthesize proteins of low nutritional quality (Rogers 1990). Transgenic tall fescue plants that express foreign genes encoding a rumen-stable protein rich in sulphur amino acids have been developed with the aim of improving protein quality (Wang et al. 2001). In this case, chimeric gene encoding sunflower seed albumin 8 (SFA8) protein which contains high levels of methionine and cysteine, resistant to rumen degradation, a sequence coding for endoplasmic reticulum retention signal (KDEL) were constructed under the control of constitutive promoter (CaMV 35S) and light-regulated (wheat cab) promoters. These chimeric gene constructs were introduced into the tall fescue genome by microprojectile bombardment of embryogenic suspension cells. The results of the study revealed that transgenes were stably integrated into genome and produced expected transcript. The corresponding sulfur-rich SFA8 protein was accumulated at levels of up to 0.20% of total soluble protein in transgenic plant leaves. Constitutive and light-regulated expression of the transgenes led to similar levels of SFA8 protein accumulation (Wang et al. 2001). The study revealed that the expression of chimeric genes to the level of 2–4% of total leaf protein is required for significant impact in ruminal diets.

Postharvest storage of forages (Ensiling, Silage) often leads to degradation of protein to nonprotein nitrogen, which is poorly utilized by ruminants. Postharvest protein degradation is especially high in alfalfa (*Medicago sativa* L.). Contrary to it, red clover (*Trifolium pratense* L.) has up to 90% less protein loss during storage due to enzyme polyphenol oxidase (PPO) that oxidizes o-diphenols to O-quinones and subsequently form complex with proteins. Such complexes are resistant to enzymatic proteolysis thus reduces protein degradation during storage (Sullivan and Hatfield 2006). Kingston-Smith et al. (2013) suggested that the inhibitory effect of PPO on proteolysis could be exploited to breed forage varieties with increased PPO activity. Sullivan et al. (2004) transferred gene encoding PPO enzyme from red clover to alfalfa (*Medicago sativa*) and reported reduction of postharvest proteolysis.

Strategies for increasing the accumulation of foreign proteins in forage plants may include use of strong promoters and development of chloroplast transformation system to recover transplastomic lines. In transgenic subterranean clover (*Trifolium subterraneum*), the accumulation of sunflower albumin increased with maturity and the old leaves of highly expressing plants contain 1.30% of total extractable protein (Khan et al. 1996). Plant transformation appears to be the most promising genetic approach to creating and identifying novel protein phenotypes of forage plants. Furthermore, because each transgenic event is inherited as a simple Mendelian dominant, it can be backcrossed relatively easily into other germplasm and different transgenic events can be combined to potentially increase the phenotypic response.

15.2.1.2 Prevention of Bloat

A further application of genetic engineering is the improvement in forage quality through introduction of genes responsible for condensed tannins (proanthocyanidins) biosynthesis in tannins negative forage crops such as alfalfa. Condensed tannins (proanthocyanidins) are polymeric phenyl-propanoid-derived compounds synthesized by the flavonoid pathway. They are agronomically important in a range of forage legumes, where they are regarded either as beneficial or detrimental. High proanthocyanidin content is not desirable but total lack of proanthocyanidins causes the problem of bloating in animals, which is characterized as loss of protein in the form of ammonia (Ammonia overflow). It is a concerning fact that bloat in animals is alone responsible for more than hundred million dollars loss globally, thus considered as an important agronomic trait (Li et al. 2016). It has been reported that moderate concentrations of proanthocyanidins (About 2–4% of dry matter) prevent bloat, enhance ruminant nutrition and reduce protein degradation during storage. Alfalfa is relatively low in fiber, but high in protein compared to other forages which results in high intake of the forage. Improving fiber digestibility, protein quality and decreasing the rate of protein degradation in rumen are the major areas of research toward alfalfa quality improvement. The rapid degradation of protein in alfalfa poses a risk of bloat to ruminants. Coupled with relatively high lignin content impeding carbohydrate degradation in rumen, alfalfa has unbalanced and asynchronous nitrogen to carbohydrate ratio. In contrast to alfalfa, some legume forages, such as sainfoin (*Onobrychis viciifolia*), birdsfoot trefoil (*Lotus corniculatus*) and big trefoil (*Lotus pedunculatus*) also have a higher content of protein but are bloat-free. This anti-bloat character is highly associated with their possession of proanthocyanidins (PAs), which are also known as condensed tannins. The first genetic transformation of alfalfa was reported in the mid-1980s. Further studies have been carried out for increasing proanthocyanidins accumulation in alfalfa. Ray et al. (2003) expressed flavonoid regulatory maize gene (Lc gene) in alfalfa and reported accumulated higher amounts of proanthocyanidins. Expression of MtPAR gene from *Medicago truncatula* enhanced proanthocyanidins accumulation in alfalfa leaves (Verdier et al. 2012). Over expression of MYB14, an R2R3-MYB transcription factor from *Trifolium arvense* in alfalfa increased the concentration of proanthocyanidins up to 1.80% of DM (Hancock et al. 2014).

15.2.1.3 Lipid Content in Forages

Forages with high energy can be developed by increasing the lipid content in vegetative tissue. Under normal circumstances, plants store neutral lipids only in seeds and pollen as a source of essential energy for subsequent germination. The majority of leaf lipid is found in membranes. Lipids have twice the caloric value of the other main sources of energy in forage leaves, namely, WSC and protein (Flowers et al. 2008). Supplementary feeding trials in sheep have demonstrated that increasing the level of dietary fat to 8% DM led to a 30% increase in feed conversion efficiency (Cosgrove et al. 2004). There is minimal genetic variation in leaf lipid content and therefore the opportunity to rapidly and substantially change the leaf lipid content through conventional breeding is limited (Hegarty et al. 2013). A technology to increase plant lipid content and plant biomass has been recently reported. A synthetic gene encoding a novel lipid-encapsulation protein, cysteine oleosin, was constitutively co-expressed with a gene encoding diacylglycerol acyltransferase (DGAT1) to increase energy density and reduce photorespiration of *Arabidopsis thaliana via* production and encapsulation of neutral lipids (Winichayakul et al. 2013). The same genes have been co-expressed in perennial rye grass under the control of light-regulated, green tissue-specific promoters derived from the rice chlorophyll a/b-binding protein and ribulose bisphosphate carboxylase genes (Jang et al. 1999). When compared with a control population in containment glasshouse conditions with nonlimiting water and nutrient supply, these transgenic plants have significant increase in leaf lipid content.

15.2.2 Engineering Metabolic Pathways

Metabolic pathways are chains of linked chemical reactions catalyzed by enzymes that convert substrates into products through a series of intermediates. Metabolic engineering could be used to modulate endogenous metabolic pathways in plants or introduce new metabolic capabilities in order to increase the production of a desirable compound or reduce the accumulation of an undesirable one. Forage quality traits are under control of many genes (Polygenic) encoding regulatory enzymes of their biosynthetic pathway. Since manipulation in one gene can affect the function of the other and can lead to an undesirable trait, therefore, deep understanding of key metabolic pathways and the regulatory genes in forages is essential for developing improved forages.

15.2.2.1 Manipulation of Lignin Biosynthesis

The quality of forages directly affects the livestock health and production and it is usually defined in terms of voluntary intake and digestibility of forage. Feeding and digestibility studies have shown that a smaller increase in forage digestibility would provide financial benefits to farmers; therefore it has been considered as a prime goal in forage crop improvement programme. Since heritability of digestibility is low, with a large number of genes involved, the potential for developing highly digestible varieties using conventional breeding is low. Dry matter digestibility of forage grasses declines with the advanced stages of development. One of the important constraints for the digestion of plant cell walls is lignin. Lignin, a major component of secondary cell wall, is a heterogeneous complex polymer resulting from oxidative coupling of hydroxylated and methoxylated phenylpropane units (Monolignols), i.e., hydroxyphenyl (H lignin), guaiacyl (G lignin) and syringyl (S lignin), synthesized from phenylpropanoid pathway. In grasses, lignin mainly consists of S and G units and ratio of these two units (S/G ratio) influences their digestibility (Christensen and Rasmussen 2019).

Lignin provides strength to cell wall and protection against pathogens and physical damage. However, these advantages of lignification in plant cell walls can represent themselves as barriers in utilization of forage crops in livestock feeding and other agro-industrial uses. Metabolic engineering of lignin biosynthetic pathway by downregulating expression of genes offers opportunities to modulate lignin content and improve digestibility of forage crops (Tu et al. 2010). Different techniques have been used for lignin manipulation such as antisense RNA, sense expression, co-suppression, and RNA interference (RNAi). However, the choice of method depends upon utility of end product (Verma and Dwivedi 2014). RNAi (RNA Interference) has been the most commonly used method for downregulation of lignin biosynthetic genes (Christensen and Rasmussen 2019).

Most of the researches are focusing on downregulating key genes involved in the biosynthesis pathway of lignin production. The first lignin gene for cinnamyl alcohol dehydrogenase (CAD) was downregulated in alfalfa by Baucher et al. (1998). However, these studies did not provide any definitive information on the factors associated with the digestibility of forage. Reddy et al. (2005) developed transgenic alfalfa lines through antisense expression of cinnamate 4-hydroxylase (C4H), coumaroyl shikimate 3-hydroxylase (CSH) and coniferaldehyde 5-hydroxylase (C5H), with altered lignin content and composition, which catalyze reactions in the formation of monolignol, G units and S units, respectively. The results of the study revealed that downregulation of expression of coumarate 3-hydroxylase *(C3H)* enzyme resulted in reduction of lignin with enriched H lignin. Downregulation of ferulate 5-hydroxylase (F5H) gene slightly increased the G lignin and reduced the S lignin content. Molecular breeding for improved digestibility by downregulating monolignol biosynthetic enzymes through transgenesis has also been explored. For example, digestibility of tall fescue has been improved by transgenic downregulation of cinnamyl alcohol dehydrogenase enzyme (Chen et al. 2003) and caffeic acid O-methyltransferase enzyme (Chen et al. 2004).

With lignin manipulation, some contradictory results have been obtained regarding the relationship between lignin content and digestibility of transgenic plants. For instance, silencing of lignin biosynthesis genes resulted in changes in lignin composition (lignin monomer ratio) with no effect on total lignin content (Boerjan et al. 2003). Vailhe et al. (1996) reported a tenfold decrease in S/G ratio with increased *in vitro* digestibility in transgenic tobacco. The lignin content was unaltered in transgenic plants. Transgenic alfalfa generated through downregulating CAD genes revealed unaltered Klason lignin content, reduced S/G ratio with increased forage digestibility (Baucher et al. 1998). Such reports reveal the importance of lignin content and lignin composition in determining the digestibility of forage crops.

Though expression of many genes regulating lignin biosynthetic pathway has been engineered in forages, but the research outcomes are still not sufficient for field demonstrations. Therefore, genes from other biosynthetic pathways affecting lignin biosynthesis must be identified and used for lignin engineering. Tricin has been reported to play a key role in lignification exclusively in monocots (Lam et al. 2017). This molecule could be an alternative target for lignin engineering (Halpin 2019). Metabolic engineering of the transcriptional regulatory network of secondary cell wall synthesis has also been found effective in reducing lignin content in plants (Wuddineh et al. 2016). The use of (1) specific proteins which have functioned as negative regulators of lignin biosynthesis; (2) enzymes producing alternative lignin monomers; (3) engineered enzymes have the ability to modify monolignols into their nonoxidizable forms, and (4) proteins mediating posttranscriptional degradation of enzymes of lignin biosynthetic pathway have also been reported as a promising strategy for spatiotemporal reduction of lignin in plants (Eudes et al. 2015).

Metabolic engineering of lignin biosynthesis pathway have demonstrated reduction in lignin content, however, there are some reports of undesirable effects in engineered transgenic plants (Shadle et al. 2007; Liu et al. 2017). For example, antisense expression of HCT constructs in transgenic *Medicago sativa* significantly reduced lignin content and composition but affected plant growth and flowering (Shadle et al. 2007). Major limitation in using the silencing approach is off targeting due to which silencing effect transports within the plant system and affect most of the plant tissues. Moreover, the phenylpropanoid pathway is linked with the other metabolic pathways associated with the synthesis of compounds derived from the same precursors of lignin (Weng and Chapple 2010). Therefore, any alteration in lignin biosynthesis affects synthesis of other metabolites. Use of tissue-specific promoters has been proposed to reduce the risk of off-targeting. However, the use of such promoters is challenging because most lignin biosynthetic genes belong to phenylpropanoid pathway (Van Holme et al. 2012) that may affect biosynthesis of other metabolites. Use of promoters of vessel-specific and fiber-specific genes expressed during secondary cell wall synthesis has raised concerns regarding integrity of plant cell walls. The recent genome editing technology offer opportunities to meet challenges associated with metabolic engineering of lignin biosynthetic pathway.

The basic effect of manipulating the expression of some enzymes in the monolignol pathway may be similar to native genes coding for decreased enzyme activity (Casler and Kaeppler 2000). Transgenic technology offer the potential for (a) producing plants with highly unusual, novel lignins; (b) higher frequencies of novel-lignin phenotypes in comparison with natural variation; (c) effective simultaneous downregulation of multiple enzymes; (d) and highly targeted downregulation of enzymes through a choice of cell type-specific and developmentally regulated promoters. Genetic manipulation of monolignol biosynthesis transgenetically to enhance herbage quality have been explored in some forage legumes (e.g., *Stylosanthes humilis*, *Medicago sativa*) and forage grasses (e.g., *Lolium perenne* and *Festuca arundinacea*) (Spangenberg et al. 2000). cDNAs and genomic clones encoding of the key enzymes COMT, 4CL, CCR, and CAD in perennial rye grass have been isolated, sequenced, characterized and used for molecular genetic dissection of this biosynthetic pathway in grasses (Heath et al. 1998). Once the suitable transformation events in forage plants with downregulation of individual enzymes or simultaneously for multiple enzymes of monolignol biosynthetic pathway are obtained, a thorough agronomic assessment of these transgenic plants particularly for vigor and stress tolerances will be required.

15.2.2.2 Manipulation of Fructan Metabolism

Fructans, polymer of fructose are the main constituent of WSC in forage grasses. Many grass species produce fructans (polyfructose molecules) as their main soluble storage carbohydrate. It has been shown that rye grass lines that accumulate higher soluble carbohydrates do not suffer from decline in digestibility during summer (Radojevic et al. 1994). The increased level of soluble carbohydrates appears to offset the decline in digestibility due to lignification. In addition, by increasing nonstructural carbohydrates in pasture plants, herbage intake, protein capture in the rumen and live weight gains may improve. Fructan synthesis in grasses involves action of at least three enzymes; sucrose: sucrose 1-fructosyltransferase (1-SST);

fructan: fructan 1-fructosyltransferase (1-FFT); and sucrose: fructan 6-fructosyltransferase (6-SFT), which synthesizes the more complex mixed-linkage fructans that prevail in grasses and cereals.

Transgenic approaches for the genetic manipulation of fructan biosynthesis to enhance herbage quality and tolerance to abiotic stresses are being explored in both forage legumes (e.g., *Trifolium repens, Medicago sativa*) and forage grasses (e.g., *Lolium perenne* and *Festuca arundinacea*) (Jenkins et al. 2000; LePage et al. 2000; Lidgett et al. 2000; Ye et al. 2001). Transgenic rye grass plants with altered fructan metabolism have been generated by the expression of chimeric levansucrase genes (Ye et al. 2001). In the study, the expression of sacB genes in transgenic rye grass and the concomitant accumulation of low levels of bacterial levan substantially distorted the native grass fructan synthesis pattern. Growth of the levan-accumulating transgenic rye grass plants slowed down with the onset of the reproductive phase (Ye et al. 2001). cDNAs encoding perennial rye grass (*L. perenne*) fructosyltransferase homologs have been isolated, characterized and are being used for the systematic molecular genetic dissection of fructan biosynthesis in transgenic grasses. A cDNA encoding 1-SST from tall fescue has been isolated and functionally characterized in transient assays with tobacco protoplasts and in methylotrophic yeast, *Pichia pastoris* (Luescher et al. 2000). Transgenic white clover (*Trifolium repens*) plants expressing chimeric *Bacillus subtilis* SacB genes for enhanced tolerance to drought have been generated (LePage et al. 2000). Transgenic lucerne (*Medicago sativa*) and white clover plants expressing a fructosyltransferase gene, derived from *Streptococcus salivarius*, have also been generated (Jenkins et al. 2000).

15.2.2.3 Manipulation of Stay-Green Trait in Forages

Chlorophyll degradation is dramatically initiated during senescence as well as in response to biotic and abiotic stresses. Mutation of a gene regulating chlorophyll degradation results in indefinite retention of greenness in senescing leaves of temperate grass *Festuca pratensis* (Vicentini et al. 1995). Chl degradation involves multiple biochemical reactions catalyzed by at least six Chl catabolic enzymes (CCEs) in the chloroplast, namely pheophytin pheophorbide hydrolyase (PPH), Non-Yellow Coloring1 (NYC1), NYC1-like (NOL), Chl a reductase (HCAR), pheide a oxidase (PAO), and red Chl catabolite reductase (RCCR) (Hörtensteiner 2006). The PPH transcript level is positively correlated with the extent of leaf senescence during the natural leaf maturation process or induced by stresses. Zhang et al. (2016a, b) carried out functional characterization and hormonal regulation of PHEOPHYTINASE gene LpPPH. A phylogenetic tree of LpPPH and six PPH orthologs of different plant species revealed that the phylogenetic relationship between the PPHs was consistent with that of the plant species, indicating that the cloned LpPPH is the functional pheophytinase-encoding gene. The expression level of LpPPH was positively related to the extent of leaf senescence. Exogenous application of abscisic acid (ABA) and ethephon (an ethylene-releasing agent) accelerate the decline in Chl content in leaves of perennial rye grass, whereas

Table 15.1 Genes controlling important traits in forage crops

Target trait	Species	Gene	Reference
Fructans	Perennial rye grass	Wheat fructosyltransferase genes	Hisano et al. (2004)
	Lolium multiflorum	*Levan sucrase (SacB)*	Ye et al. (2001)
	Trifolium repens	*Levan sucrase (SacB)*	Jenkins et al. (2000)
	Medicago sativa	*Levan sucrase (SacB)*	Jenkins et al. (2000)
Digestibility—lignin	*Lolium multiflorum*	*Lignin biosynthetic genes (CAD, COMT, 4CL CCR)*	Spangenberg et al. (2000)
	Medicago sativa	*Cinnamyl alcohol dehydrogenase*	Baucher et al. (1998)
	Festuca arundinacea	*Cinnamyl alcohol dehydrogenase*	Chen et al. (2003)
	Lolium perenne	*Cinnamoyl CoA-reductase (CCR) and caffeic acid O-methyltransferase (COMT)*	Tu et al. (2010)
Stay green trait	*Lolium perenne*	*pph-1*	Zhang et al. (2016a, b)
		LpSGR	Xu et al. (2018)

cytokinin and aminoethoxyvinylglycine treatments suppress leaf senescence, corresponding to the up or downregulation of LpPPH expression.

Manipulation of stay green or chlorophyll degradation is responsible due to the identification of NYE1 gene in *Arabidopsis,* which is generally referred to as SGR (Stay-Green) (Wei et al. 2011). Forages tend to respond to various biotic and abiotic stresses leading to degradation of chlorophyll, thereby reducing the forage yield and quality. Orthologue of AtNYE1 of Arabidopsis from tall fescue, FaNYE1 has been isolated and characterized by RACE-PCR (Wei et al. 2011). The transcript level of FaNYE1 in early senescent and senescent leaves was about 14-fold higher than green leaves. Table 15.1 summarizes the important genes exploited for forage crop development.

15.3 Genomic *In Situ* Hybridization (GISH) and DNA-Based Genetic Marker Technology

15.3.1 Genomic *In Situ* Hybridization (GISH)

Molecular cytogenetic techniques, such as *in situ* hybridization methods, are admirable tools to analyze the genomic structure and function, chromosome constituents, recombination patterns, alien gene introgression, genome evolution, aneuploidy, polyploidy and also genome constitution visualization and chromosome discrimination from different genomes in allopolyploids of different crops. *In situ* hybridization technique was developed by John et al. (1969) and Gall and Pardue (1969). Genomic *in situ* hybridization (GISH) was first used to distinguish chromosomes from different genomes in interspecific/intergeneric hybrids or allopolyploids. In GISH, the probe is easier to obtain and may be used directly with no need for amplification, because the amount of DNA is limitless (Peñaloza and Pozzobon 2007). Total genomic DNA must be broken into smaller fragments, which can be done using restriction enzymes, autoclaving, or sonicating. Although autoclaving is more often used to break DNA due to the low cost of the equipment, sonication provides greater precision as regards the size of the DNA fragments. Probe labeling may be carried out with random primers, PCR, or nick translation (Guerra 2004).

GISH technique is not useful in the improvement of forage quality unless mapping of genes governing a particular specific trait is done. It provides an insight into technology to identify introgressed genes either through cloning or through hybridization. Not much of GISH analysis has been done in forage crops with respect to forage quality improvement. *Festulolium* is an intergeneric hybrid between *Festuca* and *Lolium* species. Through crossing, the hybrids acquire agronomic benefits typically associated with both genera, such as good persistence and high tolerance to environmental stress from *Festuca* and nutritional quality and high forage yield from *Lolium* (Yamada et al. 2005). In *Festulolium*, the GISH technique could discriminate between two genomes of triploid hybrids of *Lolium multiflorum* and *Festuca pratensis* Thomas et al. (1994). The ability to discriminate between chromosomes from *Festuca* and *Lolium* in *Festulolium* using GISH was also demonstrated by Mizukami et al. (1998). Kopecky and colleagues (2008) analyzed genomic constitution of *Festulolium* varieties.

The genomic constitution of the first *Festulolium* cultivar developed in Japan, "Tohoku 1," released in 2009, was analyzed by GISH by Akiyama et al. (2010). GISH revealed numerous recombinant chromosomes with a relatively small proportion of the *Festuca* genome in Tohoku 1. The ratio of the *Festuca*-specific genome (f ratio) was tested for use as an evaluation index for estimating the genomic constitution of *Festulolium*. Although, the observed number of *Festuca* chromatins in Tohoku 1 were reduced during the mitosis, the f ratio was stable throughout mitosis, indicating that the f ratio is a useful index for characterizing the *Festulolium* genome. Reis et al. (2014) studied the homeology between genomes A and A′ and confirmed by the extensive marking of *Pennisetum glaucum* chromosomes by the probe A′B of *Pennisetum purpureum*. All 14 chromosomes from genome A of *Pennisetum glaucum* were almost completely marked, with large blocks of probe signals observed on the chromosomes. The markings by the probe of genome A′B observed in the centromeric and pericentric regions represented 63% of the genome of *Pennisetum glaucum*. These marked portions result from hybridization, both between the genomes A and A′ and, in smaller proportion, genomes A and B, observed both in karyograms of *Pennisetum purpureum* and triploid hybrids.

15.3.2 DNA-Based Marker Technology/Linkage Mapping

Molecular marker technology, which allows the development of genome-wide linkage maps, presents the possibility of identification and selection on the basis of individual QTL for forage traits. Marker-assisted selection (MAS) consists of (1) identifying putative QTL by correlating phenotypic data to markers within a linkage map, (2) determining which putative QTL explain the largest amount of phenotypic variance and have the desired effect (positive or negative) on the phenotypic trait, and (3) selecting plants on the basis of the molecular marker (s) linked to the putative QTL. Because many forage crops are complex polyploids with genomes that derive from multiple progenitors or with polysomic inheritance, the development of reliable linkage maps in most forage crops is more complicated than in simple diploids. Maker-assisted breeding with high throughput phenotyping could increase the speed and precision with which traits can be introduced and selected into and from breeding populations. This is based on the detection of DNA variation among individuals and quantitative trait locus (QTL) analysis for dissecting complex traits and assessing gene effects. Associations of QTL with valuable traits in different forage crops have been identified.

Mapping of QTLs involves screening of mapping populations with respect to particular quantitative traits. QTL mapping identifies and provides linkage maps in which genes or markers are linked to complex traits like forage quality, yield, and various biotic and abiotic stresses. The identified locus of genes is used to clone or introgress sequence into the background DNA of forage crops. Lübberstedt et al. (1997) published the first QTL analysis devoted to forage quality in maize. Quantitative trait loci for cell-wall digestibility and lignification traits in maize were also investigated in two recombinant inbred lines (RIL) progeny by Méchin et al. (2001). Cardinal et al. (2003) detected 65 QTL related to fiber and lignin content in maize.

The best options for breeding grasses for improved digestibility were assessed based on genome locations involved in forage quality traits through QTL analysis (Ralph et al. 2004). In two pearl millet × elephant grass crosses, a total of 15 putative QTL were identified for *in vitro* organic matter digestibility, NDF, and CP, six of which were associated with more than one trait suggesting either pleiotropic effects or close linkage of genes controlling two different forage nutritional value traits (Smith et al. 1993). An association between isozymes of phosphoglucose isomerase (PGI-2) and WSC concentration has proven relatively consistent in perennial rye grass. The b allele of the Pgi-2 locus has been associated with high WSC concentration in populations undergoing natural selection in a range of cultivars that differe in Pgi-2 allele frequency and WSC content (Smith et al. 1998), and marker-selected progenies of a cross between parents homozygous for different isozymes of PGI-2 (Hayward et al. 1994).

Ground herbage samples from genotypes of the p150/ 112 population of perennial rye grass (*Lolium perenne*) were analyzed for quality traits such as CP, IVDMD, NDF, metabolizable energy and WSC by NIRS analysis (Cogan et al. 2005). A total of 42 QTL were observed in six different sampling experiments varying by developmental stage (anthesis or vegetative growth), location, or year. QTL was detected on LG 3, 5, and 7. The region on LG 3 was associated with variation for all measured traits across various experimental datasets. The region on LG-7 was associated with variation for all traits except CP and was located in the vicinity of the lignin biosynthesis gene loci *xlpomt1* (Caffeic acid-O-methyltransferase), *xlpccr1* (Cinnamoyl CoA-reductase), and *xlpssrcad2.1* (Cinnamyl alcohol dehydrogenase). Water-soluble carbohydrates provide the most readily available source of energy for grazing ruminants. In the F_2 (Aurora × Perma) population, high molecular fructan, which constitutes the major part of the WSC pool, was analyzed in samples collected during spring and autumn in tiller bases and leaves with replication of data over years (i.e., collection one replicate each year for several years; Turner et al. unpublished data). Correlations between traits did not always lead to a corresponding cluster of QTL and some traits revealed no reproducible QTL. Tiller base QTL were identified on LG 1, 5, and leaf QTL on LG 2 and 6 (Humphreys and Turner 2003). Biazzi et al. (2017) studied alfalfa genome and highlighted a genomic area on chromosome 8 that was characterized by a high concentration of SNP markers linked to ADL in leaves. A linked annotated gene, namely *Medtr8g091470*, which is known to encode for a cellulose synthase interactive protein was identified in that chromosome region by the presence of two SNPs in the same exon.

15.4 Prospects of Gene Editing Techniques in Nutritional Quality Improvement of Forage Crops

Conventional plant breeding is the major crop improvement approach used for the improvement of yield and quality traits in forage crops. Through this approach, the potential way to develop improved forage crops is through selecting parental lines having superior traits and cross-hybridizing them to produce progeny with desirable and novel traits. Major limitations of forage breeding program are allogamous nature and self-incompatibility in forage crops which make conventional breeding a tedious job. The major breakthrough in crop improvement is the advancement of agricultural biotechnology that has allowed the exploration and development to tackle challenges in forage breeding program.

Biotechnological approaches like transgenic technology, gene silencing through RNAi and antisense technology offer opportunity tremendous scope in manipulating the genetic constituents of forage crops. Although conventional breeding approaches have been explored to some level, biotechnological techniques generate new horizons to create genetic variation in forage crops by the introduction of foreign genes and/or by up- or downregulation of genes. Using this technology, significant progress in forage breeding has been made. Some examples include improvement in *in vitro* dry matter digestibility (Chen et al. 2004; Reddy et al. 2005; Tu et al. 2010); greenness (Xu et al. 2018); increased stress tolerance (Dong et al. 2007); improved turf quality (Agharkar et al. 2007); increasing sulphur-rich proteins (Rafiqul et al. 1996; Wang et al. 2001); increased sugar release (Fu et al. 2011; Saathoff et al. 2011). New biotechnological tools like Cisgenesis/Intragenesis involve genetic material transfer between related species by avoiding any possibility of linkage drag. Since it doesn't involve the transfer of foreign DNA, the bred cultivar is just similar to conventionally bred cultivars. The discovery of artificially engineered nucleases (ZFNs, TALENs) has been used successfully for the improvement of economic crops.

Zinc Finger Nucleases (ZFNs) are the fusion product of zinc finger motif exclusively with nuclease domain of endonuclease FolK I (Weinthal et al. 2010). Each zinc finger domain has three to four DNA binding modules and before cleavage, the zinc finger domains dimerize and bind with nucleotide triplet, giving a recognition sequence of 18–36 bp. A double-strand break generated by the action of FolK I nuclease domain is repaired through nonhomologous end joining (NHEJ) or homologous recombination (HR) (Heyer et al. 2010). During the DNA repair process in cells, mutations such as gene deletions or insertions are introduced in target DNA (Fig. 15.1).

Similar to ZFNs, the DNA binding domain of TALE proteins, firstly discovered in *Xanthomonas*, is fused with FolK I endonuclease to obtain TALENs (Boch et al. 2011). TALENs recognize much longer sequences than ZFNs thereby less number of off-targets are produced. TALEN systems could be customized to increase specificity in binding with target DNA by modifying the amino acid repeats in TALEs and induce cleavage by FolK I enzyme between two distinct TAL array binding sites. However, these gene-editing technologies suffer from lower specificity and require customization (Figs. 15.2 and 15.3).

Gene editing technology has re-emerged in the year 2012 with the development of CRISPR/Cas system which has become a powerful tool for bringing targeted manipulations in genome through only a single protein and a guide RNA (sgRNA) (Jinek et al. 2012). CRISPR/Cas system has several desirable features, which include simplicity, efficiency, high specificity, minimal off-target effects and gene multiplexing. This system is discovered as an adaptive immune system in bacteria analogous to the immune system of animals in which foreign bacteriophage and

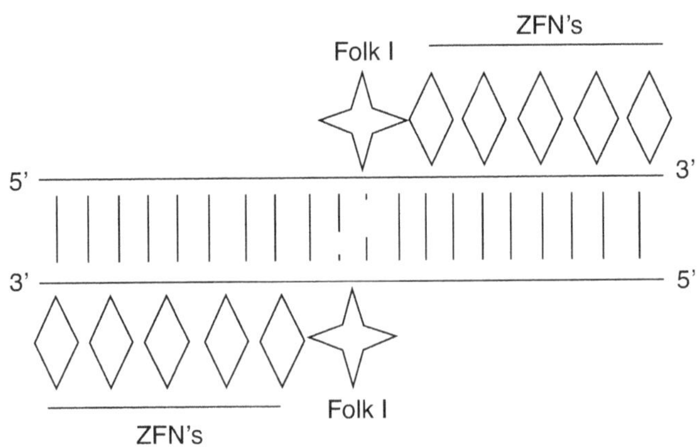

Fig. 15.1 Illustration showing the components of Zinc Finger Nucleases (ZFNs)

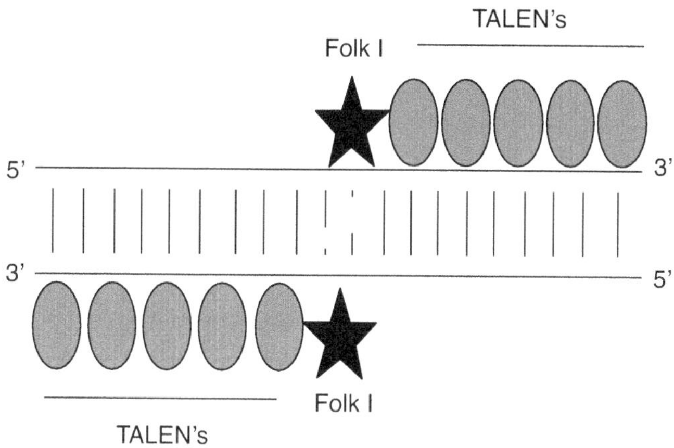

Fig. 15.2 Illustration showing the components of TALENs (Transcription activator-like effector nucleases)

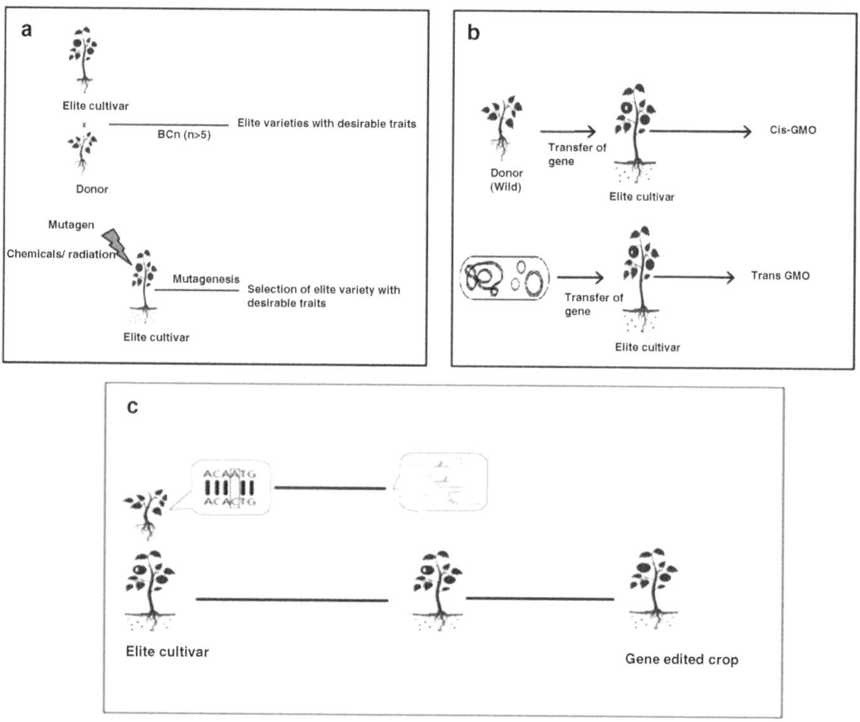

Fig. 15.3 Different approaches for genetic improvement. (**a**) Conventional breeding. (**b**) Genetic modification. (**c**) Genome editing

plasmid DNA are degraded through RNA-guided endonuclease digestion (Jinek et al. 2012). The microbial adaptive immune system is promptly accepted for genome editing in eukaryotic organisms (Jinek et al. 2012). This gene-editing technology has a common feature that utilize endonucleases, generating double-strand breaks (DSBs) and DNA repair mechanisms [Homologus end joining (HEJ) and Nonhomologous end joining(NHEJ)] for gene manipulation. CRISPR/Cas system differs from gene silencing approaches in that the CRISPR/Cas system utilizes site-specific nuclease and produces more defined alterations in targeted gene (gene knockouts) whereas in case of gene silencing or downregulation mechanism, knock-down of gene is achieved due to transcriptional repression of gene. Unlike transgene technology and RNAi, gene editing techniques target gene(s) in a more precise and predictable manner. Moreover, they generate stable and heritable mutations without having any effects on other valuable traits (Arora and Narula 2017).

Till date three most common types of CRISPR/Cas system (Type I, Type II, and Type III) (Makarova et al. 2011) and a distinct and less common type IV system (Koonin and Makarova 2013; Makarova et al. 2013) have been discovered and among them Type II system is widely accepted for genome editing. CRISPR genome editing systems allow users to design gRNA which targets their DNA sequence of interest. After cloning in binary vector and transformation in conjunction with endonuclease protein (Cas9), gRNA directs Cas9 to target sequence where it unwinds and induces site-specific cleavage after PAM (Promoter Adjacent Motif) sequence. The execution of targeted gene modification in plants using CRISPR/Cas9 system involves following steps (Fig. 15.4).

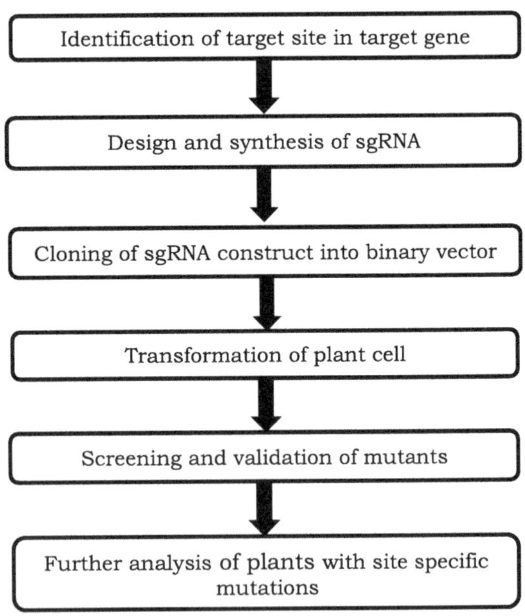

Fig. 15.4 Steps of CRISPR–Cas-mediated plant genome editing

Genome editing is a tool like an editing a video clip where desired traits in plants are precisely and quickly introgressed compared to traditional breeding approaches. Due to orthogonality, versatility, high precision in targeted gene manipulation and minimal possibility of pleiotropic effects, gene-editing techniques have provided unprecedented advancement in genome research. Since the advent of genome editing technology, different reports have been published on the improvement of nutritional quality and other agronomically desirable traits in many food crops (El-Mounadi et al. 2020). Some concept studies have suggested future implications of CRISPR/cas system for targeted manipulations in forage grasses and legumes (Park et al. 2017; Meng et al. 2017; Zhang et al. 2020).

Forages digestibility is the foremost nutritional quality trait and lignin is one of the major components contributing to the digestibility constraints in livestock. For many grasses and legumes, low digestibility reduces utilization of nutrients in animals where lignin binds with cell wall polysaccharides. Manipulation of lignin and its composition is one of the ways to increase digestibility in forages. Since gene (s) encoding regulatory enzymes of lignin biosynthetic pathway have been cloned and characterized from various forage crops, gene-editing techniques could be used to breed forages with reduced lignin content and high digestibility. Park et al. (2017) used CRISPR/Cas9 system to reduce lignin content by downregulating the expression 4-coumarate: coenzyme A ligase 1 (*Pv4CL1*) gene in switch grass. Gene editing techniques could also be used to manipulate lignin genes encoding regulatory enzymes precisely in targeted tissue using tissue-specific promoters in forage crops with aim of improving their digestibility. Gene editing techniques can also be used for the manipulation of fructan metabolism to increase nonstructural carbohydrate content, manipulate tannin synthesis, develop bloat-free forage crops, and to remove anti-nutritional factors through downregulation of gene(s) encoding regulatory enzymes of biosynthetic pathway. The forage nutritional quality traits are complex and typically intercorrelated. Almost all quality traits are under control of many genes (Polygenic) encoding regulatory enzymes of their biosynthetic pathway. Since manipulation in one gene can affect the function of the other and can lead to an undesirable trait, a deep understanding of key metabolic pathways and the regulatory genes in forages is essential for developing improved forages. CRISPR/Cas system has the ability to target multiple genes in a single event, thus offer a greater possibility of enhancing the nutritional quality of forage crops.

Though gene editing using CRISPR/Cas system has shown unprecedented success in the improvement of food crops, there are some concerns regarding the use of this technology (El-Mounadi et al. 2020) which include the generation of unwanted changes as a result of manipulation in genome at unexpected positions during DNA repair (Zhang et al. 2016a, b; Svitashev et al. 2016; Shimatani et al. 2017; Liang et al. 2018). The other issues, such as deregulation of gene-edited crops associated with CRISPR/Cas system also need consideration for commercialization (Eckerstorfer et al. 2019; El-Mounadi et al. 2020).

Biotechnological techniques have shown new horizon in crop improvement programmes. These techniques make it possible to create desirable crop cultivars in fast and more efficient ways to meet the demand for improved crops to support sustainable agricultural productivity.

Although the new biotechnological techniques have one common goal of precise, fast and efficient crop improvement, but individually they are different in approach and characteristics. CRISPR/Cas9 system is being used extensively used in most crops and is paving the way to precision trait improvements including yield, quality, biotic and abiotic stress resistance.

References

Agharkar M, Lomba P, Altpeter F, Zhang H, Kenworthy K, Lange T (2007) Stable expression of AtGA2ox1 in a low-input turfgrass (*Paspalum notatum* Flugge) reduces bioactive gibberellin levels and improves turf quality under field conditions. Plant Biotechnol J 5:791–801

Akiyama Y, Ueyama Y, Hamada S, Kubota A, Kato D, Yamada-Akiyama H, Takahara Y, Fujimori M (2010) Utilization of flow cytometry for festulolium breeding (Lolium multiflorum (2x) × Festuca arundinacea (6x)). Breed Sci 66(2):234–243

Arora L, Narula A (2017) Gene editing and crop improvement using CRISPR-Cas9 system. Front Plant Sci 8:1932

Avraham T, Badani H, Galili S, Amir R (2005) Enhanced levels of methionine and cysteine in transgenic alfalfa (*Medicago sativa* L.) plants over-expressing the Arabidopsis cystathionine γ-synthase gene. Plant Biotechnol J 3:71–79

Bagga S, Sutton D, Kemp JD, Sengupta-Gopalan C (1992) Constitutive expression of the β-phaseolin gene in different tissues of transgenic alfalfa does not ensure phaseolin accumulation in non-seed tissue. Plant Mol Biol 10:951–958

Bagga S, Armendaris A, Klypina N, Ray I, Ghoshroy S, Endress M, Sengupta-Gopalan C (2004) Genetic engineering ruminal stable high methionine protein in the foliage of alfalfa. Plant Sci 166(2):273–283

Baucher M, Bernard Vailh MA, Chabbert B, Besle JM, Opsomer C, Van Montagu M, Botterman J (1998) Down-regulation of cinnamyl alcohol dehydrogenase in transgenic alfalfa (*Medicago sativa* L.) and the effect on lignin composition and digestibility. Plant Mol Biol 39:437–447

Bellucci M, Lazzari B, Viotti A, Arcioni S (1997) Differential expression of a γ-zein gene in *Medicago sativa*, *Lotus corniculatus* and *Nicotiana tabacum*. Plant Sci 127:161–169

Bellucci M, Alpini A, Arcioni S (2002) Zein accumulation in forage species (*Lotus corniculatus* and *Medicago sativa*) and co-expression of the γ-zein: KDEL and β-zein: KDEL polypeptides in tobacco leaf. Plant Cell Rep 20:848–856

Biazzi E, Nazzicari N, Pecetti L, Brummer EC, Palmonari A, Tava A, Annicchiarico P (2017) Genome-wide association mapping and genomic selection for alfalfa (*Medicago sativa*) forage quality traits. PLoS One 12(1):e0169234

Bingham TB, Conger BV (1995) Forage crop development: application of emerging technologies. In: Barns RF, Miller DA, Nelson CJ (eds) Forages, the science of grassland agriculture. Iowa State University Press, Ames, Iowa, pp 21–27

Boch J, Scholze H, Schornack S, Landgraf A, Hahn S, Kay S, Lahaye T, Nickstadt A (2011) Boch effectors breaking the code of DNA binding specificity of TAL-type III. Science 326:1509–1512

Boerjan W, Ralph J, Baucher M (2003) Lignin biosynthesis. Ann Rev Plant Biol 54:519–546

Cardinal A, Lee M, Moore KJ (2003) Genetic mapping and analysis of quantitative trait loci (QTL) affecting fiber and lignin content in maize. Theor Appl Genet 106:866–874

Casler MD, Kaeppler HF (2000) Molecular breeding for herbage quality in forage crops. In: Spangenberg G (ed) Molecular breeding of forage crops. Kluwer Academic Publishers, Dordrecht

Chen L, Auh CK, Dowling P, Bell J, Chen F, Hopkins A, Dixon RA, Wang Z (2003) Improved forage digestibility of tall fescue (*Festuca arundinacea*) by transgenic down-regulation of cinnamyl alcohol dehydrogenase. Plant Biotechnol J 1:437–449

Chen L, Auh CK, Dowling P, Bell J, Deane L, Wang Z (2004) Transgenic down-regulation of caffeic acid O-methyltransferase (COMT) led to improved digestibility in tall fescue (*Festuca arundinacea*). Funct Plant Biol 31:235–245

Cho H, Brotherton JE, Song H, Widholm JM (2000) Increasing tryptophan synthesis in a forage legume *Astragalus sinicus* by expressing the tobacco feedback-insensitive anthranilate synthase (ASA2) gene. Plant Physiol 123:1069–1076

Christensen CSL, Rasmussen SK (2019) Low lignin mutants and reduction of lignin content in grasses for increased utilization of lignocelluloses. Agron 9:256. https://doi.org/10.3390/agronomy9050256

Cogan NOI, Smith KF, Yamada T, Francki MG, Vecchies AC, Jones ES, Spangenberg GC, Forster JW (2005) QTL analysis and comparative genomics of herbage quality traits in perennial grass (*Lolium perenne* L.). Theor Appl Genet 110:364–380

Cosgrove GP, Anderson CB, Knight TW, Roberts NJ, Waghorn GC (2004) Forage lipid concentration, fatty acid profile and lamb productivity. Proc N Z Grassl Assoc 66:251–256

Dong S, Tredway LP, Shew HD, Wang G, Sivamani E, Qu R (2007) Resistance of transgenic tall fescue to two major fungal diseases. Plant Sci 173:501–509

Eckerstorfer MF, Engelhard M, Heissenberger A, Simon S, Teichmann H (2019) Plants developed by new genetic modification techniques- comparison of existing regulatory frameworks in the EU and non-EU countries. Front Bioener Biotechnol 7:26

El-Mounadi K, Morales-Floriano ML, Garcia-Ruiz H (2020) Principles, applications, and biosafety of plant genome editing using CRISPR-Cas9. Front Plant Sci 11(56):1–16

Eudes A, Sathitsuksanoh N, Baidoo EE, George A, Liang Y, Yang F, Singh S, Keasling JD, Simmons BA, Loqué D (2015) Expression of a bacterial 3-dehydroshikimate dehydratase reduces lignin content and improves biomass saccharification efficiency. Plant Biotechnol J 13:1241–1250

Flowers G, Ibrahim SA, AbuGhazaleh AA (2008) Milk fatty acid composition of grazing dairy cows when supplemented with linseed oil. J Dairy Sci 91:722–730

Fu C, Mielenz JR, Xiao X, Yaxin Ge Choo Y, Hamilton MR Jr, Chen F, Foston M, Ragauskas A, Bouton J, Dixon RA, Wang Z-Y (2011) Genetic manipulation of lignin reduces recalcitrance and improves ethanol production from switchgrass. Proc Natl Acad Sci U S A 108:3803–3808

Gall JG, Pardue ML (1969) Formation and detection of RNA-DNA hybrid molecules in cytological preparations. Proc Natl Acad Sci U S A 63:378–383

Gao X, She MY, Yin GX, Yu Y, Qiao WH, Du LP, Ye XG (2010) Cloning and characterization of genes coding for fructan biosynthesis enzymes (FBEs) in Triticeae plants. Agric Sci China 9: 313–324

Gebril S, Seger M, Villanueva FM, Ortega JL, Bagga S, Sengupta-Gopalan C (2015) Transgenic alfalfa (*Medicago sativa*) with increased sucrose phosphate synthase activity shows enhanced growth when grown under N_2-fixing conditions. Planta 242:1009–1024

Guerra M (2004) Hybridization in situ: Princípios Básicos. In: Guerra M (ed) FISH: Conceitos e Aplicações na Citogenética. Sociedade Brasileira de Genética, Ribeirão Preto, pp 1–32

Halpin C (2019) Lignin engineering to improve saccharification and digestibility in grasses. Curr Opi Biotechnol 56:223–229

Hancock K, Collette V, Chapman E, Hanson K, Temple S, Moraga R, Caradus J (2014) Progress towards developing bloat-safe legumes for the farming industry. Crop Pasture Sci 65:1107–1113

Haverkort AJ, Boonekamp PM, Hutten R, Jacobsen E, Lotz LAP, Kessel GJT, Visser RGF, Van der Vossen AEG (2008) Societal costs of late blight in potato and prospects of durable resistance through cisgenic modification. Potato Res 51:47–57

Hayward MD, McAdam NJ, Jones JG, Evans C, Evans GM, Forster JW, Ustin A, Hossain KG, Quader B, Stammers M, Will JK (1994) Genetic markers and the selection of quantitative traits in forage grasses. Euphytica 77:269–275

Heath R, Huxley H, Stone B, Spangenberg G (1998) cDNA cloning and differential expression of three caffeic acid *O*-methyltransferase homologues from perennial ryegrass (*Lolium perenne*). J Plant Physiol 153:649–657

Hegarty M, Yadav R, Lee M, Armstead I, Sanderson R, Scollan N, Powell W, Skøt L (2013) Genotyping by RAD sequencing enables mapping of fatty acid composition traits in perennial ryegrass (*Lolium perenne*(L.)). Plant Biotechnol J 11:572–581

Heyer W-D, Ehmsen KT, Liu J (2010) Regulation of homologous recombination in eukaryotes. Ann Rev Genet 44:113–139

Higgins TJ, O'Brien PA, Spencer D, Schroeder HE, Dove H, Freer M (1988) Potential of transgenic plants for improved amino acid supply for wool growth. In: Rogers GE, Reis PJ, Ward KA, Marshall RC (eds) The biology of wool and hair. Springer, Dordrecht, pp 441–445

Hisano H, Kanazawa A, Kawakami A, Yoshida M, Shimamotoa Y, Yamada T (2004) Transgenic perennial ryegrass plants expressing wheat fructosyltransferase genes accumulate increased amounts of fructan and acquire increased tolerance on a cellular level to freezing. Plant Sci 167:861–868

Holme IB, Dionisio G, Brinch-Pedersen H, Wendt T, Madesen CK, Vincze E, Holm BP (2012) Cisgenic barley with improved phytase activity. Plant Biotechnol J 10:237–247

Hörtensteiner S (2006) Chlorophyll degradation during senescence. Annu Rev Plant Biol 57:55–77

Humphreys M, Turner L (2003) Nutritive quality QTL and marker assisted selection in rye grass. Vortrage für Pflanzenzuchtung 59:280–288

Jaganathan D, Ramasamy K, Sellamuthu G, Jayabalan S, Venkataraman G (2018) CRISPR for crop improvement: an update review. Front Plant Sci 9:985

Jang IC, Nahm B, Kim J-K (1999) Sub-cellular targeting of green fluorescent protein to plastids in transgenic rice plants provides a high-level expression system. Mol Breeding 5:453–461

Jenkins C, Simpson R, Higgins TJ, Larkin P (2000) Transgenic white clover and lucerne containing bacterial fructan as a novel carbohydrate for grazing animals. In: Abstracts 2nd International Symposium Molecular Breeding of Forage Crops. Lorne and Hamilton, Victoria, p 32

Jinek M, Chylinski K, Fonfara I, Hauer M, Doudna JA, Charpentier E (2012) A programmable dual-RNA–guided DNA endonuclease in adaptive bacterial immunity. Science 337:816–821

John HA, Birnstile ML, John KW (1969) RNA-DNA hybrids at the cytological level. Nature 223: 582–587

Khan MRI, Ceriotti A, Tabe L, Aryan A, McNabb W, Moore A, Craig S, Spencer D, Higgins TJV (1996) Accumulation of a sulphur-rich seed albumin from sunflower in the leaves of transgenic subterranean clover (*Trifolium subterraneum* L.). Transgenic Res 5:179–185

Kingston-Smith AH, Marshall AH, Moorby JM (2013) Breeding for genetic improvement of forage plants in relation to increasing animal production with reduced environmental footprint. Animal 1:79–88

Koonin EV, Makarova KS (2013) CRISPR-Cas: evolution of an RNA-based adaptive immunity system in prokaryotes. RNA Biol 10:679–686

Kopecky D, Lukaszewski AJ, Dolezel J (2008) Cytogenetics of festulolium (*Festuca* × *Lolium* hybrids). Cytogenet Genome Res 120:370–383

Lam PY, Tobimatsu Y, Takeda Y, Suzuki S, Yamamura M, Umezawa T, Lo C (2017) Disrupting flavone synthase II alters lignin and improves biomass digestibility. Plant Physiol. https://doi.org/10.1104/pp.16.01973

Lepage C, Mackin L, Lidgett A, Spangenberg G (2000) Development of transgenic white clover expressing chimeric bacterial levansucrase genes for enhanced tolerance to drought stress. In: Abstracts 2nd International Symposium Molecular Breeding of Forage Crops. Lorne and Hamilton, Victoria, p 80

Lepiniec L, Debeaujon I, Routaboul JM, Baudry A, Pourcel L, Nesi N, Caboche M (2006) Genetics and biochemistry of seed flavonoids. Ann Rev Plant Biol 57:405–430

Li JF, Norville JE, Aach J, McCormack M, Zhang D, Bush J, Church GM, Sheen J (2013) Multiplex and homologous recombination-mediated genome editing in Arabidopsis and *Nicotiana benthamiana* using guide RNA and Cas9. Nat Biotechnol 31(8):688–691

Li P, Dong Q, Ge S, He X, Verdier J, Li D, Zhao J (2016) Metabolic engineering of proanthocyanidin production by repressing the isoflavone pathways and redirecting anthocyanidin precursor flux in legume. Plant Biotechnol J 14:1604–1618

Liang Z, Chen K, Zhang Y, Liu J, Yin K, Qiu JL et al (2018) Genome editing of bread wheat using biolistic delivery of CRISPR/Cas9 in vitro transcripts or ribonucleoproteins. Nat Protoc 13:413–430

Lidgett A, Johnson X, Terdich K, Spangenberg G (2000) The manipulation of fructan biosynthesis in pasture grasses. In: Abstracts 2nd International Symposium Molecular Breeding of Forage Crops. Lorne and Hamilton, Victoria, p 31

Liu S, Fu C, Gou J, Sun L, Huhman D, Zhang Y, Wang Z-Y (2017) Simultaneous Down-regulation of MTHFR and COMT in Switch grass affects plant performance and induces lesion-mimic cell death. Front Plant Sci 8:1–15

Lorenzo CD, García-Gagliardi P, Antonietti MS, Sánchez-Lamas M, Mancini E, Dezar CA, Vazquez M, Watson G, Yanovsky MJ, Cerdán PD (2020) Improvement of alfalfa forage quality and management through the down-regulation of MsFTa1. Plant Biotechnol J 18:944–954

Lübberstedt T, Melchinger AE, Klein D, Degenhardt H, Paul C (1997) QTL mapping in testcrosses of European flint lines of maize: II. Comparison of different testers for forage quality traits. Crop Sci 37:1913–1922

Luescher M, Hochstrasser U, Vogel G, Aeschbacher R, Galati V, Nelson CJ, Boller T, Wiemken A (2000) Cloning and functional analysis of sucrose:sucrose 1-fructosyltransferase (1-SST) from tall fescue. Plant Physiol 124:1217–1227

Makarova KS, Haft DH, Barrangou R, Brouns SJ, Charpentier E, Horvath P, Moineau S, Mojica FJ, Wolf YI, Yakunin AF, Van der Oost J, Koonin EV (2011) Evolution and classification of the CRISPR-Cas systems. Nat Rev Microbiol 9:467–477

Makarova KS, Wolf YI, Koonin EV (2013) The basic building blocks and evolution of CRISPR-CAS systems. Biochem Soc Trans 41:1392–1400

McCarthy RD, Klusmeyer TH, Vicini J, Clark JH, Nelson DR (1989) Effects of source of protein and carbohydrate on ruminal fermentation and passage of nutrients to the small intestine of lactating cows. J Dairy Sci 72:2002–2016

Méchin V, Argillier O, Hébert Y, Guingo E, Moreau L, Charcosset A, Barrière Y (2001) Genetic analysis and QTL mapping of cell wall digestibility and lignification in silage maize. Crop Sci 41:690–697

Meng YY, Hou YL, Wang H, Ji RH, Liu B, Wen J, Niu L, Lin H (2017) Targeted mutagenesis by CRISPR/Cas9 system in the model legume *Medicago truncatula*. Plant Cell Rep 36:371–374

Mitchell R, Dupree P, Shewry P (2007) Novel bioinformatics approach identifies candidate genes for the synthesis and feruloylation of arabinoxylan. Plant Physiol 144:43–53

Mizukami Y, Sugita SI, Ohmido N, Fukui K (1998) Agronomic and cytological characterization of F1 hybrids between *Lolium multiflorum* Lam and *Festuka arundinaceae* var. glaucescens Boiss. Grassl Sci 44:14–21

Nocek E, Russell JB (1988) Protein and energy as an integrated system. Relationship of ruminal protein and carbohydrate availability to microbial synthesis and milk production. J Dairy Sci 71: 2070–2107

Orskov ER, Chen XB (1990) Assessment of amino acid requirements in ruminants. In: Hoshino S, Onodera R, Minato H, Itabashi H (eds) The rumen ecosystem. Japan Scientific Societies Press and Springer Verlag, pp 161–167

Park JJ, Yoo CG, Flanagan A, Pu Y, Debnath S, Ge Y, Ragauskas A, Wang Z-Y (2017) Defined tetra-allelic gene disruption of the 4-coumarate: coenzyme A ligase 1 (Pv4CL1) gene by CRISPR/Cas9 in switchgrass results in lignin reduction and improved sugar release. Biotechnol Biof 10(1):1–11

Peñaloza APS, Pozzobon MT (2007) Caracterização Citogenética de Germoplasma Vegetal. In: Nass LL (ed) Recursos Genéticos Vegetais. Embrapa Recursos Genéticos e Biotecnologia, Brasília, pp 308–342

Pickering FS, Reis PJ (1993) Effects of abomasal supplement of methionine on wool growth of grazing sheep. Aust J Exp Agric 33:7–12

Poehlman JM (ed) (1987) Breeding forage crops. Breeding field crops. Springer, Dordrecht, pp 625–679

Radojevic I, Simpson RJ, St John JA, Humphreys MO (1994) Chemical composition and in vitro digestibility of lines of Lolium perenne selected for high concentrations of water-soluble carbohydrate. Aust J Agric Res 45:901–912

Rafiqul M, Khan I, Ceriotti A, Tabe L, Aryan A, McNabb W, Moore A, Craig S, Spencer D, Higgins TJV (1996) Accumulation of a sulphur-rich seed albumin from sunflower in the leaves of transgenic subterranean clover (*Trifolium subterraneum* L.). Transgenic Res 5:179–185

Ralph J, Guillaumie S, Grabber JH, Lapierre C, Barrière Y (2004) Genetic and molecular basis of grass cell-wall biosynthesis and degradability. III. Towards a forage grass ideotype. C R Biol 327:467–479

Ray H, Yu M, Auser P, Blahut-Beatty L, McKersie B, Bowley S, Westcott N, Coulman B, Lloyd A, Gruber MY (2003) Expression of anthocyanins and proanthocyanidins after transformation of alfalfa with maize Lc. Plant Physiol 132:1448–1463

Reddy MSS, Chen F, Shadle G, Jackson L, Aljoe H, Dixon RA (2005) Targeted down- regulation of cytochrome P_{450} enzymes for forage quality improvement in alfalfa (*Medicago sativa* L.). Proc Natl Acad Sci U S A 102:16573–16578

Reis PJ (1979) Effects of amino acids on the growth and properties of wool. In: Black JL, Reis PJ (eds) Physiological and environmental limitations to wool growth. University of New England Publishing Unit, Armidale, pp 223–242

Reis GB, Mesquita AT, Torres GA, Andrade-Vieira LF, Pereira AV, Davide LC (2014) Genomic homeology between Pennisetum purpureum and Pennisetum glaucum (Poaceae). Comp Cytogenet 8:199–209

Ritsema T, Smeekens S (2003) Fructans: beneficial for plants and humans. Curr Opin Plant Biol 6: 223–230

Rogers GE (1990) Improvement of wool production through genetic engineering. TIBTECH 8:6–11

Russell AW, Sparrow R (2008) The case for regulating intragenic GMOs. J Agric Envi Ethics 21: 153–181

Saathoff AJ, Sarath G, Chow EK, Dien BS, Tobias CM (2011) Downregulation of cinnamyl-alcohol dehydrogenase in switch grass by RNA silencing results in enhanced glucose release after cellulase treatment. PLoS One 6:e16416

Salunkhe DK, Chavan JK, Kadam SS (2017) Occurrence, nature, and composition. In: Salunkhe DK (ed) Dietary tannins: consequences and remedies. CRC Press, Boca Raton, pp 54–60

Schaart JG, Van de Wiel CCM, Lotz LAP, Smulders MJM (2016) Opportunities for products of new plant breeding techniques. Trends Plant Sci 21:438–449

Schelbert S, Sylvain A, Bo B, Birgit A, Felix K, Karin K, Stefan H (2009) Pheophytin Pheophorbide Hydrolase (Pheophytinase) Is Involved in Chlorophyll Breakdown during Leaf Senescence in *Arabidopsis*. Plant Cell 21(3):13. https://doi.org/10.1105/tpc.108.064089

Shadle G, Chen F, Srinivasa Reddy MS, Jackson L, Nakashima J, Dixon RA (2007) Down-regulation of hydroxycinnamoyl CoA: Shikimate hydroxycinnamoyl transferase in transgenic alfalfa affects lignification, development and forage quality. Phytochemistry 68:1521–1529

Shan Q, Wang Y, Li J, Zhang Y, Chen K, Liang Z, Zhang K, Liu J, Xi JJ, Qiu JL, Gao C (2013) Targeted genome modification of crop plants using a CRISPR-Cas system. Nat Biotechnol 31(8):686–688

Shimatani Z, Kashojiya S, Takayama M, Terada R, Arazoe T, Ishii H, Teramura H, Yamamoto T, Komatsu H, Miura K, Ezura H, Nishida K, Ariizumi T, Kondo A (2017) Targeted base editing in rice and tomato using a CRISPR-Cas9 cytidine deaminase fusion. Nat Biotechnol 35:441–443

Smith RL, Schweder ME, Chowdhury MKU, Seib JC, Schank SC (1993) Development and application of RFLP and RAPD DNA markers in genetic improvement of Pennisetum for biomass and forage production. Biomass Bioenergy 5:51–62

Smith KF, Simpson RJ, Oram RN, Dove H, Culvenor RA, Humphreys MO, Boller B, Stadelmann FJ (1998) Increasing the nutritive value of perennial ryegrass in dairy pastures of temperate Australia. In: Boller B (ed) Breeding for a multifunctional agriculture: Proceedings of the 21st meeting of the EUCARPIA Fodder Crops and Amenity Grasses Section. Swiss Fed. Res. Stn. Agrecol. and Agric., Zurich

Spangenberg GC (2005) Transgenesis and genomics in molecular breeding of pasture grasses and legumes for forage quality and other traits. In: Makkar HPS, Viljoen G (eds) Applications of gene-based technologies for improving animal production and health in developing countries. IEE Global Telecommunications Conference, San Francisco

Spangenberg G, Kalla R, Lidgett A, Sawbridge T, Ong EK, John U (2000) Breeding forage plants in the genome era. In: Spangenberg G (ed) Molecular breeding of forage crops. Kluwer Academic Publishers, Dordrecht, pp 1–39

Sullivan ML, Hatfield RD (2006) Polyphenol oxidase and o-diphenols inhibit postharvest proteolysis in red clover and alfalfa. Crop Sci 46:662–670

Sullivan ML, Hatfield RD, Thoma SL, Samac DA (2004) Cloning and characterization of red clover polyphenol oxidase cDNAs and expression of active protein in *Escherichia coli* and transgenic alfalfa. Plant Physiol 136:3234–3244

Svitashev S, Schwartz C, Lenderts B, Young JK, Mark Cigan A (2016) Genome editing in maize directed by CRISPR-Cas9 ribonucleoprotein complexes. Nat Commun 7:13274. https://doi.org/10.1038/ncomms13274

Tabe LM, Droux M (2002) Limits to sulfur accumulation in transgenic lupin seeds expressing a foreign sulfur-rich protein. Plant Physiol 128(113):1137–1148

Tabe LM, Higgins CM, McNabb WC, Higgins TJ (1993) Genetic engineering of grain and pasture legumes for improved nutritive value. Genetica 90(2–3):181–200

Tabe L, Wardley-Richardson MT, Ceriotti A, Aryan A, McNabb W, Moore A, Higgins TJV (1995) A biotechnological approach to improving the nutritive value of alfalfa. J Anim Sci 73:2752–2759

Thomas HM, Morgan WG, Meredith MR, Humphreys MW, Leggett JM (1994) Identification of parental and recombined chromosomes in hybrid derivatives of *Lolium multiflorum* × *Festuca pratensis* by genomic in situ hybridization. Theor Appl Genet 88:909–913

Tong Z, Xie C, Ma L, Liu L, Jin Y, Dong J, Wang T (2014) Co-expression of bacterial aspartate kinase and adenylylsulfate reductase genes substantially increases sulphur amino acid levels in transgenic alfalfa (*Medicago sativa* L.). PLoS One 10:e88310

Tu Y, Rochfort S, Liu Z, Ran Y, Griffith M, Badenhorst P, Louie GV, Bowman ME, Smith KF, Noel JP, Mouradov A, Spangenberg G (2010) Functional analysis of caffeic acid *O*-methyltransferase and cinnamoyl-CoA reductase genes from perennial reygrass (*Lolium perenne*). Plant Cell 22:3357–3373

Vailhe MAB, Migne C, Cornu A, Maillot MP, Grenet E, Besle JM, Atanassova R, Martz F, Legrand M (1996) Effect of modification of the O-methyltransferase activity on cell wall composition, ultrastructure and degradability of transgenic tobacco. J Sci Food Agric 72:385–392

Van Loo EN, Dolstra O, Humphreys MO, Wolters L, Luessink W, De Riek J, Bark N (2003) Lower nitrogen losses through marker assisted selection for nitrogen use efficiency and feeding value in ryegrass (NIMGRASS). Vortrage für Pflanzenzuchtung 59:270–279

Verdier J, Zhao J, Torres-Jerez I, Ge S, Liu C, He X, Mysore KS, Udvardi MK (2012) MtPAR MYB transcription factor acts as an on switch for proanthocyanidin biosynthesis in *Medicago truncatula*. Proc Nat Acad Sci U S A 109:1766–1771

Verma RS, Dwivedi U (2014) Lignin genetic engineering for improvement of wood quality: Applications in paper and textile industries, fodder and bioenergy production. South Afr J Bot 91:107–125

Vicentini F, Hörtensteiner S, Schellenberg M, Thomas H, Matile P (1995) Chlorophyll breakdown in senescent leaves: identification of the biochemical lesion in a *stay-green* genotype of *Festuca pratensis* Huds. New Phytol 129:247–252

Vogel KP, Pedersen JF (1993) Breeding systems for cross-pollinated perennial grasses. Plant Breed Rev 11:251–274

Wang Z, Brummer C (2012) Is genetic engineering ever going to take off in forage, turf and bioenergy crop breeding? Ann Bot 110:1317–1325

Wang Z, Ge Y (2006) Recent advances in genetic transformation of forage and turf grasses. In Vitro Cell Dev Biol Plant 42:1–18

Wang Z, Ye X, Nagel J, Potrykus I (2001) Expression of a sulphur-rich sunflower albumin gene in transgenic tall fescue (*Festuca arundinacea* Schreb.) plants. Plant Cell Rep 20:213–219

Wang J, Feng J, Jia W, Chang S, Li S, Li Y (2015) Lignin engineering through laccase modification: a promising field for energy plant improvement. Biotechnol Biofuel 8:145

Weeks D, Spalding MH, Yang B (2016) Use of designer nucleases for targeted gene and genome editing in plants. Plant Biotechnol J 14:483–495

Wei Q, Guo Y, Kuai B (2011) Isolation and characterization of a chlorophyll degradation regulatory gene from tall fescue. Plant cell Rep 30:1201–1207

Weinthal D, Tovkach A, Zeevi V, Tzfira T (2010) Genome editing in plant cells by zinc finger nucleases. Trends Plant Sci 15:308–321

Weng JK, Chapple C (2010) The origin and evolution of lignin biosynthesis. New Phytol 187:273–285

Winichayakul S, Scott RW, Roldan M, Hatier J-HB, Livingston S, Cookson R, Curran AC, Roberts NJ (2013) In vivo packaging of triacylglycerols enhances *Arabidopsis* leaf biomass and energy density. Plant Physiol 162:626–639

Wuddineh WA, Mazarei M, Zhang JY, Turner GB, Sykes RW, Decker SR, Davis MF, Udvardi MK, Stewart CN (2016) Identification and overexpression of a knotted1-like transcription factor in Switch grass (*Panicum virgatum* L.) for lignocellulosic feedstock improvement. Front Plant Sci 7:1–15

Xu B, Yu G, Li H, Xie Z, Wen W, Zhang J, Huang B (2018) Knockdown of STAYGREEN in perennial ryegrass (*Lolium perenne* L.) leads to transcriptomic alterations related to suppressed leaf senescence and improved forage quality. Plant Cell Physiol 60. https://doi.org/10.1093/pcp/pcy203

Yamada T, Forster JW, Humphreys MW, Takamizo T (2005) Genetics and molecular breeding in *Lolium/Festuca* grass species complex. Grassl Sci 51:89–106

Ye XD, Wu XL, Zhao H, Frehner M, Nösberger J, Potrykus I, Spangenberg G (2001) Altered fructan accumulation in transgenic *Lolium multiflorum* plants expressing a *Bacillus subtilis* SacB gene. Plant Cell Rep 20:205–212

Zhang Y, Liang Z, Zong Y, Wang Y, Liu J, Chen K, Qiu JL, Gao C (2016a) Efficient and transgene-free genome editing in wheat through transient expression of CRISPR/Cas9 DNA or RNA. Nat Commun 7:12617. https://doi.org/10.1038/ncomms12617

Zhang J, Yu G, Wen W, Ma X, Xu B, Huang B (2016b) Functional characterization and hormonal regulation of the PHEOPHYTINASE gene LpPPH controlling leaf senescence in perennial ryegrass. J Exp Bot 67(3):935–945

Zhang Y, Ran Y, Nagy I, Lenk I, Qiu J, Asp T, Jensen CS, Gao C (2020) Targeted mutagenesis in ryegrass (*Lolium spp.*) using the CRISPR/Cas9 system. Plant Biotechnol J 18:1854–1856

16 Lignin: Possible Manipulations in Forages

Lignin, a complex phenolic compound, is the second most abundant compound on earth after cellulose (Monties 1991; Boudet 1998). It is the main noncarbohydrate component of plant cell wall where it is cross-linked with carbohydrates and forms large complex. Lignin as a structural component which provides strength and rigidity to cell wall. Lignin limits the water loss by reducing the permeability of cell wall (Zeikus 1980; Dean and Eriksson 1992). These attributes together make lignin a desirable compound in plant cell walls from the perspective of plant function and survival. However, lignin content in forages has a negative relationship with their nutritional quality intake and digestibility.

Forage grasses have higher lignin content than leguminous forages therefore, its proportions have greater impact on the amount of total digestible energy. Lignin content in forages varies between 3 and 15% depending on different plant-based and environmental factors. With increasing maturity, plant has higher proportion of stems in comparison to leaves. Plant stems contain more lignified tissues than leaves thus have low digestibility. Typically warm-season grasses have high fiber and lignin content than cool-season grasses at similar growth stages. In both grasses and legumes, lignin content increases with maturity. Lignification in plants is also affected by light conditions and moisture levels in soil. Lignification decreases under low light conditions and increases under moisture stress conditions.

16.1 Lignin and Plant Growth Development

Lignin being the structural component of plant cell wall, maintain the structure and rigidity of plants tissues and organs. Its deposition starts at cell corners and middle lamellae and proceeds through the primary walls into the secondary walls and cross-links with cellulosic and noncellulosic components. Lignin metabolism is involved in plant growth and development and it has been proved that any interference in lignin biosynthesis, often results in inhibition of plant growth and development

R. Katoch, *Nutritional Quality Management of Forages in the Himalayan Region*, https://doi.org/10.1007/978-981-16-5437-4_16

including seed propagation (*Arabidopsis*) (Liljegren et al. 2000). Role of lignin deposition in providing protection from adverse extrinsic factors has been well established. Liang et al. (2006) observed that low lignin content in *Arabidopsis thaliana* seed coat decreases the germination rate. Low lignin content in *Arabidopsis thaliana* CCR1 mutant results in stunted growth and a reduced number of seeds (Mir Derikvand et al. 2008; Vanholme et al. 2012). Schilmiller et al. (2009) reported that the mutation in the lignin biosynthetic gene (C4H) in *Arabidopsis thaliana* inhibited the growth as well as lignin accumulation. Moreover, mutation in C4H gene also resulted in male sterility in plants with loss of apical dominance. Herrero et al. (2013) reported reduction in total lignin and S units in *Arabidopsis thaliana* peroxidase AtPrx72 knock-out mutant. The mutant has stunted growth, less branches, smaller flower and stem with reduced photosynthetic efficiency.

The major functions of lignin in plant growth and development have been discussed below:

16.1.1 Lodging Resistance to Plants

Lignin plays an important role in providing strength to plant to resist lodging (Hu et al. 2017). It has been well established that resistance to lodging is associated with plant height, biomass, stem diameter, composition and characteristics of stem cell walls (Tanaka et al. 2003; Zhang and Zhou 2011). The deposition of lignin in cell wall increases mechanical strength of the plant and provides resistance to lodging (Tripathi et al. 2003). Crop density changes the morphological characteristics and lignin biosynthesis in stem and thus increases the mechanical strength of stem that ultimately provides lodging resistance to plants (Zheng et al. 2017). Application of silicon can enhance expression of CAD gene, improve lignin biosynthesis and increase stem strength, thereby enhancing lodging resistance in plants (Dorairaj et al. 2017).

16.1.2 Lignin and Plant Stress Adaptation

Lignin accumulation has a significant role in making plant adaptable to various biotic and abiotic stresses. It has been reported that expression of PAL, C4H, and pathogenesis-related 9 (PR-9) genes that are associated with lignin biosynthesis are up-regulated in brown plant hopper-infested insect-resistant rice varieties. This suggests that these genes may participate in lignin biosynthesis synergistically and increase resistance against insects (Duan et al. 2014; Jannoey et al. 2017). An insect-specific toxin peptide LqhIT2 increased the accumulation of lignin in rice and increased resistance against leaf rollers infestation (Tianpei et al. 2015).

Lignin is a direct barrier or through associated signaling pathways to increase insect resistance of plants. When plants are infected with pathogens, the increased accumulation of lignin provides barrier and reduces the infiltration of fungal enzymes and toxins into cell walls (Zhang et al. 2007; Ma et al. 2017). The expression of gene encoding lignin biosynthesis enzymes is closely related to the disease resistance in plants. Invasion of pathogens increases the expression of CAD5 gene in *Arabidopsis* roots with strong lignification (Tronchet et al. 2010). In maize, ZmCCoAOMT2 gene has been associated with resistance to multiple pathogens and regulation of programmed cell death (Yang et al. 2017).

16.1.3 Plant Drought and Salt Stress Tolerance

Drought and salt stress affect plant growth and development and resulting crop loss. Lignin reduces water penetration and transpiration. Studies have shown that lignin biosynthesis is increased under drought stress. Level of expression of lignin biosynthesis genes (CAD and COMT) and lignin content in leaves of inbred maize lines have is positively correlated with drought tolerance (Hu et al. 2009). Under drought stress, expression of CCoAOMT enzyme increases root elongation region limit water loss and promote water transportation (Yamaguchi et al. 2010). Studies have shown that lignin deposition is involved in providing salt tolerance to plants. Over expression of *Ipomoea batatas* IbLEA14 genes increased the salt and osmotic stress tolerance in transgenic callus by increasing accumulation of lignin (Park et al. 2011). Shafi et al. (2015) reported that overexpression of *Potentilla atrosanguinea* superoxide dismutase and *Rheum australe* ascorbate peroxidase gene increased transgenic *Arabidopsis thaliana* plants tolerance by increasing lignin accumulation *via* maintaining the level of hydrogen peroxide. MYBs transcription factors also play important role in both secondary cell wall biosynthesis and stress tolerance in plants.

16.1.4 Plant Temperature Stress Adaptability

High temperature increases peroxidation of membrane lipids disturbing normal plant metabolism. In contrast, low-temperature stress also results in cell membrane damage, reduces photosynthesis and respiration rates, and inhibits plant growth and development. Lignin increases process of cold acclimation in plant tissues. For example, C3H gene expression in cold-acclimated *Rhododendron* leaves has been reported than that of control leaves (Wei et al. 2006). Lignin deposition in terminal latewood tracheids of Norway spruce is positively correlated with low temperature (Gindl et al. 2000). Lignin content in *Medicago truncatula* CAD1 mutant has also been reported low than wild-type plants. Moreover, there was no significant growth difference between wild-type and mutants at normal temperature (22 °C). At 30 °C, the growth of MtCAD1 mutants was inhibited however, the phenotype was similar to wild type. Possible reason could be reduced lignin accumulation leading to vascular damage, thereby affecting transpiration and causing overheating of plant (Zhao et al. 2013).

Lignin being the most important cell wall constituent has negative relation with forage quality and digestibility. Lignin inhibits activity of ruminal microflora. Although lignin has about 30% higher gross energy than cellulose (Novaes et al.

2010), however, this energy is barely accessible to the livestock. Lignin deposition reduces feed intake due to the association of lignin with polysaccharide constituents. This forms a physical barrier and thus hinders the access of rumen microbes to cell wall polysaccharides.

16.2 Factors Affecting Lignification in Plants

In the context of forage quality, there are a number of factors influencing the nutritive value of forages. Different factors affecting lignification in plant have been discussed.

16.2.1 Species Difference

Lignin content and composition vary greatly among forages. There are large differences in lignification between grasses and legumes and effect of lignin on their quality and digestibility. Lignin content of forage legumes often appears comparable to forage grasses. However, when expressed as a proportion of fiber, legumes demonstrate a wider range of lignin content that is generally higher than forage grasses (Buxton and Russell 1988; Buxton and Brasche 1991). The differences in the digestibility of forage legumes and grasses indicate that there are qualitative differences between the two families in the way lignin interacts with other fiber components. Most dicotyledonous shrubs and trees have relatively high lignin content and concomitantly low fiber digestibility.

The ploidy of plants has been demonstrated to have an impact on lignin content in forage grasses. In *Brachiaria* grass, lignin content is higher in tetraploid in comparison to diploid plants (Valle et al. 1988). However, there were no differences in digestibility of fiber among the ploidy levels. The lignin composition on the ploidy level in plants has controversial and contradictory results. Triploid and tetraploid willow plants have lignin content in comparison to diploid plants (Serapiglia et al. 2015). Morrison (1980) reported lignin content in tetraploid temperate grasses is higher in comparison to diploids.

16.2.2 Temperature

Plant lignification tends to increase under warm temperature regimes. This is related to increased activities of lignin biosynthesis enzymes at higher temperatures (Buxton and Fales 1994). In some instances it is also related to increased partitioning of plant dry matter to the lignified tissues (Cone and Engels 1990; da Silva et al. 1987). Wilson et al. (1991) reported that growing a number of tropical and temperate forage species under high-temperature regime did not alter the proportion of various tissues, but increase the rate of lignification. Similar results have also been reported for tall fescue (*Festuca arundinacea* Schreb.) grown under different temperature regimes (Akin et al. 1987). Increasing temperature did not affect leaf anatomy, but has a negative effect on the proportion of degraded tissues.

Fales (1986) reported that growing tall fescue under increasing temperatures did not affect lignification, but increase the amount of indigestible fiber. Henderson and Robinson (1982) observed increasing growth temperatures of subtropical grass species resulted increased lignification and decreased digestibility of fiber. Despite the inconsistencies in the effect of temperature on lignification in plants, the negative effect of increasing temperature on fiber digestibility appears to be consistent. Cool-season and warm-season grasses have similar lignin and fiber compositions but the content is different (Jung and Vogel 1992). Cool- and warm-season grasses have higher lignin content than warm-season grass species (Akin 1989). Warm-season grasses partition more dry matter into lignified tissues with thickened secondary walls.

16.2.3 Moisture Stress

Moisture stress decreases lignifications in plants, although this effect occurs largely as a function of changes in plant development and morphology (Buxton and Casler 1993). In alfalfa, total cell wall content increases with increasing irrigation levels, but there is no effect on cell wall composition (Halim et al. 1989). Differences in digestibility of forages between stressed and nonstressed plants are related to delayed plant maturity caused by water stress. Excess moisture can also reduce whole plant lignin content. Growing alfalfa under conditions of excess moisture results in stunted growth and development (Buscaglia et al. 1994).

16.2.4 Light Intensity and Quality

Low light intensity decreases the rate of lignification in plants (Buxton and Fales 1994). Shaded plants have less lignin content (Buxton and Casler 1993; Buxton 1996). However, the effect of light intensity on the rate of lignification and fiber digestion is not always consistent. Henderson and Robinson (1982) reported that increasing light intensity has variable effects on lignification and fiber digestibility of subtropical forage grasses. Kephart and Buxton (1993) also reported inconsistent effects of shading on lignification of a number of temperate and warm-season grasses. However, shading consistently resulted in improvement in dry matter digestibility.

16.2.5 Plant Anatomy and Morphology

Lignification occurs during the process of cell differentiation and is often associated with the formation of secondary walls (Iiyama et al. 1993). The deposition of lignin start at the corners of cell adjacent to middle lamella and proceeds inward into the primary and secondary walls. Lignification is always preceded by deposition of structural carbohydrates. Because of this, lignin content is usually high in primary wall in comparison to secondary wall (Terashima et al. 1993). The process of lignification occurs rapidly in growing tissues. Tissues with high lignin content include epidermis (grasses), xylem, and sclerenchyma. These tissues are relatively less affected by rumen microbes and comprise a large proportion of total indigestible fraction of forages (Buxton and Redfearn 1997; Twidwell et al. 1991). Other tissues such as mesophyll and other types of parenchyma are not lignified. Forage digestibility represents the relative contributions of various plant tissues to total herbage dry matter (Magai et al. 1994). Higher proportion of structural and other lignified tissues is responsible for low digestibility of forage (Akin 1989). Lignification in plants increases with the advancement in maturity. This is largely a function of changes in the morphology of the plant and to a lesser degree, the aging of plant tissues. As plants mature, leaves to stems ratio decreases (Nelson and Moser 1994; Martiniello et al. 1997). As stems has more lignified and less metabolically active tissues, the digestibility of stems is lesers than leaves. Quantitatively, the changes in lignification that occur as plants mature have high impact on forage quality. In grasses, lignin content more than double as plants develop from vegetative to reproductive stages of growth.

16.3 Lignin Content and Forage Digestibility

Lignin is a limiting factor in forage utilization and is negatively related to digestibility (Jung and Deetz 1993). The lignin content always increases with increasing cell wall content (Jung et al. 1997). Lignin inhibits the digestion of grasses more than legumes (Buxton and Russell 1988; Engels and Jung 1998). Lignin acts as a physical barrier to the microbial enzymes reaching their target polysaccharides (Jung and Deetz 1993).

Although, forages contain a sufficient amount of energy, the feeding value of forages is lower and much more variable. The relationship of lignin content of forages is negatively correlated with digestibility (Jung and Deetz 1993). Lignin can be classified as core lignin which has highly condensed polymeric matrices and noncore lignins having low molecular weight phenolic monomers. Core lignins are covalently linked to hemicellulose in forage cell walls, with at least some of these linkages being alkali labile in grasses. Among the noncore lignins, *p*-coumaric acid is primarily associated with core lignin whereas ferulic acid is generally linked to hemicellulose fraction. Majority of noncore lignin moieties are esterified to other cell wall components, but ether linkages also exist. Lignification seems to depress digestion of grass cell walls more strongly than legumes. All forages contain core lignin and its content is higher in stem than leaf tissue (Morrison 1980). Total core lignin content of both grasses and legumes has been negatively correlated with both *in vivo* and *in vitro* digestibility.

16.4 Lignin Biosynthesis

The cell wall of plants has been characterized as cellulose microfibrils embedded in a ligno-hemicellulosic macromolecule to which acetyl and phenolic acid groups are bound (Morrison 1979). The cellulose microfibrils are bound to the hemicellulose polymers by hydrogen bonding. During plant cell development, primary cell wall contains cellulose, hemicellulose and pectins (Bacon 1979). Lignin becomes part of the cell wall during formation and thickening of the secondary cell wall (Theander and Aman 1984).

Lignin biosynthesis in plants can be divided into three major phases: (1) synthesis of monolignols in the symplastic shikimate and phenyl propanoid pathway (2) export of monolignols to the apoplast and (3) activation of monolignols by enzyme-mediated formation of monolignol radicals in the apoplast and their polymerization to form complex lignin polymers. Lignin polymer is characterized by the bonding of these units by β-O-4 ether bonds and by resistant carbon-carbon and biphenyl-ether linkages or condensed bonds (Yamamoto et al. 1989; Higuchi 1990; Terashima et al. 1993). Phenolic constituents of forages are covalently linked to the cell wall are divided into core and noncore lignin components (Hartley 1972). Core lignin is a highly condensed, high molecular weight polymer of cinnamyl alcohols (Grisebach 1981). Klason type lignin preparations such as acid detergent lignin (Van Soest 1982) are considered core lignins (Figs. 16.1 and 16.2).

Lignin is a phenolic polymer of p-hydroxyphenyl (H), guaiacyl (G), and syringyl (S) units which are derivatives of hydroxycinnamyl alcohol, also called monolignols, p-coumaryl alcohol, coniferyl alcohol and sinapyl alcohol, respectively. Lignin of grasses is primarily made up of S- and G-units. Additionally, grasses also contain H-units and significantly larger amounts of ferulic acid. Lignin deposition in cell wall appears to involve cross-linking of ferulic acid with arabinoxylan *via* ester linkage (Fig. 16.3). Ferulic acid and *p*-coumaric acid cross-links to lignocellulosic matrix and provide structural integrity of cell wall. They form covalent linkages or ether bonds between polysaccharides and lignin components. Furthermore tricin, a flavonoid, has recently been discovered in lignin polymer and designated an initiator of lignin chains.

Monolignols are coupled with recalcitrant C-C, C-O-O- (ester) and C-O-C (ether) bonds. However, the coupling of monomers differs: H- and G-units can couple *via* β-5 (from monomer–monomer and monomer–oligomer reactions) and 5–5 (from

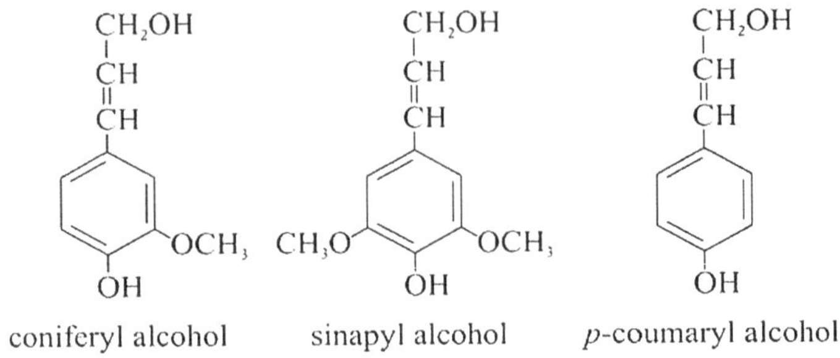

Fig. 16.1 Monolignol precursors of lignin

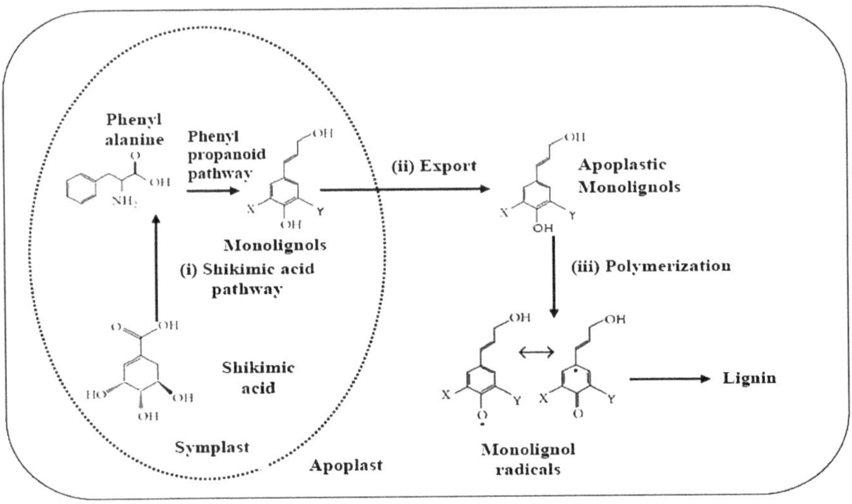

Fig. 16.2 Lignin biosynthesis phases

oligomer–oligomer reactions) coupling modes with C-C linkages, whereas S-units are linked with β-O-4 which are more easily degraded. A ratio between the monolignols (S/G ratio) is often used as a validation factor to draw conclusions about cell wall degradation ability. A higher ratio (>1.0) indicates more S-units than G-units and a low ratio (<1.0) indicates less S-unit than G-units. A high ratio favors digestibility. One hypothesis is that more S-units compared to G-units increases the number of labile β-O-4 bonds and thereby favors enzymatic digestibility positively. On the contrary, increased S-units lead to a linear structure with uncondensed (high β-O-4´) lignin, which provides high coverage and interaction with cellulose fibers and thereby lower enzymatic digestibility.

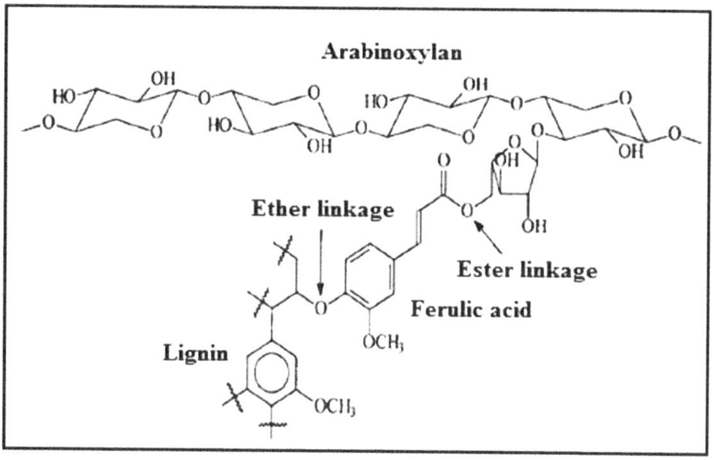

Fig. 16.3 Illustration of the cross-linking structure in grasses where ferulate esters of arabinoxylan form bridge to lignin. (Ralph et al. 1992)

16.5 Lignin Biosynthetic Pathway

The biosynthetic route of lignin synthesis involves different steps, which include (1) synthesis of aromatic amino phenylalanine from photosynthetic assimilation products (Shikimic acid pathway), (2) synthesis of different compounds from phenylalanine *via* phenylpropanoid pathway which involves enzymes phenylalanine ammonia-lyase (PAL), cinnamate-4-hydroxylase (C4H), 4-(hydroxy) cinnamoyl-CoA ligase (4CL) and (3) a specific metabolic pathway involving enzymes p-coumarate-3-hydroxylase (C3H); hydroxycinnamoyl-transferase (HCT); caffeic acid/5- hydroxyferulic acid O-methyltransferase (COMT), caffeoyl CoA O-methyltransferase (CCoAOMT), cinnamoyl CoA reductase (CCR) and cinnamoyl alcohol dehydrogenase (CAD) which leads to the synthesis of monolignols and (4) polymerization of monolignol precursors *via* oxidative coupling, which is then cross-linked with structural polysaccharides in the cell wall. The process of lignin biosynthesis in plants has been discussed:

16.5.1 Shikimic Acid Pathway

Shikimic acid pathway is the precursor pathway for lignin biosynthesis in plants. Important steps of shikimic acid pathway are localized in plastids (Vogt 2010). This biochemical pathway is a major link between primary and secondary metabolism in higher plants (Tzin and Galili 2010). The shikimate pathway involves seven enzymatic steps and begins with an aldol-type condensation of phosphoenolpyruvic acid (PEP) from the glycolytic pathway and carbohydrate D-erythrose-4-phosphate from

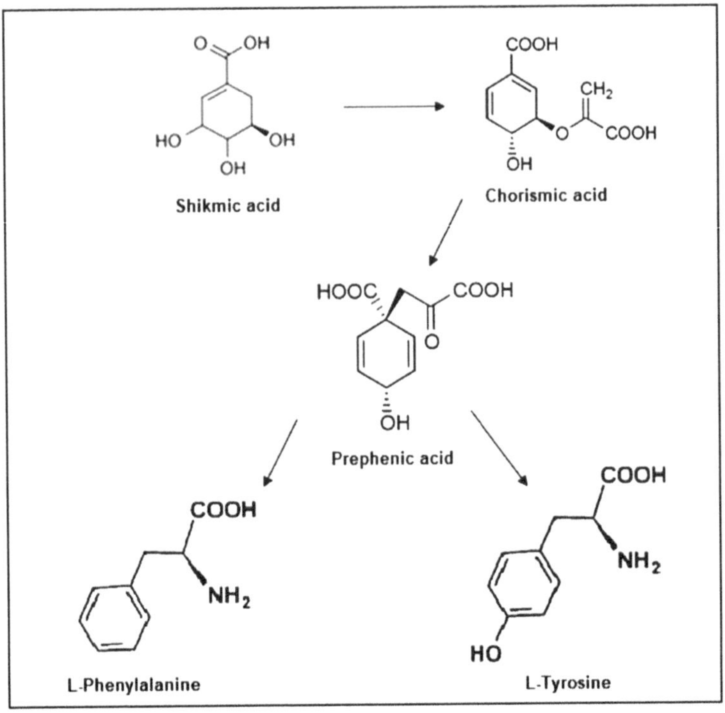

Fig. 16.4 Overview of shikimic acid pathway

pentose phosphate pathway, to give 3-deoxy-D-arabino-heptulosonic acid 7-phosphate (DAHP). Seven enzymes that catalyze the pathway are 3-deoxy-D-arabino-heptulosonate-7-phosphate synthase (DAHPS), 3-dehydroquinate synthase (DHQS), 3-dehydroquinate dehydratase/shikimate dehydrogenase (DHQ/SDH), shikimate kinase (SK), 5-enolpyruvylshikimate 3-phosphate synthase (EPSPS), and chorismate synthase (CS) (Santos-Sánchez et al. 2019). Principal aromatic phenolic compounds synthesized from shikimic acid pathway are L-phenylalanine and L-tyrosine in plants (Fig. 16.4).

16.5.2 Phenylpropanoid Pathway

Phenylpropanoid pathway is associated with the synthesis of different secondary metabolites such as flavonoids, anthocyanins, tannins, coumarin, and volatiles (Vogt 2010; Liu 2012). Shikimate pathway involves conversion of carboxylic acid shikimic to the aromatic amino acid L-phenylalanine. The deamination of L-phenylalanine to trans-cinnamate is catalyzed by enzyme phenylalanine ammonia-lyase (PAL). From here, synthesis of all phenyl propanoids is carried out. The role of different enzymes in phenylpropanoid pathway has been discussed under the following subheads:

(a) **Phenylalanine ammonia-lyase (PAL) (EC 4.3.1.24)**: PAL catalyzes the first step in phenylpropanoid metabolism in which phenylalanine undergoes deamination to yield *trans*-cinnamic acid and ammonia. Kinetic analysis of the proteins encoded by the Arabidopsis PAL gene family has shown that PAL1, 2, and 4 can catalyze phenylalanine deamination *in vitro*, whereas PAL3 shows only minimal activity (Cochrane et al. 2004).

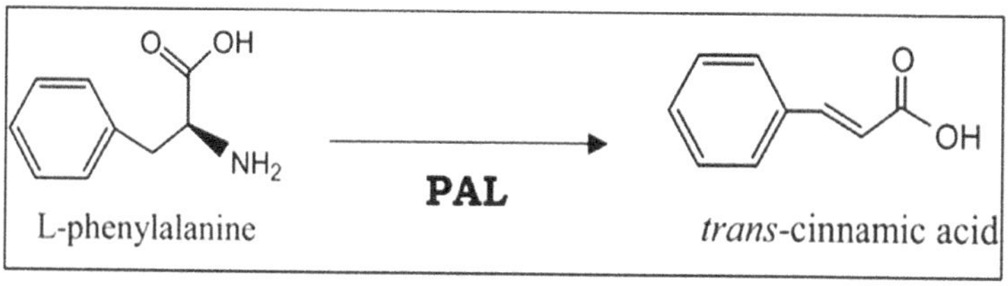

(b) **Cinnamate 4-hydroxylase (C4H) (EC 1.14.13.11)**: C4H is a cytochrome P_{450}-dependent monooxygenase (P_{450}; CYP) which catalyzes hydroxylation of cinnamate to 4-coumarate (also known as *p*-coumarate).

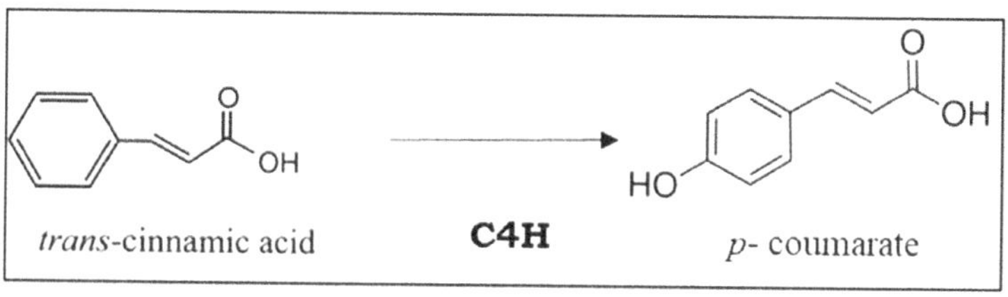

(c) **4-coumarate CoA ligase (4CL) (EC 6.2.1.12)**: 4CL catalyzes the ATP-dependent formation of CoA thioester 4-coumaroyl CoA (also known as *p*-coumaroyl CoA).

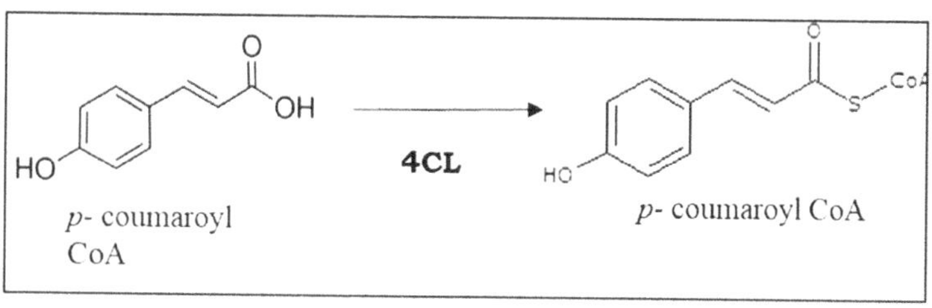

16.5.3 Monolignol Biosynthesis Pathway

Monolignols biosynthesis requires hydroxylation of aromatic rings, methylation of hydroxyl groups, and stepwise reduction of monolignol side chains from carboxylic acids to alcohols. These steps are mediated by specific enzymes and lead to the formation of major monolignols, *p*-coumaryl alcohol, coniferyl alcohol, and sinapyl alcohol. These monolignols are the building blocks of lignin and have to be transported across the plasma membrane to apoplast for polymerization. Different enzymes involved in monolignol biosynthetic pathway have been discussed:

(a) **Hydroxycinnamoyl-coenzyme A shikimate- quinate hydroxycinnamoyl-transferase (HCT) (EC 2.3.1.133):** HCT catalyzes two steps in phenylpropanoid metabolism. After the synthesis of *p*-coumaroyl CoA by 4CL, HCT catalyzes the transfer of *p*-coumaroyl group to shikimate.

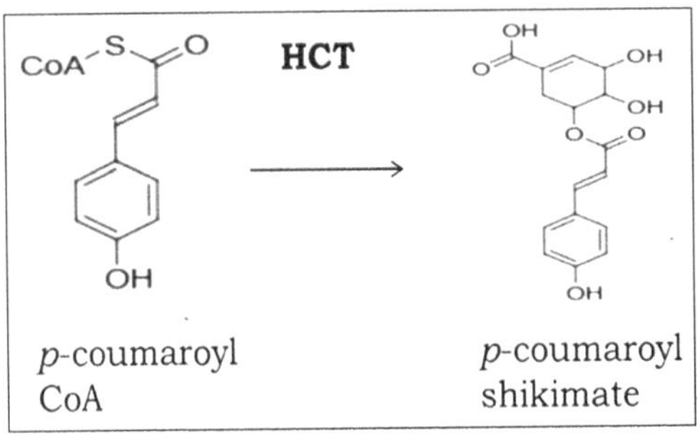

(b) **Cinnamoyl-CoA reductase (CCR) (EC 1.2.1.44):** Cinnamoyl-CoA reductase (CCR) catalyzes the synthesis of p-coumaryaldehyde which later with the reaction of CAD leads to the formation of p-coumaryl alcohol.

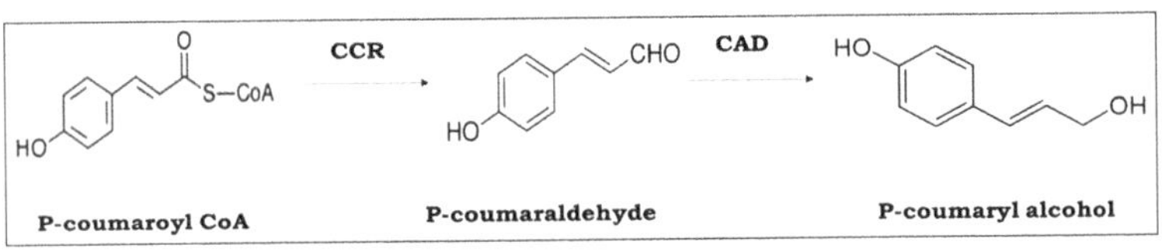

(c) ***p*-coumaroyl shikimate 3′-hydroxylase (C3′H) (EC 1.14.13.11):** p-coumaroyl shikimate 3′ hydroxylase (C3′H) catalyzes the hydroxylation of *p*-coumaroyl shikimate to yield caffeoyl shikimate.

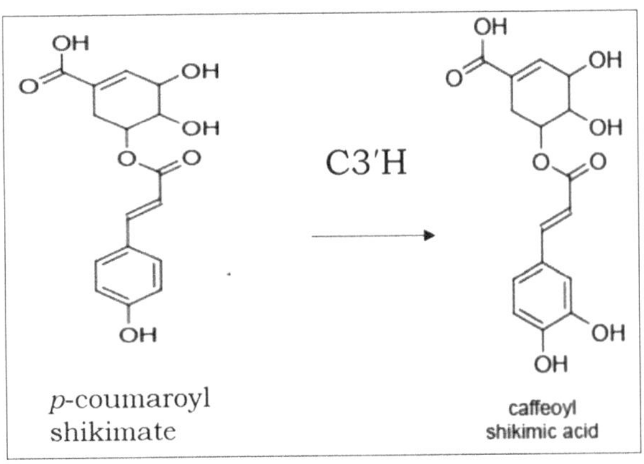

(d) **Hydroxycinnamoyl-coenzyme A shikimate- quinate hydroxycinnamoyl-transferase (HCT) (EC 2.3.1.133):** Caffeoyl shikimate in presence of HCT forms caffeoyl CoA

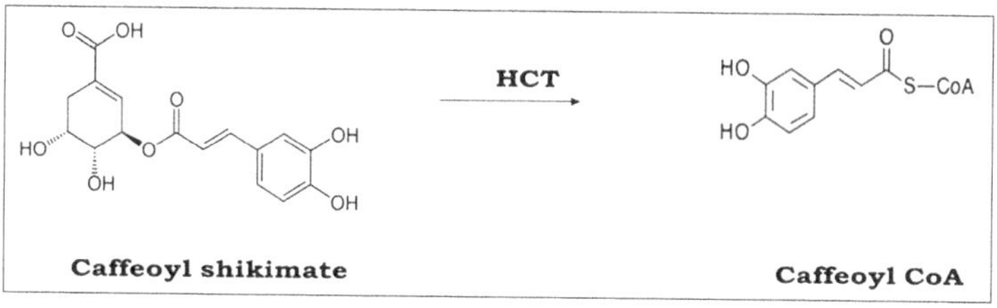

(e) **Caffeoyl CoA 3-O-methyltransferase (CCoAOMT) (EC 2.1.1.104):** CCoAOMT catalyzes first methyl-transfer reaction in phenylpropanoid metabolism, synthesizing feruloyl CoA from caffeoyl CoA.

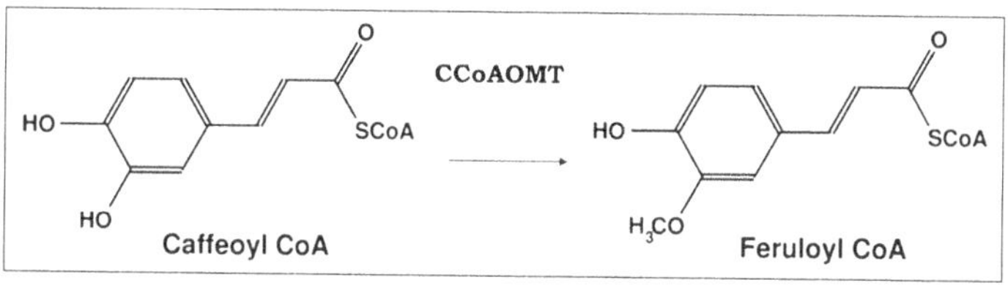

(f) **Cinnamoyl-CoA reductase (CCR) (EC 1.2.1.44):** Cinnamoyl-CoA reductase (CCR) catalyzes the synthesis of hydroxycinnamaldehydes from hydroxycinnamoyl CoA thioesters of caffeoyl CoA to form caffeylaldehyde, while feruloyl CoA produces coniferaldehyde.

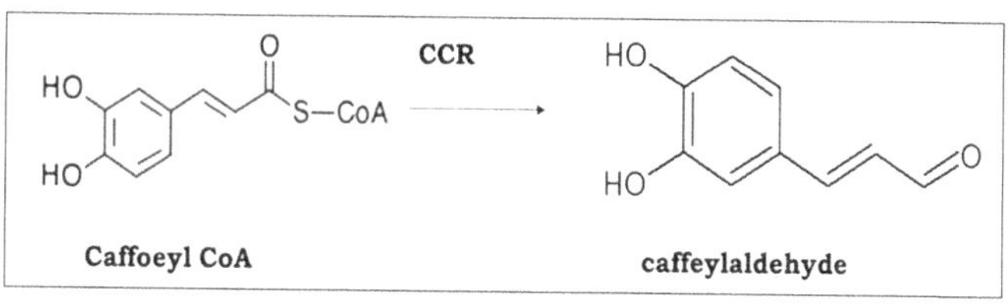

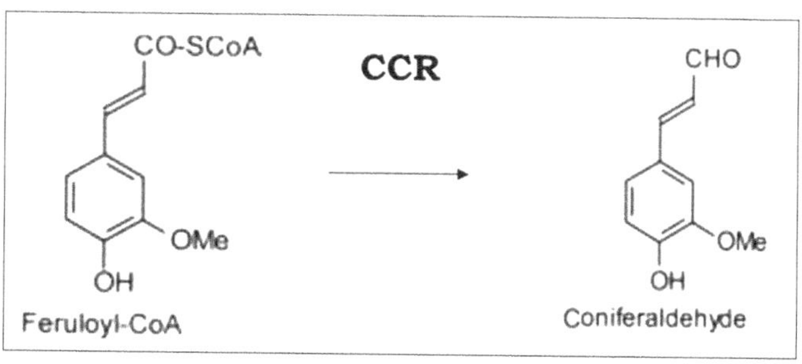

(g) **Ferulate 5-hydroxylase (F5H) (EC 1.14.14.13):** Ferulate 5-hydroxylase (F5H) is third P_{450} enzyme involved in phenylpropanoid metabolism, catalyzing NADPH and O_2 dependent hydroxylation of coniferaldehyde to yield 5-hydroxyconiferaldehyde.

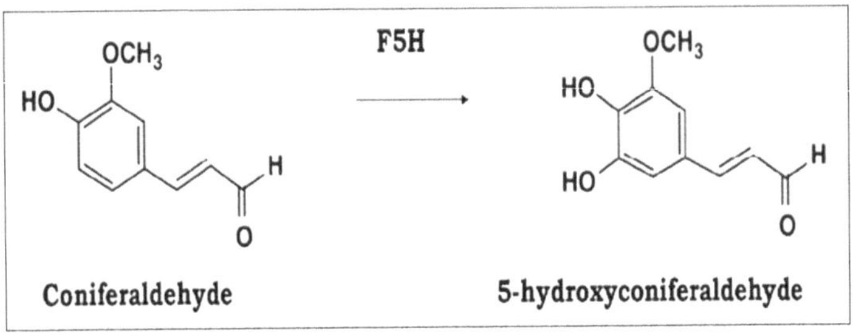

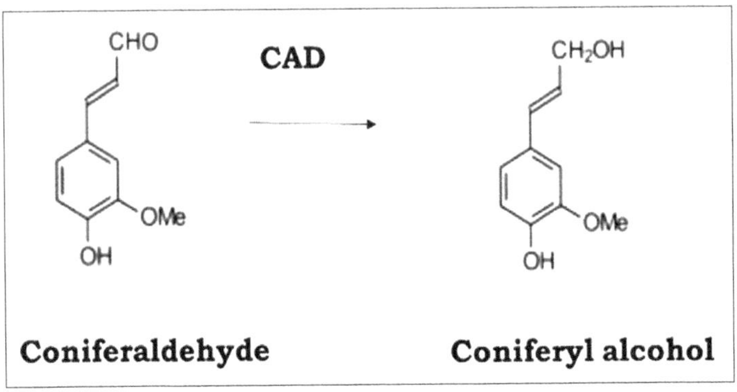

(h) **Caffeic acid/5-hydroxyferulic acid *O*-methyltransferase (COMT) (EC 2.1.1.68):** Caffeic acid/5-hydroxyferulic acid *O*-methyltransferase (COMT) act at the level of aldehyde and alcohol precursors of S lignin by methylating 5-hydroxy-coniferaldehyde and 5-hydroxyconiferyl alcohol to yield sinapaldehyde and sinapyl alcohol, respectively. Cinnamyl alcohol dehydrogenase (CAD) catalyzes the final step in biosynthesis of G and S lignin monomers: the NADPH-dependent reduction of coniferaldehyde and sinapaldehyde to monolignols coniferyl alcohol and sinapyl alcohol, respectively.

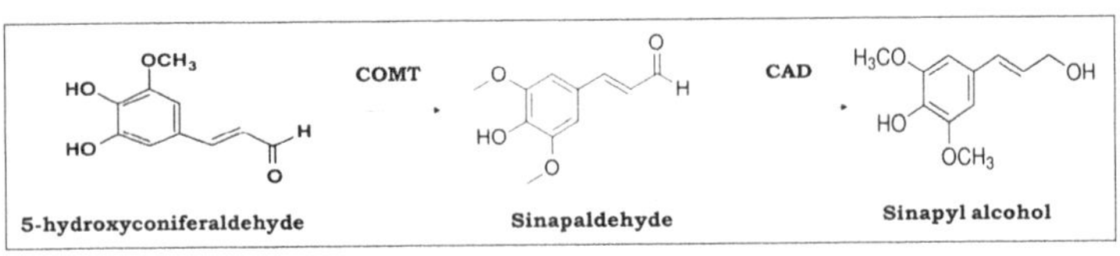

Three mechanisms have been suggested for transportation of monolignols for lignification. The vesicular secretion involves exocytosic transport *via* the endoplasmic reticulum-golgi derived vesicles. Immunocytochemical studies suggested that transport of lignin precursors is catalyzed by UDPG: coniferyl alcohol glucosyltransferase. However, recent studies suggested that endoplasmic reticulum-golgi vesicles mediate exocytosis and play a major role in monolignol transport (Kaneda et al. 2008). Passive diffusion is another proposed mechanism for monolignol transport based on the plasticity of lignin composition. The accumulation of alternative precursors in lignification infers nonspecific passive diffusion of lignin precursors across the cell membrane (Boija et al. 2007). Another possible way for the transport of lignin precursors is the ATP-binding cassette (ABC) transporters (Yazaki 2006).

16.5.4 Polymerization of Monolignols to Form H, G, and S Lignin

Activation of monolignols in the apoplast requires enzymes such as peroxidases (POX), laccases (LAC), or other polyphenol oxidases that transfer electrons from monolignols to electron receptors. These apoplastic enzymes interact with reactive oxygen species (ROS) such as hydrogen peroxide or superoxide, which act as electron receptors or modulators of POX and LAC enzymes through their signaling function. ROS are formed as by-products of many metabolic processes in plants, but can also be actively produced through enzymes such as NADPH-oxidases or class III peroxidases (also termed as guaiacol POX), and they accumulate excessively when plants encounter abiotic or biotic stress (O'Brien et al. 2012; Apel and Hirt 2004). POX uses hydrogen peroxide as an electron receptor to oxidize a variety of phenolic compounds including monolignols (Liu 2012). Laccases (LAC) are copper-containing apoplastic enzymes that oxidize phenolic compounds using molecular oxygen as an electron receptor (Liu 2012). Generally, plant genomes contain many isoforms of POX and LAC genes with presumably overlapping functions, making it difficult to link particular isoforms of these enzymes to lignin synthesis (Bonawitz and Chapple 2010). After the activation of monolignols by oxidases, the generated monolignol radicals couple onto each other to form three-dimensionally cross-linked structures. This process is called polymerization and constitutes the final step of lignin biosynthesis (Fig. 16.5).

16.6 Intricacies of Lignin Biosynthesis in Plant Cell Wall

Lignin is composed of three kinds of alcohol monomers (Monolignols): coumaryl alcohol (H-type lignin), coniferyl alcohol (G-type lignin and sinapyl alcohol (S-type lignin). The biosynthesis of monolignols is connected to the primary metabolism through shikimate/chorismate pathway which produces phenylalanine. Phenylalanine is entry compound of the carbon skeletons into the phenylpropanoid pathway. Phenylalanine follow a cascade of reactions (Hydroxylation and methylation) and lead to the formation of cinnamoyl-CoA esters which undergo two successive reductive steps. The first step specific to monolignol synthesis is catalyzed by Cinnamoyl CoA-reductase (CCR) which causes NADPH-dependent reduction of *p*-hydroxycinnamoyl CoA esters to the *p*-hydroxy cinnamaldehydes. In second step,

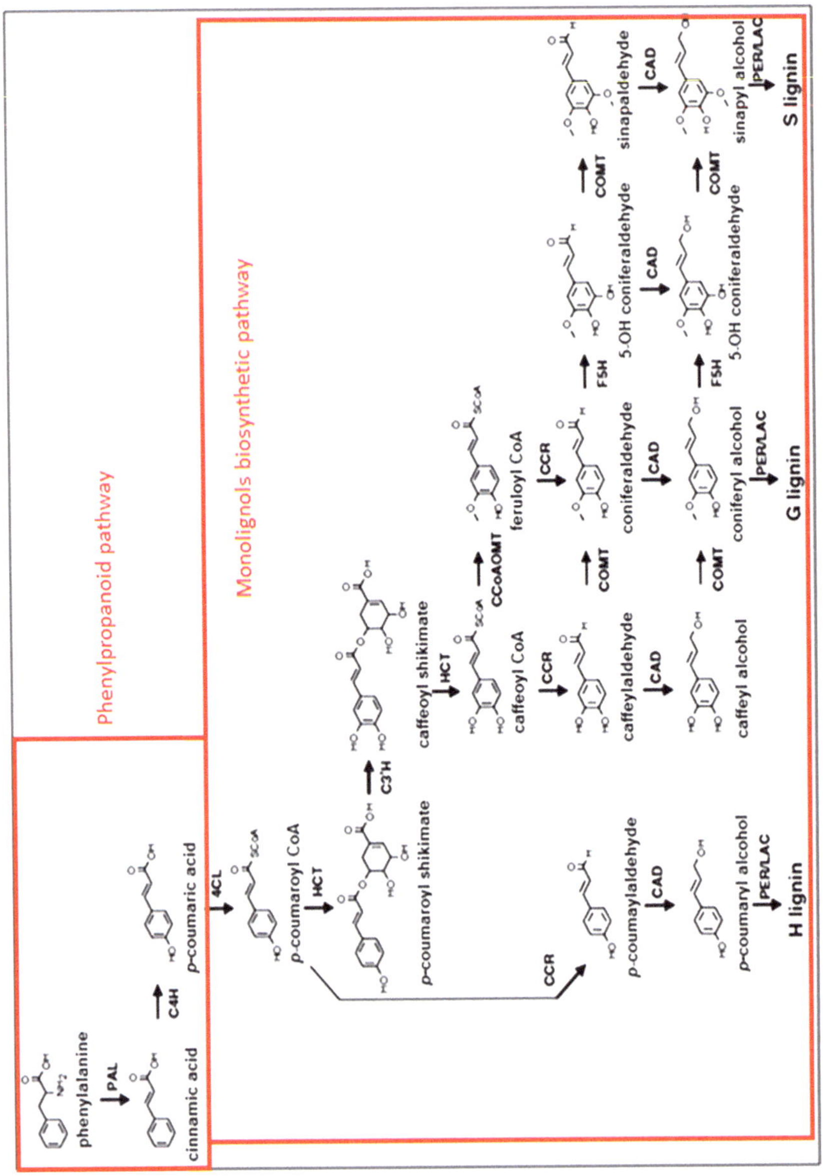

Fig. 16.5 Lignin biosynthetic pathway. (**Source:** Li et al. 2012)

these *p*-hydroxy cinnamaldehydes are reduced to their corresponding alcohols or monolignols (*p*-coumaryl, coniferyl, and sinapyl alcohol) through NADPH [Nicotinamide adenine dinucleotide phosphate (reduced)] dependent activity of cinnamyl alcohol dehydrogenase (CAD). The synthesized monolignol precursors are transported across plasma membrane and oxidized by laccase and peroxidase before polymerization into lignin polymer. Different mechanisms have been suggested for the transport of monolignols across the plasma membrane. These mechanisms are (1) Transport of monolignols through vesicles derived from golgi bodies; (2) Passive diffusion of monolignols through the plasma membrane, and (2) Active transport mediated by plasma membrane located transporters.

The polymerization of lignin monomers in secondary cell walls occurs *via* oxidative coupling of monolignols. In the first step, monolignols are oxidized (i.e., dehydrogenated) by the enzyme peroxidase and laccase to generate monolignol radicals. In the next step, coupling between two monomer radical dimers establishes a covalent bond between monolignol radicals. During the polymerization, the ratio of each possible coupling product depends on the chemical nature of monomers and conditions in the cell wall (Ralph et al. 2004). The coupled product is dehydrogenated and starts another coupling reaction with monomers. This mode of action of the addition of monomers to the growing polymer is known as endwise coupling. In each coupling reaction, two radicals are consumed at every single electron contributes to newly formed bond. Peroxidase/Laccase plays an exclusive role during the polymerization of monolignols into lignin. Peroxidases generate free radicals of monolignols, which further catalyze the coupling reactions. Peroxidases help in determining the final structure of lignin due to their substrate specificity. This opens up the possibilities for altering the structure of lignin by modifying the expression of a specific isoform of peroxidases (Fig. 16.6).

16.7 Manipulation of Lignin Biosynthesis in Forages

The content, composition and structure of lignin can be altered through genetic manipulation. The strategies for manipulation of lignin content and composition depend on end utility. Recently, the alteration of lignin in plants is accomplished through technologies including expression of antisense RNA, sense expression, co-suppression, ribozyme and RNA interference (RNAi) and mutations. Plants have the ability to tolerate large variations in lignin composition often without apparent adverse effects. Substitution of some fraction of traditional monolignols by alternative monomers through genetic engineering is a promising strategy to engineer lignin in plants (Vanholme et al. 2012).

Genes encoding enzyme(s) which are involved in monolignol biosynthesis, their transport to cell wall, their polymerization to form lignin and also the regulation of lignin biosynthesis are potential targets for lignin manipulation. The manipulation is based on alteration of monolignol biosynthetic pathway genes as a single construct

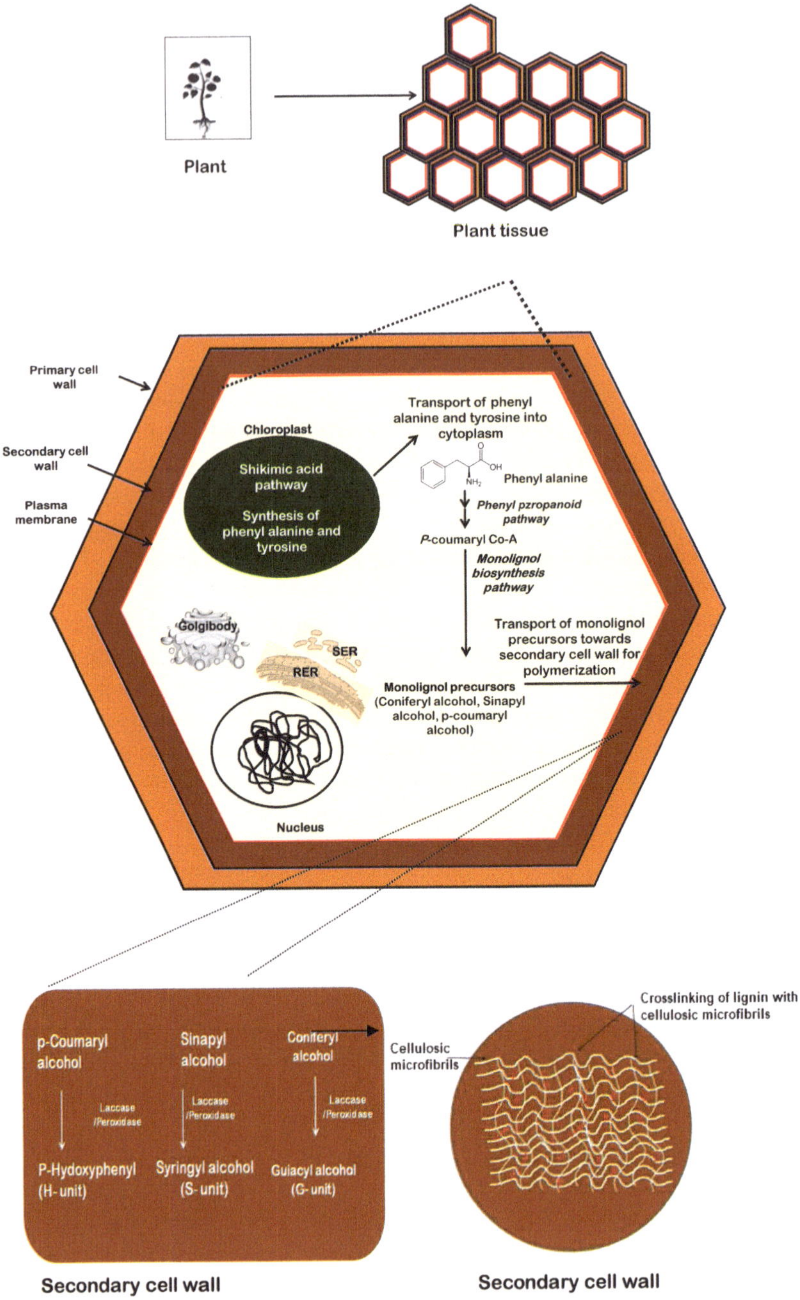

Fig. 16.6 Intricacies of lignin biosynthesis in plant cell wall

or by simultaneously targeting two or more genes involved in monolignol biosynthetic pathway (Baucher et al. 2003; Rastogi and Dwivedi 2008; Zhong et al. 1998). Several studies have reported that the incorporation of structurally related aromatic compounds, for example, cinnamaldehydes, 5-hydroxyconiferyl alcohol, dihydroconiferyl alcohol, ferulate esters, coniferyl ferulate, feruloyl tyramine, and ferulate malate into the cell walls of plants decreases the rate of lignification (Derikvand et al. 2008; Prashant et al. 2011). Transcriptional manipulation of lignin-specific genes, for example, MYB genes have also been targeted for altering lignin content and its composition in plants. Thus, by engineering the expression of transcription factors, lignification can be altered without affecting plant development. As these transcription factors bind promoters of multiple genes, the carbon flux through the pathway can be altered in a more coordinated way.

16.7.1 RNAi-Mediated Manipulation of Lignin Biosynthesis

16.7.1.1 Down-Regulation of Gene Encoding 4CL (4-Coumarate: CoA Ligase) Enzyme

The enzyme 4-Coumarate: coenzyme A ligase (4CL) is one of the key enzymes involved in the monolignol biosynthetic pathway. Xu et al. (2011) identified two homologous 4CL genes, Pv4CL1 and Pv4CL2 in switch grass (*Panicum virgatum*) through phylogenetic analysis. Gene expression patterns and enzymatic activity assays suggested that Pv4CL1 is involved in monolignol biosynthesis. Transgenic plants of switchgrass were obtained by silencing of Pv4CL1 gene. RNA interference of Pv4CL1 reduced extractable 4CL activity by 80%, which lead to reduction in lignin content with decreased guaiacyl (G) unit composition.

16.7.1.2 Down-Regulation of Gene Encoding CCR (Cinnamoyl-CoA Reductase) Enzyme

In the lignin biosynthesis pathway, CCR regulates the first committed step of the biosynthesis of monolignol precursors and is considered as a crucial regulatory point that controls the total carbon flux carbon toward lignin. CCR enzymes catalyze the conversion of hydroxycinnamoyl CoA thioesters (*p*-coumaroyl CoA, feruloyl-CoA, and sinapoyl- CoA) to their corresponding cinnamaldehydes. Studies have revealed that the synthesis of monolignol precursors is strictly associated with the activity of this enzyme. Success has been achieved for manipulation of lignin content *via* down-regulation of the activity of this enzyme. Tamasloukht et al. (2011) isolated a mutant in the *CCR1* gene from maize and named it *ZmCCR1*. *ZmCCR1* gene expression was reduced to 31% of the residual wild-type level. *Zmccr1*$^{-}$ exhibited enhanced digestibility without compromising plant growth and development. At the cellular level, alterations in lignin deposition were mainly observed in the walls of the sclerenchymatic fiber cells surrounding the vascular bundles. Lignin analysis revealed a slight decrease in lignin content and significant changes in lignin structure. *p*-Hydroxyphenyl units were strongly decreased and the syringyl/guaiacyl ratio increased. Park et al. (2012) down-regulated ZmCCR1 gene in maize with RNAi technique. In the study, among 10 out of 30 randomly tested lines, six lines revealed a significant reduction in *ZmCCR1* transcription. The second generation of these *ZmCCR1* down-regulated transgenic plants exhibited brown coloration of midribs, husk, and stems and 7.00–8.70% reduction in Klason lignin. Down-regulation of *CCR1* using an RNA interference-mediated silencing strategy resulted changes in lignin content and composition in transgenic perennial rye grass plants grown under glasshouse and field conditions (Tu et al. 2010). Giordano et al. (2014) isolated three putative *CCR1* cDNAs from *Paspalum dilatatum* and reported that their spatiotemporal expression pattern correlates with a developmental profile of lignin deposition. Reduction in *CCR1* expression levels resulted in reduction of lignin content, which largely attributed to decrease in G lignin content.

16.7.1.3 Down-Regulation of COMT (Caffeic Acid O-Methyltransferase) Enzyme

COMT catalyzes the methylation of caffeic acid, 5-hydroxy ferulic acid, and 3-O methylation of monolignol precursors. COMT enzyme uses 5-hydroxyconiferaldehyde and/or 5-hydroxyconiferyl alcohol preferentially for the synthesis of sinapaldehyde/sinapyl alcohol which leads to the synthesis of S-type lignin. In monocot species, maize *bm3* mutant is severely deficient in OMT activity, with only 10% of the activity in normal plants (Grand et al. 1985). Sequencing COMT clones obtained from the *bm3-1* and *bm3-2* maize, the *bm3-1* allele arise from an insertional event producing a COMT mRNA altered in both size and amount, whereas the *bm3-2* resulted from deletion of part of the COMT gene (Vignols et al. 1995). Transgenic maize plants generated with construct harboring maize COMT cDNA in the antisense orientation, one transgenic line displayed a significant reduction in COMT activity (15 to 30% residual activity) and barely detectable amounts of COMT protein. Fu et al. (2011a) carried out down-regulation of COMT gene in switch grass and reported a 22 to 28% reduction in COMT enzyme activity in transgenic plants.

16.7.1.4 Down-Regulation of Gene Encoding CAD (Cinnamyl Alcohol Dehydrogenase) Enzyme

CAD enzyme catalyzes the last step of biosynthesis of monolignol precursors which is reduction of hydroxyl-cinnamaldehydes to corresponding alcohols (Monolignol precursors). Since this enzyme has crucial role in determining the lignin content and composition, involvement of this enzyme in lignin biosynthesis has been investigated in several plants. Down-regulation of CAD gene alters cell wall architecture, reduces lignin level, and incorporates phenolic aldehydes into lignin (Sattler and Funnell-Harris 2013). Using particle bombardment technique, Muguerza et al. (2014) introduced CAD gene constructs in antisense and RNAi vector under the control of maize ubiquitin promoter into Bahia grass calli (*Paspalum notatum*). The ADL content in transgenic plants was 4.70% and that of the control was 5.60%, which represented 16% reduction in lignin content. In another study, transgenic switch grass plants were produced with a CAD RNAi gene construct under the control of the maize ubiquitin promoter (Fu et al. 2011b). The transgenic lines showed reduced CAD expression levels, reduced enzyme activities, reduced lignin content, and altered lignin composition. The modification of lignin biosynthesis resulted in improved sugar release and forage digestibility. Switch grass CAD (PvCAD) belongs to the first group which included other CAD proteins involved in lignin biosynthesis. At the amino acid level, our PvCAD shows high identity with predicted proteins of two recently isolated CAD sequences from switch grass, PviCAD1 (98% identity) and PviCAD2 (99% identity) (Saathoff et al. 2010). Jackson et al. (2008) reported that CAD down-regulation resulted in both reduction in lignin content and alteration in lignin composition in alfalfa. Down-regulation of CAD in dicot species such as tobacco (Bernard-Vailhe et al. 1996) and alfalfa (Baucher et al. 1999) showed reduction of CAD activity only changed lignin composition or the levels of phenolic compounds.

16.7.2 Antisense RNA-Mediated Manipulation of Lignin Biosynthesis

16.7.2.1 Down-Regulation of Gene Encoding C4H (Cinnamate-4-Hydroxylase) Enzyme

The first three biosynthetic reactions in phenylpropanoid metabolism are referred to as the general phenylpropanoid pathway because they produce *p*-coumaroyl CoA, a major branch-point metabolite between the production of flavonoids and pathway that produces monolignols, lignans and hydroxy-cinnamate conjugates. Down-regulation of upstream genes like C3H, HCT, or 4CL leads to a reduction in lignin content, while the down-regulation of F5H and COMT results in changes in S/G ratio (Weng et al. 2008). Reddy et al. (2005) developed transgenic alfalfa lines through antisense expression encoding cytochrome P_{450} enzymes, namely cinnamate 4-hydroxylase (C4H), coumaroyl shikimate 3-hydroxylase (CSH, and coniferaldehyde 5-hydroxylase (C5H) also known as ferulate 5-hydroxylase (F5H) with altered lignin content and composition which catalyze reactions in the formation of monolignol, G and S-units, respectively. In the study, down-regulation of coumarate 3-hydroxylase (C3H) resulted in reduction of lignin content with increased H lignin units, while down-regulating ferulate 5-hydroxylase (F5H) resulted in a slight increment in G lignin units and reduction in S lignin units. Down-regulation of genes early in the pathway (C4H, C3H, and HCT) is most effective at reducing lignin content, in some cases leading to plants that contain less than half the lignin in wild-type plants (Hisano et al. 2004).

16.7.2.2 Down-Regulation of Gene Encoding HCT Enzyme

4-Coumaroyl CoA, a common precursor for both lignin and flavonoid biosynthesis, is converted by HCT to 4-coumaroyl shikimate, the substrate for hydroxylation by 4-coumaroyl shikimate 3-hydroxylase, previously known as coumarate 3-hydroxylase (C3H) (Schoch et al. 2001; Hoffmann et al. 2003). The shikimate ester is then converted back to the corresponding CoA ester by HCT reacting in the reverse direction. Shadle et al. (2007) introduced antisense construct *Medicago truncatula* HCT gene and developed transgenic alfalfa *via* somatic hybridization.

To evaluate the reduction in lignin content, crude extracts of stem material from various down-regulated and control lines were analyzed for HCT enzyme activity. They reported a 15–50% reduction in HCT enzyme activity in transgenic lines.

16.7.2.3 Down-Regulation of COMT (Caffeic Acid O-Methyltransferase) Enzyme and CCoAOMT (Caffeoyl CoA 3-O-Methyltransferase) Enzyme

Guo et al. (2001a) reported the effects of down-regulation of CCoAOMT enzymes in lignin content and composition on *in vitro* and *in situ* digestibility of alfalfa. Down regulation of CCoAOMT reduced the Klason lignin content by 12–17% in transgenic alfalfa lines. Decrease in S lignin units approximately 11–33% in the CCoAOMT down-regulated lines. Maize and rice plants have been developed with manipulated CCoAOMT or F5H (Takeda et al. 2017). By comparison, COMT, the gene underlying improved digestibility in *bm3* mutants, has been an obvious target for transgenic manipulation in biomass and forage grasses and cereal crops with several evaluations progressing to the field. This demonstrate changes in COMT-suppressed transgenics consistent with those in *bm3* plants, that is, moderate reductions of lignin (6–14%), with more dramatic reductions in S units (20–46%) accompanied by a decrease in S/G ratio. Chen et al. (2004) isolated four COMT cDNA clones from tall fescue by screening a cDNA library. To confirm that the cloned sequences encode corresponding enzymes and to investigate their kinetic properties and substrate specificity, the coding region of one of the COMT cDNAs (FaCOMT1b) cloned into pET29a vector for expression in *E. coli*. The coding region of FaCOMT1b was also used to construct sense and antisense chimeric transgenes, which were introduced back to tall fescue by biolistic transformation. Biochemical and molecular analyses revealed that two co-suppressed sense transgenic tall fescue lines were identified with substantially reduced levels of transcripts, reduced enzymatic activities, decreased lignin content, apparently altered lignin composition, and significantly increased dry matter digestibility. Similar to the maize bm3 mutant, the transgenic tall fescue with down-regulated COMT also showed reduced S lignin level and reduced S/G ratio. Guo et al. (2001a) reported 13–29% reduction in Klason lignin content in COMT down-regulated lines of alfalfa.

16.7.2.4 Down-Regulation of Gene Encoding CAD (Cinnamyl Alcohol Dehydrogenase) Enzyme

Cinnamyl alcohol dehydrogenase (CAD) and caffeic acid O-methyltransferase (COMT) are the key enzymes that play important roles in lignin biosynthesis (Vailhe et al. 1998; Guo et al. 2001a, b). CAD catalyzes the last step in the biosynthesis of lignin precursors, i.e., reduction of cinnamaldehydes to cinnamyl alcohols. Transgenic manipulation of CAD by either antisense or sense strategy led to the reduction of CAD activity changing the S/G ratio or other phenolic compounds in tobacco (Vailhe et al. 1998; Baucher et al. 1999), alfalfa (Baucher et al. 1999). A slight reduction in lignin content has been observed in transgenic poplar trees (Lapierre et al. 1999; Pilate et al. 2002). Some of the changes resulted in increased lignin extractability or degradability (Baucher et al. 1999). Transgenic CAD plants with altered lignin showed normal development (Vailhe et al. 1998; Baucher et al. 1999; Pilate et al. 2002). Chen et al. (2004) cloned four CAD cDNA sequences from tall fescue. cDNA sequences cloned were highly homologous with each other and expressed very high similarity to that of other related monocot species such as perennial rye grass. Transgenic tall fescue plants carrying either sense or antisense CAD gene constructs obtained by microprojectile bombardment of single genotype-derived embryogenic suspension cells severely reduced the levels of mRNA transcripts and significantly reduced CAD enzymatic activities in two transgenic plants carrying sense and antisense CAD transgenes, respectively. Decreased lignin content and apparently altered lignin composition were observed in the transgenic lines. No significant changes in cellulose, hemicellulose, neutral sugar composition, *p*-coumaric acid, and ferulic acid levels were observed in transgenic plants. The increase in IVDMD to the extent of 9.50% was recorded for CAD down-regulated tall fescue lines.

16.7.3 Lignin Biosynthetic Pathway Editing Techniques

Plant breeding is a conventional approach to transfer agronomically important crop traits (Bradshaw 2017). The major change in breeding was during the 80s of the last century introducing genomics assisted breeding (Varshney et al. 2005; Desta and Ortiz 2014). In recent years additional tools to identify new traits *via* transcriptomics have been developed (Jiao et al. 2009; Lu et al. 2010; Zhao et al. 2012). New sequencing methods and genome assembly, metabolomics and proteomics assist breeders nowadays. The conventional breeding approaches require a lot of time and energy compared to new breeding technologies (NBT) (Schaart et al. 2016). Transgenic and genome editing technology permits the correct and precise transfer of one or a few desirable genes apart from the introgression of novel genes conferring superior desirable traits using a conventional approach. Sequence-specific nucleases (SSNs) allow for precise and targeted modifications of the complex crop genome. These SSNs currently include zinc finger nucleases (ZFN) (Kim et al. 1996), meganucleases (Smith et al. 2006), transcription activator-like effector nucleases (TALEN) (Christian et al. 2010; Morbitzer et al. 2010) and RNA-guided nuclease (CRISPR/Cas) systems (Jinek et al. 2012).

16.7.3.1 TALEN-Mediated Editing of Lignin Biosynthetic Pathway

TALEN (transcription activator-like effector nucleases) originated from effectors of phytopathogenic bacteria of the *Xanthomonas* genus (Kay et al. 2009). The discovered transcription activator-like effectors were observed to bind to host DNA sequences (Boch et al. 2009; Moscou and Bogdanove 2009). TALEN effectors tend to contain 5–30 tandem monomer repeats each consisting of 33–35 residues with each tandem repeat determines the specificity of the target sequence binding (Mak et al. 2012). A pair of these residues often referred to as repeat variable residues (RVD), is responsible for binding to the target sequence, whereas NI

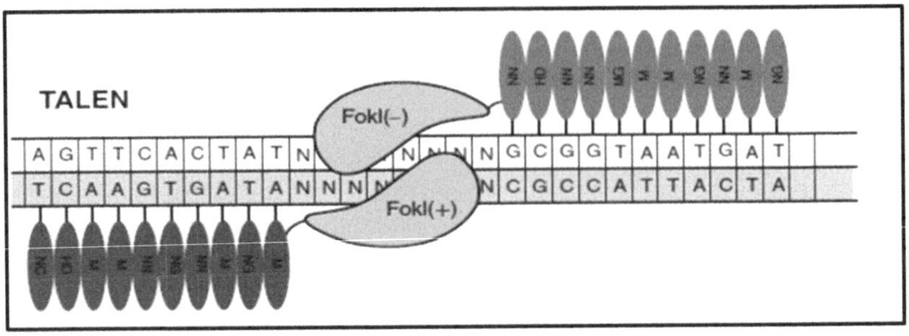

Fig. 16.7 Model of genome editing through TALENS

(Asn Ile) binds to adenine, HD (His, Asp) to cysteine, NN (Asn, Asn), NH (Asn, His) and NK (Asn, Lys) to guanine and NG (Asn Gly) to thymine (Boch et al. 2009; Moscou and Bogdanove 2009). From the three candidates binding to guanine, the highest affinity was shown for NH. Due to the pioneering work with ZFNs, TALEN can be used with numerous proteins for targeted genetic modifications (Li et al. 2012).

Jung and Altpeter (2016) selected TALEN binding sites in the conserved region of the first exon among the sugarcane COMT gene and its putative homo(eo)logs. TALEN expression cassette was introduced into sugarcane by *Agrobacterium*-mediated transformation *via* indirect embryogenesis (AIE) or biolistic mediated transformation *via* direct embryogenesis (BDE). The results revealed that the majority of the mutations were deletions (98%) and the length of deletions was in the range from 1 to 52 bp. The most frequent mutation was 5 bp deletion (10%) followed by 6 and 7 bp deletions. Insertions were relatively infrequent (2%), and the length of insertions was between 2 and 23 bp. Lignin reduction levels ranged from 11 to 32% in the mutation lines with relatively high mutation frequencies (90–99%) compared to wild type. All of the lignin-reduced COMT mutant lines have significant reduction of S subunit content by up to 54%, while G subunit content remained unchanged compared to wild type (Fig. 16.7).

16.7.3.2 ZFN (Zinc Finger Nucleases) Mediated Editing of Lignin Biosynthetic Pathway

Zinc finger nucleases (ZFNs) are artificial protein complexes resulting from the idea to generate new endonucleases with the ability to choose the target sequence. The first generated ZFN used the nonsequence-specific cleavage domain of the *Fok* I Type II endonuclease fused to a homeobox domain from *Drosophila melanogaster* (Kim and Chandrasegaran 1994). Research on ZFNs discovered favorable designs which lead to the conclusion that a construct with dimers inversely oriented and separated by six nucleotides is the most efficient way (Bibikova et al. 2001; Pabo et al. 2001). ZFNs have been widely used to induce DSBs to alter the DNA sequence of different model organisms (Mashimo et al. 2010) (Fig. 16.8).

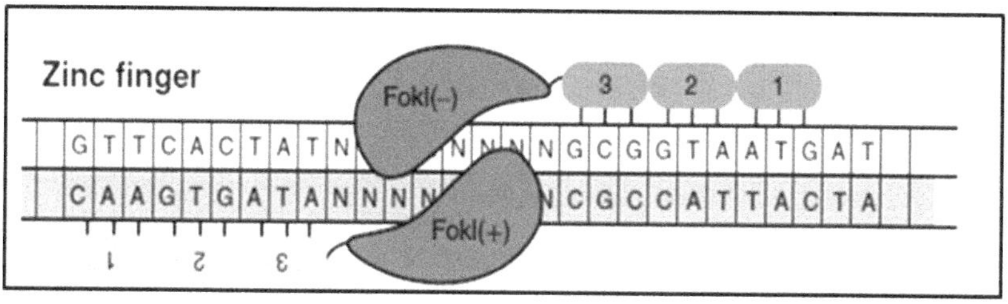

Fig. 16.8 Model of genome editing through ZFNs

16.7.3.3 CRISPR–Cas Mediated Editing of Lignin Biosynthetic Pathway

Major genome editing tools like TALEN and ZFN rely on large protein structures needed for specific binding to the target DNA. This requires large constructs with limitations for cloning and target design (Gaj et al. 2013). However, a new method for genome editing CRISPR solves the complications and challenges if the binding to target regions is impossible with large protein structures used by TALEN or ZFN. Ishino et al. (1987) were the first to have the idea of CRISPR (clustered regularly interspaced palindromic sequences) which was discovered in the gram-negative bacteria *Escherichia coli*. These sequences were later discovered in numerous other organisms from the two kingdoms of archaea and bacteria (Mojica et al. 2000). Research on these CRISPR sequences revealed their origin in acquired DNA or plasmid DNA and these sequences heavily undergo horizontal gene transfer and even closely related species could differ in their CRISPR sequences (Garneau et al. 2010). The combination of CRISPR sequences with cas enzymes was the pioneer system that quickly propagated for biotechnology purposes in phage resistance or eliminating unwanted traits of bacteria (Barrangou and Horvath 2012). The enzyme Cas9 has been identified as a large multifunctional protein, consisting of two nuclease subunit domains (Makarova et al. 2002, 2006). In its native form, this enzyme produces double-stranded DNA cleavage deleting either the HNH or the RuvC domain where single-stranded DNA cleavage occurs (nickases) (Jinek et al. 2012). In the c-terminal domain of the cas9 protein, a short protospacer adjacent motif (PAM) is identified which is needed for interrogation of the DNA to the matching PAM sequence (Jiang and Doudna 2017). In its inactive state, this region is largely disordered to prevent cleavage at the PAM site (Barrangou and Doudna 2016) (Fig. 16.9).

Only the combination of tracrRNA and crRNA leads to the activation of cas9 enzyme (Jinek et al. 2012; Perez-Rodriguez et al. 2011). Modifications of this tracrRNA:crRNA duplex leads to the construction of single-guide RNA (sgRNA) which fuse the crRNA directly to the tracrRNA with a linker loop resulting in a working CRISPR/Cas9 construct and enabling the design for genome editing constructs (Anders et al. 2014). Recently, efforts have been made for engineering better cas enzymes to increase specificity or modifying the enzyme to unlock new targeting sites with changes in PAM sequence (Tsai and Joung 2016). The difference between the downregulation and CRISPR/Cas9 technologies is that the site-specific

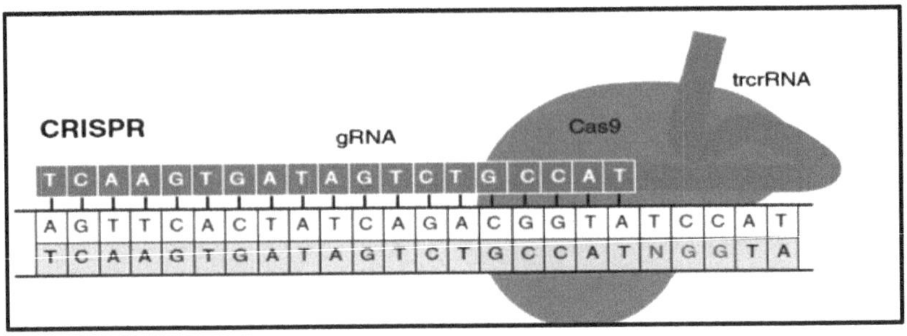

Fig. 16.9 Model of genome editing through CRISPR/Cas9

nucleases produce defined gene knock-outs by altering the genomic DNA sequence, while antisense, RNAi, and artificial microRNA techniques generate knock-downs by repressing transcription of the targeted gene. Park et al. (2017) employed the CRISPR/Cas9 system to generate low-lignin switch grass to test the system in an out-crossing species with complicated genomes by producing total knockouts of a gene previously knocked down by RNAi. The 4-coumarate: coenzyme A ligase (4CL) was targeted for genetic modification which is a key enzyme related to the early steps of the monolignol biosynthesis pathway. Three 4CL genes, Pv4CL1, Pv4CL2, and Pv4CL3 have been identified in switch grass. Expression analysis revealed that Pv4CL1 transcripts were more abundant in the stem than in the leaf, while Pv4CL2 transcripts were barely detectable and Pv4CL3 was mainly expressed in the leaf. Pv4CL1 was selected as the target for CRISPR/Cas9 editing because of its preferential expression in highly lignified stem tissues. Specific guide RNA was constructed to target Pv4CL1. After introducing the construct into switch grass calli, 39 transgenic plants were regenerated. Using two rounds of PCR screening and sequencing, four plants were confirmed to have tetra-allelic mutations simultaneously. The Pv4CL1 knock-out plants had reduced cell wall thickness, an 8–30% reduction in total lignin content, a 7–11% increase in glucose release, and a 23–32% increase in xylose release. Park et al. (2017) established a successful CRISPR/Cas9 system in switch grass with mutation efficiency reaching 10%. The system allows the precise targeting of the selected Pv4CL1 gene to create switch grass knock-out mutant plants with decreased lignin content and reduced recalcitrance (Table 16.1).

Lignin is a polymer of phenylpropanoid compounds formed through a complex biosynthesis route, represented by a metabolic grid for which most of the genes involved have been sequenced in several plants. Plants are exposed to different stresses, which may change lignin content and composition particularly during plant-microbe interactions. Understanding stress modulated expression of the genes related to lignin biosynthesis may allow us to understand the metabolic control of lignin deposition in the cell wall. Lignin deposition is considered an anti-quality trait that affect digestibility and utilization of nutrients from forage crops. Large number of enzymes are involved in the biosynthesis of lignin. The first step in lignin biosynthesis is shikimate pathway, followed by phenyl alanine pathway, third major step is monolignol biosynthetic pathway followed by polymerization leading to lignin deposition on cell wall. Studies in the manipulation of enzymes like CAD,

Table 16.1 Forage crops developed through manipulation of lignin biosynthesis

Transgenic plants and genes targeted	Genes targeted	Alterations in lignin content and composition; Production of other metabolites; Changes in plant characteristics	References
Alfalfa	C3H	Decreased lignin content; lignin-rich in H units; improved digestibility	Reddy et al. (2005)
Alfalfa	C3H	Decreased lignin content; normal G and S units; increased p-hydroxyphenyl units; higher levels of phenyl coumarans andresinols	Ralph et al. (2006)
Alfalfa	C3H	Significant structural changes had occurred to the alfalfa ball-milled lignins; increased *p*-hydroxyphenyl units; decreased guaiacyl and syringyl units; decreased methoxyl content (~55–58% in the ball-milled lignin); decreased β-O-4linkage contents; abundance of phenyl coumaran and resinol in the ball-milled lignins	Pu et al. (2009)
Alfalfa	C4H	Decreased lignin content; no significant impact on lignin composition; Improved digestibility	Reddy et al. (2005)
Alfalfa	OMT	Decreased (30%) lignin content; decreased G; new 5-OH G	Guo et al. (2001a)
Maize	OMT	Decreased (25–30%) lignin content; decreased S; lower p-coumaric acid content; new 5-OHG	Piquemal et al. (2002)
Alfalfa	CCoAOMT	Decreased (47%) lignin content; decreased G; No reduction of S	Guo et al. (2001a, b)
Alfalfa	CAD	Unchanged lignin content; decreased S	Baucher et al. (1999)
Tall fescue	COMT	Decreased lignin content; altered lignin composition, significantly increased dry matter digestibility	Chen et al. (2004)
Switch grass	4CL	Decreased lignin content by 34%	Xu et al. (2011)
Maize	CCR	Decreased lignin content by 10% with higher enzymatic digestibility	Tamasloukht et al. (2011)
Sorghum	COMT	Lignin reduced	Palmer et al. (2008)
Switch grass	CAD	Lignin reduced by 15%	Saathoff et al. (2011)

COMT, CCR, 4CL, F5H, C4H and HCT have proved the feasibility of genetic manipulation of lignin biosynthetic pathways for altering the lignin content for improving the forage quality. Lower lignin content is preferred in forages crops for digestibility and easy uptake of nutrients from the feed resources. Considering influence of lignin on the forage nutritive value, manipulation of lignin content in forage crops through different biotechnological approaches could have a tremendous effect on nutritional value of forages.

References

Akin DE (1989) Histological and physical factors affecting digestibility at forages. Agron J 81:17–25

Akin DE, Fales SL, Rigsby LL, Snook ME (1987) Temperature effects on leaf anatomy, phenolic acids, and tissue digestibility in tall fescue. Agron J 79:271–275

Anders C, Niewoehner O, Duerst A, Jinek M (2014) Structural basis of PAM-dependent target DNA recognition by the Cas9 endonuclease. Nature 513:569–573

Apel K, Hirt H (2004) Reactive oxygen species: metabolism, oxidative stress and signal transduction. Ann Rev Plant Biol 55:373–399

Bacon JSD (1979) Plant cell wall digestibility and chemical structure. Rep Rowett Inst 35:99–108

Barrangou R, Doudna JA (2016) Applications of CRISPR technologies in research and beyond. Nat Biotechnol 34:933–941

Barrangou R, Horvath P (2012) CRISPR: new horizons in phage resistance and strain identification. Ann Rev Food Sci Technol 3:143–162

Baucher M, Bernard-Vailhe MA, Chabbert B, Besle JM, Opsomer C, Van Montagu M, Botterman J (1999) Down-regulation of cinnamyl alcohol dehydrogenase in transgenic alfalfa (*Medicago sativa* L.) and the effect on lignin composition and digestibility. Plant Mol Biol 39:437–447

Baucher M, Halpin C, Petit-Conil M, Boerjan W (2003) Lignin: genetic engineering and impact on pulping. Cri Rev Biochem Mol Biol 38:305–350

Bibikova M, Carroll D, Segal DJ, Trautman JK, Smith J, Kim Y-G, Chandrasegaran S (2001) Stimulation of homologous recombination through targeted cleavage by chimeric nucleases. Mol Cell Biol 21:289–297

Boch J, Scholze H, Schornack S, Landgraf A, Hahn S, Kay S, Lahaye T, Nickstadt A, Bonas U (2009) Breaking the code of DNA binding specificity of TAL-type III effectors. Science 326: 1509–1512

Boija E, Lundquist A, Edwards K, Isaksson R, Johansson G (2007) Evaluation of bilayer disks as plant cell membrane models in partition studies. Anal Biochem 364(2):145–152

Bonawitz ND, Chapple C (2010) The genetics of lignin biosynthesis: connecting genotype to phenotype. Ann Rev Genet 44:337–363

Boudet AM (1998) A new view of lignification. Trends Plant Sci 3:67–71

Bradshaw JE (2017) Plant breeding: past, present and future. Euphytica 213:60. https://doi.org/10.1007/s10681-016-1815-y

Buscaglia HJ, Van EHM, Geohring LD, Vermeulen HCAM, Fick GW, Lucey RF (1994) Alfalfa yield and quality are affected by soil hydrologic conditions. Agron J 86:535–542

Buxton DR (1996) Quality-related characteristics of forages as influenced by plant environment and agronomic factors. Anim Feed Sci Tech 59:37–49

Buxton DR, Brasche MR (1991) Digestibility of structural carbohydrates in cool season grass and legume forages. Crop Sci 31:1338–1345

Buxton DR, Casler MD (1993) Environmental and genetic effects on cell wall composition and digestibility. In: Jung HG, Buxton DR, Hatfield RD, Ralph J (eds) Forage cell wall structure and digestibility. ASA-CSSA-SSSA, Madison, pp 685–714

Buxton DR, Fales SL (1994) Plant environment and quality. In: Fahey GC Jr, Collins M, Mertens DR, Moser LE (eds) Forage quality, evaluation and utilization. ASA, CSSA, and SSA, Madison, pp 155–199

Buxton DR, Redfearn DD (1997) Plant limitations to fiber digestion and utilization. J Nutr 127: 814–818

Buxton DR, Russell JR (1988) Lignin constituents and cell-wall digestibility of grass and legume stems. Crop Sci 28:553–558

Chen L, Auh C, Dowling P, Bell J, Lehmann D, Wang Z (2004) Transgenic down-regulation of caffeic acid O-methyltransferase (COMT) led to improved digestibility in tall fescue (*Festuca arundinacea*). Funct Plant Biol 31:235–245

Christian M, Cermak T, Doyle EL, Schmidt C, Zhang F, Hummel A, Bogdanove AJ, Voytas DF (2010) Targeting DNA double-strand breaks with TAL effector nucleases. Genetics 186:757–761

Cochrane FC, Davin LB, Lewis NG (2004) The Arabidopsis phenylalanine ammonia lyase gene family: kinetic characterization of the four PAL isoforms. Phytochem 65:1557–1564

Cone JW, Engels FM (1990) Influence of growth temperature on anatomy and *in vitro* digestibility of maize tissues. J Agric Sci (Camb.) 114:207–212

da Silva JHS, Johnson WL, Burns JC, Anderson CE (1987) Growth and environment effects on anatomy and quality of temperate and subtropical forage species. Crop Sci 27:1266–1273

Dean JFD, Eriksson KE (1992) Biotechnological modification of lignin structure and composition in forest trees. Holzforschung 46:135–147

Derikvand MM, Sierra JB, Ruel K, Pollet B, Do CT, Thevenin J, Buffard D, Jouanin L, Lapierre C (2008) Effect of shade stress on lignin biosynthesis in soybean stems. Planta 227:943–956

Desta ZA, Ortiz R (2014) Genomic selection: genome-wide prediction in plant improvement. Trends Plant Sci 19:592–601

Dorairaj D, Ismail MR, Sinniah UR, Tan KB (2017) Influence of silicon on growth, yield, and lodging resistance of MR219, a lowland rice of Malaysia. J Plant Nutr 40:1111–1124

Duan C, Yu J, Bai J, Zhu Z, Wang X (2014) Induced defense responses in rice plants against small brown planthopper infestation. Crop J 2:55–62

Engels FM, Jung HG (1998) Alfalfa stem tissues: cell-wall development and lignification. Ann Bot 82:561–568

Fales SL (1986) Effects of temperature on fiber content, composition, and in vitro digestion kinetics of tall fescue. Agro J 78:963–966

Fu C, Mielenz JR, Xiao X, Ge Y, Choo Y, Hamilton MR Jr, Chen F, Foston M, Ragauskas A, Bouton J, Dixon RA, Wang ZY (2011a) Genetic manipulation of lignin reduces recalcitrance and improves ethanol production from switchgrass. Proc Nat Acad Sci U S A 108:3803–3808

Fu C, Xiao X, Xi Y, Ge Y, Chen F, Bouton J, Dixon R, Wang ZY (2011b) Downregulation of cinnamyl alcohol dehydrogenase (CAD) leads to improved saccharification efficiency in Switch grass. BioEnergy Res 4:153–164

Gaj T, Gersbach CA, Barbas CF (2013) ZFN, TALEN, and CRISPR/Cas-based methods for genome engineering. Trends Biotechnol 31:397–405

Garneau JE, Dupuis M, Villion M, Romero DA, Barrangou R, Boyaval P, Fremaux C, Horvath P, Magadán AH, Moineau S (2010) The CRISPR/Cas bacterial immune system cleaves bacteriophage and plasmid DNA. Nature 468:67

Gindl W, Grabner M, Wimmer R (2000) The influence of temperature on latewood lignin content in treeline Norway spruce compared with maximum density and ring width. Trees 14:409–414

Giordano A, Liu Z, Panter SN, Dimech AM, Shang Y, Wijesinghe H, Fulgueras K, Ran Y, Mouradov A, Rochfort S, Patron NJ, Spangenberg GC (2014) Reduced lignin content and altered lignin composition in the warm season forage grass *Paspalum dilatatum* by down-regulation of a Cinnamoyl CoA Reductase Gene. Transgenic Res 23:503–517

Grand CP, Parmentier P, Boudet A, Boudet AM (1985) Comparison of lignins and enzymes involved in lignification in normal and brown-midrib (bm3) mutant corn seedlings. Physiol Veg 23:905–911

Grisebach H (1981) Lignins. In: Conn EE (ed) The biochemistry of plants, vol 7. Secondary plant products. Academic Press, New York, pp 457–478

Guo D, Chen F, Inoue K, Blount JW, Dixon RA (2001a) Down regulation of caffeic acid 3-O-methyltransferase and caffeoyl CoA 3-O-methyltransferase in transgenic alfalfa: impacts on lignin structure and implications for the biosynthesis of G and S lignin. Plant Cell 13:73–88

Guo D, Chen F, Wheeler J, Winder J, Selman S et al (2001b) Improvement of in-rumen digestibility of alfalfa forage by genetic manipulation of lignin O-methyltransferases. Transgenic Res 10: 457–464

Halim RA, Buxton DR, Hattendorf MJ, Carlson RE (1989) Water-stress effects on alfalfa forage quality after adjustment for maturity differences. Agron J 81:189–194

Hartley RD (1972) p-Coumaric and ferulic acid components of cell walls of ryegrass and their relationships with lignin and digestibility. J Sci Food Agric 23:1347

Henderson MS, Robinson DL (1982) Environmental influences on yield and *in vitro* true digestibility of warm-season perennial grasses and the relationships to fiber components. Agron J 74: 943–946

Herrero J, Fernándezpérez F, Yebra T, Novouzal E, Pomar F, Pedreño MÁ, Cuello J, Guéra A, Estebancarrasco A, Zapata JM (2013) Bioinformatic and functional characterization of the basic peroxidase from *Arabidopsis thaliana* involved in lignin biosynthesis. Planta 237:1599–1612

Higuchi T (1990) Lignin biochemistry: biosynthesis and biodegradation. Wood Sci Technol 24:23–63

Hisano H, Kanazawa A, Kawakami A, Yoshida M, Shimamotoa Y, Yamada T (2004) Transgenic perennial ryegrass plants expressing wheat fructosyltransferase genes accumulate increased amounts of fructan and acquire increased tolerance on a cellular level to freezing. Plant Sci 167:861–868

Hoffmann L, Maury S, Martz F, Geoffroy P, Legrand M (2003) Purification, cloning, and properties of an acyltransferase controlling shikimate and quinate ester intermediates in phenylpropanoid metabolism. J Biol Chem 278:95–103

Hu Y, Li WC, Xu YQ, Li GJ, Liao Y, Fu FL (2009) Differential expression of candidate genes for lignin biosynthesis under drought stress in maize leaves. J Appl Genet 50(3):213–223

Hu D, Liu XB, She HZ, Gao Z, Ruan RW, Wu DQ, Yi ZL (2017) The lignin synthesis related genes and lodging resistance of *Fagopyrum esculentum*. Biol Plant 61:1–9

Iiyama K, Lam TBT, Miekle PJK, Rhodes DI, Stone BA (1993) Cell wall biosynthesis and its regulation. In: Jung HG, Buxton DR, Hatfield RD, Ralph J (eds) Forage cell wall structure and digestibility. ASA-CSSA-SSSA, Madison, pp 621–683

Ishino Y, Shinagawa H, Makino K, Amemura M, Nakata A (1987) Nucleotide sequence of the iap gene, responsible for alkaline phosphatase isozyme conversion in Escherichia coli, and identification of the gene product. J Bacteriol 169:5429–5433

Jackson L, Shadle G, Zhou R, Nakashim J, Chen F, Dixon R (2008) Improving saccharification efficiency of alfalfa stems through modification of the terminal stages of monolignol biosynthesis. Bioenergy Res 1:180–192

Jannoey P, Channei D, Kotcharerk J, Pongprasert W, Nomura M (2017) Expression analysis of genes related to rice resistance against brown planthopper, *Nilaparvata lugens*. Rice Sci 24: 163–172

Jiang F, Doudna JA (2017) CRISPR–Cas9 structures and mechanisms. Ann Rev Biophys 46:505–529

Jiao Y, Tausta SL, Gandotra N, Sun N, Liu T, Clay NK, Ceserani T, Chen M, Ma L, Holford M, Zhang HY, Zhao H, Deng XW, Nelson T (2009) A transcriptome atlas of rice cell types uncovers cellular, functional and developmental hierarchies. Nat Genet 41:258–263

Jinek M, Chylinski K, Fonfara I, Hauer M, Doudna JA, Charpentier E (2012) A programmable dual-RNA–guided DNA endonuclease in adaptive bacterial immunity. Science 337:816–821

Jung JH, Altpeter F (2016) TALEN mediated targeted mutagenesis of the caffeic acid O-methyltransferase in highly polyploid sugarcane improves cell wall composition for production of bioethanol. Plant Mol Biol 92:131–142

Jung HG, Deetz DA (1993) Cell wall lignifications and degradability. In: Jung HG et al (eds) Forage cell wall structure and digestibility. ASA-CSSA-SSSA, Madison, pp 315–346

Jung HG, Vogel KP (1992) Lignification of switch grass (*Panicum virgatum*) and big bluestem (*Andropogon gerardii*) plant parts during maturation and its effect on fiber degradability. J Sci Food Agr 59:169–176

Jung HG, Mertens DR, Payne AJ (1997) Correlation of acid detergent lignin and Klason lignin with digestibility of forage dry matter and neutral detergent fiber. J Dairy Sci 80:1622–1628

Kaneda M, Rensing KH, Wong JCT, Banno B, Mansfield SD, Samuels AL (2008) Tracking monolignols during wood development in lodgepole pine. Plant Physiol 147:1750–1760

Kay S, Hahn S, Marois E, Wieduwild R, Bonas U (2009) Detailed analysis of the DNA recognition motifs of the Xanthomonas type III effectors AvrBs3 and AvrBs3Δrep16. The Plant J 59:859–871

Kephart KD, Buxton DR (1993) Forage quality responses of C3 and C4 perennial grasses to shade. Crop Sci 33:831–837

Kim YG, Chandrasegaran S (1994) Chimeric restriction endonuclease. Proc Nat Acad Sci U S A 91: 883–887

Kim YG, Cha J, Chandrasegaran S (1996) Hybrid restriction enzymes: zinc finger fusions to Fok I cleavage domain. Proc Natl Acad Sci U S A 93:1156–1160

Lapierre C, Pollet B, Petit-Conil M, Toval G, Romero J, Pilate G, Leplé JC, Boerjan W, Ferret V, Nadai VD, Jouanin L (1999) Structural alterations of lignins in transgenic poplars with depressed cinnamyl alcohol dehydrogenase or caffeic acid Omethyltransferase activity have opposite impact on the efficiency of industrial Kraft pulping. Plant Physiol 119:153–163

Li T, Liu B, Spalding MH, Weeks DP, Yang B (2012) High efficiency TALEN-based gene editing produces disease resistant rice. Nat Biotechnol 30:390–392

Liang M, Davis E, Gardner D, Cai X, Wu Y (2006) Involvement of AtLAC15 in lignin synthesis in seeds and in root elongation of *Arabidopsis*. Planta 224(5):1185–1196

Liljegren SJ, Ditta GS, Eshed Y, Savidge B, Bowman JL, Yanofsky MF (2000) SHATTERPROOF MADS-box genes control seed dispersal in *Arabidopsis*. Nature 404:766–770

Liu CJ (2012) Deciphering the enigma of lignification: precursor transport, oxidation, and the topochemistry of lignin assembly. Mol Plant 5:304–317

Lu T, Lu G, Fan D, Zhu C, Li W, Zhao Q, Feng Q, Zhao Y, Guo Y, Li W, Huang X, Han B (2010) Function annotation of the rice transcriptome at single-nucleotide resolution by RNA-seq. Genome Res 20:1238–1249

Ma QH, Zhu HH, Han JQ (2017) Wheat ROP proteins modulate defense response through lignin metabolism. Plant Sci. 262:32–38

Magai MM, Sleper DA, Beuselinck PR (1994) Degradation of three warm-season grasses in a prepared cellulase solution. Agron J 86:1049–1053

Mak AN, Bradley P, Cernadas RA, Bogdanove AJ, Stoddard BL (2012) The crystal structure of TAL effector PthXo1 bound to its DNA target. Science 335:716–719

Makarova KS, Aravind L, Grishin NV, Rogozin IB, Koonin EV (2002) A DNA repair system specific for thermophilic Archaea and bacteria predicted by genomic context analysis. Nucleic Acids Res 30:482–496

Makarova KS, Grishin NV, Shabalina SA, Wolf YI, Koonin EV (2006) A putative RNA-interference-based immune system in prokaryotes: computational analysis of the predicted enzymatic machinery, functional analogies with eukaryotic RNAi, and hypothetical mechanisms of action. Biol Direct 1:7. https://doi.org/10.1186/1745-6150-1-7

Martiniello P, Paolette R, Berardo (1997) Effect of phenological stages on dry matter and quality components in lucerne. Eur J Agron 6:79–87

Mashimo T, Takizawa A, Voigt B, Yoshimi K, Hiai H, Kuramoto T, Serikawa T (2010) Generation of knockout rats with X-linked severe combined immunodeficiency (X-SCID) using zinc-finger nucleases. PLoS One 5:e8870

Mir Derikvand M, Sierra JB, Ruel K, Pollet B, Do CT, Thévenin J, Buffard D, Jouanin L, Lapierre C (2008) Redirection of the phenylpropanoid pathway to feruloyl malate in Arabidopsis mutants deficient for cinnamoyl-CoA reductase 1. Planta 227(5):943–956

Mojica FJM, Díez-Villaseñor C, Soria E, Juez G (2000) Biological significance of a family of regularly spaced repeats in the genomes of Archaea, Bacteria and mitochondria. Mol Microbiol 36:244–246

Monties B (1991) Plant cell walls as fibrous lignocellulosic composites: Relations with lignin structure and function. Anim Feed Sci Tech 32:159–175

Morbitzer R, Römer P, Boch J, Lahaye T (2010) Regulation of selected genome loci using de novo-engineered transcription activator-like effector (TALE)-type transcription factors. Proc Natl Acad Sci U S A 107:21617–21622

Morrison IM (1979) Carbohydrate chemistry and rumen digestion. Proc Nutr Soc 38:269–274
Morrison IM (1980) Changes in the lignin and hemicellulose content of 10 varieties of temperate grasses with increasing maturity. Grass Forage Sci 35:287–293
Moscou MJ, Bogdanove AJ (2009) A simple cipher governs DNA recognition by TAL effectors. Science 326:1501. https://doi.org/10.1126/science.1178817
Muguerza M, Gondo T, Ishigaki G, Akashi R (2014) Lignin content and digestibility in transgenic bahiagrass (*Paspalum notatum* Flügge) obtained by genetic manipulation of cinnamyl alcohol dehydrogenase gene. Asian J Plant Sci 13:8–17
Nelson CJ, Moser LE (1994) Plant factors affecting forage quality. In: Fahey GC Jr (ed) Forage quality, evaluation, and utilization. ASA, CSSA, SSSA, Madison, pp 115–154
Novaes E, Kirst M, Chiang V, Winter-Sederoff H, Sederoff R (2010) Lignin and biomass: a negative correlation for wood formation and lignin content in trees. Plant Physiol 154:555–561
O'Brien JA, Daudi A, Butt VS, Bolwell GP (2012) Reactive oxygen species and their role in plant defence and cell wall metabolism. Planta 236:765–779
Pabo CO, Peisach E, Grant RA (2001) Design and selection of novel Cys2His2 zinc finger proteins. Ann Rev Biochem 70:313–340
Palmer NA, Sattler SE, Saathoff AJ, Funnell D, Pedersen JF, Sarath G (2008) Genetic background impacts soluble and cell wall-bound aromatics in brown midrib mutants of sorghum. Planta 229: 115–127
Park SC, Kim YH, Jeong JC, Kim CY, Lee HS, Bang JW, Kwak SS (2011) Sweet potato late embryogenesis abundant 14 (IbLEA14) gene influences lignification and increases osmotic- and salt stress-tolerance of transgenic calli. Planta 233(3):621–634
Park SH, Chuansheng M, Markus P, Rebecca O, Bruce D, Robab S, Alameldin, Hussien, Thang N, Sticklen M (2012) Downregulation of maize cinnamoyl-coenzyme a reductase via RNA interference technology causes brown midrib and improves ammonia fiber expansion-pretreated conversion into fermentable sugars for biofuels. Crop Sci 52:2687–2701
Park JJ, Yoo CG, Flanagan A, Pu Y, Debnath S, Ge Y, Ragauskas A, Wang ZY (2017) Defined tetra-allelic gene disruption of the 4-coumarate:coenzyme A ligase 1 (Pv4CL1) gene by CRISPR/Cas9 in switchgrass results in lignin reduction and improved sugar release. Biotech Biofuels 10(1):284. https://doi.org/10.1186/s13068-017-0972-0
Perez-Rodriguez R, Haitjema C, Huang Q, Nam KH, Bernardis S, Ke A, DeLisa MP (2011) Envelope stress is a trigger of CRISPR RNA-mediated DNA silencing in *Escherichia coli*. Mol Microbiol 79:584–599
Pilate G, Guiney E, Holt K, Petit-Conil M, Lapierre C, Leple JC, Pollet B, Mila I, Webster EA, Marstorp HG, Hopkins DW, Jouanin L, Boerjan W, Schuch W, Cornu D, Halpin C (2002) Field and pulping performancesof transgenic trees with altered lignification. Nat Biotechnol 20:607–612
Piquemal J, Chamayou S, Nadaud I, Beckert M, Barriere Y et al (2002) Down-regulation of caffeic acid Omethyltransferase in maize revisited using a transgenic approach. Plant Physiol 130: 1675–1685
Prashant S, Srilakshmi Sunita M, Pramod S et al (2011) Down-regulation of *Leucaena leucocephala* cinnamoyl CoA reductase (*LlCCR*) gene induces significant changes in phenotype, soluble phenolic pools and lignin in transgenic tobacco. Plant Cell Rep 30:2215–2231
Pu Y, Chen F, Ziebell A, Davison B, Ragauskas AJ (2009) NMR characterization of C3H and HCT down-regulated alfalfa lignin for improved fermentable sugar yields. BioEnergy Res 2:198–208
Ralph J, Helm RF, Quideau S, Hatfield RD (1992) Lignin-feruloyl ester cross-links in grasses. Part 1. Incorporation of feruloyl esters into coniferyl alcohol dehydrogenation polymers. J Chem Soc Perkin Trans 1:2961–2969
Ralph J, Lundquist K, Brunow G, Lu F, Kim H, Schatz PF, Marita JM (2004) Lignins: Natural polymers from oxidative coupling of 4-hydroxyphenyl-propanoids. Phytochem Rev 3(1):29–60
Ralph J, Akiyama T, Kim H, Lu F, Schatz PF, Marita JM, Ralph SA, Reddy MSS, Chen F, Dixon RA (2006) Effects of coumarate 3-hydroxylase down-regulation on lignin structure. J Biol Chem 281:8843–8853

Rastogi S, Dwivedi UN (2008) Manipulation of lignin in plants with special reference to O-methyltransferase. Plant Sci 174:264–277

Reddy MSS, Chen F, Shadle G, Jackson L, Aljoe H, Dixon RA (2005) Targeted down- regulation of cytochrome P_{450} enzymes for forage quality improvement in alfalfa (*Medicago sativa* L.). Proc Nat Acad Sci U S A 102:16573–16578

Saathoff AJ, Sarath G, Chow EK, Dien BS, Tobias CM (2010) Downregulation of cinnamyl-alcohol dehydrogenase in switchgrass by RNA silencing results in enhanced glucose release after cellulase treatment. PLoS One 6:e16416. https://doi.org/10.1371/journal.pone.0016416

Saathoff AJ, Sarath G, Chow EK, Dien BS, Tobias CM (2011) Downregulation of cinnamyl-alcohol dehydrogenase in switch grass by RNA silencing results in enhanced glucose release after cellulase treatment. PLoS One 6:e16416

Santos-Sánchez NF, Salas-Coronado R, Hernández-Carlos B, Villanueva-Cañongo C (2019) Shikimic acid pathway in biosynthesis of phenolic compounds. In: Plant physiological aspects of phenolic compounds. Intech Open, London

Sattler SE, Funnell-Harris DL (2013) Modifying lignin to improve bioenergy feedstocks: strengthening the barrier against pathogens? Front Plant Sci 4:70. https://doi.org/10.3389/fpls.2013.00070

Schaart JG, van de Wiel CCM, Lotz LAP, Smulders MJM (2016) Opportunities for products of new plant breeding techniques. Trends Plant Sci 21:438–449

Schilmiller AL, Stout J, Weng JK, Humphreys J, Ruegger MO, Chapple C (2009) Mutations in the cinnamate 4-hydroxylase gene impact metabolism, growth and development in Arabidopsis. Plant J 60:771–782

Schoch G, Goepfert S, Morant M, Hehn A, Meyer D, Ullmann P, Werck-Reichhart D (2001) CYP98A3 from Arabidopsis thaliana is a 3′-hydroxylase of phenolic esters, a missing link in the phenylpropanoid pathway. J Biol Chem 276:36566–36574

Serapiglia MJ, Gouker FE, Hart JF, Unda F, Mansfield SD, Stipanovic AJ, Smart LB (2015) Ploidy level affects important biomass traits of novel shrub willow (Salix) hybrids. BioEnergy Res 8: 259–269

Shadle G, Chen F, Srinivasa Reddy MS, Jackson L, Nakashima J, Dixon RA (2007) Down-regulation of hydroxycinnamoyl CoA: Shikimate hydroxycinnamoyl transferase in transgenic alfalfa affects lignification, development and forage quality. Phytochem 68:1521–1529

Shafi A, Chauhan R, Gill T, Swarnkar MK, Sreenivasulu Y, Kumar S, Kumar N, Shankar R, Ahuja PS, Singh AK (2015) Expression of SOD and APX genes positively regulates secondary cell wall biosynthesis and promotes plant growth and yield in Arabidopsis under salt stress. Plant Mol Biol 87:615–631

Smith J, Grizot S, Arnould S, Duclert A, Epinat JC, Chames P, Prieto J, Redondo P, Blanco FJ, Bravo J, Montoya G, Pâques F, Duchateau P (2006) A combinatorial approach to create artificial homing endonucleases cleaving chosen sequences. Nucleic Acids Res 34:e149

Takeda Y, Koshiba T, Tobimatsu Y, Suzuki S, Murakami S, Yamamura M, Rahman MM, Takano T, Hattori T, Sakamoto M et al (2017) Regulation of coniferaldehyde 5-hydroxylase expression to modulate cell wall lignin structure in rice. Planta 246:337–349

Tamasloukht B, Lam MSJWQ, Martinez Y, Tozo K, Barbier O, Jourda C, Jauneau A, Borderies G, Balzergue S, Renou JP, Huguet S, Martinant JP, Tatout C, Lapierre C, Barriere Y, Goffner D, Pichon M (2011) Characterization of cinnamoyl-CoA reductase 1 (CCR1) mutant in maize: effect of lignification, fibre development, and global gene expression. J Exp Bot 62:3837–3848

Tanaka K, Murata K, Yamazaki M, Onosato K, Miyao A, Hirochika H (2003) Three distinct rice cellulose synthase catalytic subunit genes required for cellulose synthesis in the secondary wall. Plant Physiol 133(1):73–83

Terashima N, Fukushima K, He L-F, Takabe K (1993) Comprehensive model of the lignified plant cell wall. In: Jung HG, Buxton DR, Hatfield RD, Ralph J (eds) Forage cell wall structure and digestibility. ASA-CSSA-SSSA, Madison, pp 247–270

Theander O, Aman P (1984) Anatomical and chemical characteristics. In: Sundstol F, Owen E (eds) Straw and other fibrous by-products as feed. Elsevier, Amsterdam, pp 45–78

Tianpei X, Li D, Qiu P, Luo J, Zhu Y, Li S (2015) Scorpion peptide LqhIT2 activates phenylpropanoid pathways via jasmonate to increase rice resistance to rice leaf rollers. Plant Sci 230:1–11

Tripathi SC, Sayre KD, Kaul JN, Narang RS (2003) Growth and morphology of spring wheat (*Triticum aestivum* L.) culms and their association with lodging: Effects of genotypes, N levels and ethephon. Field Crop Res 84:271–290

Tronchet M, Balagué C, Kroj T, Jouanin L, Roby D (2010) Cinnamyl alcohol dehydrogenases-C and D, key enzymes in lignin biosynthesis, play an essential role in disease resistance in *Arabidopsis*. Mol Plant Pathol 11:83–92

Tsai SQ, Joung JK (2016) Defining and improving the genome-wide specificities of CRISPR–Cas9 nucleases. Nat Rev Genet 17:300–312

Tu Y, Rochfort S, Liu Z, Ran Y, Griffith M, Badenhorst P, Louie GV, Bowman ME, Smith KF, Noel JP, Mouradov A, Spangenberg G (2010) Functional analysis of caffeic acid *O*-methyltransferase and cinnamoyl-CoA reductase genes from perennial ryegrass (*Lolium perenne*). Plant Cell 22:3357–3337

Twidwell EK, Johnson KD, Patterson JA, Cherney JH, Bracker CE (1991) Degradation of switchgrass anatomical tissue by rumen microorganisms. Crop Sci 30:1321–1328

Tzin V, Galili G (2010) The biosynthetic pathways for shikimate and aromatic amino acids in *Arabidopsis thaliana*. Arabidopsis Book 8:e0132

Vailhe MAB, Migne C, Cornu A, Maillot MP, Grenet E, Besle JM, Atanassova R, Martz F, Legrand M (1996) Effect of modification of the O-methyltransferase activity on cell wall composition, ultrastructure and degradability of transgenic tobacco. J Sci Food Agric 72:385–392

Vailhe MAB, Besle JM, Maillot MP, Cornu A, Halpin C, Knight M (1998) Effect of down-regulation of cinnamyl alcohol dehydrogenase on cell wall composition and on degradability of tobacco stems. J Sci Food Agric 76:505–514

Valle CBD, Moore KJ, Miller DA (1988) Effect of ploidy level on cell wall composition and digestibility of *Brachiaria ruziziensis* germain et evrard. Trop. Agr. (Trinidad) 65:16–20

Van Soest PJ (1982) Nutritional ecology of the ruminant. O and B Books, Inc., Corvalis

Vanholme R (2012) Metabolic engineering of novel lignin in biomass crops. New Phytol 196:978–1000

Varshney RK, Graner A, Sorrells ME (2005) Genomics-assisted breeding for crop improvement. Trends Plant Sci 10:621–630

Vignols F, Rigau J, Torres MA, Capellades M, Puigdomenech P (1995) The brown midrib3 (bm3) mutation in maize occurs in the gene encoding caffeic acid O-methyltransferase. Plant Cell 7: 407–416

Vogt T (2010) Phenylpropanoid biosynthesis. Mol Plant 3:2–20

Wei H, Dhanaraj AL, Arora R, Rowland LJ, Fu Y, Sun L (2006) Identification of cold acclimation-responsive Rhododendron genes for lipid metabolism, membrane transport and lignin biosynthesis: importance of moderately abundant ESTs in genomic studies. Plant Cell Environ 29(4): 558–570

Weng JK, Li X, Bonawitz ND, Chapple C (2008) Emerging strategies of lignin engineering and degradation for cellulosic biofuel production. Curr Opin Biotechnol 19(2):166–172

Wilson JR, Deinum B, Engels FM (1991) Temperature effects on anatomy and digestibility of leaf and stem of tropical and temperate forage species. Netherlands J Agr Sci 39:3–48

Xu B, Escamilla-Treviño LL, Sathitsuksanoh N, Shen Z, Shen H, Zhang YHP, Dixon RA, Zhao B (2011) Silencing of 4-coumarate:coenzyme A ligase in switch grass leads to reduced lignin content and improved fermentable sugar yields for biofuel production. New Phytol 192:611–625

Yamaguchi M, Valliyodan B, Zhang J, Lenoble ME, Yu O, Rogers EE, Nguyen HT, Sharp RE (2010) Regulation of growth response to water stress in the soybean primary root. I. Proteomic analysis reveals region-specific regulation of phenylpropanoid metabolism and control of free iron in the elongation zone. Plant Cell Environ. 33(2):223–243

Yamamoto E, Bokelman GH, Lewis NG (1989) Phenylpropanoid metabolism in cell walls. In: Lewis NG, Paice MG (eds) Plant cell wall polymers: biogenesis and biodegradation. ACS Symposium Series, 399. American Chemical Society, Washington, pp 68–88

Yang Q, He Y, Kabahuma M, Chaya T, Kelly A, Borrego E, Bian Y, El KF, Yang L, Teixeira P (2017) A gene encoding maize caffeoyl-CoA O-methyltransferase confers quantitative resistance to multiple pathogens. Nat Genet 49:1364–1372

Yazaki K (2006) ABC transporters involved in the transport of plant secondary metabolites. FEBS Lett 80(4):1183–1191

Zeikus JG (1980) Fate of lignin and related aromatic substrates in anaerobic environments. In: Kirk TK et al (eds) Lignin biodegradation: microbiology, chemistry, and potential applications. CRC Press, Boca Raton, pp 101–109

Zhang B, Zhou Y (2011) Rice brittleness mutants: a way to open the 'black box' of monocot cell wall biosynthesis. J Integr Plant Biol 53(2):136–142

Zhang S, Yang Q, Ma R (2007) *Erwinia carotovora* ssp. carotovora infection induced "defense lignin" accumulation and lignin biosynthetic gene expression in chinese cabbage (*Brassica rapa* L. ssp. pekinensis). J Integr Plant Biol 49:993–1002

Zhao X, Wang Q, Jiao Y, Huang R, Deng Y, Wang H, Du X (2012) Identification of genes potentially related to biomineralization and immunity by transcriptome analysis of pearl sac in pearl oyster *Pinctada martensii*. Mar Biotechnol 14:730–739

Zhao Q, Tobimatsu Y, Zhou R, Pattathil S, Gallego-Giraldo L, Fu C, Jackson LA, Hahn MG, Kim H, Chen F, Ralph J, Dixon RA (2013) Loss of function of cinnamyl alcohol dehydrogenase 1 leads to unconventional lignin and a temperature-sensitive growth defect in *Medicago truncatula*. Proc Natl Acad Sci U S A 110(33):13660–13665

Zheng M, Chen J, Shi Y, Li Y, Yin Y, Yang D, Luo Y, Pang D, Xu X, Li W, Ni J, Wang Y, Wang Z, Li Y (2017) Manipulation of lignin metabolism by plant densities and its relationship with lodging resistance in wheat. Sci Rep 7:41805. https://doi.org/10.1038/srep41805

Zhong R, Morrison WH III, Negrel J, Ye ZH (1998) Dual methylation pathways in lignin biosynthesis. Plant Cell 10:2033–2046

Postharvest Processing and Preservation of Forages 17

Maintaining a continuous supply of quality feed to livestock throughout the year is essential for profitable livestock farming. Preservation of forage provides a mean of storing surplus fodder that ensures feed availability during the lean period. The preparation of hay, haylage or baleage and silage are the most common methods of preservation of forages. However, care should be taken in the selection and suitability of the technique for given forage, the storage capability of processed feed products, weather conditions and the availability of material. The selected processing and preservation technique should not affect the nutrient availability from feed and minimize production costs.

17.1 Postharvest Preservation of Forages

Forage crops should be harvested at a stage when production and nutritional quality parameters are favourable (Adjolohoun et al. 2008; Olawoye and Kubkomawa 2018a). The major goal of preservation is to retain the original nutritional quality of feed during storage (Olawoye and Kubkomawa 2018b). Hay and silage preparation are the common methods for forage preservation. Haymaking and silage preparation are usually practised when green fodder is in surplus quantity (Akinlade et al. 2002; Adamu and Kubkomawa 2018).

Harvesting of the crop at the proper stage is the prerequisite for the preparation of silage of good nutritional quality (Rotz et al. 2003). Crops with a good percentage of sugar and appropriate moisture (35–40% dry matter and 65–60% moisture contents) are preferred for silage making. Crops at the pre-flowering to flowering stage should be harvested. They should not contain >75% moisture content during silage preparation. The time of harvesting is also a crucial factor for silage preparation.

R. Katoch, *Nutritional Quality Management of Forages in the Himalayan Region*, https://doi.org/10.1007/978-981-16-5437-4_17

Crops like maize, sorghum, bajra, hybrid Napier and oat are most suitable for silage making. However, berseem, lucerne and cowpea are not appropriate for ensiling, unless sprayed with molasses while filling into silo pit. Silage quality can be increased with the use of additives. The pasture should be cut when they contain the highest nutritional quality. For silage preparation, the grasses is should be allowed to wilt in the field for a few hours to reduce the moisture content to around 60–75%.

17.1.1 Haymaking

Hay is the final product of cutting and curing forage grasses or legumes. The moisture content of hay should not be >12–14% so that it can be stored safely. Hay is more nutritious and palatable than crop residues. Crops suitable for haymaking should be thin stemmed for easy drying. The thick-stemmed fodder can be dried quickly if chopped into small pieces. There are two main factors, which determine haymaking *viz.*, expected weather of season and maturity stage of the crop. Usually, the season will be predetermined by forage availability and haymaking conditions. Crops for haymaking should be harvested at the flower initiation stage. Nutritional quality of forage depends on the maturity stage. After flowering, there is a reduction in fodder quality. Total volume obtained from hay is less than crop residues but has nutritional quality better than the crop residues.

For haymaking, oats are harvested when grain kernels are in soft dough maturity stage. At this stage, heads are filled with grain or reached the near maturity stage. Hay cut at soft dough stage is usually preferred. Early harvest can be important with both dryland and irrigated oat hay plantings. Late maturing oats may fail to produce grain in exceptionally dry springs and early harvest may be of benefit when double cropping is planned on irrigated lands. Top-quality fodder can be produced when alfalfa is not cut until early bloom (25–50% flowering). Early cut hay often contains higher amount of nitrates and is low in high-quality protein. Mustard crop should be cut from the early stage just after flowering to the stage when lower leaves starting drying. For millets, the optimum harvest time is when seeds are in the milk to soft dough stage.

Types of hay: Hay has been classified based on the level of technology used in the preparation and crop from which hay is made. The main types of hay based on basic technology are:

(a) **Baled hay:** Initially hay is prepared by hand and then by stationary machines. Round bales are the simplest to make and most popular. Their shape sheds rain and resists water better than traditional bales.
(b) **Barn-dried hay:** Hay is prepared by air drying the raw material with the help of a fan. This type of hay is not widely prepared.
(c) **Chopped hay**: This type of hay is prepared where drying conditions are good and the system is highly mechanized. It is less bulky and better for mechanical handling.

(d) **Dried grass:** In this method, grass/crop residues are artificially dried at high temperatures. This type of hay making is not economic.
(e) **Hand-trussed hay:** It is widespread in manual haymaking to reduce shattering.
(f) **Long hay:** It is the traditional type of hay where crop is tied in small bundles, turned and stored in a stack form. This is the most common type of haymaking.

Depending on the crop from which hay is made, they may be classified as follows:

(a) **Legume hay**: They are made from leguminous plants like lucerne and other clovers. They are generally rich in proteins, vitamins and minerals as compared to nonleguminous ones.
(b) **Nonlegume hay**: They contain less proteins, minerals and vitamins. These types of hay are prepared from nonleguminous forages. Hay prepared from forage grasses is usually rich in carbohydrates.
(c) **Mixed hay**: These are made from a mixed crop of leguminous and nonleguminous plants. Such a crop is generally cut at an early facet of growth. The nutritive value of such hay depends on the proportion of leguminous and nonleguminous plants.

Suitable crops for haymaking: Crops with thin, soft stems are more suitable and convenient for haymaking. Oat is the best-suited crop for haymaking because of its thin and hollow stem and low leaf-to-stem ratio. Other fodder crops like berseem, lucerne and cowpea are suitable for haymaking. Hay from leguminous crops is more nutritive and usually used to supplement straws and stovers and other fibrous feeds fed to animals. Annual and perennial crops (mainly grasses) also form good hay provided they are harvested at the optimum stage of their growth. In Indian conditions, most of the hay is prepared from the crops like sorghum, maize and millets normally harvested after procuring grains.

17.1.1.1 Process of Haymaking

1. Before flowering forage crop preferably with less number of stems and a high number of leaves is cut from the fields. Since leaves become more brittle after drying and tend to shatter, therefore care should be taken.
2. Avoid drying of forage in moist and humid conditions as these conditions increase molds growth, which could be toxic to livestock. Alternatively, drying can also be done by passing hot air through forage.
3. Wilt the cut forage for 2–3 days with intermittent raking and turning. After wilting forage to about 40% moisture, it is rolled into small loose fluffy bundles "Windrows." Check moisture content in cut forage to prevent overdrying of cut forage.

4. Store dried hay in the form of bales undercover or create haystacks. The haystacks may be covered by plastic sheets to keep out rain and prevent overexposure to the sun. The most common method of storing dry fodder is heaps or stacks in open covered with plastic sheets or small branches of trees to prevent wind and rain spoilage.
5. A good quality hay should have
   (a) leafy and greenish in color
   (b) no foreign material
   (c) no foul smell
   (d) 20–23% moisture content

During haymaking some nutrients are lost in the curing process. The major losses in haymaking are

- Nutrient losses due to late cutting
- Shattering and dropping of leaves
- Dry matter losses due to fermentation
- Nutritional losses due to excessive exposure of cut forage to sun during drying and curing process
- Nutritional losses due to leaching
- High moisture content at time of storing may lead to fermentation and molding of hay.

Good hay should have green color and leaves and branches stay connected.

Factors, which affect the quality of hay, include the type of forage, harvest stage, leaf to stem ratio, nutritional composition, physical form and deterioration during storage. Some forages like alfalfa hay are harvested during the rainy time on the high plains and get damaged easily by weather. Alfalfa hay may lose leaves during curing and the product may become black and unpalatable. Excessive moisture leads to mold development. It may turn chocolate brown in the middle of big bales from stack burning and will lose energy. The green-colored alfalfa hay is not always containing high protein and high energy. If the brown hay is cured and harvested properly, it may be superior to green hay.

Millet hay sometimes may cause scouring. Feeding this type of hay to horses should be avoided. Hairy and waxy varieties of millet could be difficult to cure as hay (Fig. 17.1).

Mustard can be used as a hay crop. Mustard hay typically contains 50–55% TDN and 10–12% CP content. Mustard does not dry down rapidly and could be difficult to bale at proper moisture levels. In addition, trace minerals should be provided in sufficient quantity because long-term feeding of mustard hay has been associated with reduced absorption of trace minerals.

Mustard could also accumulate nitrate, thereby nitrate poisoning may be a potential problem. Wheat, oats, barley, triticale and rye hay can be used in beef, sheep and dairy rations. Harvesting should be performed between the heading and

Fig. 17.1 Grass hay

soft dough stages. Rye hay loses palatability and protein content rapidly after flowering. The nutritive value of these hays should be similar to brome hay when cut at the heading to soft dough stage. Energy per acre could be managed by delaying harvest until the soft dough stage. However, good proportion of protein is retained by harvesting at late boot to early heading stage. All these feeds should be checked for nitrate content if drought-stressed and/or fertilized heavily with nitrogen.

Soybeans can also be used as hay. CP content in soybean hay is generally similar to alfalfa or clover with moderate TDN. The stems of soybean are generally not preferred by cattle due to their coarse nature. Chopping or tub grinding and blending with other forages or feedstuffs could be used for improving consumption by the livestock. When stored in round bales, water penetration and spoilage are more of a problem because of the coarse stems. Hay quality does not change drastically with increasing maturity of the plant. Specific varieties of soybeans are available for use as forages to the livestock.

17.1.2 Silage Making

With the constant increase in the prices of feed concentrates and other feed ingredients green fodder is considered an economic nutrient source for livestock. While an increase in green fodder production per hectare of land has been emphasized, it is equally important to conserve green fodder to ensure a regular supply for feeding animals, especially during lean period. Conserving green fodder in the form of silage is one of the best options available to ensure regular supply of quality fodder through different periods of the year. Fodder crops like maize, sorghum, oats and pearl millet rich in water-soluble carbohydrates are suitable for ensiling forage crops. Nutritional quality of silage can be improved using additives like molasses, urea, salt, formic acid, etc. Silage is the conserved green fodder having moisture content between 65 and 70%. Fodder crops rich in soluble carbohydrates are incubated after chaffing for 45–50 days under anaerobic conditions. Sugars present in fodder are converted to lactic acid, which acts as a preservative and a good source of fermentable sugars. Under proper storage conditions, silage can be stored for up to 2 years. Good quality silage should not have any butyric acid, which often impart off flavor to silage.

Advantages of Silage Preparation

(a) It is the only means by which the entire forage plant can be preserved in a succulent form.
(b) Retains higher proportions of nutrients than hay.
(c) Silage preserves 85% of feed energy.
(d) It could be prepared from thick stem crops which are not suitable for haymaking.
(e) Could be prepared in almost any kind of weather.
(f) Silage require less storage space than hay.
(g) Less fire hazard with silage compared to hay and other dry crops.
(h) Silage is more palatable, digestible and slightly laxative.
(i) Less wastage in feeding.
(j) Nutritive value of ensiled forage could be increased by adding suitable additives.

17.1.2.1 Phases of Silage Fermentation

Fermentation is the critical process in silage making, which is completed in 3–6 weeks. The whole process of fermentation has four phases.

(a) **Aerobic phase (1-day process)**: During this phase, microbes metabolize plant carbohydrates in the presence of oxygen and generate heat. Silage temperature is increased. It is important that if anaerobic conditions are not achieved quickly, high temperature and prolonged heating occur due to unwanted aerobic bacterial growth, yeast, and molds. Therefore, it is critical to ensure good compaction, proper moisture and good sealing conditions during the preparation of silage.
(b) **Lag phase**: After the oxygen is used up in the aerobic phase, plant cells are broken down and used as a food source by bacteria. Plant enzymes break down complex carbohydrates into simpler sugars. Enzymes also break down plant proteins at this time, making the protein more soluble. The bacteria will use cell content during the lag phase to grow during the fermentation phase. Silage pH drops from 6.50 to 6.70 at ensiling to 5.50–5.70.
(c) **Fermentation (1-week to 1-month process)**: After the lag phase, once the anaerobic conditions are maintained, lactic acid bacteria and other anaerobic microbes start ferment sugars into lactic acid and other organic acids that drop the pH of silage. The decrease in pH prevents the breakdown of plant proteins in silage.
(d) **Stable phase***:* By maintaining anaerobic conditions, silage can be stable for months and up to years. However, silage should be used in a year. Slow entry of air can slowly deteriorate the material, thus silos should be checked regularly and maintained to avoid any potential break of seal integrity.

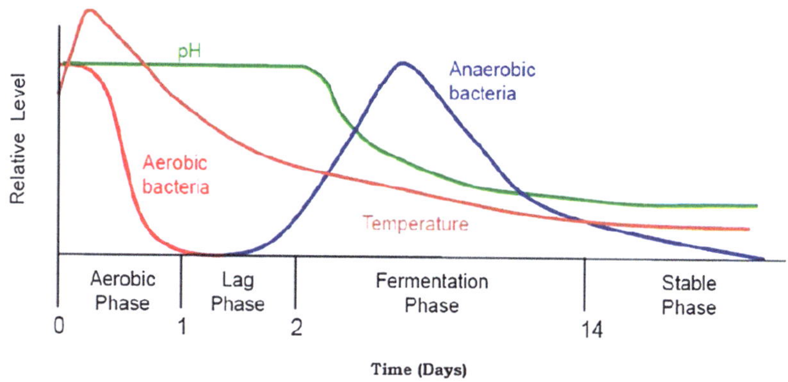

Fig. 17.2 Phases of silage fermentation

(e) **Feed out:** With the opening of silo or bale, silage should be used immediately to avoid aerobic deterioration. When oxygen becomes available in ensiled material, organic acids produced after fermentation, increase pH and restarts aerobic activity, causing silage spoilage. Silo opening should occur only after the fermentation phase has been completed (Fig. 17.2).

17.1.2.2 Forage Resources for Silage Preparation

Clovers are sometimes used for silage, usually in mixtures with grasses or other type of material fit for ensiling. Peas and vetches could be used for making silage with high protein content. These crop resources are also often mixed with spring barley or oats as arable silage mixtures. Lucerne silage contains high amount of protein (20–22%). For ensiling, lucerne and clovers should be wilted to at least 30% DM.

In the rainy season, there is abundance of green fodder which could be harvested at pre-flowering or flowering stage. If conditions are too humid for hay-making, green fodder could be preserved in the form of silage. Stover from sorghum, bajra and maize is commonly used for silage preparation. Tall varieties of paddy are susceptible to lodging. To avoid lodging, about one-third of the top portion is removed during the vegetative stage and used for ensiling. Paddy straw with 50–65% moisture at the time of harvesting is suitable for ensiling.

Fruit and vegetable wastes are a rich source of nutrients and these can be fed either after drying or ensiling with straws, without affecting palatability, nutrient utilization or livestock performance. The utilization of fruit and vegetable wastes reduces the cost of animal feeding and helping in waste management along with reduction of environmental pollution.

Water hyacinth leaves, an aquatic weed are rich in protein but plant contains 90–94% moisture. To prepare silage from water hyacinth, it is necessary to reduce moisture to about 70%.

17.1.2.3 Factors Affecting Silage Preparation

(a) **Type of crop and substrate**: During silage preparation, microorganisms need supply of carbohydrates. Corn is an ideal crop for silage preparation. The carbohydrates present in the crop material are utilized by the microbes and produce acids that results in reduction in pH. For good fermentation, it is essential that a large proportion of sugars should be converted to acid.

(b) **Moisture**: It is one of the most critical factors affecting the process of fermentation. Silage microbes require moisture to thrive and multiply, therefore curing forages before silage preparation is essential for good silage preparation. Moreover, amount of moisture determines the category of microbes, which would thrive best in the ensiling material. During silage preparation, estimation of moisture content is calculated in terms of DM content of the crop. After cutting of the material, wilting and curing has become a standard treatment during silage preparation. Optimum dry matter content in forages for silage preparation is between 25 and 50%, depending on the type of crop and storage system.

(c) **Weather**: Dry, sunny weather is generally preferred for silage making. Grass sugar levels are also high in sunny weather. Grass mown in the afternoon has maximum WSC content.

(d) **Silage bacteria**: Microorganisms particularly bacteria play an important role during silage preparation. Among them, lactic acid bacteria (LAB) are of great significance. After the crop is chopped and placed in a silo, the plant cells and bacteria respire and conditions in the silo become increasingly anaerobic. Under these anaerobic conditions, the lactic acid bacteria begin to proliferate in ensiling material.

(e) **Harvesting**: Wilted grass is usually chopped between 20 and 50 mm, whereas maize benefits from even fine chopping between 10 and 20 mm. Finely chopped material consolidates better and acquire anaerobic conditions in the clamp more quickly. Fine chopping also results in rapid release of sugars from grass for fermentation.

(f) **Silage additives: The addition of** compounds can accelerate the ensiling process and increase shelf life. There are two main categories of silage additives including fermentation inhibitors and stimulants. Silage inhibitors include aerobic and anaerobic inhibitors, which affect ensiling process. Aerobic inhibitors suppress microbial growth and prevent silage spoilage. Aerobic inhibitors include propionic acid and anhydrous ammonia. Anaerobic inhibitors tend to restrict undesirable bacteria and plant enzymes.

Fermentation stimulants include bacterial inoculants and substrate suppliers. The bacteria needed for fermentation are normally present in forage tissues. However, under certain harvest conditions, there may not be enough lactic acid-producing bacteria in chopped forage. The addition of bacterial inoculants speeds up lactic acid production and lowers pH. This low pH suppresses growth of undesirable organisms. Enzyme additives break complex carbohydrates into simple sugars or substrates. Various combination products containing both enzymes and lactic acid bacteria have been developed.

Steps in Silage Making

1. Construct silo/trench structure.
2. Harvest and wilt the crop to attain 60–75% moisture and 30–35% dry matter (DM).
3. Chop the fodder into small pieces of 2–3 cm size and fill the chopped fodder into the silo/trench.
4. Ensure even distribution of chopped fodder in the pit/trench. Press chopped material either by feet or with a tractor or bullocks as per the size of pit to exclude air as much as possible.
5. Pack the fodder up to1 meter at the top of silo or above ground level of pit.
6. Use additives during filling of fodder in the silo.
7. After filling and pressing, cover the fodder with 14–15 cm thick layer of wheat straw/ paddy straw from all sides.
8. Put mud on it and finally plaster it to seal the material preventing the entry of air and water or alternatively seal the silo with thick polyethylene sheet.
9. Open the silo for feeding, minimum after 45 days, as per need.
10. For feeding, initially add small amount of silage to diet. After few days start feeding with large quantity of silage (Fig. 17.3).

Characteristics of Good Quality Silage

(a) Light green-yellow or green-brown in color.
(b) Lactic acid odor without butyric acid and ammonical odor.
(c) Firm texture with softer material.
(d) Moisture in the range of 65–70%.
(e) Lactic acid: 3–14%.
(f) Butyric acid: <0.20%.
(g) pH: 4.0–4.2.

Factors Affecting Production of Good Quality Silage

(a) **Type of silo**: Surface silo is best due to ease of ensiling.
(b) **Dry matter of fodder**: Ideal 30–35%.
(c) **Chop length of fodder:** Ideal 2–3 cm, easy to compact.
(d) **Pressing/compaction of fodder**: As quick as possible to minimize aerobic fermentation.
(e) **Sealing of silo**: To check inflow of air and water into silo.

17.1.3 Fodder Chaffing

Cutting stems or leaves of the fodder into pieces by manual or mechanical means is known as chafing. The leafy material and tender part of fodder shoot are consumed by the animal while, the thick or hard stem is left behind which goes waste. By

Fig. 17.3 Overview of steps in silage making. (1. Green crop ready for ensiling; 2. Chopped green fodder; 3. Manual filling and pressing of fodder; 4. Silo sealing; 5. Silage)

chaffing, the whole plant or stems are fully consumed by the animal. Moreover, chaffed material can be packed in bags and require less space for storage. Feed additives or other feed supplements can be easily mixed with chaffed fodder. The leafy portion, shoots and stems pieces are mixed while chaffing, resulting more palatability of fodder.

17.1.4 Bale Silage

Grass is wilted up to 40% dry matter prior to baling in high-density round or square bales. It is easier to stack square bales, which are available in many lengths and widths, but round bales are easier to feed in self-feed ruminant systems. As soon as the grass is baled it is wrapped with polythene. The use of efficient bale wrapping machines, using polythene stretch film has made this system of covering almost universal. Four layers of wrap is the minimum, six being preferred to increase protection of the material and to ensure anaerobic conditions in the bale.

17.1.5 Baleage or Haylage

The level of initial dry matter and moisture content at the time of harvesting forages is the main difference between silage and haylage. Silage with low-moisture content is referred to as haylage. When baled and wrapped properly, it is referred to as baleage. Forage preservation as haylage requires high-density packing to maintain anaerobic conditions because microbes are less active and fermentation is low in haylage as compared to high moisture silage. Ensiling bales at 45–55% moisture provide adequate moisture for fermentation resulting in final pH between 4.70 and 5.80. Forage baled at lower moisture levels will have reduced fermentation resulting in high pH values and a greater chance of yeast and mold growth. Bales should be wrapped as soon as possible, ideally within 5–12 h of baling as spoilage increases dramatically if the time lag is longer. Bales should remain sealed for at least 20 days for ensiling process to be completed before feeding.

17.2 Packaging and Storage of Preserved Forages

17.2.1 Baling of Hay/Straw

Baling is a procedure in which harvested grasses are baled into a round or rectangular structure with compact packing. This helps in the easy storage and transportation of fodder. Optimum moisture for baling could range from 15 to 20% depending on the size of the bale. Baling at moisture levels lower than 15% will result in higher harvesting losses. On the other hand, storing straw at moisture levels greater than 20% could result in molding and heating, which translates into discoloration and greater dry matter and nutrient losses. Baling can be applied to improve the characteristics of straw for transportation and storage. The bulk density of rice straw is around 75 kg for loose straw and 100–180 kg in packed and baled form as energy feedstock. Transport of bailed straw could be up to 50% cheaper than the transportation of loose material. Bales of hay are made either by baler or hand baler.

Hand hay baler is a wooden hand-operated hay baler that makes a compressed bale of dry grass clippings/hay approximately 12″ × 15″ × 28″. After harvesting, hay is filled in the wooden construct tightly and is compacted and tied to get a bale. Typical DM losses during hay baling vary between 2 and 6% of the yield with the loss equally divided between pickup and chamber losses.

Balers are classified as stationary balers, movable and field balers. They are further classified as square balers, rectangular balers and round balers according to the bale shape produced. According to the density of bale, they could be high (200–350 kg/m^3), medium (100–200 kg/m^3), or low density (<100 kg/m^3) balers. The bale chamber contains a plunger that drives in and out, each time packing and compressing straw into the desired shape. The plunger also cuts the ends of the straw to make the bale of uniform size. The chamber feeds into a spring tension section that keeps the bale tightly compressed until enough straw has been processed to complete the bale. Similar to square balers, rectangular balers are also classified as small rectangular balers and large rectangular balers based on the dimension. Small rectangular bales of a conventional type made by a variety of machines have dimensions close to 0.40 × 0.60 × 1.20 m (0.30 m^3) and weighing approximately 32 kg (dry matter) each (110 kg/m^3). Large rectangular bales have nominal dimensions of 1.2 × 1.2 × 2.4 m (3.5 m^3) weighing approximately 600 kg (175 kg/m^3) and nominal dimensions of 0.9 × 1.2 × 2.4 m (2.6 m^3) weighing 450 kg (170 kg/m^3). Based on the form of working unit, round balers could be classified into long-belt type, short-belt type, chain type and roller type. They are also classified as inside winding type and outside winding type by their working principle. Long-belt and chain types are inside winding; short-belt and roller types are outside winding. According to the chamber size adjustment, balers are classified as expandable chamber round pick up baler, ground roll baler and fixed volume round pick up baler (Fig. 17.4).

Fig. 17.4 Bale of hay

17.2.2 Pelleting

Pelleted feeds have been defined as "agglomerated feeds formed by extruding individual ingredients or mixtures by compacting and forcing through openings by any mechanical process." Basically, the purpose of pelleting is to convert finely grinded, unpalatable and difficult-to-handle feed material to useful pallets by using heat, moisture and pressure. These larger particles are easier to handle, more palatable and usually result in improved feeding results when compared with unpelleted feed.

Livestock make better gains on pelleted feed than a meal ration. The most logical reasons are that (a) the heat generated in conditioning and pelleting makes the feedstuffs more digestible by breaking down the starches (b) the pellet simply puts the feed in a concentrated form, and (c) pelleting minimize waste during feeding. Pelleting prevents the segregation of ingredients in a mixing, handling or feeding process. Bulk density is increased, which enhances the storage capabilities of most bulk facilities. Shipping facilities are also increased, thereby reducing transportation costs (Fig. 17.5).

Types of Feed Pellets

(a) **Pure forage/grass pellet**: This type of feed pellet is made only from grass, hay, grain straw, stalk, haulms etc. They are the one of the easily made pellets for livestock feeding.

(b) **Concentrated feed pellets**: Concentrate feed pellet supplement the nutritive value, which are generally fed on coarse feed, green fodder and silage.

(c) **Complete feed pellet**: They are composed of protein component, coarse feed and different additives. By feeding livestock animals with complete feed pellets the nutritive demand of animals is fulfilled.

(d) **Premix feed pellet**: It is referred to the feed pellets that are mainly composed of feed additives. There are mainly three kinds of feed additives *viz.*, nutritive additives, drug additives and general additives.

Fig. 17.5 Feed pellets

17.2.2.1 Preparation of Feed Pellets

(a) **Procurement of raw feed material and storage**: Crop straw and grains from crop like wheat, maize and oil cake/meal, additives etc. constitute the raw material for feed pallets.

(b) **Cleaning of raw feed material**: Feed impurities are not only affect feed product quality but are also related to the safety of feed processing equipment.

(c) **Crushing of raw feed material**: It can be divided into one stage crushing process, circulation crushing or two stage-crushing processes.

(d) **Batching/proportioning of feed material**: Manual batching is mostly utilized in small and medium scale feed pellet factories. In this batching technology, all feed materials are weighed manually by workers and poured into mixing machine.

(e) **Feed mixing process:** Batch mixing and continuous mixing are two important methods for feed mixing.

(f) **Feed pelletizing and cooling**: During the feed pelletizing process, the mixed feed material is uniformly distributed and pressed into the machine. The pressed and densified material go through holes and form cylindrical granules. The finally desirable size feed pellets are cut by cutting blades and discharged from the feed pellet machine outlet. During feed pelletizing, feed is put under high temperature and humidity steam and squeezing process generates great heat, so that moisture content in feed pellets can reach to 16–18% and temperature to 75–85 °C.

(g) **Feed pellets crumbling, sieving and packaging**: For making small size feed pellets, crumbling is necessary. Feed material is often mould into certain size pellets. In small feed pellet plants, this process is realized by feed pellet cooler. However, for large capacity, a pellet-sieving machine is also needed, so that uniform feed pellets could be delivered for packaging (Fig. 17.6).

17.2.3 Densified Complete Feed Block (DCFB)

Densified complete feed block (DCFB) is composed of forage resource, concentrate and other supplementary nutrients in particular proportion that fulfil nutrient requirements of animals. Feeding of livestock with complete feed blocks stabilizes rumen fermentation, minimizes fermentation loss and ensures better ammonia utilization. DCFB could be easily transported to forage deficit regions or to the areas of natural calamities (Mattoo and Ganai 2004). Nutritive value of poor quality and bulky roughages can be improved by Densified Complete Feed Block (Salem and Nefzaoui 2003).

DCFB contains three main components namely, roughages, concentrates, and additives. Roughages part constitutes crop residues of rice, wheat, sugarcane, millets and legumes. In the Eastern Himalayan region especially, Nagaland, there is a practice of using dried forest grasses and tree foliage in feed blocks. The feed blocks are fed to Mithun in these regions (Das et al. 2009).

Fig. 17.6 Process for the preparation of feed pellets

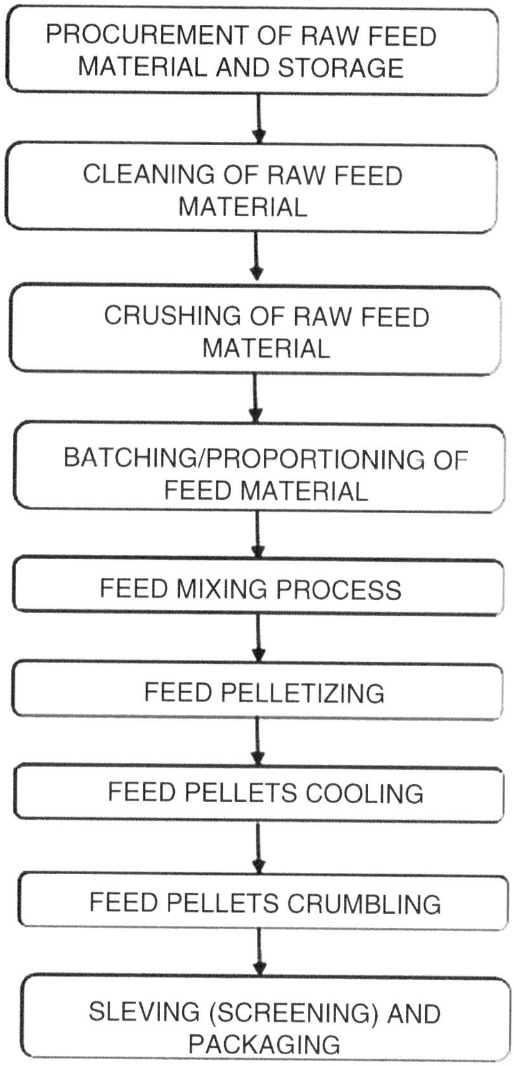

Concentrates form the second component of the densified complete feed block. Feed blocks are generally made with 86 parts of straw, 10 parts of molasses, 2 parts of the mineral mixture, 1 part of urea and 1 part of salt, which could meet maintenance requirements of animals. (Fig. 17.7).

Micronutrients and feed additives are the third component of feed blocks. They include the strategic and catalytic supplements and other feed additives like vitamins, minerals, bentonite (binder), probiotics, enzymes, antioxidants, immune-protective agents, antitoxins and herbal extracts (Table 17.1 and Fig. 17.8).

Fig. 17.7 Densified complete feed block

Table 17.1 Composition of densified complete feed blocks (DCFB) (Karangiya et al. 2016)

Ingredients	**Per 100 kg**
Mustard cake	25 kg
Barley grain	10 kg
Leucaena leaf meal	20 kg
Wheat bran	32 kg
Urea	01 kg
Molasses	10 kg
Mineral mixture	01 kg
Common salt	01 kg
Ingredients during feed shortage	**Per 100 kg**
Cereals	25 kg
Wheat/rice bran	37 kg
Oil cakes	35 kg
Mineral mixture	02 kg
Salt	01 kg

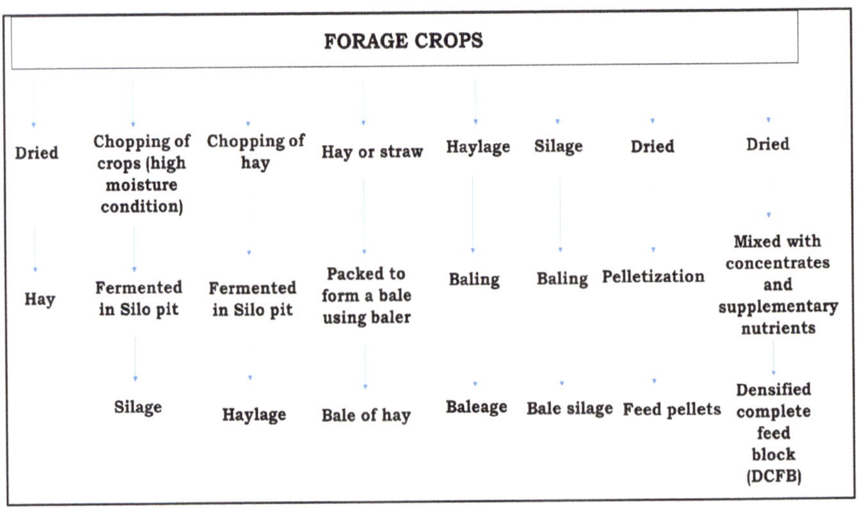

Fig. 17.8 Overview of postharvest processing and conservation of forage

17.3 Transporting and Storage Practices

Baling reduces the space required for fodder storage. Careful handling could avoid damage to bales during transportation (Fig. 17.9).

Paddy straw is generally stored on uplands in the form of stacks. The upper part of the stack is given a conical shape, which prevents the water from entering into the stack. The straw could kept well for many years. However, improper shaping of the above portion of the stack may allow water to leak into the stack. In Southern India, finger millet straw is also stored like paddy straw (Fig. 17.10).

Long, dried stover of maize, sorghum and pearl millet (Bajra) are tied in bundles, weighing about 10–15 kg each. They are stacked on a platform or elevation in such a way that their slopes are toward outside. These structures are covered with paddy straw, sugarcane leaves or coarse and long dried grasses in such a way that rainwater cannot enter into the stack. Large quantities of such stovers are stored in drought-prone areas for feeding during feed scarcity.

Fig. 17.9 Temporary storage place for hay and straw in hilly terrain

Fig. 17.10 Storing of straw on the ground (**a**) and above ground (**b**)

The practices of harvesting and conserving fodder in the Himalayan region are typical. Wheat or rice straws are stacked on the tree branches. This is the traditional practice for storing straws in Himachal Pradesh while hays are also stacked in the same manner. These conservation techniques either modern or traditional provide fodder during a scarcity period.

For postharvest preservation of forages, farmers must be aware of quality and yield losses during storage and preservation. Silage preparation, baling, pelleting and chopping of fodder are practised for forage conservation. Densified complete feed blocks provide most of the nutrients required by the animals for meeting energy requirements.

References

Adamu SM, Kubkomawa HI (2018) Beef production and marketing in tropical Africa: entrepreneurship in animal agriculture, 1st edn. Tapass Institute of Scientific Research and Development, Owerri, pp 131–145

Adjolohoun S, Bindelle J, Adandédjan C, Buldgen A (2008) Some suitable grasses and legumes for ley pastures in Sudanian Africa: the case of the Borgou region in Benin. Biotechnol Agron Soc Environ 12(4):405–419

Akinlade J, Smith JW, Larbi A, Archibong IO, Adekunle IO (2002) Forage from cropping systems as dry season supplements for sheep. Trop Grass 36(2):102–106

Das KC, Haque N, Rajkhowa C (2009) Tree leaves and forest grasses based feed block for Mithun in North Eastern region of India. In: Walli TK (ed) Proceedings of the national symposium on fodder block technology. ILDEX India, New Delhi, pp 80–83

Karangiya VK, Savsani HH, Ribadiya NK (2016) Use of densified complete feed blocks as ruminant feed for sustainable livestock production: a review. Agric Rev 37(2):141–147

Mattoo FA, Ganai AM (2004) Evaluation and improving locally available feed resources and developing feeding system for economic livestock production. Final report. NATP, SKUAST-K, Srinagar

Olawoye UH, Kubkomawa HI (2018a) Feed requirements of livestock and product processing for agricultural science and related disciplines, 1st edn. Tapass Institute of Scientific Research and Development, Owerri, pp 1–139

Olawoye UH, Kubkomawa HI (2018b) Principles of pasture and range management for agricultural science and related disciplines, 1st edn. Tapass Institute of Scientific Research and Development, Owerri, pp 1–102

Rotz CA, Ford S, Buckmaster D (2003) Silages in farming systems. In: Buxton DR, Muck RE, Harrison JH (eds) Silage science and technology edition: monograph 42 Chapter: silages in farming sytems. American Society of Agronomy, pp 505–546

Salem HB, Nefzaoui A (2003) Feed block as alternative supplements for sheep and goats. Small Ruminant Res 29:275–288

18 Effect of Harvest and Storage on Forage Quality

Forages are the primary feed resource to livestock. The nutritional quality of forages is often affected by seasonal variations as a consequence of changing climatic conditions. Seasonal variations have resulted in the periods of surplus and shortage in availability of feed. In both cases, preservation and conservation of forages play crucial role in bridging the animal feed shortages. During the periods of feed scarcity, preserved forages could meet feed and nutritional requirements of livestock.

The prime objective of forage conservation is to maintain quantity and quality of forages with minimal loss during harvesting and storage (Buskirk et al. 2003; Mackie 2004). Forages at early stages of growth have good quality but there is loss in quality during harvesting, transportation, and storage (Roberts et al. 2009; Bernués et al. 2011). There are number of factors that are associated with loss of nutritional quality during harvesting and storage. During cutting, curing and baling of forages, climatic conditions affect the nutritional quality of forages by

1. Affecting plant respiration which reduces WSC content and energy during storage
2. Increasing the rate of leaching of soluble nutrients
3. Leaf shattering and loss of digestible proteins
4. Affecting microbial activity
5. Bleaching and excessive drying (Rankin and Undersander 2004)

Hay and silage making are common practices of forage preservation (Suttie 2000). It has been observed that inappropriate harvest, storage and preservation practices result loss in quality of forages. Each step in the forage preservation incur loss in dry matter and essential nutrients. Some of these losses are because of mechanical action; others are due to biological processes. Total quality losses range from 20 to 30% in typical forage harvesting and preservation system.

R. Katoch, *Nutritional Quality Management of Forages in the Himalayan Region*, https://doi.org/10.1007/978-981-16-5437-4_18

During hay preparation, nutrient losses are mostly as a result of mechanical handling and weather conditions during harvesting in the fields. In silage preparation, nutrient losses occur during storage whereas in hay making most of the dry matter losses from forage during harvest and storage. Dry matter losses and quality changes cannot be eliminated during hay preservation, but they can be minimized by using appropriate harvest and storage practices (Rotz and Muck 1994). These changes during harvest and storage of forage legumes and grasses have been presented in Table 18.1.

18.1 Nutrient Losses During Harvest

18.1.1 Carbohydrate Losses

Carbohydrates are important source of digestible energy to livestock. The reduction in carbohydrate content in forages is important from the prospects of nutritional quality of final feed product after storage and processing. During the storage, respiration oxidizes carbohydrates to carbon dioxide, heat and moisture. Since this loss is primarily associated with the loss of readily digestible carbohydrates, respiration increases the total fiber content and reduces energy content in forages. During storage, respiration of microorganisms causes heating and results in loss of dry matter and nutrient in hay. Losses from plant respiration are relatively small when the crop is dried quickly under good conditions, but under slow drying conditions losses can be up to 7% of dry matter.

With natural drying, forages harvested in evening hours generally have higher nutrient losses than harvesting during daytime. It has been reported that, losses are greater during warm nights than during cool nights due to high respiration rate. Since starch and other storage carbohydrates are hydrolyzed and yield simpler carbohydrates for respiration, the sucrose content increases in alfalfa during night (Knapp et al. 1973). Respiration losses of total soluble carbohydrates in air-dried forages are around 3.50% during 24 h period. However, alfalfa cut during evening has overnight dry matter losses ranging from 7.20 to 11% (Knapp et al. 1973). In rye grass, fructosans and total soluble fructose content decrease throughout the drying process. High moisture, soluble sugar content, relative humidity and temperature result in carbohydrate loss. Since structural carbohydrates in cell wall are undiminished in the drying process, the proportion of the structural fraction during drying is higher because of reduction in total soluble nutrients content. Respiration during curing of forages results in loss of soluble carbohydrates.

18.1.2 Protein Losses

Protein content in forages is less affected than carbohydrate content during drying process. Loss of volatile nitrogenous compounds is small under normal drying conditions, however, when drying temperature is high, a higher loss in nitrogen

Table 18.1 Dry matter losses and quality changes during forage harvest and storage

Type of loss		Dry matter loss (%dry matter)		Change in nutrient content (% dry matter)		
		Range	Normal	CP	NDF	TDN
Forage legumes						
1. Respiration		1–7%	4%	0.90%	−1.70%	1.70%
2. Rain damage						
	5 mm	1–7%	5%	−0.40%	1.40%	−1.50%
	25 mm	3–7%	17%	−1.70%	6.00%	−7.00%
	50 mm	12–50%	30%	−3.50%	14.00%	−14.00%
3. Mowing/ conditioning		1–4%	2%	−0.70%	1.20%	−1.40%
4. Bailing						
	Small bale	2–6%	4%	−0.90%	1.50%	−2.00%
	Round bale	3–9%	6%	−1.70%	3.00%	−4.00%
5. Hay storage						
	Inside	3–9%	5%	−0.70%	2.0%	−2.00
	Outside	6–30%	15%	0	5.0%	−7.00
6. Silo storage						
	Sealed	6–14%	8%	1.40	0.70	−4.70
	Bunker	10–16%	12%	2.30	2.70	−5.60
Forage grasses						
1. Respiration		2–8%	5%	0.80%	3.20%	−1.80%
2. Rain damage						
	5 mm	1–3%	2%	−0.20%	0.90%	−0.50%
	25 mm	4–14%	8%	−1.30%	5.30	−0.30%
	50 mm	8–27%	15%	−2.70%	11.00%	−0.60%
3. Mowing/ conditioning		1–2%	1%	0%	0%	0%
4. Bailing						
	Small bale	2–6%	4%	−0.50%	0.90%	−1.00%
	Round bale	3–9%	6%	−1.00%	1.80%	−2.00%
5. Hay storage						
	Inside	3–9%	5%	−1.30%	3.20%	−1.80%
	Outside	5–22%	12%	0	8.00%	−4.80%
6. Silo storage						
	Sealed	6–14%	8%	0.80%	0.90%	−3.70%
	Bunker	10–16%	12%	1.50%	3.60%	−5.60%

Source: Rotz and Muck (1994)

content of forages can occur. Dry matter losses are often higher than N losses in forages when drying temperature is high.

In the presence of sufficient moisture, protein breakdown is very rapid and the extent of degradation is influenced by the time required for drying process. With rapid wilting, protein breakdown is negligible. NPN content increases to 6 g/kg of total *N* each hour during field wilting. A moderate increase in temperature during drying increases the rate of protein breakdown. However, high temperature normally provides faster drying; therefore the net effect on protein degradation is less. Protein degradation during wilting results in rapid accumulation of amides. Other NPN forms include free amino acids, volatile bases and peptides. Protein changes during wilting result in the decreased solubility of true protein.

18.1.3 Vitamin Losses

Vitamin A or its precursor, carotene in forages are affected by drying process, especially if drying period is prolonged. Losses could range from 90 to 95% in field cured hay, 80–90% in barn-dried hay and 40–60% in silage (Carter 1960). A major loss of carotene content in forages during drying is due to lipoxidase destruction (Sullivan 1973). Photochemical breakdown of carotene in forages occurs under bright sunlight. Green forages have high carotene, while bleached forages have low carotene content. Vitamin E content in forages is also affected by prolonged drying. B complex vitamins may be readily leached by rain since they are water-soluble. Vitamin D is formed during drying process by light radiation of its sterol precursor, ergosterol. Vitamin D content is high in fresh and sun cured hay.

18.1.4 Other Nutrient Losses

Losses of minerals, such as P and Ca, in the drying process are often quite small. However, prolonged field drying exposes forages to the weathering. As a result, leaching, leaf shattering and other indirect physical factors could increase the extent of loss of minerals from forages. Some toxic compounds present prior to harvest become nontoxic after drying. For example, toxicity of cyanide-releasing glucosides in fresh tissues of sorghum Sudan grass reduces after drying which may be due to disorganization of plant cells upon drying, releasing glucosidase, which breaks sugar from cyanogenic glucoside. Dry hay from forage legumes does not produce foamy bloat because soluble proteins, which appear to be largely responsible for bloat are denatured and lose their ability to stabilize foam. Level of estrogenic substances (Coumesterol in alfalfa which interferes with oestrus cycle and disturbs conception), decreases with drying process. Sweet clover can become toxic if molding takes place. Toxicity occurs due to the conversion of coumarin to dicoumarol (Vitamin K antagonist), which reduces blood clotting. Drying also reduces the tannins content in Lespedeza (*Lespedeza cuneata*) (Terrill et al. 1989).

18.2 Nutrient Losses Due to Weathering

During curing of forages, the incidence of rain in open field reduces yield and quality of forages. Yield or organic matter losses of up to 30% have been reported with over 50% loss with heavy rain. Excessive weathering reduces the quality of hay. Leaching nutrients is important, but not all losses from rain are due to leaching. These losses occur only from dead plant cells when protoplasm and membranes have lost their selective permeability. Nutritional loss due to leaching increases with the number of rain showers, amount of rain and with the extent of dryness of hay before rains. Prolonged exposure of forage could also result in nearly complete loss of carotene. If soluble nutrients are leached out, the forage digestibility decreases. Leaching of soluble nitrogen from tissues increases the water-insoluble protein content in forages (Rotz and Abrams 1988).

Soluble mineral elements especially potassium are leached out from rain-damaged forage. Feed value of forage left standing in the field decreases due to the rupture of cells and leaching of soluble nutrients with rain. Since leaves are the most valuable part of forage, nutrient loss is greatly affected by leaf shattering. Rain not only leaches soluble nutrients, but also causes extra handling to facilitate drying which in turn increases leaf loss. Nutrional losses due to microbial activity are another indirect effect of rain damage. When forages are rehydrated, microbial population is increased. This increase may cause additional carbohydrate loss from fungi respiration. Generally most of the microbial loss occurs after harvest when hay is stored with excessive moisture; however, under wet conditions, microbial and leaching losses may occur in the field before storage.

18.3 Losses During Mowing and Conditioning

Nutrient losses during mowing and conditioning of forages generally vary between 1 and 5% of the dry matter (Shinners et al. 1991). Stage of maturity influences mowing and conditioning loss in both forage legumes and forage grasses. Nutrient loss can double as forages mature from late vegetative to reproductive stage. The moisture content of crop decreases with the advancement of maturity. In forage legumes, higher loss is expected with leaves because leaves have high CP content and low fiber content than stem. Dry matter losses and associated nutrient changes caused by conditioning increase with crop maturity. Although, severe conditioning provides faster field curing, but harvest losses are generally higher. Normally less severe conditioning is recommended to procure forage of good nutritional quality.

Forage crops should be mown at defined maturity stage to stablize yield and nutritional quality. Nutritional quality of most forage crops declines rapidly with commencement of reproductive stage. The optimum maturity level varies among forages. Mowing at this time provides a good yield, a relatively low fiber content and adequate energy and protein in forages.

During heavy rains and poor drying conditions, it is advised to delay the mowing and conditioning. On warm, sunny days when crop is actively growing, plant readily stores total digestible carbohydrates in tissues. Therefore, forages mown in late afternoon have low fiber and higher energy contents.

18.4 Nutrient Losses During Baling and Chopping

Forages conserved as silage or hay, represent the most secure feed resource for lead periods. With the increase in availability of different baler systems, there is remarkable increase in herbage quantity stored as silage. The process of bailing involves mechanical treatments ranging from harvesting to storage in order to achieve baled feed of high quality. However, the nutritional quality of baled feed is dependent on moisture content during baling and of storage time, environmental conditions and forage species. Hay that is baled with moisture contents more than 20% can develop mold and lose dry matter and quality due to bacterial degradation. Hay baled at moisture content between 15 and 20% undergo increase in temperature during first 2–3 weeks after baling. This heat buildup is due to plant respiration and microbial activity. At a moisture level of about 30%, a bale may maintain a high temperature up to 40 days regardless of forage species or bale shape. Temperature below 130 °F has no effect on nutritional quality of hay. However, with increase in temperature, nutritional losses up to 5% in hay may be recorded.

Typically dry matter losses during hay baling vary between 2 and 6% which depends on many factors. Tabacco et al. (2006) observed that when the chopping system was used during bailing process, the dry matter losses in alfalfa range from 53.20 to 253.00 kg dry matter ha^{-1}.

18.5 Changes in Forage Quality During Storage

Changes which occur during the storage of forages in the form of hay and silage are associated with moisture content in forages. The changes in quality of forages during storage are primarily affected by microbial activity and subsequent generation of heat, oxidation of nonstructural carbohydrates, mold growth and the level of fiber components. Such changes can negatively affect the feed value of forage and result in decreased animal performance. Therefore, understanding quality changes during storage is important for better management and utilization of these resources.

18.5.1 Quality Loss During Inside Storage

The major factor influencing the nutrient loss during storage of hay is the moisture content of hay. In hay having high moisture content, microbial respiration increases the production of heat during initial period of storage. Heating and associated nutrient losses increases with moisture content in hay. Dry matter loss and heating affect the level of most nutrient constituents in hay during storage. Much of lost dry matter is nonstructural carbohydrate respired to carbon dioxide and water. Sugars are primary plant carbohydrates lost during storage respiration. During extended storage periods, microbial and fungal respiration reduces oxygen levels and produces carbon dioxide, water, and heat, causing reduction in dry matter content. Dry matter losses could exceed 10% if moisture level ranges between 20 and 30%. Due to low carbohydrate content, perennial warm-season grasses may lose less dry matter than forage legumes and cool-season forage grasses. Total nonstructural carbohydrates (TNC) content in forages decreases depending on the storage conditions. Larger losses of TNC usually occur within 10 days of baling, when most of spontaneous heating occurs and microbial activity is high. Subsequently, TNC reduces during storage, while NDF and ADF content in hay increase. These changes in forage composition reduce hay palatability, digestibility and quality. Fiber content usually increases along with spontaneous heating during first 12 days of hay storage. Storage temperatures below 120 °F generally do not alter the digestibility, but temperatures above 140 °F can lead to approximately 15% reduction in digestibility of hay.

The loss in CP content occurs due to slow volatilization of ammonia contained in hay or produced through microbial respiration. Loss of more soluble nitrogen component increases water-insoluble nitrogen and ADIN content (Rotz and Abrams 1988). Additionally, heating of high-moisture hay causes formation of ADIN through Maillard reactions which involve polymerization of sugars and other carbohydrates with amino acids.

Fiber content consistently increases during hay storage. Forage digestibility and many nutrient constituents also decreases during storage (Rotz and Abrams 1988). The improper storage conditions can lead to molding which markedly reduces the feeding value of hay. Saprophytic fungi use the readily available carbohydrates and thereby decrease the feeding value of hay and affect animal performance.

During silo filling, the dominant process affecting forage quality is plant respiration. This respiration causes dry matter loss and may increase silage temperature, which influences the rate of ensiling process. Once the silo is filled, remaining oxygen allows anaerobic fermentation to decrease silage pH. After anaerobic microbial activity has ceased due to low pH or lack of substrate, the stable phase begins. Even if silos are well sealed, slow diffusion of air occurs through the silo walls or cover. This oxygen again is used in respiration contributing to further nutrient loss. When the silo is opened, much more oxygen is present at the open face which penetrates into the silage mass. With higher oxygen availability, aerobic microbial growth and respiration increase substantially, causing heating as well as dry matter and nutrient loss.

Digestible carbohydrates are largely lost through respiration. Changes in NDF content are also dependent upon amount of enzymatic and acid hydrolysis of structural carbohydrates during storage. The change in NDF levels during silo storage may range from 1 to 4% increase, depending on respiration loss. Losses of digestible carbohydrates is more, therefore dry matter digestibility or TDN content declines by 2–7%.

18.5.2 Quality Loss During Outside Storage

Quality changes in large round bales stored outside range from 3 to 40%. Among different factors weather, length of storage and storage methods have significant effect on quality of forages. Excessive rain causes loss by knocking off leaves and leaching soluble nutrients from within the plant tissue. Since leaves have normally higher nutrient contents than stem parts, the loss of leaves reduces overall quality. Soluble carbohydrates, proteins and minerals are washed from the plant material leaving a higher fiber content and lower energy content.

Nutrient loss is less in hay stored over winter periods or in dry climatic conditions. When rain drops on an exposed bale, the moisture content of bale increases from 25 and 40% (Harrigan and Rotz 1992). The increased moisture leads to higher microbial activity. Digestible soluble carbohydrates decline in weathered hay, resulting in high fiber content and low digestibility. CP content declines if hay undergoes excessive heating and weathering.

To fill the gap between demand and supply of feed, preservation and conservation of forages is crucial. Different methods are in practice for the conservation of the forages. When forages are cut for hay or silage, physiological changes occur in the plants resulting in unavoidable nutrient losses. During harvesting and storage of forages, reduction in the level of carbohydrates, protein and other nutritional constituents result in the significant nutritional loss. Thus, use of appropriate harvest and storage practices is imperative to reduce loss in nutritional quality of forages during harvest and storage.

References

Bernués A, Ruiz R, Olaizola A, Villalba D, Casasús I (2011) Sustainability of pasture-based livestock farming systems in the European Mediterranean context: synergies and trade-offs. Livestock Sci 139(1):44–57

Buskirk DD, Zanella AJ, Harrigan TM, Van Lente JL, Gnagey LM, Kaerche MJ, Buskirk DD, Zanella AJ, Harrigan TM, Van Lente JL, Gnagey LM, Kaerche MJ (2003) Large round bale feeder design affects hay utilization and beef cow behaviour. J Anim Sci 81:109–115

Carter WRB (1960) A review of nutrient losses and efficiency of conserving forage as silage, bam dried hay and field cured hay. J Brit Grassl Soc 15:220–230

Harrigan TM, Rotz CA (1992) Net, plastic and twine wrapped large round bale storage loss. Paper 921572. Am Soc Agric Eng St. Joseph, MI

Knapp WR, Holt DA, Vough LR (1973) Diurnal variation in alfalfa (*Medicago sativa* L.) dry matter yield and overnight losses in harvested alfalfa forage. Agron J 65:413–417

Mackie PG (2004) US Patent No. 6711996. US Patent and Trademark Office, Washington. http://tinyurl.com/n7mef3e

Rankin M, Undersander D (2004) Rain damage to forage during hay and silage making. http://fyi.uwex.edu/forage/files/2014/01/Raindam.pdf

Roberts CA, Kallenbach RL, Hill NS, Rottinghaus GE, Evans TJ (2009) Ergot alkaloid concentrations in tall fescue hay during production and storage. Crop Sci 49(4):1496–1502

Rotz CA, Abrams SM (1988) Losses and quality changes during alfalfa hay harvest and storage. Trans ASAE 31:350–355

Rotz CA, Muck RE (1994) Changes in forage quality during harvest and storage. In: Fahey GC, Collins M, Mertens DR, Moser LE (eds) Forage quality, evaluation, and utilization. Univ. Nebraska, Lincoln. ASA-CSSASSSA, Madison, WI, pp 828–868

Shinners KJ, Koegel RG, Straub RJ (1991) Leaf loss and drying rate of alfalfa as affected by conditioning roll type. Appl Eng Agric 7:46–49

Sullivan JT (1973) Drying and storing herbage as hay. In: Butler GW, Bailey RW (eds) Chemistry and biochemistry of herbage, vol 3. Academic Press, New York, pp 1–31

Suttie JM (2000) Hay and straw conservation: for small-scale farming and pastoral conditions (No. 29). Food & Agriculture Organization. http://tinyurl.com/hx2dv3r

Tabacco E, Borreani G, Crovetto GM, Galassi G, Colombo D, Cavallarin L (2006) Effect of chestnut tannin on fermentation quality, proteolysis, and protein rumen degradability of alfalfa silage. J Dairy Sci 89(12):4736–4746

Terrill TH, Windham WR, Hoveland CS, Amos HE (1989) Forage preservation method influences on tannin concentration, intake, and digestibility of *Serecia lespedeza* by sheep. Agron J 81: 435–439

19 Challenges and Opportunities in Forage Production and Livestock Management in the Himalayan Region

The uninterrupted supply of feed for the nutritional requirement of livestock is one of the ultimate objectives for successful livestock production. Since the majority of cultivable land is being utilized for food crops, therefore, to meet the increasing demands of livestock products, the competition for natural feed resources is rapidly increasing. The availability of grazing land is continuously decreasing and this trend could result in an acute shortage of fodder to feed livestock in nearby future.

Livestock production and crop farming are crucial components of agriculture in the Himalayan region. Based on two core activities (Family-based agricultural production and Local consumption), the agricultural systems in the Himalayan region have been classified as subsistence-based and family farming-based systems (Shukla et al. 2018). Traditional methods of subsistence farming in the Himalayan region depend on natural resources such as natural grasslands and forests. Livestock farming, one of the important components of subsistence farming is an important vocation for the people of the Himalayan region. It plays an important role in the economy of the rural households in the Himalayan region. More than 80% of land holdings of farmers in the region are marginal and generally, it is not preferred practice to set aside agricultural land for the cultivation of fodder crops. Therefore, livestock rearing is mainly dependent upon natural grazing resources. Natural grasslands and pastures cover approximately 55% of the total area in the Himalayan region comprised of different vegetation types distributed from subtropical to the temperate and to the alpine regions. These natural feed resources meet over 50–60% of the total requirement of fodder in the Himalayan region.

Despite of importance of livestock in the Himalayan region, their productivity remains low and one of the major constraints is the scarcity of feed both in quantitative and qualitative dimensions. The increasing prices and unavailability of feed concentrate, improper management of available forage resources, overdependency on low quality feeds, felling of fodder trees, shifts in cropping patterns and lack of knowledge and information of techniques and developments in forage crop improvement programs are the major constraints in the sustenance of the

R. Katoch, *Nutritional Quality Management of Forages in the Himalayan Region*,
https://doi.org/10.1007/978-981-16-5437-4_19

livestock in the region. Knowing the importance of livestock farming in the region, there is a need to address different challenges in fodder production and livestock management and explore the opportunities for achieving sustainability in livestock rearing in the Himalayan region.

19.1 Challenges for Forage Production and Livestock Management

Livestock production plays a crucial role in the sustenance of the livelihood of the rural community in the Himalayan region. Various challenges for forage and livestock production in the Himalayan region have been discussed.

19.1.1 Land Constraints

Livelihoods of the majority of the population in the Himalayan region depend on a mixed farming system. Under such system, the land is the center of all socio-economic activities. For the majority of the small and marginal farmers, their wealth and poverty are associated with the ownership of the size of land holdings (Partap 1999). For large number of small and marginal farmers of the Himalayan region, shrinking cropland is a key concern for managing food and livelihoods (Pokhriyal and Bist 1988; Partap 1998). The fragmented and scattered nature of land holdings has also been identified as one of the main factors for decreasing agriculture productivity in the Himalayan region. The smaller landholdings have been devoted for the cultivation of food crops on high priority and cultivation of forage crops have least priority. With continuous pressure on cultivable lands and deterioration of grazing lands, it is imperative to adopt the practice of a land-use system with multiple crops in a sustainable manner (Sarker et al. 2017).

19.1.2 Uncontrolled Grazing of Natural Pasture Resources

Animals particularly yak, yak-cattle hybrids, hill cattle, Mithun, sheep, goat, buffalo, camel and horse are reared by different pastoral communities in the Himalayan region. Animal rearing plays a very important role in the food security, economy and culture of these pastoral communities in the region. These communities depend on natural pasturelands for their livelihood. They move along with their herds from one pasture land to another. Hence, grasslands are the major feed reservoirs for their animals. Besides grazing in temperate/alpine pastures, fodder trees and shrubs are also fed to the livestock. As a result of overgrazing, coupled with poor management and care, these grazing lands are continuously degrading. Therefore, there is a need for the amelioration or rehabilitation of natural pastures and grasslands *via* judicious implementation of management techniques for sustained forage production.

19.1.3 Changing Climatic Conditions

The changing climatic conditions due to global warming and its impact on crop production are important issue in recent times. The Himalayan region is comparatively more susceptible to climate change, therefore there is increasing pressure on natural resources for meeting the demand of rural households as well as livestock production systems. The adverse effect of climate change is visible in the form of increasing intensity and frequency of natural disasters. It is important to note that the production of forage crops per unit area is the consequence of interactions between genotype and environment. Therefore, varieties with regeneration potential, and high persistency under adverse climatic conditions need to be developed. Improved tillage practices, changing planting and harvest dates, adoption of crop rotations, selection of suitable crops and crop varieties for cultivation, irrigation and fertilizer management can reduce the impact of climate change on forage crop production.

19.1.4 Lack of Planting Material

Inadequate forage seed supply has been a serious drawback in animal husbandry programmes. It is important that quality seed and planting material are readily available for supply to the farmers. Indeterminate growth, uneven maturity, blank seed and seed dormancy are the contributing factors for low seed yield in forage crops. Moreover, the nature of seed production is more complex and different environmental and physiological factors have a significant impact on the seed production of forage crops.

19.1.5 Inefficient Forage Resource Management

Forage resource management practices have considerable importance in determining productivity and nutritional quality of natural pastures and grasslands. Lack of fertilizer application, absence of forage legumes, inappropriate cutting intervals, lack of knowledge of improving nutritional quality of crop residues and indiscriminate grazing are major contributing factors for poor livestock productivity in the Himalayan region. The farmers in the region follow conventional and traditional approaches to fodder production and livestock feeding system. Forage crops in the Himalayan region are harvested without using machinery. Forage mechanization has the potential to meet contemporary challenges and to increase fodder productivity in a sustainable way. Topographic constraints play a significant role in restricting farmers from using large machinery for the cultivation and harvesting of forages. Processing and conservation techniques are rarely practiced in livestock rearing in the Himalayan region. Poor infrastructure facilities for meeting the storage of fodder are also one of the serious constraints.

19.1.6 Inefficient Livestock Management Practices

Animal husbandry serves as a source of employment for the Himalayan rural households. Sometimes livestock is the only source of income for the family and provides income insurance during crop failure. Despite holding a sufficient livestock population the productivity of the livestock is quite low. Grazing on natural pastures and grasslands is the conventional way of feeding livestock. However, feeding dry grasses and crop residues leads to malnutrition, poor growth and low production and makes animals more susceptible to various diseases.

The low genetic potential of livestock and irregularity in the availability of feed superior in nutritional quality are other reasons for low productivity. To achieve the target of sustainable livestock productivity, it is imperative to supply nutritionally adequate feed to livestock in a manner that encourages consumption with minimum waste and allows high efficiency of feed utilization. Moreover, small and marginal farmers, pastoralists and their animals often live in harsh environments in the Himalayan region. Under these conditions, smallholders and pastoralists need different and diverse animal genotypes, species mix and types to make animal husbandry more productive in their specific environmental locations. The livestock breeds having the ability to survive in harsh environmental conditions and high output capacity in the region need to be developed or introduced from other similar topographical regions.

19.1.7 Lack of Strong Policy Support

The sluggish research and development on fodder and forages in the Himalayan region are associated with poor organizational structure and lack of policy support. The policy aspects are required to provide a favorable environment for research and development on forages. The focus should involve (a) collection of database of fodder production and utilization; (b) accelerated investment in the forage resource development; (c) credit facility to forage production; (d) support price for forage and seed marketing; (e) nondiversion of crop residues to other use; (f) strong policy on unregulated grazing in natural pastures and grassland; (g) policy on common property resources and legal protection of natural pastures and grasslands.

19.2 Opportunities for Improving Forage and Livestock Production

1. The adoption of Silvi-pastoral and hortipastoral systems suitable to the area could enhance the sustainability in forage availability. Silvi-pastoral system integrates pasture with trees. In this system, trees having significant fodder value are introduced and managed. Under poor soil, water and nutrient situations where cropping is not possible such systems can be useful in making forage supply more sustainable and also serve the purpose of ecosystem conservation.

In silvipastoral system, grasses provide benefits by reducing soil and water loss. The importance of grasses in silvipastoral system could be understood from the fact that the average soil loss from deforested land is higher than natural pastures and grasslands. Apart from being an important source of fodder, forage grasses and legumes are important for checking soil erosion. For example, *Dichanthium annulatum* is best in reducing soil loss from 20.50 t/ha to 1.50 t/ha on sandy loam soil having a 2% slope. Horti-pastoral system also has applicability with the farmers for utilizing their degraded lands and increasing the fodder production in the Himalayan region.

Forage crops could be grown in space between the fruit trees for economic utilization of orchard lands. Introduction of tall fescue grass in apple orchards could provide 83.50% higher fodder yield over local grasses (Sharma and Jindal 1989). It is important to note that there is a considerable area under orchards in the Himalayan region that could be utilized for fodder production. The area under the forest can also be utilized for growing fodder on partially shaded ground without affecting the establishment of fodder trees. A noncompetitive land-use system for forage production in the hills is to grow forage on terrace bunds and risers (Singh et al. 1993). Area under terrace risers, bunds and along the irrigation channels could be utilized for the cultivation of forage grasses/legumes/fodder trees. This is beneficial as the cultivation of forages on terrace risers, bunds and irrigation channels could provide a sufficient quantity of fodder to livestock and could reduce the nutrient loss in run-off water thus improving the soil's physical and textural properties as well as soil fertility. For example, two rows of Bajra Napier hybrid/ guinea grass/setaria/tall fescue could supply 7–11 q green fodder per 100 m length of boundary per year. The suitable grass and fodder legumes that can be grown under silvipasture and hortipasture system of Himachal Pradesh have been enlisted in Tables 19.1 and 19.2.

2. Achieving high herbage production per unit area should be the challenge in fodder production. Moreover, opportunities to produce fodder on arable lands must be seized by following intensive forage production systems such as multiple cropping, intercropping, overlapping cropping, and relay cropping for maximum harvest of nutritive herbage per unit area and time. Multiple cropping sequences like sorghum (multicut) + cowpea—berseem + mustard—maize+ cowpea and sorghum (multicut) + cowpea—berseem + mustard are promising for fetching maximum food and fodder harvest. Overlapping cropping systems should also be practiced to maximize fodder production. For example, overlapping cropping system developed involving seasonal and perennial forage crops like guinea grass and Napier bajra hybrid intercropped with cowpea during summer and *Kharif* season and berseem in *Rabi* season has capacity of providing round the year green fodder (200–300 t/ha) to livestock.
3. Cropping of graminaeous and leguminous forages improves the nutritional quality of feed in terms of protein and mineral balances (Kumar et al. 2014). This system of cropping also helps to maintain soil fertility due to addition of root organic matter and better utilization of nutrients from different soil depths.

Table 19.1 Forage grasses and legumes suitable under silvipastoral system in Himalayan region

Zones	Trees	Grasses	Legumes
Lower & mid-hills	*Grewia optiva, Celtis australis, Bauhinia variegate, Ficus raxburghii, Melia azedarach, Leucaena leucocephala, Albizia chinensis, A. lebbeck, Morus alba, Quercus leucotrichophora, Salix tetrasperma*	*Pasphalum notatum, Dichanthium annulatum, Chrysopogon fulvus, Lolium perenne, Festuca arundinacea, Setaria anceps, Cenchrus ciliaris, Dactylis glomerata, Chloris gayana, Apluda mutica, Pleum pretense, Panicum maximum, Heteropogon controtus*	*Trifolium alexandrium, T. repens, Stylosanthes hamata, S. scabra, Macrotyloma axillare, Neonotonia wightii*
High hills	*Ulmus wallichiana, Robinia pseudoacacia, Salix alba, Popularis ciliate, Quercus dilatata, Morus serrata, M. alba, B. variegate*	*Chrysopogon montanus, Poa annua, L. multiflorum, F. arundinacea, D. glomerata, Arundinella nepalensis, P. alpinum, Stipa* spp., *Chrysopogon gryllus*	*Medicago sativa, T. repens, T. pretense, onobrychis viciifolia*
Cold dry	*Salix* spp., *Populus* spp., *R. pseudoacacia, Ulmus* spp., *Hippophae rhamnoides*	*F. arundinacea, Avena sativa, Phalaris* spp., *D. glomerata*	*M. sativa, M. falcata, Lotus corniculatis, Astragalus* spp., *Caragana* spp., *Melilotus officinalis, Cicer microphyllum*

Table 19.2 Forage grasses and legumes suitable for Horti-pastoral system in Himalayan region

Zones	Horticulture trees	Grasses	Legumes
Lower hills	Mango, Litchi, Guava, Aonla, Loquot, Peach, plum, citrus fruits	*Bracharia brizantha*, Bajra-napier hybrids, *Setaria anceps, Paspalum notatum, Dicanthium annulatum, Chrysopogon fulvus*, Guinea grass	*Trifolium alexandrium, Stylosanthes hamata, S. scabra, Macrotyloma axillare, Nenoatonia wightii*
Mid hills	Peach, Plum, and Apricot	Perennial rye grass, Tall fescue, Setaria grass, Anjan grass, *Paspalum* spp., *Dactylis glomerata*	White clover, *Stylosanthes hamata, Macrotyloma axillare, Nenoatonia wightii*
High hills	Apple, Apricot, Peach, Plum, Pear, and other stones fruits	Tall fescue, Orchard grass, *Phalaris* spp.	Lucerne, White clover, red clover, Sainfoin
Cold dry	Apples, Prunes, Apricot, Almond, Chilgoza, Pistacio, Walnut, Hazel-nut, Hops	Tall fescue, Orachard grass, *Phalaris* spp.	Lucerne, White clover, red clover, Sainfoin

4. For sustainable production of fodder from arable lands, emphasis should also be given on conservation tillage, contingent crop planning, climate-resilient cropping system, crop diversification, soil carbon sequestration and its management, farming system research, soil micronutrient management and crop modeling approaches (Kumar et al. 2012).
5. Forage crop improvement is a multidisciplinary approach that involves several disciplines, *viz.*, plant breeding, genetics, agronomy, pathology, nematology, entomology, physiology, biotechnology and animal nutrition, etc. A thorough understanding of species relationship, chromosomal constitution, genome structure, putative parentage, and extent of possible gene exchange/recombination and nature of polyploidy is also required for the success of the forage breeding program.
6. The viable strategy need to be developed for improvement of different crops varieties as per the problems encountered in that species. Leguminous species such as berseem, lucerne, cowpea, *Dolichos,* and *Stylos* are important forages and attention is required for developing suitable lines of these forages with efficient nitrogen-fixing ability, having high phosphate response resulting in the high dry matter besides high protein content, wide tolerance for different soil conditions and diseases and pest. There are several anti-quality factors that affect the large-scale acceptance of many important and high-yielding fodder crops such as oxalates in NB hybrids, HCN in sorghum, mimosine in *Leucaena* and high fiber content in maize stover. These problems are to be resolved with improved suitable breeding techniques.
7. The development of new forage varieties excelling in production and nutritional quality is important for the acquisition, conservation and utilization of fodder biodiversity as well as for improving livestock production. The generation and identification of novel planting material is an important issue for harnessing the full potential of both cultivated and noncultivated lands for forage production in the Himalayan region. Different fodder varieties have been released in the past few decades keeping in mind the present status of fodder and livestock production in the Himalayan region. Some advances in biotechnological approaches have also been made which include such as development of interspecific crosses using embryo rescue in *Trifolium* sp., regeneration protocols through tissue culture in *Cenchrus, Dicanthium, Trifolium,* and *Medicago*, development of a protocol for transgenic plants in *Dicanthium* and *Medicago*, identification of QTL's related to nutritional quality in sorghum, development of SSR markers in berseem, molecular characterization of genes involved in the expression of components of apomixis, ploidy series in guinea grass using components of apomixis, identification of sexual plants in grasses, use of molecular markers in the improvement of *Stylosanthes* for nutritional quality and drought tolerance. Now there is a need for judicious use of FGR resources and conserve them for future needs. Breeding dual-purpose cereals and grain crops, breeding specifically for stress tolerance and degraded lands and finding new niches for fodder production are also important for sustainable improvement of fodder and livestock production in the Himalayan region. According to an estimate, pests

including pathogens, insects and nematodes incur yield losses in major forage crops. Since the chemical control of diseases may have adverse implications from the viewpoint of the environment, emphasis is required on developing forage crop varieties having resistance to pests and diseases. Transfer of desirable genes through successive backcrossing and/or evaluation of progenies/germplasm condition could result in the development of tolerant/resistant lines. Gene transfer from wild plants into fodder crops through biotechnological tools/molecular gene manipulations may also help to solve some specific problems. Moreover, there is a need to develop reliable molecular techniques for screening of the grass species for existence of sexuality as it would accelerate the breeding process in such grasses as many of these plants could be used in crossing. Plants with better agronomic traits and apomixis can be selected and advanced for developing varieties. Forage breeders must overcome the barriers in the production of seeds of high-yielding forage crops. Furthermore, emphasis should also be laid on increasing the availability of seeds of short-duration and dual-purpose crops.

8. To increase the productivity of natural pastures and grassland, it is essential to replace low-yielding annual grasses with high-yielding perennial grasses and introduce multipurpose tree species (Palsaniya et al. 2011; Palsaniya and Dhyani 2012). Introduction of forage legumes is one of the best ways to improve the quality of feed and soil health. Excessive run-off leading to soil erosion and nutrient loss could be effectively reduced by improved practices of pasture establishment, contour bunding, grazing management to reduce soil loss, make more water available and improve soil conditions development/restoration of degraded rangelands. It has been possible to increase land productivity from 0.50 to 1.50 t/ha/year. to approximately 10 t/ha/year on a rotation of 10 years on degraded rangelands (Pathak and Kumar 2004). Many studies have supported this view that silvipastoral systems provide effective land cover as well as produce nutritious fodder for livestock and beneficial environmental effects (Palsaniya et al. 2011; Rai and Palsaniya 2015). Grassland restoration and improvement techniques for forage production are based on various approaches such as protection of grassland for vegetation recovery, removal of unwanted bushes, rejuvinating grasslands with forage grasses and legumes, application of fertilizers for high productivity, and subsequently utilization of grassland either through cutting or grazing in a suitable manner (Rai and Palsaniya 2015). Ghosh et al. (2016) emphasized the need for strengthening research on grazing resources, grazing intensity, and their impact on productivity, biodiversity, and soil health for restoration of productivity of grassland and pasture.
9. Postharvest management of surplus fodder is the best mitigating strategy for abating regular phenomenon of seasonal and regional deficit of forage and during natural calamities. Farmers in the Himalayan region lose a sufficient amount of fodder every year due to the lack of technology for postharvest management of forages. Therefore adaptation of technologies for the conservation of forages is essential in order to extend feed supplies for improved animal performance. Bailing, haymaking and silage preparation are the best

conservation methods developed for the conservation of forages. Use of leaf meal from leguminous fodder, crop residues from pulses, and tree leaves could substitute costly feed concentrates in livestock feeding.

10. Although several technologies have been developed for storing and preserving forages, it is important to understand that conserved forages could rarely match the nutritive value of fresh forage because some losses of highly digestible nutrients are unavoidable during conservation and storage. When selecting a conservation method, a producer should consider the suitability of the forage for a given method, storage capability, weather conditions, and the intended use of the conserved forage. The selected conservation technique should maximize nutrient conservation efficiency and minimize production costs. The losses in the quality of hay, straw, and silage and can be reduced by proper storage of feed stuffs. Bale silage, baling of hay, pelleting and densified complete feed blocks can reduce losses and improve storage and conservation procedure.
11. Crop residues constitute the major proportion of feed in Himalayan region during lean periods where green fodder shortage is prevalent. The large-scale cultivation of cereal crops and pulses produces a tremendous amount of crop residues. Crop residues like wheat straw, rice straw, oat and barley straw, maize and sorghum stover, and legumes haulms are the mainstream crop residues in the region for livestock feeding. Main constraint in the use of crop residues is their low nutritional quality. The crop residues no doubt produces bulk of yield, but one has to compromise with their low nutritional quality. The high lignin, low protein, high silica, and high fiber are the main constraints in the efficient utilization of crop residues. Different studies have been conducted to improve the nutritive value of crop residues. Biological, chemical, physicochemical and physical treatment of crop residues are efficient methods to improve the nutritive value of crop residues. Ligno-cellulolytic microorganism, urea and sodium hydroxide treatment, ammonia treatment, steam treatment, and steam treatment in combination with chemical treatment have been recommended to improve palatability and digestibility of dry fodder. The productivity and quality of green fodder could be improved with the application of different chemicals and bio-stimulants. The foliar application of seaweed extract, urea, micronutrient, plant growth regulators and chitosan could improve the nutritive value and production of forage crops.
12. Establishment of improved veterinary facilities can increase the efficiency of livestock. The veterinary services should be focused on prevention and control of diseases, with active participation of famers in the planning and execution of animal health programs. Besides implementation of technologies for improving livestock productivity and livelihood of the resource-poor famers, pasture management, coupled with animal health programs should be implemented for sustainability of livestock production in the Himalayan region.
13. Conservation of animal genetic resources is essential for the development of animal breeds with improved productivity. Animal genetic resources can be preserved by following two approaches (a) *in-situ* conservation of live animals and (b) *ex-situ* conservation of germplasm in cryogenic storage. Live

animal conservation has the advantage that the breed can gradually respond to changing external influences. However, this approach requires high number of animals, and associated disease risks are large, which makes this approach less attractive. The second approach involves cryogenic preservation of semen and embryos. In other breeding option in the livestock is the use of nucleus breeding herds.

14. The rural entrepreneurial capacity should be build up to increase farmer's income to overcome poverty and to improve their socio-economic status. The forage resource development-related activities should conduct with policies of state and central government. The balance between credit and market linkage to forage-based livestock production system needs governmental support to enable livestock keepers for improving their income from animal husbandry. Strategies should be changed from simple Minikit programs to focused technology demonstration. More emphasis should also be given on FLD of major forage crops with the active participation of KVK network. Though multitude of technological options for sustainability of livestock in the Himalayan region has been developed their application depends on their economic attractiveness to marginal and small farmers. In addition to this, parallel development of private technical support services is also required for sustainable rural livestock development. However, systematic monitoring and critical evaluation of extension, research, development work, services, and group activities are essential.

Livestock production is an important source of livelihood for the resource-poor farmers in the Himalayan region. It is a source of food and food products and a source of income to rural households. However, livestock productivity in the region is below their genetic potential, due to many challenges such as irregularity in feed availability, rising prices of feed concentrates, improper management practices, felling of fodder trees, shifts in cropping patterns, inadequate nutrition, poor reproductive management, animal diseases and lack of support services. From a nutritional perspective, livestock productivity is restricted due to overdependence on low quality of feed such as crop residues which is exacerbated by the seasonal availability of the feed resources. Furthermore, feeding fibrous forages results in high methane production that can lead to a 5–15% loss in gross energy intake. To keep the pace of logarithmically growing demands for feed with desired nutritional quality, it is necessary to steer efforts to transfer potential technologies developed by various research organizations to farmer's field for increasing the production and productivity of fodder of good nutritional quality. To lead livestock production as economically attractive venture, their production and productivity need adequate management. In India, national institutes like IGFRI, Jhansi and state agricultural universities are actively involved in improving the present status of livestock and forage production in the Himalayan region. They are also actively engaged in disseminating information regarding the technologies for improving livestock as well as forage production among the farmers through different extension programs and services.

References

Ghosh PK, Palsaniya DR, Ramakrishnan S (2016) Forage research in India: issues and strategies. Agric Res J 53(1):1–12

Kumar S, Dev I, Agrawal RK, Dixit AK, Ram SN (2012) Agronomic research on forages in India: an overview. Indian J Agron 57:92–104

Kumar S, Palsaniya DR, Shivkumar BG, Ram SN, Ghosh PK (2014) *Stylosanthes* production and utilization in India. Technological Bulletin, Indian Grassland and Fodder Research Institute, Jhansi, p 28

Palsaniya DR, Dhyani SK (2012) Natural silvipastoral covers of India. Indian J Forest 35(2): 157–166

Palsaniya DR, Dhyani SK, Rai P (2011) Silvipasture in India: present perspectives and challenges ahead. Scientific Publishers, Jodhpur, p 207

Partap T (1998) Sloping land agriculture and resource management in semi-arid and humid Asia: perspectives and issues in perspectives on sustainable farming systems in upland areas. Asian Productivity Organization, Tokyo, pp 38–84

Partap T (1999) Sustainable land management in marginal mountain areas of the himalayan region. Mountain Res Dev 19:251–260

Pathak PS, Kumar S (2004) Forage and grazing resources in different agro-climatic regions of India. In: Kundu SS, Misra AK, Pathak PS (eds) Buffalo production under different climatic regions, vol XIV. International Book Distributing Company, Lucknow, p 534

Pokhriyal HC, Bist NS (1988) Planning for agricultural development in the Himalayan region: an environmental approach. Environmentalist 8:47–56

Rai P, Palsaniya DR (2015) Silvipasture research and development in India: looking back to move ahead. In: Dhyani SK, Ram N, Badre A, Inder D (eds) Agroforestry—present status and way forward. Biotech Books, New Delhi, pp 471–486

Sarker NR, Habib MA, Amin MR, Yeasmin S, Tabassum F, Yeasmin D (2017) Feeds and fodder dynamics in selected river basins of Bangladesh. Bang J Anim Sci 46:206–214

Sharma JR, Jindal KK (1989) Introduction of superior varieties of grasses in orchards. Paper presented in the workshop on pasture and grassland improvement at HPKV, Palampur on 12 to 13 October, 1989

Shukla R, Chakraborty A, Sachdeva K, Joshi PK (2018) Agriculture in the western Himalayas—an asset turning into a liability. Dev Prac 28:318–324

Singh KA, Prasad RN, Stapathy KK, Sharma UC (1993) Need for forage resource development in the hills of eastern Himalayas. Indian Farming 43(8):21–26

Perspective

1. The Indian Himalayan region (IHR) is the one of the oldest geographical entities on the Earth covering the mountainous ranges of major Asian countries and considered as one of the world's most important treasure of biodiversity in the world. The region has been known for supporting the life of diverse species including plants, animals and humans. Mixed farming system, which involves livestock rearing in combination with the cultivation of cereal, horticultural and cash crops is the integral component of their livelihood in the region. Livestock production system is also a cornerstone of pastoral livelihoods and is contributing significantly to subsistence and regional economy in the Himalayan region. Natural pasture and grasslands are one of the important feed resources to the livestock in the Himalayan region. Fodder trees, shrubs, and cultivated forages are among other feed resources in the region. Climatic, topography, physiographic factors, altitude and other related factors have influenced the distribution of these fodder resources in the region, which affect the total fodder as well as livestock production in the region both qualitatively and quantitatively. Moreover, anthropogenic activities such as industrialization and urbanization have also affected the fodder and livestock production in the Himalayan region.
2. Livestock rearing and crop farming are indispensable components of Indian agriculture and both components make multifaceted contributions to growth and development of the agricultural sector. Livestock farming helps to attain food and nutritional security by providing nutrient-rich animal products, generate income and employment and insurance against crop failure, provide draft power and manure inputs to crop subsector. It also contributes significantly to total foreign exchange through exports.
3. India holds significant position in the total livestock production worldwide. Despite this, the productivity of livestock is low in comparison to the countries with smaller number of livestock holdings. There are a number of factors, which affect the livestock productivity in the country. Feed scarcity has been one of the major limiting factors for better performance of livestock in India. Presently there is shortage of feed and fodder in the country and likely to continue in the near future. The heavy livestock and human population pressure, quantitative

R. Katoch, *Nutritional Quality Management of Forages in the Himalayan Region*,
https://doi.org/10.1007/978-981-16-5437-4

and qualitative deterioration in common grazing lands, lack of adoption of fodder production technologies and poor livestock management practices are the contributing factors responsible for low livestock productivity.

4. Natural grasslands and pastures have played an important role in maintaining human and animal lives and livelihoods for millennia. The area under these natural feed resources is continuously declining, which is mainly due to conversion to arable land for production of food crops and lack of management practices and land abandonment. The competition for land and other resources for agriculture will increase considerably in the near future and would be accentuated by climate change and further increasing livestock population. Therefore, practical policies and actions should be made and employed to stop the degradation of natural pastures and grasslands.
5. The Indian Himalayan region covers 12 Indian States and nearly 50 million people reside in the region alone. The livelihood of Himalayan people depends on sustenance agriculture and cattle production system. In the hills, natural pastures and grasslands, fodder trees, shrubs and open grazing are main feed resource for livestock. Increasing livestock population and diminishing feed resource availability have increased pressure on natural feed resources. The pressure of livestock grazing on forest and grazing land is much higher than normal carrying capacity of grazing areas in the Himalayan region. Due to feed-fodder deficiency, declining per capita land availability, geographic and climatic constraints and poor livestock management, the sustenance of livestock production has become a formidable challenge in the hill region. Therefore, sufficient practical measure are required for boosting fodder production and livestock productivity in the region.
6. In the Himalayan region, farmers have small land holdings and are always biased in choosing the food crops over forage crops. Due to this, the area under fodder crops is meager and is not more than one percent of the total cultivated land. One possible solution for overcoming the crises of arable land for cultivation of forages is the utilization of area under forest and fruit orchards. Moreover, the cultivation of forages on terrace risers, bunds, and along irrigation channels could also be beneficial in alleviating the feed shortage in the region. Habitat-specific studies are required to optimize forage supply from such lands. Further, analysis for introducing diversity in the components and designs for optimizing yields are needed for different types of degraded lands.
7. At the national level, there is shortage of both dry and green fodder. The deficit in green fodder is due to low priority of fodder crops in research and development efforts compared to food crops. The share of fodder crops in total cropped area has remained stagnant during the last two decades. It is expected that the policy emphasis will shift from food grains to fodder crops in view of the growing demand for livestock products. The increase in the area under forage crops particularly leguminous crops, besides alleviating fodder constraints, would make agricultural systems more sustainable by improving soil fertility and contributing toward increased production of sequential crops. Comprehensive breeding strategies are required to generate high yielding fodder cultivars

that can be profitably cultivated under field conditions in different agro-ecological conditions.

8. Enhancing genome base of forage crops is also a key aspect for forage breeding programs. The maintenance of genetic diversity through both *ex situ* and *in situ* conservation can ensure the availability of genetic material to plant breeders for forage improvement. In the recent past, NBPGR, India and IGFRI have made significant contribution in conserving the forage crops germplasm from diverse locations from the country as well as from other countries. Though, efforts have been made in the past, still there is a possibility to widen the genome resource base of forage crops from other unexploited resources. To address the need of increasing forage production, there is a need to broaden the genomic resource base and focus of research and development with the ultimate objective of sustainable livestock productivity.
9. Livestock are key components of hill agriculture farming systems and are increasingly viewed as important components for rural households to overcome poverty and socioeconomic barriers. Low availability of quality feeds is a major constraint limiting livestock productivity among smallholder farmers in the Himalayan region. At present, concerns have been raised for meeting the demand of livestock population, especially on account of climate and land use change. Fodder from trees being rich source of nutrients and metabolizable energy has been considered as an important source of alternative fodder for livestock. Fodder trees can withstand marginal growing conditions and can sustain green fodder availability round the year without extra efforts for their management. Fodder trees could fit in different cropping systems such as alley cropping systems and agroforestry systems. Farmers must be motivated for the inclusion of fodder trees in the cropping system from where they could draw multiple benefits to make the livestock production system more profitable and sustainable.
10. Forages provide the cheapest source of nutrition to livestock and their nutritional value is expressed in terms of different nutrients available to livestock. Nutritional quality of forages is affected by various factors and understanding of interrelationship of these factors with the quality of forages will help in preparation of diets for providing optimum nutrition to livestock. Generally, legumes have excellent nutritional quality than grasses, but the quality of both forage resources is affected by the stage of maturity, which is thus considered as one of the important factors in determining the nutritional quality of forages. Most of forages have better nutritional quality during their early growth stages and decrease markedly thereafter with advancement in maturity. Dry matter, NDF, ADF, and hemicellulose content increase whereas crude protein, IVDMD and oxalate content decrease as the maturity advances. *Setaria* and tall fescue are important grasses of the Himalayan region. The study has revealed that the nutritional quality of both grasses is influenced by harvesting stages. In *Setaria* grass, the nutritional components are in the optimum proportion during May and July; however, oxalate content is simultaneously high, which is not desirable despite other merits. Harvesting of *Setaria* in the month of August maintains a

better balance in nutritive composition in terms of nutritional as well as anti-nutritional components. Tall fescue grass retains high nutritive value during the months of March and April. Since harvesting stages have significant influence on nutritional quality of forages, harvesting of fodder at appropriate stages is essential.

11. Forage nutritional quality has a direct effect on livestock performance and ultimately, profitability of the livestock production system. Forage quality is generally estimated in terms of nutritional and anti-nutritional constituents and digestibility. Generally, high-quality forages also have variable amount of anti-nutritional constituents. Nutritional analysis of forage is essential for estimates of the level of different nutrients and digestibility. Nutritional analysis is useful in providing the first assessment of relative potential of forage crops on livestock performance. If the animal has the genetic potential, productivity will depends on nutritive value of forages. Different factors have been identified that have a significant impact on the nutritional quality of forages. Forage maturity is the most important factor determining the amount of available digestible energy and protein in forages. Studies have revealed that forage quality decreases with the advancement in maturity. Therefore, harvesting forages at the stage when forages have maximum nutritional quality is essential for fetching maximum output. Environment factors also have a significant impact on the quality of forage crops.
12. Nutritional quality of forages is not uniform round the year and changes with variation in seasonal conditions. In advancing months of the year, NDF, ADF, and hemicellulose content increases while crude protein, ether extract, ash, and cell content decreases in *Leucaena* foliage. *Leucaena leucocephala* has excellent nutritive value from the month of January to April and lowest during May to July. Nutritional quality of forage grasses also varies with altitude of growing region. White clover from high altitudes has better nutritional profile in comparison to those from lower altitudes. Anti-nutritional factors except HCN content have inverse relationship with altitude. White clover plants at higher altitude may have high HCN content; therefore, excessive feeding should be monitored to avoid the lethal effects of cyanide poisoning on livestock.
13. Seasonal variations also affect the nutritional quality of tree fodder. Trees often contribute substantial amount of leaf fodder during lean period through lopping, popularly known as top feed. Top feeds are also considered important in stabilization of vegetation and sustained productivity of rangelands. They also play an important role as windbreaks and by providing shade for the grazing animal. The foliage yield and nutritional quality depend on species, initial age, lopping intensity and interval as well as agro-climatic conditions. In advancing months of the year, the NDF, ADF, and hemicellulose content increases while crude protein, ether extract, ash, cell content decreases. *Morus alba* has highest nutritive value from March to April and September to December. Further, *Grewia oppositifolia* is good fodder tree from May to June, July to August, and from January to February. *Grewia oppositifolia* and *Bauhinia variegate* have low anti-nutritional content and could serve as an alternate feed stuff for

livestock. Both trees could be lopped during the period between September and December. *Grewia oppositifolia* may be lopped preferably in November, while October is good lopping season for *Bauhinia variegata*. In the month of September and November, *Morus alba* has low anti-nutrients. Delayed lopping of tree fodder may result in higher levels of anti-nutrients, which could have adverse effect on the livestock health.

14. Forage quality is also affected by incidence of diseases and insects. Insect infestation not only decrease forage yield, but also affects the quality of forage. For example, alfalfa weevil feed on mesophyll cell content due to which nutritional quality as well as forage yield decreases. Leaf chlorosis and dropping after insect attack reduce photosynthetic area and damage cells that ultimately affect the quality of forage crops. The incidence of foliar diseases increases the level of undesirable nutritional constituents and reduces desirable constituents. Diseased plants have low digestibility due to higher level of cell wall constituents. Incidence of foliar diseases and pest outbreak has adverse effects on quality attributes, *viz*. CP, NDF, ADF, Hemicellulose, Chl a, Chl b, Total Chl, Ash, K, Ca, IVDMD, ADF-N, CHO and phenolic content of cowpea fodder; therefore, the management of diseases and insect pests is of utmost importance not only for increasing yield but also for the production of quality fodder. In chemically protected crop, the desirable quality parameters were higher in comparison to unprotected crop.
15. Besides natural grazing resources and fodder trees, crop residues have due importance in livestock feeding in the Himalayan region, especially during the period of scarcity of green fodder. Paddy straw, wheat straw, and maize stover are important crop residues. Although, crop residues are important lean season forage resources, their effective utilization in livestock feeding is constrained by low digestibility and low voluntary intake. The high lignin and fiber content are responsible for low feeding value of crop residues. Different strategies have been developed for improvement of quality and digestibility of crop residues. Different treatments for improving quality and digestibility improvement of different crop residues have been developed. Treatment of maize stover with alkalis [sodium hydroxide (2%) and calcium hydroxide (2%)] and inoculation of *Trichoderma viride* and *Pseudomonas fluorescens* is promising treatments for improving the digestibility and quality of crop residues.
16. Owing to the shortage of feed in terms of quantity and quality to livestock, the future hopes for feeding of animals and safeguarding their food and nutritional security will also depend on nonconventional feed resources (NCFR), which are non-competitive in terms of human consumption. NCFR are feed resources which have not been utilized traditionally and/or commercially in livestock feeding. They could bridge the gap between demand and supply of fodder, decrease direct competition for food between humans and animals, reduce the feeding cost and contribute nutritional sufficiency on locally available feed sources. The propagation of nonconventional or alternative fodder can also lead to diversification of traditional agriculture and conservation of biodiversity through sustainable utilization of natural resources. It is therefore imperative to

investigate the advantages of inclusion of NCFR in livestock feeding to maximize livestock productivity vis-à-vis economic viability of the livestock production system. The successful inclusion of NCFR in farming system requires facilitation of knowledge intensive practices like farmer to farmer extension, different training sessions and field trial demonstrations by local and government extension agencies.

17. It is generally assumed that weeds are nutritionally inferior and could not be used for livestock feeding, however, certain weeds have better nutritive profile, and could be used in livestock feeding. *Rumex* species (*R. hastatus, R. obtusifolius, and R. nepalensis*) have considerable merits in the nutritive traits with high crude protein content. These *Rumex* spp. have prospects of utilization as a feed resource. These nutritional nonconventional forage resources also have prospects of blending in conventional feeds.
18. For bridging gap between demand and supply of feed, preservation and conservation of forages are essential. There is a concern that the nutritional quality of forages during harvesting, storage, conservation and processing may decline. Due precautions are required during harvesting and storage of forages and adoption of conservation and preservation methods to avoid nutrient losses vis-à-vis feeding value of forages.
19. Livestock are an integral part of agriculture and instrumental for future growth and development of the agriculture and allied sectors. Livestock production is a valuable source of income to complement crop production. Farmers rely heavily on livestock rearing activities which in turn provide source of savings, cash income, draft power and nutritious food. Nevertheless, scarcity and poor quality of feed are major constraints for livestock productivity in the Himalayan region. The introduction of multipurpose forages such as food-feed crops in these systems has the potential for increasing the availability of food as well as fodder resources to the farmer. Cultivation of multipurpose forages including dual-purpose legumes is economically viable and profitable than solely for fodder or feed purpose. The productivity and economics of cultivating maize varieties under different use patterns has recommended for the cultivation of maize varieties for baby corn and green fodder was a profitable venture than sole cropping of maize for green fodder. The research on ricebean has established that ricebean forage has good excellent nutritive value, mainly attributed to high bioavailability of proteins, favorable amino acid composition and appreciable level of different micronutrients.
20. Breeding of forage varieties with desirable traits has been the method of choice. Allogamous nature and self-incompatibility limits inbreeding in forage crops and turns conventional breeding a difficult venture in forage crops. Due to time consumption, limited genetic resources and exploitation of traditional landraces of most of the forage crops, there is little space for the improvement of forage crops by these methods. Advancement in agricultural biotechnology has allowed the exploration and development of technologies to tackle challenges in forage breeding program. Genetic engineering techniques are highly efficient which have enabled researchers to upregulate or downregulate

genes or to move genes between distantly related species. Using these techniques, significant advances in forage breeding could be made in the near future. Gene editing techniques have emerged as a new horizon for further improvement in forage crops. Gene editing technologies have facilitated plant scientists to insert desirable traits precisely and quickly in plants than traditional breeding. Although, these technologies have been implemented in many food crops, their potential in forage crops breeding is yet to be exploited. Gene editing techniques could be used to manipulate the expression of regulatory gene (s) encoding regulatory enzyme(s) of metabolic pathways leading to the biosynthesis of nutrients as well as anti-nutrients.

21. Recently, the CRISPR-Cas 9 genome editing system has received attention from researchers globally for genome manipulation due to its simplicity, flexibility, consistency and high efficiency. Moreover, the CRISPR/Cas9 approach enables manipulation of multiple target genes in a single event. In the last 5 years, this technique has been applied vigorously in many food crops for functional studies and combating biotic and abiotic stresses as well as for improving other important agronomic traits. The unique attributes of the CRISPR/Cas9 system make this technique a much needed tool to breed improved forages without affecting their genetic makeup and going through long breeding cycles.
22. Availability of digestible energy from forages depends upon cell wall constituents. Lignin is a major component of secondary cell wall. It is a heterogeneous complex polymer resulting from oxidative coupling of monolignols synthesized from phenylpropanoid pathway. Lignin provides strength to cell wall and protection against pathogens and physical damage. However, these advantages of lignification in plant cell wall can represent themselves as barriers in utilization of forage crops in livestock feeding and other agro-industrial uses. In the recent years, manipulation of content and composition of lignin content in forages has been accomplished through metabolic engineering of lignin biosynthetic pathway using antisense RNA, sense expression, co-suppression and RNA interference techniques. The CRISPR–Cas system for metabolic engineering is a potential area of research for manipulating lignin content in different forage species.

Conclusion

- The Himalayan region has great ethnical and cultural diversity. It harbors different rare plant and animal species, which are considered as one of the important treasure houses of biodiversity of flora and fauna. The Indian Himalayan region covers approximately 16.20% of total geographic area of India and mountainous ranges of twelve Indian states. The representativeness of floral and faunal species in the Himalayan mountain ecosystem varies from the subtropical forests of the Shivalik to alpine meadows and scrub in the high ranges of the Greater Himalayan region. Its expansion in latitude and altitude offer habitat to diverse flora and fauna, which supports biodiversity.
- The rural people and farmers of the Himalayan region depend on the natural flora and fauna for their sustenance. The rural households of the region follows mixed farming approach for the sustenance of their livelihoods in which agriculture and livestock farming are two essential components. Natural pasture, grassland and forest are the main feed resource for the livestock in the region. Besides these, shrubs and crop residues are other important feed resources. The increasing livestock population and diminishing availability of feed resources have increased the pressure on natural feed resources. It is estimated that the pressure of livestock grazing on forest and grazing land is much higher than normal carrying capacity of grazing land in the region. Due to feed-fodder deficiency, declining per capita land availability, geographic and climatic constraints and poor livestock management, livestock production in the hill region has become a formidable challenge. Factors, which affect the livestock and fodder production in the Himalayan region have been presented in Fig. A.1.
- Natural feed resources in the Himalayan region have been exploited in an unplanned manner leading to their degradation. Reckless cutting of trees, indiscriminate use of grazing areas and absence of rehabilitation programs have led to denudation of hill slopes, resulting in low availability of feed resource and thereby adverse effects on livestock production. This invite more consistent actions to achieve sustainability in fodder as well as livestock production in the Himalayan region.

R. Katoch, *Nutritional Quality Management of Forages in the Himalayan Region*,
https://doi.org/10.1007/978-981-16-5437-4

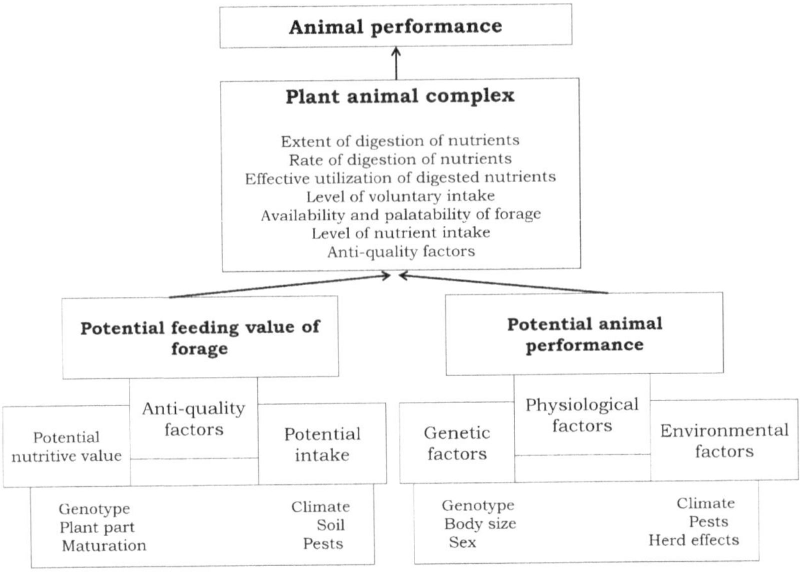

Fig. A.1 Factors affecting livestock performance on forages

- The division of families has further fragmented available agricultural land in the region. At present, land holdings are very small and the farmer is always biased in the choice of the crops. The agricultural land ratio does not permit diversion of land from food production to cultivated fodder. The large area under forest and orchards in the Himalayan region could be utilized under fodder production programmes. Moreover, the cultivation of forages on terrace risers, bunds and along the irrigation channel could be useful in alleviating fodder shortages in the Himalayan region. In general, legumes are higher in essential nutrients such as crude protein, digestible energy, calcium and magnesium compared to grasses. Forage legumes and grasses are often planted together in a farming system to improve the nutritional value of dominating forage and increase the productivity of system.
- Productivity of grazing lands has declined and therefore, it is essential to replace low yielding annual grasses with high yielding perennial grasses with the introduction of multipurpose tree species that are adaptable to prevailing conditions. Introduction of pasture legumes is one of the best ways to improve the nutritional quality of forage resource as well as soil health. Participatory watershed-based approach could also be utilized for improving the productivity of natural grassland and rangelands. The undulating topography and ragged terrain are responsible for excessive soil erosion and nutrient losses. This could also be effectively checked by development/restoration of degraded rangelands by introduction of

forage legumes and grasses in pasturelands. Introduction of grasses, legumes and trees in a silvipastoral system could also provide effective land cover in pasturelands as well as to produce nutritious fodder for livestock. Due to excessive grazing, the invasion of alien species in pasturelands has also increased. These species offer severe competition for light and nutrients to major forage grasses and legumes in pasturelands.

- In recent years, the Himalayan region has become more vulnerable to climate change due to dependency of large population on agriculture and livestock farming. Since the forage production per unit area is the consequence of interactions between genotypes and environment, varieties having resistance to any changes in the growing conditions with excellent quality and high production potential need to be developed. Emphasis should also be given on increasing the availability of seeds of short duration and multipurpose forages crops in the region.
- Different plant and environment-based factors have significant effect on the nutritional quality of forages. Maturity stage is one of the crucial plant-based factors having significant influence on the nutritional quality of forages. Generally, forages are nutritious at early stages of growth. However, as the maturity advances, nutritional quality decreases considerably due to changes in composition of cell wall, lignin deposition and other polyphenolic compounds. The presence of anti-nutritional factors also influences the nutritional quality of forages. These constituents if consumed in large quantity, not only affect the productivity of animals but may even cause death of animals.
- The alterations in quality of forages occur as a result of unfavorable environmental growing conditions during the active period of growth. The influence of various environmental factors on quality of forages has been the subject of interest for fodder researchers for many years. Among different environmental factors, temperature has significant impact on nutritive quality of forages because it alters the leaf to stem ratio, concentrations of indigestible cell wall constituents, concurrently the nonstructural carbohydrates. In contrast, shading reduces yield, but has small and variable effects on quality of forage legumes. The effect of drought on forage quality is usually low particularly if the stress on leaf mass is not severe. Besides these factors, the forage quality is also influenced by soil properties, uptake of nutrients from soil and incidence of insect and diseases.
- The information on the nutrient composition of forages is essential for offering balanced feed to livestock. Nutritional quality of forage is highly variable since it is affected by different plant-based and environmental factors. At present, different traditional and advanced techniques are available for the forage nutritional quality analysis.
- Forage breeders have the challenge to tackle different barriers in the breeding of forage crops such as genotypic and phenotypic heterogeneity, polyploid nature and polygenic control of desirable traits. Different objectives underlying forage crop improvement focus on following approaches for substantial improvement in forage and livestock production.

- **Improving nutritional quality of forage crops**: New varieties with higher crude protein, less fiber, high digestibility, better quality amino acid composition and slow protein degradation in rumen are to be developed. Additionally, most of the forages have some anti-quality constituents, which need to be eliminated or minimized through breeding strategies.
- **Utilization of plant genetic resource and broadening genetic base**: Efforts are required for enriching the genetic base either by introduction from exotic sources, collection/exploration or pre-breeding. Much emphasis is needed for interspecific or distinct intervarietal crossing programs. Use of modern biotechnological tools in the development of interspecific hybrids through embryo rescue, transgenic plants and for assessment of genetic variability is also the need of the hour.
- **Development of multicut and quick regeneration varieties**: Breeding should be directed for the development of forage varieties multicut in nature and having quick regeneration ability, in order to meet the requirement of green fodder regularly.
- **Tolerance to abiotic and biotic stress**: Changing climatic situations warrent that the breeding programme must be directed for the development of forage varieties having the potential to grow in marginal, submarginal and degraded habitats. Crop-specific biotic stress could be addressed through breeding mainly by utilizing secondary and tertiary gene pools.
- **Quality seed production**: Seed shattering and seed dormancy, indeterminate habit, uneven pod ripening and low fertility are the common problems in most of the forage legumes, making their commercial cultivation quite difficult. These factors affect the availability and production of quality seed as well as important reason for their non-adoption by farmers. These problems need to be addressed through breeding including selection and crop-specific seed production with suitable package of practices.

• Fodder tree and shrubs are important components of the livestock feeding system, as a good supplier of protein and other essential nutrients to livestock. Their importance has been well recognized during dry season when other feed sources are in limited supply and grazing is not sufficient. In other words, fodder trees provide insurance to farmers and rural people against seasonal dry spell. Fodder trees are less affected by dry conditions because of their more extensive root systems and longer life span. Since the benefits associated with fodder trees are enormous, cultivation and integration of fodder trees in farming system is essential to increase livestock production and to support the life of rural livelihoods.

• Conventional feed sources are an indispensable part in livestock feeding. The unavailability or fluctuating year round feed supply of quality feed is one of the critical reasons for low livestock productivity in the Himalayan region. Therefore, for the sustainable development of the livestock sector, it is important to look into all possibilities to overcome feed shortages. The inclusion of nonconventional feed resources (NCFR) could be an alternative for bridging the gap between supply and demand for animal feeds, for reducing competition between humans and animals for food and for providing nutritional sufficiency on available feed resources. This would also be an effective way of diversification of traditional agriculture and biodiversity conservation. NCFR are an excellent source of nutrients. However, some of them are low in nutritive quality and contain high concentrations of aversive factors, which reduce nutrient utilization and affect different metabolic processes. Different strategies and methods have been developed for managing the level of anti-nutrients in nonconventional feed resources, however advance studies are required for the safe use of these resources.

- Conventional breeding has been a method of choice for the breeding of forage crops for different desirable traits. Using this approach, forage breeders have developed and released different varieties of forage crops. However, conventional breeding is a time consuming process and forage breeders have to overcome barriers in forage crop breeding such as genotypic and phenotypic heterogeneity, polyploid nature and polygenic control of desirable traits. Due to these reasons, the success in forage breeding program is comparatively less in comparison to the food crops. Biotechnology and molecular biology techniques provide enormous opportunities and novel possibilities to generate variation in forage crops when sufficient genetic variability is not available. Advancing molecular biology techniques allows unlimited gene modifications in gene pool, which holds promise for improvements in forage crops. Though success has been achieved in improving nutritional quality and other desirable traits in some forage crops through genetic engineering, however this approach is confronted some bottlenecks which would be resolved with further research. The advancement in biotechnological tools such as development of gene editing techniques has equipped scientists with enormous options to tailor forage crops for almost any given desirable trait. Although, developed recently, the CRISPR/Cas system has been vigorously used for the manipulation of desirable traits in different economic crops, but its implication in forage breeding has not been fully recognized. The expertise for use of this technology in forage improvement could provide new dimensions to the forage breeding program.
- The scarcity of green as well as dry fodder has been reported, but in the pocket where surplus fodder is available, due to the lack of knowledge on harvesting methods, storage and value addition, a major part of this surplus fodder goes waste. Different technologies have been developed for the post-harvest management and value addition of feed resource like bailing and silage making, preparation of densified feed block, use of leaf meal technology and improvement in quality of crop residues with physical, chemical and biological treatments.
- Improving the entrepreneurial capacity of rural households need emphasis. The forage resource development-related activities should be modified in harmony with the policies of government for poverty reduction and livelihood promotion. Credit and market linkages to forage-based livestock production need support from central and state governments. Strategies should be changed from simple program on cultivated fodder to focused technology demonstration. Involvement of different non-government organizations along with government policies is also required to create a supportive environment.

- In conclusion, the forage and livestock production situation in the Himalayan region needs attention for sustainability. The growth scenario of the livestock and forage sector and their implications for farmers and economy of the region necessitate efforts for improving the nutritional status of forage as well as livestock production in the Himalayan region. Looking at the vast gap between the demand and supply, it becomes necessary for "lab to land" transfer the potential technologies developed by various research organization to farmer's field in order to increase the production and productivity of high-quality fodder.

To increase the economic feasibility of the livestock production system, availability of nutritional feed and fodder in adequate quantity is required. Therefore, emphasis is needed on intensive and diversified cropping systems, alternate land use, integration of agro-forestry in farming, alternate feed resources and management of natural grassland and pastures. The increased degradation of the environment, combined with the increased population pressure, requires reconsideration of existing livestock development programs on priorities.

In the past few decades, research attempts have been made for improvement in nutritional components of forage crops, however a number of technological issues must be addressed for breeding high-quality forages. The existing information on forage germplasm should be cataloged, updated accurately and disseminated among forage breeders for variability in germplasm.

Though different technologies have been developed for improving fodder and livestock productivity, the decision to adopt a particular technology depends to a great extent on the farmer's perceptions about the technology attributes and their socioeconomic relevance. Therefore, inclusion of farmer's perceptions in generation and adoption of technology is crucial.

Lack of coordination among various agencies has often resulted in delay or failure in the accomplishment of the set goals. There is a need to work out possible feed options with feed resources available to farmers in different agro-ecological zones of the country to match the nutrient requirements of livestock of that region. Socio-economic analysis and interventions for optimizing pasture quality and productivity and sustenance through participatory management on common property resources need further attention (Fig. A.2).

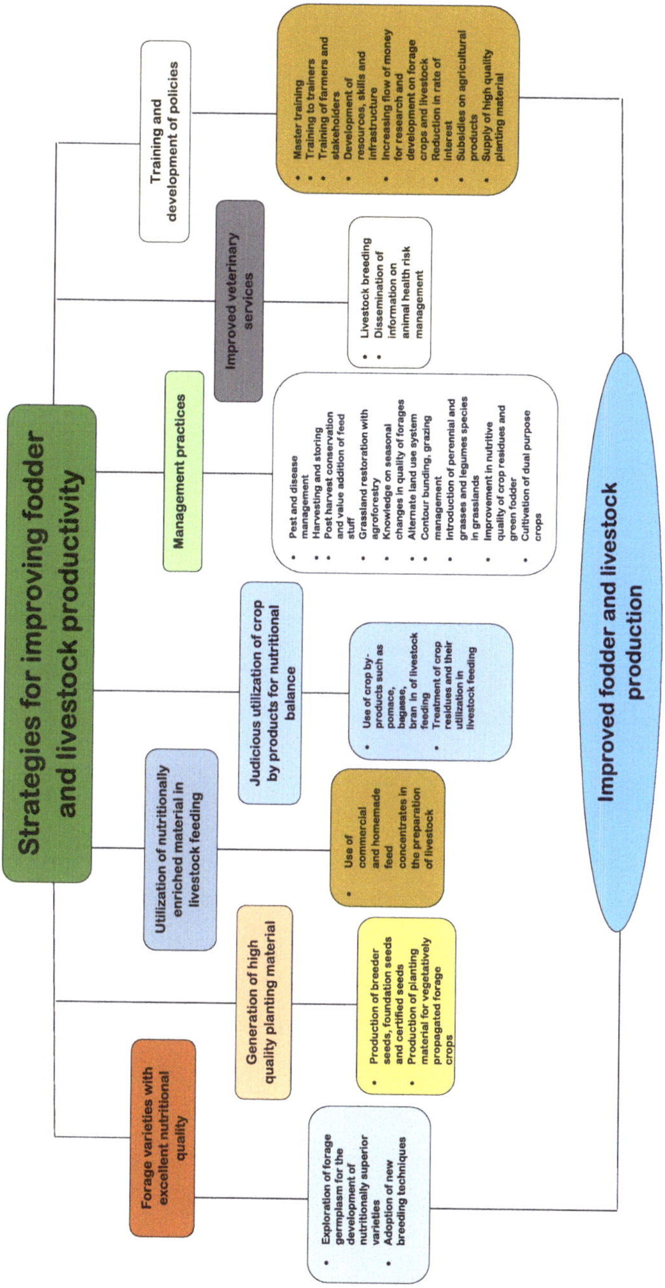

Fig. A.2 Strategies for improving fodder and livestock production in the Himalayan region

Glossary

Acid detergent fiber (ADF) The amount of residue remains after boiling forage sample in acid detergent solution. ADF contains cellulose, lignin, and silica, but does not exclude hemicelluloses. This term is inversely related to digestible energy content.

Acid detergent lignin (ADL) Percentage proportion of plant material that is insoluble in 72% sulfuric acid. ADL analysis is performed sequentially with ADF residue.

Agrisilvopastoral system Land use system in which woody perennials are grown with agricultural crops, forage crops and livestock farming system.

Agro-forestry Agro-forestry combines agriculture and forestry to create integrated and sustainable land use systems. Agro-forestry takes the advantage of interactive benefits from combining the trees and shrubs with crops and/or livestock.

Agrostology It is the study of forage crops and their management and utilization.

Annual forages Forages complete their life cycle in 1 year.

Anti-nutrients Compounds cause problems in livestock health and production. Even though often present in small amounts, they decrease the nutritional quality of fodder.

Ash The residue represents the total inorganic mineral material of a feed sample. Ash is determined in a laboratory by burning the sample at a high temperature and weighing the residue.

Biennial forage Forages complete their life cycle in 2 years, normally flowering only in the second year.

Bloat A livestock disorder which results from gas accumulation in the rumen or intestine.

Browse Leaves, shoots, and sprouts of woody plants which are eaten by domestic and wild animals.

Bunch grass A grass having the characteristic growth habit of forming a bunch; lacking stolons or rhizomes, e.g., Tall fescue.

Carrying capacity Average number of livestock and/or wildlife that may be sustained on a management unit compatible with management objectives for the unit.

R. Katoch, *Nutritional Quality Management of Forages in the Himalayan Region*, https://doi.org/10.1007/978-981-16-5437-4

Cellulose A structural carbohydrate; a long-chain polymer of glucose that is the main constituent of plant cell wall. It is the most abundant carbohydrate in nature and is slowly and partially digestible.

Cool season forages Forage crops that make most of their growth and flowers during spring or early summer.

Concentrate Any feed that is low in fiber and high in TDN.

Crop residue The portion of crop remaining after harvest of economic part. It may be managed for grazing and/or ground cover and to replenish soil organic matter levels.

Crude fat An estimate of the fat content of feeds that is measured by ether extraction. Crude fat contains true fat (triglycerides) as well as alcohols, waxes, terpenes, steroids, pigments, ester, aldehydes and other lipids.

Crude fiber The coarse, fibrous portions of plants that are partially digestible and relatively low in nutritional quality. In chemical analysis, it is the residue obtained after boiling plant material with dilute acid and then with dilute alkali.

Crude protein Amount of protein which is calculated from percent nitrogen content of a feedstuff, using Kjedahl digestion analysis. Crude protein content represents proteins and designated true proteins as well as non-protein nitrogen compounds.

Digestibility Extent of digestion of nutrients in the rumen and intestine of livestock. It has inverse relationship with ADL and fiber content in forages.

Digestible Dry Matter (DDM) Percentage proportion of dry matter, organic matter, or nutrients in feed which are destined to digest and absorbed in the digestive tract. Digestibility estimated from ADF. Higher the ADF content, lower the digestibility.

Dry matter Dry matter represents everything contained in a feed sample excluding moisture. It is determined by drying the feed sample until a constant weight is achieved.

Dry matter basis The expression of percent proportion of nutrients in feed with respect to dry matter content. In other words, dry matter basis indicates the nutrient levels in a feed sample based on its dry matter content. This is also referred to as dry basis, dry results, or moisture-free basis.

Dry matter intake The amount of forage consumed by an animal on a dry matter basis.

Ensiling The process by which silage is produced through fermentation of forages.

Feeding value Some consider this term synonymous with nutritive value. The characteristics that make a particular feed a valuable source of nutrients to animals. It can also be understood as the combination of chemical, biochemical, physical and organoleptic characteristics of forage that determines its potential to produce animal products.

Fodder Any plant that is cut before being feeding to animals in green stage or after converting hay and silage.

Forage It is the edible part of plants other than grains which is harvested for livestock feeding.

Forage quality The characteristics making forage a valuable nutrient source to livestock. In other words, the combination of the chemical and biological characteristics of forage that determines its potential to produce animal products.

Forb It includes herbaceous broad-leaved plant other than grasses.

Grassland It is an area in which the vegetation is dominated by a nearly continuous grass cover. Grasslands occur in environments conducive to the growth of this plant cover but not to that of taller plants, particularly trees and shrubs.

Grazing When animals eat or partially defoliate any kind of standing vegetation.

Grazing season The period when grazing can normally be practiced each year or portion of the year.

Habitat The environmental conditions or kind of place where a plant naturally grows.

Hay The herbage of grasses, legumes, or comparatively fine-stemmed forbs cut and cured to preserve forage.

Haylage It is called no moisture silage, hay crop silage, or drylage. It is the combination of hay and silage in which moisture in grass or forage is reduced 40–60% before it is chopped and stored in the silage.

Hortipastoral system Agroforestry system that involves cropping of grasses in fruit orchards.

***In sacco* degradability** It is the same as *in situ* degradability, wherein a ground feed sample is incubated in a porous nylon bag placed within the rumen for a fixed time period; and loss in dry matter or nitrogen is measured and this loss is taken as degradability.

***In situ* digestibility** *In situ* digestibility is determined by incubating a ground forage sample in a porous nylon bag placed within the rumen *via* a fistula or port in the animal's side (*in situ*) for a fixed time period.

***In vitro* dry matter digestibility (IVDMD)** Digestibility determined by incubation of a ground forage sample with rumen fluid in beaker or test tube for 24–48 h, followed either by addition of acid and pepsin and further incubation for 24 h (IVDM or IVDMD) or by boiling in neutral detergent fiber solution.

Lignin Lignin is a complex strengthening material in the cell walls of plants. Lignin reduces the digestibility of plant tissues; as lignin increases, the digestibility of the forage decreases.

Macrominerals Macrominerals that are required in relatively large amounts by livestock include calcium (Ca), phosphorous (P), sodium (Na), potassium (K), magnesium (Mg), and sulfur (S).

Meadow An area covered with grasses or succulent fodder legumes grown primarily for hay and silage, rather than for grazing.

Metabolizable energy (ME) The energy available for metabolism by an animal (energy in the food minus fecal energy, urinary energy, and gaseous energy losses).

Microminerals Minerals which are required in relatively small amounts by livestock. For example, manganese (Mn), iron (Fe), copper (Cu), zinc (Zn), and molybdenum (Mo).

Net energy (NE) Metabolizable energy minus energy lost as heat of fermentation and heat of nutrient metabolism. The net increase in useful animal product expressed per unit increase in food intake.

Net energy for lactation Change in lactation energy per unit of feed intake measured under conditions in which retained energy remains constant.

Net energy for maintenance Change in retained energy per unit of feed intake measured between an intake of zero and intake when retained energy equals zero.

Neutral detergent fiber (NDF) Residue left after boiling a sample in neutral detergent solution. NDF content in forages represents indigestible and slowly digestible components in plant cell walls (cellulose, hemicellulose, lignin, and ash).

Nomadic systems Systems based on extensive movement of herds and flocks in search of forage, led by human family units.

Non-fibrous carbohydrate (NFC) An estimate of the rapidly available carbohydrates in forage (primarily starch and sugars). This NFC content is calculated as: NFC = 100% – (CP% + NDF% + CF% + Ash %).

Non-protein nitrogen (NPN) The portion of total nitrogen that is not in protein. If high, NPN is an indicator of potential for nitrate toxicity.

Noxious weed A plant which is undesirable, troublesome, and difficult to control. It grows and spreads in places where it interferes with the growth and production of the desired crop.

Organic matter intake The amount of forage consumed by an animal on an organic matter basis.

Palatability The degree to which feed is agreeable to an animal's taste is savory and acceptable.

Pastoralism A livelihood strategy based on moving livestock to seasonal pastures primarily in order to convert grasses, forbs, tree leaves, or crop residues into human food.

Pasture land Land devoted to the production of indigenous (native) or to introduce forage species that are harvested primarily by grazing.

Pastures Land covered with grass and other low plants suitable for grazing animals, especially cattle or sheep.

Permanent pasture Pasture composed of perennial or self-seeding annual plants that are grazed annually, generally for 10 years or more successive years.

Rangelands Any extensive area of land that is occupied by native herbaceous or shrubby vegetation which is grazed by domestic or wild herbivores. The vegetation of ranges may include tallgrass prairies, steppes (shortgrass prairies), desert lands, shrub woodlands, savannas, chaparrals and tundras.

Relative feed value (RFV) Index for ranking forage grasses and legume forages based on intake of digestible energy, is calculated from acid detergent fiber (ADF) and neutral-detergent fiber (NDF) concentrations in forage.

Relative forage quality (RFQ) Index for ranking all forages based on intake of total digestible nutrients calculated by summative equations after estimating digestible portions of protein, fatty acids, fiber and non-fibrous carbohydrate.

Roughages Roughages are bulky field stuff that is relatively high in fiber and low in digestible nutrient. They usually contain more than 80% fiber.

Sedentary systems Grazing systems which are managed at a particular location (s) by resident management and may be managed by either or both extensive and intensive grazing managements and can include rangeland, pastureland, cropland, and forestland within the grazing system.

Semi-sedentary systems Systems based on a village permanently occupied by women and children from which herds and flocks, usually tended by men and boys, are absent for extended periods of time in search of forage.

Senescence It is the process of remobilization and transfer of soluble constituents from mature to immature plant tissues with the advancement of maturity. It is usually accompanied by chlorosis and subsequent death of mature tissue.

Shrub A woody plant having multiple stems arising at or near the base. Mature height is generally less than 5–6 m.

Silage Forage preserved in a succulent condition by organic acids produced by partial anaerobic fermentation of carbohydrates in the forage.

Silage additives Material which is added to forage at the time of ensiling during silage preparation to enhance the fermentation process.

Silvopastoral system Type of agroforestry system that involves cropping of grasses in between the trees.

Sod grass Perennial grasses which form a mass of individuals by means of rhizomes or stolons.

Steppe Semi-arid grassland which is characterized by grasses occurring in scattered bunches with other herbaceous vegetation and occasional woody species.

Stubble The basal portion of herbaceous plants remaining after harvesting of top portion either manually or by grazing animals.

Supplement A nutritional additive intended to improve nutrition balance and remedy deficiencies in livestock diets.

Sward A population of herbaceous plants, characterized by a relatively short habit of growth and relatively continuous ground cover, including both above and belowground parts.

Total digestible nutrients (TDN) A unit that represents sum total of digestibility of organic components of forage (e.g., crude protein + nitrogen-free extract + crude fiber + fat).

Transhumance systems Systems that differ from semi-sedentary systems in that the grazing is cyclical beginning at the end of winter with flocks and herds leaving lowland grazing areas where permanent villages are located and moving to mountain pastures for grazing during summer.

Vegetation type The type of existing plant community with distinguishable characteristics described in terms of present dominating vegetation or physiognomy of area.

Voluntary intake Amount of forage consumed by an animal unrestricted by the quantity available.

Warm season forages Forage crops which make most or all of its growth during the late spring and summer and flowers in the summer.

Water soluble carbohydrates (WSC) Carbohydrates solubilized and extracted in water and include monosaccharides, disaccharides, and some polysaccharides (mainly fructan). Fructan is a major storage carbohydrate in grasses.

The manufacturer's authorised representative in the EU is Springer Nature Customer Service Centre GmbH, Europaplatz 3, 69115 Heidelberg, Germany. If you have any concerns regarding our products, please contact ProductSafety@springernature.com

Printed and bound by CPI Group (UK) Ltd, Croydon, CR0 4YY
30/09/2026
02225248-0019